Monographs in Theoretical Computer Science
An EATCS Series

Manfred Droste • Werner Kuich • Heiko Vogler
Editors

Handbook of Weighted Automata

Editors

Prof. Dr. Manfred Droste
Universität Leipzig
Institut für Informatik
04009 Leipzig, Germany
droste@informatik.uni-leipzig.de

Prof. Dr. Werner Kuich
Technische Universität Wien
Institut für Diskrete Mathematik und Geometrie
1040 Wien, Austria
kuich@tuwien.ac.at

Prof. Dr.-Ing. Heiko Vogler
Technische Universität Dresden
Institut für Theoretische Informatik
01062 Dresden, Germany
heiko.vogler@tu-dresden.de

Series Editors

Prof. Dr. Wilfried Brauer
Institut für Informatik der TUM
Boltzmannstr. 3
85748 Garching, Germany
brauer@informatik.tu-muenchen.de

Prof. Dr. Juraj Hromkovič
ETH Zentrum
Department of Computer Science
Swiss Federal Institute of Technology
8092 Zürich, Switzerland
juraj.hromkovic@inf.ethz.ch

Prof. Dr. Grzegorz Rozenberg
Leiden Institute of Advanced
Computer Science
University of Leiden
Niels Bohrweg 1
2333 CA Leiden, The Netherlands
rozenber@liacs.nl

Prof. Dr. Arto Salomaa
Turku Centre of
Computer Science
Lemminkäisenkatu 14 A
20520 Turku, Finland
asalomaa@utu.fi

ISBN 978-3-642-26049-0 e-ISBN 978-3-642-01492-5

Monographs in Theoretical Computer Science. An EATCS Series. ISSN 1431-2654

ACM Computing Classification (1998): F4.3, F1.1, F1.2
Mathematics Subjects Classification (2000): 68Q45, 68Q70, 03B50, 03B70, 03D05, 68R99, 68Q85

Softcover reprint of the hardcover 1st edition 2009

Cover Design: KünkelLopka GmbH

Printed on acid-free paper

9 8 7 6 5 4 3 2 1

springer.com

Preface

The purpose of this Handbook is to highlight both theory and applications of weighted automata.

Weighted finite automata are classical nondeterministic finite automata in which the transitions carry weights. These weights may model, e.g., the cost involved when executing a transition, the amount of resources or time needed for this, or the probability or reliability of its successful execution. The behavior of weighted finite automata can then be considered as the function (suitably defined) associating with each word the weight of its execution. Clearly, weights can also be added to classical automata with infinite state sets like pushdown automata; this extension constitutes the general concept of weighted automata.

To illustrate the diversity of weighted automata, let us consider the following scenarios. Assume that a quantitative system is modeled by a classical automaton in which the transitions carry as weights the amount of resources needed for their execution. Then the amount of resources needed for a path in this weighted automaton is obtained simply as the sum of the weights of its transitions. Given a word, we might be interested in the minimal amount of resources needed for its execution, i.e., for the successful paths realizing the given word. In this example, we could also replace the "resources" by "profit" and then be interested in the maximal profit realized, correspondingly, by a given word. Furthermore, if the transitions carry probabilities as weights, the reliability of a path can be formalized as the product of the probabilities of its transitions, and the reliability of a word could be defined again as the maximum of the reliabilities of its successful paths. As another example, we may obtain the multiplicity of a word, defined as the number of paths realizing it, as follows: let each transition have weight 1; for paths take again the product of the weights of its transitions (which equals 1); then the multiplicity of a word equals the sum of the weights of its successful paths. Finally, if in the latter example we replace sum by "maximum," weight 1 is associated to a word if and only if it is accepted by the given classical automaton.

In all of these examples, the algebraic structure underlying the computations with the weights is that of a semiring. Therefore, we obtain a uniform and powerful automaton model if the weights are taken from an abstract semiring. Here the multiplication of the semiring is used for determining the weight of a path, and the weight of a word is then obtained by the sum of the weights of its successful paths. In particular, classical automata are obtained as weighted automata over the Boolean semiring. Many constructions and algorithms known from classical automata theory can be performed very generally for such weighted automata over large classes of semirings. For particular properties, sometimes additional assumptions on the underlying semiring are needed.

Another dimension of diversity evolves by considering weighted automata over discrete structures other than finite words, e.g., infinite words, trees, traces, series-parallel posets, or pictures. Alternatively, in a weighted automaton, the state set needs not to be finite, so we can consider, e.g., weighted pushdown automata with states being pairs of states (in the usual meaning) and the contents of the pushdown tape. Moreover, weighted context-free grammars and algebraic systems arise from weighted automata over trees by using the well-known equivalence between frontier sets of recognizable tree languages and context-free languages.

For the definition of weighted automata and their behaviors, matrices and formal power series are used. This makes it possible to use methods of linear algebra over semirings for more succinct, elegant, and convincing proofs.

Weighted finite automata and weighted context-free grammars were first introduced in the seminal papers of Marcel-Paul Schützenberger (1961) and Noam Chomsky and Marcel-Paul Schützenberger (1963), respectively. These general models have found much interest in Computer Science due to their importance both in theory as well as in current practical applications. For instance, the theory of weighted finite automata and weighted context-free grammars was essential for the solution of classical automata theoretic problems like the decidability of the equivalence: of unambiguous context-free languages and regular languages; of deterministic finite multitape automata; and of deterministic pushdown automata. For the variety of theoretical results discovered, we refer the reader to the indispensable monographs by Samuel Eilenberg (1974), Arto Salomaa and Matti Soittola (1978), Wolfgang Wechler (1978), Jean Berstel and Christophe Reutenauer (1984), Werner Kuich and Arto Salomaa (1986), and Jacques Sakarovitch (2003). (See Chap. 1 for precise references.) On the other hand, weighted automata and weighted context-free grammars have been used as basic concepts in natural language processing and speech recognition, and recently, weighted automata have been used in algorithms for digital image compression.

Since the publication of the mentioned monographs, the field of weighted automata has further developed both in depth and breadth.[1] The editors of this Handbook are very happy that international experts of the different areas agreed to write survey articles on the present shape of their respective field. The chapters of this Handbook were written such that a basic knowledge of automata and formal language theory suffices for their understanding.

Next, we give a short overview of the contents of this Handbook. Part I provides foundations. More specifically, in Chap. 1, Manfred Droste and Werner Kuich present basic foundations for the theory of weighted automata, in particular, semirings, formal power series, and matrices. As is well known, regular and context-free languages can be obtained as least solutions of suitable fixed point equations. In Chap. 2, Zoltán Ésik provides an introduction to that part of the theory of fixed points that has applications to weighted automata and their behaviors, and to weighted context-free grammars in the shape of algebraic systems.

Part II of this Handbook investigates different concepts of weighted recognizability. In Chap. 3, Zoltán Ésik and Werner Kuich develop the theory of finite automata starting from ideas based on linear algebra over semirings. In particular, they derive the fundamental Kleene–Schützenberger characterization of the behaviors of weighted automata over Conway semirings. In Chap. 4, Jacques Sakarovitch presents the theory of rational and recognizable formal power series over arbitrary semirings and graded monoids. As a consequence, he derives that the equivalence of deterministic multitape transducers is decidable. A seminal theorem of J. Richard Büchi (1960) and Calvin C. Elgot (1961) shows the equivalence in expressive power between classical finite automata (over finite and infinite words) and monadic second-order logic. In Chap. 5, Manfred Droste and Paul Gastin present a weighted version of monadic second-order logic and derive corresponding equivalence results for weighted automata. In Chap. 6, Mehryar Mohri presents several fundamental algorithms for weighted graphs, weighted automata, and regulated transducers as, e.g., algorithms for shortest-distance computation, ε-removal, determinization, minimization, and composition.

In Part III of this Handbook, alternative types of weighted automata and various discrete structures other than words are considered. In Chap. 7, Ion Petre and Arto Salomaa present the core aspects of the theory of algebraic power series in noncommuting variables, weighted pushdown automata, and their relationship to formal languages. In Chap. 8, Juha Honkala extends the theory of algebraic power series by considering Lindenmayerian algebraic systems and several restricted such systems. The following two chapters consider weighted automata acting on extensions of finite words. In Chap. 9, Zoltán Fülöp and Heiko Vogler survey the theory of weighted tree automata and weighted tree transducers. This combines classical results of weighted au-

[1] For instance, see the biennial workshops on "Weighted Automata: Theory and Applications" (WATA) since 2002.

tomata and transducers on words and of unweighted tree automata and tree transducers. In Chap. 10, Ina Fichtner, Dietrich Kuske, and Ingmar Meinecke present different weighted automata models for concurrent processes, formalized by traces and series-parallel posets, and analyze their relationships. They also consider two-dimensional extensions of words, namely pictures.

Part IV deals with applications of weighted automata. In Chap. 11, Jürgen Albert and Jarkko Kari present the use of weighted automata and transducers for digital image compression and give comparisons with the image compression standard JPEG. In Chap. 12, George Rahonis describes the theory of fuzzy recognizable languages. This theory arises by considering weighted automata over particular semirings, namely bounded distributive lattices. In Chap. 13, Christel Baier, Marcus Größer, and Frank Ciesinski present the main concepts of Markov decision processes as an operational model for probabilistic systems, and basic steps for the (qualitative and quantitative) analysis against linear-time properties. In Chap. 14, Kevin Knight and Jonathan May address the reawakened interest in string and tree automata among computational linguists. The chapter surveys tasks occurring in natural language processing and shows their solutions by using weighted automata.

Some of the chapters contain open problems. We hope that this will stimulate further research.

Finally, we would like to express our thanks to all authors of this Handbook and to the referees for their careful work. Moreover, warm thanks go to Carmen Heger for her support in the technical compilation of the chapters.

Manfred Droste	Werner Kuich	Heiko Vogler
Leipzig	Wien	Dresden

May 13, 2009

List of Contributors

Jürgen Albert
Universität Würzburg
Informatik II
97074 Würzburg, Germany
albert@informatik.
uni-wuerzburg.de

Christel Baier
Technische Universität Dresden
Faculty of Computer Science
01062 Dresden, Germany
baier@tcs.inf.tu-dresden.de

Frank Ciesinski
Technische Universität Dresden
Faculty of Computer Science
01062 Dresden, Germany
ciesinsk@tcs.inf.
tu-dresden.de

Manfred Droste
Universität Leipzig
Institut für Informatik
04009 Leipzig, Germany
droste@informatik.
uni-leipzig.de

Zoltán Ésik
University of Szeged,
Szeged, Hungary, and
Rovira i Virgili University
Tarragona, Spain

Ina Fichtner
Universität Leipzig
Institut für Informatik
04009 Leipzig, Germany
fichtner@informatik.
uni-leipzig.de

Zoltán Fülöp
University of Szeged
Department of Computer Science
6720 Szeged, Hungary
fulop@inf.u-szeged.hu

Paul Gastin
LSV, ENS Cachan, INRIA, CNRS
94230 Cachan, France
Paul.Gastin@lsv.ens-cachan.fr

Marcus Größer
Technische Universität Dresden
Faculty of Computer Science
01062 Dresden, Germany
groesser@tcs.inf.tu-dresden.de

Juha Honkala
University of Turku
Department of Mathematics
20014 Turku, Finland
juha.honkala@utu.fi

Jarkko Kari
University of Turku
Department of Mathematics
20014 Turku, Finland
jkari@utu.fi

Kevin Knight
USC Information Sciences Institute
Marina del Rey, CA 90292
USA
knight@isi.edu

Werner Kuich
Institut für Diskrete Mathematik und Geometrie
Technische Universität Wien
1040 Wien, Austria
kuich@tuwien.ac.at
www.dmg.tuwien.ac.at/kuich

Dietrich Kuske
Universität Leipzig
Institut für Informatik
04009 Leipzig, Germany
kuske@informatik.uni-leipzig.de

Jonathan May
USC Information Sciences Institute
Marina del Rey, CA 90292
USA
jonmay@isi.edu

Ingmar Meinecke
Universität Leipzig
Institut für Informatik
04009 Leipzig, Germany
meinecke@informatik.uni-leipzig.de

Mehryar Mohri
Courant Institute of Mathematical Sciences
New York, NY 10012
mohri@cs.nyu.edu
Google Research
New York, NY 10011
USA
mohri@google.com

Ion Petre
Academy of Finland and
Åbo Akademi University
Turku 20520, Finland
ion.petre@abo.fi

George Rahonis
Aristotle University of Thessaloniki
Department of Mathematics
54124 Thessaloniki, Greece
grahonis@math.auth.gr

Jacques Sakarovitch
CNRS and Ecole Nationale Supérieure des Télécommunications
75634 Paris Cedex 13, France
sakarovitch@enst.fr

Arto Salomaa
Turku Centre for Computer Science
Turku 20520, Finland
arto.salomaa@utu.fi

Heiko Vogler
Technische Universität Dresden
Faculty of Computer Science
01062 Dresden, Germany
vogler@tcs.inf.tu-dresden.de

Contents

Part I Foundations

Part II Concepts of Weighted Recognizability

Part I

Foundations

Chapter 1: Semirings and Formal Power Series

Manfred Droste[1] and Werner Kuich[2]

[1] Institut für Informatik, Universität Leipzig, 04009 Leipzig, Germany
droste@informatik.uni-leipzig.de
[2] Institut für Diskrete Mathematik und Geometrie, Technische Universität Wien, 1040 Wien, Austria
kuich@tuwien.ac.at
www.dmg.tuwien.ac.at/kuich

1 Introduction

It is the goal of this chapter to present basic foundations for the theory of weighted automata: semirings and formal power series.

Weighted automata are classical automata in which the transitions carry weights. These weights may model, e.g., the cost involved when executing the transition, the amount of resources or time needed for this, or the probability or reliability of its successful execution. In order to obtain a uniform model of weighted automata for different realizations of weights and their computations, the weight structures are often modeled as semirings. A semiring consists of a set with two operations addition and multiplication satisfying certain natural axioms like associativity, commutativity, and distributivity, just like the natural numbers with their laws for sums and products. The behavior of weighted automata can then be defined as a function associating to each word the total weight of its execution; see Chaps. 3 and 4 of this handbook [12, 38].

Any function from the free monoid Σ^* of all words over a given alphabet Σ into a semiring S is called a formal power series. It is important to notice that

M. Droste, W. Kuich, H. Vogler (eds.), *Handbook of Weighted Automata*, Monographs in Theoretical Computer Science. An EATCS Series,
DOI 10.1007/978-3-642-01492-5_1, © Springer-Verlag Berlin Heidelberg 2009

each language over Σ can be viewed as a formal power series over the Boolean semiring $\mathbb{B}$ and Σ^* (by identifying the language with its characteristic series). Therefore, formal power series form a generalization of formal languages, and similarly, weighted automata generalize classical automata. For other semirings (like the natural or real numbers), formal power series can be viewed as weighted, multivalued or quantified languages in which each word is assigned a weight, a number, or some quantity.

In this chapter, we will present the basics of the theory of semirings and formal power series as far as they are used in the forthcoming chapters of this handbook. Now, we give a summary of the contents of this chapter.

First, we consider various particular monoids and semirings. Many semirings (like the natural numbers) carry a natural order. Also, when generalizing the star operation (= Kleene iteration) from languages to formal power series, important questions on the existence of infinite sums arise. This leads to the notions of ordered, complete or continuous monoids and semirings. Besides these, we will consider the related concepts of star semirings and Conway semirings, and also locally finite semirings.

Next, we introduce formal power series, especially locally finite families of power series and cycle-free power series. It is a basic result that the collection of all formal power series over a given semiring and an alphabet can be endowed with addition and Cauchy multiplication yielding again the structure of a semiring, as well as with several further useful operations like the Hadamard product or the Hurwitz (shuffle) product. We prove that, under suitable assumptions, certain equalities involving the Kleene-star of elements are valid. Moreover, various important properties of the underlying semiring transfer to the semiring of formal power series. In particular, this includes properties like being ordered, complete, continuous, or Conway. We also consider morphisms between semirings of formal power series.

As is well known, the set of transitions of a classical finite automaton can be uniformly represented by matrices with entries 0 or 1. A similar representation is also easily possible for the transitions of a weighted automaton: here the matrices have entries from the underlying semiring, namely the weights of the transitions. This yields very compact representations of weighted automata and often very concise algebraic proofs about their behaviors. We prove a theorem on (infinite) matrices central for automata theory: In a complete star semiring, the blocks of the star of a matrix can be represented by applying rational operations to the blocks of the matrix. Moreover, the Kronecker (tensor) product of matrices is considered.

Finally, we consider cycle-free equations. They have a unique solution and can be used to show that two expressions represent the same formal power series. Again, we obtain results on how to compute the blocks of the star of a matrix, but now for arbitrary semirings, by imposing restrictions on the matrix.

In the literature, a number of authors have dealt with the interplay between semirings, formal power series and automata theory. The following books and surveys deal with this topic: Berstel [2], Berstel and Reutenauer [3], Bloom and Ésik [4], Carré [5], Conway [6], Eilenberg [8], Ésik and Kuich [9], Kuich [28], Kuich and Salomaa [29], Sakarovitch [37], Salomaa and Soittola [39], Wechler [40].

Further books on semirings and formal power series are Golan [15] and Hebisch and Weinert [20]. Głazek [13] is a bibliography on semirings and formal power series.

Some ideas and formulations of this presentation originate from Kuich and Salomaa [29] and Ésik and Kuich [11].

2 Monoids and Semirings

In this section, we consider monoids and semirings. The definitions and results on monoids and semirings are mainly due to Bloom and Ésik [4], Eilenberg [8], Goldstern [16], Karner [22, 23], Krob [25, 26], Kuich [27, 28], Kuich and Salomaa [29], Manes and Arbib [31], and Sakarovitch [36]. Our notion of continuous monoids and semirings is a specialization of the continuous algebras as defined, e.g., in Guessarian [17], Goguen, Thatcher, Wagner, and Wright [14], Adámek, Nelson, and Reiterman [1].

A *monoid* consists of a non-empty set M, an associative binary operation $\cdot$ on M and a neutral element 1 such that $m \cdot 1 = 1 \cdot m = m$ for every $m \in M$. A monoid M is called *commutative* if $m_1 \cdot m_2 = m_2 \cdot m_1$ for every $m_1, m_2 \in M$. The binary operation is usually denoted by juxtaposition and often called product.

If the operation and the neutral element of M are understood, then we denote the monoid simply by M. Otherwise, we use the triple notation $\langle M, \cdot, 1\rangle$. A commutative monoid M is often denoted by $\langle M, +, 0\rangle$.

The most important type of a monoid in our considerations is the *free monoid* Σ^* generated by a nonempty set Σ. It has all the (*finite*) *words over* Σ

$$x_1 \dots x_n, \qquad \text{with } x_i \in \Sigma,\ 1 \le i \le n,\ n \ge 0,$$

as its elements, and the product $w_1 \cdot w_2$ is formed by writing the string w_2 immediately after the string w_1. The neutral element of Σ^* (the case $n = 0$), also referred to as the *empty word*, is denoted by ε.

The elements of Σ are called *letters* or *symbols*. The set Σ itself is called an *alphabet*. The *length* of a word $w = x_1 \dots x_n$, $n \ge 0$, in symbols $|w|$, is defined to be n.

A *morphism* h of a monoid M into a monoid M' is a mapping $h : M \to M'$ compatible with the neutral elements and operations in $\langle M, \cdot, 1\rangle$ and $\langle M', \circ, 1'\rangle$, i.e., $h(1) = 1'$ and $h(m_1 \cdot m_2) = h(m_1) \circ h(m_2)$ for all $m_1, m_2 \in M$.

If Σ is an alphabet and $\langle M, \cdot, 1\rangle$ is any monoid, then every mapping $h : \Sigma \to M$ can be uniquely extended to a morphism $h^\sharp : \Sigma^* \to M$ by putting $h^\sharp(\varepsilon) = 1$ and $h^\sharp(x_1 x_2 \dots x_n) = h(x_1) \cdot h(x_2) \cdot \dots \cdot h(x_n)$ for any $x_1, \dots, x_n \in \Sigma$, $n \ge 1$. Usually, $h^\sharp$ is again denoted by h.

Next, we consider monoids with particular properties, like carrying an order or having an infinite sum operation. For our purposes, it suffices to consider commutative monoids. A commutative monoid $\langle M, +, 0\rangle$ is called *idempotent*, if $m + m = m$ for all $m \in M$, and it is called *ordered* if it is equipped with a partial order $\leq$ preserved by the $+$ operation. An ordered monoid M is *positively ordered*, if $m \geq 0$ for each $m \in M$. A commutative monoid $\langle M, +, 0\rangle$ is called *naturally ordered* if the relation $\sqsubseteq$ defined by: $m_1 \sqsubseteq m_2$ if there exists an m such that $m_1 + m = m_2$, is a partial order. Clearly, this is the case, i.e., $\sqsubseteq$ is antisymmetric, iff whenever $m, m', m'' \in M$ with $m + m' + m'' = m$, then $m + m' = m$. Then in particular M is positively ordered. We note that if $\langle M, +, 0\rangle$ is idempotent, then M is naturally ordered and for any $m_1, m_2 \in M$ we have $m_1 + m_2 = \mathsf{sup}\{m_1, m_2\}$ in $\langle M, \sqsubseteq\rangle$. Further, $m_1 \sqsubseteq m_2$ iff $m_1 + m_2 = m_2$. Morphisms of ordered monoids are monoid morphisms which preserve the order.

If I is an index set, an infinitary sum operation $\sum_I : M^I \to M$ associates with every family $(m_i \mid i \in I)$ of elements of M an element $\sum_{i \in I} m_i$ of M. A monoid $\langle M, +, 0\rangle$ is called *complete* if it has infinitary sum operations $\sum_I$ (for any index set I) such that the following conditions are satisfied:

(i) $\sum_{i \in \emptyset} m_i = 0$, $\sum_{i \in \{j\}} m_i = m_j$, $\sum_{i \in \{j,k\}} m_i = m_j + m_k$, for $j \neq k$.
(ii) $\sum_{j \in J}(\sum_{i \in I_j} m_i) = \sum_{i \in I} m_i$, if $\bigcup_{j \in J} I_j = I$ and $I_j \cap I_{j'} = \emptyset$ for $j \neq j'$.

A morphism of complete monoids is a monoid morphism preserving all sums. Note that any complete monoid is commutative.

Recall that a non-empty subset D of a partially ordered set P is called *directed* if each pair of elements of D has an upper bound in D.

A positively ordered commutative monoid $\langle M, +, 0\rangle$ is called a *continuous monoid* if each directed subset of M has a least upper bound and the $+$ operation preserves the least upper bound of directed sets, i.e., when

$$m + \mathsf{sup}\, D = \mathsf{sup}(m + D),$$

for all directed sets $D \subseteq M$ and for all $m \in M$. Here, $m + D$ is the set $\{m + d \mid d \in D\}$.

It is known that a positively ordered commutative monoid M is continuous iff each chain in M has a least upper bound and the $+$ operation preserves least upper bounds of chains, i.e., when $m + \mathsf{sup}\, C = \mathsf{sup}(m + C)$ holds for all non-empty chains C in M. (See Markowsky [32].)

Proposition 2.1. *Any continuous monoid $\langle M, +, 0\rangle$ is a complete monoid equipped with the following sum operation:*

$$\sum_{i \in I} m_i = \mathsf{sup}\left\{\sum_{i \in F} m_i \mid F \subseteq I,\ F \textit{ finite}\right\},$$

for all index sets I and all families $(m_i \mid i \in I)$ in M.

A function $f : P \to Q$ between partially ordered sets is *continuous* if it preserves the least upper bound of any directed set, i.e., when $f(\sup D) = \sup f(D)$, for all directed sets $D \subseteq P$ such that $\sup D$ exists. It follows that any continuous function preserves the order. A morphism of continuous monoids is defined to be a monoid morphism which is a continuous function. Clearly, any morphism between continuous monoids is a complete monoid morphism.

A *semiring* is a set S together with two binary operations $+$ and $\cdot$ and two constant elements 0 and 1 such that:

(i) $\langle S, +, 0\rangle$ is a commutative monoid,
(ii) $\langle S, \cdot, 1\rangle$ is a monoid,
(iii) the distributivity laws $(a+b)\cdot c = a\cdot c + b\cdot c$ and $c\cdot(a+b) = c\cdot a + c\cdot b$ hold for every a, b, $c \in S$,
(iv) $0\cdot a = a\cdot 0 = 0$ for every $a \in S$.

A semiring S is called *commutative* if $a\cdot b = b\cdot a$ for every $a, b \in S$. Further, S is called *idempotent* if $\langle S, +, 0\rangle$ is an idempotent monoid. By the distributivity law, this holds iff $1+1=1$.

If the operations and the constant elements of S are understood, then we denote the semiring simply by S. Otherwise, we use the notation $\langle S, +, \cdot, 0, 1\rangle$. In the sequel, S will denote a semiring.

Intuitively, a semiring is a ring (with unity) without subtraction. A typical example is the semiring of nonnegative integers $\mathbb{N}$. A very important semiring in connection with language theory is the *Boolean* semiring $\mathbb{B} = \{0, 1\}$ where $1+1 = 1\cdot 1 = 1$. Clearly, all rings (with unity), as well as all fields, are semirings, e.g., the integers $\mathbb{Z}$, rationals $\mathbb{Q}$, reals $\mathbb{R}$, complex numbers $\mathbb{C}$, etc.

Let $\mathbb{N}^\infty = \mathbb{N} \cup \{\infty\}$ and $\overline{\mathbb{N}} = \mathbb{N} \cup \{-\infty, \infty\}$. Then $\langle \mathbb{N}^\infty, +, \cdot, 0, 1\rangle$, $\langle \mathbb{N}^\infty, \min, +, \infty, 0\rangle$ and $\langle \overline{\mathbb{N}}, \max, +, -\infty, 0\rangle$, where $+$, $\cdot$, $\min$ and $\max$ are defined in the obvious fashion (observe that $0\cdot\infty = \infty\cdot 0 = 0$ and $(-\infty)+\infty = -\infty$), are semirings.

Let $\mathbb{R}_+ = \{a \in \mathbb{R} \mid a \geq 0\}$, $\mathbb{R}_+^\infty = \mathbb{R}_+ \cup \{\infty\}$ and $\overline{\mathbb{R}}_+ = \mathbb{R}_+ \cup \{-\infty, \infty\}$. Then $\langle \mathbb{R}_+, +, \cdot, 0, 1\rangle$, $\langle \mathbb{R}_+^\infty, +, \cdot, 0, 1\rangle$ and $\langle \mathbb{R}_+^\infty, \min, +, \infty, 0\rangle$ are semirings. The semirings $\langle \mathbb{N}_+^\infty, \min, +, \infty, 0\rangle$ and $\langle \mathbb{R}_+^\infty, \min, +, \infty, 0\rangle$ are called *tropical semirings* or *min-plus semirings.* Similarly, the semirings $\langle \overline{\mathbb{N}}, \max, +, -\infty, 0\rangle$ and $\langle \overline{\mathbb{R}}_+, \max, +, -\infty, 0\rangle$ are called *max-plus semirings* or *arctic semirings.* A further example is provided by the semiring $\langle [0,1], \max, \cdot, 0, 1\rangle$, called the *Viterbi semiring* in probabilistic parsing.

We note that the tropical and the arctic semirings are very often employed in optimization problems of networks, cf., e.g., Heidergott, Olsder, and van der Woude [21].

Let Σ be a finite alphabet. Then each subset of Σ^* is called a *formal language over* Σ. We define, for formal languages $L_1, L_2 \subseteq \Sigma^*$, the *product* of L_1 and L_2 by

$$L_1 \cdot L_2 = \{w_1 w_2 \mid w_1 \in L_1,\ w_2 \in L_2\}.$$

Then $\langle 2^{\Sigma^*}, \cup, \cdot, \emptyset, \{\varepsilon\}\rangle$ is a semiring, called the *semiring of formal languages over* Σ. Here, 2^U denotes the power set of a set U and $\emptyset$ denotes the empty set.

If U is a set, $2^{U\times U}$ is the set of binary relations over U. Define, for two relations R_1 and R_2, the product $R_1 \cdot R_2 \subseteq U \times U$ by

$$R_1 \cdot R_2 = \{(u_1, u_2) \mid \text{there exists } u \in U \text{ such that } (u_1, u) \in R_1 \text{ and } (u, u_2) \in R_2\}$$

and furthermore, define

$$\Delta = \{(u, u) \mid u \in U\}.$$

Then $\langle 2^{U\times U}, \cup, \cdot, \emptyset, \Delta\rangle$ is a semiring, called the *semiring of binary relations over* U.

Further semirings are the chain of nonnegative reals $\langle \mathbb{R}_+^\infty, \mathsf{max}, \mathsf{min}, 0, \infty\rangle$ and any Boolean algebra, in particular the power set Boolean algebras $\langle 2^U, \cup, \cap, \emptyset, U\rangle$ where U is any set. These examples can be generalized as follows. Recall that a partially ordered set $\langle L, \leq\rangle$ is a *lattice* if for any two elements $a, b \in L$, the least upper bound $a \vee b = \mathsf{sup}\{a, b\}$ and the greatest lower bound $a \wedge b = \mathsf{inf}\{a, b\}$ exist in $\langle L, \leq\rangle$. A lattice $\langle L, \leq\rangle$ is *distributive*, if $a \wedge (b \vee c) = (a \wedge b) \vee (a \wedge c)$ for all $a, b, c \in L$; and *bounded*, if L contains a smallest element, denoted 0, and a greatest element, denoted 1. Now, let $\langle L, \leq\rangle$ be any bounded distributive lattice. Then $\langle L, \vee, \wedge, 0, 1\rangle$ is a semiring. Since any distributive lattice L also satisfies the dual law $a\vee(b\wedge c) = (a\vee b)\wedge(a\vee c)$ for all $a, b, c \in L$, the structure $\langle L, \wedge, \vee, 1, 0\rangle$ is also a semiring. Such semirings are often used for fuzzy automata; see Chap. 12 [35] of this book. Another semiring is the *Łukasiewicz semiring* $\langle [0, 1], \mathsf{max}, \otimes, 0, 1\rangle$ where $a \otimes b = \mathsf{max}\{0, a+b-1\}$ which occurs in multivalued logic (see Hájek [18]).

Recall that in formal language theory, the Kleene-iteration L^* of a language $L \subseteq \Sigma^*$ is defined by $L^* = \bigcup_{n\geq 0} L^n$. Later on, we wish to extend this star operation to formal power series (i.e., functions) $r : \Sigma^* \to S$ where S is a semiring. For this, it will be useful to know which semirings carry such a star operation like the semiring of formal languages. We will call a *star semiring* any semiring equipped with an additional unary operation *. The following semirings are star semirings:

(i) The Boolean semiring $\langle \mathbb{B}, +, \cdot, ^*, 0, 1\rangle$ with $0^* = 1^* = 1$.
(ii) The semiring $\langle \mathbb{N}^\infty, +, \cdot, ^*, 0, 1\rangle$ with $0^* = 1$ and $a^* = \infty$ for $a \neq 0$.
(iii) The semiring $\langle \mathbb{R}_+^\infty, +, \cdot, ^*, 0, 1\rangle$ with $a^* = 1/(1-a)$ for $0 \leq a < 1$ and $a^* = \infty$ for $a \geq 1$.
(iv) The tropical semirings $\langle \mathbb{R}_+^\infty, \mathsf{min}, +, ^*, \infty, 0\rangle$ and $\langle \mathbb{N}^\infty, \mathsf{min}, +, ^*, \infty, 0\rangle$ with $a^* = 0$ for all $a \in \mathbb{R}_+^\infty$ resp. all $a \in \mathbb{N}^\infty$.
(v) The arctic semirings $\langle \overline{\mathbb{R}}_+, \mathsf{max}, +, ^*, -\infty, 0\rangle$ and $\langle \overline{\mathbb{N}}, \mathsf{max}, +, ^*, -\infty, 0\rangle$ with $(-\infty)^* = 0^* = 0$ and $a^* = \infty$ for $a > 0$.
(vi) The semiring $\langle 2^{\Sigma^*}, \cup, \cdot, ^*, \emptyset, \{\varepsilon\}\rangle$ of formal languages over a finite alphabet Σ, as noted before, with $L^* = \bigcup_{n\geq 0} L^n$ for all $L \subseteq \Sigma^*$.

(vii) The semiring $\langle 2^{U\times U}, \cup, \cdot, ^*, \emptyset, \Delta\rangle$ of binary relations over U with $R^* = \bigcup_{n\geq 0} R^n$ for all $R \subseteq U \times U$. The relation R^* is called the *reflexive and transitive closure* of R, i.e., the smallest reflexive and transitive binary relation over S containing R.

(viii) The Łukasiewicz semiring $\langle [0,1], \mathsf{max}, \otimes, ^*, 0, 1\rangle$ with $a^* = 1$ for all $a \in [0,1]$.

(ix) The idempotent naturally ordered commutative semiring $\langle \{0, 1, a, \infty\}, +, \cdot, ^*, 0, 1\rangle$, with $0 \sqsubseteq 1 \sqsubseteq a \sqsubseteq \infty$, $a \cdot a = a$, $0^* = 1^* = 1$, $a^* = \infty^* = \infty$.

(x) The bounded distributive lattice semiring $\langle L, \vee, \wedge, ^*, 0, 1\rangle$ with $a^* = 1$ for all $a \in L$.

The semirings (i)–(v) and (viii)–(x) are commutative. The semirings (i), (iv)–(x) are idempotent.

A semiring $\langle S, +, \cdot, 0, 1\rangle$ is called *ordered* if $\langle S, +, 0\rangle$ is an ordered monoid and multiplication with elements $s \geq 0$ preserves the order; it is *positively ordered*, if furthermore, $\langle S, +, 0\rangle$ is positively ordered. When the order on S is the natural order, $\langle S, +, \cdot, 0, 1\rangle$ is automatically a positively ordered semiring.

A semiring $\langle S, +, \cdot, 0, 1\rangle$ is called *complete* if $\langle S, +, 0\rangle$ is a complete monoid and the following distributivity laws are satisfied (see Bloom and Ésik [4], Conway [6], Eilenberg [8], Kuich [28]):

$$\sum_{i\in I}(a \cdot a_i) = a \cdot \left(\sum_{i\in I} a_i\right), \qquad \sum_{i\in I}(a_i \cdot a) = \left(\sum_{i\in I} a_i\right) \cdot a.$$

This means that a semiring S is complete if it is possible to define "infinite sums" (i) that are an extension of the finite sums, (ii) that are associative and commutative and (iii) that satisfy the distributivity laws.

In complete semirings for each element a, we can define the *star* a^* of a by

$$a^* = \sum_{j\geq 0} a^j,$$

where $a^0 = 1$ and $a^{j+1} = a \cdot a^j = a^j \cdot a$ for $j \geq 0$. Hence, with this star operation, each complete semiring is a star semiring called a *complete star semiring*. The semirings (i)–(viii) are complete star semirings. The semiring (ix) is complete, but it violates the above equation for the element a, hence it is not a complete star semiring. The distributive lattice semiring L satisfies $a^* = \bigvee_{j\geq 0} a^j$ for each $a \in L$, but is not necessarily complete. It is a complete semiring iff $(L, \vee, \wedge)$ is a join-continuous complete lattice, i.e., any subset of L has a supremum in L and $a \wedge \bigvee_{i\in I} a_i = \bigvee_{i\in I}(a \wedge a_i)$ for any subset $\{a_i \mid i \in I\}$ of L.

A semiring $\langle S, +, \cdot, 0, 1\rangle$ is called *continuous* if $\langle S, +, 0\rangle$ is a continuous monoid and if multiplication is continuous, i.e.,

$$a \cdot (\mathsf{sup}_{i\in I}\, a_i) = \mathsf{sup}_{i\in I}(a \cdot a_i) \quad \text{and} \quad (\mathsf{sup}_{i\in I}\, a_i) \cdot a = \mathsf{sup}_{i\in I}(a_i \cdot a)$$

for all directed sets $\{a_i \mid i \in I\}$ and $a \in S$ (see Bloom and Ésik [4]). It follows that the distributivity laws hold for infinite sums:

$$a \cdot \left(\sum_{i\in I} a_i\right) = \sum_{i\in I}(a \cdot a_i) \quad \text{and} \quad \left(\sum_{i\in I} a_i\right) \cdot a = \sum_{i\in I}(a_i \cdot a)$$

for all families $(a_i \mid i \in I)$.

Proposition 2.2. *Any continuous semiring is complete.*

All the semirings in (i)–(ix) are continuous.

We now consider two equations that are important in automata theory. Let S be a star semiring. Then for $a, b \in S$:

(i) The *sum star identity* is valid for a and b if $(a+b)^* = (a^*b)^*a^*$.
(ii) The *product star identity* is valid for a and b if $(ab)^* = 1 + a(ba)^*b$.

If the sum star identity (resp. the product star identity) is valid for all $a, b \in S$, then we say that the *sum star identity* (resp. the *product star identity*) is *valid* (in the star semiring S).

A *Conway semiring* is now a star semiring in which the sum star identity and the product star identity are valid (see Conway [6], Bloom and Ésik [4]). All the star semirings in (i)–(x) are Conway semirings. The semiring $\langle \mathbb{Q}_+^\infty, +, \cdot, ^*, 0, 1\rangle$, with $\mathbb{Q}_+^\infty = \mathbb{R}_+^\infty \cap (\mathbb{Q} \cup \{\infty\})$ and operations defined as in (iii), is a Conway semiring (since the sum star identity and product star identity hold in $\mathbb{R}_+^\infty$) but is not complete. Now we have the following proposition.

Proposition 2.3. *Let S be a star semiring. Then S is a Conway semiring iff, for all $a, b \in S$:*

(i) $(a+b)^* = (a^*b)^*a^*$.
(ii) $(ab)^*a = a(ba)^*$.
(iii) $a^* = 1 + aa^* = 1 + a^*a$.

Proof. If S is a Conway semiring, we obtain (iii) from the product star identity with $b = 1$, resp. $a = 1$. Then (ii) follows from the product star identity, distributivity, and (iii). Conversely, for the product star identity compute $(ab)^*$ by using (iii) and then (ii). □

Next we introduce conditions which often simplify the definition or the calculation of the star of elements. A semiring S is *k-closed*, where $k \geq 0$, if for each $a \in S$,

$$1 + a + \cdots + a^k = 1 + a + \cdots + a^k + a^{k+1}.$$

It is called *locally closed*, if for each $a \in S$, there is an integer $k \geq 0$ such that the above equality is valid. (See Carré [5], Mohri [33], Ésik and Kuich [10], Zhao [41], Zimmermann [42].) If $\langle S, +, \cdot, 0, 1\rangle$ is a k-closed semiring, then define the star of $a \in S$ by

$$a^* = 1 + a + \cdots + a^k.$$

An analogous equality defines the star in a locally closed semiring. With this star operation, each k-closed (resp. locally closed) semiring is a star semiring called a *k-closed* (resp. *locally closed*) *star semiring.* The semirings (i), (iv), (viii), and (x) are 0-closed star semirings; the semiring (ix) is a 1-closed semiring, but not a 1-closed star semiring. In [10, 41], the following was shown.

Theorem 2.4. *Any locally closed star semiring is a Conway semiring.*

Next we consider morphisms between semirings. Let S and S' be semirings. Then a mapping $h : S \to S'$ is a *morphism* from S into S' if $h(0) = 0$, $h(1) = 1$, $h(a + b) = h(a) + h(b)$ and $h(a \cdot b) = h(a) \cdot h(b)$ for all $a, b \in S$. That is, a morphism of semirings is a mapping that preserves the semiring operations and constants. A bijective morphism is called an *isomorphism.* For instance, the semirings $\langle \mathbb{R}_+^\infty, \mathsf{min}, +, \infty, 0\rangle$ and $\langle [0,1], \mathsf{max}, \cdot, 0, 1\rangle$ are isomorphic via the mapping $x \mapsto e^{-x}$, and the semiring $\langle \mathbb{R}_+^\infty, \mathsf{max}, \mathsf{min}, 0, \infty\rangle$ is isomorphic to $\langle [0,1], \mathsf{max}, \mathsf{min}, 0, 1\rangle$ via the mapping $x \mapsto 1 - e^{-x}$. A morphism h of star semirings is a semiring morphism that preserves additionally the star operation, i.e., $h(a^*) = h(a)^*$ for all $a \in S$. Similarly, a morphism of ordered (resp. complete, continuous) semirings is a semiring morphism that preserves the order (resp. all sums, all suprema of directed subsets). Note that every continuous semiring is an ordered semiring and every continuous semiring morphism is an ordered semiring morphism.

Complete and continuous semirings are typically infinite. For results on weighted automata, sometimes it is assumed that the underlying semiring is finite or "close" to being finite. A large class of such semirings can be obtained by the notion of local finiteness (which stems from group theory where it is well known).

A semiring S is *locally finite* (see Wechler [40], Droste and Gastin [7]) if each finitely generated subsemiring is finite. We note that a semiring $\langle S, +, \cdot, 0, 1\rangle$ is locally finite iff both monoids $\langle S, +, 0\rangle$ and $\langle S, \cdot, 1\rangle$ are locally finite. Indeed, if $\langle S, +, 0\rangle$ and $\langle S, \cdot, 1\rangle$ are locally finite and U is a finite subset of S, then the submonoid V of $\langle S, \cdot, 1\rangle$ generated by U is finite and the submonoid W of $\langle S, +, 0\rangle$ generated by V is also finite. Now, it is easy to check that $W \cdot W \subseteq W$ and we deduce that the subsemiring of $\langle S, +, \cdot, 0, 1\rangle$ generated by U is the finite set W.

For instance, if both sum and product are commutative and idempotent, then the semiring is locally finite. Consequently, any bounded distributive lattice $\langle L, \vee, \wedge, 0, 1\rangle$ is a locally finite semiring. In particular, the chain $\langle [0,1], \mathsf{max}, \mathsf{min}, 0, 1\rangle$ and any Boolean algebra are locally finite. Further, the Łukasiewicz semiring $\langle [0,1], \mathsf{max}, \otimes, 0, 1\rangle$ is locally finite, since its additive and multiplicative monoid are commutative and locally finite. Moreover, each positively ordered locally finite semiring is locally closed, and each positively ordered finite semiring is k-closed where k is less than the number of elements of the semiring.

Examples of infinite but locally finite fields are provided by the algebraic closures of the finite fields $\mathbb{Z}/p\mathbb{Z}$ for any prime p.

3 Formal Power Series

In this section, we define and investigate formal power series (for expositions, see Salomaa and Soittola [39], Kuich and Salomaa [29], Berstel and Reutenauer [3], Sakarovitch [37]). Let Σ be an alphabet and S a semiring. Mappings r from Σ^* into S are called *(formal) power series*. The values of r are denoted by (r, w), where $w \in \Sigma^*$, and r itself is written as a formal sum

$$r = \sum_{w \in \Sigma^*} (r, w)w.$$

The values (r, w) are also referred to as the *coefficients* of the series. The collection of all power series r as defined above is denoted by $S\langle\langle \Sigma^* \rangle\rangle$.

This terminology reflects the intuitive ideas connected with power series. We call the power series "formal" to indicate that we are not interested in summing up the series but rather, for instance, in various operations defined for series.

Given $r \in S\langle\langle \Sigma^* \rangle\rangle$, the *support* of r is the set

$$\text{supp}(r) = \{w \in \Sigma^* \mid (r, w) \neq 0\}.$$

A series $r \in S\langle\langle \Sigma^* \rangle\rangle$ where every coefficient equals 0 or 1 is termed the *characteristic series* of its support L, in symbols, $r = \text{char}(L)$ or $r = \mathbb{1}_L$. The subset of $S\langle\langle \Sigma^* \rangle\rangle$ consisting of all series with a finite support is denoted by $S\langle \Sigma^* \rangle$. Series of $S\langle \Sigma^* \rangle$ are referred to as *polynomials*. It will be convenient to use the notations $S\langle \Sigma \cup \{\varepsilon\} \rangle$, $S\langle \Sigma \rangle$ and $S\langle \{\varepsilon\} \rangle$ for the collection of polynomials having their supports in $\Sigma \cup \{\varepsilon\}$, Σ and $\{\varepsilon\}$, respectively.

Examples of polynomials belonging to $S\langle \Sigma^* \rangle$ are 0 and aw, where $a \in S$ and $w \in \Sigma^*$, defined by:

$$(0, w) = 0 \quad \text{for all } w,$$
$$(aw, w) = a \quad \text{and} \quad (aw, w') = 0 \quad \text{for } w \neq w'.$$

Often, $1w$ is denoted by w or $\mathbb{1}_{\{w\}}$.

Next, we introduce several operations on power series. For $r_1, r_2, r \in S\langle\langle \Sigma^* \rangle\rangle$ and $a \in S$, we define the *sum* $r_1 + r_2$, the *(Cauchy) product* $r_1 \cdot r_2$, the *Hadamard product* $r_1 \odot r_2$, and *scalar products* ar, ra, each as a series belonging to $S\langle\langle \Sigma^* \rangle\rangle$, as follows:

- $(r_1 + r_2, w) = (r_1, w) + (r_2, w)$
- $(r_1 \cdot r_2, w) = \sum_{w_1 w_2 = w} (r_1, w_1)(r_2, w_2)$
- $(r_1 \odot r_2, w) = (r_1, w)(r_2, w)$
- $(ar, w) = a(r, w)$
- $(ra, w) = (r, w)a$

for all $w \in \Sigma^*$.

It can be checked that $\langle S\langle\langle \Sigma^* \rangle\rangle, +, \cdot, 0, \varepsilon \rangle$ and $\langle S\langle \Sigma^* \rangle, +, \cdot, 0, \varepsilon \rangle$ are semirings, the semirings of formal power series resp. of polynomials over Σ and S.

We just note that the structure $\langle S\langle\langle\Sigma^*\rangle\rangle, +, \odot, 0, \text{char}(\Sigma^*)\rangle$ is also a semiring (the full Cartesian product of Σ^* copies of the semiring $\langle S, +, \cdot, 0, 1\rangle$).

Clearly, the formal language semiring $\langle 2^{\Sigma^*}, \cup, \cdot, \emptyset, \{\varepsilon\}\rangle$ is isomorphic to $\langle \mathbb{B}\langle\langle\Sigma^*\rangle\rangle, +, \cdot, 0, \varepsilon\rangle$. Essentially, a transition from 2^{Σ^*} to $\mathbb{B}\langle\langle\Sigma^*\rangle\rangle$ and vice versa means a transition from L to $\text{char}(L)$ and from r to $\text{supp}(r)$, respectively. Furthermore, the operation corresponding to the Hadamard product is the intersection of languages. If r_1 and r_2 are the characteristic series of the languages L_1 and L_2, then $r_1 \odot r_2$ is the characteristic series of $L_1 \cap L_2$.

This basic transition between 2^{Σ^*} and $\mathbb{B}\langle\langle\Sigma^*\rangle\rangle$ will be very important in all of the following as it often gives a hint how to generalize classical results from formal language theory into the realm of formal power series (with an arbitrary or suitable semiring S replacing $\mathbb{B}$).

Let $r_i \in S\langle\langle\Sigma^*\rangle\rangle$ $(i \in I)$, where I is an arbitrary index set. Then for $w \in \Sigma^*$ let $I_w = \{i \mid (r_i, w) \neq 0\}$. Assume now that for all $w \in \Sigma^*$, I_w is finite. Then we call the family of power series $\{r_i \mid i \in I\}$ *locally finite*. In this case, we can define the sum $\sum_{i\in I} r_i$ by

$$\left(\sum_{i\in I} r_i, w\right) = \sum_{i\in I_w} (r_i, w)$$

for all $w \in \Sigma^*$. Also, in this case for each $r \in S\langle\langle\Sigma^*\rangle\rangle$, the families $\{r \cdot r_i \mid i \in I\}$ and $\{r_i \cdot r \mid i \in I\}$ are also locally finite, and $r \cdot \sum_{i\in I} r_i = \sum_{i\in I} r \cdot r_i$ and $(\sum_{i\in I} r_i) \cdot r = \sum_{i\in I} r_i \cdot r$. Indeed, let $w \in \Sigma^*$ and put $J = \bigcup_{w=uv} I_v$, a finite set. Then

$$\begin{aligned}\left(r \cdot \sum_{i\in I} r_i, w\right) &= \sum_{w=uv} (r, u)\left(\sum_{i\in J} r_i, v\right) = \sum_{w=uv}\sum_{i\in J} (r, u) \cdot (r_i, v) \\ &= \sum_{i\in J}\sum_{w=uv} (r, u) \cdot (r_i, v) = \sum_{i\in J} (r \cdot r_i, w) = \left(\sum_{i\in I} r \cdot r_i, w\right),\end{aligned}$$

as $(r \cdot r_i, w) \neq 0$ implies $i \in J$. This proves the first equation, and the second one follows similarly.

A power series $r \in S\langle\langle\Sigma^*\rangle\rangle$ is called *proper* or *quasiregular* if $(r, \varepsilon) = 0$. The *star* r^* of a proper power series $r \in S\langle\langle\Sigma^*\rangle\rangle$ is defined by

$$r^* = \sum_{n\geq 0} r^n.$$

Since r is proper, we infer $(r^n, w) = 0$ for each $n > |w|$. Hence, $\{r^n \mid n \geq 0\}$ is locally finite, $(r^*, w) = \sum_{0\leq n\leq |w|} (r^n, w)$, and the star of a proper power series is well-defined.

We generalize this result to cycle-free power series. A power series $r \in S\langle\langle\Sigma^*\rangle\rangle$ is called *cycle-free of index* $k > 0$ if $(r, \varepsilon)^k = 0$. It is called *cycle-free* if there exists a $k \geq 1$ such that r is cycle-free of index k. Again, we define the *star* of a cycle-free power series $r \in S\langle\langle \Sigma^* \rangle\rangle$ by

$$r^* = \sum_{n \geq 0} r^n.$$

Since r is cycle-free of some index $k \geq 1$, an easy proof by induction on the length of $w \in \Sigma^*$ yields $(r^n, w) = 0$ for $n \geq k \cdot (|w| + 1)$. Hence, $\{r^n \mid n \geq 0\}$ is locally finite, $(r^*, w) = \sum_{0 \leq n < k(|w|+1)} (r^n, w)$, and the star of a cycle-free power series is well-defined.

Next, we wish to consider identities that are valid for a cycle-free power series r, like, e.g., $rr^* + \varepsilon = r^* r + \varepsilon = r^*$. Using the distributivity laws given above for locally finite families, this follows from:

$$rr^* + \varepsilon = r \cdot \sum_{n \geq 0} r^n + \varepsilon = \sum_{n \geq 0} r^{n+1} + \varepsilon = r^*.$$

Theorem 3.1. *Let r be a cycle-free power series. Then, for each $n \geq 0$,*

$$r^* = r^{n+1} r^* + \sum_{0 \leq j \leq n} r^j = r^* r^{n+1} + \sum_{0 \leq j \leq n} r^j.$$

Proof. We obtain by substitutions

$$r^* = rr^* + \varepsilon = r(rr^* + \varepsilon) + \varepsilon = r^2 r^* + r + \varepsilon = \cdots .$$

The proof of the second equality is analogous. □

Theorem 3.2. *Let $r, s \in S\langle\langle \Sigma^* \rangle\rangle$ and assume that rs is cycle-free. Then sr is cycle-free and*

$$(rs)^* r = r(sr)^*.$$

Proof. Since rs is cycle-free, $((rs)^k, \varepsilon) = 0$ for some $k > 0$. Hence,

$$((sr)^{k+1}, \varepsilon) = (s, \varepsilon)((rs)^k, \varepsilon)(r, \varepsilon) = 0$$

and sr is cycle-free. It follows that $(rs)^* r = \sum_{n \geq 0}((rs)^n \cdot r) = \sum_{n \geq 0} r \cdot (sr)^n = r \cdot (sr)^*$. □

The *Hurwitz product* (also called *shuffle product*) is defined as follows. For $w_1, w_2 \in \Sigma^*$ and $x_1, x_2 \in \Sigma$, we define $w_1 \sqcup\!\sqcup\, w_2 \in S\langle\langle \Sigma^* \rangle\rangle$ by

$$w_1 \sqcup\!\sqcup\, \varepsilon = w_1, \qquad \varepsilon \sqcup\!\sqcup\, w_2 = w_2,$$

and

$$w_1 x_1 \sqcup\!\sqcup\, w_2 x_2 = (w_1 x_1 \sqcup\!\sqcup\, w_2) x_2 + (w_1 \sqcup\!\sqcup\, w_2 x_2) x_1.$$

For $r_1, r_2 \in S\langle\langle \Sigma^* \rangle\rangle$, the Hurwitz product $r_1 \sqcup\!\sqcup\, r_2 \in S\langle\langle \Sigma^* \rangle\rangle$ of r_1 and r_2 is then defined by

$$r_1 \sqcup\!\sqcup\, r_2 = \sum_{w_1, w_2 \in \Sigma^*} (r_1, w_1)(r_2, w_2)(w_1 \sqcup\!\sqcup\, w_2).$$

Observe that

$$(r_1 \sqcup\!\sqcup\, r_2, w) = \sum_{|w_1|+|w_2|=|w|} (r_1, w_1)(r_2, w_2)(w_1 \sqcup\!\sqcup\, w_2, w)$$

is a finite sum for all $w \in \Sigma^*$. Hence, $\{\sum_{w_1 w_2 = w}(r_1, w_1)(r_2, w_2) w_1 \sqcup\!\sqcup\, w_2 \mid w \in \Sigma^*\}$ is locally finite and the Hurwitz product of two power series is well-defined.

In language theory, the shuffle product is customarily defined for languages L and L' by

$$L \sqcup\!\sqcup\, L' = \{w_1 w_1' \dots w_n w_n' \mid w_1 \dots w_n \in L,\ w_1' \dots w_n' \in L',\ n \geq 1\}.$$

If $r_1, r_2 \in \mathbb{B}\langle\langle \Sigma^* \rangle\rangle$, then this definition is "isomorphic" to that given above for formal power series.

When the semiring S is ordered by $\leq$, we may order $S\langle\langle \Sigma^* \rangle\rangle$, and thus $S\langle \Sigma^* \rangle$ by the pointwise order: We define $r \leq r'$ for $r, r' \in S\langle\langle \Sigma^* \rangle\rangle$ iff $(r, w) \leq (r', w)$ for all $w \in \Sigma^*$. Equipped with this order, clearly both $S\langle\langle \Sigma^* \rangle\rangle$ and $S\langle \Sigma^* \rangle$ are ordered semirings.

If $\langle S, +, \cdot, 0, 1 \rangle$ is a complete semiring, we can define an infinitary sum operation on $S\langle\langle \Sigma^* \rangle\rangle$ as follows: If $r_i \in S\langle\langle \Sigma^* \rangle\rangle$ for $i \in I$, then $\sum_{i \in I} r_i = \sum_{w \in \Sigma^*} (\sum_{i \in I} (r_i, w)) w$. By arguing elementwise for each word $w \in \Sigma^*$, we obtain the following proposition.

Proposition 3.3. *Let S be a semiring.*

(a) If S is complete, $S\langle\langle \Sigma^* \rangle\rangle$ is also complete.
(b) If S is continuous, $S\langle\langle \Sigma^* \rangle\rangle$ is also continuous.

We just note here that an analogous result holds if S is a Conway semiring, with an appropriate definition of the star operation in $S\langle\langle \Sigma^* \rangle\rangle$; see Chap. 3, Theorem 2.8 [12] of this book.

Proposition 3.3 and the Hurwitz product are now used to prove that each complete star semiring is a Conway semiring (see Kuich [27], Hebisch [19]).

Theorem 3.4. *Each complete star semiring is a Conway semiring.*

Proof. Let S be a complete star semiring and let $a, b \in S$. Let $\bar{a}, \bar{b}$ be letters. Note that to each word $\bar{w} = \bar{c_1}\bar{c_2} \dots \bar{c_n}$, with $\bar{c_i} \in \{\bar{a}, \bar{b}\}$ for $1 \leq i \leq n$, there corresponds the element $w = c_1 c_2 \dots c_n \in S$. Let S' be the complete star semiring generated by 1. Then S' is commutative. By Proposition 3.3, $\langle S'\langle\langle \{\bar{a}, \bar{b}\} \rangle\rangle, +, \cdot, 0, \varepsilon \rangle$ is a complete semiring. Also, observe that $\bar{a} \mapsto a$, $\bar{b} \mapsto b$ induces a complete star semiring morphism from the complete star semiring $S'\langle\langle \{\bar{a}, \bar{b}\}^* \rangle\rangle$ to S.

Using induction, the following equalities can be shown for all $n, m \geq 0$:

$$(\bar{a}+\bar{b})^n = \sum_{0 \leq j \leq n} \bar{a}^j \sqcup\!\sqcup \bar{b}^{n-j},$$

$$\bar{a}^n \sqcup\!\sqcup \bar{b}^m = \sum_{0 \leq j \leq n} (\bar{a}^j \sqcup\!\sqcup \bar{b}^{m-1})\bar{b}\bar{a}^{n-j}$$

and

$$\bar{a}^* \sqcup\!\sqcup \bar{b}^n = \sum_{j \geq 0} \bar{a}^j \sqcup\!\sqcup \bar{b}^n = (\bar{a}^*\bar{b})^n \bar{a}^*.$$

Hence, we infer the equality

$$(\bar{a}+\bar{b})^* = \sum_{n \geq 0} \sum_{j \geq 0} \bar{a}^j \sqcup\!\sqcup \bar{b}^n,$$

which implies immediately

$$(\bar{a}+\bar{b})^* = (\bar{a}^*\bar{b})^*\bar{a}^*.$$

Applying the complete star semiring morphism defined above, we obtain the sum star identity in S:

$$(a+b)^* = (a^*b)^*a^*.$$

The product star identity is clear by

$$\begin{aligned}(ab)^* &= 1 + \sum_{n \geq 1} (ab)^n = 1 + a\Big(\sum_{n \geq 0} (ba)^n\Big)b \\ &= 1 + a(ba)^*b. \quad \square\end{aligned}$$

Finally, we show that morphisms between two semirings and also particular morphisms between free monoids induce morphisms between the associated semirings of formal power series.

First, let Σ be an alphabet, S, S' two semirings and $h : S \to S'$ a morphism. We define $\bar{h} : S\langle\langle \Sigma^* \rangle\rangle \to S'\langle\langle \Sigma^* \rangle\rangle$ by $\bar{h}(r) = h \circ r$ for each $r \in S\langle\langle \Sigma^* \rangle\rangle$, i.e., $(\bar{h}(r), w) = h((r, w))$ for each $w \in \Sigma^*$. Often, $\bar{h}$ is again denoted by h. The following is straightforward by elementary calculations.

Proposition 3.5. *Let Σ be an alphabet, S, S' two semirings and $h : S \to S'$ a semiring morphism. Then $h : S\langle\langle \Sigma^* \rangle\rangle \to S'\langle\langle \Sigma^* \rangle\rangle$ is again a semiring morphism. Moreover, if r is cycle-free, so is $h(r)$ and $h(r^*) = (h(r))^*$.*

Second, let S be a semiring, Σ, Σ' two alphabets and $h : \Sigma^* \to \Sigma'^*$ a morphism. We define $h^{-1} : S\langle\langle \Sigma'^* \rangle\rangle \to S\langle\langle \Sigma^* \rangle\rangle$ by $h^{-1}(r') = r' \circ h$ for each $r' \in S\langle\langle \Sigma'^* \rangle\rangle$, that is, $(h^{-1}(r'), v) = (r', h(v))$ for each $v \in \Sigma^*$. We call $h : \Sigma^* \to \Sigma'^*$ *length-preserving*, if $|v| = |h(v)|$ for each $v \in \Sigma^*$; equivalently,

$h(x) \in \Sigma'$ for each $x \in \Sigma$. Further, h is *non-deleting*, if $h(x) \neq \varepsilon$ for each $x \in \Sigma$; equivalently, $|v| \leq |h(v)|$ for each $v \in \Sigma^*$. If h is non-deleting or if S is complete, we define $\bar{h} : S\langle\langle \Sigma^* \rangle\rangle \to S\langle\langle \Sigma'^* \rangle\rangle$ by letting $(\bar{h}(r), w) = \sum_{v \in \Sigma^*, h(v)=w} (r, v)$ for each $r \in S\langle\langle \Sigma^* \rangle\rangle$ and $w \in \Sigma'^*$. Observe that if h is non-deleting, $h^{-1}(w)$ is a finite set for each $w \in \Sigma^*$, and hence $\bar{h}(r)$ is well defined.

Proposition 3.6. *Let S be a semiring, Σ, Σ' two alphabets and $h : \Sigma^* \to \Sigma'^*$ a morphism.*

(i) Let h be length-preserving. Then the mapping $h^{-1} : S\langle\langle \Sigma'^* \rangle\rangle \to S\langle\langle \Sigma^* \rangle\rangle$ is a semiring morphism. Moreover, if $r' \in S\langle\langle \Sigma'^* \rangle\rangle$ is cycle-free, then so is $h^{-1}(r')$, and $h^{-1}(r'^*) = (h^{-1}(r'))^*$.
(ii) Let h be nondeleting, or assume that S is complete. Then $\bar{h} : S\langle\langle \Sigma^* \rangle\rangle \to S\langle\langle \Sigma'^* \rangle\rangle$ is a semiring morphism. Moreover, if h is non-deleting and $r \in S\langle\langle \Sigma^* \rangle\rangle$ is cycle-free, then so is $\bar{h}(r)$, and $\bar{h}(r^*) = (\bar{h}(r))^*$.

Proof. This can be shown again by elementary calculations. For (i), note that if $v \in \Sigma^*$ and $h(v) = w_1 w_2$ with $w_1, w_2 \in \Sigma'^*$, then since h is length-preserving, there are $v_1, v_2 \in \Sigma^*$ with $v = v_1 v_2$ and $h(v_1) = w_1, h(v_2) = w_2$. This implies that h^{-1} preserves the Cauchy product. □

4 Matrices

In this section, we introduce and investigate (possibly infinite) matrices. These are important here since the structure and the behavior of weighted automata can often be compactly described using matrices (see Chaps. 3, 4, and 7 of this book [12, 38, 34]), and hence results from matrix algebra can be used to derive results for weighted automata.

Consider two nonempty index sets I and I' and a set U. A mapping $A : I \times I' \to U$ is called a *matrix*. The values of A are denoted by $A_{i,i'}$, where $i \in I$ and $i' \in I'$. The values $A_{i,i'}$ are also referred to as the *entries* of the matrix A. In particular, $A_{i,i'}$ is called the *(i, i')-entry* of A. The collection of all matrices as defined above is denoted by $U^{I \times I'}$.

If both I and I' are finite, then A is called a *finite matrix*. If I or I' is a singleton, then $A^{I \times I'}$ is denoted by $A^{1 \times I'}$ or $A^{I \times 1}$, and A is called a *row* or *column vector*, respectively. If $A \in U^{I \times 1}$ (resp. $A \in U^{1 \times I'}$), then we often denote the ith entry of A for $i \in I$ (resp. $i \in I'$), by A_i instead of $A_{i,1}$ (resp. $A_{1,i}$). If $I = \{1, \ldots, m\}$ and $I' = \{1, \ldots, n\}$, the set $U^{I \times I'}$ is denoted by $U^{m \times n}$.

As before, we introduce some operations and special matrices inducing a monoid or semiring structure to matrices. Let S be a semiring. For $A, B \in S^{I \times I'}$, we define the *sum* $A + B \in S^{I \times I'}$ by $(A + B)_{i,i'} = A_{i,i'} + B_{i,i'}$ for all $i \in I$, $i' \in I'$. Furthermore, we introduce the *zero matrix* $0 \in S^{I \times I'}$. All entries of the zero matrix 0 are 0. By these definitions, $\langle S^{I \times I'}, +, 0 \rangle$ is a commutative monoid.

Let $A \in S^{I \times I'}$. Consider, for $i \in I$, the set of indices $\{j \mid A_{ij} \neq 0\}$. Then A is called a *row finite matrix* if these sets are finite for all $i \in I$. Similarly, consider, for $i' \in I'$, the set of indices $\{j \mid A_{ji'} \neq 0\}$. Then A is called a *column finite matrix* if these sets are finite for all $i' \in I'$.

If A is row finite or B is column finite, or if S is complete, then for $A \in S^{I_1 \times I_2}$ and $B \in S^{I_2 \times I_3}$, we define the *product* $AB \in S^{I_1 \times I_3}$ by

$$(AB)_{i_1,i_3} = \sum_{i_2 \in I_2} A_{i_1,i_2} B_{i_2,i_3} \quad \text{for all } i_1 \in I_1,\ i_3 \in I_3.$$

Furthermore, we introduce the *matrix of unity* $E \in S^{I \times I}$. The diagonal entries $E_{i,i}$ of E are equal to 1, the off-diagonal entries E_{i_1,i_2} $(i_1 \neq i_2)$ of E are equal to 0, for $i, i_1, i_2 \in I$.

It is easily shown that matrix multiplication is associative, the distributivity laws are valid for matrix addition and multiplication, E is a multiplicative unit, and 0 is a multiplicative zero. So, we infer that $\langle S^{I \times I}, +, \cdot, 0, E \rangle$ is a semiring if I is finite or if S is complete. Moreover, the row finite matrices in $S^{I \times I}$ and the column finite matrices in $S^{I \times I}$ form semirings.

If S is complete, infinite sums can be extended to matrices. Consider $S^{I \times I'}$ and define, for $A_j \in S^{I \times I'}$, $j \in J$, where J is an index set, $\sum_{j \in J} A_j$ by its entries:

$$\left(\sum_{j \in J} A_j \right)_{i,i'} = \sum_{j \in J} (A_j)_{i,i'}, \quad \text{for all } i \in I,\ i' \in I'.$$

By this definition, $S^{I \times I}$ is a complete semiring.

If S is ordered, the order on S is extended pointwise to matrices A and B in $S^{I \times I'}$:

$$A \leq B \quad \text{if} \quad A_{i,i'} \leq B_{i,i'} \text{ for all } i \in I,\ i' \in I'.$$

If S is continuous, then so is $S^{I \times I}$.

Eventually, if S is a locally closed star semiring, then $S^{n \times n}$, $n \geq 1$, is again a locally closed star semiring (see Ésik and Kuich [10], Zhao [41]); and if S is a Conway semiring, then $S^{n \times n}$, $n \geq 1$, is again a Conway semiring (see Conway [6], Bloom and Ésik [4], Ésik and Kuich [9]). Clearly, if S is locally finite, then so is $S^{n \times n}$ for each $n \geq 1$ (cf. [7]).

For the remainder of this chapter, I (resp. Q), possibly provided with indices, denotes an *arbitrary* (resp. *finite*) *index set*. For the rest of this section, we assume that all products of matrices are well-defined.

We now introduce blocks of matrices. Consider a matrix A in $S^{I \times I}$. Assume that we have a decomposition $I = \bigcup_{j \in J} I_j$ where J and all I_j $(j \in J)$ are non-empty index sets such that $I_{j_1} \cap I_{j_2} = \emptyset$ for $j_1 \neq j_2$. The mapping A, restricted to the domain $I_{j_1} \times I_{j_2}$, i.e., $A \restriction_{I_{j_1} \times I_{j_2}} \colon I_{j_1} \times I_{j_2} \to S$ is, of course, a matrix in $S^{I_{j_1} \times I_{j_2}}$. We denote it by $A(I_{j_1}, I_{j_2})$ and call it the *(I_{j_1}, I_{j_2})-block* of A.

We can compute the blocks of the sum and the product of matrices A and B from the blocks of A and B in the usual way:

$$(A+B)(I_{j_1}, I_{j_2}) = A(I_{j_1}, I_{j_2}) + B(I_{j_1}, I_{j_2}),$$
$$(AB)(I_{j_1}, I_{j_2}) = \sum_{j \in J} A(I_{j_1}, I_j) B(I_j, I_{j_2}).$$

In a similar manner, the matrices of $S^{I \times I'}$ can be partitioned into blocks. This yields the computational rule

$$(A+B)(I_j, I'_{j'}) = A(I_j, I'_{j'}) + B(I_j, I'_{j'}).$$

If we consider matrices $A \in S^{I \times I'}$ and $B \in S^{I' \times I''}$ partitioned into compatible blocks, i.e., I' is partitioned into the same index sets for both matrices, then we obtain the computational rule

$$(AB)(I_j, I''_{j''}) = \sum_{j' \in J'} A(I_j, I'_{j'}) B(I'_{j'}, I''_{j''}).$$

Now let us assume that I and I' are finite, or that S is complete. In the sequel, the following isomorphisms are needed:

(i) The semirings

$$\left(S^{I' \times I'}\right)^{I \times I}, \; S^{(I \times I') \times (I \times I')}, \; S^{(I' \times I) \times (I' \times I)}, \; \left(S^{I \times I}\right)^{I' \times I'}$$

are isomorphic by the correspondences between

$$(A_{i_1, i_2})_{i'_1, i'_2}, \; A_{(i_1, i'_1),(i_2, i'_2)}, \; A_{(i'_1, i_1),(i'_2, i_2)}, \; (A_{i'_1, i'_2})_{i_1, i_2}$$

for all $i_1, i_2 \in I$, $i'_1, i'_2 \in I'$.

(ii) The semirings $S^{I \times I}\langle\langle \Sigma^* \rangle\rangle$ and $(S\langle\langle \Sigma^* \rangle\rangle)^{I \times I}$ are isomorphic by the correspondence between $(A, w)_{i_1, i_2}$ and (A_{i_1, i_2}, w) for all $i_1, i_2 \in I$, $w \in \Sigma^*$.

Moreover, analogous isomorphisms are valid if the semirings of row finite or column finite matrices are considered. Observe that, in case S is complete, these correspondences are isomorphisms of complete semirings, i.e., they respect infinite sums. These isomorphisms are used without further mention. Moreover, the notation A_{i_1, i_2}, where $A \in S^{I_1 \times I_2}\langle\langle \Sigma^* \rangle\rangle$ and $i_1 \in I_1, i_2 \in I_2$, is used: A_{i_1, i_2} is the power series in $S\langle\langle \Sigma^* \rangle\rangle$ such that the coefficient (A_{i_1, i_2}, w) of $w \in \Sigma^*$ is equal to $(A, w)_{i_1, i_2}$. Similarly, the notation (A, w), where $A \in (S\langle\langle \Sigma^* \rangle\rangle)^{I_1 \times I_2}$ and $w \in \Sigma^*$, is used: (A, w) is the matrix in $S^{I_1 \times I_2}$ whose (i_1, i_2)-entry $(A, w)_{i_1, i_2}$ is equal to (A_{i_1, i_2}, w) for each $i_1 \in I_1, i_2 \in I_2$.

For the proof of the next theorem, we need a lemma.

Lemma 4.1. *Let S be a complete star semiring. Then for all $a, b \in S$,*

$$(a+b)^* = (a + ba^*b)^*(1 + ba^*).$$

Proof. Using Theorem 3.4, we have

$$\begin{aligned}(a+ba^*b)^*(1+ba^*) &= (a^*ba^*b)^*a^*(1+ba^*)\\ &= \sum_{n\geq 0}(a^*b)^{2n}a^* + \sum_{n\geq 0}(a^*b)^{2n+1}a^*\\ &= (a^*b)^*a^* = (a+b)^*. \quad \square\end{aligned}$$

The next theorem is central for automata theory (see Conway [6], Lehmann [30], Kuich and Salomaa [29], Kuich [28], Bloom and Ésik [4], Kozen [24]). It allows us to compute the blocks of the star of a matrix A by sum, product, and star of the blocks of A.

For notational convenience, we will denote in Theorem 4.2 and in Corollaries 4.3 and 4.4 the matrices $A(I_i, I_j)$ by $A_{i,j}$, for $1 \leq i, j \leq 3$.

Theorem 4.2. *Let S be a complete star semiring. Let $A \in S^{I\times I}$ and $I = I_1 \cup I_2$ with $I_1, I_2 \neq \emptyset$ and $I_1 \cap I_2 = \emptyset$. Then*

$$\begin{aligned}A^*(I_1,I_1) &= (A_{1,1} + A_{1,2}A_{2,2}^*A_{2,1})^*,\\ A^*(I_1,I_2) &= (A_{1,1} + A_{1,2}A_{2,2}^*A_{2,1})^*A_{1,2}A_{2,2}^*,\\ A^*(I_2,I_1) &= (A_{2,2} + A_{2,1}A_{1,1}^*A_{1,2})^*A_{2,1}A_{1,1}^*,\\ A^*(I_2,I_2) &= (A_{2,2} + A_{2,1}A_{1,1}^*A_{1,2})^*.\end{aligned}$$

Proof. Consider the matrices

$$A_1 = \begin{pmatrix} A_{1,1} & 0 \\ 0 & A_{2,2}\end{pmatrix} \quad \text{and} \quad A_2 = \begin{pmatrix} 0 & A_{1,2} \\ A_{2,1} & 0\end{pmatrix}.$$

The computation of $(A_1 + A_2A_1^*A_2)^*(E + A_2A_1^*)$ and application of Lemma 4.1 prove our theorem. □

Corollary 4.3. *If $A_{2,1} = 0$, then*

$$A^* = \begin{pmatrix} A_{1,1}^* & A_{1,1}^*A_{1,2}A_{2,2}^* \\ 0 & A_{2,2}^*\end{pmatrix}.$$

Corollary 4.4. *Let $I = I_1 \cup I_2 \cup I_3$ be a decomposition into pairwise disjoint nonempty subsets. If $A_{2,1} = 0$, $A_{3,1} = 0$, and $A_{3,2} = 0$, then*

$$A^* = \begin{pmatrix} A_{1,1}^* & A_{1,1}^*A_{1,2}A_{2,2}^* & A_{1,1}^*A_{1,2}A_{2,2}^*A_{2,3}A_{3,3}^* + A_{1,1}^*A_{1,3}A_{3,3}^* \\ 0 & A_{2,2}^* & A_{2,2}^*A_{2,3}A_{3,3}^* \\ 0 & 0 & A_{3,3}^*\end{pmatrix}.$$

Next, we consider an arbitrary partition of the index set I.

Theorem 4.5. *Let S be a complete star semiring, and let $I = \bigcup_{j \in J} I_j$ be a decomposition into pairwise disjoint nonempty subsets. Fix $j_0 \in J$. Assume that the only non-null blocks of the matrix $A \in S^{I \times I}$ are $A(I_j, I_{j_0})$, $A(I_{j_0}, I_j)$ and $A(I_j, I_j)$, for all $j \in J$. Then*

$$A^*(I_{j_0}, I_{j_0}) = \Big(A(I_{j_0}, I_{j_0}) + \sum_{j \in J,\, j \neq j_0} A(I_{j_0}, I_j) A(I_j, I_j)^* A(I_j, I_{j_0}) \Big)^* .$$

Proof. We partition I into I_{j_0} and $I' = I - I_{j_0}$. Then $A(I', I')$ is a block-diagonal matrix and $(A(I', I')^*)(I_j, I_j) = A(I_j, I_j)^*$ for all $j \in J - \{j_0\}$. By Theorem 4.2, we obtain

$$A^*(I_{j_0}, I_{j_0}) = \big(A(I_{j_0}, I_{j_0}) + A(I_{j_0}, I') A(I', I')^* A(I', I_{j_0}) \big)^* .$$

The computation of the right-hand side of this equality proves our theorem. □

We now introduce the *Kronecker product* (also called *tensor product*) $A \otimes B \in S^{(I_1 \times I_2) \times (I'_1 \times I'_2)}$ for the matrices $A \in S^{I_1 \times I'_1}$ and $B \in S^{I_2 \times I'_2}$, by defining its entries:

$$(A \otimes B)_{(i_1, i_2), (i'_1, i'_2)} = A_{i_1, i'_1} B_{i_2, i'_2}, \quad \text{for all } i_1 \in I_1,\ i'_1 \in I'_1,\ i_2 \in I_2,\ i'_2 \in I'_2.$$

Sometimes, the Kronecker product $A \otimes B$ is defined to be in $(S^{I_2 \times I'_2})^{I_1 \times I'_1}$ with

$$\big((A \otimes B)_{i_1, i'_1} \big)_{i_2, i'_2} = A_{i_1, i'_1} B_{i_2, i'_2}, \quad \text{for all } i_1 \in I_1,\ i'_1 \in I'_1,\ i_2 \in I_2,\ i'_2 \in I'_2.$$

Since the semirings $S^{(I_1 \times I_2) \times (I_1 \times I_2)}$ and $(S^{I_2 \times I_2})^{I_1 \times I_1}$ are isomorphic, this will not make any difference in the computations.

Easy proofs show the following computational rules for Kronecker products.

Theorem 4.6. *Let $A, A' \in S^{I_1 \times I'_1}$, $B, B' \in S^{I_2 \times I'_2}$, $C \in S^{I_3 \times I'_3}$. Then:*

(i) $(A + A') \otimes B = A \otimes B + A' \otimes B$.
(ii) $A \otimes (B + B') = A \otimes B + A \otimes B'$.
(iii) $A \otimes 0 = 0$ *and* $0 \otimes B = 0$.
(iv) $A \otimes (B \otimes C) = (A \otimes B) \otimes C$.

Theorem 4.7. *Let $A \in S^{I_1 \times I_2}\langle\{\varepsilon\}\rangle$, $B \in S^{I_2 \times I_3}\langle\{\varepsilon\}\rangle$, $C \in S^{I_4 \times I_5}\langle\langle \Sigma^* \rangle\rangle$ and $D \in S^{I_5 \times I_6}\langle\langle \Sigma^* \rangle\rangle$. Assume that S is complete or that (A, ε) and (C, w) are row finite for all $w \in \Sigma^*$, or that (B, ε) and (D, w) are column finite for all $w \in \Sigma^*$. Furthermore, assume that all entries of (B, ε) commute with those of (C, w) for all $w \in \Sigma^*$. Then*

$$(AB) \otimes (CD) = (A \otimes C)(B \otimes D).$$

Proof. Let $i_j \in I_j$ for $j = 1, 3, 4, 6$. Then we obtain

$$\begin{aligned}
&((AB) \otimes (CD))_{(i_1,i_3),(i_4,i_6)} \\
&= (AB)_{i_1,i_3}(CD)_{i_4,i_6} \\
&= \sum_{i_2 \in I_2} \sum_{i_5 \in I_5} A_{i_1,i_2} B_{i_2,i_3} C_{i_4,i_5} D_{i_5,i_6} \\
&= \sum_{i_2 \in I_2} \sum_{i_5 \in I_5} A_{i_1,i_2} C_{i_4,i_5} B_{i_2,i_3} D_{i_5,i_6} \\
&= \sum_{(i_2,i_5) \in I_2 \times I_5} (A \otimes C)_{(i_1,i_4),(i_2,i_5)} (B \otimes D)_{(i_2,i_5),(i_3,i_6)} \\
&= ((A \otimes C)(B \otimes D))_{(i_1,i_4),(i_3,i_6)}. \qquad \square
\end{aligned}$$

The Kronecker product is useful for investigating the Hadamard product of formal power series, cf., e.g., Chap. 4, Sect. 4.2 [38] of this book.

5 Cycle-Free Linear Equations

Let Σ be an alphabet and S any semiring. Cycle-free linear equations over $S\langle\langle \Sigma^* \rangle\rangle$ are a useful tool for proving identities in $S\langle\langle \Sigma^* \rangle\rangle$. Assume that two expressions are shown to be solutions of such an equation. Then the uniqueness of the solution (shown below) implies that these two expressions represent the same formal power series in $S\langle\langle \Sigma^* \rangle\rangle$.

A *cycle-free linear equation (over* $S\langle\langle \Sigma^* \rangle\rangle$*)* has the form

$$y = ry + s,$$

where $r, s \in S\langle\langle \Sigma^* \rangle\rangle$ and r is cycle-free. A *solution* to this equation is given by a power series $\sigma \in S\langle\langle \Sigma^* \rangle\rangle$ such that $\sigma = r\sigma + s$.

Theorem 5.1. *The cycle-free equation* $y = ry + s$ *with* $r, s \in S\langle\langle \Sigma^* \rangle\rangle$, r *cycle-free, has the unique solution* $\sigma = r^* s$.

Proof. By Theorem 3.1, we obtain

$$r\sigma + s = rr^*s + s = (rr^* + \varepsilon)s = r^*s = \sigma.$$

Hence, σ is a solution.

Assume that r is cycle-free of index k, i.e., $(r, \varepsilon)^k = 0$, and that ϱ is a solution. Then by substitution, we obtain for all $n \geq 0$,

$$\varrho = r\varrho + s = \cdots = r^n \varrho + \sum_{0 \leq j < n} r^j s.$$

We now compute the coefficients (ϱ, w) for each $w \in \Sigma^*$:

$$(\varrho, w) = (r^{k(|w|+1)}\varrho, w) + \sum_{0 \le j < k(|w|+1)} (r^j s, w) = (r^* s, w) = (\sigma, w).$$

Hence, $\varrho = \sigma$. □

For power series over arbitrary semirings, the sum star identity and the product star identity are valid only for some cycle-free power series.

Theorem 5.2. *Let $r, s \in S\langle\langle \Sigma^* \rangle\rangle$ and assume that r, $r^* s$ and $r + s$ are cycle-free. Then the sum star identity is valid for r and s.*

Proof. We show that $(r^* s)^* r^*$ is a solution of the cycle-free equation $y = (r + s)y + \varepsilon$:

Indeed, by Theorem 3.2, we have

$$\begin{aligned}(r + s)(r^* s)^* r^* + \varepsilon &= rr^*(sr^*)^* + (sr^*)(sr^*)^* + \varepsilon \\ &= rr^*(sr^*)^* + (sr^*)^* = r^*(sr^*)^* = (r^* s)^* r^*.\end{aligned}$$

We now apply Theorem 5.1. □

Theorem 5.3. *Let $r, s \in S\langle\langle \Sigma^* \rangle\rangle$ and assume that rs is cycle-free. Then the product star identity is valid for r and s.*

Proof. By Theorems 3.2 and 3.1, we obtain

$$\varepsilon + r(sr)^* s = \varepsilon + rs(rs)^* = (rs)^*. \quad □$$

Corollary 5.4. *Let $r, s \in S\langle\langle \Sigma^* \rangle\rangle$ and assume that r is cycle-free and s is proper. Then the sum star identity and the product star identity are valid for r and s.*

Compare the next lemma with Lemma 4.1.

Lemma 5.5. *Let $r, s \in S\langle\langle \Sigma^* \rangle\rangle$ and assume that r, $r + s$ and $r + sr^* s$ are cycle-free. Then*

$$(r + s)^* = (r + sr^* s)^*(\varepsilon + sr^*).$$

Proof. By our assumptions, the power series r^*, $(r + s)^*$ and $(r + sr^* s)^*$ exist. By Theorem 3.1, we have

$$(r + s)^* = r(r + s)^* + s(r + s)^* + \varepsilon.$$

Hence, $(r + s)^*$ is a solution of the equation $y = ry + s(r + s)^* + \varepsilon$. By Theorem 5.1 and the cycle-freeness of r, another representation of this unique solution is $r^* s(r + s)^* + r^*$. Substituting $r^* s(r + s)^* + r^*$ into the third occurrence in the above equality yields

$$(r + s)^* = (r + sr^* s)(r + s)^* + sr^* + \varepsilon.$$

This shows that $(r+s)^*$ is a solution of the equation

$$y = (r + sr^*s)y + sr^* + \varepsilon.$$

By Theorem 5.1 and the cycle-freeness of $r + sr^*s$, another representation for the unique solution of this equation is

$$(r + sr^*s)^*(\varepsilon + sr^*). \quad \square$$

Consider matrices $A \in S^{I_1 \times I_2}\langle\langle \Sigma^* \rangle\rangle$ and $B \in S^{I_2 \times I_3}\langle\langle \Sigma^* \rangle\rangle$ such that either the matrices $(A, w) \in S^{I_1 \times I_2}$ are row finite for all $w \in \Sigma^*$ or the matrices $(B, w) \in S^{I_2 \times I_3}$ are column finite for all $w \in \Sigma^*$. Then $AB \in S^{I_1 \times I_3}\langle\langle \Sigma^* \rangle\rangle$ is well-defined. Hence, for a matrix $A \in S^{I \times I}\langle\langle \Sigma^* \rangle\rangle$, such that the matrices $(A, w) \in S^{I \times I}$ are row (resp. column) finite for all $w \in \Sigma^*$, all powers $A^n \in S^{I \times I}\langle\langle \Sigma^* \rangle\rangle$ are well-defined. If, furthermore, A is cycle-free then $A^* \in S^{I \times I}\langle\langle \Sigma^* \rangle\rangle$ is well-defined. If (A, w) is row and column finite for all $w \in \Sigma^*$, then so is (A^n, w) for all $n \in \mathbb{N}$ and $w \in \Sigma^*$.

Lemma 4.1 is the main tool for proving the matrix identities of Theorem 4.2. In an analogous manner, Lemma 5.5 is a main tool for proving—under different assumptions—the same matrix identities in the next theorem.

For the rest of the section, let $I = I_1 \cup I_2$ with $I_1, I_2 \neq \emptyset$ and $I_1 \cap I_2 = \emptyset$. The notation is similar to that in Theorem 4.2, but with $A \in S^{I \times I}\langle\langle \Sigma^* \rangle\rangle$ instead of $A \in S^{I \times I}$. For notational convenience, we will denote in Theorems 5.6 and 5.7 and in Corollaries 5.8 and 5.9 the matrices $A(I_i, I_j)$ by $A_{i,j}$, for $1 \leq i, j \leq 3$.

Theorem 5.6. *Assume that $A \in S^{I \times I}\langle\langle \Sigma^* \rangle\rangle$ is cycle-free and (A, w) is row and column finite for all $w \in \Sigma^*$. Furthermore, assume that $A_{1,1}$, $A_{2,2}$, $A_{1,1} + A_{1,2}A_{2,2}^*A_{2,1}$ and $A_{2,2} + A_{2,1}A_{1,1}^*A_{1,2}$ are cycle-free. Then*

$$\begin{aligned}
A^*(I_1, I_1) &= (A_{1,1} + A_{1,2}A_{2,2}^*A_{2,1})^*, \\
A^*(I_1, I_2) &= (A_{1,1} + A_{1,2}A_{2,2}^*A_{2,1})^*A_{1,2}A_{2,2}^*, \\
A^*(I_2, I_1) &= (A_{2,2} + A_{2,1}A_{1,1}^*A_{1,2})^*A_{2,1}A_{1,1}^*, \\
A^*(I_2, I_2) &= (A_{2,2} + A_{2,1}A_{1,1}^*A_{1,2})^*.
\end{aligned}$$

Proof. Consider the matrices

$$A_1 = \begin{pmatrix} A_{1,1} & 0 \\ 0 & A_{2,2} \end{pmatrix} \quad \text{and} \quad A_2 = \begin{pmatrix} 0 & A_{1,2} \\ A_{2,1} & 0 \end{pmatrix}.$$

Since the blocks of the block-diagonal matrix A_1 are cycle-free, the matrix A_1^* exists and equals

$$A_1^* = \begin{pmatrix} A_{1,1}^* & 0 \\ 0 & A_{2,2}^* \end{pmatrix}.$$

This implies that

$$A_1 + A_2A_1^*A_2 = \begin{pmatrix} A_{1,1} + A_{1,2}A_{2,2}^*A_{2,1} & 0 \\ 0 & A_{2,2} + A_{2,1}A_{1,1}^*A_{1,2} \end{pmatrix}.$$

Since the blocks of the block-diagonal matrix $A_1 + A_2 A_1^* A_2$ are cycle-free, the matrix $(A_1 + A_2 A_1^* A_2)^*$ exists and equals

$$(A_1 + A_2 A_1^* A_2)^* = \begin{pmatrix} (A_{1,1} + A_{1,2} A_{2,2}^* A_{2,1})^* & 0 \\ 0 & (A_{2,2} + A_{2,1} A_{1,1}^* A_{1,2})^* \end{pmatrix}.$$

We now apply Lemma 5.5 with $r = A_1$ and $s = A_2$. The computation of

$$(A_1 + A_2 A_1^* A_2)^* (E + A_2 A_1^*)$$

proves the theorem. □

Theorem 5.7. *Consider $A \in S^{I \times I}\langle\langle \Sigma^* \rangle\rangle$ such that (A, w) is row and column finite for all $w \in \Sigma^*$. Furthermore, assume that $A_{1,1}$ and $A_{2,2}$ are cycle-free, and $A_{1,2}$ or $A_{2,1}$ is proper. Then A is cycle-free and*

$$\begin{aligned} A^*(I_1, I_1) &= (A_{1,1} + A_{1,2} A_{2,2}^* A_{2,1})^*, \\ A^*(I_1, I_2) &= (A_{1,1} + A_{1,2} A_{2,2}^* A_{2,1})^* A_{1,2} A_{2,2}^*, \\ A^*(I_2, I_1) &= (A_{2,2} + A_{2,1} A_{1,1}^* A_{1,2})^* A_{2,1} A_{1,1}^*, \\ A^*(I_2, I_2) &= (A_{2,2} + A_{2,1} A_{1,1}^* A_{1,2})^*. \end{aligned}$$

Proof. We only prove the case where $A_{2,1}$ is proper. The proof of the other case is similar. An easy proof by induction on $j \geq 1$ shows that

$$(A, \varepsilon)^j = \begin{pmatrix} (A_{1,1}, \varepsilon)^j & \sum_{j_1 + j_2 = j-1} (A_{1,1}, \varepsilon)^{j_1} (A_{1,2}, \varepsilon)(A_{2,2}, \varepsilon)^{j_2} \\ 0 & (A_{2,2}, \varepsilon)^j \end{pmatrix}.$$

Now let $A_{1,1}$ and $A_{2,2}$ be cycle-free of index k. Then $(A, \varepsilon)^{2k} = 0$ and A is cycle-free. Furthermore $(A_{1,1} + A_{1,2} A_{2,2}^* A_{2,1}, \varepsilon) = (A_{1,1}, \varepsilon)$ and $(A_{2,2} + A_{2,1} A_{1,1}^* A_{1,2}, \varepsilon) = (A_{2,2}, \varepsilon)$. Hence, the assumptions of Theorem 5.6 are satisfied and our theorem is proved. □

Corollary 5.8. *Consider $A \in S^{I \times I}\langle\langle \Sigma^* \rangle\rangle$ such that (A, w) is row and column finite for all $w \in \Sigma^*$. Furthermore, assume that $A_{1,1}$ and $A_{2,2}$ are cycle-free and that $A_{2,1} = 0$. Then A is cycle-free and*

$$A^* = \begin{pmatrix} A_{1,1}^* & A_{1,1}^* A_{1,2} A_{2,2}^* \\ 0 & A_{2,2}^* \end{pmatrix}.$$

Observe that for finite matrices, the row and column finiteness of (A, w) for all $w \in \Sigma^*$ is satisfied and is not needed as assumption in Theorem 5.7. If A is finite and proper, all assumptions of Theorem 5.7 are satisfied.

Corollary 5.9. *Let I be finite and $A \in S^{I \times I}\langle\langle \Sigma^* \rangle\rangle$ be proper. Then*

$$\begin{aligned} A^*(I_1, I_1) &= (A_{1,1} + A_{1,2} A_{2,2}^* A_{2,1})^*, \\ A^*(I_1, I_2) &= (A_{1,1} + A_{1,2} A_{2,2}^* A_{2,1})^* A_{1,2} A_{2,2}^*, \\ A^*(I_2, I_1) &= (A_{2,2} + A_{2,1} A_{1,1}^* A_{1,2})^* A_{2,1} A_{1,1}^*, \\ A^*(I_2, I_2) &= (A_{2,2} + A_{2,1} A_{1,1}^* A_{1,2})^*. \end{aligned}$$

References

1. J. Adámek, E. Nelson, and J. Reiterman. Tree constructions of free continuous algebras. *Journal of Computer and System Sciences*, 24:114–146, 1982.
2. J. Berstel, editor. *Séries formelles en variables non commutatives et applications*. Laboratoire d'Informatique Théorique et Programmation, Ecole Nationale Supérieure de Techniques Avancées, Paris, 1978.
3. J. Berstel and C. Reutenauer. *Les séries rationelles et leurs langages*. Masson, Paris, 1984. English translation: *Rational Series and Their Languages*, volume 12 of *Monographs in Theoretical Computer Science. An EATCS Series*. Springer, Berlin, 1988.
4. S.L. Bloom and Z. Ésik. *Iteration Theories, Monographs in Theoretical Computer Science. An EATCS Series*. Springer, Berlin, 1993.
5. G. Carré. *Graphs and Networks*. Clarendon, Oxford, 1979.
6. J.H. Conway. *Regular Algebra and Finite Machines*. Chapman & Hall, London, 1971.
7. M. Droste and P. Gastin. On aperiodic and star-free formal power series in partially commuting variables. *Theory of Computing Systems*, 42:608–631, 2008. Extended abstract in: D. Krob, A.A. Milchalev, and A.V. Milchalev, editors, *Formal Power Series and Algebraic Combinatorics, 12th Int. Conf., Moscow*, pages 158–169. Springer, Berlin, 2000.
8. S. Eilenberg. *Automata, Languages and Machines, volume A*. Academic Press, San Diego, 1974.
9. Z. Ésik and W. Kuich. Modern automata theory. www.dmg.tuwien.ac.at/kuich.
10. Z. Ésik and W. Kuich. Locally closed semirings. *Monatshefte für Mathematik*, 137:21–29, 2002.
11. Z. Ésik and W. Kuich. Equational axioms for a theory of automata. In C. Martin-Vide, V. Mitrana, and G. Paun, editors, *Formal Languages and Applications*, volume 148 of *Studies in Fuzziness and Soft Computing*, pages 183–196. Springer, Berlin, 2004.
12. Z. Ésik and W. Kuich. Finite automata. In M. Droste, W. Kuich, and H. Vogler, editors, *Handbook of Weighted Automata*. Chapter 3. Springer, Berlin, 2009.
13. K. Głazek. *A Guide to the Literature on Semirings and Their Applications in Mathematics and Information Science*. Kluwer Academic, Dordrecht, 2002.
14. J.A. Goguen, J.W. Thatcher, E.G. Wagner, and J.B. Wright. Initial algebra semantics and continuous algebras. *Journal of the Association for Computing Machinery*, 24:68–95, 1977.
15. J. Golan. *Semirings and Their Applications*. Kluwer Academic, Dordrecht, 1999.
16. M. Goldstern. Vervollständigung von Halbringen. Diplomarbeit, Technische Universität Wien, 1985.

17. I. Guessarian. *Algebraic Semantics*, volume 99 of *Lecture Notes in Computer Science*. Springer, Berlin, 1981.
18. P. Hájek. *Metamathematics of Fuzzy Logic*. Kluwer Academic, Dordrecht, 1998.
19. U. Hebisch. The Kleene theorem in countably complete semirings. *Bayreuther Mathematische Schriften*, 31:55–66, 1990.
20. U. Hebisch and H.J. Weinert. *Halbringe—Algebraische Theorie und Anwendungen in der Informatik*. Teubner, Leipzig, 1993. English translation: *Semirings—Algebraic Theory and Applications in Computer Science*. World Scientific, Singapore, 1998.
21. B. Heidergott, G.J. Olsder, and J. van der Woude. *Max Plus at Work*. Princeton University Press, Princeton, 2006.
22. G. Karner. On limits in complete semirings. *Semigroup Forum*, 45:148–165, 1992.
23. G. Karner. Continuous monoids and semirings. *Theoretical Computer Science*, 318:355–372, 2004.
24. D. Kozen. A completeness theorem for Kleene algebras and the algebra of regular events. *Information and Computation*, 110:366–390, 1994.
25. D. Krob. Monoides et semi-anneaux complets. *Semigroup Forum*, 36:323–339, 1987.
26. D. Krob. Monoides et semi-anneaux continus. *Semigroup Forum*, 37:59–78, 1988.
27. W. Kuich. The Kleene and the Parikh theorem in complete semirings. In *ICALP'87*, volume 267 of *Lecture Notes in Computer Science*, pages 212–225. Springer, Berlin, 1987.
28. W. Kuich. Semirings and formal power series: Their relevance to formal languages and automata theory. In G. Rozenberg and A. Salomaa, editors, *Handbook of Formal Languages, volume 1*, Chapter 9, pages 609–677. Springer, Berlin, 1997.
29. W. Kuich and A. Salomaa. *Semirings, Automata, Languages*, volume 5 of *Monographs in Theoretical Computer Science. An EATCS Series*. Springer, Berlin, 1986.
30. D.J. Lehmann. Algebraic structures for transitive closure. *Theoretical Computer Science*, 4:59–76, 1977.
31. E.G. Manes and M.A. Arbib. *Algebraic Approaches to Program Semantics*. Springer, Berlin, 1986.
32. G. Markowsky. Chain-complete posets and directed sets with applications. *Algebra Universalis*, 6:53–68, 1976.
33. M. Mohri. Semiring frameworks and algorithms for shortest-distance problems. *Journal of Automata, Languages and Combinatorics*, 7:321–350, 2002.
34. I. Petre and A. Salomaa. Algebraic systems and pushdown automata. In M. Droste, W. Kuich, and H. Vogler, editors, *Handbook of Weighted Automata*. Chapter 7. Springer, Berlin, 2009.

35. G. Rahonis. Fuzzy languages. In M. Droste, W. Kuich, and H. Vogler, editors, *Handbook of Weighted Automata.* Chapter 12. Springer, Berlin, 2009.
36. J. Sakarovitch. Kleene's theorem revisited. In *Trends, Techniques, and Problems in Theoretical Computer Science, 4th International Meeting of Young Computer Scientists*, volume 281 of *Lecture Notes in Computer Science*, pages 39–50. Springer, Berlin, 1987.
37. J. Sakarovitch. *Éléments de Théorie des Automates.* Vuibert, Paris, 2003.
38. J. Sakarovitch. Rational and recognisable power series. In M. Droste, W. Kuich, and H. Vogler, editors, *Handbook of Weighted Automata.* Chapter 4. Springer, Berlin, 2009.
39. A. Salomaa and M. Soittola. *Automata-Theoretic Aspects of Formal Power Series.* Springer, Berlin, 1978.
40. W. Wechler. *The Concept of Fuzziness in Automata and Language Theory.* Akademie Verlag, Berlin, 1978.
41. X. Zhao. Locally closed semirings and iteration semirings. *Monatshefte für Mathematik*, 144:157–167, 2005.
42. U. Zimmermann. *Linear and Combinatorial Optimization in Ordered Algebraic Structures*, volume 10 of *Annals of Discrete Mathematics.* North-Holland, Amsterdam, 1981.

Chapter 2: Fixed Point Theory

Zoltán Ésik[1,2,*]

[1] University of Szeged, Szeged, Hungary
[2] Rovira i Virgili University, Tarragona, Spain

1 Introduction

Fixed points and fixed point computations occur in just about every field of computer science. Their widespread use is due to the fact that the semantics of recursion can be described by fixed points of functions or functionals, or more generally, functors or morphisms. Of course, the treatment of fixed points in mathematics goes well back before their first use in computer science: They frequently occur in analysis, algebra, geometry, and logic. One

* The author was partially supported by grant no. MTM2007-63422 from the Ministry of Education and Science of Spain.

M. Droste, W. Kuich, H. Vogler (eds.), *Handbook of Weighted Automata*, Monographs in Theoretical Computer Science. An EATCS Series, DOI 10.1007/978-3-642-01492-5_2,

of the first occurrences of fixed points in the theory of automata and formal languages were probably the equational characterizations of regular and context-free languages as least solutions to right-linear and polynomial fixed point equations. Kleene's theorem for regular languages follows from the fixed point characterization just by a few simple equational properties of the fixed point operation. Many results in the theory of automata and languages can be derived from basic properties of fixed points.

The aim of this paper is to provide an introduction to that part of the theory of fixed points that has applications to weighted automata and weighted languages. We start with a treatment of fixed points in the ordered setting and review some basic theorems guaranteeing the existence of least (or greatest) fixed points. Then we establish several (equational) properties of the least fixed point operation including the Bekić identity asserting that systems of fixed point equations can be solved by the technique of successive elimination. Then we use the Bekić identity and some other basic laws to introduce the axiomatic frameworks of Conway and iteration theories. We provide several axiomatizations of these notions and review some completeness results showing that iteration theories capture the equational properties of the fixed point operation in a large class of models. In the last two sections, we treat fixed points of linear functions and affine functions over semirings and semimodules. The main results show that for such functions, the fixed point operation can be characterized by a star operation possibly in conjunction with an omega operation. We show that the equational properties of the fixed point operation are reflected by corresponding properties of the star and omega operations.

Some Notation

The composition of functions $f : A \to B$ and $g : B \to C$ is written $g \circ f$, $x \mapsto g(f(x))$. The identity function $A \to A$ will be denoted $\mathbf{id}_A$. When $f : A \to B$ and $g : A \to C$, the *target pairing* (or just *pairing*) of f and g is the function $\langle f, g \rangle : A \to B \times C$, $x \mapsto (f(x), g(x))$, $x \in A$. In the same way, one defines the *(target) tupling* $f = \langle f_1, \ldots, f_n \rangle : A \to B_1 \times \cdots \times B_n$ of $n \geq 0$ functions $f_i : A \to B_i$. When $n = 0$, the Cartesian product $B_1 \times \cdots \times B_n$ is a singleton set and f is the unique function from A to this set. The ith *projection function* $A_1 \times \cdots \times A_n \to A_i$ will be denoted $\mathbf{pr}_{A_i}^{A_1 \times \cdots \times A_n}$, or $\mathbf{pr}_i^{A_1 \times \cdots \times A_n}$, or just $\mathbf{pr}_i$. A *base function* is any tupling of projections. When $f : A \to A'$ and $g : B \to B'$, $f \times g$ is the function $A \times B \to A' \times B'$ mapping each pair $(x, y) \in A \times B$ to $(f(x), g(y))$. Clearly, $f \times g = \langle f \circ \mathbf{pr}_A^{A \times B}, g \circ \mathbf{pr}_B^{A \times B} \rangle$.

2 Least Fixed Points

When A is a set, an *endofunction* over A is a function $A \to A$. We say that $a \in A$ is a *fixed point* of f if $f(a) = a$. We also say that a is a *solution of* or *solves the fixed point equation* $x = f(x)$. When A is partially ordered by a relation $\leq$, we also define *prefixed points* of f as those elements $a \in A$ with $f(a) \leq a$. Dually, we call $a \in A$ a *post-fixed point* of f if $a \leq f(a)$. Thus, a fixed point is both a prefixed point and a post-fixed point. A *least fixed point* of f is least among the fixed points of f, and a *least prefixed point* is least among all prefixed points of f. *Greatest fixed points* and *greatest post-fixed points* are defined dually. It is clear that the extremal (i.e., least or greatest) fixed points, prefixed points and post-fixed points are unique whenever they exist.

Least prefixed points give rise to the following *fixed point induction* principle. When P is a poset and $f : P \to P$ has a least prefixed point x, then we have $x \leq y$ whenever $f(y) \leq y$. As an application of the principle, we establish a simple fact.

Proposition 2.1. *Let P be a partially ordered set and let $f : P \to P$ be monotone. If f has a least prefixed point, then it is the least fixed point of f. Dually, if f has a greatest post-fixed point, then it is the greatest fixed point of f.*

Proof. We only prove the first claim since the second follows by reversing the order. Suppose that p is the least prefixed point of f. Then $f(p) \leq p$, and since f is monotone, $f(f(p)) \leq f(p)$. This shows that $f(p)$ is a prefixed point. Thus, by fixed point induction, $p \leq f(p)$. Since p is both a prefixed point and a post-fixed point, it is a fixed point. □

Next, we provide conditions guaranteeing the existence of fixed points. Recall that a *directed set* in a partially ordered set P is a nonempty subset D of P such that any two elements of D have an upper bound in D. A *chain* in P is a linearly ordered subset of P. Note that every nonempty chain is a directed set.

Definition 2.2. *A* complete partial order, *or cpo is a partially ordered set $P = (P, \leq)$ which has a least element denoted $\bot_P$ or just $\bot$ such that each directed set $D \subseteq P$ has a supremum $\bigvee D$.*

It is known that a partially ordered set P is a cpo iff it has suprema of all chains, or suprema of well-ordered chains, cf. [43]. See also [16].

Definition 2.3. *Suppose that P, Q are partially ordered sets and $f : P \to Q$. We say that f is* continuous *if it preserves all existing suprema of directed sets: For all directed sets $D \subseteq P$, if $\bigvee D$ exists, then so does $\bigvee f(D)$, and*

$$f\Big(\bigvee D\Big) = \bigvee f(D).$$

Every continuous function $P \to Q$ is monotone, since for all $x, y \in P$ with $x \leq y$, $f(y) = f(\bigvee\{x, y\}) = \bigvee\{f(x), f(y)\}$, i.e., $f(x) \leq f(y)$. From [43], it is also known that a function $f : P \to Q$ is continuous iff it preserves suprema of nonempty chains, or suprema of nonempty well-ordered chains.

Remark 2.4. Suppose that P, Q are partially ordered sets and $f : P \to Q$ and $g : Q \to P$ are monotone functions. We say that (f, g) is a *Galois connection* if for all $x \in P$ and $y \in Q$, $f(x) \leq y$ iff $x \leq g(y)$. It is known that when (f, g) is a Galois connection then f preserves all existing suprema, and g preserves all existing infima. In particular, f is continuous, and when P has a least element $\bot_P$ then Q also has a least element $\bot_Q$, and $f(\bot_P) = \bot_Q$.

Theorem 2.5. *Suppose that P is a cpo and $f : P \to P$ is monotone. Then f possesses a least prefixed point (which is the least fixed point of f).*

Proof. Define $x_\alpha = f(x_\beta)$, if α is the successor of the ordinal β, and $x_\alpha = \bigvee\{x_\beta : \beta < \alpha\}$ if α is a limit ordinal. In particular, x_0 is the least element $\bot$. It is a routine matter to verify that $x_\alpha \leq x_\beta$ whenever $\alpha \leq \beta$. Thus, there is a (least) ordinal α with $x_\alpha = x_{\alpha+1}$. This element x_α is the least prefixed point of f. □

A partial converse of Theorem 2.5 is proved in [43]: If P is a partially ordered set such that any monotone endofunction $P \to P$ has a least fixed point, then P is a cpo.

The above rather straightforward argument makes use of the axiom of choice. An alternative proof which avoids using this axiom is presented in [19]. A special case of the theorem is the *Knaster–Tarski fixed point theorem* [50, 19] asserting that a monotone endofunction of a complete lattice L has a least (and by duality, also a greatest) fixed point.

When the endofunction f in Theorem 2.5 is continuous, the least fixed point can be constructed in ω steps.

Corollary 2.6. *Suppose that P is a cpo and $f : P \to P$ is continuous. Then the least prefixed point of f is $\bigvee\{f^n(\bot) : n \geq 0\}$ (which is the least fixed point of f).*

Proof. Using the above notation, we have by continuity that $f(x_\omega) = x_\omega$, where $x_\omega = \bigvee\{f^n(\bot) : n \geq 0\}$. □

Note that the same result holds if we only assume that P is a *countably complete* or *ω-complete* partially ordered set, i.e., when it has a least element and suprema of ω-chains, or equivalently, suprema of countable directed sets, and if f is *ω-continuous*, i.e., it preserves suprema of ω-chains, or suprema of countable directed sets.

Dually, if P is a partially ordered set which has infima of all chains, and thus a greatest element $\top$, and if $f : P \to P$ is monotone, then f has a greatest post-fixed point which is the greatest fixed point of f. This greatest fixed point can be constructed as the first x_γ with $x_\gamma = x_{\gamma+1}$, where $x_\alpha = f(x_\beta)$ if α is the successor of the ordinal β, and $x_\alpha = \bigwedge\{x_\beta : \beta < \alpha\}$ if α is a limit ordinal, the infimum of the set $\{x_\beta : \beta < \alpha\}$. Thus, $x_0 = \top$. If, in addition, f preserves infima of nonempty chains, then the greatest post-fixed point is $\bigwedge\{f^n(\top) : n \geq 0\}$.

Besides single fixed point equations $x = f(x)$, we will consider finite *systems of fixed point equations*:

$$x_1 = f_1(x_1, \ldots, x_n),$$
$$\vdots$$
$$x_n = f_n(x_1, \ldots, x_n).$$

Each component equation $x_i = f_i(x_1, \ldots, x_n)$ of such a system may be considered as a fixed point equation in the unknown x_i and the *parameters* $x_1, \ldots, x_{i-1}, x_{i+1}, \ldots, x_n$. This leads to *parametric fixed point equations* of the sort $x = f(x, y)$, where f is a function $P \times Q \to P$.

Note that when P and Q are cpo's and A is a set, then $P \times Q$ and Q^A, equipped with the pointwise order, are cpo's. Moreover, the set of all continuous functions $P \to Q$ is also a cpo denoted $(P \to Q)$. For any partially ordered sets, P_1, P_2, Q and function $f : P_1 \times P_2 \to Q$, f is monotone or continuous iff it is *separately* monotone or continuous in either argument, i.e., when the functions ${}_{p_1}f : P_2 \to Q$ and $f_{p_2} : P_1 \to Q$, ${}_{p_1}f(y) = f(p_1, y)$, $f_{p_2}(x) = f(x, p_2)$, $p_1 \in P_1$, $p_2 \in P_2$ have the appropriate property. Moreover, a function $f = \langle f_1, f_2 \rangle : Q \to P_1 \times P_2$ is monotone or continuous iff both functions $f_i = \mathbf{pr}_i \circ f$, $i = 1, 2$ are monotone or continuous, where $\mathbf{pr}_1$ and $\mathbf{pr}_2$ denote the first and second projection functions $P_1 \times P_2 \to P_1$ and $P_1 \times P_2 \to P_2$.

Definition 2.7. *Suppose that P, Q are partially ordered sets such that $f : P \times Q \to P$ is monotone and for each $y \in P$ the endofunction $f_y : P \to P$, $f_y(x) = f(x, y)$ has a least prefixed point. Then we define $f^\dagger : Q \to P$ as the function mapping each $y \in Q$ to the least prefixed point of f_y.*

In a similar fashion, one could define a greatest (post)fixed point operation. Since the properties of this operation follow from the properties of the least (pre)fixed point operation by simple duality, below we will consider only the least fixed point operation. Nested least and greatest fixed points are considered in the μ-calculus, cf. Arnold and Niwinski [1]. It is known that over complete lattices, the alternation hierarchy obtained by nesting least and greatest fixed points is infinite.

Notice the pointwise nature of the above definition. For each $y \in Q$, $f^\dagger(y)$ is $(f_y)^\dagger$, the least prefixed point of the function $f_y : P \to P$ (which may be identified with a function $P \times R \to P$, where R has a single element).

The above definition of the *dagger operation* is usually applied in the case when P, Q are cpo's. In that case, the existence of the least prefixed point is guaranteed by Theorem 2.5.

Proposition 2.8. *Suppose that $f : P \times Q \to P$ is monotone. Then $f^\dagger$ is also monotone. Moreover, when P and Q are cpo's and f is continuous, so is $f^\dagger$.*

Proof. Assume that $y \leq z$ in Q. If x is a prefixed point of f_z, then $f_y(x) = f(x,y) \leq f(x,z) = f_z(x) \leq x$, so that x is also a pre-fixed point of f_y. Thus, the least prefixed point of f_y is below the least prefixed point of f_z, i.e., $f^\dagger(y) \leq f^\dagger(z)$.

Assume now that P and Q are cpo's and f is continuous. Let D denote a directed subset of Q. We have

$$f^\dagger(y) = \bigvee\{f_y^n(\bot_P) : n \geq 0\} = \bigvee\{f^n(\bot_P, y) : n \geq 0\}, \quad \text{for all } y \in Q,$$

where we define $f^0(x,y) = x$ and $f^{n+1}(x,y) = f(f^n(x,y),y)$, for all $n \geq 0$. Now,

$$\begin{aligned}\bigvee\Big\{f^n\Big(\bot_P, \bigvee D\Big) : n \geq 0\Big\} &= \bigvee\Big\{\bigvee\{f^n(\bot_P, y) : y \in D\} : n \geq 0\Big\} \\ &= \bigvee\{f^n(\bot_P, y) : n \geq 0,\ y \in D\} \\ &= \bigvee\Big\{\bigvee\{f^n(\bot_P, y) : n \geq 0\} : y \in D\Big\} \\ &= \bigvee\{f^\dagger(y) : y \in D\}.\end{aligned}$$

□

It is also known that for cpo's P, Q, the function $((P \times Q) \to P) \to (Q \to P)$ which maps each continuous $f : P \times Q \to P$ to the continuous function $f^\dagger : Q \to P$ is itself continuous; see, e.g., [19].

The dagger operation satisfies several nontrivial equational properties. We list a few below. Let P, Q, R denote cpo's and $f, g, \ldots$ monotone or continuous functions whose sources and targets are specified below.

FIXED POINT IDENTITY

$$f^\dagger = f \circ \langle f^\dagger, \mathbf{id}_Q\rangle \tag{1}$$

where $f : P \times Q \to P$.

PARAMETER IDENTITY

$$\big(f \circ (\mathbf{id}_P \times g)\big)^\dagger = f^\dagger \circ g \tag{2}$$

where $f : P \times Q \to P$ and $g : R \to Q$.

COMPOSITION IDENTITY

$$\big(f \circ \langle g, \mathbf{pr}_R^{P\times R}\rangle\big)^\dagger = f \circ \big\langle \big(g \circ \langle f, \mathbf{pr}_R^{Q\times R}\rangle\big)^\dagger, \mathbf{id}_R\big\rangle \tag{3}$$

where $f : Q \times R \to P$, $g : P \times R \to Q$ and $\mathbf{pr}_R^{P\times R} : P \times R \to P$ and $\mathbf{pr}_R^{Q\times R} : Q \times R \to Q$ are projection functions.

DOUBLE DAGGER IDENTITY or DIAGONAL IDENTITY

$$\big(f^\dagger\big)^\dagger = \big(f \circ (\langle \mathbf{id}_P, \mathbf{id}_P\rangle \times \mathbf{id}_Q)\big)^\dagger, \tag{4}$$

where $f : P \times P \times Q \to P$.

PAIRING IDENTITY or BEKIĆ IDENTITY

$$\langle f, g\rangle^\dagger = \langle f^\dagger \circ \langle h^\dagger, \mathbf{id}_R\rangle, h^\dagger\rangle \tag{5}$$

where $f : P \times Q \times R \to P$, $g : P \times Q \times R \to Q$ and

$$h = g \circ \langle f^\dagger, \mathbf{id}_{Q\times R}\rangle. \tag{6}$$

PERMUTATION IDENTITY

$$\left(\pi \circ f \circ (\pi^{-1} \times \mathbf{id}_Q)\right)^\dagger = \pi \circ f^\dagger, \tag{7}$$

where

$$f : P_1 \times \cdots \times P_n \times Q \to P_1 \times \cdots \times P_n \quad \text{and}$$
$$\pi = \langle \mathbf{pr}_{i_1}^{P_1\times\cdots\times P_n}, \ldots, \mathbf{pr}_{i_n}^{P_1\times\cdots\times P_n}\rangle$$

for some permutation $(i_1, \ldots, i_n)$ of the first n positive integers, and where π^{-1} is the inverse of π, i.e., $\pi^{-1} = \langle \mathbf{pr}_{j_1}^{P_1\times\cdots\times P_n}, \ldots, \mathbf{pr}_{j_n}^{P_1\times\cdots\times P_n}\rangle$ where $(j_1, \ldots, j_n)$ is the inverse of $(i_1, \ldots, i_n)$.

For these identities, we refer to [4, 20, 44, 45, 47, 51] and [9]. Each of the above identities can be explained using an ordinary functional language. For example, the fixed point identity (1) says that $f^\dagger(y)$ is a solution of the fixed point equation $x = f(x, y)$ in the unknown x and parameter y. It is customary to write this least solution as $\mu x.f(x, y)$. Using this μ-notation, the fixed point identity reads $f(\mu x.f(x, y), y) = \mu x.f(x, y)$. The parameter identity (2) is implicit in the μ-notation. It is due to the pointwise nature of the definition of dagger, and it says that solving $x = f(x, y)$ and then substituting $g(z)$ for y gives the same result as first substituting $g(z)$ for y and then solving $x = f(x, g(z))$. In the composition identity (3), one considers the equations $x = f(g(x, z), z)$ and $y = g(f(y, z), z)$, with least solutions $\mu x.f(g(x, z), z)$ and $\mu y.g(f(y, z), z)$. The composition identity asserts that these are related: $\mu x.f(g(x, z), z) = f(\mu y.g(f(y, z), z), z)$. The double dagger identity (4) asserts that the least solution of $x = f(x, x, z)$ is the same as the least solution of $y = f^\dagger(y, z)$, where $f^\dagger(y, z)$ is in turn the least solution of $x = f(x, y, z)$. In the μ-notation, $\mu x.\mu y.f(x, y, z) = \mu x.f(x, x, z)$. The Bekić identity (5) asserts that systems

$$x = f(x, y, z), \tag{8}$$
$$y = g(x, y, z) \tag{9}$$

can be solved by *Gaussian elimination* (or *successive elimination*). To find the least solution of the above system, where f, g are appropriate functions, one can proceed as follows. First, solve the first equation to obtain $x = f^\dagger(y, z)$, then substitute this solution for x in the second equation to obtain $y = g(f^\dagger(y, z), y, z) = h(y, z)$. The identity asserts that the second component of the least solution of the above system is the least solution of $y =$

$h(y,z)$, i.e., $h^\dagger(z)$. Moreover, it asserts that the first component is $f^\dagger(h^\dagger(z),z)$, which is obtained by back substituting $h^\dagger(z)$ for y in $f^\dagger(y,z)$, the solution of just the first equation. In the μ-notation, $\mu(x,y).(f(x,y,z),g(x,y,z)) = (\mu x.f(\mu y.h(y,z),z),\mu y.h(y,z))$, where $h(y,z) = g(\mu x.f(x,y,z),y,z)$.

We still want to illustrate the Bekić identity over semirings. So, suppose that S is a continuous semiring, cf. [21]. It will be shown later that there is a star operation $^* : S \to S$ such that least solutions of fixed point equations $x = ax+b$ can be expressed as a^*b. Suppose now that $f,g : S^2 \to S$, $f(x,y) = ax + by + e$ and $g(x,y) = cx + dy + f$ and consider the system of fixed point equations

$$\begin{aligned} x &= ax + by + e, \\ y &= cx + dy + f. \end{aligned}$$

Then $f^\dagger(y)$, the least solution of just the first equation is $a^*(by + e) = a^*by + a^*e$. Thus, $h(y) = g(f^\dagger(y),y)$ is $(d + ca^*b)y + ca^*e + f$, and $h^\dagger = (d + ca^*b)^*(ca^*e + f)$ is the second component of the least solution of the above system. The first component is $f^\dagger(h^\dagger) = a^*b(d+ca^*b)^*(ca^*e+f)+a^*e$. Using the matrix notation, the least solution of

$$\begin{pmatrix} x \\ y \end{pmatrix} = \begin{pmatrix} a & b \\ c & d \end{pmatrix}\begin{pmatrix} x \\ y \end{pmatrix} + \begin{pmatrix} e \\ f \end{pmatrix}$$

is

$$\begin{pmatrix} x \\ y \end{pmatrix} = \begin{pmatrix} a^*b(d + ca^*b)^*ca^* + a^* & a^*b(d + ca^*b)^* \\ (d + ca^*b)^*ca^* & (d + ca^*b)^* \end{pmatrix}\begin{pmatrix} e \\ f \end{pmatrix}.$$

We leave it to the reader to express the permutation identity (7) in the μ-notation.

A special case of the fixed point identity is

$$(f \circ \mathbf{pr}_Q^{P\times Q})^\dagger = f, \tag{10}$$

where $f : Q \to P$, and a special case of the parameter identity is

$$(f \circ \mathbf{pr}_{P\times Q}^{P\times Q\times R})^\dagger = f^\dagger \circ \mathbf{pr}_Q^{Q\times R}, \tag{11}$$

where $f : P \times Q \to P$. A special case of the permutation identity (7) is

Transposition identity

$$(\pi_{Q,P}^{P,Q} \circ \langle f,g\rangle \circ (\pi_{P,Q}^{Q,P} \times \mathbf{id}_R))^\dagger = \pi_{Q,P}^{P,Q} \circ \langle f,g\rangle^\dagger, \tag{12}$$

where $f : P \times Q \times R \to P$ and $g : P \times Q \times R \to Q$, and where $\pi_{Q,P}^{P,Q} = \langle \mathbf{pr}_Q^{P\times Q}, \mathbf{pr}_P^{P\times Q}\rangle$ and $\pi_{P,Q}^{Q,P} = \langle \mathbf{pr}_P^{Q\times P}, \mathbf{pr}_Q^{Q\times P}\rangle$.

In the μ-notation, (10) can be written as $\mu x.f(y) = f(y)$, while the transposition identity (12) asserts that $\mu(x,y).(f(x,y,z), g(x,y,z))$ is the transposition of $\mu(y,x).(g(x,y,z), f(x,y,z))$. Equation (11), being a special case of the parameter identity, is implicit in the μ-notation.

Theorem 2.9. *All of the above identities hold for the least prefixed point operation.*

Proof. It is clear that the fixed point (1), parameter (2), and permutation (7) identities hold. We now establish the pairing identity (5). Suppose that $f : P \times Q \times R \to P$, $g : P \times Q \times R \to Q$ such that $f^\dagger$ and $h^\dagger$ exist. This means that for all $y \in Q$ and $z \in R$, $f^\dagger(y,z)$ is the least prefixed point solution of the single equation (8), and for all $z \in R$, $h^\dagger(z)$ is the least prefixed point solution of the equation

$$y = h(y,z).$$

We want to show that for all z, $(f^\dagger(h^\dagger(z),z), h^\dagger(z))$ is the least prefixed point solution of the system consisting of (8) and (9). But

$$f(f^\dagger(h^\dagger(z),z), h^\dagger(z), z) = f^\dagger(h^\dagger(z), z)$$

and

$$g(f^\dagger(h^\dagger(z),z), h^\dagger(z), z) = h(h^\dagger(z), z) = h^\dagger(z),$$

showing that $(f^\dagger(h^\dagger(z),z), h^\dagger(z))$ is a solution. Suppose that (x_0, y_0) is any prefixed point solution, so that $f(x_0, y_0, z) \le x_0$ and $g(x_0, y_0, z) \le y_0$. Then $f^\dagger(y_0, z) \le x_0$, and thus

$$h(y_0, z) = g(f^\dagger(y_0, z), z) \le g(x_0, y_0, z) \le y_0.$$

Thus, by fixed point induction, $h^\dagger(z) \le y_0$ and $f^\dagger(h^\dagger(z), z) \le f^\dagger(y_0, z) \le x_0$.

The double dagger and composition identities may be established directly using fixed point induction. Below, we show that these are already implied by (10), (11) and the pairing (5), and transposition (12) identities. First, note that by the pairing and transposition identities, we also have the following version of the pairing identity:

$$\langle f, g\rangle^\dagger = \langle k^\dagger, (g \circ (\pi^{Q,P}_{P,Q} \times \mathbf{id}_R))^\dagger \circ \langle k^\dagger, \mathbf{id}_R\rangle\rangle \tag{13}$$

where

$$k = f \circ \langle \mathbf{pr}^{P\times R}_P, (g \circ (\pi^{Q,P}_{P,Q} \times \mathbf{id}_R))^\dagger, \mathbf{pr}^{P\times R}_R\rangle. \tag{14}$$

Now, for the double dagger identity (4), assume that $f : P \times P \times Q \to P$. Let $g = \mathbf{pr}^{P\times P\times Q}_1$ and consider the function $\langle f, g\rangle : P \times P \times Q \to P \times P$. We can compute the second component of $\langle f, g\rangle^\dagger$ in two ways using the two versions of the pairing identity. The first version gives $f^{\dagger\dagger}$, while the second gives, using (10) and (11), $(f \circ (\langle \mathbf{id}_P, \mathbf{id}_P\rangle \times \mathbf{id}_Q))^\dagger$.

As for the composition identity (3), assume that $f : Q \times R \to P$, $g : P \times R \to P$. Then define $f' = f \circ \langle \mathbf{pr}_Q^{P\times Q\times R}, \mathbf{pr}_R^{P\times Q\times R}\rangle$ and $g' = g \circ \langle \mathbf{pr}_P^{P\times Q\times R}, \mathbf{pr}_R^{P\times Q\times R}\rangle$, and use the two versions of the pairing identity (and (10)) to compute the first component of $\langle f', g'\rangle^\dagger$ in two different ways. □

As already noted, the above identities are not all independent, (10), (11), (12) are instances of (1), (2), and (7), respectively. By the proof of Theorem 2.9, (10), (11) and the pairing (5), and transposition (12) identities imply (in conjunction with the usual laws of function composition and the Cartesian structure) the double dagger (4) and composition (3) identities. The fixed point identity is a particular instance of the composition identity (take $P = Q$ and $g = \mathbf{pr}_P^{P\times Q}$). In fact, the following systems are all equivalent, cf. [9]:

1. The system consisting of (10), (11), and the pairing (5), and transposition (12) (or permutation (7)) identities.
2. The system consisting of (10), (11) and the two versions of the pairing identity, (5) and (13).
3. The system consisting of the parameter (2), double dagger (4), and composition (3) identities.

Several other identities follow. For example, the following "symmetric version" of the Bekić identity follows. For all f, g as in the Bekić identity,

$$\langle f, g\rangle^\dagger = \langle k^\dagger, h^\dagger\rangle \tag{15}$$

where h and k are defined in (6) and (14). In the μ-notation, (15) can be written as

$$\begin{aligned}&\mu(x,y).(f(x,y,z), g(x,y,z))\\ &\quad = \big(\mu x.f(x, \mu y.g(x,y,z), z), \mu y.g(\mu x.f(x,y,z), y, z)\big).\end{aligned}$$

3 Conway Theories

In most applications of fixed point theory, one considers a collection T of functions $f : A^n \to A^m$, for a *fixed set* A, sometimes equipped with additional structure, where n, m are nonnegative integers. For example, T may consist of the monotone, or continuous functions $P^n \to P^m$, where P is a cpo. When T contains the projection functions and is closed under composition and tupling, we call T a *Lawvere theory of functions*, or just a *theory of functions*. The collection of all functions $A^n \to A$ of a theory of functions is a *function clone*, cf. [16].

There is a more abstract notion due to Lawvere [42]. We may think of a theory T of functions over a set A as a category whose objects are not the sets A^n, but rather the nonnegative integers n. A morphism $n \to m$ in T is a function $A^n \to A^m$, subject to certain conditions. As such, T is a *category* with all finite products in the categorical sense (cf., e.g., [3]), with $n+m$ being the *product* of n and m, and 0 being the *terminal object*.

Definition 3.1. *A theory is a small category whose objects are the nonnegative integers such that each integer n is the n-fold product of object 1 with itself.*

Morphisms between theories are defined in the natural way. They preserve objects, composition, and the projections. It follows that morphisms also preserve tupling (and thus pairing) and the identity morphisms. Below, when T is a theory, we denote by $T(m,n)$ the set of morphisms $n \to m$ in T. (Note the reversal of the source and the target.)

Below, we will assume that each theory T comes with given *projection morphisms* $\mathbf{pr}_i^n : n \to 1$, $i = 1, \ldots, n$ making object n the n-fold product of 1 with itself. In a similar way, we write $\mathbf{pr}_n^{n,m}$ and $\mathbf{pr}_m^{n,m}$ for the projections $n+m \to n$ and $n+m \to m$, given by $\langle \mathbf{pr}_1^{n+m}, \ldots, \mathbf{pr}_n^{n+m} \rangle$ and $\langle \mathbf{pr}_{n+1}^{n+m}, \ldots, \mathbf{pr}_{n+m}^{n+m} \rangle$, respectively, and $\mathbf{id}_n = \langle \mathbf{pr}_1^n, \ldots, \mathbf{pr}_n^n \rangle$ for the identity morphism $n \to n$. Without loss of generality, we will assume that $\mathbf{id}_1 = \mathbf{pr}_1^1$. Since 0 is a terminal object, for each n, there is a unique morphism $n \to 0$. In any theory, a *base morphism* is a tupling of projection morphisms. For example, the identity morphisms and the morphisms with target 0 are base morphisms. Note that there is a base morphism $n \to m$ corresponding to each function $\rho : \{1, \ldots, m\} \to \{1, \ldots, n\}$, namely the morphism $\langle \mathbf{pr}_{\rho(1)}^n, \ldots, \mathbf{pr}_{\rho(m)}^n \rangle$. When ρ is bijective, injective, etc. we will also say that the corresponding base morphism has the appropriate property. When $f : p \to m$ and $g : q \to n$, $f \times g : p+q \to m+n$ is $\langle f \circ \mathbf{pr}_p^{p,q}, g \circ \mathbf{pr}_q^{p,q} \rangle$. When T is understood, we will just write $f : n \to m$ for $f \in T(m,n)$.

There is a representation theorem for theories (see, e.g., [9]) by which each theory is isomorphic to a theory of functions. But very often there are more natural ways of representing the morphisms of a theory (e.g., as matrices over a semiring).

When a theory T is equipped with a dagger operation $^\dagger : T(n, n+p) \to T(n,p)$, $n, p \geq 0$, we define when the fixed point identity and the other identities given above hold in T in the natural and expected way. For example, the fixed point identity (1) is given by

$$f^\dagger = f \circ \langle f^\dagger, \mathbf{id}_p \rangle, \tag{16}$$

where $f : n+p \to n$. As another example, the pairing identity (5) is understood in the form

$$\langle f, g \rangle^\dagger = \langle f^\dagger \circ \langle h^\dagger, \mathbf{id}_p \rangle, h^\dagger \rangle \tag{17}$$

where $f : n+m+p \to n$, $g : n+m+p \to m$ and

$$h = g \circ \langle f^\dagger, \mathbf{id}_{m+p} \rangle.$$

Definition 3.2. *A* Conway theory *is a theory T equipped with a dagger operation $^\dagger : T(n, n+p) \to T(n,p)$ which satisfies (10), (11), the pairing (5), and transposition (12) (or permutation (7)) identities.*

Morphisms of Conway theories, or theories equipped with a dagger operation, also preserve dagger. Two alternative axiomatizations of Conway theories are given below. By the discussion at the end of the preceding section, we have the following theorem.

Theorem 3.3. *Let T be a theory equipped with a dagger operation. The following are equivalent:*

1. *T is a Conway theory.*
2. *T satisfies (10), (11), and the two versions of the pairing identity, (5) and (13).*
3. *T satisfies the parameter (2), double dagger (4), and composition (3) identities.*

Corollary 3.4. *Any Conway theory satisfies all of the identities defined above.*

Yet another axiomatization can be derived from the following result.

Theorem 3.5. *Suppose that T is a theory equipped with a* scalar dagger operation $^\dagger : T(1, 1+p) \to T(1,p)$, $p \geq 0$ *satisfying the* scalar parameter *(18),* scalar composition *(19) and* scalar double dagger *(20) identities below.*

SCALAR PARAMETER IDENTITY

$$(f \circ (\mathbf{id}_1 \times g))^\dagger = f^\dagger \circ g, \tag{18}$$

for all $f : 1+p \to 1$ and $g : q \to p$.

SCALAR COMPOSITION IDENTITY

$$(f \circ \langle g, \mathbf{pr}_p^{1,p} \rangle)^\dagger = f \circ \langle (g \circ \langle f, \mathbf{pr}_p^{1,p} \rangle)^\dagger, \mathbf{id}_p \rangle, \tag{19}$$

for all $f, g : 1+p \to 1$.

SCALAR DOUBLE DAGGER IDENTITY

$$f^{\dagger\dagger} = (f \circ (\langle \mathbf{id}_1, \mathbf{id}_1 \rangle \times \mathbf{id}_p))^\dagger, \tag{20}$$

for all $f : 2+p \to 1$.

Then there is a unique way to extend the dagger operation to all morphisms $n+p \to n$ for all $n, p \geq 0$ such that T becomes a Conway theory.

Proof. The unique extension is given by induction on n. When $n = 0$, $^\dagger : T(0,p) \to T(0,p)$ is the identity function on the singleton set $T(0,p)$. On morphisms in $T(1, 1+p)$, the dagger is already defined. Suppose that $n > 1$ and $f \in T(n, n+p)$. Then let $m = n-1$ and write f as $f = \langle f_1, f_2 \rangle$ where $f_1 : m+1+p \to m$, $f_2 = m+1+p \to 1$. Then define $f^\dagger$ as $\langle f_1^\dagger \circ \langle h^\dagger, \mathbf{id}_p \rangle, h^\dagger \rangle$ where $h = g \circ \langle f_1^\dagger, \mathbf{pr}_1^{1+p} \rangle$. □

Corollary 3.6. *A theory T equipped with a dagger operation is a Conway theory iff T satisfies the scalar versions of the parameter, composition, and double dagger identities and the* scalar version of the pairing identity *(17), where f is arbitrary but g is scalar (i.e., $m = 1$).*

A detailed study of the identities true of all Conway theories is given in [5]. It is shown that there is an algorithm to decide whether an identity holds in all Conway theories, and that this decision problem is complete for PSPACE. The proof is based on a description of the structure of the free Conway theories using "aperiodic congruences" of flowchart schemes.

Remark 3.7. In any theory T equipped with a dagger operation, one may define a *feedback operation* $\uparrow : T(n+p, n+q) \to T(p,q)$, $n,p,q \geq 0$: Given $\langle f,g \rangle : n+q \to n+p$ with $f : n+q \to n$ and $g : n+q \to p$, we *define* $\uparrow\langle f,g \rangle = g \circ \langle f^\dagger, \mathbf{id}_q \rangle$. Then T, equipped with the feedback operation and the operation $\times$ as "tensor product" is a *traced monoidal category* [38]. The same notion was earlier defined under a different name in connection with flowcharts; see [49]. In fact, Conway theories correspond to traced monoidal categories whose tensor product is a (Cartesian) product. Another aspect of the connection is that traced monoidal categories are axiomatized by the identities that hold for flowchart schemes, and Conway theories by those that hold for flowchart schemes modulo aperiodic simulations (and the iteration theories defined in the next section are axiomatized by the identities that hold for flowchart schemes with respect to arbitrary simulations, or strong behavioral equivalence). Flowchart schemes were first axiomatized in [8]. For more information on the connection between Conway theories and traced monoidal categories, we refer to [36, 49]. See also Chap. 6, Sect. 8 in [9].

4 Iteration Theories

The Conway identities do not capture all equational properties of the least (pre)fixed point operation. In order to achieve completeness, we now introduce the *commutative identity* in any theory T equipped with a dagger operation:

$$\mathbf{pr}_1 \circ \langle f \circ (\rho_1 \times \mathbf{id}_p), \ldots, f \circ (\rho_n \times \mathbf{id}_p) \rangle^\dagger = \left(f \circ (\rho \times \mathbf{id}_p)\right)^\dagger, \qquad (21)$$

where $f : n+p \to 1$, $n \geq 1$, each $\rho_i : n \to n$ is a base morphism (i.e., a tupling of projections), and ρ is the unique base morphism $1 \to n$, i.e., ρ is the *diagonal* $\langle \mathbf{id}_1, \ldots, \mathbf{id}_1 \rangle$. Particular instances of the commutative identity are the *group identities*. Suppose that G is a finite group of order n with group operation denoted. Moreover, suppose for simplicity that the carrier of G is the set $\{1, \ldots, n\}$ with 1 being the unit element of G. For each i, define ρ_i as the tupling of the n projection morphisms $\mathbf{pr}^n_{i \cdot j}$, so that $\rho_i = \langle \mathbf{pr}^n_{i1}, \ldots, \mathbf{pr}^n_{in} \rangle$. Then the commutative identity above is called the group identity associated with G. (When the permutation identity holds, as will be the case below, it does not matter how the elements of the group G are enumerated.)

The commutative identity (21) can be explained in theories of continuous or monotone functions over cpo's as follows. Suppose that P is a cpo and $f : P^{n+p} \to P$ is continuous. Moreover, suppose that each $\rho_i : P^n \to P^n$ is a tupling of projections, i.e., a base function. Then consider the system of equations in n unknowns and p parameters:

$$x_1 = f(x_{\rho_1(1)}, \ldots, x_{\rho_1(n)}, y_1, \ldots, y_p),$$

$$\vdots$$

$$x_n = f(x_{\rho_n(1)}, \ldots, x_{\rho_n(n)}, y_1, \ldots, y_p).$$

The commutative identity asserts that the first component of the least solution of this parametric system is just the least solution of the single parametric equation

$$x = f(x, \ldots, x, y_1, \ldots, y_p).$$

When the permutation identity holds, the same is true for all other components.

Definition 4.1. *An iteration theory is a Conway theory satisfying the group identities.*

Morphisms of iteration theories are Conway theory morphisms. Iteration theories were defined in [6, 7] and independently in [24]. The axiomatization in [24] used the Conway theory identities and the "vector form" of the commutative identity; see below. The completeness of the group identities in conjunction with the Conway theory identities was proved in [27].

Theorem 4.2. *An identity involving the dagger operation holds in all theories of continuous functions on cpo's iff it holds in all theories of monotone functions on cpo's iff it holds in iteration theories.*

The proof is based on a concrete description of the free iteration theories as theories of *regular trees*, cf. [9], which are the unfoldings of finite flowchart schemes [8]. By this concrete description, it is known that there is a P-time algorithm to decide whether an identity holds in all iteration theories; see [18].

Theorem 4.2 can be generalized to a great extent. The following result was proved in [26].

Theorem 4.3. *The iteration theory identities are complete for the class of all theories T equipped with a partial order $\leq$ on each hom-set $T(n, m)$ and a dagger operation such that the operations of composition and tupling are monotone. Moreover, the fixed point identity (1), the parameter identity (2), and the fixed point induction axiom hold, so that*

$$f \circ \langle g, \mathbf{id}_p \rangle \leq g \implies f^\dagger \leq g, \tag{22}$$

for all $f : n + p \to n$ and $g : p \to n$.

Thus, in such theories, called *Park theories* in [26], the fixed point equation $\xi = f \circ \langle \xi, \mathbf{id}_p \rangle$ has a least solution, namely $f^\dagger$. With the same argument as in the proof of Proposition 2.8, it follows that the dagger operation is also monotone. Equivalently, one may define Park theories as ordered theories as above satisfying the scalar parameter identity (18), the scalar versions of the fixed point and pairing identities, i.e., (16) with $n = 1$ and (17) with $m = 1$, and the fixed point induction axiom (22) for $n = 1$. See the proof of the Bekić identity. Moreover, the fixed point identity may be replaced by the inequality $f \circ \langle f^\dagger, \mathbf{id}_p \rangle \leq f^\dagger$, for all appropriate f. Instances of Park theories are the *continuous theories* and *rational theories*, cf. [9, 51]. In a continuous theory T, each $T(m, n)$ is a cpo and composition is continuous. The dagger operation is defined by least fixed points. In particular, the theory of continuous functions over a cpo is a continuous theory. A rational theory T is also ordered, but not all directed sets in $T(m, n)$ have suprema. But there are enough suprema to have least solutions of fixed point equations. It is known that each rational theory embeds in a continuous theory.

More generally, one often considers certain 2-categories, called *2-theories*, such that for each $f : n + p \to n$ there is an *initial solution* of the fixed point equation $\xi = f \circ \langle \xi, \mathbf{id}_p \rangle$. The identities satisfied by such 2-theories are again those of iteration theories, cf. [34].

An essential feature of iteration theories is that the "vector form" of each identity true of iteration theories holds in all iteration theories. In a semantic setting, this means the following. Given a theory T and an integer k, we can form a new theory Tk whose morphisms $m \to n$ are the morphisms $mk \to nk$ of T. The composition operation in Tk is that inherited from T, and the ith projection morphism $n \to 1$ in Tk is $\langle \mathbf{pr}^{nk}_{(i-1)k+1}, \ldots, \mathbf{pr}^{nk}_{ik} \rangle$. If T is equipped with a dagger operation, then Tk is equipped with the dagger operation inherited from T, since if $f : n+p \to n$ in Tk, then f is a morphism $nk + pk \to nk$ in T and we may define $f^\dagger$ in Tk as the morphism $f^\dagger$ in T. For details, see [27].

Theorem 4.4. *When T is a Conway or iteration theory, so is Tk for each k.*

Proof. The claim is clear for Conway theories, since the vector form of each defining identity of Conway theories is also a defining identity. As for iteration theories, by the completeness of the iteration theory identities for the least fixed point operation on continuous functions on cpo's (Theorem 4.2), it suffices to prove that if T is the theory of continuous functions on a cpo P, equipped with the least fixed point operation, then each Tk is an iteration theory. But Tk is isomorphic to the theory of continuous functions over P^k, which is an iteration theory. □

The commutative identity and the group identities seem to be extremely difficult to verify in practice. But in most cases, this is not so. The commutative identity, and thus the group identities are implied by certain quasi-identities, which are usually easy to establish.

Definition 4.5. *Let C be a set of morphisms in a theory T equipped with a dagger operation. We say that T has a* functorial dagger *with respect to C if*

$$f \circ (\rho \times \mathbf{id}_p) = \rho \circ g \quad \Longrightarrow \quad f^\dagger = \rho \circ g^\dagger, \tag{23}$$

for all $f : n + p \to n$, $g : m + p \to m$ in T and $\rho : m \to n$ is in C.

When T has a functorial dagger with respect to the set of all base morphisms (all morphisms, respectively), we also say that T has a *weak* (*strong*, respectively) functorial dagger. It is known that every Conway theory has a functorial dagger with respect to the set of injective base morphisms. Moreover, if T has a strong functorial dagger, then it has a unique morphism $0 \to 1$. In [25], it is proved that if a Conway theory has a functorial dagger with respect to the set of base morphisms $1 \to n$, $n \geq 2$, then it has a weak functorial dagger.

Proposition 4.6. *If a Conway theory T has a weak functorial dagger, then T is an iteration theory.*

Proof. We show that under the assumptions, the commutative identity (21) holds. So, let $f : n + p \to 1$ and let $\rho_1, \ldots, \rho_n$ be base morphisms $n \to n$, and let ρ denote the unique base morphism $1 \to n$. Define $g = f \circ (\rho \times \mathbf{id}_p)$ and $h = \langle f \circ (\rho_1 \times \mathbf{id}_p), \ldots, f \circ (\rho_n \times \mathbf{id}_p)\rangle$. Then $h \circ (\rho \times \mathbf{id}_p) = \rho \circ g$, so that $h^\dagger = \rho \circ g^\dagger$, completing the proof. □

For other quasi-identities implying the commutative identity, we refer to [9, 11]. It is known that there exist iteration theories which do not have a weak functorial dagger.

Simpson and Plotkin [48] proved the following equational completeness result for iteration theories. Suppose that T is a nontrivial iteration theory equipped with a dagger operation, so that T has at least two morphisms $2 \to 1$, or equivalently, $\mathbf{pr}_1^2 \neq \mathbf{pr}_2^2$ in T. Then there are two cases. Either an identity holds in T iff it holds in all iteration theories, or it holds in all iteration theories with a unique morphism $0 \to 1$. It was argued in [9, 11] that all fixed point models satisfy at least the iteration theory identities. Thus, by the Plotkin–Simpson result, all nontrivial fixed point models either satisfy exactly the iteration theory identities, or the identities that hold in all iteration theories with a single "constant." Such iteration theories are, for example, the matrix theories over nontrivial iteration semirings defined below. Iteration theories of Boolean functions are described in [28].

5 Unique Fixed Points

Suppose that T is a theory. We say that a morphism $f = \langle f_1, \ldots, f_m\rangle : n \to m$ in T is *ideal* if none of the morphisms $f_i : n \to 1$ is a projection. Following Elgot [22], we call T an *ideal theory* if whenever f is ideal, then for all g in T with appropriate target, $f \circ g$ is ideal.

An important example of an ideal theory can be constructed over *complete metric spaces* $M = (M, d)$, where d denotes a distance function. It is clear that when (M, d) is complete, so is any finite power M^n of M equipped with the distance function d_n defined by $d_n((x_1, \ldots, x_n), (y_1, \ldots, y_n)) = \max\{d(x_i, y_i) : i = 1, \ldots, n\}$. Now, a function $f : M \to M'$ between metric spaces $M = (M, d)$ and $M' = (M', d')$ is called a *proper contraction* if there is a constant $0 < c < 1$ such that $d'(f(x), f(y)) \leq cd(x, y)$ for all $x, y \in M$. The following simple but important fact is Banach's fixed point theorem [2].

Theorem 5.1. *When M is a complete metric space and $f : M \to M$ is a proper contraction, then f has a unique fixed point.*

Proof. If x, y are both fixed points, then $d(x, y) = d(f(x), f(y)) \leq cd(x, y)$ for some $0 < c < 1$. It follows that $d(x, y) = 0$, i.e., $x = y$.

To show that there is at least one fixed point, let $x_0 \in M$ and define $x_{n+1} = f(x_n)$ for all $n \geq 0$. Since f is a proper contraction, the sequence $(x_n)_n$ is a Cauchy sequence, and since M is complete, it has a limit x. Since f is a proper contraction, it follows that $f(x) = x$. □

Let M be a complete metric space. Consider the collection T_M of all functions $M^n \to M^m$, $n, m \geq 0$ of the form $f = \langle f_1, \ldots, f_m \rangle$ such that each $f_i : M^m \to M$ is a proper contraction *or* a projection. It is clear that T_M is closed under composition and tupling, so that it is a theory of functions over M. Moreover, T_M is an ideal theory, since if M is nontrivial then a function $f = \langle f_1, \ldots, f_m \rangle$ is an ideal morphism iff each f_i is a proper contraction which implies that each component function of $f \circ g$ is also a proper contraction for any g in T_M with appropriate target.

Definition 5.2. *An iterative theory (cf. Elgot [22]) is an ideal theory T equipped with a dagger operation defined on ideal morphisms in $T(n, n + p)$, $n, p \geq 0$ such that for each ideal $f : n + p \to n$, the morphism $f^\dagger : p \to n$ is the unique solution of the fixed point equation $\xi = f \circ \langle \xi, \mathbf{id}_p \rangle$.*

Thus, the fixed point identity (16) and the *unique fixed point rule*

$$f \circ \langle g, \mathbf{id}_p \rangle = g \quad \Longrightarrow \quad g = f^\dagger$$

hold for all ideal $f : n + p \to n$ and all $g : p \to n$ in T.

Remark 5.3. Let T be an ideal theory. We say that $f : n + p \to n$ in T is a *power ideal* morphism if for some $k \geq 1$, f^k is ideal. When $f : n + p \to n$ is a power ideal morphism in an iterative theory T, then the fixed point equation $\xi = f \circ \langle \xi, \mathbf{id}_p \rangle$ has a unique solution, namely the solution of $\xi = f^k \circ \langle \xi, \mathbf{id}_p \rangle$, where f^k is ideal. See [22]. (Here, $f^0 = \mathbf{pr}_n^{n,p}$ and $f^{k+1} = f \circ \langle f^k, \mathbf{pr}_p^{n,p} \rangle$.)

The following result is from [13]; see also [9].

Theorem 5.4. *An ideal theory T is an iterative theory iff for each ideal morphism $f : 1 + p \to 1$ there is a unique solution of the equation $\xi = f \circ \langle \xi, \mathbf{id}_p \rangle$.*

Proof. The proof is based on a version of the pairing identity. One argues by induction. In the induction step, one shows that if the fixed point equation for ideal morphisms $n + q \to n$ and $m + q \to m$ have unique solutions, then the same holds for ideal morphisms $n + m + p \to n + m$. □

In an iterative theory, the dagger operation is only partially defined. In order to be able to solve all fixed point equations over an iterative theory T, there must be at least one morphism $0 \to 1$ in T.

Theorem 5.5. *Suppose that T is an iterative theory with at least one morphism $0 \to 1$. Then for each $\bot : 0 \to 1$, the dagger operation on T has a unique extension to all morphisms $n+p \to n$, $n, p \geq 0$ such that T becomes a Conway theory with $\mathbf{id}_1^\dagger = \bot$. Moreover, equipped with this unique extension, T is an iteration theory having a weak functorial dagger.*

This result was proved in [6, 7] and [24]. Iteration theories arising from Theorem 5.5 are called *pointed iterative theories.* One application of the theorem is the following.

Corollary 5.6. *Suppose that M is a complete metric space and consider the theory T_M defined above. Let x_0 be a point in M. Then there is a unique way to define a dagger operation on T_M such that T_M becomes a Conway theory with $\mathbf{id}_M^\dagger = x_0$. This unique Conway theory is an iteration theory with a weak functorial dagger.*

Without proof, we mention the following theorem.

Theorem 5.7. *An identity holds in all pointed iterative theories iff it holds in iteration theories.*

See [9, 24]. Thus, the equational properties of the least fixed point operation are the same as the equational properties of the unique fixed point operation.

6 Fixed Points of Linear Functions

Let S be a semiring. A function $S^n \to S$ is called *linear* if it is of the form

$$f(x_1, \ldots, x_n) = s_1 x_1 + \cdots + s_n x_n$$

for some $s_1, \ldots, s_n \in S$. A linear function $S^n \to S^m$ is a tupling of linear functions $S^n \to S$. Since any composition of linear functions is linear, it follows that linear functions over S determine a theory T_S.

The linear function f given above may be represented by the n-dimensional row matrix $(s_1, \ldots, s_n)$. More generally, any linear function $S^n \to S^m$ may be represented by an $m \times n$ matrix $M = (s_{ij})_{ij}$ over S: The linear function determined by M maps $x \in S^n$, an n-dimensional column vector to Mx, an m-dimensional column. It follows that T_S can be represented as the theory T with $T(m,n) = S^{m\times n}$, $m,n \geq 0$, the set of all $m \times n$ matrices over S, whose composition operation is matrix product. The projections are the row matrices with an entry equal to 1 and all other entries equal to 0. The identity morphism $\mathbf{id}_n$, $n \geq 0$ is the $n \times n$ unit matrix E_n. We denote this theory by $\mathbf{MAT}_S$ and call it the *matrix theory* over S.

Proposition 6.1. *T_S is isomorphic to $\mathbf{MAT}_S$.*

Note that in $\mathbf{MAT}_S$, a base morphism, also called a *base* or *functional matrix*, is a 0–1 matrix with a single occurrence of 1 in each row (at least when S is nontrivial). In particular, every *permutation matrix* is a base matrix. Note that the inverse of a permutation matrix π is its transpose, π^T. It is educational to see that for all $A \in \mathbf{MAT}_S(p,n)$ and $B \in \mathbf{MAT}_S(q,n)$,

$$\langle A, B\rangle = \begin{pmatrix} A \\ B \end{pmatrix},$$

and if $A \in \mathbf{MAT}_S(p,n)$ and $B \in \mathbf{MAT}_S(q,m)$ then

$$A \times B = \begin{pmatrix} A & 0_{pm} \\ 0_{qn} & B \end{pmatrix},$$

where 0_{pn} and 0_{qm} are zero matrices of appropriate dimension.

Below, we will show that any dagger operation on $\mathbf{MAT}_S$ satisfying the parameter identity determines and is determined by a star operation on $\mathbf{MAT}_S$, and, in fact, on the semiring S. Moreover, we show how to express the Conway identities and the commutative and group identities in terms of the star operation giving rise to Conway matrix theories, matrix iteration theories, and Conway and iteration semirings.

Proposition 6.2. *Suppose that $\mathbf{MAT}_S$ is equipped with a dagger operation such that the parameter identity holds. Then there is a unique* star operation *$A \mapsto A^*$ defined on the square matrices $A \in \mathbf{MAT}_S(n,n)$, $n \geq 0$ such that for all $(A,B) \in \mathbf{MAT}(n, n+p)$ with $A \in \mathbf{MAT}_S(n,n)$ and $B \in \mathbf{MAT}_S(n,p)$*

$$(A, B)^\dagger = A^* B. \tag{24}$$

If $\mathbf{MAT}_S$ is equipped with a star operation and if we define dagger by (24), then the parameter identity holds.

Proof. If the parameter identity (2) holds, then

$$(A \quad B)^\dagger = \left((A \quad E_n)\begin{pmatrix} E_n & 0 \\ 0 & B \end{pmatrix}\right)^\dagger = (A \quad E_n)^\dagger B.$$

Thus, we define $A^* = (A, E_n)^\dagger$. With this definition, (24) holds. Moreover, if $\mathbf{MAT}_S$ is equipped with a star operation and if we define dagger by (24), then the parameter identity holds. □

Theorem 6.3. *Suppose that* $\mathbf{MAT}_S$ *is equipped with both a star and a dagger operation which are related by (24).*

1. *The fixed point identity (1) holds iff the* star fixed point identity *holds:*
$$A^* = AA^* + E_n \tag{25}$$
for all $A \in \mathbf{MAT}_S(n,n)$, $n \geq 0$.
2. *The double dagger identity (4) holds iff the* sum star identity *holds:*
$$(A+B)^* = (A^*B)^*A^* \tag{26}$$
for all $A, B \in \mathbf{MAT}_S(n,n)$, $n \geq 0$.
3. *The composition identity (3) holds iff the* product star identity *holds:*
$$(AB)^* = E_n + A(BA)^*B \tag{27}$$
for all $A \in \mathbf{MAT}_S(n,m)$, $B \in \mathbf{MAT}_S(m,n)$, $m, n \geq 0$.
4. *The identity (10) holds iff the* zero star identity *holds:*
$$0_{nn}^* = E_n, \tag{28}$$
where the entries of the $n \times n$ *matrix* 0_{nn} *are all* 0.
5. *The pairing identity (5) holds iff the* matrix star identity *holds:*
$$\begin{pmatrix} A & B \\ C & D \end{pmatrix}^* = \begin{pmatrix} \alpha & \beta \\ \gamma & \delta \end{pmatrix} \tag{29}$$
where $A \in \mathbf{MAT}_S(n,n)$, $B \in \mathbf{MAT}_S(n,m)$, $C \in \mathbf{MAT}_S(m,n)$, *and* $D \in \mathbf{MAT}_S(m,m)$, *and where*
$$\alpha = A^*B\delta CA^* + A^*, \qquad \beta = A^*B\delta,$$
$$\gamma = \delta CA^*, \qquad \delta = (D + CA^*B)^*.$$
6. *The permutation identity (7) holds iff the* star permutation identity *holds:*
$$\left(\pi A \pi^T\right)^* = \pi A^* \pi^T, \tag{30}$$
where $A \in \mathbf{MAT}_S(n,n)$ *and where* $\pi \in \mathbf{MAT}_S(n,n)$ *is a permutation matrix with transpose* π^T.
7. *The transposition identity (12) holds iff the* star transposition identity *holds, i.e., the identity (30) when* $n = p + q$ *and* $\pi = \left(\begin{smallmatrix} 0 & E_p \\ E_q & 0 \end{smallmatrix}\right)$.
8. *The group identity associated with a finite group* G *of order* n *holds iff* star group identity *associated with* G *holds:*
$$e_1 M_G^* u_n = (a_1 + \cdots + a_n)^* \tag{31}$$
where M_G *is the* $n \times n$ *matrix whose* (i,j)*th entry is* $a_{i^{-1}j}$, *for all* $1 \leq i, j \leq n$, *and* $e_1 = \mathbf{pr}_1^n$ *is the* $1 \times n$ *0–1 matrix whose first entry is* 1 *and whose other entries are* 0. *Finally,* u_n *is the* $n \times 1$ *matrix all of whose entries are* 1.

Definition 6.4. *A* Conway matrix theory (matrix iteration theory) *is a matrix theory* $\mathbf{MAT}_S$ *equipped with a star operation defined on square matrices such that when dagger is defined by (24) then it is a Conway theory (iteration theory, respectively). A morphism of Conway matrix theories or matrix iteration theories is a theory morphism which preserves star.*

It follows that morphisms also preserve the additive structure.

Note that $A^* = A^*A + E_n$ holds for all $n \times n$ matrices A in any Conway matrix theory. As an immediate corollary to Theorem 3.3, Theorem 3.5, Corollary 3.6, and Theorem 6.3 we obtain the following corollary.

Corollary 6.5. *A matrix theory T equipped with a star operation is a Conway theory iff one of the following three groups of identities holds in T.*

1. *The zero star (28), matrix star (29), and star transposition (or star permutation (30)) identities.*
2. *The product star (27) and sum star (26) identities.*
3. *The scalar versions of the product star, sum star, and matrix star identities, i.e., (27) with $m = n = 1$, (26) with $n = 1$, and (29) with $m = 1$.*

Thus, all identities (25)–(30) hold in Conway matrix theories. By adding to the axioms, the star group identities (31) associated with the finite groups, one obtains three sets of equational axioms for matrix iteration theories.

We note the following version of the matrix star identity:

$$\begin{pmatrix} A & B \\ C & D \end{pmatrix}^* = \begin{pmatrix} (A + BD^*C)^* & (A + BD^*C)^*BD^* \\ (D + CA^*B)^*CA^* & (D + CA^*B)^* \end{pmatrix}.$$

When $n = m = 1$, the product star and sum star identities only involve elements of the semiring S. This consideration gives rise to the following definitions; see also [21].

Definition 6.6. *A* *-semiring *is a semiring S equipped with a unary star operation $^* : S \to S$. A Conway semiring [9] is a *-semiring S which satisfies the (scalar) product star and sum star identities, i.e.,*

$$(ab)^* = a(ba)^*b + 1, \tag{32}$$

$$(a + b)^* = (a^*b)^*a^*, \quad a, b \in S. \tag{33}$$

An iteration semiring [9, 27] is a Conway semiring, which when the star of a square matrix is inductively defined by the scalar version of the matrix star identity (i.e., (29) with $m = 1$), satisfies each star group identity (31) associated with a finite group G. A morphism of *-semirings also preserves the star operation. A morphism of Conway or iteration semirings is a *-semiring morphism.

Corollary 6.7. *When* $\mathbf{MAT}_S$ *is a Conway matrix theory or a matrix iteration theory, then* S *is a Conway semiring or an iteration semiring. Suppose that* S *is a Conway semiring (iteration semiring, resp.). Then there is a unique way of extending the star operation on* S *to all square matrices over* S *such that* $\mathbf{MAT}_S$ *becomes a Conway matrix theory, or a matrix iteration theory.*

Proof. This follows from Theorems 3.5 and 6.3. □

Of course, the unique extension is given by the scalar version of the matrix star identity (29) with $m = 1$. Using the above results, one can show that the category of Conway semirings is equivalent to the category of Conway matrix theories, and that the category of iteration semirings is equivalent to the category of matrix iteration theories.

In any Conway matrix theory, the group identities follow from the functorial star conditions defined below.

Definition 6.8. *Suppose that* $\mathbf{MAT}_S$ *is equipped with a star operation. Let* $\mathcal{C}$ *be a set of matrices in* $\mathbf{MAT}_S$*. We say that* $\mathbf{MAT}_S$ *satisfies the* functorial star *implication for* $\mathcal{C}$*, or that* $\mathbf{MAT}_S$ *has a functorial star with respect to* $\mathcal{C}$*, if for all* $A \in \mathbf{MAT}_S(n,n)$ *and* $B \in \mathbf{MAT}_S(m,m)$ *and all* $C \in \mathbf{MAT}_S(n,m)$ *in* $\mathcal{C}$*,*

$$AC = CB \quad \Longrightarrow \quad A^*C = CB^*.$$

When $\mathbf{MAT}_S$ *has a functorial star with respect to the set of all matrices in* $\mathbf{MAT}_S$*, then* $\mathbf{MAT}_S$ *is said to have a strong functorial star. And when* $\mathbf{MAT}_S$ *has a functorial star with respect to the set of all base matrices,* $\mathbf{MAT}_S$ *is said to have a weak functorial star.*

Proposition 6.9. *A Conway matrix theory* $\mathbf{MAT}_S$ *has a functorial star with respect to* $\mathcal{C}$ *iff it has functorial dagger with respect to* $\mathcal{C}$ *when dagger is defined by (24).*

Thus, $\mathbf{MAT}_S$ has a strong or weak functorial star iff it has a strong or weak functorial dagger. Moreover, $\mathbf{MAT}_S$ has a weak functorial star iff it has a weak functorial star with respect to all $n \times 1$ base matrices, $n \geq 2$.

Corollary 6.10. *If* $\mathbf{MAT}_S$ *is a Conway matrix theory with a weak functorial star, then* $\mathbf{MAT}_S$ *is a matrix iteration theory.*

We mention one more property of Conway and iteration semirings. The *dual* of a *-semiring S is equipped with the same sum and star operation and constants as S, but multiplication, denoted $\circ$, is defined by $a \circ b = ba$, the product of a and b in S in the reverse order.

Proposition 6.11. *The dual of a Conway or iteration semiring is also a Conway or iteration semiring.*

See [27]. In the rest of this section, we will exhibit three subclasses of iteration semirings.

6.1 Inductive *-Semirings

Recall from [21] that an *ordered monoid* is a commutative monoid $(M, +, 0)$ such that the sum operation is monotone. An *ordered semiring* is a semiring which is an ordered monoid such that the product operation is also monotone. Moreover, an ordered semiring S is *positively ordered* if $0 \leq s$ for all $s \in S$. A morphism of ordered semirings is a monotone semiring morphism. This section is based on [31].

Definition 6.12. *An* inductive *-semiring *is an ordered semiring S which is a *-semiring such that for any $a, b \in S$, a^*b is the least prefixed point of the function $S \to S$, $x \mapsto ax + b$. A morphism of inductive *-semirings is a morphism of ordered semirings which is a *-semiring morphism.*

Proposition 6.13. *Any inductive *-semiring S is positively ordered.*

Proof. The least solution of the equation $x = x$ is $1^* \cdot 0 = 0$. Since any element of S is a solution, it follows that 0 is the least element of S. □

Proposition 6.14. *When S is an inductive *-semiring, the star operation is monotone.*

Proof. This follows from Proposition 2.8. □

The dual of an inductive *-semiring is not necessarily an inductive *-semiring.

Definition 6.15. *A* symmetric inductive *-semiring *is an inductive *-semiring whose dual is also an inductive *-semiring.*

Proposition 6.16. *An inductive *-semiring S is symmetric iff for all $a, b, x \in S$, if $xa + b \leq x$, then $ba^* \leq x$.*

If S is an ordered semiring, $\mathbf{MAT}_S$ is equipped with the pointwise partial order. It is clear that the theory operations are monotone as is the sum operation on matrices.

Theorem 6.17. *Let S be an ordered semiring which is a *-semiring. Then S is an inductive *-semiring iff $\mathbf{MAT}_S$ is a Park theory when the dagger is defined by (24).*

Proof. This follows from Theorem 6.3 and the Bekić identity (5). □

Corollary 6.18. *Thus, when S is an inductive *-semiring, then for each $A \in \mathbf{MAT}_S(n, n)$ and $B \in \mathbf{MAT}_S(n, p)$, $(A, B)^\dagger = A^*B$ is the least prefixed point solution of the equation $X = AX + B$.*

Corollary 6.19. *Any inductive *-semiring S is an iteration semiring, so that $\mathbf{MAT}_S$ is a matrix iteration theory.*

Proof. By Theorem 4.3. □

Corollary 6.20. *If S is an inductive *-semiring, so is $S^{n\times n}$, for each $n \geq 0$.*

Recall from [21] that an ordered semiring S is *continuous* if S is a cpo with least element 0 and the sum and product operations are continuous.

Proposition 6.21. *Every continuous semiring is a symmetric inductive *-semiring where $a^* = \bigvee\{\sum_{i=1}^n a^i : n \geq 0\}$.*

Proof. This follows from Corollary 2.6. When S is a continuous semiring, then for each $a, b \in S$, the function $f(x) = ax + b$ is continuous with least prefixed point $\bigvee f^n(0, b)$. But for each n, $f^n(0, b) = \sum_{i=1}^n a^i b = (\sum_{i=1}^n a^i)b$, so that by continuity, $\bigvee f^n(0, b) = (\bigvee\{\sum_{i=1}^n a^i : n \geq 0\})b$. Since the dual of a continuous semiring is also continuous, it follows now that any continuous semiring is a symmetric inductive *-semiring. □

Kozen [40] defines a *Kleene algebra* as an idempotent symmetric inductive *-semiring. In [39], it is shown that there is an idempotent inductive *-semiring which is not a Kleene algebra.

Remark 6.22. Kozen showed in [40] that for each alphabet A, the semiring of regular languages over A, equipped with the partial order of set inclusion is the free Kleene algebra on A. Krob [41] proved that the same semiring is the free iteration semiring on A satisfying the identity $1^* = 1$, and thus also the free idempotent inductive *-semiring on A. See also [14, 15] and [10]. For recent extensions of these results, see [12, 29]. It is shown in [12] that for each alphabet A, the *-semiring of rational power series [33] over the semiring $\mathbb{N}^\infty$ is the free iteration semiring over A satisfying three additional simple identities. Moreover, an identity holds in these semirings iff it holds in all continuous (or complete, see below) semirings. And the same semirings, equipped with the sum order, are the free symmetric inductive *-semirings. The paper [12] also contains a characterization of the semirings of rational power series over the semiring $\mathbb{N}$ as the free "*partial iteration semirings.*"

6.2 Complete Semirings

For the definition of *complete semirings* and their morphisms, we refer to [21] where original references can be found. When S is a complete semiring, then we may equip each hom-set $\mathbf{MAT}_S(n, m) = S^{n\times m}$ with the pointwise sum operation, so that the straightforward generalizations of the defining axioms of complete semirings hold. In particular,

$$\sum_{j\in J}\sum_{i\in I_j} A_i = \sum_{i\in\bigcup_{j\in J} I_j} A_i,$$

$$B\Big(\sum_{i\in I} A_i\Big) = \sum_{i\in I} BA_i,$$

$$\Big(\sum_{i\in I} A_i\Big)C = \sum_{i\in I} A_iC,$$

where I is any set which is the disjoint union of sets I_j, $j \in J$, and where A_i, $i \in I$ is a family of matrices in $\mathbf{MAT}_S(m,n)$, and $B \in \mathbf{MAT}_S(p,m)$ and $C \in \mathbf{MAT}_S(n,q)$.

Now, any complete semiring S can be turned into a *-semiring by defining $s^* = \sum s^k$. By the above remark, if S is complete, then each semiring $S^{n\times n}$ is also complete and is thus a *-semiring with $A^* = \sum A^k$, for each $A \in S^{n\times n}$. We can use the star operation on S and the scalar version of the matrix star identity to define another star operation on $S^{n\times n}$. However, the two star operations coincide as noticed in [17, 37]. We have the following result, cf. [9].

Theorem 6.23. *When S is a complete semiring, then S is an iteration semiring with a strong functorial star. Thus, S is an iteration semiring and* $\mathbf{MAT}_S$ *is a matrix iteration theory.*

In a *rationally additive semiring* S, only certain sums are required to exist including the geometric sums $s^* = \sum_{n\geq 0} s^n$, for all $s \in S$. It is shown in [30] that they are also iteration semirings with a strong functorial star.

6.3 Iterative Semirings

An *ideal* of a semiring S is a set $I \subseteq S$ which is closed under the sum operation and contains 0. Moreover, $SI = IS = I$. Let I be an ideal of S and S_0 a subsemiring of S. Below, we will say that S is the *direct sum* of S_0 and I if each $s \in S$ can be written in a unique way in the form $s_0 + a$, where $s_0 \in S_0$ and $a \in I$. The following result is from [9].

Theorem 6.24. *Suppose that S is the direct sum of S_0 and I, where S_0 is a subsemiring of S and I is an ideal. Moreover, suppose that each fixed point equation $x = ax + b$ with $a \in I$ has a unique solution in S. If S_0 is a Conway semiring, then there is a unique way to extend the star operation on S_0 to the whole semiring S such that S becomes a Conway semiring. Moreover, when S_0 is an iteration semiring, then so is S.*

Proof. First, we define a^* for all $a \in I$ as the unique solution of the equation $x = ax + 1$. When $a \in I \cap S_0$, the star fixed point identity guarantees that this unique solution is just a^* taken in the Conway semiring S_0. Moreover, it follows that the unique solution of $x = ax + b$ is a^*b, for all b. Then the star operation on S is defined as follows. Given $s \in S$, write s in the unique way as a sum $s_0 + a$ with $s_0 \in S_0$ and $a \in I$. Then by the sum star identity we are forced to define $s^* = (s_0^* a)^* s_0^*$, where s_0^* is taken in S_0 and $s_0^* a$ is in I, since I is an ideal. For more details, the reader is referred to [9]. □

Thus, under the above assumptions, if S_0 is a Conway semiring, then $\mathbf{MAT}_S$ is a Conway matrix theory and if S_0 is an iteration semiring then $\mathbf{MAT}_S$ is a matrix iteration theory. In either case, we have the following proposition.

Proposition 6.25. *Under the assumptions of Theorem 6.24, for any matrices $A \in \mathbf{MAT}_S(n,n)$ and $B \in \mathbf{MAT}_S(n,p)$ such that each entry of A is in I, A^*B is the unique solution of the fixed point equation $X = AX + B$.*

An application of Theorem 6.24 is that if S is a Conway semiring or an iteration semiring, then so is the power series semiring $S\langle\langle A^* \rangle\rangle$ (for the definition of power series semirings, see [21]) for any set A. This follows since $S\langle\langle A^* \rangle\rangle$ is the direct sum of S and the ideal I of proper power series, and when s is a proper power series and r is any power series in $S\langle\langle A^* \rangle\rangle$, the function $x \mapsto sx + r$ is a proper contraction with respect to the complete metric on $S\langle\langle A^* \rangle\rangle$ defined by $d(s, s') = 2^{-n}$, where n is the length of the shortest word w with $(s, w) \neq (s', w)$, for all distinct series s, s'.

Definition 6.26. *We call a semiring S an iterative semiring if S is the direct sum of an iteration semiring S_0 generated by 1 and an ideal I such that each equation $x = ax + b$ with $a \in I$ has a unique solution.*

Corollary 6.27. *Each iterative semiring is an iteration semiring.*

7 Fixed Points of Affine Functions

In this section, we will consider pairs (S, V) consisting of a semiring S and a (left) S-semimodule V. An *affine function* is a function $f : V^n \to V$ of the form

$$f(x_1, \ldots, x_n) = s_1 x_1 + \cdots + s_n x_n + v,$$

where each s_i is in S and v is in V. An affine function $V^n \to V^m$ is a target tupling of affine functions $V^n \to V$. The collection of all affine functions is a theory of functions over V denoted T_V.

Each affine function $V^n \to V^m$ may be represented by a pair (A, v) consisting of a matrix $A \in S^{m \times n}$ and a column vector $v \in V^m$. This representation gives rise to the following definition.

Definition 7.1. *Let (S, V) be a semiring–semimodule pair. The* matricial theory $\mathbf{Matr}_{S,V}$ *[23] over (S, V) has as morphisms $n \to m$ all pairs (A, v) where $A \in \mathbf{MAT}_S(m, n)$ and $v \in V^m$. Composition is defined by*

$$(A, v) \circ (B, w) = (AB, v + Aw),$$

where AB is the usual matrix product and Aw is the action of A on w, i.e., $(Aw)_i = \sum_{j=1}^{n} A_{ij} w_j$. The projection morphism $\mathbf{pr}_i^n$ is the pair $(e_i, 0)$, where e_i is the ith n-dimensional unit row vector considered as a row matrix. Morphisms between matricial theories are theory morphisms which preserve the additive structure.

It can be seen that a morphism $\mathbf{Matr}_{S,V} \to \mathbf{Matr}_{S',V'}$ is completely determined by a semiring morphism $h_S : S \to S'$ and a semimodule morphism $h_V : V \to V'$ such that $(sv)h_V = (sh_S)(vh_V)$ for all $s \in S$ and $v \in V$. Thus, the category of matricial theories is equivalent to the category of semiring–semimodule pairs.

Proposition 7.2. *The theory T_V is a quotient of $\mathbf{Matr}_{S,V}$. A surjective theory morphism $\mathbf{Matr}_{S,V} \to T_V$ maps $(A, v) \in \mathbf{Matr}_{S,V}(m, n)$ to the function $\langle f_1, \ldots, f_m \rangle : V^n \to V^m$ with $f_i(u_1, \ldots, u_n) = A_{i1}u_1 + \cdots + A_{in}u_n + v_i$, for all i.*

The above morphism is usually not injective. To get a faithful representation, one can use $(A, v) \in \mathbf{Matr}_{S,V}(m, n)$ to induce a function $(S \times V)^n \to (S \times V)^m$. Indeed, we can map $(A, v) \in \mathbf{Matr}_{S,V}(m, n)$ to the function $g = \langle g_1, \ldots, g_m \rangle : (S \times V)^n \to (S \times V)^m$, $g_i((x_1, u_1), \ldots, (x_n, u_n)) = (A_{i1}x_1 + \cdots + A_{in}x_n, A_{i1}u_1 + \cdots + A_{in}u_n + v_i)$. This mapping $(A, v) \mapsto g$ is always injective.

Below, we will show that when $\mathbf{Matr}_{S,V}$ is equipped with a dagger operation such that it is a Conway or an iteration theory, then the dagger operation determines and is determined by a star and an omega operation satisfying certain natural axioms. For all omitted details we refer to [9]. Each matricial theory $\mathbf{Matr}_{S,V}$ has $\mathbf{MAT}_S$ as its *underlying matrix theory.*

Suppose that $\mathbf{Matr}_{S,V}$ is equipped with a dagger operation. Hence the dagger operation applied to $(A, v) \in \mathbf{Matr}_{S,V}(n, n + p)$ produces $f^\dagger = (C, z)$ where $C \in \mathbf{MAT}_S$ and $z \in V^n$. Two operations are implicitly defined by the dagger operation. For each $A \in \mathbf{MAT}_S$, consider $((A, E_n), 0^n)$, where all entries of $0^n \in V^n$ are 0. Then we define A^* and A^ω by

$$\big((A, E_n), 0^n\big)^\dagger = \big(A^*, A^\omega\big). \tag{34}$$

Thus, $A \mapsto A^*$ is a map $\mathbf{MAT}_S(n, n) \to \mathbf{MAT}_S(n, n)$, and $A \mapsto A^\omega$ is a map from $\mathbf{MAT}_S(n, n)$ to V^n, for each $n \geq 0$.

Theorem 7.3. *Suppose that $\mathbf{Matr}_{S,V}$ is equipped with a dagger operation and that the star and omega operations are defined as above. Then the parameter identity holds in T if and only if the dagger operation is determined by the star and omega operations:*

$$\big((A, B), v\big)^\dagger = \big(A^*B, A^*v + A^\omega\big), \tag{35}$$

for all $((A, B), v) \in \mathbf{Matr}_{S,V}(n, n+p)$ *with* $A \in \mathbf{MAT}_S(n, n)$, $B \in \mathbf{MAT}_S(n, p)$ *and* $v \in V^n$.

Suppose that the dagger, star, and omega operations are related by (35).

- *The star fixed point identity (25) and the* omega fixed point identity
$$AA^\omega = A^\omega, \quad A \in \mathbf{MAT}_S(n, n), \tag{36}$$
hold if and only if the fixed point identity holds.
- *The product star identity (27) and the* product omega identity
$$(AB)^\omega = A(BA)^\omega, \tag{37}$$
$A \in \mathbf{MAT}_S(n, m)$, $B \in \mathbf{MAT}_S(m, n)$, *hold if and only if the composition identity holds.*
- *The sum star identity and the* sum omega identity
$$(A + B)^\omega = (A^*B)^\omega + (A^*B)^*A^\omega, \tag{38}$$
$A, B \in \mathbf{MAT}_S(n, n)$, *hold if and only if the double dagger identity holds.*
- *The zero star identity and the* zero omega identity
$$0_{nn}^\omega = 0^n \tag{39}$$
hold if and only if (10) holds.
- *The matrix star identity (29) and the* matrix omega identity *(40) hold if and only if the pairing identity holds.*
$$\begin{pmatrix} A & B \\ C & D \end{pmatrix}^\omega = \begin{pmatrix} A^*B(D + CA^*B)^\omega + A^*B(D + CA^*B)^*CA^\omega + A^\omega \\ (D + CA^*B)^\omega + (D + CA^*B)^*CA^\omega \end{pmatrix} \tag{40}$$
for all $A \in \mathbf{MAT}_S(n, n)$, $B \in \mathbf{MAT}_S(n, m)$, $C \in \mathbf{MAT}_S(m, n)$, $D \in \mathbf{MAT}_S(m, m)$.
- *The star permutation identity (30) and the* omega permutation identity *(41) hold if and only if the permutation identity holds.*
$$(\pi A \pi^T)^\omega = \pi A^\omega, \tag{41}$$
where $\pi \in \mathbf{MAT}_S(n, n)$ *is a permutation matrix and* $A \in \mathbf{MAT}_S(n, n)$.
- *The star transposition identity and the* omega transposition identity *hold iff the transposition identity holds, where the omega transposition identity is (41) with* π *restricted to matrices of the form* $\begin{pmatrix} 0 & E_p \\ E_q & 0 \end{pmatrix}$.
- *The star group identity (31) and the* omega group identity *(42) associated with a finite group G hold if and only if the group identity associated with G holds.*
$$e_1 M_G^\omega = (a_1 + \cdots + a_n)^\omega \tag{42}$$
where $a_1, \ldots, a_n \in S$ *and* M_G *is defined above.*

Conversely, if $\mathbf{Matr}_{S,V}$ *is a matricial theory equipped with star and omega operations defined for all square matrices in* $\mathbf{MAT}_S$*, and if the dagger operation is defined by (35), then* $\mathbf{Matr}_{S,V}$ *satisfies the parameter identity and all of the above equivalences hold.*

If the star and omega sum and product identities hold, then the omega pairing identity can be expressed in either of the following two forms:

$$\begin{pmatrix} A & B \\ C & D \end{pmatrix}^\omega = \begin{pmatrix} (A+BD^*C)^\omega + (A+BD^*C)^*BD^\omega \\ (D+CA^*B)^\omega + (D+CA^*B)^*CA^\omega \end{pmatrix},$$

$$\begin{pmatrix} A & B \\ C & D \end{pmatrix}^\omega = \begin{pmatrix} (A^*BD^*C)^*A^\omega + (A^*BD^*C)^*A^*BD^\omega + (A^*BD^*C)^\omega \\ (D^*CA^*B)^*D^*CA^\omega + (D^*CA^*B)^*D^\omega + (D^*CA^*B)^\omega \end{pmatrix}.$$

Definition 7.4. *A* matricial iteration theory *is a matricial theory which is also an iteration theory. A* Conway matricial theory *is a matricial theory which is a Conway theory. Morphisms of matricial iteration theories and Conway matricial theories are matricial theory morphism which preserve dagger and thus star and omega.*

The following results follow from Theorem 7.3, and the axiomatization results in Sect. 3 (Theorems 3.3 and 3.5).

Corollary 7.5. *Let* $\mathbf{Matr}_{S,V}$ *be a matricial theory. Suppose that either* $\mathbf{Matr}_{S,V}$ *is equipped with a star and an omega operation and dagger is defined by (35), or that* $\mathbf{Matr}_{S,V}$ *is equipped with a dagger operation satisfying the parameter identity in which the star and omega operations are defined by (34). Then* T *is a Conway matricial theory if and only if* T *satisfies either of the following groups of equational axioms:*

1. *The zero star (28) and zero omega (39) identities, the matrix star (29) and matrix omega (40) identities, and the star and omega transposition identities.*
2. *The sum and product star and omega identities, (26), (27), (37), (38).*
3. *The scalar versions of the sum and product star and omega identities (i.e., the identities (26), (38), (27), and (37) with* $n = m = 1$*), and the scalar version of the matrix star and matrix omega identities, i.e., (29) and (40) with* $m = 1$.

Moreover, $\mathbf{Matr}_{S,V}$ *is a matricial iteration theory iff it is a Conway matricial theory satisfying the star and omega group identities (31), (42) associated with finite groups. In either case, the dagger, star, and omega operations are related by (34) and (35).*

Note that the star and omega fixed point identities hold in any Conway matricial theory. Also, if $\mathbf{Matr}_{S,V}$ is a Conway matricial theory, then $\mathbf{MAT}_S$ is a Conway matrix theory, and if $\mathbf{Matr}_{S,V}$ is a matricial iteration theory then $\mathbf{MAT}_S$ is a matrix iteration theory.

If $\mathbf{MAT}_S$ is equipped with a star operation, we may equip $\mathbf{Matr}_{S,V}$ with an omega operation such that $A^\omega = 0^n$, for all $A \in \mathbf{MAT}_S(n,n)$. When $\mathbf{MAT}_S$ is a matrix iteration theory, $\mathbf{Matr}_{S,V}$ is a matricial iteration theory. Similarly, if $\mathbf{MAT}_S$ is a Conway matrix theory, $\mathbf{Matr}_{S,V}$ is a Conway matricial theory. In particular, any matrix iteration theory $\mathbf{MAT}_S$ may be viewed as the matricial iteration theory $\mathbf{Matr}_{S,V}$ where $V = \{0\}$ is the trivial S-semimodule.

In a matricial theory $\mathbf{Matr}_{S,V}$, any morphism $0 \to 1$ may be identified with an element of V. Similarly, each morphism $1 \to 1$ in the underlying matrix theory $\mathbf{MAT}_S$ may be considered to be an element of the semiring S. Thus, when the matrix star and omega identities hold, the star and omega operations are determined by operations $^* : S \to S$ and $^\omega : S \to V$.

Definition 7.6. *A* Conway semiring–semimodule pair *consists of a Conway semiring S, an S semimodule V and an operation $^\omega : S \to V$ which satisfies the sum and product omega identities*

$$(a+b)^\omega = (a^*b)^*a^\omega + (a^*b)^\omega, \tag{43}$$

$$(ab)^\omega = a(ba)^\omega \tag{44}$$

for all a,b in S. An iteration semiring–semimodule pair *is a Conway semiring–semimodule pair such that S is an iteration semiring, which when star and omega on matrices are defined by the matrix star and matrix omega identities (29) and (40) with $m = 1$, satisfies the omega group identity associated with any finite group. Morphisms of Conway and iteration semiring–semimodule pairs are morphisms of semiring–semimodule pairs which preserve star and omega.*

Proposition 7.7.

- *When* $\mathbf{Matr}_{S,V}$ *is a matricial iteration theory, (S,V) is an iteration semiring–semimodule pair, and when* $\mathbf{Matr}_{S,V}$ *is a Conway matricial theory, (S,V) is a Conway semiring–semimodule pair.*
- *Let (S,V) be an iteration (or Conway) semiring–semimodule pair. There is a unique way to extend the star and omega operations on S to all square matrices in* $\mathbf{MAT}_S$ *so that* $\mathbf{Matr}_{S,V}$ *becomes a matricial iteration theory (or Conway matricial theory, respectively).*

In fact, the category of Conway matricial theories is equivalent to the category of Conway semiring–semimodule pairs, and the category of matricial iteration theories is equivalent to the category of iteration semiring–semimodule pairs.

In any Conway matricial theory, the group identities follow from the functorial star and omega conditions.

Definition 7.8. *Suppose that* $\mathbf{Matr}_{S,V}$ *is equipped with a star and omega operation. Let $\mathcal{C}$ be a set of matrices in* $\mathbf{MAT}_S$*. We say that* $\mathbf{Matr}_{S,V}$ *satisfies the functorial star implication for $\mathcal{C}$, or has a functorial star with respect to $\mathcal{C}$ if*

$\mathbf{MAT}_S$ *does. We say that* $\mathbf{Matr}_{S,V}$ *satisfies the functorial omega implication for* $\mathcal{C}$, *or that* $\mathbf{Matr}_{S,V}$ *has a functorial omega with respect to* $\mathcal{C}$, *if for all* $A \in \mathbf{MAT}_S(n,n)$ *and* $B \in \mathbf{MAT}_S(m,m)$ *and all* $C \in \mathbf{MAT}_S(n,m)$ *in* $\mathcal{C}$,

$$AC = CB \quad \Longrightarrow \quad A^\omega = CB^\omega.$$

When $\mathbf{Matr}_{S,V}$ *has a functorial star and omega with respect to the set of all matrices (all base matrices, respectively) in* $\mathbf{MAT}_S$, *then* $\mathbf{Matr}_{S,V}$ *is said to have a strong functorial star and omega (*weak functorial star and omega, *respectively).*

Proposition 7.9. *Suppose that* $\mathbf{Matr}_{S,V}$ *is a Conway matricial theory.*

- *For any set* $\mathcal{C} \subseteq \mathbf{MAT}_S$, $\mathbf{Matr}_{S,V}$ *has a functorial dagger with respect to* $\mathcal{C}$ *if and only if* $\mathbf{Matr}_{S,V}$ *has a functorial star and omega with respect to* $\mathcal{C}$.
- $\mathbf{Matr}_{S,V}$ *has a functorial star and omega with respect to all injective base matrices.*
- *If* $\mathbf{Matr}_{S,V}$ *has a functorial star and omega with respect to all* $n \times 1$ *base matrices,* $n \geq 2$, *then the star and omega group identities hold in* $\mathbf{Matr}_{S,V}$.
- $\mathbf{Matr}_{S,V}$ *has a weak functorial star and omega if and only if* $\mathbf{Matr}_{S,V}$ *has a functorial star and omega with respect to all* $n \times 1$ *base matrices,* $n \geq 2$.

Corollary 7.10. *Any Conway matricial theory with a weak functorial star and omega is a matricial iteration theory.*

We end this section by exhibiting two classes of iteration semiring–semimodule pairs.

7.1 Complete Semiring–Semimodule Pairs

This section is based on [32]. Recall the definition of a complete monoid and that of a complete semiring. We call a semiring–semimodule pair (S,V) a *complete semiring–semimodule pair* if S is a complete semiring, V is a complete monoid, and the action is *completely distributive*, so that $(\sum_{i\in I} s_i)(\sum_{j\in J} v_j) = \sum_{(i,j)\in I\times J} s_i v_j$. Moreover, we require that an *infinite product operation* $S \times S \times \cdots \to S$,

$$(s_1, s_2, \ldots) \mapsto \prod_{j\geq 1} s_j$$

is given mapping infinite sequences over S to V subject to the following conditions:

$$\prod_{j\geq 1} s_j = \prod_{j\geq 1} (s_{n_{j-1}+1} \cdots s_{n_j}), \tag{45}$$

$$s_1 \cdot \prod_{i\geq 1} s_{i+1} = \prod_{i\geq 1} s_i, \tag{46}$$

$$\prod_{j\geq 1} \sum_{i_j \in I_j} s_{i_j} = \sum_{(i_1,i_2,\ldots)\in I_1\times I_2\times\cdots} \prod_{j\geq 1} s_{i_j}, \tag{47}$$

where in the first equation $0 = n_0 \leq n_1 \leq n_2 \leq \cdots$ and $I_1, I_2, \ldots$ are sets. (Complete semimodules of complete semirings without an infinitary product operation on the semiring are studied in Chap. 23 of [35]. When (S, V) is a complete semiring–semimodule pair, then equipped only with the binary and infinitary multiplication operations, (S, V) is an *ω-semigroup* [46].)

Suppose that (S, V) is complete. Then we define

$$s^* = \sum_{i \geq 0} s^i \quad \text{and} \quad s^\omega = \prod_{i \geq 1} s,$$

for all $s \in S$.

Theorem 7.11. *Every complete semiring–semimodule pair* (S, V) *is an iteration semiring–semimodule pair.*

Proof. We already know that S is an iteration semiring. We establish the sum omega and product omega identities and leave the proof of the group identities to the reader. So, suppose that $a, b \in S$. We also consider the set $\{a, b\}$ as an alphabet Σ. When w is a finite or infinite word over this alphabet, we let $\overline{w}$ denote the corresponding product over S which is either an element of S (finite product) or an element in V (infinite product). Our proof of the sum omega identity uses the fact that $\{a, b\}^\omega = (\{a\}^*\{b\})^\omega \cup (\{a\}^*\{b\})^*\{a\}^\omega = K \cup L$ holds over the alphabet Σ.[3]

$$\begin{aligned}
(a+b)^\omega &= \prod_{j \geq 1} (a+b) \\
&= \sum_{w \in \{a,b\}^\omega} \overline{w} \\
&= \sum_{u \in K} \overline{u} + \sum_{v \in L} \overline{v} \\
&= \prod_{j \geq 1} \sum_{u \in \{a\}^*\{b\}} \overline{u} + \Big(\sum_{v \in (\{a\}^*\{b\})^*} \overline{v} \Big) \prod_{j \geq 1} a \\
&= (a^*b)^\omega + (a^*b)^* a^\omega .
\end{aligned}$$

As for the product omega identity, let $c_j = a$ if $j \geq 1$ is odd, and let $c_j = b$ if $j \geq 1$ is even. Then

$$(ab)^\omega = \prod_{j \geq 1} (ab) = \prod_{j \geq 1} c_j = a \prod_{j \geq 2} c_j = a(ba)^\omega . \quad \square$$

Thus, $\mathbf{Matr}_{S,V}$ is a matricial iteration theory, so that when the dagger is defined by (24), then all iteration theory identities hold over any complete semiring–semimodule pair. Without proof, we mention the following proposition.

[3] Here, for any language X of nonempty finite words, we denote by X^ω the set $\{x_1 x_2 \ldots : x_i \in X\}$ of ω-words.

Proposition 7.12. *When (S, V) is a complete semiring–semimodule pair then for each n, $(S^{n\times n}, V^n)$ is also a complete semimodule pair with infinitary product such that for each $A_1, A_2, \ldots \in S^{n\times n}$, and for each i, the ith entry of $A_1 \cdot A_2 \cdots$ is the sum of all elements of the form $(A_1)_{i,j_1} \cdot (A_2)_{j_1,j_2} \cdots$. Moreover, for each $A \in \mathbf{MAT}_S(n,n)$, A^ω in $\mathbf{Matr}_{S,V}$ is the same as A^ω in the complete semiring–semimodule pair $(S^{n\times n}, V^n)$.*

7.2 Bi-inductive Semiring–Semimodule Pairs

This section is based on [32]. We call a semiring–semimodule pair (S, V) ordered if S is an ordered semiring and V is an ordered monoid, ordered by $\leq$, such that $sv \leq s'v'$ whenever $s \leq s'$ in S and $v \leq v'$ in V.

Definition 7.13. *Suppose that (S, V) is an ordered semiring–semimodule pair equipped with a star operation $^* : S \to S$ and an omega operation $^\omega : S \to V$ such that*

$$aa^* + 1 \leq a^* \tag{48}$$

$$ax + y \leq x \quad \Longrightarrow \quad a^*y \leq x, \tag{49}$$

for all $a \in S$ and $x, y \in S$ or $x, y \in V$, and

$$aa^\omega \geq a^\omega \tag{50}$$

$$ax + y \geq x \quad \Longrightarrow \quad a^\omega + a^*y \geq x, \tag{51}$$

for all $a \in S$ and $x, y \in V$. Then we call (S, V) a bi-inductive semiring–semimodule pair. *A morphism of bi-inductive semiring–semimodule pairs is a morphism of semiring–semimodule pairs which preserves the order and the star and omega operations.*

The terminology is due to the fact that bi-inductive semiring–semimodule pairs satisfy both an induction axiom (49) and a coinduction axiom (51). Affine functions $x \mapsto ax + v$ over V have both a least prefixed point and a greatest post-fixed point, namely a^*v and $a^\omega + a^*v$, where a^* is the least prefixed point solution of $x = ax + 1$ over S and a^ω is the greatest post-fixed point solution of $x = ax$ over V. Note that if (S, V) is a bi-inductive semiring–semimodule pair then S is an inductive *-semiring.

Proposition 7.14. *If (S, V) is bi-inductive, then 0 is the least and 1^ω is the greatest element of V.*

Proof. The fact that 0 is least follows from Proposition 6.13. The fact that 1^ω is the greatest element of V follows by noting that any element of V solves the equation $x = 1x$. □

Theorem 7.15. *Every bi-inductive semiring–semimodule pair (S, V) is an iteration semiring–semimodule pair. Moreover, the star and omega operations are monotone.*

Proof. We already know that S is an iteration semiring and that * is monotone. The fact that $^\omega$ is monotone follows from (the dual of) Proposition 2.8.

We prove that the product omega identity holds. Indeed, if $a, b \in S$, then $aba(ba)^\omega = a(ba)^\omega$, thus $(ab)^\omega \geq a(ba)^\omega$. Thus, $(ab)^\omega \geq a(ba)^\omega \geq ab(ab)^\omega = (ab)^\omega$, proving $(ab)^\omega = a(ba)^\omega$.

Next, we prove that the sum omega identity holds. Given $a, b \in S$,

$$\begin{aligned}
&(a+b)[(a^*b)^*a^\omega + (a^*b)^\omega] \\
&= a(a^*b)^*a^\omega + a(a^*b)^\omega + b(a^*b)^*a^\omega + b(a^*b)^\omega \\
&= a[a^*(ba^*)^*b + 1]a^\omega + aa^*(ba^*)^\omega + (ba^*)^*ba^\omega + (ba^*)^\omega \\
&= aa^*(ba^*)^*ba^\omega + aa^\omega + aa^*(ba^*)^\omega + (ba^*)^*ba^\omega + (ba^*)^\omega \\
&= (aa^* + 1)(ba^*)^*ba^\omega + (aa^* + 1)(ba^*)^\omega + a^\omega \\
&= [a^*(ba^*)^*b + 1]a^\omega + a^*(ba^*)^\omega \\
&= (a^*b)^*a^\omega + (a^*b)^\omega.
\end{aligned}$$

It follows by (51) that $(a+b)^\omega \geq (a^*b)^*a^\omega + (a^*b)^\omega$. As for the reverse inequality, note that for all $x \in V$, if $(a+b)x = ax + bx \geq x$, then $a^\omega + a^*bx \geq x$, so that $(a^*b)^\omega + (a^*b)^*a^\omega \geq x$. Taking $x = (a+b)^\omega$, we have $(a^*b)^\omega + (a^*b)^*a^\omega \geq (a+b)^\omega$.

We omit the verification of the group identities. □

Thus, when (S, V) is a bi-inductive semiring–semimodule pair, then $\mathbf{Matr}_{S,V}$ is an iteration semiring–semimodule pair. Thus, $\mathbf{Matr}_{S,V}$ is a matricial iteration theory, so that when dagger is defined by (24), then all iteration theory identities hold.

Theorem 7.16. *Suppose that (S, V) is a bi-inductive semiring–semimodule pair. Then for any $(A, v) : n \to n$ in $\mathbf{Matr}_{S,V}$, A^*v is the least prefixed point solution and $A^\omega + A^*v$ is the greatest post-fixed point solution of $x = Ax + v$. Thus, each $(S^{n \times n}, V^n)$ is also a bi-inductive semiring–semimodule pair.*

References

1. A. Arnold and D. Niwinski. *Rudiments of μ-Calculus.* Elsevier, Amsterdam, 2001.
2. S. Banach. Sur les operations dans les ensembles abstraits et leur applications aux équations intégrales. *Fundamenta Mathematicae*, 22:133–181, 1922.
3. M. Barr and C. Wells. *Category Theory for Computing Science.* Prentice Hall International, Englewood Cliffs, 1990.
4. H. Bekić. Definable operations in general algebras, and the theory of automata and flowcharts. Technical Report, IBM Laboratory, Vienna, 1969.

5. L. Bernátsky and Z. Ésik. Semantics of flowchart programs and the free Conway theories. *Theoretical Informatics and Applications, RAIRO*, 32:35–78, 1998.
6. S.L. Bloom, C.C. Elgot, and J.B. Wright. Solutions of the iteration equation and extension of the scalar iteration operation. *SIAM Journal on Computing*, 9:26–45, 1980.
7. S.L. Bloom, C.C. Elgot, and J.B. Wright. Vector iteration of pointed iterative theories. *SIAM Journal on Computing*, 9:525–540, 1980.
8. S.L. Bloom and Z. Ésik. Axiomatizing schemes and their behaviors. *Journal of Computer and System Sciences*, 31:375–393, 1985.
9. S.L. Bloom and Z. Ésik. *Iteration Theories: The Equational Logic of Iterative Processes. Monographs in Theoretical Computer Science. An EATCS Series*, Springer, Berlin, 1993.
10. S.L. Bloom and Z. Ésik. Two axiomatizations of a star semiring quasi-variety. *Bulletin of the European Association for Theoretical Computer Science*, 59:150–152, 1996.
11. S.L. Bloom and Z. Ésik. The equational logic of fixed points. *Theoretical Computer Science*, 179:1–60, 1997.
12. S.L. Bloom and Z. Ésik. Axiomatizing rational power series. *Information and Computation*, 207:793–811, 2009.
13. S.L. Bloom S. Ginali, and J. Rutledge. Scalar and vector iteration. *Journal of Computer and System Sciences*, 14:251–256, 1977.
14. M. Boffa. Une remarque sur les systèmes complets d'identités rationnelles. (A remark on complete systems of rational identities). *Theoretical Informatics and Applications, RAIRO*, 24:419–423, 1990 (in French).
15. M. Boffa. Une condition impliquant toutes les identités rationnelles (A condition implying all rational identities). *Theoretical Informatics and Applications, RAIRO*, 29:515–518, 1995 (in French).
16. P.M. Cohn. *Universal Algebra*. Harper & Row, New York, 1965.
17. J.C. Conway. *Regular Algebra and Finite Machines*. Chapman & Hall, London, 1971.
18. B. Courcelle, G. Kahn, and J. Vuillemin. Algorithmes d'équivalence et de réduction á des expressions minimales dans une classe d'équations récursives simples. In *Proc. ICALP 74, Saarbrücken*, volume 14 of *Lecture Notes in Computer Science*, pages 200–213. Springer, Berlin, 1974.
19. B.A. Davey and H.A. Priestley. *Introduction to Lattices and Order*, 2nd edition, Cambridge University Press, Cambridge, 2002.
20. J.W. De Bakker and D. Scott. A theory of programs. Technical Report, IBM Vienna, 1969.
21. M. Droste and W. Kuich. Semirings and formal power series. In W. Kuich, M. Droste, and H. Vogler, editors. *Handbook of Weighted Automata*. Chapter 1. Springer, Berlin, 2009.
22. C.C. Elgot. Monadic computation and iterative algebraic theories. In J.C. Shepherdson, editor, *Logic Colloquium 1973*, volume 80 of *Studies in Logic*, pages 175–230. North-Holland, Amsterdam, 1975.

23. C.C. Elgot. Matricial theories. *Journal of Algebra*, 42:391–421, 1976.
24. Z. Ésik. Identities in iterative algebraic theories. *Computational Linguistics and Computer Languages*, 14:183–207, 1980.
25. Z. Ésik. A note on the axiomatization of iteration theories. *Acta Cybernetica*, 9:375–384, 1990.
26. Z. Ésik. Completeness of Park induction. *Theoretical Computer Science*, 177:217–283, 1997.
27. Z. Ésik. Group axioms for iteration. *Information and Computation*, 148:131–180, 1999.
28. Z. Ésik. Iteration theories of boolean functions. In *Math. Found. of Computer Science, 2000*, volume 1893 of *Lecture Notes in Computer Science*, pages 343–352. Springer, Berlin, 2000.
29. Z. Ésik. Iteration semirings. In *Proc. DLT 08*, volume 5257 of *Lecture Notes in Computer Science*, pages 1–20. Springer, Berlin, 2008.
30. Z. Ésik and W. Kuich. Rationally additive semirings. *Journal of Universal Computer Science*, 2(8):173–183, 2002.
31. Z. Ésik and W. Kuich. Inductive *-semirings. *Theoretical Computer Science*, 324:3–33, 2004.
32. Z. Ésik and W. Kuich. On iteration semiring–semimodule pairs. *Semigroup Forum*, 75:129–159, 2007.
33. Z. Ésik and W. Kuich. Finite automata. In W. Kuich, M. Droste, and H. Vogler, editors, *Handbook of Weighted Automata.* Chapter 3. Springer, Berlin, 2009.
34. Z. Ésik and A. Labella. Equational properties of iteration in algebraically complete categories. *Theoretical Computer Science*, 195:61–89, 1998.
35. J.S. Golan. *The Theory of Semirings with Applications in Computer Science.* Longman, Harlow, 1993.
36. M. Hasegawa. Recursion from cyclic sharing: Traced monoidal categories and models of cyclic lambda calculi. In *Typed Lambda Calculi and Applications, Nancy, 1997*, volume 1210 of *Lecture Notes in Computer Science*, pages 196–213. Springer, Berlin, 1997.
37. U. Hebisch. The Kleene theorem in countably complete semirings. *Bayreuther Mathematische Schriften*, 31:55–66, 1990.
38. A. Joyal, R. Street, and D. Verity. Traced monoidal categories. *Mathematical Proceedings of the Cambridge Philosophical Society*, 119:447–468, 1996.
39. D. Kozen. On Kleene algebras and closed semirings. In *Proc. MFCS'90*, volume 452 of *Lecture Notes in Computer Science*, pages 26–47. Springer, Berlin, 1990.
40. D. Kozen. A completeness theorem for Kleene algebras and the algebra of regular events. *Information and Computation*, 110:366–390, 1994.
41. D. Krob. Complete systems of B-rational identities. *Theoretical Computer Science*, 89:207–343, 1991.

42. F.W. Lawvere. Functorial semantics of algebraic theories. *Proceedings of the National Academy of Sciences of the United States of America*, 50:869–873, 1963.
43. G. Markowsky. Chain-complete posets and directed sets with applications. *Algebra Universalis*, 6:53–68, 1976.
44. D. Niwinski. Equational μ-calculus. In *Computation Theory, Zaborów, 1984*, volume 208 of *Lecture Notes in Computer Science*, pages 169–176. Springer, Berlin, 1985.
45. D. Niwinski. On fixed-point clones. In *Proc. ICALP'86*, volume 226 of *Lecture Notes in Computer Science*, pages 464–473. Springer, Berlin, 1986.
46. D. Perrin and J.-E. Pin. *Infinite Words*, volume 141 of *Pure and Applied Mathematics*. Elsevier, Amsterdam, 2004.
47. G.D. Plotkin. Domains. Lecture Notes, Department of Computer Science, University of Edinburgh, 1983.
48. A. Simpson and G. Plotkin. Complete axioms for categorical fixed-point operators. In *15th Annual IEEE Symposium on Logic in Computer Science, Santa Barbara, CA, 2000*, pages 30–41. IEEE Computer Society, Los Alamitos, 2000.
49. Gh. Stefanescu. *Network Algebra*. Springer, Berlin, 2000.
50. A. Tarski. A lattice theoretical fixpoint theorem and its applications. *Pacific Journal of Mathematics*, 5:285–309, 1955.
51. J.B. Wright, J.W. Thatcher, J. Goguen, and E.G. Wagner. Rational algebraic theories and fixed-point solutions. In *Proceedings 17th IEEE Symposium on Foundations of Computing, Houston, Texas*, pages 147–158, 1976.

Part II

Concepts of Weighted Recognizability

Chapter 3:
Finite Automata

Zoltán Ésik[1,2,*] and Werner Kuich[3]

[1] Department of Computer Science, University of Szeged, Szeged, Hungary
[2] GRLMC, Rovira i Virgili University, Tarragona, Spain
[3] Institut für Diskrete Mathematik und Geometrie, Technische Universität Wien, 1040 Wien, Austria
kuich@tuwien.ac.at
www.dmg.tuwien.ac.at/kuich

1 Introduction

The purpose of this chapter is to develop the theory of finite automata starting from ideas based on linear algebra over semirings. Many results in the theory of automata and languages depend only on a few equational axioms. For example, Conway [4] has shown that Kleene's fundamental theorem equating the recognizable languages with the regular ones follows from a few simple identities defining Conway semirings. Such semirings are equipped with a star operation subject to the sum star identity and product star identity.

The use of equations has several advantages. Proofs can be separated into two parts, where the first part establishes the equational axioms, and the sec-

* The first author was partially supported by grant no. MTM2007-63422 from the Ministry of Education and Science of Spain.

M. Droste, W. Kuich, H. Vogler (eds.), *Handbook of Weighted Automata*, Monographs in Theoretical Computer Science. An EATCS Series, DOI 10.1007/978-3-642-01492-5_3,

ond is based on simple equational reasoning. Such proofs have a transparent structure and are usually very easy to understand, since manipulating equations is one of the most common ways of mathematical reasoning. Moreover, since many results depend on the same equations, the first part of such proofs usually provides a basis to several results. Finally, the results obtained by equational reasoning have a much broader scope, since many models share the same equations.

This chapter consists of this and two more sections. Section 2 constitutes a generalization of the theory of regular languages over finite words. In this section, we define finite automata over power series semirings and over Conway semirings, and prove theorems of the Kleene–Schützenberger type. Moreover, we introduce finite linear systems and show the coincidence of the set of components of the solutions of such finite linear systems with the set of behaviors of finite automata.

Section 3 constitutes a generalization of the Büchi theory on languages over infinite words. In this section, we first define the algebraic structures needed for this generalization: semiring–semimodule pairs and quemirings. Then we define finite automata over quemirings and prove theorems of the Kleene–Büchi type. Moreover, we consider linear systems over quemirings as a generalization of regular grammars with finite and infinite derivations and show the coincidence of the set of components of the solutions of such linear systems with the set of behaviors of finite automata over quemirings.

The presentation of this chapter is influenced by the ideas presented in Bloom and Ésik [2], and Ésik and Kuich [10, 11].

2 Finite Automata over Semirings

In this section, we deal with finite automata over semirings. Here, semirings constitute a generalization of formal languages with finite words. The main results of this section are various generalizations of the theorem of Kleene–Schützenberger [24].

This section consists of three subsections. In Sect. 2.1, we define finite automata over a semiring S and an alphabet Σ whose behavior is a formal power series in $S\langle\langle \Sigma^* \rangle\rangle$. The main result of this subsection is the theorem of Kleene–Schützenberger (Theorem 2.5): A power series of $S\langle\langle \Sigma^* \rangle\rangle$ is rational iff it is the behavior of a cycle-free finite automaton over the semiring S and the alphabet Σ.

In Sect. 2.2, we introduce finite S'-automata over a Conway semiring S, where S' is a subset of S. The main result of this subsection is the following generalization of the theorem of Kleene–Schützenberger (Theorem 2.12): The star semiring generated by S' equals the set of behaviors of finite S'-automata over semirings. The results in Sects. 2.1 and 2.2 are similar; but in Sect. 2.1 we consider cycle-free finite automata over an arbitrary semiring, while in Sect. 2.2, we consider arbitrary finite automata over a Conway semiring.

In Sect. 2.3, we consider finite linear systems as a generalization of right linear context-free grammars. The main result of this subsection is that a semiring element is a component of the solution of a finite linear system iff it is the behavior of a finite automaton (Theorems 2.18 and 2.23).

We now give a typical example which will be helpful for readers with some background in semiring theory (as given in Droste and Kuich [5]) and automata theory. Readers without this background should consult it when finite automata are defined in Sect. 2.2.

Example 2.1. Let $\mathcal{A} = (Q, \Sigma, \delta, 1, \{1\})$ be a finite automaton (a definition is given at the end of Sect. 2.2), where $Q = \{1, 2\}$, $\Sigma = \{a, b, c, d\}$, and $\delta(1, a) = \{1\}$, $\delta(1, b) = \{2\}$, $\delta(2, c) = \{1\}$, $\delta(2, d) = \{2\}$ are the only nonempty images of δ.

The graph of $\mathcal{A}$ is

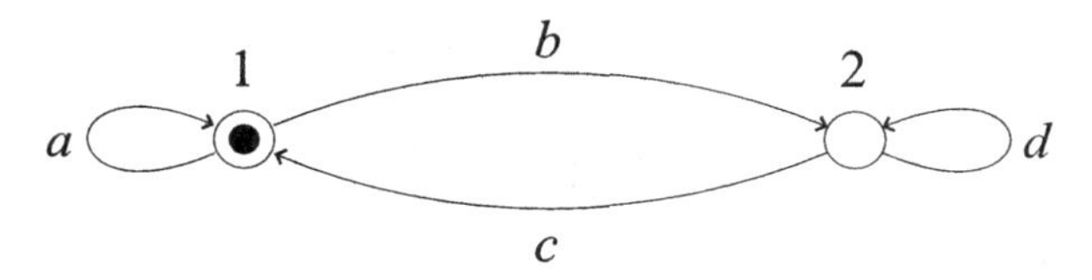

and the adjacency matrix of this graph is

$$A = \begin{pmatrix} \{a\} & \{b\} \\ \{c\} & \{d\} \end{pmatrix} .$$

(Whenever we use matrices in this example, they are 2×2-matrices and their entries are formal languages over Σ, i.e., elements of the semiring 2^{Σ^*}.)

Consider the powers of A, e.g.,

$$A^2 = \begin{pmatrix} \{aa, bc\} & \{ab, bd\} \\ \{ca, dc\} & \{cb, dd\} \end{pmatrix} ,$$

$$A^3 = \begin{pmatrix} \{aaa, abc, bca, bdc\} & \{aab, abd, bcb, bdd\} \\ \{caa, cbc, dca, ddc\} & \{cab, cbd, dcb, ddd\} \end{pmatrix} .$$

It is easily proved by induction on k that $(A^k)_{ij}$ is the language of inscriptions of the paths of length k from state i to state j, $k \geq 0$, $1 \leq i, j \leq 2$. Define A^* by $(A^*)_{ij} = \bigcup_{k \geq 0} (A^k)_{ij}$, $1 \leq i, j \leq 2$. Then $(A^*)_{ij}$ is the language of inscriptions of all the paths from state i to state j, $1 \leq i, j \leq 2$.

We now construct regular expressions for the entries of A^*. Consider the inscriptions of paths from 1 to 1 not passing 1: they are a and $bd^n c$. Hence, the language of inscriptions of these paths is $\{a\} \cup \{b\}\{d\}^*\{c\}$. Consider now the language of inscriptions of paths from 1 to 1: it is $(\{a\} \cup \{b\}\{d\}^*\{c\})^*$. Hence, $(A^*)_{11} = (\{a\} \cup \{b\}\{d\}^*\{c\})^*$. We obtain $(A^*)_{12}$, if we concatenate $(A^*)_{11}$ with the language of inscriptions of all paths from 1 to 2 not passing through 1: $\{b\}\{d\}^*$. Hence, $(A^*)_{12} = (A^*)_{11}\{b\}\{d\}^*$. By symmetry, we obtain $(A^*)_{22} = (\{d\} \cup \{c\}\{a\}^*\{b\})^*$ and $(A^*)_{21} = (A^*)_{22}\{c\}\{a\}^*$. Hence,

$$A^* = \begin{pmatrix} (\{a\} \cup \{b\}\{d\}^*\{c\})^* & (\{a\} \cup \{b\}\{d\}^*\{c\})^*\{b\}\{d\}^* \\ (\{d\} \cup \{c\}\{a\}^*\{b\})^*\{c\}\{a\}^* & (\{d\} \cup \{c\}\{a\}^*\{b\})^* \end{pmatrix}.$$

The language $\|\mathcal{A}\|$ accepted by the finite automaton $\mathcal{A}$ is the language of inscriptions of all paths from the initial state 1 to the final state 1, i.e., $\|\mathcal{A}\| = (A^*)_{11} = (\{a\} \cup \{b\}\{d\}^*\{c\})^*$.

2.1 Finite Automata over Arbitrary Power Series Semirings

In this subsection, S denotes an arbitrary semiring and Σ a finite alphabet.

A *finite automaton*

$$\mathfrak{A} = (Q, R, A, P)$$

(*over the semiring S and the alphabet Σ*) is given by:

(i) A finite nonempty *set of states* Q
(ii) A *transition matrix* $A \in (S\langle \Sigma \cup \{\varepsilon\}\rangle)^{Q \times Q}$
(iii) An *initial state vector* $R \in (S\langle \Sigma \cup \{\varepsilon\}\rangle)^{1 \times Q}$
(iv) A *final state vector* $P \in (S\langle \Sigma \cup \{\varepsilon\}\rangle)^{Q \times 1}$

The finite automaton $\mathfrak{A}$ is called *cycle-free* if the transition matrix A is cycle-free, cf. Droste and Kuich [5]. The *behavior* $\|\mathfrak{A}\|$ of the cycle-free finite automaton $\mathfrak{A} = (Q, R, A, P)$ is defined by

$$\|\mathfrak{A}\| = \sum_{q_1, q_2 \in Q} R_{q_1} (A^*)_{q_1, q_2} P_{q_2} = RA^*P.$$

Since $\mathfrak{A}$ is cycle-free, A^* and $\|\mathfrak{A}\|$ are well defined.

The *(directed) graph* of the cycle-free finite automaton $\mathfrak{A} = (Q, R, A, P)$, where $Q = \{q_1, \ldots, q_n\}$, is constructed in the usual manner. It has nodes $q_1, \ldots, q_n$ and an edge from node q_i to node q_j if $A_{q_i, q_j} \neq 0$. The *weight* of this edge is $A_{q_i, q_j} \in S\langle \Sigma \cup \{\varepsilon\}\rangle$. The *initial* (resp. *final*) *weight* of a node q_i is given by R_i (resp. P_i). A node is called *initial* (resp. *final*) if its initial (resp. final) weight is unequal to 0. The *weight* of a path is the product of the weights of its edges. It is easily shown that $(A^k)_{q_i, q_j}$ is the sum of the weights of paths of length k from node q_i to node q_j. Since $(A^*)_{q_i, q_j} = \sum_{k \geq 0} (A^k)_{q_i, q_j}$, $(A^*)_{q_i, q_j}$ is the sum of the weights of the paths from node q_i to node q_j. Hence, $R_{q_1} (A^*)_{q_i, q_j} P_{q_2}$ is this sum for nodes q_1 and q_2, multiplied on the left and right by the initial weight of node q_1 and the final weight of node q_2, respectively. Eventually, the behavior of $\mathfrak{A}$ is the sum of all these terms with summation over all initial states q_1 and all final states q_2.

Theorem 2.2. *If $\mathfrak{A}$ is a cycle-free finite automaton then $\|\mathfrak{A}\|$ is the sum of the weights of all paths from an initial state to a final state multiplied by the initial and final weights of these states.*

Two cycle-free finite automata $\mathfrak{A}$ and $\mathfrak{A}'$ are *equivalent* if $\|\mathfrak{A}\| = \|\mathfrak{A}'\|$. A finite automaton $\mathfrak{A} = (Q, R, A, P)$ is called *normalized* if $|Q| \geq 2$ and:

(i) $R_{q_1} = \varepsilon$ for some $q_1 \in Q$; $R_q = 0$, for all $q \in Q$, $q \neq q_1$.
(ii) $P_{q_n} = \varepsilon$ for some $q_n \in Q$, $q_n \neq q_1$; $P_q = 0$, for all $q \in Q$, $q \neq q_n$.
(iii) $A_{q,q_1} = A_{q_n,q} = 0$ for all $q \in Q$.

Hence, the directed graph of a normalized finite automaton has the unique initial node q_1 and the unique final node q_n, both with weight ε; moreover, no edges are leading to the initial node and no edges are leaving the final node.

Theorem 2.3. *Each cycle-free finite automaton is equivalent to a normalized cycle-free finite automaton.*

Proof. Let $\mathfrak{A} = (Q, R, A, P)$ be a cycle-free finite automaton. Define the finite automaton $\mathfrak{A}'$ by

$$\mathfrak{A}' = \left(\{q_0\} \cup Q \cup \{q_f\}, (\varepsilon\ 0\ 0), \begin{pmatrix} 0 & R & 0 \\ 0 & A & P \\ 0 & 0 & 0 \end{pmatrix}, \begin{pmatrix} 0 \\ 0 \\ \varepsilon \end{pmatrix} \right).$$

Here, q_0 and q_f are new states. Then $\mathfrak{A}'$ is normalized. Moreover, Corollary 5.8 of Droste and Kuich [5] implies that $\mathfrak{A}'$ is cycle-free and $\|\mathfrak{A}'\| = \|\mathfrak{A}\|$. □

Remark. Usually, a finite automaton $\mathfrak{A} = (Q, R, A, P)$ is defined as above with the exception that the entries of R and P are in $S\langle\{\varepsilon\}\rangle$. We have chosen the more general definitions for two reasons:

(i) This definition is compatible with the definitions of finite S'-automata given in Sects. 2.2 and 3.2.
(ii) This definition is in correspondence with the definition of finite linear systems generalizing right linear grammars. (See Theorems 2.18, 3.20 and 3.21, and Corollaries 2.20 and 3.24.)

Moreover, by Theorem 2.3, our finite automata are exactly as powerful as the usual finite automata.

The three operations sum, product, and star are customarily referred to as *rational* operations. A power series r belonging to $S\langle\langle\Sigma^*\rangle\rangle$ is termed *rational* (over S and Σ) if r can be obtained from the polynomials of $S\langle\Sigma^*\rangle$ by finitely many applications of the rational operations, where the star is applied only to *proper* power series. The formula telling how a given rational series in $S\langle\langle\Sigma^*\rangle\rangle$ is obtained from the polynomials of $S\langle\Sigma^*\rangle$ by rational operations is referred to as *regular expression* (see Salomaa [22], Kuich and Salomaa [19]). The family of rational power series (over S and Σ) is denoted by $S^{\mathrm{rat}}\langle\langle\Sigma^*\rangle\rangle$.

Observe that $S^{\mathrm{rat}}\langle\langle\Sigma^*\rangle\rangle$ can equivalently be defined as follows: a power series r is in $S^{\mathrm{rat}}\langle\langle\Sigma^*\rangle\rangle$ if r can be obtained from the polynomials of $S\langle\Sigma^*\rangle$ by finitely many applications of the rational operations, where the star is applied only to *cycle-free* power series, see Droste and Kuich [5]. Let $r = r_0 + r_1$ be cycle-free, where $r_0 = (r, \varepsilon)\varepsilon$ and $r_1 = \sum_{w \in \Sigma^+}(r, w)w$. Then by the sum star identity (Corollary 5.4 of Droste and Kuich [5]) we obtain $r^* = (r_0^* r_1)^* r_0^*$. Hence, $r^* \in S^{\mathrm{rat}}\langle\langle\Sigma^*\rangle\rangle$ according to the original definition.

The collection of all behaviors of cycle-free finite automata (over S and Σ) is denoted by $S^{\mathrm{rec}}\langle\langle\Sigma^*\rangle\rangle$. The classical theorem of Kleene essentially states that $\mathbb{B}^{\mathrm{rat}}\langle\langle\Sigma^*\rangle\rangle$ and $\mathbb{B}^{\mathrm{rec}}\langle\langle\Sigma^*\rangle\rangle$ coincide.

As a generalization of this theorem of Kleene, we now prove a variant of the theorem of Kleene–Schützenberger. (See Schützenberger [24], Conway [4], Eilenberg [6], Salomaa and Soittola [23], Kuich and Salomaa [19], Kuich [18], Berstel and Reutenauer [1], Ésik and Kuich [9]).

Before proving this generalization, we show that each cycle-free finite automaton is equivalent to one with a proper transition matrix. In the following proof and then without mention, we use the isomorphism between $(S\langle\langle\Sigma^*\rangle\rangle)^{Q\times Q}$ and $S^{Q\times Q}\langle\langle\Sigma^*\rangle\rangle$ to simplify our notation (see Droste and Kuich [5], before Lemma 4.1).

Theorem 2.4. *Each cycle-free finite automaton is equivalent to a cycle-free finite automaton* $\mathfrak{A}' = (Q', R', A', P')$, *where* $A' \in (S\langle\Sigma\rangle)^{Q\times Q}$, $P' \in (S\langle\{\varepsilon\}\rangle)^{Q\times 1}$, *and there exists a* $q_0 \in Q'$ *such that* $R'_{q_0} = \varepsilon$, $R'_q = 0$ *for all* $q \neq q_0$.

Proof. For each cycle-free finite automaton, there exists by Theorem 2.3, an equivalent normalized cycle-free finite automaton $\mathfrak{A} = (Q, R, A, P)$. Let $A_0 = (A, \varepsilon)\varepsilon$ and $A_1 = \sum_{x\in\Sigma}(A, x)x$, and define the finite automaton $\mathfrak{A}'$ by $Q' = Q$, $A' = A_0^* A_1$, $R' = R$, $P' = A_0^* P$. Then

$$\|\mathfrak{A}'\| = R(A_0^* A_1)^* A_0^* P = R(A_0 + A_1)^* P = RA^* P = \|\mathfrak{A}\|.$$

Here, we have applied in the second equality the sum star identity (Corollary 5.4 of Droste and Kuich [5]). □

We define, for finite automata $\mathfrak{A} = (Q, R, A, P)$ and $\mathfrak{A}' = (Q', R', A', P')$ with $Q \cap Q' = \emptyset$, the finite automata $\mathfrak{A} + \mathfrak{A}'$, $\mathfrak{A} \cdot \mathfrak{A}'$ and $\mathfrak{A}^+$:

$$\mathfrak{A} + \mathfrak{A}' = \left(Q \cup Q', (R\ R'), \begin{pmatrix} A & 0 \\ 0 & A' \end{pmatrix}, \begin{pmatrix} P \\ P' \end{pmatrix}\right),$$
$$\mathfrak{A} \cdot \mathfrak{A}' = \left(Q \cup Q', (R\ 0), \begin{pmatrix} A & PR' \\ 0 & A' \end{pmatrix}, \begin{pmatrix} 0 \\ P' \end{pmatrix}\right),$$
$$\mathfrak{A}^+ = (Q, R, A + PR, P).$$

In the construction of $\mathfrak{A} \cdot \mathfrak{A}'$ (resp. $\mathfrak{A}^+$), we assume that the entries of PR' (resp. PR) are in $S\langle\Sigma \cup \{\varepsilon\}\rangle$ or else $\mathfrak{A}$ or $\mathfrak{A}'$ to be normalized.

Theorem 2.5 (Theorem of Kleene–Schützenberger). $S^{\mathrm{rat}}\langle\langle\Sigma^*\rangle\rangle = S^{\mathrm{rec}}\langle\langle\Sigma^*\rangle\rangle$.

Proof. (i) An easy proof by induction on $|Q|$ using Corollary 5.9 of Droste and Kuich [5] shows that $A^* \in (S^{\mathrm{rat}}\langle\langle\Sigma^*\rangle\rangle)^{Q\times Q}$ if $A \in (S\langle\Sigma\rangle)^{Q\times Q}$. This implies by Theorem 2.4 immediately that $S^{\mathrm{rec}}\langle\langle\Sigma^*\rangle\rangle \subseteq S^{\mathrm{rat}}\langle\langle\Sigma^*\rangle\rangle$.

(ii) Let $r, r' \in S^{\mathrm{rec}}\langle\langle \Sigma^* \rangle\rangle$ and assume that $r = \|\mathfrak{A}\|$ and $r' = \|\mathfrak{A}'\|$ for cycle-free finite automata $\mathfrak{A}$ and $\mathfrak{A}'$.

Application of Corollary 5.8 of Droste and Kuich [5] shows that $\mathfrak{A} + \mathfrak{A}'$ and $\mathfrak{A} \cdot \mathfrak{A}'$ are cycle-free and that the equations $\|\mathfrak{A} + \mathfrak{A}'\| = \|\mathfrak{A}\| + \|\mathfrak{A}'\|$ and $\|\mathfrak{A} \cdot \mathfrak{A}'\| = \|\mathfrak{A}\| \cdot \|\mathfrak{A}'\|$ are valid. Hence, $r + r' = \|\mathfrak{A} + \mathfrak{A}'\|$ and $r \cdot r' = \|\mathfrak{A} \cdot \mathfrak{A}'\|$ are in $S^{\mathrm{rec}}\langle\langle \Sigma^* \rangle\rangle$.

Let now $r \in S^{\mathrm{rec}}\langle\langle \Sigma^* \rangle\rangle$ be a proper power series. By Theorem 2.4, there exists a finite automaton $\mathfrak{A} = (Q, R, A, P)$ with $A \in (S\langle \Sigma \rangle)^{Q \times Q}$, $R \in (S\langle \{\varepsilon\} \rangle)^{1 \times Q}$ and $P \in (S\langle \{\varepsilon\} \rangle)^{Q \times 1}$ such that $r = \|\mathfrak{A}\|$. Consider now the finite automaton $\mathfrak{A}^+$. Since $(r, \varepsilon) = (R, \varepsilon)(P, \varepsilon) = 0$, we obtain $((A + PR)^2, \varepsilon) = ((PR)^2, \varepsilon) = (P, \varepsilon)(R, \varepsilon)(P, \varepsilon)(R, \varepsilon) = (P, \varepsilon)(r, \varepsilon)(R, \varepsilon) = 0$ and $\mathfrak{A}^+$ is cycle-free.

We now compute the behavior of $\mathfrak{A}^+$ and obtain

$$\|\mathfrak{A}^+\| = R(A + PR)^* P = R(A^* PR)^* A^* P = (RA^* P)(RA^* P)^* = \|\mathfrak{A}\| \cdot \|\mathfrak{A}\|^*.$$

Here, we have applied in the second equality Theorem 5.2 (the sum star identity) and in the third equality Theorem 3.2 of Droste and Kuich [5].

Easy constructions yield $a\varepsilon, x \in S^{\mathrm{rec}}\langle\langle \Sigma^* \rangle\rangle$, $a \in S$, $x \in \Sigma$. Hence, $S\langle \Sigma \cup \{\varepsilon\} \rangle \subseteq S^{\mathrm{rec}}\langle\langle \Sigma^* \rangle\rangle$. Moreover, for a proper power series $r \in S^{\mathrm{rec}}\langle\langle \Sigma^* \rangle\rangle$, $r^* = \varepsilon + rr^*$ by Theorem 3.1 of Droste and Kuich [5]. Hence, $r^* \in S^{\mathrm{rec}}\langle\langle \Sigma^* \rangle\rangle$ and we have proved $S^{\mathrm{rat}}\langle\langle \Sigma^* \rangle\rangle \subseteq S^{\mathrm{rec}}\langle\langle \Sigma^* \rangle\rangle$. This implies our theorem. □

2.2 Finite Automata over Conway Semirings

Recall from Droste and Kuich [5] and Ésik [8] that a *Conway semiring* is a star semiring that satisfies the *sum star identity*

$$(a + b)^* = (a^* b)^* a^*$$

and the *product star identity*

$$(ab)^* = 1 + a(ba)^* b$$

for all semiring elements a, b. It then follows that the *star fixed point identity*

$$a^* = 1 + aa^*$$

and the *simplified product star identity*

$$a(ba)^* = (ab)^* a$$

hold for all semiring elements a, b.

Let S be a star semiring. Then for $A \in S^{n \times n}$, we define $A^* \in S^{n \times n}$ inductively as follows:

(i) For $n = 1$ and $A = (a)$, $a \in S$, we define $A^* = (a^*)$.

(ii) For $n > 1$, we partition A into blocks $A = \begin{pmatrix} a & b \\ c & d \end{pmatrix}$ and define $A^* = \begin{pmatrix} \alpha & \beta \\ \gamma & \delta \end{pmatrix}$ with $a, \alpha \in S^{1\times 1}$, $b, \beta \in S^{1\times(n-1)}$, $c, \gamma \in S^{(n-1)\times 1}$, $d, \delta \in S^{(n-1)\times(n-1)}$ by

$$\alpha = (a + bd^*c)^*, \qquad \beta = \alpha bd^*, \qquad \gamma = \delta ca^*, \qquad \delta = (d + ca^*b)^*.$$

(See Theorem 3.3 of Conway [4], Theorem 4.21 of Kuich and Salomaa [19], Theorem 2.5 of Kuich [18], Theorem 5.7 of Droste and Kuich [5], Bloom and Ésik [2], Sect. 6 of Ésik [8], and Example 2.1.) If $\langle S, +, \cdot, ^*, 0, 1\rangle$ is a star semiring, then the *star operation* in the star semiring $\langle S^{n\times n}, +, \cdot, ^*, 0, E\rangle$ will always be defined as above.

Theorem 2.6 (Conway [4], Bloom and Ésik [2], Ésik and Kuich [12], Ésik [8]). *If S is a Conway semiring, then for $n \geq 1$, $S^{n\times n}$ again is a Conway semiring.*

Let A and A^* be given as in (ii) of the definition of A^* above, but with $a, \alpha \in S^{n_1\times n_1}$, $b, \beta \in S^{n_1\times n_2}$, $c, \gamma \in S^{n_2\times n_1}$, $d, \delta \in S^{n_2\times n_2}$, $n_1 + n_2 = n$. Then the *matrix star identity* is valid in the star semiring S if A^* is independent of the partition of n in summands.

Theorem 2.7 (Conway [4], Bloom and Ésik [2], Ésik and Kuich [12], Ésik [8]). *If S is a Conway semiring, then the matrix star identity holds.*

Let S be a star semiring. Then for $r \in S\langle\langle \Sigma^*\rangle\rangle$, we define the star $r^* \in S\langle\langle \Sigma^*\rangle\rangle$ of r inductively as follows:

$$(r^*, \varepsilon) = (r, \varepsilon)^*, \quad (r^*, w) = (r, \varepsilon)^* \sum_{uv=w,\ u\neq\varepsilon} (r, u)(r^*, v), \ w \in \Sigma^*, \ w \neq \varepsilon.$$

(See Bloom and Ésik [2], and Theorem 3.5 of Kuich and Salomaa [19].) If $\langle S, +, \cdot, ^*, 0, 1\rangle$ is a star semiring, then the *star operation* in the star semiring $\langle S\langle\langle \Sigma^*\rangle\rangle, +, \cdot, ^*, 0, \varepsilon\rangle$ will be always defined as above.

Theorem 2.8 (Bloom and Ésik [2], Ésik and Kuich [12]). *If S is a Conway semiring and Σ is an alphabet then $S\langle\langle \Sigma^*\rangle\rangle$ is again a Conway semiring.*

Corollary 2.9. *If S is a Conway semiring, Σ is an alphabet and $n \geq 1$, then $(S\langle\langle \Sigma^*\rangle\rangle)^{n\times n}$ is again a Conway semiring.*

Each complete semiring is a Conway semiring (Kuich [17], Hebisch [14], Bloom and Ésik [2], see also Droste and Kuich [5] and Ésik [8]). Moreover, for a complete semiring S, the star operations in the complete semirings $S^{n\times n}$ and $S\langle\langle \Sigma^*\rangle\rangle$ are the same as the star operations in the Conway semirings $S^{n\times n}$ and $S\langle\langle \Sigma^*\rangle\rangle$, respectively.

For the rest of this subsection, S denotes a Conway semiring and S' denotes a subset of S. We now generalize the finite automata of Sect. 2.1 to finite S'-automata over a Conway semiring S.

A *finite S'-automaton*

$$\mathfrak{A} = (n, R, A, P), \quad n \geq 1$$

(*over the Conway semiring S*) is given by:

(i) A *transition matrix* $A \in (S' \cup \{0,1\})^{n \times n}$
(ii) An *initial state vector* $R \in (S' \cup \{0,1\})^{1 \times n}$
(iii) A *final state vector* $P \in (S' \cup \{0,1\})^{n \times 1}$

The *behavior* $\|\mathfrak{A}\|$ of $\mathfrak{A}$ is defined by

$$\|\mathfrak{A}\| = \sum_{1 \leq i_1, i_2 \leq n} R_{i_1} (A^*)_{i_1, i_2} P_{i_2} = RA^*P.$$

The *(directed) graph of* $\mathfrak{A}$ is constructed analogous to that in Sect. 2.1. It has nodes $1, \ldots, n$ and an edge from node i to node j if $A_{ij} \neq 0$. The *weight* of this edge is $A_{ij} \in S' \cup \{1\}$. The *initial* (resp. *final*) *weight* of a node i is given by R_i (resp. P_i). A node is called *initial* (resp. *final*) if its initial (resp. final) weight is unequal to 0. The *weight* of a path is the product of the weights of its edges. It is easily shown that $(A^k)_{ij}$ is the sum of the weights of paths of length k from node i to node j. If S is a complete semiring, and hence $(A^*)_{ij} = \sum_{k \geq 0} (A^k)_{ij}$, then $(A^*)_{ij}$ is the sum of the weights of the paths from node i to node j. Hence, $S_{i_1} (A^*)_{i_1, i_2} P_{i_2}$ is this sum for nodes i_1 and i_2, multiplied on the left and right by the initial weight of node i_1 and the final weight of node i_2, respectively. Eventually, the behavior of $\mathfrak{A}$ is the sum of all these terms with summation over all initial states i_1 and all final states i_2.

Theorem 2.10. *Let S be a complete semiring and $S' \subseteq S$. If $\mathfrak{A}$ is a finite S'-automaton then $\|\mathfrak{A}\|$ is the sum of the weights of all paths from an initial state to a final state multiplied by the initial and final weights of these states.*

Two finite S'-automata $\mathfrak{A}$ and $\mathfrak{A}'$ are *equivalent* if $\|\mathfrak{A}\| = \|\mathfrak{A}'\|$. A finite S'-automaton $\mathfrak{A} = (n, R, A, P)$ is called *normalized* if $n \geq 2$ and:

(i) $R_1 = 1$, $R_i = 0$, for all $2 \leq i \leq n$.
(ii) $P_n = 1$, $P_i = 0$, for all $1 \leq i \leq n-1$.
(iii) $A_{i,1} = A_{n,i} = 0$, for all $1 \leq i \leq n$.

Hence, the directed graph of a normalized finite S'-automaton has the unique initial node 1 and the unique final node n, both with weight 1; moreover, no edges are leading to the initial node and no edges are leaving the final node.

Theorem 2.11. *Let S be a Conway semiring and $S' \subseteq S$. Then each finite S'-automaton is equivalent to a normalized finite S'-automaton.*

Proof. Let $\mathfrak{A} = (n, R, A, P)$ be a finite S'-automaton. Define the finite S'-automaton $\mathfrak{A}'$ by

$$\mathfrak{A}' = \left(1+n+1, (1\ 0\ 0), \begin{pmatrix} 0 & R & 0 \\ 0 & A & P \\ 0 & 0 & 0 \end{pmatrix}, \begin{pmatrix} 0 \\ 0 \\ 1 \end{pmatrix}\right).$$

Then $\mathfrak{A}'$ is normalized. Applying the matrix star identity yields the proof that $\|\mathfrak{A}'\| = \|\mathfrak{A}\|$. □

The substar semiring of S that is generated by S' is denoted by $\mathfrak{Rat}(S')$. The collection of all behaviors of finite S'-automata is denoted by $\mathfrak{Rec}(S')$. The classical theorem of Kleene essentially states that $\mathfrak{Rat}(\mathbb{B}\langle\Sigma\rangle)$ and $\mathfrak{Rec}(\mathbb{B}\langle\Sigma\rangle)$ coincide. As a generalization of the theorem of Kleene–Schützenberger, we show that $\mathfrak{Rat}(S') = \mathfrak{Rec}(S')$. (See Conway [4], Bloom and Ésik [2], Kuich [17, 18], Ésik and Kuich [9]).

We now define, for given finite S'-automata $\mathfrak{A} = (n, R, A, P)$ and $\mathfrak{A}' = (n', R', A', P')$, the finite S-automata $\mathfrak{A} + \mathfrak{A}'$, $\mathfrak{A} \cdot \mathfrak{A}'$ and $\mathfrak{A}^*$:

$$\mathfrak{A} + \mathfrak{A}' = \left(n+n', (R\ R'), \begin{pmatrix} A & 0 \\ 0 & A' \end{pmatrix}, \begin{pmatrix} P \\ P' \end{pmatrix}\right),$$

$$\mathfrak{A} \cdot \mathfrak{A}' = \left(n+n', (R\ 0), \begin{pmatrix} A & PR' \\ 0 & A' \end{pmatrix}, \begin{pmatrix} 0 \\ P' \end{pmatrix}\right),$$

$$\mathfrak{A}^* = \left(1+n, (1\ 0), \begin{pmatrix} 0 & R \\ P & A \end{pmatrix}, \begin{pmatrix} 1 \\ 0 \end{pmatrix}\right).$$

Theorem 2.12. *Let S be a Conway semiring and $S' \subseteq S$. Then $\mathfrak{Rat}(S') = \mathfrak{Rec}(S')$.*

Proof. (i) An easy proof by induction on n using the matrix star identity shows that $A^* \in \mathfrak{Rat}(S')^{n\times n}$ if $A \in (S' \cup \{0,1\})^{n\times n}$. This implies immediately $\mathfrak{Rec}(S') \subseteq \mathfrak{Rat}(S')$.

(ii) Easy constructions yield $S' \cup \{0,1\} \subseteq \mathfrak{Rec}(S')$. Consider now a and a' in $\mathfrak{Rec}(S')$. Then there exist finite S'-automata $\mathfrak{A} = (n, R, A, P)$ and $\mathfrak{A}' = (n', R', A', P')$ such that $\|\mathfrak{A}\| = a$ and $\|\mathfrak{A}'\| = a'$. Clearly, $\mathfrak{A} + \mathfrak{A}'$ and $\mathfrak{A}^*$ are finite S'-automata. If PR' is in $S' \cup \{0,1\}$, then also $\mathfrak{A} \cdot \mathfrak{A}'$ is a finite S'-automaton. If PR' is not in $S' \cup \{0,1\}$, choose $\mathfrak{A}$ or $\mathfrak{A}'$ to be normalized. This is, by Theorem 2.11, no loss of generality. Then again $\mathfrak{A} \cdot \mathfrak{A}'$ is a finite S'-automaton. Application of the matrix star identity shows that the equations $\|\mathfrak{A} + \mathfrak{A}'\| = \|\mathfrak{A}\| + \|\mathfrak{A}'\| = a + a'$, $\|\mathfrak{A} \cdot \mathfrak{A}'\| = \|\mathfrak{A}\| \cdot \|\mathfrak{A}'\| = a \cdot a'$ and $\|\mathfrak{A}^*\| = \|\mathfrak{A}\|^* = a^*$ are valid. □

We now turn to the power series semiring $S\langle\langle\Sigma^*\rangle\rangle$. A finite $S\langle\Sigma \cup \{\varepsilon\}\rangle$-automaton $\mathfrak{A} = (n, R, A, P)$ is called a *finite automaton (over S and Σ) without ε-moves* if $A \in (S\langle\Sigma\rangle)^{n\times n}$, $R \in (S\langle\{\varepsilon\}\rangle)^{1\times n}$ with $R_1 = \varepsilon$, $R_j = 0$ for $2 \le j \le n$, $P \in (S\langle\{\varepsilon\}\rangle)^{n\times 1}$. (This definition holds also for arbitrary semirings.) For $S = \mathbb{B}$, this is the usual definition, i. e., such a finite $\mathbb{B}\langle\Sigma\rangle$-automaton is a nondeterministic finite automaton without ε-moves in the classical sense.

We now show that each finite $S\langle \Sigma \cup \{\varepsilon\}\rangle$-automaton is equivalent to a finite automaton without ε-moves.

Theorem 2.13. *Let S be a Conway semiring. Then each finite $S\langle \Sigma \cup \{\varepsilon\}\rangle$-automaton is equivalent to a finite automaton over A and Σ without ε-moves.*

Proof. For each finite $S\langle \Sigma \cup \{\varepsilon\}\rangle$-automaton there exits, by Theorem 2.11, an equivalent normalized finite $S\langle \Sigma \cup \{\varepsilon\}\rangle$-automaton. Let $\mathfrak{A} = (n, R, A, P)$ be such a normalized finite $S\langle \Sigma \cup \{\varepsilon\}\rangle$-automaton. Let $A_0 = (A, \varepsilon)\varepsilon$ and $A_1 = \sum_{x\in\Sigma}(A, x)x$ and define the finite automaton without ε-moves $\mathfrak{A}' = (n, R, A_0^* A_1, A_0^* P)$. Then

$$\|\mathfrak{A}'\| = R(A_0^* A_1)^* A_0^* P = R(A_0 + A_1)^* P = RA^* P = \|\mathfrak{A}\|.$$

Here, we have applied in the second equality the sum star identity. □

Observe that, in case of a Conway semiring S, we have $\mathfrak{Rat}(S\langle \Sigma \cup \{\varepsilon\}\rangle) = S^{\mathrm{rat}}\langle\langle \Sigma^* \rangle\rangle$. Indeed, it is clear that $S^{\mathrm{rat}}\langle\langle \Sigma^* \rangle\rangle \subseteq \mathfrak{Rat}(S\langle \Sigma \cup \{\varepsilon\}\rangle)$. The reverse inclusion is consequence of the fact that if $r \in S^{\mathrm{rat}}\langle\langle \Sigma^* \rangle\rangle$ then $r^* \in S^{\mathrm{rat}}\langle\langle \Sigma^* \rangle\rangle$, which is shown as follows. Given r, write $r = r_0 + r_1$, where $r_0 = (r, \varepsilon)\varepsilon$ and $r_1 = \sum_{w\in\Sigma^+}(r, w)w$. Now, one can easily see that r_1 is also in $S^{\mathrm{rat}}\langle\langle \Sigma^* \rangle\rangle$, and by the sum star identity we obtain $r^* = (r_0^* r_1)^* r_0^* \in S^{\mathrm{rat}}\langle\langle \Sigma^* \rangle\rangle$.

Analogously, in case of a Conway semiring, by Theorem 2.13, we have $\mathfrak{Rec}(S\langle \Sigma \cup \{\varepsilon\}\rangle) = S^{\mathrm{rec}}\langle\langle \Sigma^* \rangle\rangle$.

Corollary 2.14. $S^{\mathrm{rat}}\langle\langle \Sigma^* \rangle\rangle = S^{\mathrm{rec}}\langle\langle \Sigma^* \rangle\rangle = \{\|\mathfrak{A}\| \mid \mathfrak{A}$ *is a finite automaton over S and Σ without ε-moves*$\}$.

The classical theorem of Kleene in terms of formal power series essentially states that $\mathbb{B}^{\mathrm{rat}}\langle\langle \Sigma^* \rangle\rangle$ and $\mathbb{B}^{\mathrm{rec}}\langle\langle \Sigma^* \rangle\rangle$ coincide.

Corollary 2.15. $\mathbb{B}^{\mathrm{rat}}\langle\langle \Sigma^* \rangle\rangle = \mathbb{B}^{\mathrm{rec}}\langle\langle \Sigma^* \rangle\rangle = \{\|\mathfrak{A}\| \mid \mathfrak{A}$ *is a finite automaton over $\mathbb{B}$ and Σ without ε-moves*$\}$.

Usually, a nondeterministic finite automaton without ε-moves is defined as follows (see Hopcroft and Ullman [15]). A *nondeterministic finite automaton* (in the classical sense)

$$\mathcal{A} = (Q, \Sigma, \delta, q_1, F)$$

is given by:

(i) A finite nonempty *set of states* Q
(ii) An *input alphabet* Σ
(iii) A *transition function* $\delta : Q \times \Sigma \to 2^Q$
(iv) An *initial state* $q_1 \in Q$
(v) A *set of final states* $F \subseteq Q$

The transition function δ is extended to a mapping $\hat{\delta} : Q \times \Sigma^* \to 2^Q$ by

$$\hat{\delta}(q, \varepsilon) = \{q\}, \qquad \hat{\delta}(q, wx) = \{p \mid p \in \delta(r, x) \text{ for some } r \in \hat{\delta}(q, w)\},$$

for $q \in Q$, $w \in \Sigma^*$ and $x \in \Sigma$.

A word $w \in \Sigma^*$ is *accepted* by $\mathcal{A}$ if $\hat{\delta}(q_1, w) \cap F \neq \emptyset$. The *language* $\|\mathcal{A}\|$ *accepted by* $\mathcal{A}$, is defined by

$$\|\mathcal{A}\| = \{w \in \Sigma^* \mid \hat{\delta}(q_1, w) \cap F \neq \emptyset\}.$$

We now connect the notion of a finite automaton $\mathfrak{A}$ over 2^{Σ^*} without ε-moves with the notion of a nondeterministic finite automaton $\mathcal{A}$ as defined above.

Assume that $\mathfrak{A} = (n, R, A, P)$ and $\mathcal{A} = (Q, \Sigma, \delta, q_1, F)$. Then $\mathfrak{A}$ and $\mathcal{A}$ *correspond to each other* if the following conditions are satisfied:

(i) $|Q| = n$; so we may assume $Q = \{q_1, \dots, q_n\}$, where i corresponds to q_i, $1 \leq i \leq n$.
(ii) $x \in A_{ij} \Leftrightarrow q_j \in \delta(q_i, x)$, $1 \leq i, j \leq n$, $x \in \Sigma$.
(iii) $R_{q_1} = \{\varepsilon\}$, $R_{q_i} = \emptyset$, $2 \leq i \leq n$.
(iv) $P_i = \{\varepsilon\} \Leftrightarrow q_i \in F$, $P_i = \emptyset \Leftrightarrow q_i \notin F$.

It is easily seen that $\|\mathfrak{A}\| = \|\mathcal{A}\|$ if $\mathfrak{A}$ and $\mathcal{A}$ correspond to each other. This is due to the fact that

$$w \in (A^k)_{ij} \iff q_j \in \hat{\delta}(q_i, w), \quad 1 \leq i, j \leq n, \; k \geq 0, \; w \in \Sigma^*, \; |w| = k,$$

and

$$w \in (A^*)_{ij} \iff q_j \in \hat{\delta}(q_i, w), \quad 1 \leq i, j \leq n, \; w \in \Sigma^*.$$

(In the complete star semiring $(2^{\Sigma^*})^{n \times n}$, we have $A^* = \bigcup_{n \geq 0} A^n$.) Hence,

$$\begin{aligned} \|\mathfrak{A}\| = RA^*P &= \bigcup_{1 \leq i,j \leq n} R_i (A^*)_{ij} P_j = \bigcup_{q_j \in F} (A^*)_{1j} \\ &= \bigcup_{q_j \in F} \{w \mid q_j \in \hat{\delta}(q_1, w)\} = \{w \mid \hat{\delta}(q_1, w) \cap F \neq \emptyset\} = \|\mathcal{A}\|. \end{aligned}$$

It is clear that $\mathfrak{Rec}(\Sigma)$ coincides with the collection of all languages over Σ accepted by a nondeterministic finite automaton, and $\mathfrak{Rat}(\Sigma)$ coincides with the set of all languages that can be constructed from the finite subsets of Σ^* by the rational operations of union, concatenation, and Kleene-iteration.

Corollary 2.16 (Kleene's theorem [16]). *In the semiring 2^{Σ^*} of formal languages over Σ, $\mathfrak{Rat}(\Sigma) = \mathfrak{Rec}(\Sigma)$.*

2.3 Finite Linear Systems

A *finite linear system* (*over the semiring* S) is of the form

$$y = Ay + P.$$

Here, $y = \begin{pmatrix} y_1 \\ \vdots \\ y_n \end{pmatrix}$ is a column vector of variables $y_1, \ldots, y_n$, $A \in S^{n\times n}$ and $P \in S^{n\times 1}$. A *solution* to the finite linear system $y = Ay + P$ is given by a column vector $\sigma \in S^{n\times 1}$ such that $\sigma = A\sigma + P$.

A finite linear system $y = Ay + P$ over the semiring $S\langle\langle \Sigma^* \rangle\rangle$ is called *cycle-free* if A is cycle-free.

Theorem 2.17. *The cycle-free finite linear system* $y = Ay + P$ *over the semiring* $S\langle\langle \Sigma^* \rangle\rangle$ *has the unique solution* $\sigma = A^* P$.

Proof. The proof is analogous to the proof of Theorem 5.1 of Droste and Kuich [5]. □

The next theorem connects finite automata and finite linear systems. The special form of the cycle-free finite linear system in Theorem 2.18(i) is needed in connection with regular grammars in Theorem 2.20. In the sequel, e_k denotes the kth row vector of unity.

Theorem 2.18. *Let* $r \in S\langle\langle \Sigma^* \rangle\rangle$. *Then the following statements are equivalent:*

(i) r *is a component of the solution of a cycle-free finite linear system* $y = Ay + P$ *over the semiring* $S\langle\langle \Sigma^* \rangle\rangle$, *where* $A \in (S\langle \Sigma \rangle)^{n\times n}$, $A_{i,1} = 0$ *for all* $1 \le i \le n$, *and* $P_1 \in S\langle \Sigma \cup \{\varepsilon\}\rangle$, $P_i \in S\langle \Sigma \rangle$, $2 \le i \le n$.
(ii) r *is in* $S^{\mathrm{rec}}\langle\langle \Sigma^* \rangle\rangle$.

Proof. (i) ⇒ (ii): The solution of $y = Ay + P$ is $\sigma = A^* P$. We now construct cycle-free finite automata $\mathfrak{A}_k$ such that $\|\mathfrak{A}_k\| = \sigma_k$, $1 \le k \le n$: $\mathfrak{A}_k = (\{1, \ldots, n\}, e_k, A, P)$. Hence, by Theorem 2.3, $\sigma_k \in S^{\mathrm{rec}}\langle\langle \Sigma^* \rangle\rangle$.

(ii) ⇒ (i): Let $r \in S^{\mathrm{rec}}\langle\langle \Sigma^* \rangle\rangle$. Then by Theorem 2.4, there exists a cycle-free finite automaton $\mathfrak{A} = (\{1, \ldots, n\}, e_1, A, P)$, $A \in (S\langle \Sigma \rangle)^{n\times n}$, $P \in (S\langle \{\varepsilon\} \rangle)^{n\times 1}$, such that $\|\mathfrak{A}\| = r$. Consider now the finite linear system

$$\begin{pmatrix} y_0 \\ y \end{pmatrix} = \begin{pmatrix} 0 & e_1 A \\ 0 & A \end{pmatrix} \begin{pmatrix} y_0 \\ y \end{pmatrix} + \begin{pmatrix} e_1 P + e_1 AP \\ AP \end{pmatrix}.$$

Here, $y = \begin{pmatrix} y_1 \\ \vdots \\ y_n \end{pmatrix}$. By Corollary 5.8 of Droste and Kuich [5], this finite linear system is cycle-free and its solution is given by

$$\sigma = \begin{pmatrix} \varepsilon & e_1 A A^* \\ 0 & A^* \end{pmatrix} \begin{pmatrix} e_1 P + e_1 AP \\ AP \end{pmatrix}.$$

Hence, $\sigma_0 = e_1 P + e_1 AP + e_1 AA^* AP = e_1 P + e_1 AP + e_1 AAA^* P = e_1 A^* P = \|\mathfrak{A}\| = r$. Here, we have applied Theorems 3.1 and 3.2 of Droste and Kuich [5]. □

Let $G = (\{y_1, \dots, y_n\}, \Sigma, \Pi, y_1)$ be a context-free grammar. Then G is called *right linear* if Π contains only productions of the form $y_i \to xy_j$ or $y_i \to x$, where $x \in \Sigma \cup \{\varepsilon\}$ and $1 \le i, j \le n$. And G is called *regular* if Π contains only productions of the form $y_i \to xy_j$, $y_i \to x$, where $x \in \Sigma$, or $y_1 \to \varepsilon$; but if $y_1 \to \varepsilon \in \Pi$ then y_1 must not appear on the right side of any production of the grammar.

Consider the right linear grammars $G_i = (\{y_1, \dots, y_n\}, \Sigma, \Pi, y_i)$, $1 \le i \le n$. Then define the finite linear system $y = Ay + P$ over the semiring 2^{Σ^*} by $A_{i,j} = \{x \mid y_i \to xy_j \in \Pi\}$ and $P_i = \{x \mid y_i \to x \in \Pi\}$, $1 \le i, j \le n$. Conversely, given a finite linear system $y = Ay + P$ with $A_{i,j}, P_i \in \Sigma \cup \{\varepsilon\}$, $1 \le i, j \le n$, over the semiring 2^{Σ^*}, define the right linear grammars $G_i = (\{y_1, \dots, y_n\}, \Sigma, \Pi, y_i)$ by $y_i \to xy_j \in \Pi$ iff $x \in A_{i,j}$ and $y_i \to x \in \Pi$ iff $x \in P_i$. Whenever we speak of right linear grammars *corresponding* to a finite linear system, or vice versa, then we mean the correspondence in the sense of the above definition.

The next theorem is a special case of a result referred to in Petre and Salomaa [21].

Theorem 2.19. *Let $G_i = (\{y_1, \dots, y_n\}, \Sigma, \Pi, y_i)$, $1 \le i \le n$, be regular grammars and consider the corresponding finite linear system over 2^{Σ^*} with unique solution σ. Then $\sigma_i = L(G_i)$, $1 \le i \le n$.*

Consider a cycle-free finite linear system of the form of Theorem 2.18(i) and the corresponding right linear grammars $G_1, \dots, G_n$. Then G_1 is regular.

Corollary 2.20. *A language is generated by a regular grammar iff it is accepted by a nondeterministic finite automaton.*

More statements on the correspondence of finite automata and right linear grammars can be found in the forthcoming Theorems 3.20 and 3.21, Corollary 3.24 with $k = 0$, and Corollary 3.25.

We now turn to Conway semirings S.

Theorem 2.21. *Let S be a Conway semiring. Then $\sigma = A^*P$ is a solution to the finite linear system $y = Ay + P$ over S.*

Proof. $A\sigma + P = AA^*P + P = (AA^* + E)P = A^*P = \sigma$. Here, we have applied Theorem 2.6 and the star fixed point identity, cf. Ésik [8]. □

We call the above solution of the system $y = Ay + P$ the *canonical solution*. Next, we consider inductive star semirings, cf. Ésik and Kuich [9], Ésik [8]. Recall that every inductive star semiring is a Conway semiring (and an iteration semiring).

Theorem 2.22. *Let S be an inductive star semiring. Then the canonical solution is the least solution to the finite linear system $y = Ay + P$ over S.*

Proof. By Corollary 6.18 of Ésik [8]. □

Let $S' \subseteq S$. A finite linear system $y = Ay + P$ over the semiring S is called *finite S'-linear system* if the entries of A and P are in $S' \cup \{0, 1\}$.

The next theorem connects finite S'-linear systems and finite S'-automata.

Theorem 2.23. *Let S be a Conway semiring and $S' \subseteq S$. Then for $a \in S$, the following statements are equivalent:*

(i) a is a component of the canonical solution of a finite S'-linear system.
(ii) a is the behavior $\|\mathfrak{A}\|$ of a finite S'-automaton $\mathfrak{A}$.

Proof. (i) $\Rightarrow$ (ii): Consider the finite S'-linear system $y = Ay + P$ with n variables and the kth component $e_k A^* P$ of its canonical solution, $1 \leq k \leq n$. Then the behavior of the finite S'-automaton $\mathfrak{A}_k = (n, e_k, A, P)$ equals $e_k A^* P$ for all $1 \leq k \leq n$.

(ii) $\Rightarrow$ (i): By Theorem 2.11, a is, without loss of generality, the behavior of a normalized finite S'-automaton $\mathfrak{A} = (n, e_1, A, P)$. The behavior $\|\mathfrak{A}\| = e_1 A^* P$ is then the first component of the canonical solution of the finite S'-linear system $y = Ay + P$. □

Corollary 2.24. *Let S be an inductive star semiring and $S' \subseteq S$. Then for $a \in S$, the following statements are equivalent:*

(i) a is a component of the least solution of a finite S'-linear system.
(ii) a is the behavior $\|\mathfrak{A}\|$ of a finite S'-automaton $\mathfrak{A}$.

Since every continuous semiring is an inductive star semiring, the last two results also apply for continuous semirings.

3 Finite Automata over Quemirings

In this section, we deal with semiring–semimodule pairs and finite automata over quemirings. Here, semiring–semimodule pairs constitute a generalization of formal languages with finite and infinite words. The semiring models formal languages with finite words while the semimodule models formal languages with infinite words. The main result of this section is a generalization of the Kleene theorem of Büchi [3] in the setting of semiring–semimodule pairs. (See also Perrin and Pin [20].)

This section consists of three subsections. In Sect. 3.1 we introduce the algebraic structures used in this section: semiring–semimodule pairs and quemirings.

In Sect. 3.2, we define finite automata over quemirings. Given a Conway semiring–semimodule pair (S, V), cf. Bloom and Ésik [2], Ésik [8], we prove a Kleene theorem for S'-finite automata, where S' is a subset of S: the collection of all behaviors of S'-finite automata coincides with the generalized star quemiring generated by S' (Theorem 3.14). A special case of this Kleene theorem is the result of Büchi [3].

In Sect. 3.3, we consider finite linear systems over quemirings as a generalization of regular grammars with finite and infinite derivations. We show a connection between certain solutions of these finite linear systems, the weights of finite and infinite derivations with respect to regular grammars and the behaviors of finite automata over quemirings.

We now give a typical example on Büchi automata over infinite words.

Example 3.1. A *(finite) Büchi automaton*

$$\mathcal{A} = (Q, \Sigma, \delta, q, G)$$

is given by:

(i) A finite *set of states* $Q = \{q_1, \ldots, q_n\}$, $n \geq 1$
(ii) An *input alphabet* Σ
(iii) A *transition function* $\delta : Q \times \Sigma \to 2^Q$
(iv) An *initial state* $q \in Q$
(v) A *set of repeated states* $G = \{q_1, \ldots, q_k\}$, $k \geq 0$

A *run of* $\mathcal{A}$ on an infinite word $w \in \Sigma^\omega$, $w = a_1 a_2 a_3 \ldots$, is an infinite sequence of states $q(0), q(1), q(2), q(3), \ldots$ such that the following conditions are satisfied:

(i) $q(0) = q$.
(ii) $q(i) \in \delta(q(i-1), a_i)$ for $i \geq 1$.

A word $w \in \Sigma^\omega$ is *Büchi accepted by* $\mathcal{A}$ if there exists a run ρ of $\mathcal{A}$ on w and a repeated state in G occurring infinitely often in ρ.

The *behavior* $\|\mathcal{A}\| \subseteq \Sigma^\omega$ of $\mathcal{A}$ is defined to be the set of infinite words that are Büchi accepted by $\mathcal{A}$ (see Büchi [3]).

Let now $\mathcal{A} = (Q, \Sigma, \delta, 2, \{1\})$ be a Büchi automaton, where $Q = \{1, 2\}$, $\Sigma = \{a, b, c, d\}$, and $\delta(1, a) = \{1\}$, $\delta(1, b) = \{2\}$, $\delta(2, c) = \{1\}$, $\delta(2, d) = \{2\}$ are the only nonempty images of δ. The graph of $\mathcal{A}$ is

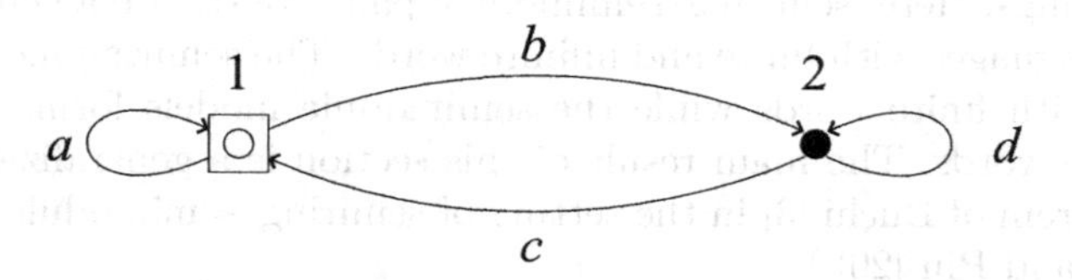

and the adjacency matrix of this graph is

$$A = \begin{pmatrix} \{a\} & \{b\} \\ \{c\} & \{d\} \end{pmatrix}.$$

(See Example 2.1.)

The language of inscriptions of paths from 1 to 1 not passing 1 is given by $\{a\} \cup \{b\}\{d\}^*\{c\}$. Hence, the ω-language of inscriptions of infinite paths starting in 1 and passing infinitely often through 1 is $(\{a\} \cup \{b\}\{d\}^*\{c\})^\omega$ and the ω-language of inscriptions of infinite paths starting in 1, passing finitely often through 1 and infinitely often through 2 is $(\{a\} \cup \{b\}\{d\}^*\{c\})^*\{b\}\{d\}^\omega$. By symmetry, the ω-language of inscriptions of infinite paths starting in 2 and passing infinitely often through 2 (resp. finitely often through 2 and infinitely often through 1) is $(\{d\} \cup \{c\}\{a\}^*\{b\})^\omega$ (resp. $(\{d\} \cup \{c\}\{a\}^*\{b\})^*\{c\}\{a\}^\omega$).

We now define a column vector A^ω by

$$A^\omega = \begin{pmatrix} (\{a\} \cup \{b\}\{d\}^*\{c\})^\omega \cup (\{a\} \cup \{b\}\{d\}^*\{c\})^*\{b\}\{d\}^\omega \\ (\{d\} \cup \{c\}\{a\}^*\{b\})^\omega \cup (\{d\} \cup \{c\}\{a\}^*\{b\})^*\{c\}\{a\}^\omega \end{pmatrix},$$

where $(A^\omega)_1$ (resp. $(A^\omega)_2$) is the ω-language of inscriptions of all infinite paths starting in 1 (resp. 2). Observe that

$$(\{a\} \cup \{b\}\{d\}^*\{c\})^\omega \cap (\{a\} \cup \{b\}\{d\}^*\{c\})^*\{b\}\{d\}^\omega = \emptyset$$

and

$$(\{d\} \cup \{c\}\{a\}^*\{b\})^\omega \cap (\{d\} \cup \{c\}\{a\}^*\{b\})^*\{c\}\{a\}^\omega = \emptyset.$$

The ω-language of inscriptions of infinite paths starting in 2 and passing infinitely often through 1 is $\{d\}^*\{c\}(\{a\} \cup \{b\}\{d\}^*\{c\})^\omega$. We define a column vector $A^{\omega,1}$ by

$$A^{\omega,1} = \begin{pmatrix} (\{a\} \cup \{b\}\{d\}^*\{c\})^\omega \\ \{d\}^*\{c\}(\{a\} \cup \{b\}\{d\}^*\{c\})^\omega \end{pmatrix},$$

where $(A^{\omega,1})_1$ (resp. $(A^{\omega,1})_2$) is the ω-language of inscriptions of all infinite paths starting in 1 (resp. 2) and passing infinitely often through 1.

The ω-language $\|\mathcal{A}\|$ is the ω-language of all inscriptions of infinite paths starting in 2 and passing infinitely often through 1, i.e., $\|\mathcal{A}\| = (A^{\omega,1})_2$.

3.1 Semiring–Semimodule Pairs and Quemirings

Suppose that S is a semiring and V is a commutative monoid written additively. We call V a (left) *S-semimodule* if V is equipped with a (left) action

$$S \times V \to V,$$
$$(s, v) \mapsto sv$$

subject to the following rules:

$$\begin{aligned} s(s'v) &= (ss')v, \\ (s+s')v &= sv + s'v, \\ s(v+v') &= sv + sv', \\ 1v &= v, \\ 0v &= 0, \\ s0 &= 0, \end{aligned}$$

for all $s, s' \in S$ and $v, v' \in V$. When V is an S-semimodule, we call (S, V) a *semiring–semimodule pair.*

Suppose that (S, V) is a semiring–semimodule pair such that S is a star semiring, and S and V are equipped with an omega operation $^\omega : S \to V$. Then we call (S, V) a *star semiring–omega semimodule pair.* Following Bloom and Ésik [2], see also Ésik [8], we call a star semiring–omega semimodule pair (S, V) a *Conway semiring–semimodule pair* if S is a Conway semiring and if the omega operation satisfies the *sum omega identity* and the *product omega identity*:

$$\begin{aligned}(a+b)^\omega &= (a^*b)^\omega + (a^*b)^*a^\omega\\ (ab)^\omega &= a(ba)^\omega,\end{aligned}$$

for all $a, b \in S$. It then follows that the *omega fixed point identity* holds, i.e.,

$$aa^\omega = a^\omega,$$

for all $a \in S$.

Theorem 3.2 (Bloom and Ésik [2], Ésik and Kuich [12], Ésik [8]). *If (S, V) is a Conway semiring–semimodule pair, then for $n \geq 1$, $(S^{n\times n}, V^{n\times 1})$ again is a Conway semiring–semimodule pair.*

Ésik and Kuich [13] define a *complete semiring–semimodule pair* to be a semiring–semimodule pair (S, V) such that S is a complete semiring, V is a complete monoid with

$$\begin{aligned}s\Big(\sum_{i\in I} v_i\Big) &= \sum_{i\in I} sv_i,\\ \Big(\sum_{i\in I} s_i\Big)v &= \sum_{i\in I} s_i v,\end{aligned}$$

for all $s \in S$, $v \in V$, and for all families s_i, $i \in I$ over S and v_i, $i \in I$ over V. Moreover, it is required that an *infinite product operation*

$$(s_1, s_2, \ldots) \mapsto \prod_{j\geq 1} s_j$$

is given mapping infinite sequences over S to V subject to the following three conditions:

$$\begin{aligned}\prod_{i\geq 1} s_i &= \prod_{i\geq 1}(s_{n_{i-1}+1}\cdot \cdots \cdot s_{n_i}),\\ s_1\Big(\prod_{i\geq 1} s_{i+1}\Big) &= \prod_{i\geq 1} s_i,\\ \prod_{j\geq 1}\sum_{i_j\in I_j} s_{i_j} &= \sum_{(i_1,i_2,\ldots)\in I_1\times I_2\times\cdots}\ \prod_{j\geq 1} s_{i_j},\end{aligned}$$

where in the first equation $0 = n_0 \leq n_1 \leq n_2 \leq \cdots$ and $I_1, I_2, \ldots$ are arbitrary index sets. Suppose that (S, V) is complete. Then we define

$$s^* = \sum_{i \geq 0} s^i,$$
$$s^\omega = \prod_{i \geq 1} s,$$

for all $s \in S$. This turns (S, V) into a star semiring–omega semimodule pair. By Ésik and Kuich [13], see also Ésik [8], each complete semiring–semimodule pair is a Conway semiring–semimodule pair.

A *star–omega semiring* is a semiring S equipped with unary operations * and $^\omega : S \to S$. A star–omega semiring S is called *complete* if (S, S) is a complete semiring–semimodule pair, i.e., if S is complete and is equipped with an infinite product operation that satisfies the three conditions stated above and the omega operation is determined by the infinite product as above.

Example 3.3. Suppose that Σ is an alphabet. Let Σ^* denote the set of all finite words over Σ including the empty word ε, and let Σ^ω denote the set of all ω-words over Σ. The set 2^{Σ^*} of all subsets of Σ^*, equipped with the operations of set union as sum and concatenation as product is a semiring, where 0 is the empty set $\emptyset$ and 1 is the set $\{\varepsilon\}$. Moreover, equipped with the usual star operation, 2^{Σ^*} is a Conway semiring. Also, 2^{Σ^ω}, equipped with union as the sum operation and the empty set as 0 is a commutative idempotent monoid. Define an action of 2^{Σ^*} on 2^{Σ^ω} by $KL = \{uv \mid u \in K,\ v \in L\}$, for all $K \subseteq \Sigma^*$ and $L \subseteq \Sigma^\omega$. Moreover, for each sequence $(K_0, K_1, \ldots)$ over 2^{Σ^*}, let $\prod_{j \geq 0} K_j = \{u_0 u_1 \ldots \in \Sigma^\omega \mid u_i \in K_i,\ i \geq 0\}$. Then $(2^{\Sigma^*}, 2^{\Sigma^\omega})$ is a complete semiring–semimodule pair with idempotent module 2^{Σ^ω}. Note that in this example, $1^\omega = 0$, where $1 = \{\varepsilon\}$ and $0 = \emptyset$.

Example 3.4. Consider the semiring $\mathbb{N}^\infty = N \cup \{\infty\}$, obtained by adjoining a top element ∞ to the semiring of the natural numbers. Note that $\mathbb{N}^\infty$ is a complete semiring where an infinite sum is ∞ iff either a summand is ∞ or the number of nonzero summands is infinite. Moreover, ∞ multiplied with a nonzero element on either side gives ∞. Define an infinite product

$$(n_1, n_2, \ldots) \mapsto \prod_{j \geq 1} n_j$$

on $\mathbb{N}^\infty$ as follows. If some n_j is 0, then so is the product. Otherwise, if all but a finite number of the n_j are 1s, then the infinite product is the product of those n_j with $n_j > 1$. In all remaining cases, the infinite product is ∞. Then $\mathbb{N}^\infty$ is a complete star–omega semiring, where * and $^\omega$ are defined as above.

Let Σ denote an alphabet. The semiring $S = \mathbb{N}^\infty\langle\langle \Sigma^* \rangle\rangle$ of all power series over Σ^* with coefficients in $\mathbb{N}^\infty$ is a complete and continuous semiring. Now let $V = \mathbb{N}^\infty\langle\langle\Sigma^\omega\rangle\rangle$ be the collection of all formal power series over Σ^ω with coefficients in $\mathbb{N}^\infty$. Thus, the elements of V are formal sums of the sort

$$s = \sum_{w\in\Sigma^\omega} (s,w)w,$$

where each coefficient (s,w) belongs to $\mathbb{N}^\infty$. Now V can be turned into an S-semimodule by the pointwise sum operation and the action $(r,s) \mapsto rs$ defined by

$$(rs,w) = \sum_{u\in\Sigma^*,\ v\in\Sigma^\omega,\ uv=w} (r,u)(s,v),$$

where the infinite sum on the right hand side exists since $\mathbb{N}^\infty$ is complete. We may also define an infinite product taking sequences over S to series in V. Given $s_1, s_2, \ldots$ in S, we define $\prod_{j\geq 1} s_j$ to be the series r in V with

$$(r,w) = \sum_{w=w_1w_2\ldots\in\Sigma^\omega} \prod_{j\geq 1}(s_j,w_j).$$

Then $(\mathbb{N}^\infty\langle\langle\Sigma^*\rangle\rangle, \mathbb{N}^\infty\langle\langle\Sigma^\omega\rangle\rangle)$ is a complete semiring–semimodule pair, and thus a Conway semiring–semimodule pair.

This can be generalized to a large extent. Suppose that S is a complete star–omega semiring. If Σ is a set, consider the complete semiring $S\langle\langle\Sigma^*\rangle\rangle$ and the complete monoid $S\langle\langle\Sigma^\omega\rangle\rangle$ of all series over Σ^ω with coefficients in S equipped with the pointwise sum operation. If we define the action sr of $s \in S\langle\langle\Sigma^*\rangle\rangle$ on $r \in S\langle\langle\Sigma^\omega\rangle\rangle$ by

$$(sr,w) = \sum_{w=uv} (s,u)(r,v),$$

then $(S\langle\langle\Sigma^*\rangle\rangle, S\langle\langle\Sigma^\omega\rangle\rangle)$ becomes a semiring–semimodule pair. Now, $S\langle\langle\Sigma^*\rangle\rangle$ is a star semiring, and if we define the infinite product operation

$$(s_1, s_2, \ldots) \mapsto \prod_{j\geq 1} s_j \in S\langle\langle\Sigma^\omega\rangle\rangle$$

by

$$\left(\prod_{j\geq 1} s_j, w\right) = \sum_{w=w_1w_2\ldots\in\Sigma^\omega} \prod_{j\geq 1}(s_j,w_j),$$

then $(S\langle\langle\Sigma^*\rangle\rangle, S\langle\langle\Sigma^\omega\rangle\rangle)$ becomes a complete semiring–semimodule pair, hence a Conway semiring–semimodule pair, satisfying $s^\omega = 0$, where each $s \in S$ is identified with a series in the usual way.

Consider a star semiring–omega semimodule pair (S,V). Following Bloom and Ésik [2], see also Ésik [8] and Example 3.1, we define a matrix operation

${}^{\omega} : S^{n\times n} \to V^{n\times 1}$ on a star semiring–omega semimodule pair (S, V) as follows. When $n = 0$, then A^ω is the unique element of $V^{0\times 1}$. When $n = 1$, so that $A = (a)$, for some $a \in S$, $A^\omega = (a^\omega)$. For $n > 1$, we partition A into blocks

$$A = \begin{pmatrix} a & b \\ c & d \end{pmatrix}, \tag{1}$$

where $a \in S^{1\times 1}$, $b \in S^{1\times(n-1)}$, $c \in S^{(n-1)\times 1}$, $d \in S^{(n-1)\times(n-1)}$, and define

$$A^\omega = \begin{pmatrix} (a + bd^*c)^\omega + (a + bd^*c)^*bd^\omega \\ (d + ca^*b)^\omega + (d + ca^*b)^*ca^\omega \end{pmatrix}. \tag{2}$$

Let A and A^ω be given as in the definition above, but with $a \in S^{n_1\times n_1}$, $b \in S^{n_1\times n_2}$, $c \in S^{n_2\times n_1}$, $d \in S^{n_2\times n_2}$, where $n_1 + n_2 = n$. Then the *matrix omega identity* is valid in the star semiring–omega semimodule pair if A^ω is independent of the partition of n into summands.

Theorem 3.5 (Bloom and Ésik [2], Ésik and Kuich [12], Ésik [8]). *If (S, V) is a Conway semiring–semimodule pair, then the matrix omega identity holds in the Conway semiring–semimodule pair $(S^{n\times n}, V^{n\times 1})$ for all $n \geq 1$.*

Following Ésik and Kuich [11] (see also Example 3.1), we define matrix operations ${}^{\omega,k} : S^{n\times n} \to V^{n\times 1}$, $0 \leq k \leq n$, as follows. Assume that $A \in S^{n\times n}$ is decomposed into blocks a, b, c, d as in (1), but with a of dimension $k \times k$ and d of dimension $(n - k) \times (n - k)$. Then

$$A^{\omega,k} = \begin{pmatrix} (a + bd^*c)^\omega \\ d^*c(a + bd^*c)^\omega \end{pmatrix}. \tag{3}$$

Observe that $A^{\omega,0} = 0$ and $A^{\omega,n} = A^\omega$.

Suppose that (S, V) is a semiring–semimodule pair and consider $T = S\times V$. Define on T the operations

$$(s, u) \cdot (s', v) = (ss', u + sv),$$
$$(s, u) + (s', v) = (s + s', u + v)$$

and constants $0 = (0, 0)$ and $1 = (1, 0)$. Equipped with these operations and constants, T satisfies the identities

$$(x + y) + z = x + (y + z), \tag{4}$$
$$x + y = y + x, \tag{5}$$
$$x + 0 = x, \tag{6}$$
$$(x \cdot y) \cdot z = x \cdot (y \cdot z), \tag{7}$$
$$x \cdot 1 = x, \tag{8}$$
$$1 \cdot x = x, \tag{9}$$
$$(x + y) \cdot z = (x \cdot z) + (y \cdot z), \tag{10}$$
$$0 \cdot x = 0. \tag{11}$$

Define also the unary operation ¶ on T: $(s,u)\P = (s,0)$. Thus, ¶ selects the "first component" of the pair (s,u), while multiplication with 0 on the right selects the "second component," for $(s,u)\cdot 0 = (0,u)$, for all $u \in V$. The new operation satisfies:

$$x\P \cdot (y+z) = (x\P \cdot y) + (x\P \cdot z), \tag{12}$$

$$x = x\P + (x \cdot 0), \tag{13}$$

$$x\P \cdot 0 = 0, \tag{14}$$

$$(x+y)\P = x\P + y\P, \tag{15}$$

$$(x \cdot y)\P = x\P \cdot y\P. \tag{16}$$

Note that when V is idempotent, also

$$x \cdot (y+z) = x \cdot y + x \cdot z$$

holds.

Elgot [7] defined a *quemiring* to be an algebraic structure T equipped with the above operations $\cdot, +, \P$ and constants $0, 1$ satisfying the identities (4)–(11) and (12)–(16). A morphism of quemirings is a function preserving the operations and constants. It follows that $x\P\P = x\P$, for all x in a quemiring T. Moreover, $x\P = x$ iff $x \cdot 0 = 0$.

When T is a quemiring, $S = T\P = \{x\P \mid x \in T\}$ is easily seen to be a semiring. Moreover, $V = T0 = \{x \cdot 0 \mid x \in T\}$ contains 0 and is closed under $+$, and furthermore, $sx \in V$ for all $s \in S$ and $x \in V$. Each $x \in T$ may be written in a unique way as the sum of an element of $T\P$ and of an element of $T0$ as $x = x\P + x \cdot 0$. Sometimes, we will identify $S \times \{0\}$ with S and $\{0\} \times V$ with V. It is shown in Elgot [7] that T is isomorphic to the quemiring $S \times V$ determined by the semiring–semimodule pair (S, V).

Suppose now that (S, V) is a star semiring–omega semimodule pair. Then we define on $T = S \times V$ a *generalized star operation*:

$$(s,v)^{\otimes} = (s^*, s^{\omega} + s^* v) \tag{17}$$

for all $(s,v) \in T$. Note that the star and omega operations can be recovered from the generalized star operation, since s^* is the first component of $(s,0)^{\otimes}$ and s^{ω} is the second component. Thus,

$$(s^*, 0) = (s,0)^{\otimes}\P,$$

$$(0, s^{\omega}) = (s,0)^{\otimes} \cdot 0.$$

Observe that, for $(s,0) \in S \times \{0\}$, $(s,0)^{\otimes} = (s^*,0) + (0, s^{\omega})$.

Suppose now that T is an (abstract) quemiring equipped with a generalized star operation $^{\otimes}$. As explained above, T as a quemiring is isomorphic to the quemiring $S \times V$ associated with the semiring–semimodule pair (S, V), where $S = T\P$ and $V = T0$, an isomorphism being the map $x \mapsto (x\P, x \cdot 0)$. It is clear that a generalized star operation $^{\otimes} : T \to T$ is determined by a star operation $^{*} : S \to S$ and an omega operation $^{\omega} : S \to V$ by (17) iff

$$x^{\otimes}\P = (x\P)^{\otimes}\P, \tag{18}$$
$$x^{\otimes} \cdot 0 = (x\P)^{\otimes} \cdot 0 + x^{\otimes}\P \cdot x \cdot 0 \tag{19}$$

hold. Indeed, these conditions are clearly necessary. Conversely, if (18) and (19) hold, then for any $x\P \in T\P$ we may define

$$(x\P)^{*} = (x\P)^{\otimes}\P, \tag{20}$$
$$(x\P)^{\omega} = (x\P)^{\otimes} \cdot 0. \tag{21}$$

It follows that (17) holds. The definition of star and omega was forced.

Let us call a quemiring equipped with a generalized star operation $^{\otimes}$ satisfying (18) and (19) a *generalized star quemiring.*

3.2 Finite Automata over Quemirings and a Kleene Theorem

In this subsection, we consider finite automata over quemirings and prove a Kleene theorem. Throughout this subsection, (S, V) denotes a Conway semiring–semimodule pair and T denotes the generalized star quemiring $S \times V$. Moreover, S' denotes a subset of S.

A *finite S'-automaton (over the quemiring T)*

$$\mathfrak{A} = (n, R, A, P, k)$$

is given by:

(i) A finite *set of states* $\{1, \ldots, n\}$, $n \geq 1$
(ii) A *transition matrix* $A \in (S' \cup \{0,1\})^{n \times n}$
(iii) An *initial state vector* $R \in (S' \cup \{0,1\})^{1 \times n}$
(iv) A *final state vector* $P \in (S' \cup \{0,1\})^{n \times 1}$
(v) a *set of repeated states* $\{1, \ldots, k\}$, $k \geq 0$

The *behavior* of $\mathfrak{A}$ is an element of T and is defined by

$$\|\mathfrak{A}\| = RA^{*}P + RA^{\omega,k}.$$

If $\mathfrak{A} = \left(n, (i_1\ i_2), \left(\begin{smallmatrix} a & b \\ c & d \end{smallmatrix}\right), \left(\begin{smallmatrix} p_1 \\ p_2 \end{smallmatrix}\right), k\right)$, where

$$\begin{aligned} &i_1 \in (S' \cup \{0,1\})^{1 \times k}, && i_2 \in (S' \cup \{0,1\})^{1 \times (n-k)}, \\ &a \in (S' \cup \{0,1\})^{k \times k}, && b \in (S' \cup \{0,1\})^{k \times (n-k)}, \\ &c \in (S' \cup \{0,1\})^{(n-k) \times k}, && d \in (S' \cup \{0,1\})^{(n-k) \times (n-k)}, \\ &p_1 \in (S' \cup \{0,1\})^{k \times 1}, && p_2 \in (S' \cup \{0,1\})^{(n-k) \times 1}, \end{aligned}$$

we write also

$$\mathfrak{A} = (n; i_1, i_2; a, b, c, d; p_1, p_2; k).$$

Theorem 3.6. *If (S, V) is a complete semiring–semimodule pair and $\mathfrak{A}$ is a finite S'-automaton, then $\|\mathfrak{A}\| = F + I$, where F is the sum of the weights of all finite paths from an initial state to a final state, multiplied by the initial and final weights of these states, and where I is the sum of the weights of all infinite paths starting at an initial state, passing infinitely often through repeated states, and multiplied by the initial weight of this initial state.*

Proof. By formalization of the considerations in Example 3.1 (see the proof of Theorem 3.5 of Ésik and Kuich [11]).

Note that the finite S'-automata $\mathfrak{A} = (n, R, A, P)$ and $\mathfrak{A}' = (n, R, A, P, 0)$ over a Conway semiring S and a star semiring–omega semimodule pair (S, V), respectively, have the same behavior $\|\mathfrak{A}\| = \|\mathfrak{A}'\| = RA^*P$.

By definition, $\omega\text{-}\mathfrak{Rat}(S')$ is the Conway semiring–semimodule pair generated by S'.

We now will prove a Kleene theorem: Let $a \in S \times V$. Then $a \in \omega\text{-}\mathfrak{Rat}(S')$ iff a is the behavior of a finite S'-automaton. To achieve this result, we need a few theorems and corollaries. In our arguments, we will use Theorem 3.5.

Let $\mathfrak{A} = (n, R, A, P, k)$ be a finite S'-automaton. It is called *normalized* if:

(i) $n \geq 2$ and $k \leq n - 2$.
(ii) $R_{n-1} = 1$ and $R_j = 0$ for $j \neq n - 1$.
(iii) $P_n = 1$ and $P_j = 0$ for $j \neq n$.
(iv) $A_{i,n-1} = 0$ and $A_{n,i} = 0$ for all $1 \leq i \leq n$.

Two finite S'-automata $\mathfrak{A}$ and $\mathfrak{A}'$ are *equivalent* if $\|\mathfrak{A}\| = \|\mathfrak{A}'\|$.

Theorem 3.7. *Each finite S'-automaton $\mathfrak{A} = (n, R, A, P, k)$ is equivalent to a normalized finite S'-automaton $\mathfrak{A}' = (n + 1 + 1, R', A', P', k)$.*

Proof. We define $R' = (0\ 1\ 0)$, $A' = \begin{pmatrix} A & 0 & P \\ R & 0 & 0 \\ 0 & 0 & 0 \end{pmatrix}$ and $P' = \begin{pmatrix} 0 \\ 0 \\ 1 \end{pmatrix}$. Let now $\mathfrak{A} = (n; i_1, i_2; a, b, c, d; p_1, p_2; k)$. Then

$$A' = \begin{pmatrix} a & b & 0 & p_1 \\ c & d & 0 & p_2 \\ i_1 & i_2 & 0 & 0 \\ 0 & 0 & 0 & 0 \end{pmatrix}$$

and the first k entries of $A'^{\omega,k}$ are equal to

$$\left(a + (b\ 0\ p_1) \begin{pmatrix} d & 0 & p_2 \\ i_2 & 0 & 0 \\ 0 & 0 & 0 \end{pmatrix}^* \begin{pmatrix} c \\ i_1 \\ 0 \end{pmatrix} \right)^\omega$$
$$= \left(a + (b\ 0\ p_1) \begin{pmatrix} d^* & 0 & d^*p_2 \\ i_2 d^* & 1 & i_2 d^* p_2 \\ 0 & 0 & 1 \end{pmatrix} \begin{pmatrix} c \\ i_1 \\ 0 \end{pmatrix} \right)^\omega$$
$$= (a + bd^*c)^\omega.$$

Hence, the last $n-k+2$ entries of $A'^{\omega,k}$ are equal to

$$\begin{pmatrix} d & 0 & p_2 \\ i_2 & 0 & 0 \\ 0 & 0 & 0 \end{pmatrix}^* \begin{pmatrix} c \\ i_1 \\ 0 \end{pmatrix} (a+bd^*c)^\omega = \begin{pmatrix} d^*c \\ i_2d^*c+i_1 \\ 0 \end{pmatrix} (a+bd^*c)^\omega$$

and we obtain

$$\begin{aligned} \|\mathfrak{A}'\| &= R'A'^*P' + R'A'^{\omega,k} = (A'^*)_{n+1,n+2} + \left(A'^{\omega,k}\right)_{n+1} \\ &= RA^*P + (i_2d^*c+i_1)(a+bd^*c)^\omega = RA^*P + RA^{\omega,k} = \|\mathfrak{A}\|. \qquad \square \end{aligned}$$

Lemma 3.8. *If* $\mathfrak{A} = (n; i_1, i_2; a, b, c, d; p_1, p_2; k)$ *is a finite* S'*-automaton then*

$$\begin{aligned} \|\mathfrak{A}\| &= i_1(a+bd^*c)^*(p_1+bd^*p_2) + i_2d^*c(a+bd^*c)^*(p_1+bd^*p_2) \\ &\quad + i_2d^*p_2 + i_1(a+bd^*c)^\omega + i_2d^*c(a+bd^*c)^\omega. \end{aligned}$$

Let $\mathfrak{A} = (n; i_1, i_2; a, b, c, d; f, g; m)$ and $\mathfrak{A}' = (n'; h, i; a', b', c', d'; p_1, p_2; k)$ be finite S'-automata. Then we define the finite S'-automata $\mathfrak{A}+\mathfrak{A}'$ and $\mathfrak{A}\cdot\mathfrak{A}'$ to be

$$\begin{aligned} \mathfrak{A}+\mathfrak{A}' = \Bigg(& n+n'; (i_1\ h), (i_2\ i); \\ & \begin{pmatrix} a & 0 \\ 0 & a' \end{pmatrix}, \begin{pmatrix} b & 0 \\ 0 & b' \end{pmatrix}, \begin{pmatrix} c & 0 \\ 0 & c' \end{pmatrix}, \begin{pmatrix} d & 0 \\ 0 & d' \end{pmatrix}; \\ & \begin{pmatrix} f \\ p_1 \end{pmatrix}, \begin{pmatrix} g \\ p_2 \end{pmatrix}, m+k \Bigg) \end{aligned}$$

and

$$\begin{aligned} \mathfrak{A}\cdot\mathfrak{A}' = \Bigg(& n+n'; (i_1\ 0), (i_2\ 0); \\ & \begin{pmatrix} a & fh \\ 0 & a' \end{pmatrix}, \begin{pmatrix} b & fi \\ 0 & b' \end{pmatrix}, \begin{pmatrix} c & gh \\ 0 & c' \end{pmatrix}, \begin{pmatrix} d & gi \\ 0 & d' \end{pmatrix}; \\ & \begin{pmatrix} 0 \\ p_1 \end{pmatrix}, \begin{pmatrix} 0 \\ p_2 \end{pmatrix}, m+k \Bigg). \end{aligned}$$

For the definition of $\mathfrak{A}\cdot\mathfrak{A}'$, we assume that either $\binom{f}{g}(h\ i) \in (S' \cup \{0,1\})^{n\times n'}$ or $\mathfrak{A}'$ is normalized. Observe that the definitions of $\mathfrak{A}+\mathfrak{A}'$ and $\mathfrak{A}\cdot\mathfrak{A}'$ (and of $\mathfrak{A}^\otimes$ which is defined below) are the usual ones except that certain rows and columns are permuted. These permutations are needed since the set of repeated states of a finite S'-automaton is always a set $\{1,\ldots,k\}$.

Theorem 3.9. *Let* $\mathfrak{A}$ *and* $\mathfrak{A}'$ *be finite* S'*-automata. Then* $\|\mathfrak{A}+\mathfrak{A}'\| = \|\mathfrak{A}\| + \|\mathfrak{A}'\|$ *and* $\|\mathfrak{A}\cdot\mathfrak{A}'\| = \|\mathfrak{A}\|\cdot\|\mathfrak{A}'\|$.

Proof. Let $\mathfrak{A}$ and $\mathfrak{A}'$ be defined as above. We first show $\|\mathfrak{A}+\mathfrak{A}'\| = \|\mathfrak{A}\| + \|\mathfrak{A}'\|$ and compute $\|\mathfrak{A}+\mathfrak{A}'\| \cdot 0$. The transition matrix of $\mathfrak{A}+\mathfrak{A}'$ is given by

$$A = \begin{pmatrix} a & 0 & b & 0 \\ 0 & a' & 0 & b' \\ c & 0 & d & 0 \\ 0 & c' & 0 & d' \end{pmatrix}.$$

We now compute the first $m+k$ entries of $A^{\omega,m+k}$. This column vector of dimension $m+k$ is given by

$$\left(\begin{pmatrix} a & 0 \\ 0 & a' \end{pmatrix} + \begin{pmatrix} b & 0 \\ 0 & b' \end{pmatrix}\begin{pmatrix} d & 0 \\ 0 & d' \end{pmatrix}^* \begin{pmatrix} c & 0 \\ 0 & c' \end{pmatrix}\right)^\omega$$
$$= \begin{pmatrix} a+bd^*c & 0 \\ 0 & a'+b'd'^*c' \end{pmatrix}^\omega = \begin{pmatrix} (a+bd^*c)^\omega \\ (a'+b'd'^*c')^\omega \end{pmatrix}.$$

The last $n+n'-(m+k)$ entries of $A^{\omega,m+k}$ are given by the product of

$$\begin{pmatrix} d & 0 \\ 0 & d' \end{pmatrix}^* \begin{pmatrix} c & 0 \\ 0 & c' \end{pmatrix} = \begin{pmatrix} d^*c & 0 \\ 0 & d'^*c' \end{pmatrix}$$

with the column vector computed above. Hence, we obtain by Lemma 3.8

$$\begin{aligned} \|\mathfrak{A}+\mathfrak{A}'\| \cdot 0 &= (i_1\ h\ i_2\ i)A^{\omega,m+k} \\ &= i_1(a+bd^*c)^\omega + h(a'+b'd'^*c')^\omega + i_2 d^*c(a+bd^*c)^\omega \\ &\quad + id'^*c'^*(a'+b'd'^*c')^\omega \\ &= (\|\mathfrak{A}\| + \|\mathfrak{A}'\|) \cdot 0. \end{aligned}$$

We now compute $\|\mathfrak{A}+\mathfrak{A}'\|\P$. If, in the transition matrix A of $\mathfrak{A}+\mathfrak{A}'$ we commute the $m+1,\ldots,m+k$ row and column with the $m+k+1,\ldots,n+k$ row and column, and do the same with the initial and final vector we obtain by the star permutation identity (see Conway [4], Bloom and Ésik [2], Ésik and Kuich [9], and Ésik [8]).

$$\begin{aligned} \|\mathfrak{A}+\mathfrak{A}'\|\P &= (i_1\ i_2\ h\ i)\begin{pmatrix} a & b & 0 & 0 \\ c & d & 0 & 0 \\ 0 & 0 & a' & b' \\ 0 & 0 & c' & d' \end{pmatrix}^* \begin{pmatrix} f \\ g \\ p_1 \\ p_2 \end{pmatrix} \\ &= (i_1\ i_2)\begin{pmatrix} a & b \\ c & d \end{pmatrix}^* \begin{pmatrix} f \\ g \end{pmatrix} + (h\ i)\begin{pmatrix} a' & b' \\ c' & d' \end{pmatrix}^* \begin{pmatrix} p_1 \\ p_2 \end{pmatrix} \\ &= (\|\mathfrak{A}\| + \|\mathfrak{A}'\|)\P. \end{aligned}$$

Hence, $\|\mathfrak{A}+\mathfrak{A}'\| = \|\mathfrak{A}\| + \|\mathfrak{A}'\|$.

We now show $\|\mathfrak{A}\cdot\mathfrak{A}'\| = \|\mathfrak{A}\| \cdot \|\mathfrak{A}'\|$ and compute $\|\mathfrak{A}\cdot\mathfrak{A}'\| \cdot 0$. The transition matrix of $\mathfrak{A}\cdot\mathfrak{A}'$ is given by

$$A = \begin{pmatrix} a & fh & b & fi \\ 0 & a' & 0 & b' \\ c & gh & d & gi \\ 0 & c' & 0 & d' \end{pmatrix}.$$

We now compute the first $m+k$ entries of $A^{\omega,m+k}$. This column vector of dimension $m+k$ is given by

$$\begin{aligned}
&\left(\begin{pmatrix} a & fh \\ 0 & a' \end{pmatrix} + \begin{pmatrix} b & fi \\ 0 & b' \end{pmatrix}\begin{pmatrix} d^* & d^*gid'^* \\ 0 & d'^* \end{pmatrix}\begin{pmatrix} c & gh \\ 0 & c' \end{pmatrix}\right)^\omega \\
&= \begin{pmatrix} a+bd^*c & (f+bd^*g)(h+id'^*c') \\ 0 & a'+b'd'^*c' \end{pmatrix}^\omega \\
&= \begin{pmatrix} (a+bd^*c)^\omega + (a+bd^*c)^*(f+bd^*g)(h+id'^*c')(a'+b'd'^*c')^\omega \\ (a'+b'd'^*c')^\omega \end{pmatrix}.
\end{aligned}$$

The last $n+n'-(m+k)$ entries of $A^{\omega,m+k}$ are given by the product of

$$\begin{pmatrix} d & gi \\ 0 & d' \end{pmatrix}^* \begin{pmatrix} c & gh \\ 0 & c' \end{pmatrix} = \begin{pmatrix} d^*c & d^*g(h+id'^*c') \\ 0 & d'^*c' \end{pmatrix}$$

with the column vector computed above. Hence, we obtain

$$\begin{aligned}
&\|\mathfrak{A}\cdot\mathfrak{A}'\|\cdot 0 \\
&= (i_1\ 0\ i_2\ 0)A^{\omega,m+k} \\
&= i_1(a+bd^*c)^\omega + i_1(a+bd^*c)^*(f+bd^*g)(h+id'^*c)(a'+b'd'^*c')^\omega \\
&\quad + i_2d^*c(a+bd^*c)^\omega + i_2d^*c(a+bd^*c)^*(f+bd^*g)(h+id'^*c') \\
&\quad \times (a'+b'd'^*c')^\omega + i_2d^*g(h+id'^*c')(a'+b'd'^*c')^\omega.
\end{aligned}$$

On the other side, we obtain by Lemma 3.8

$$\begin{aligned}
&\|\mathfrak{A}\|\cdot\|\mathfrak{A}'\|\cdot 0 \\
&= \|\mathfrak{A}\|\cdot 0 + \|\mathfrak{A}\|\P\cdot\|\mathfrak{A}'\|\cdot 0 \\
&= i_1(a+bd^*c)^\omega + i_2d^*c(a+bd^*c)^\omega + (i_1(a+bd^*c)^*(f+bd^*g) \\
&\quad + i_2d^*c(a+bd^*c)^*(f+bd^*g) + i_2d^*g)(h+id'^*c')(a'+b'd'^*c')^\omega.
\end{aligned}$$

Hence, $\|\mathfrak{A}\cdot\mathfrak{A}'\|\cdot 0 = \|\mathfrak{A}\|\cdot\|\mathfrak{A}'\|\cdot 0$.

We now compute $\|\mathfrak{A}\cdot\mathfrak{A}'\|\P$. If, in the transition matrix A of $\mathfrak{A}\cdot\mathfrak{A}'$, we commute the $m+1,\ldots,m+k$ row and column with the $m+k+1,\ldots,n+k$ row and column, and do the same with the initial and final vector we obtain by the star permutation identity (see Conway [4], Bloom and Ésik [2], Ésik and Kuich [9], and Ésik [8]),

$$\|\mathfrak{A}\cdot\mathfrak{A}'\|\P = (i_1\ i_2\ 0\ 0)\begin{pmatrix} a & b & fh & fi \\ c & d & gh & gi \\ 0 & 0 & a' & b' \\ 0 & 0 & c' & d' \end{pmatrix}^* \begin{pmatrix} 0 \\ 0 \\ p_1 \\ p_2 \end{pmatrix}$$

$$= (i_1\ i_2)\begin{pmatrix} a & b \\ c & d \end{pmatrix}^* \begin{pmatrix} f \\ g \end{pmatrix} (h\ i) \begin{pmatrix} a' & b' \\ c' & d' \end{pmatrix}^* \begin{pmatrix} p_1 \\ p_2 \end{pmatrix}$$

$$= \|\mathfrak{A}\|\P \cdot \|\mathfrak{A}'\|\P = \|\mathfrak{A}\| \cdot \|\mathfrak{A}'\|\P .$$

Hence, $\|\mathfrak{A}\cdot\mathfrak{A}'\| = \|\mathfrak{A}\|\cdot\|\mathfrak{A}'\|$. □

Let $\mathfrak{A} = (n; h, i; a, b, c, d; f, g; k)$ be a finite S'-automaton and write $R = (h\ \ i)$, $A = \left(\begin{smallmatrix} a & b \\ c & d \end{smallmatrix}\right)$ and $P = \left(\begin{smallmatrix} f \\ g \end{smallmatrix}\right)$. Then we define the finite S'-automaton $\mathfrak{A}^{\otimes}$ to be

$$\mathfrak{A}^{\otimes} = \Bigg(1+n+n; (1\ 0), (0\ 0); \begin{pmatrix} 0 & h \\ 0 & a \end{pmatrix}, \begin{pmatrix} i & R \\ b & 0 \end{pmatrix}, \begin{pmatrix} 0 & c \\ P & 0 \end{pmatrix}, \begin{pmatrix} d & 0 \\ 0 & A \end{pmatrix}; \begin{pmatrix} 1 \\ 0 \end{pmatrix}, \begin{pmatrix} 0 \\ 0 \end{pmatrix}; 1+k\Bigg).$$

Theorem 3.10. *Let $\mathfrak{A}$ be a finite S'-automaton. Then $\|\mathfrak{A}^{\otimes}\| = \|\mathfrak{A}\|^{\otimes}$.*

Proof. Let $\mathfrak{A}$ be defined as above. Let

$$A' = \begin{pmatrix} 0 & h & i & R \\ 0 & a & b & 0 \\ 0 & c & d & 0 \\ P & 0 & 0 & A \end{pmatrix}.$$

We first compute $\|\mathfrak{A}^{\otimes}\|\P$. Observe that A' can be written as

$$A' = \begin{pmatrix} 0 & R & R \\ 0 & A & 0 \\ P & 0 & A \end{pmatrix}$$

and that $\|\mathfrak{A}^{\otimes}\|\P = (A'^*)_{11}$. We obtain

$$(A'^*)_{11} = \left((R\ R)\begin{pmatrix} A & 0 \\ 0 & A \end{pmatrix}^* \begin{pmatrix} 0 \\ P \end{pmatrix}\right)^*$$

$$= (RA^*P)^* = (\|\mathfrak{A}\|\P)^* = \|\mathfrak{A}\|^{\otimes}\P .$$

We now compute the first $1+k$ entries of $A'^{\omega,1+k}$. This column vector of dimension $1+k$ is given by

$$\left(\begin{pmatrix} 0 & h \\ 0 & a \end{pmatrix} + \begin{pmatrix} i & R \\ b & 0 \end{pmatrix}\begin{pmatrix} d & 0 \\ 0 & A \end{pmatrix}^* \begin{pmatrix} 0 & c \\ P & 0 \end{pmatrix}\right)^{\omega}$$

$$= \begin{pmatrix} RA^*P & h + id^*c \\ 0 & a + bd^*c \end{pmatrix}^{\omega}.$$

Hence,

$$\|\mathfrak{A}^{\otimes}\| \cdot 0 = \left(A'^{\omega,1+k}\right)_1 = (RA^*P)^\omega + (RA^*P)^*(h + id^*c)(a + bd^*c)^\omega .$$

By definition, $\|\mathfrak{A}\|^{\otimes} \cdot 0 = (\|\mathfrak{A}\|\P)^\omega + (\|\mathfrak{A}\|\P)^* \|\mathfrak{A}\| \cdot 0$. Thus,

$$\|\mathfrak{A}\|^{\otimes} \cdot 0 = (RA^*P)^\omega + (RA^*P)^*(h + id^*c)(a + bd^*c)^\omega = \|\mathfrak{A}^{\otimes}\| \cdot 0$$

and we obtain $\|\mathfrak{A}^{\otimes}\| = \|\mathfrak{A}\|^{\otimes}$. □

Theorem 3.11. *Let $\mathfrak{A} = (n, R, A, P, k)$ be a finite S'-automaton. Then there exists a finite S'-automaton $\mathfrak{A}\P$ such that $\|\mathfrak{A}\P\| = \|\mathfrak{A}\|\P$.*

Proof. $\|\mathfrak{A}\|\P = (n, R, A, P, 0)$. □

Theorem 3.12. *Let $a \in S' \cup \{0, 1\}$. Then there exists a finite S'-automaton $\mathfrak{A}_a$ such that $\|\mathfrak{A}_a\| = a$.*

Proof. Let $\mathfrak{A}_a = \left(2, (1\ 0), \left(\begin{smallmatrix} 0 & a \\ 0 & 0 \end{smallmatrix}\right), \left(\begin{smallmatrix} 0 \\ 1 \end{smallmatrix}\right), 0\right)$. Then

$$\|\mathfrak{A}_a\| = (1\ 0) \begin{pmatrix} 1 & a \\ 0 & 1 \end{pmatrix} \begin{pmatrix} 0 \\ 1 \end{pmatrix} = a.$$ □

Corollary 3.13. *The behaviors of finite S'-automata form a generalized star quemiring that contains S'.*

Theorem 3.14 (Kleene theorem). *Let (S, V) be a Conway semiring–semimodule pair. Then the following statements are equivalent for $(s, v) \in S \times V$:*

(i) $(s, v) = \|\mathfrak{A}\|$, where $\mathfrak{A}$ is a finite S'-automaton.
(ii) $(s, v) \in \omega\text{-}\mathfrak{Rat}(S')$.
(iii) $s \in \mathfrak{Rat}(S')$ and v is of the form $\sum_{1 \le i \le m} s_i t_i^\omega$ with $s_i, t_i \in \mathfrak{Rat}(S')$.

Proof. (i) ⇒ (iii): Each entry in A^* and $A^{\omega,k}$ is of the form s and $\sum_{1 \le i \le m} s_i t_i^\omega$, respectively, with $s, s_i, t_i \in \mathfrak{Rat}(S')$.

(iii) ⇒ (ii): $(s, v) = (s, 0) + (0, v)$. Since $(s, 0)$ is in $\mathfrak{Rat}(S') \subseteq \omega - \mathfrak{Rat}(S')$ and $(0, v) = (0, \sum_{1 \le i \le m} s_i t_i^\omega)$ is in $\omega - \mathfrak{Rat}(S')$, (s, v) is in $\omega - \mathfrak{Rat}(S')$.

(ii) ⇒ (i): By Corollary 3.13. □

We now consider finite $S\langle \Sigma \cup \{\varepsilon\}\rangle$-automata over the Conway semiring–semimodule pair $(S\langle\langle \Sigma^* \rangle\rangle, S\langle\langle \Sigma^\omega \rangle\rangle)$, where S is a complete star–omega semiring, and will delete ε-moves in these finite $S\langle \Sigma \cup \{\varepsilon\}\rangle$-automata without changing their behavior.

Theorem 3.15. *Let $(S\langle\langle \Sigma^* \rangle\rangle, S\langle\langle \Sigma^\omega \rangle\rangle)$ be a Conway semiring–semimodule pair satisfying $(s\varepsilon)^\omega = 0$ for all $s \in S$. Let $\mathfrak{A} = (n, R, A, P, k)$ be a finite $S\langle \Sigma \cup \{\varepsilon\}\rangle$-automaton. Then there exists a finite $S\langle \Sigma \cup \{\varepsilon\}\rangle$-automaton $\mathfrak{A}' = (n, R', A', P', k)$ with $\|\mathfrak{A}'\| = \|\mathfrak{A}\|$ satisfying the following conditions:*

(i) $A' \in (S\langle\Sigma\rangle)^{n\times n}$.
(ii) $R' \in (S\langle\{\varepsilon\}\rangle)^{1\times n}$.
(iii) $P' \in (S\langle\{\varepsilon\}\rangle)^{n\times 1}$.

Proof. Without loss of generality, we assume by Theorem 3.7 that $R \in (S\langle\{\varepsilon\}\rangle)^{1\times n}$ and $P \in (S\langle\{\varepsilon\}\rangle)^{n\times 1}$. Let $A = \left(\begin{smallmatrix} a & b \\ c & d \end{smallmatrix}\right)$, where a is a $k \times k$-matrix and d is a $(n-k)\times(n-k)$-matrix. Let $a = a_0 + a_1$, $b = b_0 + b_1$, $c = c_0 + c_1$, $d = d_0 + d_1$, such that the supports of the entries of a_0, b_0, c_0, d_0 (resp. a_1, b_1, c_1, d_1) are subsets of $\{\varepsilon\}$ (resp. Σ). By assumption, $(a_0 + b_0 d_0^* c_0)^\omega = 0$.

Define the matrices A_{01}, A_{02}, and A_1 to be $A_{01} = \left(\begin{smallmatrix} 0 & b_0 \\ 0 & d_0 \end{smallmatrix}\right)$, $A_{02} = \left(\begin{smallmatrix} a_0 & 0 \\ c_0 & 0 \end{smallmatrix}\right)$ and $A_1 = \left(\begin{smallmatrix} a_1 & b_1 \\ c_1 & d_1 \end{smallmatrix}\right)$. We now specify the finite $S\langle\Sigma\cup\{\varepsilon\}\rangle$-automaton $\mathfrak{A}'$: $R' = R(A_{01}^* A_{02})^*$, $A' = A_{01}^* A_1 (A_{01}^* A_{02})^*$ and $P' = A_{01}^* P$. The behavior of $\mathfrak{A}'$ is then given by

$$\begin{aligned}
\|\mathfrak{A}'\| &= R'A'^*P' + R'A'^{\omega,k} \\
&= R(A_{01}^*A_{02})^*\left(A_{01}^*A_1(A_{01}^*A_{02})^*\right)^* A_{01}^*P \\
&\quad + R(A_{01}^*A_{02})^*\left(A_{01}^*A_1(A_{01}^*A_{02})^*\right)^{\omega,k} \\
&= R(A_{01} + A_{02} + A_1)^*P + R(A_{01} + A_{02} + A_1)^{\omega,k} \\
&= RA^*P + RA^{\omega,k} = \|\mathfrak{A}\|.
\end{aligned}$$

Here, we have applied Theorem 2.10 of Ésik and Kuich [11] in the third equality. □

In the next theorem, we construct a finite $S\langle\Sigma\cup\{\varepsilon\}\rangle$-automaton without ε-moves with a unique initial state of initial weight ε.

Theorem 3.16. *Let* $(S\langle\langle\Sigma^*\rangle\rangle, S\langle\langle\Sigma^\omega\rangle\rangle)$ *be a Conway semiring–semimodule pair as in Theorem 3.15, and consider a finite* $S\langle\Sigma\cup\{\varepsilon\}\rangle$*-automaton* $\mathfrak{A} = (n, R, A, P, k)$. *Then there exists a finite* $S\langle\Sigma\cup\{\varepsilon\}\rangle$*-automaton* $\mathfrak{A}' = (n+1, R', A', P', k)$ *with* $\|\mathfrak{A}'\| = \|\mathfrak{A}\|$ *satisfying the following conditions:*

(i) $A' \in (S\langle\Sigma\rangle)^{(n+1)\times(n+1)}$.
(ii) $R'_j = 0$, $1 \le j \le n$, *and* $R'_{n+1} = \varepsilon$.
(iii) $P' \in (S\langle\{\varepsilon\}\rangle)^{(n+1)\times 1}$.

Proof. We assume that $\mathfrak{A}$ satisfies the conditions of Theorem 3.15. We specify $\mathfrak{A}'$ by $R' = (0\ \varepsilon)$, $A' = \left(\begin{smallmatrix} A & 0 \\ RA & 0 \end{smallmatrix}\right)$ and $P' = \left(\begin{smallmatrix} P \\ RP \end{smallmatrix}\right)$. We compute $A'^* = \left(\begin{smallmatrix} A^* & 0 \\ RAA^* & \varepsilon \end{smallmatrix}\right)$ and, for $A = \left(\begin{smallmatrix} a & b \\ c & d \end{smallmatrix}\right)$, $R = (i_1\ i_2)$,

$$\begin{aligned}
A'^{\omega,k} &= \begin{pmatrix} a & b & 0 \\ c & d & 0 \\ i_1a + i_2c & i_1b + i_2d & 0 \end{pmatrix}^{\omega,k} \\
&= \begin{pmatrix} (a+bd^*c)^\omega \\ d^*c(a+bd^*c)^\omega \\ (i_1(a+bd^*c) + i_2d^*c)(a+bd^*c)^\omega \end{pmatrix} \\
&= \begin{pmatrix} A^{\omega,k} \\ RA^{\omega,k} \end{pmatrix}.
\end{aligned}$$

Hence, $\|\mathfrak{A}'\| = RAA^*P + RP + RA^{\omega,k} = \|\mathfrak{A}\|$. □

In the case of the Boolean semiring, the finite $\mathbb{B}\langle\Sigma \cup \{\varepsilon\}\rangle$-automata of Theorem 3.16 with $P' = 0$ are nothing else than the finite automata introduced by Büchi [3].

In the case of the semiring $\mathbb{N}^\infty$, we get the following result.

Theorem 3.17. *The constructions of Theorems 3.15 and 3.16 do not change, for $w \in \Sigma^*$ (resp. for $w \in \Sigma^\omega$), in the digraphs of the finite automata, the number of finite paths with label w from an initial state to a final state (resp. the number of infinite paths with label w starting in an initial state and passing infinitely often through repeated states).*

3.3 Finite Linear Systems over Quemirings

In this subsection, we consider finite linear systems over quemirings as a generalization of regular grammars with finite and infinite derivations.

A *finite S'-linear system (with variables $y_1, \ldots, y_n$, over the quemiring $S \times V$)* is a system of equations

$$y = Ay + P \tag{22}$$

where $A \in (S' \cup \{0,1\})^{n\times n}$, $P \in (S' \cup \{0,1\})^{n\times 1}$, $y = \begin{pmatrix} y_1 \\ \vdots \\ y_n \end{pmatrix}$. A column vector $\sigma \in (S \times V)^{n\times 1}$ is called a *solution* to the system (22) if

$$\sigma = A\sigma + P.$$

Theorem 3.18. *Let (S, V) be a Conway semiring–semimodule pair. Consider a finite S'-linear system*

$$y = Ay + P,$$

where $A \in (S' \cup \{0,1\})^{n\times n}$, $P \in (S' \cup \{0,1\})^{n\times 1}$, and $y = \begin{pmatrix} y_1 \\ \vdots \\ y_n \end{pmatrix}$ is a column vector of variables. Then for each $0 \leq k \leq n$, $A^{\omega,k} + A^*P$ is a solution of $y = Ay + P$.

Proof. We obtain by Theorem 3.2 and the star and omega fixed point identities (cf. Ésik [8]), for each $0 \leq k \leq n$,

$$A(A^{\omega,k} + A^*P) + P = A^{\omega,k} + A^*P. \quad \square$$

Let $k \in \{0, \ldots, n\}$ and $\mathfrak{A}_i = (n, e_i, A, P, k)$, $1 \leq i \leq n$, be finite S'-automata, where e_i is the ith vector of unity. Then $\|\mathfrak{A}_i\|$ is the ith component of a solution given in Theorem 3.18 of the finite S'-linear system $y = Ay + P$. Therefore, we call the solution $\begin{pmatrix} \|\mathfrak{A}_1\| \\ \vdots \\ \|\mathfrak{A}_n\| \end{pmatrix} = A^{\omega,k} + A^*P$ of $y = Ay + P$ the *kth automata-theoretic solution* of $y = Ay + P$.

Theorem 3.19. *Let (S, V) be a Conway semiring–semimodule pair and $S' \subseteq S$. Let $\mathfrak{A} = (n, R, A, P, k)$ be a finite S'-automaton. Then $\|\mathfrak{A}\| = R\sigma$, where σ is the kth automata-theoretic solution of the finite S'-linear system $y = Ay + P$.*

Bi-inductive semiring–semimodule pairs were defined in Ésik and Kuich [13], see also Ésik [8]. As shown in these papers, every bi-inductive semiring–semimodule pair is a Conway semiring–semimodule pair. Suppose that (S, V) is a bi-inductive semiring–semimodule pair and consider the quemiring $S \times V$, which is equipped with the pointwise partial order inherited from the orders on S and V. Moreover, consider the above finite S'-linear system $y = Ay + P$ and write it in more detail as

$$y' = ay' + by'' + p', \tag{23}$$
$$y'' = cy' + dy'' + p'' \tag{24}$$

where y' denotes the matrix of the first k entries of y, etc. We point out that the first k components of the above kth automata theoretic solution of $y = Ay + P$ over $S \times V$ can be obtained by first taking the least solution $d^*(cy' + p'')$ of (24) and substituting this least solution into (23), and then by taking the greatest solution $(a+bd^*c)^\omega + (a+bd^*c)^*(p'+bd^*p'')$ of the resulting equation. The last $n - k$ components of the kth automata theoretic solution are then obtained by substituting this greatest solution into $d^*(cy' + p'')$, the least solution of (24). Thus, we obtain the kth automata theoretic solution

$$\begin{pmatrix} a & b \\ c & d \end{pmatrix}^{\omega,k} + \begin{pmatrix} a & b \\ c & d \end{pmatrix}^{*} \begin{pmatrix} p' \\ p'' \end{pmatrix}.$$

Let S be a complete star–omega semiring and consider a finite S'-linear system $y = Ay + P$ over the quemiring $S\langle\langle \Sigma^* \rangle\rangle \times S\langle\langle \Sigma^\omega \rangle\rangle$ as defined before Theorem 3.18 for $S' = S\langle \Sigma \cup \{\varepsilon\} \rangle$. Write this system in the form

$$y_i = \sum_{1 \le j \le n} \sum_{x \in \Sigma \cup \{\varepsilon\}} (A_{ij}, x) x y_j + \sum_{x \in \Sigma \cup \{\varepsilon\}} (P_i, x) x, \quad 1 \le i \le n.$$

Analogous to the correspondence stated below Theorem 2.18, the right linear grammars $G_i = (\{y_1, \ldots, y_n\}, \Sigma, \Pi, y_i)$, $1 \le i \le n$, with weights in the semiring S, where $\Pi = \{y_i \to (A_{ij}, x) x y_j \mid 1 \le j \le n,\ x \in \Sigma \cup \{\varepsilon\}\} \cup \{y_i \to (P_i, x) x \mid x \in \Sigma \cup \{\varepsilon\}\}$ *correspond to* this finite $S\langle \Sigma \cup \{\varepsilon\} \rangle$-linear system. Here, (A_{ij}, x) and (P_i, x) are the weights of the productions $y_i \to x y_j$ and $y_i \to x$, respectively. Furthermore, let $\mathfrak{A}_i^k = (n, e_i, A, P, k)$ be finite S'-automata, $1 \le i \le n$, for some fixed $k \in \{0, \ldots, n\}$, where e_i is the ith row vector of unity.

Consider now a finite derivation with respect to G_i:

$$\begin{aligned} y_i &\Rightarrow (A_{i,i_1}, x_1) x_1 y_{i_1} \Rightarrow \cdots \Rightarrow (A_{i,i_1}, x_1) \ldots (A_{i_{m-1},i_m}, x_m) x_1 \ldots x_m y_{i_m} \\ &\Rightarrow (A_{i,i_1}, x_1) \ldots (A_{i_{m-1},i_m}, x_m)(P_{i_m}, x_{m+1}) x_1 \ldots x_m x_{m+1} \end{aligned}$$

generating the word $x_1 \dots x_m x_{m+1}$ with weight

$$(A_{i,i_1}, x_1) \dots (A_{i_{m-1},i_m}, x_m)(P_{i_m}, x_{m+1}).$$

This finite derivation corresponds to the following finite path in the directed graph of $\mathfrak{A}_i^k$:

$$(y_i, x_1, y_{i_1}), \dots, (y_{i_{m-1}}, x_m, y_{i_m})$$

with weight

$$(A_{i,i_1}, x_1) \dots (A_{i_{m-1},i_m}, x_m),$$

initial weight 1 and final weight $(P_{i_m}, x_{m+1})x_{m+1}$.

Consider now an infinite derivation with respect to G_i:

$$y_i \Rightarrow (A_{i,i_1}, x_1)x_1 y_{i_1} \Rightarrow \dots \Rightarrow (A_{i,i_1}, x_1) \dots (A_{i_{m-1},i_m}, x_m)x_1 \dots x_m y_{i_m} \Rightarrow \cdots$$

generating the infinite word $x_1 x_2 \dots x_m \dots$ with weight

$$(A_{i,i_1}, x_1) \dots (A_{i_{m-1},i_m}, x_m) \dots .$$

This infinite derivation corresponds to the following infinite path in the directed graph of $\mathfrak{A}_i^k$:

$$(y_i, x_1, y_{i_1}), \dots, (y_{i_{m-1}}, x_m, y_{i_m}), \dots$$

with weight $(A_{i,i_1}, x_1) \dots (A_{i_{m-1},i_m}, x_m) \dots$ and initial weight 1.

If S is a complete star–omega semiring, then $(S\langle\langle \Sigma^* \rangle\rangle, S\langle\langle \Sigma^\omega \rangle\rangle)$ is a complete semiring–semimodule pair by Ésik and Kuich [13], see also Ésik [8]. Hence, we obtain by Theorem 3.6, the following result for G_i and $\mathfrak{A}_i^k$ as defined above.

Theorem 3.20. *If S is a complete star–omega semiring and $1 \le i \le n$, $0 \le k \le n$, then for $w \in \Sigma^*$, $(\|\mathfrak{A}_i^k\|, w) = ((A^*P)_i, w)$ is the sum of the weights of all finite derivations of w with respect to G_i; and for $w \in \Sigma^\omega$, $(\|\mathfrak{A}_i^k\|, w) = ((A^{\omega,k})_i, w)$ is the sum of the weights of all infinite derivations of w with respect to G_i such that at least one of the variables of $\{y_1, \dots, y_k\}$ appears infinitely often in these infinite derivations.*

In particular, if $S = \mathbb{N}^\infty$ and $(A_{ij}, x), (P_i, x) \in \{0, 1\}$, $x \in \Sigma \cup \{\varepsilon\}$, $1 \le i, j \le n$, then we get the following result.

Theorem 3.21. *For $w \in \Sigma^*$, $(\|\mathfrak{A}_i^k\|, w) = ((A^*P)_i, w)$ is the number of finite derivations of w with respect to G_i; and for $w \in \Sigma^\omega$, $(\|\mathfrak{A}_i^k\|, w) = ((A^{\omega,k})_i, w)$ is the number of all infinite derivations of w with respect to G_i such that at least one of the variables of $\{y_1, \dots, y_k\}$ appears infinitely often in these infinite derivations.*

Given a right linear grammar $G_i = (\{y_1, \ldots, y_n\}, \Sigma, \Pi, y_i)$, $1 \le i \le n$, with weights as above, and $k \in \{0, \ldots, n\}$, $L(G_i)_k$ is defined to be the weighted language

$$L(G_i)_k = \{((A^*P)_i, w)w \mid w \in \Sigma^*\} \cup \{((A^{\omega,k})_i, w)w \mid w \in \Sigma^\omega\}.$$

The next theorem, Theorem 3.22, shows that such weighted languages can be generated by right linear grammars with weights in the semiring S which have only two types of productions:

$$y_i \to axy_j \quad \text{and} \quad y_i \to a\varepsilon,$$

where $a \in S$ and $x \in \Sigma$. Hence, in such right linear grammars, there are no productions $y_i \to ay_j$. Corollary 3.23 shows then that the two types of productions can be chosen as

$$y_i \to axy_j \quad \text{and} \quad y_i \to ax,$$

where $a \in S$ and $x \in \Sigma$. (Of course, ε is no longer derived. Compare this with the definition of a regular grammar below Theorem 2.18. Moreover, compare the forthcoming Corollary 3.23 with Theorem 2.18.)

Theorem 3.22. *Let $(S\langle\langle\Sigma^*\rangle\rangle, S\langle\langle\Sigma^\omega\rangle\rangle)$ be a Conway semiring–semimodule pair where $(s\varepsilon)^\omega = 0$ for all $s \in S$, consider a finite $S\langle\Sigma \cup \{\varepsilon\}\rangle$-linear system $y = Ay + P$, where $A \in (S\langle\Sigma \cup \{\varepsilon\}\rangle)^{n\times n}$, $P \in (S\langle\Sigma \cup \{\varepsilon\}\rangle)^{n\times 1}$, and $y = \begin{pmatrix} y_1 \\ \vdots \\ y_n \end{pmatrix}$ and let $i \in \{1, \ldots, n\}$. Then there exists a finite $S\langle\Sigma \cup \{\varepsilon\}\rangle$-linear system $y' = A'y' + P'$, where $A' \in (S\langle\Sigma\rangle)^{(n+1)\times(n+1)}$, $P' \in (S\langle\{\varepsilon\}\rangle)^{(n+1)\times 1}$, and $y' = \begin{pmatrix} y \\ y_{n+1} \end{pmatrix}$ such that, for all $0 \le k \le n$, the $(n+1)$st component of the kth automata-theoretic solution of $y = Ay + P$ coincides with the ith component of the kth automata theoretic solution of $y' = A'y' + P'$.*

Proof. Consider the finite $S\langle\Sigma \cup \{\varepsilon\}\rangle$-automaton $\mathfrak{A}_i^k = (n, e_i, A, P, k)$, whose behavior is $\|\mathfrak{A}_i^k\| = (A^*P)_i + (A^{\omega,k})_i$. Starting with $\mathfrak{A}_i^k$, perform the constructions of Theorems 3.15 and 3.16. This yields a finite $S\langle\Sigma \cup \{\varepsilon\}\rangle$-automaton $\mathfrak{A}' = (n+1, e_{n+1}, A', P', k)$ with behavior $\|\mathfrak{A}'\| = (A'^*P')_{n+1} + (A'^{\omega,k})_{n+1} = \|\mathfrak{A}_i^k\|$. □

Corollary 3.23. *Let $(S\langle\langle\Sigma^*\rangle\rangle, S\langle\langle\Sigma^\omega\rangle\rangle)$ be a Conway semiring–semimodule pair where $(s\varepsilon)^\omega = 0$ for all $s \in S$, consider a finite $S\langle\Sigma \cup \{\varepsilon\}\rangle$-linear system $y = Ay + P$, where $A \in (S\langle\Sigma \cup \{\varepsilon\}\rangle)^{n\times n}$, $P \in (S\langle\Sigma \cup \{\varepsilon\}\rangle)^{n\times 1}$, and $y = \begin{pmatrix} y_1 \\ \vdots \\ y_n \end{pmatrix}$ and let $i \in \{1, \ldots, n\}$. Then there exists a finite $S\langle\Sigma \cup \{\varepsilon\}\rangle$-linear system $y' = A'y' + P'$, where $A' \in (S\langle\Sigma\rangle)^{(n+1)\times(n+1)}$, $P' \in (S\langle\Sigma\rangle)^{(n+1)\times 1}$, and $y' = \begin{pmatrix} y \\ y_{n+1} \end{pmatrix}$ such that $(A'^{\omega,k} + A'^*A'P'')_{n+1} = (A^{\omega,k} + AA^*P)_i$.*

Proof. Let $y' = A'y' + P''$ be the finite $S\langle \Sigma \cup \{\varepsilon\}\rangle$-linear system constructed according to Theorem 3.22 from $y = Ay + P$. Consider the finite $S\langle \Sigma \cup \{\varepsilon\}\rangle$-linear system $y' = A'y' + P'$, where $P' = A'P''$. Then $(A'^{\omega,k} + A'^{*}P')_{n+1} = (A'^{\omega,k} + A'^{*}A'P'')_{n+1} = (A^{\omega,k} + AA^{*}P)_i$. □

Corollary 3.24. *Let S be a complete star–omega semiring and consider, for some $i \in \{1, \ldots, n\}$, the right linear grammar $G_i = (\{y_1, \ldots, y_n\}, \Sigma, \Pi, y_i)$ with weights in S.*

Then there exists a right linear grammar $G^{(i)} = (\{y_1, \ldots, y_n, y_{n+1}\}, \Sigma, \Pi^{(i)}, y_{n+1})$ with weights, which has only the two types of productions

$$y_i \to axy_i \quad \text{and} \quad y_i \to a\varepsilon \quad (\text{resp.} \quad y_i \to axy_j \quad \text{and} \quad y_i \to ax),$$

$a \in S$, $x \in \Sigma$, such that, for all $0 \le k \le n$,

$$L\big(G^{(i)}_{n+1}\big)_k = L(G_i)_k \quad \Big(\text{resp.} \quad L\big(G^{(i)}_{n+1}\big)_k = L(G_i)_k - \big\{\big(L(G_i)_k, \varepsilon\big)\varepsilon\big\}\Big).$$

If we consider finite $\mathbb{N}^{\infty}\langle \Sigma \cup \{\varepsilon\}\rangle$-linear systems, we obtain the following result about the derivations with respect to the right linear grammars G_i defined above.

Corollary 3.25. *The constructions of Theorem 3.22 and Corollary 3.23 do not change, for $w \in \Sigma^{+}$ (resp. for $w \in \Sigma^{\omega}$), the number of finite derivations of w with respect to G_i (resp. the number of infinite derivations of w with respect to G_i such that at least one of the variables of $\{y_1, \ldots, y_n\}$ appears infinitely often in these infinite derivations).*

Hence, the constructions transform unambiguous grammars into unambiguous grammars.

References

1. J. Berstel and C. Reutenauer. *Les séries rationelles et leurs langages.* Masson, Paris, 1984. English translation: *Rational Series and Their Languages*, volume 12 of *Monographs in Theoretical Computer Science. An EATCS Series.* Springer, 1988.
2. S.L. Bloom and Z. Ésik. *Iteration Theories*, *Monographs in Theoretical Computer Science. An EATCS Series.* Springer, Berlin, 1993.
3. J.R. Büchi. On a decision method in restricted second order arithmetic. In *Proc. Int. Congr. Logic, Methodology and Philosophy of Science, 1960.* pages 1–11. Stanford University Press, Stanford, 1962.
4. J.H. Conway. *Regular Algebra and Finite Machines.* Chapman & Hall, London, 1971.
5. M. Droste and W. Kuich. Semirings and formal power series. In this *Handbook*, chapter 1. Springer, Berlin, 2009

6. S. Eilenberg. *Automata, Languages and Machines, volume C.* Draft of Sects. I–III, 1978.
7. C. Elgot. Matricial theories. *Journal of Algebra*, 42:391–422, 1976.
8. Z. Ésik. Fixed point theory. In this *Handbook*, chapter 2. Springer, Berlin, 2009
9. Z. Ésik and W. Kuich. Inductive *-semirings. *Theoretical Computer Science*, 324:3–33, 2004.
10. Z. Ésik and W. Kuich. Equational axioms for a theory of automata. In C. Martin-Vide, V. Mitrana, and G. Paun, editors, *Formal Languages and Applications*, volume 148 of *Studies in Fuzziness and Soft Computing*, pages 183–196. Springer, Berlin, 2004.
11. Z. Ésik and W. Kuich. A semiring–semimodule generalization of ω-regular languages II. *Journal of Automata, Languages and Combinatorics*, 10:243–264, 2005.
12. Z. Ésik and W. Kuich. Modern Automata Theory. www.dmg.tuwien.ac.at/kuich, 2007
13. Z. Ésik and W. Kuich. On iteration semiring–semimodule pairs. *Semigroup Forum*, 75:129–159, 2007.
14. U. Hebisch. The Kleene theorem in countably complete semirings. *Bayreuther Mathematische Schriften*, 31:55–66, 1990.
15. J.E. Hopcroft and J.D. Ullman. *Introduction to Automata Theory, Languages, and Computation.* Addison–Wesley, Reading, 1979.
16. St.C. Kleene. Representation of events in nerve nets and finite automata. In C.E. Shannon and J. McCarthy, editors, *Automata Studies*, pages 3–41. Princeton University Press, Princeton, 1956.
17. W. Kuich. The Kleene and the Parikh theorem in complete semirings. In *ICALP87*, volume 267 of *Lecture Notes in Computer Science*, pages 212–225. Springer, Berlin, 1987.
18. W. Kuich. Semirings and formal power series: Their relevance to formal languages and automata theory. In G. Rozenberg and A. Salomaa, editors, *Handbook of Formal Languages, volume 1*, Chapter 9, pages 609–677. Springer, Berlin, 1997.
19. W. Kuich and A. Salomaa. *Semirings, Automata, Languages*, volume 5 of *Monographs in Theoretical Computer Science. An EATCS Series.* Springer, Berlin, 1986.
20. D. Perrin and J.-É. Pin, *Infinite Words*, Elsevier, 2004
21. I. Petre and A. Salomaa. Algebraic systems and pushdown automata. In this *Handbook*, chapter 7. Springer, Berlin, 2009
22. A. Salomaa. *Formal Languages.* Academic Press, San Diego, 1973.
23. A. Salomaa and M. Soittola. *Automata-Theoretic Aspects of Formal Power Series.* Springer, Berlin, 1978.
24. M.P. Schützenberger. On the definition of a family of automata. *Information and Control*, 4:245–270, 1961.

Chapter 4: Rational and Recognisable Power Series*

Jacques Sakarovitch

LTCI, ENST/CNRS,
46 rue Barrault, 75634 Paris Cedex 13, France
sakarovitch@enst.fr

* This chapter is adapted from Chaps. III and IV of the book *Elements of Automata Theory*, Jacques Sakarovitch, 2009, ©Cambridge University Press, where missing proofs, detailed examples and further developments can be found.

M. Droste, W. Kuich, H. Vogler (eds.), *Handbook of Weighted Automata*, Monographs in Theoretical Computer Science. An EATCS Series,
DOI 10.1007/978-3-642-01492-5_4, Springer-Verlag Berlin Heidelberg 2009

1 Introduction

Weighted automata realise power series—in contrast to 'classical' automata which accept languages. There are many good reasons that make power series worth an interest compared to languages, beyond the raw appeal to generalisation that inhabits every mathematician.

First, power series provide a more powerful mean for modelisation, replacing a pure acceptance/rejection mode by a quantification process. Second, by putting automata theory in a seemingly more complicated framework, one benefits from the strength of mathematical structures thus involved and some results and constructions become simpler, both conceptually, and on the complexity level. Let us also mention, as a third example, that in the beginning of the theory, weighted automata were probably considered for their ability of defining languages—via the supports of realised power series—rather than for the power series themselves. In all these instances, what matters is that the choice of the semiring S of multiplicity be as wide as possible and our first aim is to develop as far as possible a theory with a priori no assumption at all on S.

With this in mind, I have chosen as the main thread of this chapter to lay comprehensive bases for the proof of the decidability of the equivalence of deterministic k-tape transducers which is, at least in my opinion, one of the most striking examples of the application of algebra to "machine theory." To that end, I develop in particular the following points:

(a) The definition of rational series over *graded monoids* (in order to deal with direct product of free monoids) and not over free monoids only. A side benefit of the definition of series over arbitrary (graded) monoids is that it makes clearer the distinction between the *rational* and the *recognisable* series.
(b) The *reduction theory* of series over a free monoid and with coefficients in a (skew) field that leads to a procedure for the decidability of equivalence (with a cubic complexity).
(c) As it is natural for series with coefficients in a field, and since the topological machinery is set anyway, the star of series is defined in a slightly more general setting than cycle-free series.
(d) The basics for rational relations with multiplicity, for the weighted generalisation of the often called Kleene–Schützenberger theorem on transducers as well as of the Myhill theorem (on recognisable sets in a product of monoids) or of McKnight theorem (on the inclusion of recognisable set in rational ones in finitely generated monoids).

The core of this chapter pertains to a now classical part of automata theory, originating in the seminal paper of M.P. Schützenberger [46] and having been exposed in several treatises already quoted in Chap. 1: Eilenberg [14], Salomaa and Soittola [45], Berstel and Reutenauer [5], and Kuich and Salomaa [29]. I have not resisted though to include some more recent developments which are the result of my own work with my colleagues M.-P. Béal and S. Lombardy: the derivation of weighted expressions [33], and the connection between conjugacy and equivalence [3, 4].

The presentation given here (but for the last quoted result that is too recent) is adapted from Chaps. III and IV of my book *Elements of Automata Theory* [43], where missing proofs, detailed examples, and further developments can be found. I am grateful to Reuben Thomas who has translated this book from French to English and to Cambridge University Press for allowing me to use the material for this chapter. Finally, I want to acknowledge the always inspiring discussions I have had in the last 10 years with Sylvain Lombardy.

2 Rational Series and Weighted Rational Expressions

In the preceding chapters, the formal power series that have been considered are series over a free monoid with coefficients in a semiring S that is almost always supposed to be *complete* or *continuous*, opening the way to straightforward generalisations of results and methods developed for languages, that are series with multiplicity in the Boolean semiring, and classical automata.

Our first purpose is to build a theory where no assumptions are made on the semiring of coefficients, and as few as possible on the base monoid. There will be some redundancy with Chaps. 1 and 3, but I have preferred to write a comprehensive text that naturally flows rather than to interrupt it with references to results that are always stated under slightly different hypotheses.

In what follows, M is a monoid and S a semiring, a priori arbitrary.

2.1 Series over a Graded Monoid

For any set E, the set of maps from E to S is usually written S^E and canonically inherits from S a structure of semiring when equipped with *pointwise* addition and multiplication. When E is a monoid M, we equip S^M with another multiplication which derives from the *monoid structure* of M, and we thus use different notation and terminology for these maps together with this other semiring structure—indeed, the ones set up in Chap. 1, Sect. 3.

Any map from M to S is a *formal power series* over M with coefficients in S—abbreviated as S-series over M, or even as *series* if there is ambiguity neither on S nor on M. The set of these series is written $S\langle\langle M \rangle\rangle$. If r is a series, the image of an element m of M under r is written (r, m) rather than $(m)r$ and is called the *coefficient of* m *in* r.

The *support* of a series r is the subset of elements of M whose coefficient in r is not 0_S. A series with finite support is a *polynomial*; the set of polynomials over M with coefficients in S is written $S\langle M\rangle$.

For all r and r', and all s in S, the following operations on $S\langle\langle M\rangle\rangle$ are defined:

(i) The (left and right) *'exterior' multiplications*[1]:

$$sr \quad \text{and} \quad rs \quad \text{by} \quad \forall m \in M \quad (sr, m) = s(r, m) \text{ and } (rs, m) = (r, m)s.$$

(ii) The pointwise *addition*:

$$r + r' \quad \text{by} \quad \forall m \in M \quad (r + r', m) = (r, m) + (r', m).$$

(iii) The *Cauchy product*:

$$rr' \quad \text{by} \quad \forall m \in M \quad (rr', m) = \sum_{\substack{u,v \in M \\ uv = m}} (r, u)(r', v). \qquad (*)$$

Addition makes $S\langle\langle M\rangle\rangle$ a commutative monoid, whatever S and M; together with the two exterior multiplications, it makes $S\langle\langle M\rangle\rangle$ a left and right *semimodule*[2] on S.

The Cauchy product raises a problem for there could very well exist elements m in M such that the set of pairs (u, v) satisfying $uv = m$ is infinite, and hence there could exist series such that the sum on the right-hand side of $(*)$ is *not defined*. Thus, we cannot ensure, without further assumptions, that the Cauchy product is a binary operation totally defined on $S\langle\langle M\rangle\rangle$. This difficulty can be overcome in at least three ways.

The first is to retreat: we no longer consider $S\langle\langle M\rangle\rangle$ but only the set $S\langle M\rangle$ of polynomials. If r and r' are polynomials, the sum in $(*)$ is infinite but only a finite number of terms are non-zero; the Cauchy product is defined on $S\langle M\rangle$ and makes it indeed a semiring (a semi-algebra on S), a subsemi-algebra of $S\langle\langle M\rangle\rangle$ when that is defined.

The second is to assume that S is *complete*: every sum, even if infinite, is defined on S, and the Cauchy product of two series is defined for any M. This is the case, for example, if S is equal to $\mathbb{B}$, $\mathbb{B}\langle\langle M\rangle\rangle$, $\langle\mathbb{N}^\infty, +, \cdot\rangle$ or $\langle\mathbb{N}^\infty, \mathsf{min}, +\rangle$. The theory of finite automata over a free monoid and with multiplicity in a complete semiring has been developed in Chap. 3 of this book.

The third way, which is ours, aims at being able to define weighted automata, and hence series, *without restriction on* S, and we are led in this case

[1] Which are called *scalar products* in Chap. 1.

[2] For sake of uniformity in this book, I use the terms 'semimodule' and 'semialgebra' whereas in [43] and other publications, I follow the convention of Berstel and Reutenauer [5] and speak of 'module' and 'algebra' (over a semiring).

to make assumptions about M: we suppose for the rest of this chapter that the monoids are *graded*, a condition that we shall describe in the next paragraph and which allows the natural generalisation of the standard construction of formal power series of a single variable.[3] This somewhat different assumption makes it necessary to restate, and sometimes to reprove again, some of the statements already established when S is *complete*.

2.1.1 Graded Monoid

For the Cauchy product to be always defined on $S\langle\langle M\rangle\rangle$, independently of S, it is necessary (and sufficient) that, for all m in M, the set of pairs (u, v) such that $uv = m$ is finite—we will say that m is *finitely decomposable*. However, making $S\langle\langle M\rangle\rangle$ a semiring is not an end in itself: the development of the theory to come is the characterisation of the behaviour of finite automata by means of rational operations—a fundamental theorem—and then not only must sum and product be defined on the series, but so must the star operation, which implies an *infinite sum*. This forces us to have some sort of *topology* on $S\langle\langle M\rangle\rangle$, to which we shall return in the next paragraph.

The construction of series on Σ^*, which generalises that of series of one variable, shows that it is from the *length* of words in Σ^* that we build a topology on $S\langle\langle\Sigma^*\rangle\rangle$. The existence of an *additive length* is the main assumption that we shall make about M. Returning to the initial problem, we then seek an additional condition that ensures that every element is finitely decomposable. For reasons of simplicity, we assume that M is *finitely generated.* This solves the problem, while allowing us to deal with the cases that interest us.

Definition 2.1. *A function* $\varphi\colon M \to \mathbb{N}$ *is a* length *on* M *if:*

(i) $\varphi(m)$ *is* strictly *positive for all* m *other than* 1_M
(ii) $\forall m, n \in M \;\; \varphi(mn) \leq \varphi(m) + \varphi(n)$

We shall say that a length is a gradation *if it is* additive*; that is, if:*

(iii) $\forall m, n \in M \;\; \varphi(mn) = \varphi(m) + \varphi(n)$

and that M *is* graded *if it is equipped with a gradation.*

Every free monoid and every Cartesian product of free monoids is graded. The definition implies that $\varphi(1_M) = 0$ and that a finite monoid, more generally a monoid that contains an idempotent other than the identity (for example, a zero), cannot be equipped with a gradation.

Proposition 2.2. *In a finitely generated graded monoid, the number of elements whose length is less than an arbitrary given integer* n *is finite.*

[3] A fourth method exists that takes out of both the first and the third. It involves making an assumption about M (we require it to be an *ordered group*) and considering only a subset of $S\langle\langle M\rangle\rangle$ (those series whose support is well ordered). A reference to that set of series will be made in Sect. 5.3.

In other words, every element of a graded monoid M can only be written in a finite number of different ways as the product of elements of M other than 1_M. We can deduce in particular the following corollary.

Corollary 2.3. *In a finitely generated graded monoid, every element is finitely decomposable.*

Note that a finite monoid is not graded, but that every element in it is nonetheless finitely decomposable. From Corollary 2.3, we deduce the proposition aimed at by Definition 2.1:

Proposition 2.4. *Let M be a finitely generated graded monoid and S a semiring. Then $S\langle\langle M\rangle\rangle$, equipped with the Cauchy product, is a semiring, and what is more, a (left and right) semi-algebra*[4] *on S.*

In the following, M is a graded monoid that is implicitly assumed to be finitely generated. To simplify the notation and in imitation of the free monoid, we will write the length function as a pair of vertical bars, that is, $|m|$ rather than $\varphi(m)$.

From the semiring $S\langle\langle M\rangle\rangle$, one then builds other semirings, by means of classical constructions; let us quote in particular and for further reference the following fundamental isomorphism.

Lemma 2.5. *Let S be a semiring, M a graded monoid, and Q a finite set; then the set of square matrices of dimension Q and with entries in the semiring $S\langle\langle M\rangle\rangle$ is a semiring, isomorphic to that of series over M with coefficient in the semiring of square matrices of dimension Q and with entries in S; that is, $S\langle\langle M\rangle\rangle^{Q\times Q} \cong S^{Q\times Q}\langle\langle M\rangle\rangle$.*

Remark 2.6. A notion that is often considered in relationship with gradation is *equidivisibility.* A monoid M is *equidivisible* if whenever $mn = pq$ with m, n, p, and q in M, there exists u such that $mu = p$ and $n = uq$ or $m = pu$ and $un = q$. There is then a theorem by F.W. Levi which states that *a graded equidivisible monoid is free* (cf. [30]). This notion is also to be compared with the one of *equisubtractivity* that is considered below.

2.1.2 Topology on $S\langle\langle M\rangle\rangle$

The definition to come of the *star operation*, an infinite sum, calls for the definition of a *topology* on $S\langle\langle M\rangle\rangle$.

Since $S\langle\langle M\rangle\rangle = S^M$ is the *set of maps* from M to S, it is naturally equipped with the *product topology* of the topology on S. If this topology on S is defined by a *distance*, the product topology on $S\langle\langle M\rangle\rangle$ coincides, as M is countable, with the *simple convergence topology*:

[4] If S is a ring, $S\langle\langle M\rangle\rangle$ is even what is classically called a *graded algebra*, which is the origin of the terminology chosen for graded monoids.

$$r_n \text{ converges to } r, \quad \text{if and only if,}$$
$$\text{for all } m \text{ in } M, \quad (r_n, m) \text{ converges to } (r, m).$$

We shall reexamine the topology question using only the notion of distance, more in line with intuition and explain how to define a distance between two series under the assumption that M is graded. The foregoing reference to simple convergence topology was nevertheless worthwhile, as it made clear that *the basis of the topology on* $S\langle\langle M\rangle\rangle$ *is the topology on* S.

Distance on $S\langle\langle M\rangle\rangle$

A *distance* on a set E is a map $\mathbf{d}$ which relates to every pair (x, y) of elements of E a *positive real number* $\mathbf{d}(x, y)$, called the *distance from* x *to* y (or *between* x and y), which satisfies the following properties:

- Symmetry: $\mathbf{d}(x, y) = \mathbf{d}(y, x)$
- Positivity: $\mathbf{d}(x, y) > 0 \quad \text{if } x \neq y \text{ and } \mathbf{d}(x, x) = 0$
- Triangular inequality: $\mathbf{d}(x, y) \leq \mathbf{d}(x, z) + \mathbf{d}(y, z)$

When this triangular inequality can be replaced by

- $\forall x, y, z \in E \qquad \mathbf{d}(x, y) \leq \max\{\mathbf{d}(x, z), \mathbf{d}(y, z)\}$

the distance $\mathbf{d}$ is called *ultrametric*.

A sequence $\{x_n\}_{n\in\mathbb{N}}$ of elements of E *converges* to x if the distance between x_n and x becomes arbitrarily small as n grows; that is, more formally,

$$\forall \eta > 0 \ \exists N \in \mathbb{N} \ \forall n \geq N \quad \mathbf{d}(x_n, x) \leq \eta.$$

Such an element x is *unique*; it is called the *limit* of the sequence $\{x_n\}_{n\in\mathbb{N}}$ and we write $x = \lim_{n\to+\infty} x_n$, or simply $x = \lim x_n$ if there is no ambiguity. We say that $\mathbf{d}$ *equips* E *with a topology.*

Remark 2.7. We can always assume that a distance is a real number less than or equal to 1. If that is not the case, then by taking

$$\mathbf{f}(x, y) = \inf\{\mathbf{d}(x, y), 1\},$$

we obtain a distance $\mathbf{f}$ on E that defines *the same topology*; that is, a distance for which *the same sequences* will converge to *the same limits*.

Remark 2.8. Whatever E is, we can choose a trivial distance function which is 1 for every pair of distinct elements. This is equivalent to saying that two distinct elements are never 'close' to each other, and that the only convergent sequences are those that are eventually *stationary*. We then say that E is equipped with the *discrete topology*.

We are confronted with two situations which seem fundamentally different. The first is that of a semiring S such as $\mathbb{B}$, $\mathbb{N}$, $\mathbb{Z}$, $\mathbb{N}^\infty$, etc., whose elements are 'detached' from each other. The natural topology on these semirings is the discrete topology. The second is that of semirings such as $\mathbb{Q}$, $\mathbb{Q}_+$, $\mathbb{R}$, etc., or even later $S\langle\langle M\rangle\rangle$ itself, which can act as a semiring of coefficients for series on another monoid; that is, semirings on which there is a priori a distance which can be arbitrarily small. On these semirings as well, we can choose a discrete topology, but it is more satisfactory to preserve their 'native' topology. By means of the definition of a distance and the topological notions derived from it, we treat these two situations in the same way.

We first assume that S is equipped with a distance $\mathbf{c}$ which is bounded by 1. The length function on M allows us *to put an ordering* on the elements of M and we set

$$\mathbf{d}(r,r') = \frac{1}{2}\sum_{n\in\mathbb{N}}\left(\frac{1}{2^n}\max\{\mathbf{c}\big((r,m),(r',m)\big) \mid |m| = n\}\right).$$

We then verify that $\mathbf{d}$ is indeed a distance on $S\langle\langle M\rangle\rangle$, ultrametric when $\mathbf{c}$ is, and that the topology defined on $S\langle\langle M\rangle\rangle$ by $\mathbf{d}$ is, as stated, the *simple convergence* topology; that is, the following property.

Property 2.9. A sequence $\{r_n\}_{n\in\mathbb{N}}$ of series of $S\langle\langle M\rangle\rangle$ converges to r, if and only if, for all m in M the sequence of coefficients (r_n, m) converges to (r, m).

Furthermore, choosing a topology on a semiring only really makes sense if the constituent operations of the semiring, addition and multiplication, are consistent with the topology—we say they are *continuous*—that is, if the limit of a sum (resp. of a product) is the sum (resp. the product) of the limits. We say in this case that not only is the semiring equipped with a topology, but that it is a *topological semiring.* We easily verify that if S is topological, then so is $S\langle\langle M\rangle\rangle$. In other words, if $\{r_n\}_{n\in\mathbb{N}}$ and $\{r'_n\}_{n\in\mathbb{N}}$ are two convergent sequences of elements of $S\langle\langle M\rangle\rangle$, we have

$$\lim(r_n + r'_n) = (\lim r_n) + (\lim r'_n) \quad \text{and} \quad \lim(r_n r'_n) = (\lim r_n)(\lim r'_n).$$

Note that conversely the fact that the sequence $\{r_n + r'_n\}_{n\in\mathbb{N}}$ or $\{r_n r'_n\}_{n\in\mathbb{N}}$ converges *says nothing* about whether $\{r_n\}_{n\in\mathbb{N}}$ or $\{r'_n\}_{n\in\mathbb{N}}$ converges or not.

If S is a topological semiring, then so is $S^{Q\times Q}$ and the isomorphism quoted in Lemma 2.5 is moreover a *bi-continuous* bijection.

Summable Families

Let T be a semiring[5] equipped with a distance which makes it a topological semiring. We thus know precisely what means that an infinite sequence $\{t_n\}_{n\in\mathbb{N}}$

[5] We have temporarily changed the symbol we use for a semiring on purpose: T will not only play the role of S but also of $S\langle\langle M\rangle\rangle$ in what follows.

converges to a limit t when n tends to infinity. We must now give an equally precise meaning to the sum of an infinite family $\{t_i\}_{i\in I}$ and it turns out to be somewhat harder. The difficulty arises from the fact that we want a sort of *associativity–commutativity* extended 'to infinity', and hence to ensure that the result and its existence does not depend on an arbitrary order put on the set I of indices.

We shall therefore define an 'absolute' method of summability, and a family will be described as 'summable' if we can find an increasing sequence of finite sets of indices, a sort of 'kernels', such that not only do partial sums on these sets tend to a limit, but above all that any sum on a finite set containing one of these kernels stays close to this limit. More precisely, we take the following definition.

Definition 2.10. *A family $\{t_i\}_{i\in I}$ of elements of T indexed by an arbitrary set I is called* summable *if there exists t in T such that, for all positive η, there exists a finite subset J_η of I such that, for all finite subsets L of I which contain J_η, the distance between t and the sum of $\{t_i\}$ for i in L is less than η; that is,*

$$\exists t \in T,\ \forall \eta > 0,\ \exists J_\eta \textit{ finite},\ J_\eta \subset I,\ \forall L \textit{ finite},\ J_\eta \subseteq L \subset I$$
$$\mathbf{d}\Bigl(\sum_{i\in L} t_i, t\Bigr) \leq \eta.$$

The element t thus defined is unique *and is called the* sum *of the family $\{t_i\}_{i\in I}$.*

The sum just defined is obviously equal to the usual sum if I is finite, and we write

$$t = \sum_{i\in I} t_i.$$

From the definition of a summable family, we easily deduce an associativity property restricted to *finite groupings*, but that repeats infinitely.

Property 2.11. Let $\{t_i\}_{i\in I}$ be a summable family with sum t in T. Let K be a set of indices and $\{J_k\}_{k\in K}$ a *partition* of I where all the J_k are *finite* (that is, $I = \bigcup_{k\in K} J_k$ and the J_k are pairwise disjoint). Set $s_k = \sum_{i\in J_k} t_i$ for every k in K. Then the family $\{s_k\}_{k\in K}$ is summable with sum t.

As in the preceding chapters, we say that a family of series $\{r_i\}_{i\in I}$ is *locally finite* if for every m in M there is only a finite number of indices i such that (r_i, m) is different from 0_S.

Property 2.12. A locally finite family of power series is summable.

This simple property is a good example of what the topological structure placed on $S\langle\langle M\rangle\rangle$ imposes and adds. That we can *define a sum* for a locally finite family of series is trivial: pointwise addition is defined for each m, independently of any assumption about M. To say that the family is *summable* is to add extra information: it ensures that partial sums converge to the result of pointwise addition.

For every series r, the family of series $\{(r,m)m \mid m \in M\}$, where m is identified with its characteristic series, is locally finite, and we have

$$r = \sum_{m \in M} (r,m)m,$$

which is the usual notation that is thus justified. We also deduce from this notation that $S\langle M\rangle$ is *dense* in $S\langle\langle M\rangle\rangle$. Property 2.12 extends beyond locally finite families and generalises to a proposition which links the summability of a family of series and that of families of coefficients.

Property 2.13. A family $\{r_i\}_{i \in I}$ of $S\langle\langle M\rangle\rangle$ is summable with sum r if and only if for each m in M, the family $\{(r_i,m)\}_{i \in I}$ of elements of S is summable with sum (r,m).

2.2 Rational Series

We are now ready to define the star operation on a series. We must nevertheless introduce here an assumption on the semiring, somehow an axiom of *infinite distributivity.* After that, the definition of rational series comes easily, the double definition indeed, one as a closure under rational operations and one by means of rational expressions which opens the way to effective computations.

2.2.1 Star of a Series

We start by considering the problem in arbitrary semirings and not only in the semirings of series.

Let t be an element of a topological semiring T; it is possible for the family $\{t^n\}_{n \in \mathbb{N}}$ to be, or not to be summable. If it is summable, we call its sum the 'star of t' and write it t^*:

$$t^* = \sum_{n \in \mathbb{N}} t^n.$$

Whether t^* is defined depends on t, on T, on the distance on T, or on a combination of all these elements. For example, $(0_T)^* = 1_T$ is defined for all T; if $T = \mathbb{Q}$, we have $(\frac{1}{2})^* = 2$ if $\mathbb{Q}$ is equipped with the natural topology, or undefined if the chosen topology is the discrete topology, while 1^* is not defined in either case.

Lemma 2.14. *Let T be a topological semiring and t an element of T whose star is defined. We have the double equality*

$$t^* = 1_T + tt^* = 1_T + t^*t. \qquad (1)$$

Proof. We obviously have $t^{\leq n} = 1_T + tt^{<n} = 1_T + t^{<n}t$. As $\lim t^{<n} = \lim t^{\leq n} = t^*$, and as *addition and multiplication are continuous operations* on T, we obtain (1) by taking the limit of each side of the above equation. □

Remark 2.15. If T is a topological *ring*, and if the star of t is defined, (1) can be written $t^* - tt^* = t^* - t^*t = 1$ or $(1-t)t^* = t^*(1-t) = 1$ and so t^* is the *inverse* of $1-t$. Hence, the classic identity

$$t^* = \frac{1}{1-t} = 1 + t + t^2 + \cdots , \tag{2}$$

is justified in full generality. It also means that forming the star can be considered as a substitute of taking the inverse in poor structure that has no inverse.

Star of a Proper Series

By reference to polynomials and to series in one variable, we call the *constant term* of a series r of $S\langle\langle M\rangle\rangle$ the coefficient of the neutral element of M in r: $\mathsf{c}(r) = (r, 1_M)$. A power series is called *proper* if its constant term is zero. The sum of two proper series is a proper series; the product of a proper series with any other series is a proper series, *since M is graded.*

If r is proper, the family $\{r^n \mid n \in \mathbb{N}\}$ is locally finite, and thus the star of a proper series of $S\langle\langle M\rangle\rangle$ is defined.

Lemma 2.16 (Arden). *Let r and u be two series of $S\langle\langle M\rangle\rangle$; if r is a proper series, each of the equations*

$$X = rX + u \quad \textit{and} \tag{3}$$

$$X = Xr + u \tag{4}$$

has a unique solution*: the series r^*u and ur^*, respectively.*

Proof. In (1), we replace t by r and multiply on the left (resp. on the right) by u and we obtain that r^*u (resp. ur^*) is a solution of (3) (resp. of (4)). Conversely, if v is a solution of the equation $X = u + rX$, we have

$$v = u + rv \quad \Longrightarrow \quad v = u + ru + r^2v = \cdots = r^{<n}u + r^n v,$$

for all integers n. Since r is proper, and multiplication continuous, we have $\lim r^n = \lim r^n v = 0$, from which follows $v = \lim(r^{<n}u) = (\lim r^{<n})u = r^*u$. □

From which, we deduce the following proposition.

Proposition 2.17. *Let r and u be two proper series of $S\langle\langle M\rangle\rangle$; the following equalities (or identities) hold:*

$$(r+u)^* = r^*(ur^*)^* = (r^*u)^*r^*, \tag{$\boldsymbol{S}$}$$

$$(ru)^* = 1 + r(ur)^*u, \tag{$\boldsymbol{P}$}$$

$$\forall n \in \mathbb{N} \quad r^* = r^{<n}(r^n)^*. \tag{$\boldsymbol{Z}_n$}$$

Following [12], the identity ($\boldsymbol{S}$) is called the *sum star identity* in Chap. 1, ($\boldsymbol{P}$) the *product star identity.*

Remark 2.18. It follows by Lemma 2.5 that a square matrix m of dimension Q with elements in $S\langle\langle M\rangle\rangle$ is a proper series of $S^{Q\times Q}\langle\langle M\rangle\rangle$ if all its elements are proper series; (we say in this case that m is proper), and hence that the identities ($\boldsymbol{S}$), ($\boldsymbol{P}$), and ($\boldsymbol{Z}_n$) are satisfied by proper matrices.

Strong Semirings and Star of an Arbitrary Series

The star of an arbitrary series, not necessarily proper, may or may not be defined. The following proposition allows us to tell the difference between the two cases. First, we make a timely definition to avoid a difficulty.

Definition 2.19. *A topological semiring is* strong *if the product of two summable families is a summable family; that is, if the two families $\{r_i \mid i \in I\}$ and $\{u_j \mid j \in J\}$ are summable with sum s and t, respectively, then the family $\{r_i u_j \mid (i,j) \in I \times J\}$ is summable with sum st.*

All the semirings which we shall consider are strong: semirings equipped with the discrete topology, the sub-semirings of $\mathbb{C}^n$ (equipped with the natural topology), and the positive semirings. We then easily verify the following property.

Property 2.20. The semirings of matrices and the semirings of series on a graded monoid, with coefficients in a strong semiring are strong.

Let r be a series of $S\langle\langle M\rangle\rangle$; the *proper part* of r is the proper series that coincides with r for all the elements m of M other than 1_M. It is convenient to write $r_0 = \mathsf{c}(r)$ for the constant term of r, and r_p for the proper part of r:

$$(r_\mathsf{p} 1_M) = 0_S \qquad \text{and} \qquad \forall m \in M\backslash 1_M \quad (r_\mathsf{p}, m) = (r, m),$$

and we write $r = r_0 + r_\mathsf{p}$ (rather than $r = r_0 1_M + r_\mathsf{p}$). These definitions and notations are taken in view of the following, which generalises to a series with coefficients in an arbitrary strong semiring, a result already established for series with coefficients in a continuous semiring.

Proposition 2.21. *Let S be a strong topological semiring and M a graded monoid. Let r be a series of $S\langle\langle M\rangle\rangle$, r_0 its constant term and r_p its proper part. Then r^* is defined if and only if r_0^* is defined and in this case we have*

$$r^* = (r_0^* r_\mathsf{p})^* r_0^* = r_0^* (r_\mathsf{p} r_0^*)^* . \tag{5}$$

Proof. The condition is necessary since $(r^n, 1_M) = r_0^n$ and, if r^* is defined, the coefficients of 1_M in $\{r^n\}_{n\in\mathbb{N}}$ form a summable family.

Conversely, assume that $\{r_0^n\}_{n\in\mathbb{N}}$ is summable, with sum r_0^*. For all pairs of integers k and l, set

$$P_{k,l} = \sum_{\substack{i_0,i_1,\ldots,i_k \in \mathbb{N} \\ i_0+i_1+\cdots+i_k=l}} r_0^{i_0} r_{\mathsf{p}} r_0^{i_1} r_{\mathsf{p}} \cdots r_0^{i_{k-1}} r_{\mathsf{p}} r_0^{i_k}.$$

By convention, set $P_{0,l} = r_0^l$ and $P_{k,0} = r_{\mathsf{p}}^k$. We verify by inspection that, for all integers n,

$$r^n = (r_0 + r_{\mathsf{p}})^n = \sum_{l=0}^{l=n} P_{n-l,l}. \tag{6}$$

By induction on k, we will show that the family

$$F_k = \left\{ r_0^{i_0} r_{\mathsf{p}} r_0^{i_1} r_{\mathsf{p}} \cdots r_0^{i_{k-1}} r_{\mathsf{p}} r_0^{i_k} \;\middle|\; i_0, i_1, \ldots, i_k \in \mathbb{N} \right\}$$

is summable in $S\langle\langle M \rangle\rangle$, with sum

$$Q_k = (r_0^* r_{\mathsf{p}})^k r_0^* = r_0^* (r_{\mathsf{p}} r_0^*)^k.$$

The ingredients of the proof are depicted in Fig. 1.

In fact, the hypothesis on r_0 ensures the property for $k = 0$, and also that the family $G = \{ {r_0}^n r_{\mathsf{p}} \mid n \in \mathbb{N} \}$ is summable in $S\langle\langle M \rangle\rangle$, with sum ${r_0}^* r_{\mathsf{p}}$. The family F_{k+1} is the product of the families G and F_k and the assumption that S, and hence $S\langle\langle M \rangle\rangle$ is strong gives us the conclusion.

Hence, we deduce that, for each k, the family $\{ P_{k,l} \mid l \in \mathbb{N} \}$ is summable, with sum Q_k. The family $\{ Q_k \mid k \in \mathbb{N} \}$ is locally finite, hence summable, with sum

$$u = \sum_{k=0}^{\infty} Q_k = (r_0^* r_{\mathsf{p}})^* r_0^* = r_0^* (r_{\mathsf{p}} r_0^*)^*.$$

We can now easily finish the proof by showing that the 'doubly indexed' family $\{ P_{k,l} \mid k, l \in \mathbb{N} \}$ is summable, with sum u. Equation (6) and Property 2.11 then ensure that the family $\{ r^n \mid n \in \mathbb{N} \}$ is summable with sum u. □

The case of *cycle-free series* (see Chap. 1 and 3) falls in the scope of Proposition 2.21. In the same spirit as Remark 2.18, we note that (5) holds for every matrix m such that the star of its matrix of constant terms is defined. A particularly interesting case of this is where the matrix of constant terms is a strict upper triangular, another case of cycle-free series.

Proposition 2.22 (Bloom–Ésik [7]). *Let S be a strong topological semiring and M a graded monoid. Let r and u be series of $S\langle\langle M \rangle\rangle$ with constant terms r_0 and u_0, respectively, and such that r_0^*, u_0^*, and $(r_0 + u_0)^*$ are defined. Then the identities $(\boldsymbol{S})$, $(\boldsymbol{P})$, and $(\boldsymbol{Z}_n)$ hold for r and u.*

In other words, with the terminology of Chap. 1, and if one skips the question of the definition of star, if S is a Conway semiring, so is $S\langle\langle M \rangle\rangle$.

Remark 2.23. Along the line of Remark 2.15, it holds that if S is a ring, a series of $S\langle\langle M \rangle\rangle$ is invertible, if and only if its constant term is invertible.

For the rest of the chapter, S is a strong topological semiring.

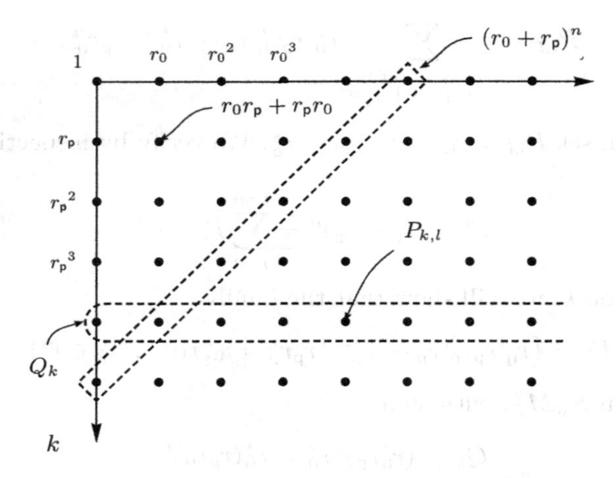

Fig. 1. A graphical representation of Proposition 2.21

2.2.2 The Family of Rational Series

We first characterise rational series 'from above' with the definition of rational operations and of closed families, and then inductively 'from below', with the definition of weighted rational expressions.

S-Rational Operations

The *rational operations* on $S\langle\langle M\rangle\rangle$ are:

(i) The *S-algebra operations*, that is:
- The two *exterior multiplications* by the elements of S
- The *addition*
- The *product*

(ii) The *star* operation, which is not defined everywhere.

Point (ii) leads us to tighten the notion of closure: a subset $\mathcal{E}$ of $S\langle\langle M\rangle\rangle$ is *closed under star* if s^* belongs to $\mathcal{E}$ for every series s in $\mathcal{E}$ such that s^* is defined.

A subset of $S\langle\langle M\rangle\rangle$ is *rationally closed* if it is closed under the rational operations; that is, if it is a subsemi-algebra of $S\langle\langle M\rangle\rangle$ closed under the star operation. The intersection of any family of rationally closed subsets is rationally closed, and thus the *rational closure* of a set $\mathcal{E}$ is the *smallest* rationally closed subset which contains $\mathcal{E}$, written $S\mathrm{Rat}\,\mathcal{E}$.

Definition 2.24. *A series of $S\langle\langle M\rangle\rangle$ is* S-rational *if it belongs to the rational closure of $S\langle M\rangle$, the set of polynomials on M with coefficients in S. The set of S-rational series (over M with coefficients in S) is written $S\mathrm{Rat}\,M$.*

If the monoid M is implied by the context, we shall say *S-rational series*, or just *rational series*, if S is also understood.

Example 2.25.

(i) Let M be the one-generator free monoid $\{x\}^*$ and S be a field $\mathbb{F}$. Then $\mathbb{F}\mathrm{Rat}\, x^*$ is exactly the set of series developments of ($\mathbb{F}$-)rational functions (that is, quotients of two polynomials) and this is where the name *rational*—rather the more common *regular* (for expressions and languages)—comes from.
(ii) If $S = \mathbb{B}$, we simply write $\mathrm{Rat}\, M$ for $\mathbb{B}\mathrm{Rat}\, M$ and its elements are the *rational subsets* of M.
(iii) If $S = \mathbb{N}$ and $M = \Sigma^* \times \Delta^*$, $\mathbb{N}\mathrm{Rat}\, \Sigma^* \times \Delta^*$ is the set of rational relations from Σ^* to Δ^* with multiplicity in $\mathbb{N}$, which we shall consider later.

Characteristic Series and Unambiguous Rational Sets

The notions introduced so far allow for a precise definition of unambiguity[6] (for rational sets) and some illustrative computations. For brevity, let us denote by $\underline{P}$ the *characteristic series* of a subset P of M (rather than by $\mathrm{char}(P)$ as in Chap. 1).

Definition 2.26. *Set $S = \mathbb{N}$ and let P and Q be two subsets of M.*

(i) The union $P \cup Q$ *is* unambiguous *if and only if* $\underline{(P \cup Q)} = \underline{P} + \underline{Q}$.
(ii) The product PQ *is* unambiguous *if and only if* $\underline{(PQ)} = \underline{P}\,\underline{Q}$.
(iii) The star *of P is* unambiguous *if and only if* $\underline{P^*} = (\underline{P})^*$.

A subset of M is unambiguously rational *if it belongs to the* unambiguous rational closure of finite subsets *of M. The family of unambiguous rational subsets of M is written* $\mathrm{URat}\, M$.

Then $P \in \mathrm{URat}\, M$ if, and only if $\underline{P} \in \mathbb{N}\mathrm{Rat}\, M$ and then $\underline{P} \in S\mathrm{Rat}\, M$ for any S. It is well known for instance that $\mathrm{URat}\, \Sigma^* = \mathrm{Rat}\, \Sigma^*$ and that $\mathrm{URat}(\Sigma^* \times \Delta^*)$ is strictly contained in $\mathrm{Rat}(\Sigma^* \times \Delta^*)$.

As Σ freely generates Σ^*, we have $(\underline{\Sigma})^* = \underline{\Sigma^*}$, and thus $\underline{\Sigma^*} = \underline{\varepsilon} + \underline{\Sigma\Sigma^*} = \underline{\varepsilon} + \underline{\Sigma^*\Sigma}$ which gives $(\underline{\varepsilon} - \underline{\Sigma})\underline{\Sigma^*} = \underline{\Sigma^*}(\underline{\varepsilon} - \underline{\Sigma}) = \underline{\varepsilon}$, and thus $\underline{\Sigma^*} = (\underline{\varepsilon} - \underline{\Sigma})^{-1}$ if $S = \mathbb{Z}$.

If P is a non-empty prefix-closed subset of Σ^*, the *border of P* is the set:

$$C = P\Sigma \backslash P.$$

As an example, Fig. 2 shows the prefix-closed subset $\{\varepsilon, b, ba\}$ and its border $\{a, bb, baa, bab\}$.

Let P is a non-empty prefix-closed subset of Σ^* and let $h = pa$ with p in P and a in Σ (this is the unique expression of h in this form). There are two, mutually exclusive, possible cases: h is in C or h is in P. Conversely, every word of $P \cup C$ can be written in this way, except ε. Hence, we deduce the equality *between characteristic series*:

[6] A more or less folklore notion; an early reference for unambiguous rational sets is [15].

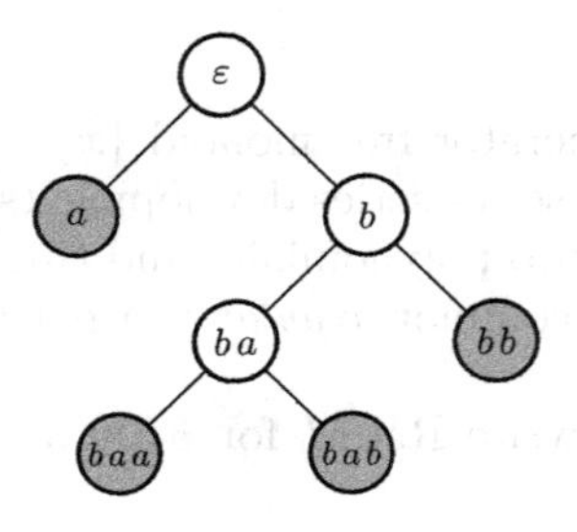

Fig. 2. A prefix-closed subset and its border

$$\underline{C} + \underline{P} = \underline{P\Sigma} + \underline{\varepsilon},$$

which we first rewrite as $\underline{\varepsilon} - \underline{C} = \underline{P}(\underline{\varepsilon} - \underline{\Sigma})$ then by right multiplication by $\underline{\Sigma^*} = (\underline{\varepsilon} - \underline{\Sigma})^{-1}$, as $\underline{\Sigma^*} - \underline{C\Sigma^*} = \underline{P}$. We thus have proved the following lemma.

Lemma 2.27. *Let P be a non-empty prefix-closed subset and $C = P\Sigma \backslash P$ its border. Every word f of $\Sigma^* \backslash P$ can be written uniquely as $f = cg$ with c in C and g in Σ^*.*

Rational S-Expressions

The definition of expressions will provide useful tools and handier ways to deal with rational series. Let $\{0, 1, +, \cdot, *\}$ be five function symbols. Naturally, the functions $+$ and $\cdot$ are binary, $*$ is unary, and 0 and 1 are nullary (they represent constants). We define, for each s in S, *two unary functions*, also written s.

Definition 2.28. *A* weighted rational expression *over M with weight in S, or rational S-expression over M, is obtained inductively in the following manner:*

(i) 0, 1, and m, for all m in M, are rational expressions (the atomic *expressions).*
(ii) If E is a rational expression and s is in S, then $(s\mathsf{E})$ and $(\mathsf{E}s)$ are rational expressions.
(iii) If E and F are rational expressions, then so are $(\mathsf{E}+\mathsf{F})$, $(\mathsf{E}\cdot\mathsf{F})$, and (E^*).

We write $S\mathsf{RatE}\,M$ for the set of rational S-expressions over M.

Remark 2.29.

(i) We can restrict the atomic expressions, other than 0 and 1, to be elements g of any given generating set G of M without reducing the power of the definition. That is what we usually do when M is a free monoid Σ^*.
(ii) We could have considered the elements of S to be atoms and not operators, again without changing the power of the definition, and that would simplify somewhat some upcoming equations. The chosen way is, however, more consistent with the upcoming definition of the *derivation* of S-expressions over Σ^*.

We define the *depth* of an expression E, $\mathsf{d}(\mathsf{E})$, as the height of the syntactical tree of the corresponding expression:

$$\mathsf{d}(0) = \mathsf{d}(1) = \mathsf{d}(m) = 0, \quad \text{for all } m \text{ in } M,$$
$$\mathsf{d}((s\mathsf{E})) = \mathsf{d}((\mathsf{E}s)) = \mathsf{d}((\mathsf{E}^*)) = 1 + \mathsf{d}(\mathsf{E}),$$
$$\mathsf{d}((\mathsf{E}+\mathsf{F})) = \mathsf{d}((\mathsf{E}\cdot\mathsf{F})) = 1 + \max\big(\mathsf{d}(\mathsf{E}), \mathsf{d}(\mathsf{F})\big).$$

The *constant term* of an expression E, $\mathsf{c}(\mathsf{E})$, is defined by induction on the depth of E; it is an element of S, computed by the following equations:

$$\mathsf{c}(1) = 1_S, \qquad \mathsf{c}(0) = \mathsf{c}(m) = 0_S \quad \text{for all } m \text{ in } M,$$
$$\mathsf{c}((s\mathsf{E})) = s\mathsf{c}(\mathsf{E}), \qquad \mathsf{c}((\mathsf{E}s)) = \mathsf{c}(\mathsf{E})s,$$
$$\mathsf{c}((\mathsf{E}+\mathsf{F})) = \mathsf{c}(\mathsf{E}) + \mathsf{c}(\mathsf{F}), \qquad \mathsf{c}((\mathsf{E}\cdot\mathsf{F})) = \mathsf{c}(\mathsf{E})\mathsf{c}(\mathsf{F}), \qquad \text{and}$$
$$\mathsf{c}((\mathsf{E}^*)) = \mathsf{c}(\mathsf{E})^* \quad \text{if the right-hand side is defined in } S.$$

A rational S-expression may represent an element of $S\langle\langle M\rangle\rangle$ or not, the distinction between the two cases being made by the constant term, exactly as for the star of an arbitrary series and using that result. We shall say that an expression in $S\mathsf{RatE}\,M$ is *valid* if its constant term is defined. The series *denoted by a valid expression* E, which we write $|\mathsf{E}|$, is defined by induction on the depth of E by the equations

$$|0| = 0_S, \qquad |1| = 1_M, \qquad |m| = m \quad \text{for all } m \text{ in } M,$$
$$|(s\mathsf{E})| = s|\mathsf{E}|, \qquad |(\mathsf{E}s)| = |\mathsf{E}|s,$$
$$|(\mathsf{E}+\mathsf{F})| = |\mathsf{E}| + |\mathsf{F}|, \qquad |(\mathsf{E}\cdot\mathsf{F})| = |\mathsf{E}||\mathsf{F}|, \qquad \text{and} \qquad |(\mathsf{E}^*)| = |\mathsf{E}|^*.$$

We verify both that these equations are well defined and that they are consistent, in the sense that the *constant term of the expression* E is the *constant term of the series* $|\mathsf{E}|$, *in parallel*, and *in the same induction*, using Proposition 2.21. In other words, and in order to define $|\mathsf{E}|$, we shall also have proved the following.

Property 2.30. For all valid S-expressions E in $S\mathsf{RatE}\,M$, $\mathsf{c}(\mathsf{E}) = (|\mathsf{E}|, 1_M)$.

Example 2.31. Take $M = \{a, b\}^*$ and $S = \mathbb{Q}$. The $\mathbb{Q}$-expression $(a^* + (-1b^*))^*$ is valid, as is $\mathsf{E}_1 = (\frac{1}{6}a^* + \frac{1}{3}b^*)^*$ since $\mathsf{c}(\frac{1}{6}a^* + \frac{1}{3}b^*) = \frac{1}{2}$, and hence $\mathsf{c}(\mathsf{E}_1) = 2$ is defined; $(a^* + b^*)^*$ is not valid.

The set of series denoted by valid S-expressions is rationally closed, and every rationally closed subset of $S\langle\langle M\rangle\rangle$ that contains every element of M (and thus $S\langle M\rangle$) contains every series denoted by a valid S-expression, which proves the following proposition.

Proposition 2.32. *A series of* $S\langle\langle M\rangle\rangle$ *is* S*-rational if and only if it is denoted by a valid rational* S*-expression over* M.

3 Weighted Automata

An *automaton over* M *with weight (or* with multiplicity*) in* S, or S*-automaton*[7] *over* M is a *graph* labelled with elements of $S\langle\langle M\rangle\rangle$, associated with two maps from the set of vertices to $S\langle\langle M\rangle\rangle$. We develop and complete this definition. We build on the identification of a graph with its incidence matrix and the proofs will be performed systematically with matrix computations. The essence of an automaton, however, remains that of a graph and the behaviour of on automaton is defined in the language of graphs. We also continue to use the graph representation and its vocabulary to aid intuition.

3.1 The Behaviour of a Weighted Automaton

An automaton $\mathfrak{A}$ over M with weights in S is specified by the choice of the following:[8]

- A non-empty set Q of *states* of $\mathfrak{A}$, also called the *dimension* of $\mathfrak{A}$.
- An element E of $S\langle\langle M\rangle\rangle^{Q\times Q}$, a square matrix of dimension Q with entries in $S\langle\langle M\rangle\rangle$, called the *transition matrix* of $\mathfrak{A}$; we can view each entry $E_{p,q}$ different from 0_S as the label of a unique edge which goes from state p to state q in the graph with vertices Q and we write $p \xrightarrow{x} q$, or $p \xrightarrow[\mathfrak{A}]{x} q$, if $x = E_{p,q}$. (If $E_{p,q} = 0_S$, we consider there to be *no* edge from p to q.)
- Two elements I and T of $S\langle\langle M\rangle\rangle^{Q}$; that is, two functions I and T from Q to $S\langle\langle M\rangle\rangle$: I is the *initial function* and T the *final function* of $\mathfrak{A}$; they can also be seen as vectors of dimension Q: I is a *row vector* and T a *column vector*, called respectively the *initial vector* and *final vector* of $\mathfrak{A}$.

The S-automaton $\mathfrak{A}$ is written, naturally enough,

$$\mathfrak{A} = \langle I, E, T\rangle.$$

We use the familiar conventions to represent S-automata graphically (see figures below); the values of I labelling the incoming arrows and those of T the outgoing arrows.

A *path* in $\mathfrak{A}$ is a sequence of transitions such that the source of each is the destination of the previous one; it can be written

$$c := p_0 \xrightarrow{x_1} p_1 \xrightarrow{x_2} p_2 \xrightarrow{x_3} \cdots \xrightarrow{x_n} p_n.$$

The *label*, or *result* of c, written $|c|$, is the *product* of the labels of the transitions of c. In the above case, $|c| = x_1 x_2 \cdots x_n$.

A *computation* in $\mathfrak{A}$ is a path to which is added an arrow arriving at the source and one leaving from the destination, with their respective labels. The computation corresponding to the above path is hence

[7] Or *weighted automaton* if S is understood or immaterial.

[8] This definition is a priori more general than the one given in Chap. 3; the two will coincide for finite automata.

$$d := \xrightarrow{I_{p_0}} p_0 \xrightarrow{x_1} p_1 \xrightarrow{x_2} p_2 \xrightarrow{x_3} \cdots \xrightarrow{x_n} p_n \xrightarrow{T_{p_n}} .$$

The *label* or *result* of d, still written $|d|$, is the product of the label of the incoming arrow, that of the path, and that of the outgoing arrow, in that order; in our case: $|d| = I_{p_0} x_1 x_2 \cdots x_n T_{p_n}$.

The definitions we have made for weighted automata are indeed a generalisation of the classical definitions:

(i) An automaton over Σ is a $\mathbb{B}$-automaton over Σ^*; an automaton over M is a $\mathbb{B}$-automaton over M.

(ii) The distinction between *path* and *computation*, which are often used as synonyms, may seem useless. But apart from the fact that it is consistent with our terminology—'path' refers to 'graph' while 'computation' refers to 'automaton', and what distinguishes an automaton from a graph is precisely that initial and final states are taken into account—it was only introduced in order to make precise definitions that incorporate the generality that we have now allowed for I and T. In the majority of cases, the non-zero elements of I and T will be scalar (that is, elements of S), usually equal to 1_S and the two notions will coincide.

(iii) Along the same lines, the disappearance of the notion of a *successful computation* is merely apparent. A state p such that the component I_p is *non-zero* (that is, different from $0_{S\langle\langle M\rangle\rangle}$) can be called *initial*, and a state where T_p is non-zero can be called *final*. We can then say that a computation is successful if its source is an initial state and its destination is a final state.

Definition 3.1. *The* behaviour *of an automaton* $\mathfrak{A} = \langle I, E, T\rangle$ *of finite dimension* Q *is defined if and only if for all* p *and* q *in* Q *the family of labels of paths with source* p *and destination* q *is summable. In this case, the family of labels of computations of* $\mathfrak{A}$ *is summable and its sum is the behaviour of* $\mathfrak{A}$, *written*[9] $|\mathfrak{A}|$. *We also say that* $\mathfrak{A}$ accepts *or* realises *the series* $|\mathfrak{A}|$.

The description of the transitions of an automaton by a matrix is justified by the fact that a walk over a graph corresponds to a matrix multiplication. This is expressed by the following proposition.

Lemma 3.2. *Let* $\mathfrak{A} = \langle I, E, T\rangle$ *be an* S*-automaton over* M *of finite dimension. For every integer* n, E^n *is the matrix of the sums of the labels of paths of length* n.

Proof. By induction on n. The assertion is true for $n = 1$ (and also for $n = 0$ by convention). The definition of the $(n+1)$st power of E is

$$\forall p, q \in Q \quad \left(E^{n+1}\right)_{p,q} = \sum_{r\in Q} \left(E^n\right)_{p,r} E_{r,q}.$$

[9] Written $\|\mathfrak{A}\|$ in Chap. 3.

Every path of length $n+1$ is the concatenation of a path of length n with a path of length 1, that is, a single transition. We can therefore write[10]

$$\{c \mid c := p \underset{\mathfrak{A}}{\rightarrow} q,\ l(c) = n+1\} = \bigcup_{r\in Q} \{(d,e) \mid d := p \underset{\mathfrak{A}}{\rightarrow} r,\ l(d) = n,\ e := r \underset{\mathfrak{A}}{\rightarrow} q \in E\},$$

and hence

$$\begin{aligned}\sum\{|c| \mid c := p \underset{\mathfrak{A}}{\rightarrow} q,\ l(c) = n+1\} \\ &= \sum_{r\in Q} \left(\{|d||e| \mid d := p \underset{\mathfrak{A}}{\rightarrow} r,\ l(d) = n,\ e := r \underset{\mathfrak{A}}{\rightarrow} q \in E\}\right) \\ &= \sum_{r\in Q} \left[\left(\sum\{|d| \mid d := p \underset{\mathfrak{A}}{\rightarrow} r,\ l(d) = n\}\right) E_{r,q}\right].\end{aligned}$$

As $\sum\{|d| \mid d = p \underset{\mathfrak{A}}{\rightarrow} r,\ l(d) = n\} = (E^n)_{p,r}$ by the induction hypothesis, the lemma is proved. □

Since the sum of the results of the *computations* of length n is equal by definition to the product $I \cdot E^n \cdot T$, and since the behaviour of $\mathfrak{A}$ is equal to the sum of the results of the computations of all the lengths, the following statement holds.

Corollary 3.3. *Let* $\mathfrak{A} = \langle I, E, T\rangle$ *be a* S*-automaton of finite dimension whose behaviour is defined, then* E^* *is defined and we have* $|\mathfrak{A}| = I \cdot E^* \cdot T$.

Example 3.4. The $\mathbb{N}$-automaton over $\{a,b\}^*$ defined by

$$\mathfrak{B}_1 = \left\langle (1\ 0), \begin{pmatrix} a+b & b \\ 0 & a+b \end{pmatrix}, \begin{pmatrix} 0 \\ 1 \end{pmatrix} \right\rangle$$

is shown in Fig. 3 (left). A simple calculation allows us to determine its behaviour:

$$\forall f \in \Sigma^* \quad (|\mathfrak{B}_1|, f) = |f|_b; \qquad \text{that is} \qquad |\mathfrak{B}_1| = \sum_{f\in\Sigma^*} |f|_b f = u_1.$$

Another $\mathbb{N}$-automaton is shown in Fig. 3 (right)

$$\mathfrak{C}_1 = \left\langle (1\ 0), \begin{pmatrix} a+b & b \\ 0 & 2a+2b \end{pmatrix}, \begin{pmatrix} 0 \\ 1 \end{pmatrix} \right\rangle.$$

If we use the convention that each word f of Σ^* is considered as a number written in binary, interpreting a as the digit 0 and b as the digit 1, and if we

[10] The length of a path c is here written $l(c)$.

Fig. 3. The $\mathbb{N}$-automata $\mathfrak{B}_1$ and $\mathfrak{C}_1$

Fig. 4. The $\mathbb{M}$-automaton $\mathfrak{S}_1$

write $\bar{f}$ for the integer represented by the word f, it is easy to verify that $\bar{f}$ is computed by $\mathfrak{C}_1$ in the sense that

$$\forall f \in \Sigma^* \quad (|\mathfrak{C}_1|, f) = \bar{f}; \qquad \text{that is,} \qquad |\mathfrak{C}_1| = \sum_{f \in \Sigma^*} \bar{f} f.$$

Example 3.5. To illustrate the case where S is *different* from $\mathbb{N}$: let $\mathbb{M} = \langle \mathbb{N}^\infty, \mathsf{min}, +, \infty, 0 \rangle$ be the 'tropical' semiring (cf. Chap. 1, Sect. 2). The $\mathbb{M}$-automaton $\mathfrak{S}_1$ over $\{a, b\}^*$ and defined by

$$\mathfrak{S}_1 = \left\langle (0\ 0), \begin{pmatrix} 0a + 1b & \infty \\ \infty & 1a + 0b \end{pmatrix}, \begin{pmatrix} 0 \\ 0 \end{pmatrix} \right\rangle$$

is shown in Fig. 4. Clearly, the support of $|\mathfrak{S}_1|$ is all of $\{a, b\}^*$ and the coefficient in $|\mathfrak{S}_1|$ of an arbitrary word f of $\{a, b\}^*$ is $\mathsf{min}\{|f|_a, |f|_b\}$.

Remark 3.6. The behaviour of an automaton was defined by returning to the essence of an 'automaton': a procedure for describing computations. With this definition, the behaviour of the two automata in Fig. 5(a), (b) are not defined although in the first case the family $\{I \cdot E^n \cdot T\}_{n \in \mathbb{N}}$ is summable since all its terms are zero, and in the second E^* is defined since $E^2 = 0$.

Such a definition of the behaviour is more 'robust' than one that would be based on the transition matrix and its star only. For instance, it is invariant under the decomposition of a transition into a strictly longer path. Figure 6

Fig. 5. Two $\mathbb{Z}$-automata with behavioural problems

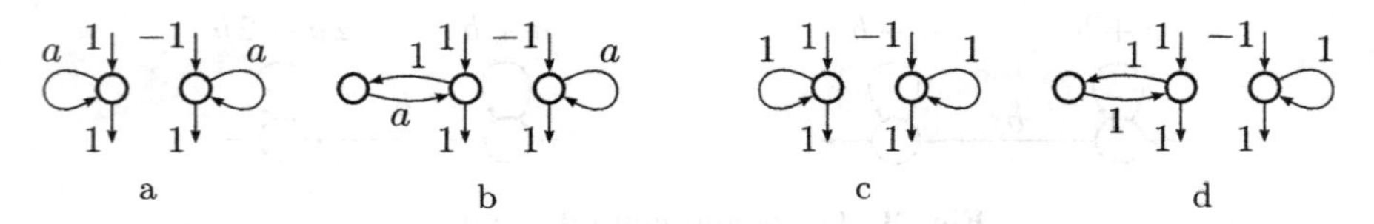

Fig. 6. Advocating for a path-based definition of the behaviour of automata

illustrates this point: as the automaton in (a) is obviously equivalent to the one in (b), those in (c) and (d) should also be equivalent.

In Chap. 3, the behaviour of an automaton is defined under the assumption that the automaton is cycle-free. Under the same assumption, the behaviour—as defined here—is always well defined and, by Corollary 3.3, equal to the one defined in Chap. 3.

Remark 3.7. On the other hand, these examples also lead us to note that the transition between each pair of states p and q must be *unique*, and labelled $E_{p,q}$; otherwise, we would be able to 'decompose' these entries in such a way that the family of labels of paths would no longer be summable.

From Lemma 2.5 and Proposition 2.21, we deduce a sort of generalisation of the same Proposition 2.21.

Proposition 3.8. *Let S be a strong topological semiring and M a graded monoid. The behaviour of an S-automaton over M, $\mathfrak{A} = \langle I, E, T \rangle$ is defined if and only if the behaviour of the S-automaton $\mathfrak{A}_0 = \langle I, E_0, T \rangle$ is defined, where E_0 is the matrix of constant terms of entries of E, and in that case we have*

$$|\mathfrak{A}| = I \cdot ({E_0}^* \cdot E_{\mathrm{p}})^* \cdot {E_0}^* \cdot T.$$

The example of Fig. 5(b) shows that it is not sufficient that ${E_0}^*$ be defined, nor even that E_0 be nilpotent[11] for the behaviour of $\mathfrak{A}$ be defined. On the other hand, the behaviour of $\mathfrak{A}$ is defined when E_0 is *strict upper triangular* since in this case the number of computations in $\mathfrak{A}_0$ is *finite*. And this is the case (up to a renaming of the states) if the automaton is cycle-free.

Definition 3.9. *A S-automaton over M, $\mathfrak{A} = \langle I, E, T \rangle$, is* finite *if:*

(i) The dimension of $\mathfrak{A}$ is finite.
(ii) The coefficients of E, I and T are polynomials; that is, have finite support.

3.2 The Fundamental Theorem of Automata

One hesitates to say of a proposition, 'here is *the fundamental theorem*'. However, this seems justified for the one that follows: it states completely generally,

[11] That is, there exists an n such that ${E_0}^n = 0$.

at least under the current assumption that M is a finitely generated graded monoid and S a strong topological semiring that what one can 'do' with a finite automaton is precisely what one can 'do' with rational operations.

Theorem 3.10. *A series of $S\langle\langle M\rangle\rangle$ is rational if and only if it is the behaviour of some finite S-automaton over M.*

Remark 3.11. Theorem 3.10 is usually called Kleene's theorem, and again in this handbook (cf. Chap. 3). When M is a free monoid Σ^*, there is no possibility to distinguish between *rational* and *recognisable* sets or series, but at the level of speech. When M is not free, recognisable sets or series take their own quality and become a distinct family from the one of rational sets, or series. We thus have two distinct results: the first one (Theorem 3.10) that states that in any graded monoid the elements of *one* certain family—for which there is no reason to coin *two* different names—may have two distinct characterisations: by rational expressions and by finite automata and another one (Theorem 4.6 below) that states that two families of sets or series, which are distinct in general, coincide in the case of free monoids.

Since every language of Σ^* is the behaviour of an *unambiguous* automaton (of a deterministic one indeed)—we quoted above that $\mathrm{URat}\,\Sigma^* = \mathrm{Rat}\,\Sigma^*$—we then have the following.

Proposition 3.12. *The characteristic series of a rational language of Σ^* is a S-rational series, for any semiring S.*

3.2.1 Proper Automata

We can make Theorem 3.10 both more precise and more general, closer to the properties used in the proof. For this, we need to define a restricted class of S-automata.

Definition 3.13. *An S-automaton over M, $\mathfrak{A} = \langle I, E, R\rangle$, is* proper *if:*

(i) The matrix E is proper.
(ii) The entries of I and T are scalar*; that is, $I \in S^{1\times Q}$ and $T \in S^{Q\times 1}$.*

It follows from Proposition 3.8 that the behaviour of a proper automaton is well defined; the following result adds the converse.

Proposition 3.14. *Every S-automaton $\mathfrak{A}$ over M whose behaviour is defined is equivalent to a proper automaton whose entries, other than the scalar entries of the initial and final vectors, are linear combinations of proper parts of the entries of $\mathfrak{A}$.*

Proof. We first show that $\mathfrak{A} = \langle I, E, T\rangle$ is equivalent to an automaton $\mathfrak{B} = \langle J, F, U\rangle$ where the entries of J and U are scalar. We set

$$J = \begin{pmatrix} 1 & 0 & 0 \end{pmatrix}, \qquad F = \begin{pmatrix} 0 & I & 0 \\ 0 & E & T \\ 0 & 0 & 0 \end{pmatrix}, \qquad U = \begin{pmatrix} 0 \\ 0 \\ 1 \end{pmatrix}. \tag{7}$$

Every path in $\mathfrak{B}$ is a path or a computation in $\mathfrak{A}$ and the behaviour of $\mathfrak{B}$ is defined if and only if that of $\mathfrak{A}$ is, and in that case E^* is defined.[12] We verify by induction that, for every integer n greater than or equal to 2,

$$F^n = \begin{pmatrix} 0 & I \cdot E^{n-1} & I \cdot E^{n-2} \cdot T \\ 0 & E^n & E^{n-1} \cdot T \\ 0 & 0 & 0 \end{pmatrix}. \tag{8}$$

We have $J \cdot U = J \cdot F \cdot U = 0$, $J \cdot F^{n+2} \cdot U = I \cdot E^n \cdot T$, hence $J \cdot F^* \cdot U = I \cdot E^* \cdot T$ and $\langle J, F, U \rangle$ is equivalent to $\mathfrak{A}$.

Next, starting from an automaton $\mathfrak{B} = \langle J, F, U \rangle$ whose initial and final vectors are scalar, we set

$$F = F_0 + F_{\mathfrak{p}}.$$

The behaviour of $\mathfrak{B}$ is defined if and only if the behaviour of the automaton $\langle J, F_0, U \rangle$ is defined, and in this case F_0^* is defined, also. We then have

$$|\mathfrak{B}| = J \cdot F^* \cdot U = J \cdot H^* \cdot V,$$

with $H = F_0^* \cdot F_{\mathfrak{p}}$ and $V = F_0^* \cdot U$. Since F_0^* is an element of $S^{Q \times Q}$, the entries of H are linear combinations (with coefficients in S) of entries of $F_{\mathfrak{p}}$ and the entries of V are scalar. □

3.2.2 Standard Automata

It is convenient to define an even more restricted class of automata and to show that an automaton of that class can be canonically associated with every S-expression.

Definition 3.15. *An S-automaton $\mathfrak{A} = \langle I, E, T \rangle$ is* standard *if the initial vector I has a single non-zero coordinate i, equal to 1_S, and if this unique initial state i is not the destination of any transition whose label is non-zero.*

In matrix terms, this means that $\mathfrak{A}$ can be written

$$\mathfrak{A} = \left\langle \begin{pmatrix} 1 & 0 \end{pmatrix}, \begin{pmatrix} 0 & K \\ 0 & F \end{pmatrix}, \begin{pmatrix} c \\ U \end{pmatrix} \right\rangle. \tag{9}$$

[12] The automaton $\mathfrak{B}$ is the *normalised automaton* $\mathfrak{A}'$ built in Chap. 3 (proof of Theorem 2.11).

The definition does not forbid the initial state i from also being final, that is, the scalar c is not necessarily zero. If $\mathfrak{A}$ is not only standard but also *proper*, c is the *constant term* of $|\mathfrak{A}|$. The proof of Proposition 3.14 itself proves the following proposition.

Proposition 3.16. *Every S-automaton $\mathfrak{A}$ over M whose behaviour is defined is equivalent to a* standard *proper automaton whose entries, other than the scalar entries of the initial and final vectors, are linear combinations of proper parts of the entries of $\mathfrak{A}$.*

We now define *operations* on standard automata (as in Chap. 3, Sect. 2.2) that are parallel to the *rational operations*. Let $\mathfrak{A}$ (as in (9)) and $\mathfrak{A}'$ (with obvious translation) be two proper standard automata; the following standard S-automata are defined:

- $$s\mathfrak{A} = \left\langle \begin{pmatrix} 1 & 0 \end{pmatrix}, \begin{pmatrix} 0 & sK \\ 0 & F \end{pmatrix}, \begin{pmatrix} sc \\ U \end{pmatrix} \right\rangle \quad \text{and}$$

  $$\mathfrak{A}s = \left\langle \begin{pmatrix} 1 & 0 \end{pmatrix}, \begin{pmatrix} 0 & K \\ 0 & F \end{pmatrix}, \begin{pmatrix} cs \\ Us \end{pmatrix} \right\rangle$$

- $$\mathfrak{A} + \mathfrak{A}' = \left\langle \begin{pmatrix} 1 & 0 & 0 \end{pmatrix}, \begin{pmatrix} 0 & K & K' \\ 0 & F & 0 \\ 0 & 0 & F' \end{pmatrix}, \begin{pmatrix} c + c' \\ U \\ U' \end{pmatrix} \right\rangle$$

- $$\mathfrak{A} \cdot \mathfrak{A}' = \left\langle \begin{pmatrix} 1 & 0 & 0 \end{pmatrix}, \begin{pmatrix} 0 & K & cK' \\ 0 & F & H \\ 0 & 0 & F' \end{pmatrix}, \begin{pmatrix} cc' \\ V \\ U' \end{pmatrix} \right\rangle$$

  where $H = (U \cdot K') \cdot F'$ and $V = Uc' + (U \cdot K') \cdot U'$

- $$\mathfrak{A}^* = \left\langle \begin{pmatrix} 1 & 0 \end{pmatrix}, \begin{pmatrix} 0 & c^*K \\ 0 & G \end{pmatrix}, \begin{pmatrix} c^* \\ Uc^* \end{pmatrix} \right\rangle$$

  which is defined if and only if c^* is defined, and where $G = U \cdot c^*K + F$.

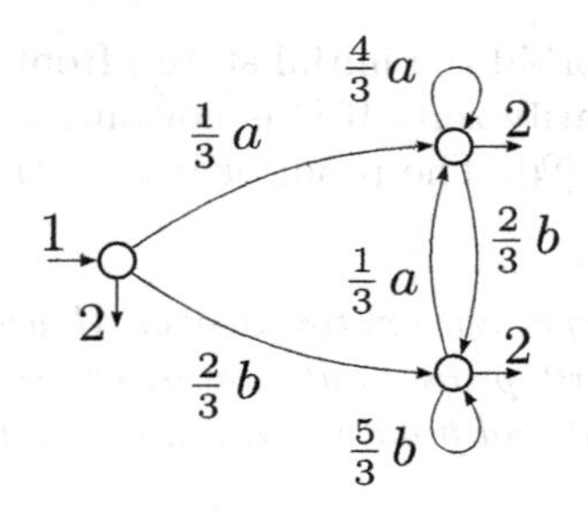

Fig. 7. The $\mathbb{Q}$-automaton $\mathfrak{S}_{\mathsf{E}_1}$

By construction, $s\mathfrak{A}$, $\mathfrak{A}s$, $\mathfrak{A}+\mathfrak{A}'$, $\mathfrak{A}\cdot\mathfrak{A}'$, and $\mathfrak{A}^*$ are all proper. Straightforward computations show that $|s\mathfrak{A}| = s|\mathfrak{A}|$, $|\mathfrak{A}s| = |\mathfrak{A}|s$, $|\mathfrak{A}+\mathfrak{A}'| = |\mathfrak{A}| + |\mathfrak{A}'|$, $|\mathfrak{A}\cdot\mathfrak{A}'| = |\mathfrak{A}||\mathfrak{A}'|$ and $|\mathfrak{A}^*| = |\mathfrak{A}^*|$.

With every valid rational S-expression E, we thus canonically associate, by induction on the depth of E, a proper standard S-automaton $\mathfrak{S}_{\mathsf{E}}$ that we call *the* standard automaton of E. Let $\ell(\mathsf{E})$ denote the *literal length* of E, that is, the number of atoms different from 0 and 1 in E. The following proposition holds.

Proposition 3.17. *If* E *is a valid rational* S*-expression, then* $|\mathfrak{S}_{\mathsf{E}}| = |\mathsf{E}|$ *and the dimension of* $\mathfrak{S}_{\mathsf{E}}$ *is* $\ell(\mathsf{E}) + 1$.

Example 3.18 (Example 2.31 continued). Figure 7 shows the $\mathbb{Q}$-automaton $\mathfrak{S}_{\mathsf{E}_1}$ associated with the rational expression $\mathsf{E}_1 = (\frac{1}{6}a^* + \frac{1}{3}b^*)^*$ by the construction described above.

3.2.3 Statement and Proof of the Fundamental Theorem

Definition 3.19. *We will say that a family of series is* proper *if it contains the proper part of each of its elements.*[13]

In particular, the polynomials form a proper family of $S\langle\langle M\rangle\rangle$.

Theorem 3.20. *Let* $\mathcal{C}$ *be a proper family of series of* $S\langle\langle M\rangle\rangle$*. A series* s *of* $S\langle\langle M\rangle\rangle$ *belongs to* $S\mathrm{Rat}\,\mathcal{C}$ *if and only if* s *is the behaviour of a* proper standard S*-automaton over* M *of finite dimension whose (non-scalar) entries are finite linear combinations of elements of* $\mathcal{C}$.

Proof. The proof of Theorem 3.20 splits in the "if" and "only if" parts, which by Proposition 2.32, essentially amount to show respectively that given a proper automaton we can compute an equivalent valid rational expression and conversely that given a valid rational expression we can compute an equivalent automaton.

[13] As opposed to all the series in the family being proper.

We write $\mathcal{D}$ for the family of behaviours of proper standard S-automata whose entries are linear combinations of elements of $\mathcal{C}$. We first show that $\mathcal{D}$ contains 0_S, behaviour of the standard automaton $\langle 1_S, 0_S, 0_S \rangle$ of dimension 1 and 1_S, behaviour of $\langle 1_S, 0_S, 1_S \rangle$, as well as every element in $\mathcal{C}$: for r in $\mathcal{C}$, r_{p} is in $\mathcal{C}$ since $\mathcal{C}$ is a proper family and it holds:

$$r = \begin{pmatrix} 1_S & 0_S \end{pmatrix} \cdot \begin{pmatrix} 0_S & r_{\mathsf{p}} \\ 0_S & 0_S \end{pmatrix}^* \cdot \begin{pmatrix} r_0 \\ 1_S \end{pmatrix}.$$

If $\mathfrak{A}$ and $\mathfrak{A}'$ are two proper standard S-automata whose entries are linear combinations of elements of $\mathcal{C}$, the above constructions $s\mathfrak{A}$, $\mathfrak{A}s$, $\mathfrak{A} + \mathfrak{A}'$, $\mathfrak{A} \cdot \mathfrak{A}'$ and $\mathfrak{A}^*$ show that $\mathcal{D}$ is rationally closed.

Conversely, we start from a proper automaton $\mathfrak{A} = \langle I, E, T \rangle$ whose behaviour is thus defined and equal to $|\mathfrak{A}| = I \cdot E^* \cdot T$. This part then amounts to prove that the entries of the star of a proper matrix E belong to the rational closure of the entries of E, a classical statement established in general under different hypotheses (e.g. [12]). Since we have to reprove it anyway, we choose a slightly different method. We write $|\mathfrak{A}| = I \cdot V$ with $V = E^* \cdot T$. Since E is proper and by Lemmas 2.5 and 2.16, V is *the unique solution* of

$$X = E \cdot X + T \tag{10}$$

and we have to prove that all entries of the vector V belong to the rational closure of the entries of E. Lemma 2.16 already states that the property holds if $\mathfrak{A}$ is of dimension 1. For $\mathfrak{A}$ of dimension Q, we write (10) as a system of $\|Q\|$ equations:

$$\forall p \in Q \quad V_p = \sum_{q \in Q} E_{p,q} V_q + T_p. \tag{11}$$

We choose (arbitrarily) one element q in Q and by Lemma 2.16 again, it comes:

$$V_q = E^*_{q,q} \left[\sum_{p \in Q \setminus \{q\}} E_{q,p} V_p + T_q \right],$$

an expression for V_q that can be substituted in every other equation of the system (11), giving a new system

$$\forall p \in Q \setminus \{q\} \quad V_p = \sum_{r \in Q \setminus \{q\}} [E_{p,r} + E_{p,q} E^*_{q,q} E_{q,r}] V_r + E_{p,q} E^*_{q,q} T_q + T_p.$$

And the property is proved by induction hypothesis. □

The fundamental theorem states the equality of two families of series (infinite objects), but its proof is better understood as the description of two algorithms. Here, we have chosen on one hand the construction of the standard automaton of an expression and on the other hand the algorithm known as the *state elimination method* for the computation of an expression denoting the behaviour of an automaton. In the latter case, the result depends on the order of elimination (the choice of the state q in (11)). The relationship between the possible different results is given by the following Proposition 3.21. We shall say that two (S-)expressions E and F are *equivalent modulo an identity* $\boldsymbol{I}$ if E can be transformed into F by using instances of $\boldsymbol{I}$ and of the so-called 'natural identities' which express that the expressions are interpreted in a semiring (associativity, distributivity of $\cdot$ over $+$, commutativity of $+$).

Proposition 3.21. *Let $\mathfrak{A}$ be an S-automaton of dimension Q. The expressions denoting $|\mathfrak{A}|$ and obtained by the state elimination method with distinct orders on Q are all equivalent modulo the identities $\boldsymbol{S}$ and $\boldsymbol{P}$.*

3.3 Conjugacy and Covering of Automata

After the definition of any structure, one looks for *morphisms* between objects of that structure, and weighted automata are no exception. Moreover, morphisms of graphs and, therefore, of classical Boolean automata, are not less classical, and one waits for their generalisation to weighted automata. Taking into account multiplicity proves, however, to be not so simple. In the sequel, all automata are supposed to be of finite dimension.

3.3.1 From Conjugacy to Covering

We choose to describe the morphisms of weighted automata, which we call *coverings*, via the notion of *conjugacy*, borrowed from the theory of symbolic dynamical systems.

Definition 3.22. *An S-automaton $\mathfrak{A} = \langle I, E, T\rangle$* is conjugate to *an S-automaton $\mathfrak{B} = \langle J, F, U\rangle$ if there exists a matrix X with entries in S such that*

$$IX = J, \qquad EX = XF, \qquad \textit{and} \qquad T = XU.$$

The matrix X is the transfer matrix *of the conjugacy and we write $\mathfrak{A} \stackrel{X}{\Longrightarrow} \mathfrak{B}$.*

In spite of the idea conveyed by the terminology, the conjugacy relation *is not an equivalence* but a *pre-order* relation. Suppose that $\mathfrak{A} \stackrel{X}{\Longrightarrow} \mathfrak{C}$ holds; if $\mathfrak{C} \stackrel{Y}{\Longrightarrow} \mathfrak{B}$, then $\mathfrak{A} \stackrel{XY}{\Longrightarrow} \mathfrak{B}$, but if $\mathfrak{B} \stackrel{Y}{\Longrightarrow} \mathfrak{C}$ then $\mathfrak{A}$ is not necessarily conjugate to $\mathfrak{B}$, and we write $\mathfrak{A} \stackrel{X}{\Longrightarrow} \mathfrak{C} \stackrel{Y}{\Longleftarrow} \mathfrak{B}$ or even $\mathfrak{A} \stackrel{X}{\Longrightarrow}\stackrel{Y}{\Longleftarrow} \mathfrak{B}$. This being well understood, we shall speak of "conjugate automata" when the orientation does not matter.

As $JF^nU = IXF^nU = IEXF^{n-1}U = \cdots = IE^nXU = IE^nT$ for every integer n, the following proposition holds.

Proposition 3.23. *Two conjugate automata are equivalent.*

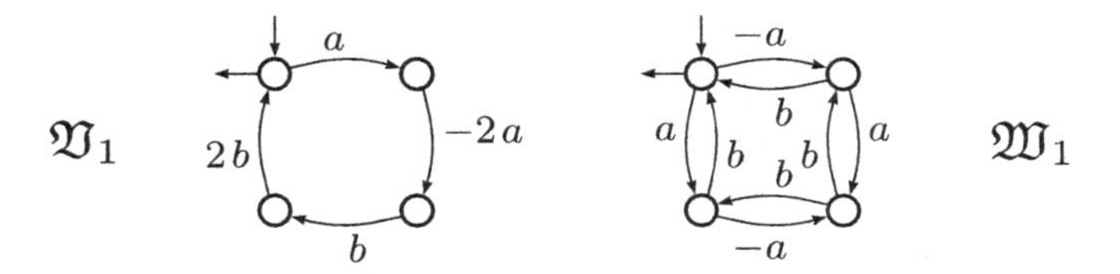

Fig. 8. Two conjugate $\mathbb{Z}$-automata

Example 3.24. It is easily checked that the $\mathbb{Z}$-automaton $\mathfrak{V}_1$ of Fig. 8 is conjugate to the $\mathbb{Z}$-automaton $\mathfrak{W}_1$ of the same figure with the transfer matrix X_1:

$$X_1 = \begin{pmatrix} 1 & 0 & 0 & 0 \\ 0 & -1 & 1 & 0 \\ 0 & 1 & 1 & 0 \\ 0 & 0 & 0 & 1 \end{pmatrix}.$$

Let φ be an *equivalence* relation on Q or what is the same, let $\varphi\colon Q \to R$ be a surjective map and H_φ the $Q \times R$-matrix where the (q,r) entry is 1 if $\varphi(q) = r$, 0, otherwise. Since φ is a map, each row of H_φ contains exactly one 1 and since φ is surjective, each column of H_φ contains at least one 1. Such a matrix is called an *amalgamation matrix* [31, Definition 8.2.4].

Definition 3.25. *Let $\mathfrak{A}$ and $\mathfrak{B}$ be two S-automata of dimension Q and R, respectively. We say that $\mathfrak{B}$ is a* S-quotient *of $\mathfrak{A}$ and conversely that $\mathfrak{A}$ is a* S-covering *of $\mathfrak{B}$ if there exists a surjective map $\varphi\colon Q \to R$ such that $\mathfrak{A}$ is conjugate to $\mathfrak{B}$ by H_φ.*

The notion of S-quotient is *lateralised* since the conjugacy relation is not symmetric. Somehow, it is the price we pay for extending the notion of morphism to S-automata. Therefore, the *dual* notions *co-S-quotient* and *co-S-covering* are defined in a natural way.

Definition 3.26. *With the above notation, we say that $\mathfrak{B}$ is a* co-S-quotient *of $\mathfrak{A}$ and conversely that $\mathfrak{A}$ is a* co-S-covering *of $\mathfrak{B}$ if there exists a surjective map $\varphi\colon Q \to R$ such that $\mathfrak{B}$ is conjugate to $\mathfrak{A}$ by ${}^{t}H_\varphi$.*

We also write $\varphi\colon \mathfrak{A} \to \mathfrak{B}$ and call φ, by way of metonymy, *a S-covering, or a co-S-covering from $\mathfrak{A}$ onto $\mathfrak{B}$.*

Example 3.27. Consider the $\mathbb{N}$-automaton $\mathfrak{C}_2$ of Fig. 9 and the map φ_2 from $\{j, r, s, u\}$ to $\{i, q, t\}$ such that $j\varphi_2 = i$, $u\varphi_2 = t$ and $r\varphi_2 = s\varphi_2 = q$, then

$$H_{\varphi_2} = \begin{pmatrix} 1 & 0 & 0 \\ 0 & 1 & 0 \\ 0 & 1 & 0 \\ 0 & 0 & 1 \end{pmatrix}$$

and φ_2 is an $\mathbb{N}$-covering from $\mathfrak{C}_2$ onto $\mathfrak{V}_2$ and a co-$\mathbb{N}$-covering from $\mathfrak{C}_2$ onto $\mathfrak{V}'_2$.

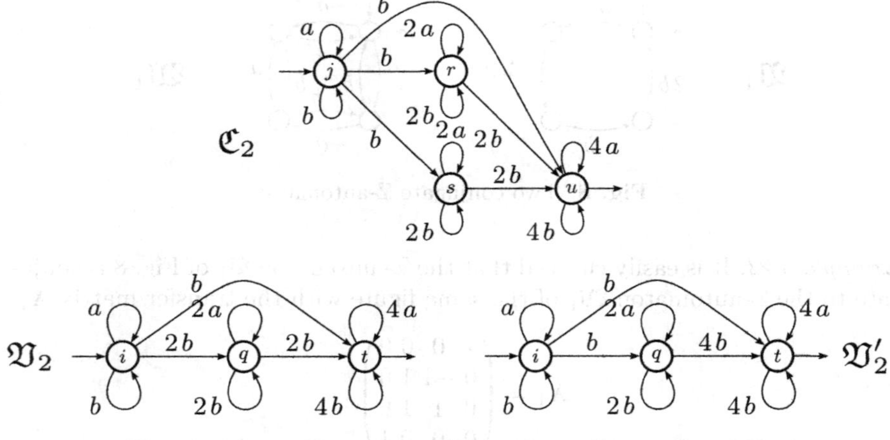

Fig. 9. $\mathfrak{C}_2$ is an $\mathbb{N}$-covering of $\mathfrak{V}_2$ and a co-$\mathbb{N}$-covering of $\mathfrak{V}_2$

3.3.2 Minimal S-Quotient

Let us first express that in a S-covering $\varphi\colon \mathfrak{A} \to \mathfrak{B}$ the image is somewhat immaterial and only counts the map equivalence of φ. From any amalgamation matrix H_φ, we construct a matrix K_φ by transposing H_φ and by arbitrarily cancelling certain entries in such a way that K_φ is row monomial (with exactly one 1 per row); K_φ is not uniquely determined by φ, but also depends on the choice of a 'representative' in each class for the map equivalence of φ. Whatever K_φ, the product $K_\varphi H_\varphi$ is the identity matrix of dimension R (as the matrix representing $\varphi^{-1}\varphi$). Easy matrix computations establish the following.

Proposition 3.28. *Let* $\mathfrak{A} = \langle I, E, T\rangle$ *and* $\mathfrak{B} = \langle J, F, U\rangle$ *be two* S*-automata of dimension* Q *and* R*, respectively. A surjective map* $\varphi\colon Q \to R$ *is* a S-covering *if and only if* $\mathfrak{A}$ *satisfies the two equations:*

$$H_\varphi \cdot K_\varphi \cdot E \cdot H_\varphi = E \cdot H_\varphi, \tag{12}$$

and

$$H_\varphi \cdot K_\varphi \cdot T = T. \tag{13}$$

In which case, $\mathfrak{B}$ *satisfies*

$$F = K_\varphi \cdot E \cdot H_\varphi, \qquad J = I \cdot H_\varphi \qquad \textit{and} \qquad U = K_\varphi \cdot T. \tag{14}$$

Theorem 3.29. *Let* $\mathfrak{A}$ *be a* S*-automaton of finite dimension over* M*. Among all the* S*-quotients of* $\mathfrak{A}$ *(resp. among all the co-*S*-quotients of* $\mathfrak{A}$*), there exists one, unique up to isomorphism and effectively computable from* $\mathfrak{A}$*, which has a minimal number of states and of which all these* S*-automata are* S*-coverings (resp. co-*S*-coverings).*

Proof. A surjective map $\varphi\colon Q \to R$ defines a S-covering $\varphi\colon \mathfrak{A} \to \mathfrak{B}$ if (12) and (13) (which do not involve $\mathfrak{B}$) are satisfied.

To prove the existence of a minimal S-quotient, it suffices to show that if $\varphi\colon Q \to R$ and $\psi\colon Q \to P$ are two maps that define S-coverings, the map $\omega\colon Q \to V$ also defines a S-covering, where $\omega = \varphi \vee \psi$ is the map whose map equivalence is the upper bound of those of φ and ψ; that is, the finest equivalence which is coarser than the map equivalences of φ and ψ. In other words, there exist $\varphi'\colon R \to V$ and $\psi'\colon P \to V$ such that $\omega = \varphi\varphi' = \psi\psi'$ and each class modulo $\omega = \varphi \vee \psi$ can be seen at the same time as a union of classes modulo φ and as a union of classes modulo ψ. It follows that

$$E \cdot H_\omega = E \cdot H_\varphi \cdot H_{\varphi'} = E \cdot H_\psi \cdot H_{\psi'}; \tag{15}$$

and if two states p and r of Q are congruent modulo ω, there exists q such that $p\varphi = q\varphi$ and $q\psi = r\psi$ (in fact, a sequence of states q_i, etc.). The rows p and q of $E \cdot H_\varphi$ are equal, and the rows q and r of $E \cdot H_\psi$ are equal; hence, by (15), the rows p and r of $E \cdot H_\omega$ are equal, also.

To compute this minimal S-quotient, we can proceed by successive refinements of partitions, exactly as for the computation of the minimal automaton of a language from a deterministic automaton which recognises the language.

In what follows, the maps φ_i are identified with their map equivalences; the image is irrelevant. A state r of Q is identified with the row vector of dimension Q, characteristic of r, and treated as such. For example, $r\varphi = s\varphi$ can be written $r \cdot H_\varphi = s \cdot H_\varphi$.

The maps φ_0 have the same map equivalence as T, that is,

$$r \cdot H_{\varphi_0} = s \cdot H_{\varphi_0} \quad\Longleftrightarrow\quad r \cdot T = s \cdot T,$$

which can also be written

$$H_{\varphi_0} \cdot K_{\varphi_0} \cdot T = T, \tag{16}$$

and the same equation holds for every map finer than φ_0. For each i, φ_{i+1} is finer than φ_i and, by definition, r and s are joint in φ_i (that is, $r \cdot H_{\varphi_i} = s \cdot H_{\varphi_i}$) and disjoint in φ_{i+1} if $r \cdot E \cdot H_{\varphi_i} \neq s \cdot E \cdot H_{\varphi_i}$. Let j be the index such that $\varphi_{j+1} = \varphi_j$, that is, such that

$$r \cdot H_{\varphi_j} = s \cdot H_{\varphi_j} \quad\Longrightarrow\quad r \cdot E \cdot H_{\varphi_j} = s \cdot E \cdot H_{\varphi_j}, \tag{17}$$

which can be rewritten

$$H_{\varphi_j} \cdot K_{\varphi_j} \cdot E \cdot H_{\varphi_j} = E \cdot H_{\varphi_j}. \tag{18}$$

By (16) and (18), φ_j is a S-covering.

Conversely, every S-covering ψ satisfies (13) and is hence finer than φ_0. Then for all i, if ψ is finer than φ_i, it must also be finer than φ_{i+1}. In fact, if r and s are joint in ψ, it follows that $r \cdot H_\psi = s \cdot H_\psi$, and hence also $r \cdot H_{\varphi_i} = s \cdot H_{\varphi_i}$ since φ_i is coarser than ψ, and hence r and s are joint in φ_{i+1}: ψ is finer than φ_j, which is thus the coarsest S-covering. □

Remark 3.30. Even if the minimal S-quotient of a S-automaton and the minimal automaton of a language are computed with the *same* algorithm, they are nevertheless fundamentally different: the second automaton is canonically associated with the language, whereas the first is associated with the S-automaton we started from, and not with its behaviour.

Remark 3.31. The above construction applies of course if $S = \mathbb{B}$, and thus shows that the notion of minimal ($\mathbb{B}$-)quotient is well defined even for a *non-deterministic automaton* (as we just wrote, this minimal quotient is not associated with the recognised language anymore). Moreover, it can be checked that two Boolean automata are *bisimilar* if and only if their minimal $\mathbb{B}$-quotients are isomorphic (cf. [2]).

3.3.3 From Covering to Conjugacy

We have defined quotients (and co-quotients) as a special case of conjugacy. Under some supplementary hypothesis—that is naturally met in cases that are important to us: $\mathbb{N}$, $\mathbb{Z}$, etc.—it can be established that a kind of converse holds and that any conjugacy can basically be realised by the composition of an inverse co-covering and a covering.

In order to state these results, we need two further definitions. A matrix is *non-degenerate* if it contains no zero row nor zero column. We call *a circulation matrix* a diagonal invertible matrix.

Theorem 3.32 ([3]). *Let $\mathfrak{A}$ be a $\mathbb{Z}$-automaton conjugate to a $\mathbb{Z}$-automaton $\mathfrak{B}$ by a non-negative and non-degenerate transfer matrix X. Then there exists a $\mathbb{Z}$-automaton $\mathfrak{C}$ that is a co-$\mathbb{Z}$-covering of $\mathfrak{A}$ and a $\mathbb{Z}$-covering of $\mathfrak{B}$.*

We can free ourselves from the two hypotheses on the transfer matrix if we allow a further conjugacy by a circulation matrix.

Theorem 3.33 ([3]). *Let $\mathfrak{A}$ be a $\mathbb{Z}$-automaton conjugate to a $\mathbb{Z}$-automaton $\mathfrak{B}$ by a transfer matrix X. Then there exists two $\mathbb{Z}$-automata $\mathfrak{C}$ and $\mathfrak{D}$ and a circulation matrix D, such that $\mathfrak{C}$ is a co-$\mathbb{Z}$-covering of $\mathfrak{A}$, $\mathfrak{D}$ a $\mathbb{Z}$-covering of $\mathfrak{B}$ and $\mathfrak{C}$ is conjugate to $\mathfrak{D}$ by D.*

Example 3.34 (Example 3.24 continued). The $\mathbb{Z}$-automata $\mathfrak{X}_1$ of Fig. 10 is a co-$\mathbb{Z}$-covering of $\mathfrak{V}_1$, $\mathfrak{Y}_1$ is a $\mathbb{Z}$-covering of $\mathfrak{W}_1$, and $\mathfrak{X}_1$ is conjugate to $\mathfrak{Y}_1$ by the circulation matrix where the only -1 entry is at state 1.

The proof of Theorem 3.33 involves indeed two properties. Let us say first that a semiring *has property* (*SU*) if *every element is a sum of units.* The semiring $\mathbb{N}$, the ring $\mathbb{Z}$, and all fields have property (SU). In any semiring with (SU), every matrix X can be written as $X = CDR$ where C is a co-amalgamation, R an amalgamation, and D a circulation matrix. In $\mathbb{Z}$, the dimension of D will be the sum of the absolute value of the entries of X.

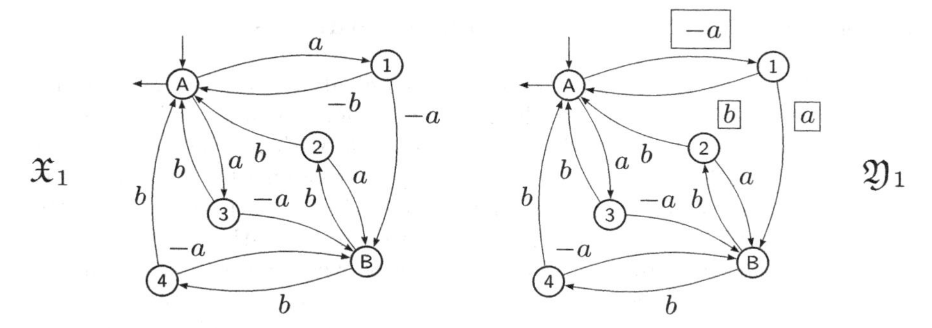

Fig. 10. The co-covering and covering of $\mathfrak{V}_1$ and $\mathfrak{W}_1$

Having secured the existence of C, D, and R, the second step consists in building the automata $\mathfrak{C}$ and $\mathfrak{D}$ that will fit in. To that end, we say that a commutative monoid is *equisubtractive* if for all p, q, r, and s such that $p+q=r+s$ there exist x, y, z, and t such that $p=x+y$, $q=z+t$, $r=x+z$ and $s=y+t$. A semiring is equisubtractive if it is so as a monoid for addition.

The semirings $\mathbb{N}$ and $\mathbb{Z}$ are equisubtractive, and if S is equisubtractive, then so are $S\langle\Sigma^*\rangle$ and $S\langle\langle\Sigma^*\rangle\rangle$. The construction of $\mathfrak{C}$ and $\mathfrak{D}$ will then follow from the following property.

Lemma 3.35. *Let S be an equisubtractive semiring and let $t_1, t_2, \ldots, t_n, s_1, s_2, \ldots, s_m$ be elements of S such that*

$$t_1 + t_2 + \cdots + t_n = s_1 + s_2 + \cdots + s_m.$$

There exists an $n \times m$ matrix G with entries in S such that the sum of the entries of each row i is equal to t_i and the sum of the entries of each column j is equal to s_j.

Another consequence of the definition of equi-subtractive semiring and of Lemma 3.35 is to allow a sort of converse to Theorem 3.29. The existence of a minimal S-covering implies a kind of Church–Rosser property: if we have two diverging arrows, that is, the *upper part* of a commutative diagram, we can construct the lower part of it. The following proposition states that it is possible to complete a commutative diagram when the *lower part* of it is known.

Proposition 3.36 ([43, 3]). *Let S be an* equisubtractive *semiring and let $\mathfrak{A}$, $\mathfrak{B}$ and $\mathfrak{C}$ be three S-automata.*

(a) If $\mathfrak{A}$ and $\mathfrak{B}$ are S-coverings of $\mathfrak{C}$ (resp. co-S-coverings of $\mathfrak{C}$), there exists a S-automaton $\mathfrak{D}$ which is a S-covering (resp. a co-S-covering) of both $\mathfrak{A}$ and $\mathfrak{B}$.

(b) If $\mathfrak{A}$ is a S-covering of $\mathfrak{C}$ and $\mathfrak{B}$ is a co-S-covering of $\mathfrak{C}$, there exists a S-automaton $\mathfrak{D}$ which is both a co-S-covering of $\mathfrak{A}$ and a S-covering of $\mathfrak{B}$.

4 Recognisable Series and Representations

As in the last section, S denotes a strong topological semiring and M a graded monoid, a priori arbitrary. We shall now consider another family of series of $S\langle\langle M\rangle\rangle$, other than $S\mathrm{Rat}\, M$, but that coincide with it when M is a free monoid Σ^*: this is the Kleene–Schützenberger theorem (Theorem 4.6). We first define these series by means of representations. We then consider the Hadamard product of series, which is a weighted generalisation of intersection. In a third subsection, by considering the series over a Cartesian product of monoids, we briefly sketch the prolegomena to a theory of weighted relations. This allows us, among other things, to establish the weighted generalisation of results on the morphic image of rational sets (Theorem 4.35).

4.1 The Family of Recognisable Series

An *S-representation of M of dimension Q* is a *morphism* μ from M to the semiring of square matrices of dimension Q with entries in S. By definition, in fact so that we can multiply the matrices, the dimension Q is *finite*. An S-representation of M (of dimension Q) is also the name we give a *triple* (λ, μ, ν) where, as before,

$$\mu : M \to S^{Q\times Q}$$

is a morphism and where λ and ν are two vectors:

$$\lambda \in S^{1\times Q} \quad \text{and} \quad \nu \in S^{Q\times 1};$$

that is, λ is a *row* vector and ν a *column* vector of dimension Q, with entries in S. Such a representation defines a map from M to S by

$$\forall m \in M \quad m \mapsto \lambda \cdot m\mu \cdot \nu;$$

that is, the *series* r:

$$r = \sum_{m\in M} (\lambda \cdot m\mu \cdot \nu)m.$$

A series r of $S\langle\langle M\rangle\rangle$ is *realised* or *recognised* by the representation (λ, μ, ν). We also say that (λ, μ, ν) *realises* or *recognises* the series r.

Definition 4.1. *A series of $S\langle\langle M\rangle\rangle$ is* S-recognisable *if it is recognised by an S-representation. The set of S-recognisable series over M is written* $S\mathrm{Rec}\, M$.

Example 4.2 (Example 3.4 continued). Take $S = \mathbb{N}$ and $M = \{a, b\}^*$. Let $(\lambda_1, \mu_1, \nu_1)$ be the representation defined by

$$a\mu_1 = \begin{pmatrix} 1 & 0 \\ 0 & 1 \end{pmatrix}, \qquad b\mu_1 = \begin{pmatrix} 1 & 1 \\ 0 & 1 \end{pmatrix}, \qquad \lambda_1 = \begin{pmatrix} 1 & 0 \end{pmatrix} \quad \text{and} \quad \nu_1 = \begin{pmatrix} 0 \\ 1 \end{pmatrix}.$$

For all f in $\{a, b\}^*$, we verify that $\lambda_1 \cdot f\mu_1 \cdot \nu_1 = |f|_b$, hence the series $u_1 = \sum_{f\in\Sigma^*} |f|_b f$ is $\mathbb{N}$-recognisable.

Remark 4.3. It is not difficult to check that Definition 4.1 coincides, for $S = \mathbb{B}$, with the definition of the *recognisable subsets* of a monoid *as the sets that are saturated by a congruence of finite index* [14]. If r is a $\mathbb{B}$-recognisable series over M, realised by the representation (λ, μ, ν), then $\mu : M \to \mathbb{B}^{Q \times Q}$ is a morphism from M *to a finite monoid.* The series r of $\mathbb{B}\langle\langle M\rangle\rangle$, $r = \sum_{m \in M}(\lambda \cdot m\mu \cdot \nu)m$, can be seen as the *subset* $r = P\mu^{-1}$ of M where $P = \{p \in \mathbb{B}^{Q \times Q} \mid \lambda \cdot p \cdot \nu = 1_{\mathbb{B}}\}$. Conversely, a morphism α from M into a finite monoid N is a morphism from M into the monoid of Boolean matrices of dimension N (the representation of N by right translations over itself) and the $\mathbb{B}$-representation that realises any subset recognised by α easily follows.

These definitions and the following two properties of $S\mathrm{Rec}\, M$ do not involve multiplication in $S\langle\langle M\rangle\rangle$, and are hence valid without even requiring that M be graded.

Proposition 4.4. *Every finite linear combination, with coefficients in S, of S-recognisable series over M is an S-recognisable series.*

Proof. Let r and u be two S-recognisable series over M, respectively recognised by the S-representations (λ, μ, ν) and (η, κ, ζ). For all s in S, the series sr is recognised by the representation $(s\lambda, \mu, \nu)$, the series rs by the representation $(\lambda, \mu, \nu s)$, and the series $r + u$ by the representation (δ, π, ξ) defined by the following block decomposition:

$$\delta = \begin{pmatrix}\lambda & \eta\end{pmatrix}, \qquad m\pi = \begin{pmatrix} m\mu & 0 \\ 0 & m\kappa \end{pmatrix}, \qquad \xi = \begin{pmatrix}\nu \\ \zeta\end{pmatrix}. \quad \square$$

Let $\varphi\colon S \to T$ be a morphism of semirings which extends to a morphism $\varphi\colon S\langle\langle M\rangle\rangle \to T\langle\langle M\rangle\rangle$ by $(r\varphi, m) = (r, m)\varphi$ for all r in $S\langle\langle M\rangle\rangle$ and all m in M. If (λ, μ, ν) is a representation of the series r of $S\langle\langle M\rangle\rangle$, then $(\lambda\varphi, \mu\varphi, \nu\varphi)$ is a representation of $r\varphi$. That is:

Proposition 4.5. *Let $\varphi\colon S \to T$ be a morphism of semirings. The image under φ of an S-recognisable series over M is a T-recognisable series over M.*

We can now get to our main point.

Theorem 4.6 (Kleene–Schützenberger). *Let S be a strong topological semiring, and Σ a finite alphabet. A series of $S\langle\langle \Sigma^*\rangle\rangle$ is S-rational if and only if it is S-recognisable. That is,*

$$S\mathrm{Rec}\, \Sigma^* = S\mathrm{Rat}\, \Sigma^* .$$

We prove the two inclusions one at a time:

$$S\mathrm{Rec}\, \Sigma^* \subseteq S\mathrm{Rat}\, \Sigma^* \quad \text{and} \quad S\mathrm{Rat}\, \Sigma^* \subseteq S\mathrm{Rec}\, \Sigma^* . \tag{19}$$

Each of the inclusions is obtained from the Fundamental Theorem together with the freeness of Σ^* and the finiteness of Σ. This is used in both cases by means of the following result.

Lemma 4.7. *Let S be a semiring and Σ a finite alphabet. Let Q be a finite set and $\mu\colon \Sigma^* \to S^{Q\times Q}$ a morphism. We set*

$$X = \sum_{a\in\Sigma} (a\mu)a.$$

Then for all f in Σ^*, we have $(X^*, f) = f\mu$.

Proof. The matrix X is a proper series of $S^{Q\times Q}\langle\langle\Sigma^*\rangle\rangle$, and hence X^* is defined. We first prove, by induction on the integer n, that

$$X^n = \sum_{f\in\Sigma^n} (f\mu)f,$$

an equality trivially verified for $n = 0$, and true by definition for $n = 1$. It follows that

$$\begin{aligned} X^{n+1} = X^n \cdot X &= \Big(\sum_{f\in\Sigma^n} (f\mu)f\Big)\cdot\Big(\sum_{a\in\Sigma}(a\mu)a\Big) = \sum_{(f,a)\in\Sigma^n\times\Sigma} (f\mu\cdot a\mu)fa \\ &= \sum_{(f,a)\in\Sigma^n\times\Sigma} (fa)\mu fa = \sum_{g\in\Sigma^{n+1}} (g\mu)g, \end{aligned}$$

since, for each integer n, Σ^{n+1} is in bijection with $\Sigma^n \times \Sigma$ as Σ^* is freely generated by Σ. For the same reason, Σ^* is the *disjoint* union of the Σ^n, for n in $\mathbb{N}$, and it follows, for all f in Σ^*, that

$$(X^*, f) = (X^{|f|}, f) = f\mu. \quad \square$$

Proof (of Theorem 4.6). Each of the two inclusions (19) is proved in the form of a property.

Property 4.8. If Σ is finite, S-recognisable series on Σ^* are S-rational.

Proof. Let (λ, μ, ν) be a representation which recognises a series r; that is, $(r, f) = \lambda \cdot f\mu \cdot \nu$, for all f in Σ^*. Let $\langle\lambda, X, \nu\rangle$ be the automaton defined by

$$X = \sum_{a\in\Sigma} (a\mu)a.$$

By Lemma 4.7, we have

$$r = \sum_{f\in\Sigma^*} (\lambda\cdot f\mu\cdot\nu)f = \lambda\cdot\Big(\sum_{f\in\Sigma^*}(f\mu)f\Big)\cdot\nu = \lambda\cdot X^*\cdot\nu.$$

By the Fundamental Theorem, the series r belongs to the rational closure of the entries of X. These entries are finite linear combinations of elements of Σ since Σ *is finite*: r belongs to $S\mathrm{Rat}\,\Sigma^*$. $\square$

Property 4.9. The S-rational series on Σ^* are S-recognisable.

Proof. By Theorem 3.20, the series r is the behaviour of a *proper* finite S-automaton $\langle I, X, T\rangle$, such that the entries of X are finite linear combinations of elements of Σ (and those of I and T are scalar). We can therefore write $X = \sum_{a\in\Sigma}(a\mu)a$ where $a\mu$ is the matrix of coefficients of the letter a in X. By Lemma 4.7, we have

$$\forall f \in \Sigma^* \quad (r, f) = (I \cdot X^* \cdot T, f) = I \cdot f\mu \cdot T,$$

and the series r is recognised by the *representation* (I, μ, T). □

The two inclusions (19) prove the theorem. □

4.2 Other Products on Recognisable Series

The two products that we shall now consider, the Hadamard and shuffle products are defined on general series—the second one for series on a free monoid—but it is their effect on recognisable series which will interest us, and we first define a product on *representations.*

4.2.1 Tensor Product of S-Representations

The *tensor product* of matrices has been defined in Chap. 1. Let A be a matrix of dimension $P \times P'$ and B a matrix of dimension $R \times R'$ (with entries in the same semiring S); the tensor product of A by B written $A \otimes B$ is a matrix of dimension $(P \times R) \times (P' \times R')$ defined by

$$\forall p \in P,\ \forall p' \in P',\ \forall r \in R,\ \forall r' \in R' \quad A \otimes B_{(p,r),(p',r')} = A_{p,p'} B_{r,r'}.$$

If S is *commutative*, the tensor product is also. We shall need the tensor product to be commutative under more general assumptions. We shall say that two sub-semirings U and V of a non-commutative semiring S are *commutable* if every element of U commutes with every element of V. For example, the *centre* of S and any sub-semiring of S are commutable. As another example, $1_T \times T$ and $T \times 1_T$ are two commutable sub-semirings[14] in $T \times T$. The following result has already been quoted (Chap. 1, Theorem 4.7).

Lemma 4.10. *Let A, B, C, and D be four matrices with entries in S, respectively of dimension $P \times Q$, $P' \times Q'$, $Q \times R$, and $Q' \times R'$, and such that all the entries of B commute with those of C. Then*

$$(A \otimes B) \cdot (C \otimes D) = (A \cdot C) \otimes (B \cdot D).$$

It then follows:

[14] On the other hand, we shall not say that two matrices A and B are *commutable* to mean that all the entries of A commute with those of B; this would be too easily confused with the fact that the two matrices *commute*, that is, $AB = BA$.

Proposition 4.11 (Tensor product of representations). *Let U and V be two commutable sub-semirings of S. Let M and N be two arbitrary monoids and $\mu : M \to U^{Q\times Q}$ and $\kappa : N \to V^{R\times R}$ two representations. The map $\mu \otimes \kappa$, defined for all (m,n) in $M \times N$ by*

$$(m,n)\,[\mu \otimes \kappa] = m\mu \otimes n\kappa$$

is a representation of $M \times N$ in $S^{(Q\times R)\times(Q\times R)}$.

Proof. For all (m,n) and (m',n') in $M \times N$, we have

$$\begin{aligned}\big((m,n)[\mu\otimes\kappa]\big)\cdot\big((m',n')[\mu\otimes\kappa]\big) &= (m\mu\otimes n\kappa)\cdot(m'\mu\otimes n'\kappa)\\ &= (m\mu\cdot m'\mu)\otimes(n\kappa\cdot n'\kappa)\\ &= (mm')\mu\otimes(nn')\kappa = (mm',nn')[\mu\otimes\kappa],\end{aligned}$$

since under the proposition's assumptions, all the entries of $m'\mu$ commute with those of $n\kappa$. □

4.2.2 Hadamard Product

The Hadamard product is to series (sets with multiplicity) what intersection is to sets, which only really makes sense if the semiring of coefficients is commutative. In the same way that the recognisable subsets of an arbitrary monoid are closed under intersection, we have the following.

Theorem 4.12. *Let S be a commutative semiring and M an arbitrary monoid. Then $S\mathrm{Rec}\,M$ is closed under the Hadamard product.*

Under the more precise assumptions of Proposition 4.11, we can state a more general result.

Theorem 4.13 (Schützenberger). *Let U and V be two commutable sub-semirings of S and M a monoid. The Hadamard product of a U-recognisable series over M and a V-recognisable series over M is an S-recognisable series over M.*

More precisely, if (λ,μ,ν) recognises r and (η,κ,ζ) recognises u, then $r \odot u$ is recognised by $(\lambda\otimes\eta, \mu\otimes\kappa, \nu\otimes\zeta)$.

Proof. First note that, since the map $m \mapsto (m,m)$ is a morphism from M to $M \times M$, Proposition 4.11 implies that the map $m \mapsto m\mu \otimes m\kappa$ is also a morphism, and we also write it $\mu \otimes \kappa$.

Let r be a series over M recognised by the U-representation (λ,μ,ν) and u be a series over M recognised by the V-representation (η,κ,ζ). By definition, we have for all m in M,

$$(r\odot u, m) = (\lambda\cdot m\mu\cdot\nu)(\eta\cdot m\kappa\cdot\zeta) = (\lambda\cdot m\mu\cdot\nu)\otimes(\eta\cdot m\kappa\cdot\zeta),$$

the second equality expressing the product of two coefficients of S as the tensor product of two 1×1 matrices. Under the assumptions of the theorem, we can apply Lemma 4.10 (three times) and obtain

$$(r \odot u, m) = (\lambda \otimes \eta) \cdot (m\mu \otimes m\kappa) \cdot (\nu \otimes \zeta) = (\lambda \otimes \eta) \cdot (m[\mu \otimes \kappa]) \cdot (\nu \otimes \zeta) .$$

Again, according to these assumptions, $\mu \otimes \kappa$ is an S-representation, the series $r \odot u$ is recognisable, and is recognised by the stated representation. □

As a consequence of Theorem 4.6, *the Hadamard product of two S-rational series on Σ^* is an S-rational series* (if S is a commutative semiring). Moreover, the tensor product of representations of Σ^* translates directly into a construction on S-automata over Σ^* whose labels are linear combinations of letters of Σ, which is the natural generalisation of the Cartesian product of automata, and which we can call the *Hadamard product* of S-automata.

Example 4.14. The $\mathbb{N}$-automaton $\mathfrak{C}_2$ of Fig. 9 is the Hadamard product of the $\mathbb{N}$-automaton $\mathfrak{C}_1$ of Fig. 3 by itself. Therefore, for every f in Σ^*, it holds $f|\mathfrak{C}_2| = \overline{f}^2$.

4.2.3 Shuffle Product

We now suppose that M is a free monoid Σ^* and that S is commutative (usually $S = \mathbb{N}$ but that is not required). The shuffle product (or Hurwitz product) of two words of Σ^*, and then by linearity of two series in $S\langle\langle\Sigma^*\rangle\rangle$, has been defined at Chap. 1, mostly for ancillary purposes. Let us recall this definition as the interest of which goes far beyond the computations it was used for so far.

Definition 4.15. *For all f and g in Σ^*, the* shuffle *of f and g, written $f \between g$, is an homogeneous polynomial of $S\langle\Sigma^*\rangle$ defined by induction on $|f| + |g|$ by*

$$\forall f \in \Sigma^* \quad f \between \varepsilon = \varepsilon \between f = f,$$
$$\forall f, g \in \Sigma^*,\ \forall a, b \in A \quad fa \between gb = (fa \between g)b + (f \between gb)a,$$

The shuffle is extended 'by linearity' to $S\langle\langle\Sigma^*\rangle\rangle$, that is,

$$\forall r, u \in S\langle\langle\Sigma^*\rangle\rangle \quad r \between u = \sum_{f,g \in \Sigma^*} (r, f)(u, g) f \between g,$$

which is defined since the family of polynomials $f \between g$ for f and g in Σ^* is locally finite.

Example 4.16.

$$ab \between ab = 4aabb + 2abab,$$
$$ab \between ba = abab + 2abba + 2baab + baba \quad \text{and}$$
$$(\varepsilon + a) \between a^* = [a^*]^2.$$

Shuffle is an associative, commutative, and continuous product and makes of $S\langle\!\langle \Sigma^* \rangle\!\rangle$ a commutative S-algebra. The shuffle of two words is characterised by the following.

Proposition 4.17. *Let $\chi \colon \Sigma^* \to S\langle \Sigma^* \times \Sigma^* \rangle$ be the morphism (of monoids) defined by $a\chi = (a, \varepsilon) + (\varepsilon, a)$, for all a in Σ^*. It then follows that*

$$\forall h \in \Sigma^* \quad h\chi = \sum_{f,g \in \Sigma^*} (f \between g, h)(f, g).$$

Theorem 4.18. *Let S be a commutative semiring. The shuffle of two S-recognisable series on Σ^* is an S-recognisable series.*

Proof. Let r and u be S-recognisable series on Σ^*, respectively recognised by the S-representations (λ, μ, ν) and (η, κ, ζ). For all h in Σ^*, the definition yields

$$\begin{aligned}
(r \between u, h) &= \sum_{f,g \in \Sigma^*} \big((r, f)(u, g)\big)(f \between g, h) \\
&= \sum_{f,g \in \Sigma^*} \big((\lambda \cdot f\mu \cdot \nu)(\eta \cdot g\kappa \cdot \zeta)\big)(f \between g, h) \\
&= \sum_{f,g \in \Sigma^*} \big((\lambda \otimes \eta) \cdot ((f, g)[\mu \otimes \kappa]) \cdot (\nu \otimes \zeta)\big)(f \between g, h) \\
&= (\lambda \otimes \eta) \cdot ((h\chi)[\mu \otimes \kappa]) \cdot (\nu \otimes \zeta) \quad \text{by Proposition 4.17.}
\end{aligned}$$

By the theorem's assumptions, $\chi \circ [\mu \otimes \kappa]$ is an S-representation; the series $r \between u$ is recognisable. □

A consequence of Theorem 4.6 again, *the shuffle of two S-rational series on Σ^* is an S-rational series* (if S is a commutative semiring). As for the Hadamard product, the construction on representations that underlies the proof of Theorem 4.18 translates into a construction on S-automata over Σ^*, which we can call the *shuffle product* of S-automata.

Formally, if $\mathfrak{A}' = \langle Q', \Sigma, E', I', T' \rangle$ and $\mathfrak{A}'' = \langle Q'', \Sigma, E'', I'', T'' \rangle$ are two proper S-automata over Σ^* whose labels are linear combinations of letters of Σ, the shuffle of $|\mathfrak{A}'|$ and $|\mathfrak{A}''|$ is realised by the S-automaton written $\mathfrak{A}' \between \mathfrak{A}''$ and defined by

$$\mathfrak{A}' \between \mathfrak{A}'' = \langle Q' \times Q'', \Sigma, E, I' \otimes I'', T' \otimes T'' \rangle ,$$

where the set E of transitions is described by

$$\begin{aligned}
E = &\big\{ ((p', p''), k'a, (q', p'')) \mid (p', k'a, q') \in E' \text{ and } p'' \in Q'' \big\} \\
&\cup \big\{ ((p', p''), k''a, (p', q'')) \mid p' \in Q' \text{ and } (p'', k''a, q'') \in E'' \big\}.
\end{aligned}$$

Example 4.19. The $\mathbb{Z}$-automaton $\mathfrak{W}_1$ of Fig. 8 is the shuffle product of the obvious two state $\mathbb{Z}$-automata that respectively accept $(ab)^*$ and $(-ab)^*$. The equivalence with $\mathfrak{V}_1$ in the same figure yields the identity[15]

$$(ab)^* \between (-ab)^* = \left(-4a^2b^2\right)^*. \tag{20}$$

4.3 Series on a Product of Monoids

Series on a (Cartesian) product of monoids is a major subject in itself and their study could occupy a whole chapter of this book: they are the behaviour of *transducers with multiplicity,* of interest both from a theoretical and applications point of view (cf. Chaps. 7, 11, and 14, for instance). Here, we confine ourselves to few definitions and results stemming from the canonical isomorphisms between several semirings of series and with the aim of being able to state (and to prove) results about the image of series under morphisms and of comparing the families of rational and recognisable series.

4.3.1 The Canonical Isomorphisms

Polynomials or series in several (commutative) variables can be ordered with respect to one or another variable. It is a purely formal exercise to verify that these manipulations generalise to polynomials, or to series, over a product of monoids.

The semialgebras $S\langle\langle M\rangle\rangle$ and $S\langle\langle N\rangle\rangle$ are canonically isomorphic to two sub-S-semi-algebras of $S\langle\langle M\times N\rangle\rangle$: we identify m with $(m, 1_N)$ and n with $(1_M, n)$. This identification enables us to build the two canonical isomorphisms.

Proposition 4.20. *The three S-semi-algebras*

$$S\langle\langle M \times N\rangle\rangle, \qquad [S\langle\langle M\rangle\rangle]\,\langle\langle N\rangle\rangle \qquad \textit{and} \qquad [S\langle\langle N\rangle\rangle]\,\langle\langle M\rangle\rangle$$

are isomorphic. Under these isomorphisms, the three sub-S-semi-algebras

$$S\langle M \times N\rangle, \qquad S\langle M\rangle\langle N\rangle \qquad \textit{and} \qquad [S\langle N\rangle]\,\langle M\rangle$$

correspond.

Remark 4.21. Modulo this canonical embedding *and if S is commutative,* then every element of $S\langle\langle M\rangle\rangle$ commutes with every element of $S\langle\langle N\rangle\rangle$ in $S\langle\langle M\times N\rangle\rangle$.

Definition 4.22. *Let r be in $S\langle\langle M\rangle\rangle$ and u be in $S\langle\langle N\rangle\rangle$. The* tensor product *of r and u, written $r \otimes u$, is the series of $S\langle\langle M \times N\rangle\rangle$ defined by*

$$\forall (m,n) \in M \times N \quad \big(r \otimes u, (m,n)\big) = (r,m)(u,n).$$

[15] Due to M. Petitot (see Sect. 7).

This definition allows the weighted generalisation of a result and is usually credited to Myhill.

Proposition 4.23. *Suppose that S is commutative. A series r of $S\langle\langle M \times N\rangle\rangle$ is recognisable if and only if there exists a finite family $\{j_i\}_{i\in I}$ of series of $S\mathrm{Rec}\, M$ and a finite family $\{u_i\}_{i\in I}$ of series of $S\mathrm{Rec}\, N$ such that*

$$r = \sum_{i\in I} j_i \otimes u_i .$$

Proof. If j is in $S\mathrm{Rec}\, M$, that is, if j is recognised by the representation (λ, μ, ν), the map $(m, n) \mapsto m\mu$ is also a morphism and the series j' of $S\langle\langle M \times N\rangle\rangle$ defined by $(j', (m, n)) = \lambda \cdot m\mu \cdot \nu = (j, m)$ is recognisable. Likewise, if $u \in S\mathrm{Rec}\, N$, the series u' of $S\langle\langle M \times N\rangle\rangle$ defined by $(u', (m, n)) = (u, n)$ is recognisable. Definition 4.22 shows that

$$j \otimes u = j' \odot u' ,$$

which is thus recognisable and Proposition 4.4—hence, we need S to be commutative—implies that the condition is sufficient.

Conversely, suppose that r is recognised by (λ, μ, ν), a representation of $M \times N$ of dimension Q. By definition of a representation, for all (m, n) in $M \times N$, it holds $(m, n)\mu = (m, 1_N)\mu(1_M, n)\mu$. The map $\mu' \colon M \to S^{Q\times Q}$ defined by $m\mu' = (m, 1_N)\mu$ is a morphism. For each q in Q, let j_q be the series defined by

$$\forall m \in M \quad (j_q, m) = [\lambda \cdot m\mu']_q ,$$

which is a recognisable series of $S\langle\langle M\rangle\rangle$. Likewise, $\mu'' \colon N \to S^{Q\times Q}$ defined by $n\mu'' = (1_M, n)\mu$ is a morphism and u_q defined by

$$\forall n \in N \quad (u_q, n) = [n\mu'' \cdot \nu]_q ,$$

is a recognisable series of $S\langle\langle N\rangle\rangle$. Since for all (m, n) of $M \times N$, we have

$$\lambda \cdot (m, n)\mu \cdot \nu = \sum_{q\in Q} [\lambda \cdot m\mu']_q [n\mu'' \cdot \nu]_q ,$$

it follows that

$$r = \sum_{q\in Q} j_q \otimes u_q . \quad \square$$

4.3.2 Rational Series in a Product

The Fundamental Theorem of (S-)automata for series in $S\langle\langle M \times N\rangle\rangle$ directly yields (weighted and generalised version of a theorem by Elgot and Mezei [16]) the following.

Proposition 4.24. *Let G and H be generating sets of M and N, respectively. A series of $S\langle\langle M \times N\rangle\rangle$ is rational if and only if it is the behaviour of a proper finite S-automaton whose coefficients are S-linear combinations of elements of $(G \times 1_N) \cup (1_M \times H)$.*

Proposition 4.25. *The canonical isomorphism from $S\langle\langle M \times N\rangle\rangle$ to $[S\langle\langle N\rangle\rangle]\langle\langle M\rangle\rangle$ sends $S\mathrm{Rat}(M \times N)$ to $[S\mathrm{Rat}\, N]\mathrm{Rat}\, M$.*

Proof. From the inclusion

$$S\langle N\rangle \subseteq S\langle M\rangle N \subseteq S\mathrm{Rat}(M \times N),$$

we deduce successively, by liberal use of the canonical embeddings,

$$\begin{aligned} S\mathrm{Rat}\, N &\subseteq S\mathrm{Rat}(M \times N), \\ [S\mathrm{Rat}\, N]\langle M\rangle &\subseteq S\mathrm{Rat}(M \times N), \\ [S\mathrm{Rat}\, N]\mathrm{Rat}\, M &\subseteq S\mathrm{Rat}(M \times N). \end{aligned}$$

Conversely, let r be in $S\mathrm{Rat}(M \times N)$. There exists a proper S-automaton $\langle I, X, T\rangle$ such that $r = I \cdot X^* \cdot T$ and such that the coefficients of X are finite S-linear combinations of elements of $(M \times 1) \cup (1 \times N)$. We write $X = Y + Z$, in such a way that the coefficients of Y are linear combinations of elements of $M \times 1$ and those of Z are linear combinations of elements of $1 \times N$ (with coefficients in S). The series r is the result of the automaton $\langle I, Z^* \cdot Y, Z^* \cdot T\rangle$ whose coefficients are linear combinations of elements of $M \times 1$, with coefficients in $1 \times S\mathrm{Rat}\, N$. □

The specialisation of this proposition when M is a free monoid gives the weighted version of what is often known as the 'Kleene–Schützenberger theorem for rational relations' (cf. Corollary 4.29). We shall state it after the definition of *weighted relations.*

4.3.3 Weighted Relations

We first need a few more definitions and notation. We write S_{c} for the *centre* of S, that is, the set of elements of S which commute with every element of S—S_{c} is a sub-semiring of S. In any case, 1_S belongs to S_{c}, which is thus never empty.

The *scalar product*[16] of two series r and u in $S\langle\langle M\rangle\rangle$, written (r, u) is defined by

$$(r, u) = \sum_{m \in M} (r, m)(u, m),$$

which may or may not be defined since the family $\{(r, m)(u, m) \mid m \in M\}$ is not necessarily summable. It is defined if r or u is a polynomial. The identification of m with its *characteristic series* $\underline{m}$ makes this notation consistent

[16] Different from what is called the *scalar product* in Chap. 1.

with the notation (r, m) for the coefficient of m in r. Even *if S is not commutative*, but if r *or* u belong to $S_\mathsf{c}\langle\langle M\rangle\rangle$, we have $(r, u) = (u, r)$. In this case, the scalar product is even compatible with left and right multiplication by arbitrary elements of S:

$$k(r, u) = (kr, u),$$
$$(r, u)k = (u, r)k = (u, rk) = (rk, u).$$

Definition 4.26. *An S-*relation *from M to N, written $\theta\colon S\langle\langle M\rangle\rangle \to S\langle\langle N\rangle\rangle$, or more often $\theta\colon M \to N$, is any series θ of $[S_\mathsf{c}\langle\langle N\rangle\rangle]\,\langle\langle M\rangle\rangle$.*

The image *of every m in M under θ is the series (θ, m) of $S\langle\langle N\rangle\rangle$, written* more simply *$m\theta$.*

The image of every r in $S\langle\langle M\rangle\rangle$ under θ, denoted $r\theta$, is then obtained 'by linearity'. It is defined if and only if the family $\{(r, m)(\theta, m) \mid m \in M\}$ is a summable family *of series of $S\langle\langle N\rangle\rangle$ and is its sum.*

The *graph* $\widehat{\theta}$ of an S-relation θ is the series of $S_\mathsf{c}\langle\langle M \times N\rangle\rangle$ which corresponds to θ under the canonical isomorphism. The *inverse* of θ, namely θ^{-1}, is the S-relation from N to M, and hence a series of $[S_\mathsf{c}\langle\langle M\rangle\rangle]\,\langle\langle N\rangle\rangle$, which has the same graph $\widehat{\theta}$ as θ. It then holds

$$\forall (m, n) \in M \times N \quad (m\theta, n) = \big(\widehat{\theta}, (m, n)\big) = \big(m, n\theta^{-1}\big). \tag{21}$$

Remark 4.27. Instead of assuming that the semiring of coefficients is commutative, we have 'only' imposed the condition that the coefficients of the relation, $\widehat{\theta}$, belong to the *centre* of this semiring. This could seem a rather weak generalisation; in fact, it allows us first and foremost to consider, as S-relations from M to N, the *characteristic relations* of relations from M to N, even if S is not commutative.

Example 4.28. For every series u in $S_\mathsf{c}\langle\langle M\rangle\rangle$, and in particular for every characteristic series u, the Hadamard product with u (or *S-intersection with u*) is an S-relation from M to itself, written ι_u: $r\iota_u = r \odot u$ and $r\iota_u$ is defined for all r in $S\langle\langle M\rangle\rangle$.

It is then natural to say that an S-relation from M to N is *rational* if its graph is a S_c-rational series of $S\langle\langle M \times N\rangle\rangle$. And the announced specialisation of Proposition 4.25 then reads as the following corollary.

Corollary 4.29 (Kleene–Schützenberger). *An S-relation θ from Σ^* to N is rational if and only if there exists an $(S_\mathsf{c}\mathrm{Rat}\,N)$-representation of Σ^*, namely (λ, μ, ν), such that for all f in Σ^*, $f\theta = \lambda \cdot f\mu \cdot \nu$, that is,*

$$S_\mathsf{c}\mathrm{Rat}(\Sigma^* \times N) \cong [S_\mathsf{c}\mathrm{Rat}\,N]\mathrm{Rec}\,\Sigma^*.$$

Example 4.30. The rational $\mathbb{B}$-relation from $\Sigma^* = \{a, b\}^*$ into itself realised by the transducer of Fig. 11 is also realised by the $[\mathbb{B}\mathrm{Rat}\,\Sigma^*]$-representation of Σ^* of dimension 1 $(1, \mu, 1)$ with $a\mu = aa^*$ and $b\mu = bb^*$.

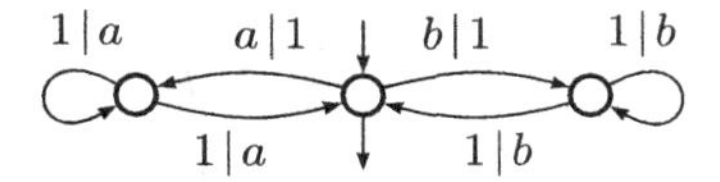

Fig. 11. A transducer to be transformed into a representation

It follows from (21) that the image $r\theta$ of a series r in $S\langle\langle M\rangle\rangle$ by an S-relation θ from M to N is defined if and only if $(r, n\theta^{-1})$ is defined for every n in N, and we have

$$(r\theta, n) = \big(r, n\theta^{-1}\big).$$

Hence, we have the following definition.

Definition 4.31. *We say that an S-relation $\theta\colon M \to N$ is* of finite co-image *if $n\theta^{-1}$ is a* polynomial *for all n.*

The image of any series by a relation of finite co-image is always defined, and this is the case that we shall only consider here. *Regulated relations* which were defined by Jacob starting from their representations as in Corollary 4.29 are relations of finite co-image; they were popularised by a number of authors inspired by Jacob's work (cf. Chap. 7, Sect. 4).

Proposition 4.32. *Let M and N be two graded monoids. An S-relation θ: $M \to N$ with finite co-image is continuous.*

4.3.4 Morphic Image of Recognisable and Rational Series

An S-relation $\theta\colon M \to N$ is *multiplicative* if its restriction to M is a morphism to $S\langle\langle N\rangle\rangle$, viewed as a multiplicative monoid. The definition of S-relations implies in fact that θ is a morphism from M to $S_{\mathsf{c}}\langle\langle N\rangle\rangle$. In particular, the characteristic relation $\underline{\theta}$ of a morphism θ from M to N is a multiplicative S-relation.

We begin with a weighted generalisation of a theorem on recognisable sets.

Proposition 4.33. *Let $\theta\colon M \to N$ be a morphism of monoids and u an S-recognisable series on N. Then $u\underline{\theta}^{-1}$ is an S-recognisable series on M.*

Proof. By assumption, there exists (λ, μ, ν), an S-representation of N, such that for all n in N, $(u, n) = (\lambda \cdot n\mu \cdot \nu)$. Whence, for all m in M,

$$\big(u\underline{\theta}^{-1}, m\big) = (u, m\theta) = \lambda \cdot (m\theta)\mu \cdot \nu.$$

Thus, the S-representation of M $(\lambda, \theta\mu, \nu)$ recognises the series $u\underline{\theta}^{-1}$. □

The hypothesis that the coefficients of an S-relation are taken in S_{c} allows us to establish the following.

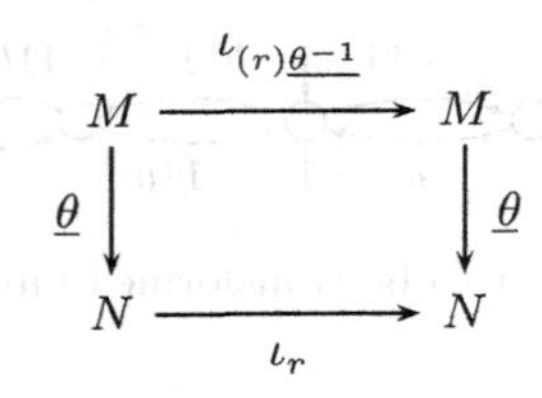

Fig. 12. Lifting of S-intersection with r

Proposition 4.34. *If $\theta\colon M \to N$ is a multiplicative S-relation, then θ is a morphism of S-semialgebras, from $S\langle M\rangle$ to $S\langle\langle N\rangle\rangle$.*

Let M and N be two graded monoids. Let $\theta\colon M \to N$ be a multiplicative S-relation; if for all m in M, $m\theta$ is a *proper series* of $S\langle\langle N\rangle\rangle$, then θ is of finite co-image, hence is defined on all of $S\langle\langle M\rangle\rangle$ and is continuous. In particular, a monoid morphism $\theta\colon M \to N$ is *continuous* if $m\theta \neq 1_N$ for all m in M and then the S-relation $\underline{\theta}$ is a continuous morphism of S-semi-algebras from $S\langle\langle M\rangle\rangle$ to $S\langle\langle N\rangle\rangle$. It follows that if r is in $S\langle\langle M\rangle\rangle$, r^* is defined if and only if $(r\underline{\theta})^*$ is defined and we have $(r^*)\underline{\theta} = (r\underline{\theta})^*$. And the following theorem then holds.

Theorem 4.35. *Let M and N be graded monoids and $\theta\colon M \to N$ a* continuous *morphism of monoids.*

(i) If $r \in S\mathrm{Rat}\,M$, then $r\underline{\theta} \in S\mathrm{Rat}\,N$.
(ii) If θ is surjective and $u \in S\mathrm{Rat}\,N$, then there exists $r \in S\mathrm{Rat}\,M$ such that $r\underline{\theta} = u$.

Example 4.36. Let $\alpha\colon \Sigma^* \to M$ be a surjective morphism; a set R of Σ^* is a *cross-section* of Σ^* for α if α is injective over R and $R\alpha = M$, that is, if $\underline{M} = (\underline{R})\underline{\alpha}$. A monoid M is *rationally enumerable* if such an R exists that is a rational subset of Σ^*.

It easily comes that M is rationally enumerable if and only if it is an unambiguous rational subset of itself: $M \in \mathrm{URat}\,M$, that is, $\underline{M} \in \mathbb{N}\mathrm{Rat}\,M$ and then $\underline{M} \in S\mathrm{Rat}\,M$ for any S.

We prove a last lemma before the result we are aiming at.

Lemma 4.37. *Let $\theta\colon M \to N$ be a function and r a S-series on N. We have (cf. diagram in Fig. 12)*

$$\underline{\theta}\iota_r = \iota_{r\underline{\theta}^{-1}}\underline{\theta}.$$

Proof. For every m in M, we have

$$\begin{aligned}(m\iota_{r\underline{\theta}^{-1}})\underline{\theta} &= (r\underline{\theta}^{-1} \odot m)\underline{\theta} = ((r\underline{\theta}^{-1}, m)m)\underline{\theta} \\ &= (r, m\underline{\theta})m\underline{\theta} = r \odot m\underline{\theta} = (m\underline{\theta})\iota_r. \quad \square\end{aligned}$$

Theorem 4.38. *Let U and V be two commutable sub-semirings of S, u in $V\mathrm{Rat}\,N$ and r in $U\mathrm{Rec}\,N$. Then the Hadamard product of u and r is an S-rational series on N.*

Proof. As N is finitely generated there exists a finite alphabet Σ and a surjective continuous morphism $\theta\colon \Sigma^* \to N$. By Theorem 4.35(ii), there exists a series u in $V\mathrm{Rat}\,\Sigma^*$ such that

$$u\underline{\theta} = u.$$

The coefficients of u commute with those of r, and hence with those of $r\underline{\theta^{-1}}$. Lemma 4.37 allows us to rewrite the equality $r \odot u = r \odot u\underline{\theta}$ as

$$r \odot u = \left[r\underline{\theta^{-1}} \odot u\right]\underline{\theta}. \tag{22}$$

Proposition 4.33 ensures that $r\underline{\theta^{-1}}$ is U-recognisable (on Σ^*), Theorem 4.13 that $r\underline{\theta^{-1}} \odot u$ is S-recognisable, hence S-rational, and finally (22) and Theorem 4.35(i) that $r \odot u$ is S-rational on N. □

Corollary 4.39. *If M is rationally enumerable, then $S\mathrm{Rec}\,M \subseteq S\mathrm{Rat}\,M$.*

Proof. By hypothesis (cf. Example 4.36), $\underline{M} \in S\mathrm{Rat}\,M$ for any S. We have $r \odot \underline{M} = r$ for all r in $S\langle\langle M\rangle\rangle$ and we apply Theorem 4.38. □

Corollary 4.39 is the weighted generalisation of a theorem by McKnight [35], Theorem 4.38, the one of a classical result on subsets of a monoid. As for subsets also, the morphic image of a recognisable series is not necessarily recognisable, the inverse morphic image of a rational series is not necessarily rational.

We stop here with the theory of weighted relations, which could, of course, be further developed. In particular, the composition and evaluation theorems hold for weighted rational relations (cf. [24, 45, 43]). But our aim here was just to set the framework in which we could establish Theorems 4.35, 4.38, and Corollary 4.39, and in the next section, Corollary 5.32.

5 Series over a Free Monoid

So far, we have developed the theory of rational series under the assumption that M is *graded* (so that we knew how to define star). In our presentation, the Kleene–Schützenberger theorem and recognisable series appeared as a last touch added to the fundamental theorem of automata in the case of free monoids. We now require M to be a *free monoid* and change our point of view: rational and recognisable series coincide and somehow recognisable series and their representations become the main subject.

The whole thing takes an algebraic turn. We first give another characterisation of recognisable series, and then under the hypothesis that the semiring of weights is a field, we develop the theory of reduction (that is, minimisation) of representations. In a third subsection, we review a number of applications of this reduction theory—and first of all, the decidability of equivalence—which in many instances, do not apply only to the case of weights in a field, but also in *any sub-semiring* of a field.

5.1 The Representability Theorem

Representations define recognisable series; we first show how, by means of the *quotient operation*, we can recover a representation from a series when it is recognisable. This is an abstract view since a series is an infinite object; we then give an effective implementation of this result, starting from a rational expression that denotes a rational series; this is another proof of one direction of the Kleene–Schützenberger theorem.

5.1.1 Characterisation of Recognisable Series

The (left) quotient of a series is the generalisation to series of the (left) quotient of a subset of a monoid (a free monoid in this case).

The free monoid Σ^* *acts by quotient* on $S\langle\langle\Sigma^*\rangle\rangle$: for all f in Σ^* and all series r in $S\langle\langle\Sigma^*\rangle\rangle$, the series $f^{-1}r$ is defined by

$$f^{-1}r = \sum_{g\in\Sigma^*} (r, fg)g, \qquad \text{that is,} \qquad \forall g \in \Sigma^* \quad (f^{-1}r, g) = (r, fg),$$

and in particular

$$\forall f \in \Sigma^* \quad (f^{-1}r, \varepsilon) = (r, f). \tag{23}$$

As the definition says, the quotient is an *action*, that is,

$$\forall f, g \in \Sigma^* \quad (fg)^{-1}r = g^{-1}[f^{-1}r],$$

and for every given f, the operation $r \mapsto f^{-1}r$ is an *endomorphism* of the S-semi-module $S\langle\langle\Sigma^*\rangle\rangle$: it is *additive*:

$$f^{-1}(r + u) = f^{-1}r + f^{-1}u,$$

and *commutes with the exterior multiplications* of S on $S\langle\langle\Sigma^*\rangle\rangle$:

$$f^{-1}(kr) = k(f^{-1}r) \quad \text{and} \quad f^{-1}(rk) = (f^{-1}r)k.$$

Moreover, it is *continuous*. These three properties ensure that the operation of quotient by f is entirely defined on $S\langle\langle\Sigma^*\rangle\rangle$ by its values on Σ^*.

Example 5.1. Let $r_2 = (\underline{a^*})^2 = \sum_{k\in\mathbb{N}}(k+1)a^k$ in $\mathbb{N}\mathrm{Rat}\, a^*$. For every integer n, we have

$$(a^n)^{-1}r_2 = \sum_{k\in\mathbb{N}} (n + k + 1)a^k = r_2 + n\underline{a^*}.$$

All quotients of r_2 are distinct.

Example 5.1 shows that, in general, and unlike the case for (recognisable) languages, the family of quotients of a rational, and thus recognisable series is not necessarily finite. On the other hand, and despite its simplicity, it exhibits the property that we seek: of course, there are infinitely many quotients, but they can all be expressed as the linear combination of a *finite number* of suitably chosen series.

Definition 5.2. *A subset U of $S\langle\langle\Sigma^*\rangle\rangle$ is called* stable *if it is closed under quotient; that is, for all r in U and all f in Σ^*, $f^{-1}r$ is still in U.*

Theorem 5.3. *A series on Σ^* with coefficients in S is S-recognisable if and only if it is contained in a finitely generated stable subsemimodule of $S\langle\langle\Sigma^*\rangle\rangle$.*

To allow later references to parts of the proof of this result, it is split into more precise properties and definitions. Let us begin with a notation: Lemma 4.7 shows how close automata and representations are. We shall thus denote the latter in the same way as the former by uppercase gothic letters.

Definition 5.4. *With every S-representation $\mathfrak{A} = (\lambda, \mu, \nu)$ of dimension Q we associate a morphism of S-semimodules $\Phi_{\mathfrak{A}} : S^Q \to S\langle\langle\Sigma^*\rangle\rangle$ by*

$$\forall x \in S^Q \quad (x)\Phi_{\mathfrak{A}} = |(x, \mu, \nu)| = \sum_{f\in\Sigma^*} (x \cdot f\mu \cdot \nu) f .$$

Proposition 5.5. *If r is a series realised by $\mathfrak{A} = (\lambda, \mu, \nu)$, then $\operatorname{Im}\Phi_{\mathfrak{A}}$ is a stable (finitely generated) subsemi-module of $S\langle\langle\Sigma^*\rangle\rangle$ that contains r.*

Proof. The subsemimodule $\operatorname{Im}\Phi_{\mathfrak{A}}$ is finitely generated since S^Q is, and it is stable since for all f in Σ^* and all x in S^Q we have

$$f^{-1}\left[(x)\Phi_{\mathfrak{A}}\right] = (x \cdot f\mu)\Phi_{\mathfrak{A}} ,$$

and contains $r = (\lambda)\Phi_{\mathfrak{A}}$. □

Proposition 5.6. *Let U be a stable subsemimodule of $S\langle\langle\Sigma^*\rangle\rangle$ generated by $G = \{g^{(1)}, g^{(2)}, \ldots, g^{(n)}\}$. Then every series in U is an S-recognisable series of $S\langle\langle\Sigma^*\rangle\rangle$, realised by a representation of dimension n.*

Proof. The set G canonically defines a linear map from S^n onto U:

$$x = (x_1, x_2, \ldots, x_n) \longmapsto x \cdot G = x_1 g^{(1)} + x_2 g^{(2)} + \cdots + x_n g^{(n)} .$$

A series u belongs to U means that there exists at least one x in S^n such that $u = x \cdot G$.

If U is stable, for every a in Σ, and every i, $a^{-1}g^{(i)}$ belongs to U and there exists a vector $m^{(i)}$ in S (at least one) such that $a^{-1}g^{(i)} = m^{(i)} \cdot G$. Let $a\mu$ be the $n \times n$-matrix whose ith row is $m^{(i)}$. As the quotient by a is a linear map, for any u in U, $u = x \cdot G$ it holds $a^{-1}u = (x \cdot a\mu) \cdot G$. These matrices $a\mu$, for a

in Σ, define a representation of Σ^* and as the quotient is an action of Σ^*, for every f in Σ^*, it holds $f^{-1}u = (x \cdot f\mu) \cdot G$.

From (23), follows then $(u, f) = (f^{-1}u, \varepsilon) = ((x \cdot f\mu) \cdot G, \varepsilon)$ and u is realised by the representation $(x, \mu, (G, \varepsilon))$ where (G, ε) denotes the (column) vector $((g^{(1)}, \varepsilon), (g^{(2)}, \varepsilon), \ldots, (g^{(n)}, \varepsilon))$. □

Propositions 5.5 and 5.6 together prove Theorem 5.3.

5.1.2 Derivation of Rational S-Expressions

The *derivation* of rational S-expressions is the lifting to the level of expressions of the quotient of series and will enable us to effectively implement Theorem 5.3: the derived terms of an expression denote a set of generators of a stable subsemimodule that contains the series denoted by the expression. It will give us the weighted generalisation of Antimirov's construction on rational expressions [1]; this is another example where taking multiplicities into account yield better understanding of constructions and results on languages.

S-Derivatives

For the rest of this subsection, addition in S is written $\oplus$ to distinguish it from the $+$ operator in expressions. The addition induced on $S\langle\langle \Sigma^* \rangle\rangle$ is also written $\oplus$. The set of left linear combinations of S-expressions with coefficients in S, or polynomials of $S\langle S\,\mathsf{RatE}\,\Sigma^* \rangle$, is a left S-semi-module on S:

$$k\mathsf{E} \oplus k'\mathsf{E}' \equiv k'\mathsf{E}' \oplus k\mathsf{E} \quad \text{and} \quad k\mathsf{E} \oplus k'\mathsf{E} \equiv [k \oplus k']\,\mathsf{E}. \qquad (\boldsymbol{B_K})$$

In the following, $[k\,\mathsf{E}]$ or $k\,\mathsf{E}$ is a monomial in $S\langle S\,\mathsf{RatE}\,\Sigma^* \rangle$ whereas $(k\,\mathsf{E})$ is an expression in $S\,\mathsf{RatE}\,\Sigma^*$.

As it is the case in general for semi-modules, there is no multiplication defined on $S\langle S\,\mathsf{RatE}\,\Sigma^* \rangle$. However, an external right multiplication of an element of $S\langle S\,\mathsf{RatE}\,\Sigma^* \rangle$ by an expression and by a scalar is needed. This operation is first defined on monomials and then extended to polynomials by linearity:

$$([k\,\mathsf{E}] \cdot \mathsf{F}) \equiv k\,(\mathsf{E} \cdot \mathsf{F}), \qquad ([k\,\mathsf{E}]\,k') \equiv k\,(\mathsf{E}\,k'),$$
$$([\mathsf{E} \oplus \mathsf{E}'] \cdot \mathsf{F}) \equiv (\mathsf{E} \cdot \mathsf{F}) \oplus (\mathsf{E}' \cdot \mathsf{F}), \qquad ([\mathsf{E} \oplus \mathsf{E}']\,k) \equiv (\mathsf{E}\,k) \oplus (\mathsf{E}'\,k).$$

This multiplication on $S\langle S\,\mathsf{RatE}\,\Sigma^* \rangle$ is *not associative*—since the product operator in expression is not—but is consistent with interpretation: the series denoted by the left-hand sides and right-hand sides are equal.

Definition 5.7. *Let* E *be in* $S\,\mathsf{RatE}\,\Sigma^*$ *and let* a *be in* Σ. *The* S-derivative *of* E *with respect to* a, *denoted by* $\frac{\partial}{\partial a}\,\mathsf{E}$, *is a polynomial of rational expressions with coefficients in* S, *defined inductively by the following formulas.*

$$\frac{\partial}{\partial a} 0 = \frac{\partial}{\partial a} 1 = 0, \qquad \frac{\partial}{\partial a} b = \begin{cases} 1 & \text{if } b = a, \\ 0 & \text{otherwise,} \end{cases}$$

$$\frac{\partial}{\partial a}(k\,\mathsf{E}) = k\,\frac{\partial}{\partial a}\,\mathsf{E}, \qquad \frac{\partial}{\partial a}(\mathsf{E}\,k) = \left(\left[\frac{\partial}{\partial a}\,\mathsf{E}\right]\,k\right),$$

$$\frac{\partial}{\partial a}(\mathsf{E}{+}\mathsf{F}) = \frac{\partial}{\partial a}\,\mathsf{E} \oplus \frac{\partial}{\partial a}\,\mathsf{F}, \tag{24}$$

$$\frac{\partial}{\partial a}(\mathsf{E}\cdot\mathsf{F}) = \left(\left[\frac{\partial}{\partial a}\,\mathsf{E}\right]\cdot\mathsf{F}\right) \oplus \mathsf{c}(\mathsf{E})\frac{\partial}{\partial a}\,\mathsf{F}, \tag{25}$$

$$\frac{\partial}{\partial a}(\mathsf{E}^*) = \mathsf{c}(\mathsf{E})^*\left(\left[\frac{\partial}{\partial a}\,\mathsf{E}\right]\cdot(\mathsf{E}^*)\right). \tag{26}$$

The derivative of a polynomial of expressions is defined by linearity:

$$\frac{\partial}{\partial a}\left(\bigoplus_{i\in I} k_i\,\mathsf{E}_i\right) = \bigoplus_{i\in I} k_i\,\frac{\partial}{\partial a}\,\mathsf{E}_i. \tag{27}$$

Implicitly, the (polynomials of) expressions are reduced with trivial identities, for instance,

$$\frac{\partial}{\partial a}\,\mathsf{E} = 1 \quad\Longrightarrow\quad \frac{\partial}{\partial a}(\mathsf{E}\cdot\mathsf{F}) = \mathsf{F} \oplus \mathsf{c}(E)\frac{\partial}{\partial a}\,\mathsf{F}.$$

Notice that (26) is defined only if (E^*) is a valid expression. The S-derivative of an expression with respect to a *word* f is defined by induction on the length of f:

$$\forall f \in \Sigma^+,\ \forall a \in \Sigma \quad \frac{\partial}{\partial fa}\,\mathsf{E} = \frac{\partial}{\partial a}\left(\frac{\partial}{\partial f}\,\mathsf{E}\right). \tag{28}$$

The definition of S-derivatives of S-expressions is consistent with that of quotient of series, as expressed by the following.

Proposition 5.8. $\forall \mathsf{E} \in S\,\mathsf{RatE}\,\Sigma^*,\ \forall f \in \Sigma^+$

$$\left|\frac{\partial}{\partial f}(\mathsf{E})\right| = f^{-1}|\mathsf{E}|.$$

The Derived Term Automaton

Definition 5.9. *The set* $\mathrm{TD}(\mathsf{E})$ *of* true derived terms *of an expression* E *in* $S\,\mathsf{RatE}\,\Sigma^*$ *is inductively defined by the following rules:*

$$\mathrm{TD}(0) = \mathrm{TD}(1) = \emptyset, \qquad \forall a \in \Sigma \quad \mathrm{TD}(a) = \{1\},$$

$$\forall k \in S,\quad \mathrm{TD}(k\,\mathsf{E}) = \mathrm{TD}(\mathsf{E}),\quad \mathrm{TD}(\mathsf{E}\,k) = \bigcup_{\mathsf{K}\in\mathrm{TD}(\mathsf{E})} (\mathsf{K}\,k),$$

$$\mathrm{TD}(\mathsf{E}+\mathsf{F}) = \mathrm{TD}(\mathsf{E}) \cup \mathrm{TD}(\mathsf{F}),$$

$$\mathrm{TD}(\mathsf{E}\cdot\mathsf{F}) = \left[\bigcup_{\mathsf{K}\in\mathrm{TD}(\mathsf{E})} (\mathsf{K}\cdot\mathsf{F})\right] \cup \mathrm{TD}(\mathsf{F}),$$

$$\mathrm{TD}(\mathsf{E}^*) = \bigcup_{\mathsf{K}\in\mathrm{TD}(\mathsf{E})} (\mathsf{K}\cdot(\mathsf{E}^*)).$$

It follows from the definition that $\mathrm{TD}(\mathsf{E})$ is a *finite* set of monomials of $S\langle S\,\mathsf{RatE}\,\Sigma^*\rangle$, whose cardinal is smaller than or equal to $\ell(\mathsf{E})$. The reason for the two distinct definitions (Definitions 5.7 and 5.9), which may look redundant will be explained below.

The expression E itself does not belong necessarily to $\mathrm{TD}(\mathsf{E})$ and we define the set of *derived terms* of E to be: $\mathrm{D}(\mathsf{E}) = \mathrm{TD}(\mathsf{E}) \cup \{\mathsf{E}\}$. A mechanical induction on the depth of the expressions establishes then the following.

Theorem 5.10. *Let* $\mathrm{D}(\mathsf{E}) = \{\mathsf{K}_1, \ldots, \mathsf{K}_n\}$ *be the set of derived terms of an expression* E *in* $S\,\mathsf{RatE}\,\Sigma^*$. *For every letter* a *in* Σ, *there exists an* $n \times n$-*matrix* $a\mu$ *with entries in* S *such that*

$$\forall i \in [n] \qquad \frac{\partial}{\partial a}\mathsf{K}_i = \bigoplus_{j\in[n]} a\mu_{i,j}\mathsf{K}_j .$$

From (28), it then follows, by induction on the length of words.

Corollary 5.11. *For every word* f *in* Σ^*, *the* S-*derivative of any expression* E *in* $S\,\mathsf{RatE}\,\Sigma^*$ *with respect to* f *is a linear combination of derived terms of* E.

The statement of Theorem 5.10 is in itself the definition of an S-representation $\mathfrak{A}_{\mathsf{E}} = (\lambda, \mu, \nu)$ of dimension $\mathrm{D}(\mathsf{E})$ if we add

$$\lambda_i = \begin{cases} 1_S & \text{if } \mathsf{K}_i = \mathsf{E}, \\ 0_S & \text{otherwise,} \end{cases} \qquad \text{and} \qquad \nu_j = \mathsf{c}(\mathsf{K}_j).$$

We also write $\mathfrak{A}_{\mathsf{E}}$ for the S-automaton $\langle \lambda, X, \nu \rangle$ where $X = \bigoplus_{a\in\Sigma} a\mu a$ and call it *the derived term automaton* of E.

Proposition 5.12. *Let* E *be in* $S\,\mathsf{RatE}\,\Sigma^*$. *Then* $|\mathfrak{A}_{\mathsf{E}}| = |\mathsf{E}|$.

Derivation is thus another means to build an automaton from an expression, different from the one we have seen in the course of the proof of Theorem 3.20 which yielded the *standard automaton* of the expression. The two constructions are related by the following, which is the weighted generalisation of a theorem by Champarnaud and Ziadi [10].

Theorem 5.13 ([33]). *Let* E *be in* $S\,\mathsf{RatE}\,\Sigma^*$. *Then* $\mathfrak{S}_{\mathsf{E}}$ *is an* S-*covering of* $\mathfrak{A}_{\mathsf{E}}$.

Remark 5.14. Definitions 5.7 and 5.9 are both based on an induction on the depth of the expression and then reunited by Theorem 5.10 and Corollary 5.11. It seems that it could be possible, and more natural, to define the derived terms of E as the monomials that appear in the S-derivatives of E.

The problem is that this is not always true if S is not a positive semiring: some derived terms may never appear in an S-derivative—as it can be observed for instance with the $\mathbb{Z}$-expression $\mathsf{E}_5 = (1 - a)a^*$ (cf. Fig. 13). And with such a definition of derived terms, more utilitarian than structural, Theorem 5.13 would not hold anymore.

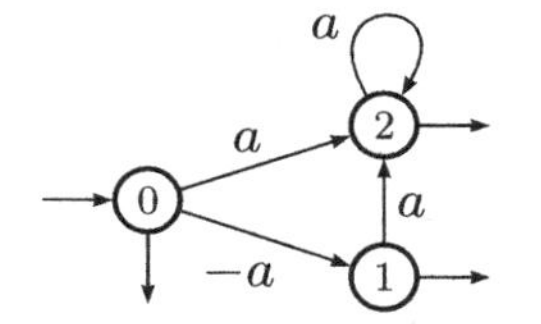

a The standard automaton: $\mathfrak{S}_{\mathsf{E}_5}$

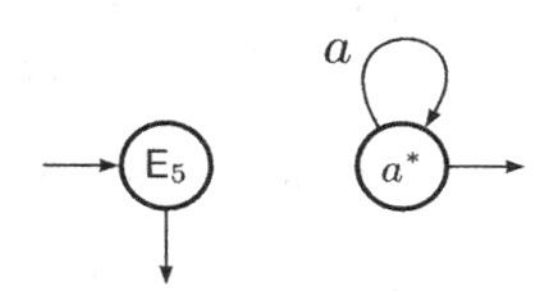

b The derived term automaton: $\mathfrak{A}_{\mathsf{E}_5}$

Fig. 13. Two $\mathbb{Z}$-automata for E_5

5.2 Reduced Representations

We now suppose that S is a *field*, not necessarily commutative, hence a *skew field*, or *division ring*. The preceding considerations about quotients of series will take on, we might say, a new dimension since the ring of series $S\langle\langle\Sigma^*\rangle\rangle$ is not only an S-algebra, but a left and right S-*vector space*, and the notion of *dimension* of subspaces will give us a new invariant.

5.2.1 Rank of a Series

Definition 5.15. *Let S be a division ring. The* rank *of a series r of $S\langle\langle\Sigma^*\rangle\rangle$ is the dimension of the subspace of $S\langle\langle\Sigma^*\rangle\rangle$ generated by the (left) quotients of r.*

In this setting, and with no further ado, Theorem 5.3 becomes the following theorem.

Theorem 5.16. *A series r over Σ^* with coefficients in a division ring is recognisable if and only its rank is finite.*

From Definition 5.4 and Proposition 5.5, it follows that if r is a series realised by an S-representation $\mathfrak{A} = (\lambda, \mu, \nu)$ of dimension n, the rank of r is smaller than or equal to $\mathsf{dim}(\mathsf{Im}\,\Phi_{\mathfrak{A}})$ which is smaller than or equal to n, that is, the rank of a recognisable series r is smaller than, or equal to, the dimension of any representation that realises it.

Definition 5.17. *A representation of a recognisable series r is* reduced *if its dimension is minimal, equal to the rank of r.*

From Proposition 5.6, it follows that with every base of the subspace generated by the quotients of r is associated a reduced representation. The reduced representations will be characterised by means of the following definition. With every S-representation $\mathfrak{A} = (\lambda, \mu, \nu)$ of dimension Q, we associate the morphism of S-semi-modules $\Psi_{\mathfrak{A}} : S\langle\Sigma^*\rangle \to S^Q$ defined by

$$\forall f \in \Sigma^* \quad (f)\Psi_{\mathfrak{A}} = \lambda \cdot f\mu .$$

Theorem 5.18. *An S-representation $\mathfrak{A} = (\lambda, \mu, \nu)$ is reduced if and only if $\Psi_{\mathfrak{A}}$ is surjective and $\Phi_{\mathfrak{A}}$ injective.*

Proof. Let r be the series realised by $\mathfrak{A}$. The morphism

$$\Psi_{\mathfrak{A}} \circ \Phi_{\mathfrak{A}} : S\langle\Sigma^*\rangle \to S\langle\langle\Sigma^*\rangle\rangle \quad \text{is such that} \quad (f)[\Psi_{\mathfrak{A}} \circ \Phi_{\mathfrak{A}}] = f^{-1}r$$

for every f in Σ^* and $\mathsf{Im}\,\Psi_{\mathfrak{A}} \circ \Phi_{\mathfrak{A}}$ is the subspace generated by the quotients of r. For the dimension of $\mathsf{Im}\,\Psi_{\mathfrak{A}} \circ \Phi_{\mathfrak{A}}$ be equal to n, the dimension of $\mathfrak{A}$, it is necessary and sufficient that the dimension of both $\mathsf{Im}\,\Psi_{\mathfrak{A}}$ and $\mathsf{Im}\,\Phi_{\mathfrak{A}}$ be equal to n. The second equality holds if and only if the dimension of $\mathsf{Ker}\,\Phi_{\mathfrak{A}}$ is zero. □

Remark 5.19. The significance of the map $\Psi_{\mathfrak{A}}$ goes beyond the case of weights taken in a field. Without linearisation, $(\Sigma^*)\Psi_{\mathfrak{A}}$ is the *reachability set* of $\mathfrak{A}$. If $S = \mathbb{B}$, $(\Sigma^*)\Psi_{\mathfrak{A}}$ is a set of subsets of states of $\mathfrak{A}$, namely the set of states of the determinisation of $\mathfrak{A}$ (by the so-called *subset construction*).

5.2.2 The Reduction Algorithm

It is not enough to know that reduced representations exist and to characterise them. We want to be able to effectively compute them and establish the following.

Theorem 5.20. *A reduced representation of a recognisable series r is effectively computable from any representation that realises r with a procedure whose complexity is cubic in the dimension of the representation.*

For the rest of this section, let $\mathfrak{A} = (\lambda, \mu, \nu)$ be a S-representation of Σ^* of dimension n (that realises the series $r = |\mathfrak{A}|$).

Word Base

The effective computation from $\mathfrak{A}$ of a reduced representation of r is based on one definition and two propositions that are related but whose scope and aim are nevertheless rather different.

Definition 5.21. *We call* word base *for $\mathfrak{A}$ a prefix-closed subset P of Σ^* such that the set $(P)\Psi_{\mathfrak{A}} = \{\lambda \cdot p\mu \mid p \in P\}$ is a base of $\mathsf{Im}\,\Psi_{\mathfrak{A}}$.*

Proposition 5.22. *Word bases for $\mathfrak{A}$ do exist.*

Proof. If $\lambda = 0$, $\mathsf{Im}\,\Psi_{\mathfrak{A}}$ is the null vector space of dimension 0 and the empty set (which is prefix-closed!) is a word base. Assuming that λ is non-zero, the family of prefix-closed subsets P of Σ^* such that $\{\lambda \cdot p\mu \mid p \in P\}$ is a free subset of S^n is not empty since it contains at least the singleton $\{\varepsilon\}$. Every such subset contains at most $k = \dim(\mathsf{Im}\,\Psi_{\mathfrak{A}})$ elements and there exist thus maximal elements (for the inclusion order) in that family.

It remains to show that such a maximal element P is a word base, that is, $(P)\Psi_{\mathfrak{A}}$ generates $\mathsf{Im}\,\Psi_{\mathfrak{A}}$. By way of contradiction, let f in Σ^* such that $\lambda \cdot f\mu$ does not belong to $\langle (P)\Psi_{\mathfrak{A}} \rangle$; the word f factorises in $f = pg$, with p in P, and we choose f in such a way that g is of minimal length. The word g is not empty: $g = ah$, with a in Σ, and $\lambda \cdot f\mu = \lambda \cdot (pa)\mu \cdot h\mu$. As P is maximal, $\lambda \cdot (pa)\mu$ belongs to $\langle (P)\Psi_{\mathfrak{A}} \rangle$, that is, $\lambda \cdot (pa)\mu = \sum_{p_i \in P} x_i(\lambda \cdot p_i\mu)$. It then follows

$$\lambda \cdot f\mu = \Big(\sum_{p_i \in P} x_i(\lambda \cdot p_i\mu)\Big) \cdot h\mu = \sum_{p_i \in P} x_i\big(\lambda \cdot (p_i h)\mu\big).$$

By the minimality of g, every $\lambda \cdot (p_i h)\mu$ belongs to $\langle (P)\Psi_{\mathfrak{A}} \rangle$: contradiction. □

In the sequel, we do not consider the trivial case $\lambda = 0$ anymore.

Proposition 5.23. *With every word base P for $\mathfrak{A}$ of cardinal m is associated a representation $\mathfrak{A}' = (\lambda', \mu', \nu')$ of dimension m—effectively computable from P and $\mathfrak{A}$—which is conjugate to $\mathfrak{A}$ and with the property that $\Psi_{\mathfrak{A}'}$ is surjective. Moreover, if $\Phi_{\mathfrak{A}}$ is injective, then so is $\Phi_{\mathfrak{A}'}$.*

Proof. Let $P = \{p_1 = \varepsilon, p_2, \ldots, p_m\}$ be a word base for $\mathfrak{A}$ and X the $m \times n$-matrix (with entries in S) whose i-th row is $\lambda \cdot (p_i)\mu$. Let us denote $\nu' = X \cdot \nu$ and by λ' the (row) m-vector whose entries are all 0 but the first one which is 1—thus $\lambda' \cdot X = \lambda$.

For every a in Σ, let $a\mu'$ be the $m \times m$-matrix (with entries in S) whose ith row is the vector of coordinates of $\lambda \cdot (p_i a)\mu$ in the base $\lambda \cdot (P)\mu$, that is,

$$\lambda \cdot (p_i a)\mu = \sum_{j=1}^{j=m} (a\mu')_{i,j}(\lambda \cdot p_j\mu). \tag{29}$$

Since $\lambda \cdot (p_i a)\mu = (\lambda \cdot p_i\mu) \cdot a\mu$, the set of equations (29) for all i may be rewritten in a more compact way as

$$a\mu' \cdot X = X \cdot a\mu$$

and $\mathfrak{A}'$ is conjugated to $\mathfrak{A}$ by X.

If P is not a word base for $\mathfrak{A}'$, there exist m coefficients α_i such that $\sum_{i=1}^{i=m} \alpha_i(\lambda' \cdot p_i\mu') = 0$, but multiplying this equality on the right by X yields $\sum_{i=1}^{i=m} \alpha_i(\lambda \cdot p_i\mu) = 0$, a contradiction (with the fact that P is a word base for $\mathfrak{A}$).

If $\Phi_{\mathfrak{A}'}$ is not injective, there exists a non-zero vector y in S^m such that $y \cdot f\mu' \cdot \nu' = 0$, and thus $(y \cdot X) \cdot f\mu \cdot \nu = 0$ for every f in Σ^*. If $\Phi_{\mathfrak{A}}$ is injective, then $y \cdot X = 0$, and thus $y = 0$ for the same reason as above, a contradiction. □

Remark 5.24 (Remark 5.19 continued). Let $\mathfrak{D}$ be the determinisation of a classical automaton $\mathfrak{A}$ (that is, an automaton with weight in $\mathbb{B}$) of dimension Q by the subset construction. If we form the (Boolean) matrix X whose rows are the states of $\mathfrak{D}$ (Boolean vectors of dimension Q), then $\mathfrak{D}$ is conjugate to $\mathfrak{A}$ by X.

Demonstration of the Reduction Theorem (Theorem 5.20)

We first observe that Proposition 5.23 has obviously a dual formulation, which we rather state on the transpose of the representation $\mathfrak{A}$, ${}^{t}\mathfrak{A} = ({}^{t}\nu, {}^{t}\mu, {}^{t}\lambda)$ where $a\,{}^{t}\mu = {}^{t}(a\mu)$ for every a in Σ and it comes $f\,{}^{t}\mu = {}^{t}({}^{t}f\mu)$ for every f in Σ^*. We then have the following connection between $\mathfrak{A}$ and ${}^{t}\mathfrak{A}$.

Lemma 5.25. *If $\Psi_{{}^{t}\mathfrak{A}}$ is surjective, then $\Phi_{\mathfrak{A}}$ is injective.*

Proof. If $x\Phi_{\mathfrak{A}} = 0$ then $x \cdot f\mu \cdot \nu = 0$ for every f in Σ^* and x belongs to the orthogonal of the subspace generated by the vectors $\{f\mu \cdot \nu \mid f \in \Sigma^*\}$ which is of dimension n by hypothesis: thus $x = 0$. □

Starting from a representation $\mathfrak{A}$, we first compute a word base for ${}^{t}\mathfrak{A}$ which determines a representation ${}^{t}\mathfrak{A}'$ such that $\Psi_{{}^{t}\mathfrak{A}'}$ is surjective, and thus by Lemma 5.25, $\Phi_{\mathfrak{A}'}$ is injective. We then compute a word base for $\mathfrak{A}'$ which determines a representation $\mathfrak{A}''$ such that $\Psi_{\mathfrak{A}''}$ is surjective and $\Phi_{\mathfrak{A}''}$ is injective: $\mathfrak{A}''$ is reduced. The proof of Theorem 5.20 will be complete when we have proved that word bases are effectively computable (with the ascribed complexity).

The foregoing proofs all correspond to effective computations, assuming of course that the operations in S (addition, multiplication, taking the inverse) are effective. All the complexities that follow are calculated assuming that each operation in S has a fixed constant cost, independent of its operands. Computations in S^n are based on the *Gaussian elimination* procedure.

Definition 5.26. *A sequence of k vectors $(v^1, v^2, \ldots, v^k)$ of S^n is an* echelon system *if, for all i in $[k]$:*

(i) $v^i{}_i = 1_S$.
(ii) $\forall j < i \;\; v^i{}_j = 0_S$.

An echelon system is free, and hence $k \leqslant n$.The following proposition is classic, at least for commutative fields, and its proof is not really different for division rings.

Proposition 5.27 (Gaussian elimination). *Let S be a skew field and let us view S^n as a left vector space over S. Let $U = (v^1, v^2, \ldots, v^k)$ be an echelon system and let w be a vector in S^n.*

(i) We can decide whether w is in $\langle U \rangle$, the subspace generated by U, and in this case, compute effectively the coordinates of w in U.

(ii) If w is not in $\langle U \rangle$, we can compute effectively w' such that $U' = U \cup \{w'\}$ is echelon and generates the same subspace as $U \cup \{w\}$.

The complexity of these operations (deciding whether w is in $\langle U \rangle$ and computing the coordinates of either w or w') is $O(kn)$.

From this proposition, we deduce the effective nature of the assertions, constructions, and specifications used in the proofs of this section. More precisely, the corollary follows.

Corollary 5.28. *Let U be a finite set of vectors of S^n and let w be in S^n.*

(i) We can decide whether w belongs to $\langle U \rangle$.
(ii) We can extract effectively from U a basis V of $\langle U \rangle$.
(iii) We can compute effectively the coordinates in V of an (explicitly given) vector of $\langle U \rangle$.

The following proposition and its proof exhibit the computation underlying Proposition 5.23 (remember, we have defined the *border* of a prefix-closed subset at Sect. 2.2.2).

Proposition 5.29. *Word bases for $\mathfrak{A}$ are effectively computable, with complexity $O(dn^3)$, where d is the cardinal of Σ.*

Proof. We set $P_0 = \{\varepsilon\}$ and $C_0 = \emptyset$. The algorithm to compute a word base P can be written in the following manner.

If $E_k = (P_k\Sigma \backslash P_k)\backslash C_k$ is non-empty, choose an arbitrary f in E_k and decide whether $\lambda \cdot f\mu$ belongs to $\langle \lambda \cdot P_k\mu \rangle$.

(i) If not, then $P_{k+1} = P_k \cup \{f\}$ and $C_{k+1} = C_k$.
(ii) If so, then $P_{k+1} = P_k$ and $C_{k+1} = C_k \cup \{f\}$.

Set $k = k + 1$ and start again.

The algorithm terminates when E_k is empty and at that moment $C_k = P_k\Sigma\backslash P_k$ is the border of P_k. The algorithm must terminate since P_k has at most n elements, so $P_k \cup C_k$ has at most $\|\Sigma\|n+1$ elements and this set grows by 1 at each step of the algorithm.

By construction, P_k is prefix-closed, and each element f of C_k is such that $\lambda \cdot f\mu$ belongs to $\langle \lambda \cdot P_k\mu \rangle$: when E_k is empty, P_k is maximal. □

5.3 Applications of the Reduction of Representations

We consider here three applications: the decidability of equivalence of S-automata (for certain S), the generalisation of the recurrence relation on the coefficients of a rational series over non-commuting variables, and a *structural interpretation* of equivalence of S-automata in terms of conjugacy and covering (again for certain S).

5.3.1 Equivalence Decidability

Even if a series has not a unique reduced representation (they are all *similar*), the existence of reduced representations implies the decidability of equivalence for automata with weights in a field.

Theorem 5.30. *The equivalence of recognisable series over Σ^* with coefficients in a (sub-semiring of a) skew field—and thus of rational series—is decidable, with a procedure which is cubic in the dimension of the representation of the series.*

Proof. Let S be a sub-semiring of a skew field $\mathbb{F}$. Two series r_1 and r_2 of $S\mathrm{Rec}\,\Sigma^*$ are also in $\mathbb{F}\mathrm{Rec}\,\Sigma^*$ and $r_1 = r_2$ holds if, and only if, $(r_1 - r_2)$ is a series of $\mathbb{F}\mathrm{Rec}\,\Sigma^*$ of rank 0, and the rank of $(r_1 - r_2)$ can be computed effectively. □

This result, together with the well-known decidability of equivalence of classical Boolean automata, should not let us think that this is the universal status. For instance, the following holds.

Theorem 5.31 ([28]). *The equivalence of recognisable series over Σ^* with coefficients in the semiring $\mathbb{M} = \langle \mathbb{N}^\infty, \mathsf{min}, + \rangle$ is undecidable.*

Theorem 5.30 has however far reaching and to some extent 'unexpected' consequences, as the following one, discovered by T. Harju and J. Karhumäki.

Corollary 5.32 ([22]). *The equivalence of rational series over $\Sigma_1^* \times \Sigma_2^* \times \cdots \times \Sigma_k^*$ with coefficients in $\mathbb{N}$ is decidable.*

Proof. By Proposition 4.25, a series in $\mathbb{N}\mathrm{Rat}\,\Sigma_1^* \times \Sigma_2^* \times \cdots \times \Sigma_k^*$ is a series in $[\mathbb{N}\mathrm{Rat}\,\Sigma_2^* \times \cdots \times \Sigma_k^*]\mathrm{Rat}\,\Sigma_1^*$. By Corollary 4.29, the latter family is isomorphic to $[\mathbb{N}\mathrm{Rat}\,\Sigma_2^* \times \cdots \times \Sigma_k^*]\mathrm{Rec}\,\Sigma_1^*$ and the decidability of equivalence follows from Theorem 5.33. □

Theorem 5.33. $\mathbb{N}\mathrm{Rat}\,\Sigma_2^* \times \cdots \times \Sigma_k^*$ *is a sub-semiring of a skew field.*

This result is the direct consequence of a series of classical results in mathematics which we shall not prove here (cf. for instance [11]) but simply state.

Definition 5.34 (Hahn–Malcev–Neumann). *Let S be a semiring and G an ordered group. We write $S_{\mathsf{wo}}\langle\langle G \rangle\rangle$ to denote the set of series on G with coefficients in S whose support is a well-ordered subset of G.*

Theorem 5.35 (Birkhoff–Tarski–Neumann–Iwazawa[17]). *A finite direct product of free groups is ordered.*

Theorem 5.36 (Malcev–Neumann). *If S is a skew field and G an ordered group, then $S_{\mathsf{wo}}\langle\langle G \rangle\rangle$ is a skew field.*

---

[17] And possibly others.

Theorems 5.35 and 5.36 imply that $S_{\mathsf{wo}}\langle\langle F(\Sigma_2) \times \cdots \times F(\Sigma_k)\rangle\rangle$ is a skew field (here $F(\Sigma)$ is the free group generated by Σ). To deduce Theorem 5.33, we must also ensure that $S\mathrm{Rat}\,\Sigma^*$—in fact $S\langle\langle\Sigma^*\rangle\rangle$—is included in $S_{\mathsf{wo}}\langle\langle F(\Sigma)\rangle\rangle$, respectively that

$$S\langle\langle\Sigma_2^* \times \cdots \times \Sigma_k^*\rangle\rangle \subseteq S_{\mathsf{wo}}\langle\langle F(\Sigma_2) \times \cdots \times F(\Sigma_k)\rangle\rangle;$$

that is, to be more precise, that we can order $F(\Sigma_2) \times \cdots \times F(\Sigma_k)$ in such a way that the above inclusion is true and this is not difficult either.

Now, by straightforward computations, 1-way k-tape Turing machines are faithfully modelised by automata over $\Sigma_1^* \times \Sigma_2^* \times \cdots \times \Sigma_k^*$ and two *deterministic* such machines are equivalent if and only if the corresponding automata are equivalent as automata over $\Sigma_1^* \times \Sigma_2^* \times \cdots \times \Sigma_k^*$ *with multiplicity in* $\mathbb{N}$.

Corollary 5.37 ([22]). *The equivalence of* 1*-way* k*-tape deterministic Turing machines is decidable.*

5.3.2 Recurrence Relations

Another consequence of Theorem 5.16 is the generalisation to series over non-commuting variables of the characterisation by linear recurrences of coefficients of rational series over one variable (recall also Lemma 2.27).

Theorem 5.38 ([46]). *A series r of $S\langle\langle\Sigma^*\rangle\rangle$ is recognisable if and only if there exists a finite prefix-closed subset P and its border $C = P\Sigma\backslash P$, such that for each pair (c,p) in $C \times P$, there exists a coefficient $s_{c,p}$ in S such that*

$$\forall g \in \Sigma^*,\ \forall c \in C \quad (r, cg) = \sum_{p \in P} s_{c,p}(r, pg). \tag{30}$$

Proof. Let P be a word base for an S-representation $\mathfrak{A} = (\lambda, \mu, \nu)$ that recognises r and (λ', μ', ν') the S-representation computed as in Proposition 5.23. For each $c = pa$ in C and all q in P, we set $s_{c,q} = (a\mu')_{p,q}$. From (29) follows that, for all g in Σ^*, it holds:

$$(r, cg) = \lambda \cdot p\mu \cdot a\mu \cdot g\mu \cdot \nu = \sum_{q \in P} a\mu'_{p,q}\lambda \cdot q\mu \cdot g\mu \cdot \nu = \sum_{q \in P} s_{c,q}(r, qg).$$

Conversely, (30) implies that every quotient $f^{-1}r$ belongs to the subspace T generated by $p^{-1}r$ for p in P. This last property is trivially verified if f is in P and (30) can be rewritten as

$$\forall c \in C \quad c^{-1}r = \sum_{p \in P} s_{c,p}p^{-1}r;$$

that is, the property is verified for f in C. *A contrario*, suppose that $f^{-1}r$ does not belong to T; by Lemma 2.27, we have $f = cg$ and choose f such that g is of minimal length. By (30), we have, for all h in Σ^*,

$$(r, cgh) = \sum_{p \in P} s_{c,p}(r, pgh) \qquad \text{that is,} \qquad f^{-1}r = \sum_{p \in P} s_{c,p}(pg)^{-1}r.$$

For each p in P, either pg is in P, or $pg = c'g'$ with c' in C, then $|c'| > |p|$; hence, $|g'| < |g|$ and $(pg)^{-1}r$ is in T by the assumption of minimality of g. Hence, $f^{-1}r$ belongs to T, which is a contradiction. Also, r is recognisable by Theorem 5.16. □

Remark 5.39. If $\Sigma = \{a\}$, every prefix-closed subset of Σ^* has the form $P = \{\varepsilon, a, \ldots, a^{j-1}\}$ for some integer j, and C is a singleton: $C = \{a^j\}$. Equation (30) becomes

$$\forall n \in \mathbb{N} \quad (r, a^{n+j}) = s_{j-1}(r, a^{n+j-1}) + s_{j-2}(r, a^{n+j-2}) + \cdots + s_0(r, a^n);$$

that is, a linear recurrence in its standard form.

Another way to exploit Proposition 5.23, is by 'computing' the coefficients of a *reduced representation* of a recognisable series as a function of the coefficients of the series itself. Going from the series back to the representation does not so much correspond to an effective procedure like those described in Proposition 5.23 and Theorem 5.38, as it expresses a fundamental property of recognisable series on a field (see an application with Theorem 6.4).

Proposition 5.40 ([46]). *Let S be a skew field, r an S-recognisable series of rank n, and (λ, μ, ν) a reduced representation of r. There exist two sets of n words: $P = \{p_1, p_2, \ldots, p_n\}$ and $Q = \{q_1, q_2, \ldots, q_n\}$ (which we can choose to be respectively prefix-closed and suffix-closed) and two $n \times n$ matrices α_P and β_Q such that*

$$\forall f \in \Sigma^* \quad f\mu = \alpha_P \cdot ((r, p_i f q_j)) \cdot \beta_Q,$$

where $((r, p_i f q_j))$ denote the $n \times n$ matrix whose entry (i, j) is $(r, p_i f q_j)$.

5.3.3 From Equivalence to Conjugacy

At Section 3.3, we have seen that it directly follows from the definition that two conjugate automata are equivalent (Proposition 3.23). For certain semirings S, this statement can be given a kind of converse, which reads as follows.

Theorem 5.41 ([4]). *Let S be $\mathbb{B}$, $\mathbb{N}$, $\mathbb{Z}$, or any (skew) field. Two S-automata are equivalent if and only if there exists a third S-automaton that is conjugate to both of them.*

The proof of Theorem 5.41 relies on the idea of *joint reduction* which is defined by means of the notion of *representation*. Let $\mathfrak{A} = \langle \lambda, \mu, \nu \rangle$ be an S-representation of dimension Q and the associated map $\Psi_{\mathfrak{A}} \colon \Sigma^* \to S^Q$. We have already seen (Proposition 5.23 and Remark 5.24) that, in the two contrasting cases of the Boolean semiring and of a field, we can choose a word base P such that:

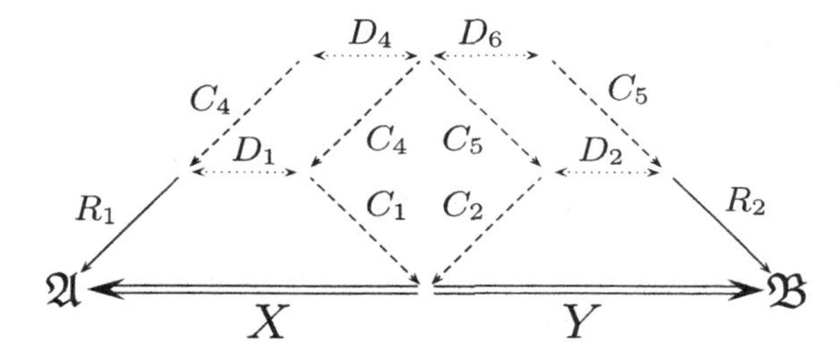

Fig. 14. Structural decomposition of the equivalence of two S-automata

(i) $\{\lambda \cdot p\mu \mid p \in P\}$ is a set of vectors, which is equal to $(\Sigma^*)\Psi_{\mathfrak{A}}$ in the Boolean case, which generates the same S-vector space in the field case.
(ii) There exists an automaton $\mathfrak{R}$ which is conjugate to $\mathfrak{A}$ by the transfer matrix X whose rows are the vectors $\{\lambda \cdot p\mu \mid p \in P\}$.

Let now $\mathfrak{A} = \langle \lambda, \mu, \nu \rangle$ and $\mathfrak{B} = \langle \eta, \kappa, \chi \rangle$ be two S-representations of dimension Q and R, respectively, and let $\mathfrak{C}$ be the *sum* of $\mathfrak{A}$ and $\mathfrak{B}$: $\mathfrak{C} = \langle \zeta, \pi, \omega \rangle$ is an S-representation of dimension $Q \cup R$, $\zeta = [\lambda \ \eta]$ is the horizontal concatenation of λ and η, $\omega = \left[\begin{smallmatrix} \nu \\ \chi \end{smallmatrix}\right]$ the vertical concatenation of ν and χ, and $\pi = \left[\begin{smallmatrix} \mu & 0 \\ 0 & \kappa \end{smallmatrix}\right]$ is the representation whose diagonal blocs are μ and κ. We perform the same construction as before on $\mathfrak{C}$; we consider the set of vectors $(\Sigma^*)\Psi_{\mathfrak{C}} = \{[\lambda \cdot f\mu \ \eta \cdot f\kappa] \mid f \in \Sigma^*\}$ and look for a *finite* set V of vectors $[x \ y]$ which, roughly speaking, generates the same S-semi-module as $(\Sigma^*)\Psi_{\mathfrak{C}}$.

The computation of V provides indeed at the same time an automaton $\mathfrak{Z}$ which is conjugate to $\mathfrak{C}$ by the transfer matrix Z whose rows are the vectors in V. *If* $\mathfrak{A}$ *and* $\mathfrak{B}$ *are equivalent*, then $\mathfrak{Z}$, or a slight modification of it (depending on which semiring S the computations are currently done), is conjugate to both $\mathfrak{A}$ and $\mathfrak{B}$ by the transfer matrices X and Y, respectively, where X and Y are respectively the 'left' and 'right' parts of the matrix Z. In every case listed in Theorem 5.41, the finite set V is effectively computable, a proof that has to be done separately for each case (cf. [4]).

Together with the result of decomposition of conjugacy by means of a sequence of co-covering, circulation, and covering (Theorem 3.33), and Proposition 3.36 that allows us to build diagrams upward; this result yields a structural decomposition of the equivalence of two S-automata as shown in Fig. 14. In the case $S = \mathbb{N}$, this decomposition takes the following form.

Corollary 5.42. *Two equivalent* $\mathbb{N}$*-automata can be transformed, one into the other, by a chain of two state-splittings (in- and out-) and two state-mergings (out- and in-).*

6 Support of Rational Series

It follows directly from Proposition 4.5 that for any (graded) monoid M, we have the following corollary.

Corollary 6.1. *If S is a positive semiring, the support of an S-recognisable series over M is a recognisable subset of M.*

The assumption on S is necessary, even in the case where M is a free monoid Σ^*, as shown by the following example.

Example 6.2 (Example 4.2 continued). We have seen that $u_1 = \sum_{f \in \Sigma^*} |f|_b f$ is a $\mathbb{Z}$-rational series, and thus so is $r_1 = \sum_{f \in \Sigma^*} |f|_a f$. The series $z_1 = r_1 - u_1 = \sum_{f \in \Sigma^*} (|f|_a - |f|_b) f$ is a $\mathbb{Z}$-rational series. The complement of $\mathsf{supp}\, z_1 = \{f \in \Sigma^* \mid |f|_a \neq |f|_b\}$ is the language $Z_1 = \{f \in \Sigma^* \mid |f|_a = |f|_b\}$, which we know is not rational.

In this short section, we study certain conditions which ensure the rationality of the support of a series, and some closure properties of the family of languages thus defined. We end with several undecidable properties for $\mathbb{Z}$-rational series, somewhat surprising in this context where properties seem to be all decidable and effective.

Recall that a series r of $S\langle\langle \Sigma^* \rangle\rangle$ is fundamentally a map from Σ^* to S. It is therefore natural to write, for every subset U of S, Ur^{-1} for the set of words of Σ^* whose coefficient in r belongs to U:

$$Ur^{-1} = \{f \in \Sigma^* \mid (r, f) \in U\}.$$

The first result concerns *locally finite* semirings (defined in Chap. 1).

Proposition 6.3. *Let S be a* locally finite *semiring and let r be an S-rational series over Σ^*. For all subsets U of S, Ur^{-1} is rational.*

Proof. Since r is also recognisable, r is recognised by a S-representation (λ, μ, ν), of finite dimension Q, that is, $\mu \colon \Sigma^* \to S^{Q \times Q}$ is a morphism. Since S is locally finite, the image $(\Sigma^*)\mu = M$ is a *finite submonoid* of $S^{Q \times Q}$. The language Ur^{-1} is recognised by the morphism $\mu \colon \Sigma^* \to M$, a well-known characterisation of rational (or recognisable) languages of Σ^*. □

Another way to state (and to prove indeed) Proposition 6.3 is to remark that if S is locally finite, then the *reachability set* $(\Sigma^*)\Psi_{\mathfrak{A}}$ of any S-representation $\mathfrak{A}$ is finite—opening the way to the immediate construction for equivalent deterministic or minimal automata, a basic fact that seems to have been often overlooked, and thus often rediscovered (cf. also Chap. 12). To express it in another way again: *Counting in a (locally) finite semiring is not counting.*

Proposition 6.3 generalises in a remarkable way if S is a field. But it is not a trivial remark anymore; it follows from the whole algebraic theory we have built in this case.

Theorem 6.4 ([46]). *Let S be a (skew) field. If r is an S-rational series over Σ^* with a finite image, then kr^{-1} is rational for all k in S.*

Proof. Let (λ, μ, ν) be a reduced representation that recognises r. By Proposition 5.40, the image $(\Sigma^*)\mu$ is a *finite sub-monoid* of $S^{Q \times Q}$ if r has a finite image and the conclusion follows as in Proposition 6.3. □

Since the family of supports of S-rational series over Σ^* strictly contains $\mathrm{Rat}\,\Sigma^*$, a natural question is to ask under which operations this family is closed. The answer certainly depends on S; a fairly complete one can be given for sub-semirings of $\mathbb{R}$.

Proposition 6.5 ([46]). *Let S be a sub-semiring of $\mathbb{R}$. The set of supports of S-rational series on Σ^* contains* $\mathrm{Rat}\,\Sigma^*$ *and is closed under union, product, star, and intersection.*

Proof. The first assertion is a restatement of Proposition 3.12. Since $S\mathrm{Rat}\,\Sigma^*$ is closed under the Hadamard product, we deduce first the closure by intersection, then because r and $r \odot r$ have the same support, it follows that every support of an S-rational series is the support of an S-rational series with *non-negative coefficients.* Then for such series, we clearly have

$$\mathsf{supp}\,(r + r') = \mathsf{supp}\;r \cup \mathsf{supp}\;r', \qquad \mathsf{supp}\,(rr') = \mathsf{supp}\;r\,\mathsf{supp}\;r' \qquad \text{and}$$
$$\mathsf{supp}\,(r^*) = (\mathsf{supp}\;r)^*. \quad \square$$

The closure under morphisms and inverse morphisms is somewhat more difficult to establish.

Proposition 6.6 (Fliess [17]). *Let S be a sub-semiring of $\mathbb{R}$. The set of supports of S-rational series over Σ^* is closed under morphisms and inverse morphisms.*

The set $\mathrm{Rat}\,\Sigma^*$ is also closed under complement, but if S is not positive, the set of supports of S-rational series can strictly contain $\mathrm{Rat}\,\Sigma^*$. The closure under complement is precisely characteristic of membership of $\mathrm{Rat}\,\Sigma^*$ as stated in the following result. Besides the reduction theory, its proof is based upon the strongest version of the iteration theorem (or pumping lemma) for rational languages, due to A. Ehrenfeucht, R. Parikh, and G. Rozenberg [13], and itself is based on Ramsey's theorem.

Theorem 6.7 (Restivo–Reutenauer [38]). *Let S be a (sub-semiring of a) skew field. If a language and its complement are each the support of an S-rational series over Σ^*, then this language is rational.*

We then construct, with this simple model of finite weighted automata, some series for which we cannot answer some elementary questions, as soon as the semiring of coefficients contains $\mathbb{Z}$.

Theorem 6.8. *It is undecidable if the support of a $\mathbb{Z}$-rational series over Σ^* is all of Σ^*.*

Proof. Let $\Delta = \{x, y\}$; the morphism $\alpha\colon \Delta^* \to \mathbb{N}^{2\times 2}$ defined by

$$x\alpha = \begin{pmatrix} 1 & 0 \\ 0 & 2 \end{pmatrix} \quad \text{and} \quad y\alpha = \begin{pmatrix} 1 & 1 \\ 0 & 2 \end{pmatrix}$$

is *injective* (cf. the automaton $\mathfrak{C}_1$ at Example 3.4).

Then let $\theta\colon \Sigma^* \to \Delta^*$ and $\theta\colon \Sigma^* \to \Delta^*$ be two morphisms. For i and j equal to 1 or 2, the series $r_{i,j}$ defined by

$$\forall f \in \Sigma^* \quad (r_{i,j}, f) = \big((f\theta)\alpha\big)_{i,j} - \big((f\mu)\alpha\big)_{i,j}$$

are $\mathbb{Z}$-rational, hence so are the series $u_{i,j} = r_{i,j} \odot r_{i,j}$, and the series

$$u = \sum_{i,j} u_{i,j}.$$

The support of u is not all of Σ^* if and only if there exists f such that $(u, f) = 0$; that is, since α is *injective*, if and only if $f\theta = f\mu$, which we know to be undecidable (Post Correspondence Problem). □

With the same construction, we easily obtain the following corollary.

Corollary 6.9. *Let r be a $\mathbb{Z}$-rational series over Σ^*. It is undecidable whether:*

(i) r has infinitely many coefficients equal to zero.
(ii) r has at least one positive coefficient.
(iii) r has infinitely many positive coefficients.

Corollary 6.10. *It is undecidable whether the supports of two $\mathbb{Z}$-rational series over Σ^* are equal.*

7 Notes

I am grateful to M. Droste, Ch. Reutenauer, and W. Kuich who pointed out some interesting references to me.

7.1 General Sources

As already mentioned in the Introduction, this chapter is essentially an epitome of Chap. III and of a part of Chap. IV of [43] where more details, proofs, and examples are to be found. More precise references to some of them are given below.

A classical, and above all pioneering, reference on the subject is the treatise by S. Eilenberg [14] whose influence is willingly acknowledged. Each of the references quoted in the Introduction [45, 29] or in Chaps. 1 and 3 develop a particular point of view that is worth interest. But the most advanced one is the book of Jean Berstel and Christophe Reutenauer [5], a new revised edition of which is now available and anyone really interested in weighted automata should certainly not miss this work.

7.2 Notes to Sect. 2: Rational Series

One can say that it is equation (2) that justifies the choice of Eilenberg, op. cit., to call *rational* what was called *regular* in the foregoing literature. Schützenberger and his school, to which I acknowledge membership, followed him but one must recognise it has not been a universal move. If the terminology is still disputable for languages, and expressions, I do not think the question may even be asked when it comes to series. On the other hand and in the same work, Eilenberg calls a monoid with the property that every element is finitely decomposable a *locally finite monoid.* This terminology inconveniently conflicts with another accepted meaning of the phrase: a monoid such that every finitely generated submonoid is finite (cf. [49]).

I was led to define *strong semirings*, a terminology suggested to me by J. van der Hoeven, to be able to prove the equivalence between the existence of the star of an arbitrary series and that of the star of its constant term.

7.3 Notes to Sect. 3: Weighted Automata

The construction of $\mathfrak{S}_{\mathsf{E}}$ is the version given in [33] of the generalisation to weighted automata of the construction of the *Glushkov automaton* or *position automaton* first given by Caron and Flouret [9].

In a sense the Fundamental Theorem is what Kleene showed for automata over Σ [26], or its usual weighted generalisation (often called the Kleene–Schützenberger theorem). However, because these results apply to automata over free monoids, their standard form—cf. Theorem 2.12, Chap. 3—states the identity between rational and recognisable languages or series, which *no longer holds* for automata (weighted or otherwise) over arbitrary monoids. Kleene's theorem was therefore split in two, as it were: one part which holds for automata over arbitrary monoids and which, considering what the proof involves, concentrates the substance of the theorem; and one part which holds only for automata over free monoids and which is nearly a formality; this distinction seems to appear for the first time in [42].

Proposition 3.21 can be credited to Conway [12] and Krob [27]; an elementary proof is given in [43, 44].

The matter of Sect. 3.3 is taken from [43] and [3]. Conjugacy of $\mathfrak{A}$ to $\mathfrak{B}$ by X is called *simulation* from $\mathfrak{A}$ into $\mathfrak{B}$ in [7]. In a different setting, this kind of mapping was called *morphism of 'modules sériels'* by Fliess in [18]. The definition of S-covering as conjugacy by an amalgamation matrix is a hint for similarity between S-coverings and *state amalgamation* in *symbolic dynamical systems* [31, Sect. 2.4]. If $\mathfrak{B}$ is obtained from $\mathfrak{A}$ by an In-amalgamation, then $\mathfrak{A}$ is an $\mathbb{N}$-covering of $\mathfrak{B}$. But the converse is not true. Roughly speaking, and with the notations of Proposition 3.28, $\mathfrak{A} = \langle I, E, T\rangle$ is an S-covering of $\mathfrak{B}$ if the rows with 'equivalent' indices of the matrix $E{\cdot}H_{\varphi}$ are equal while $\mathfrak{B}$ is obtained by amalgamation from $\mathfrak{A}$ if the rows with 'equivalent' indices of the matrix E are equal. The notion of 'equisubtractivity' used in Sect. 3.3.3 in order to express conjugacy in terms of coverings and co-coverings is very similar to a property introduced by Tarski in [51] where an extension of Lemma 3.35 to infinite sums is established.

The presentation of the minimal S-quotient is taken from [43], whereas the notion itself probably exists in many other works; for instance, two S-automata are *in bisimulation* if and only if their minimal S-quotients are isomorphic.

7.4 Notes to Sect. 4: Recognisable Series

The definition of representations in the form (λ, μ, ν) is due to Fliess [18]. Lemma 4.10 is a classic statement in matrix theory and can be found already in Gröbner [21] (cf. also [29, Theorem 4.33]). Theorem 4.12 is due to Schützenberger [48], including the more general formulation of Theorem 4.13. Theorem 4.18 is also due to Fliess [19]; the proof given here is that of [43].

The 'shuffle identity' (20) is an unpublished result of M. Petitot and was indicated to me by M. Waldschmidt (personal communication); the proof I gave for it in [43] was the starting point of [3].

The matter of Sect. 4.3, and especially the definition of weighted relations, is taken from Chap. IV of [43]. Another theory of weighted relations, slightly different from what I have very briefly sketched here, is that of Jacob [24, 25]. It consists of defining with *regulated rational transductions* the largest possible family of relations which satisfy the evaluation and composition theorems and which correspond to total maps (and hence maps whose composition is also always defined), and to do that *independently of the semiring of coefficients.* This point of view was adopted in related works [45, 29] which popularised the work of Jacob.

7.5 Notes to Sect. 5: Series over a Free Monoid

Some authors speak of the *translation* of a series instead of quotient; I have preferred to use the same term as for languages.

The original work is due to Schützenberger [46, 47]. The characterisation of recognisable series (Theorem 5.3) is a generalisation, due to Jacob [24], of the property stated by Fliess for the case of series on a field [18].

The derivation of weighted expressions is a generalisation of V. Antimirov's work [1] (where derived terms were called *partial derivatives*). We note once more that the introduction of weights clarifies and structures a result on languages, even if having to take into account that not necessarily positive semirings adds a certain complexity. This presentation is taken from [32]. With somewhat different techniques, Rutten [40, 41] also proved Theorem 5.10 and Proposition 5.12.

The original work for reduction of representations is again from Schützenberger [46]. The presentation here follows roughly [5] but as in a background and owes much to my discussions with S. Lombardy. It keeps the Hankel matrix of a series—which could be given the central role as M. Fliess did in [18]—as a subliminal object. It is important for the sequel that the theory is generalised to non-commutative fields. In [20], it was also observed that Schützenberger's reduction algorithm applies to the case of series on a skew field, but with a reference to a previous theory of non-commutative determinants [39]. The cubic complexity of the reduction algorithm was already established in [8].

The problem of the decidability of equivalence of *deterministic* k-tape automata was posed in [37] and was solved for $k = 2$ by M. Bird [6] by an *ad hoc* method, then by L. Valiant [52] as a corollary of the decidability of the equivalence of 'finite-turn' deterministic pushdown automata. The problem remained open for $k \geqslant 3$ until the solution in [22]. The material for Theorems 5.35 and 5.36 is standard if not elementary algebra, and is explained in sufficiently comprehensive treatises such as [11]. A self-contained presentation and proof of this is given in [43, IV.7]. The original proof of Theorem 5.36 by Neumann [36] has been greatly simplified by Higman [23] where he proved what is often known as 'Higman's lemma'. The Russian version of the same result was proved in [34].

Section 5.3.2 is adapted from [5] and Sect. 5.3.3 from [4]. A result analogous to Theorem 5.41 holds for functional transducers as well, but this, its proof, and its consequences somewhat fall out of the scope of this chapter (cf. [4]).

7.6 Notes to Sect. 6: Support of Rational Series

The subject is hardly touched there and the reader is referred once again to [45] or to [5]. Theorem 6.4 has been generalised to commutative rings by Sontag [50]. The proof of Theorem 6.8 is taken from [14].

References

1. V. Antimirov. Partial derivatives of regular expressions and finite automaton constructions. *Theoretical Computer Science*, 155:291–319, 1996.
2. A. Arnold. *Systèmes de transitions finis et sémantique des processus communiquants.* Masson, Paris, 1992. Translation: *Finite Transitions Systems.* Prentice–Hall, New York, 1994.
3. M.-P. Béal, S. Lombardy, and J. Sakarovitch. On the equivalence of $\mathbb{Z}$-automata. In *ICALP 2005*, volume 3580 of *Lecture Notes in Computer Science*, pages 397–409. Springer, Berlin, 2005.
4. M.-P. Béal, S. Lombardy, and J. Sakarovitch. Conjugacy and equivalence of weighted automata and functional transducers. In *CSR 2006*, volume 3967 of *Lecture Notes in Computer Science*, pages 58–69. Springer, Berlin, 2006.

5. J. Berstel and C. Reutenauer. *Les séries rationnelles et leurs langages.* Masson, Paris, 1984. Translation: *Rational Series and Their Languages.* Springer, Berlin, 1988. New revised English edition available from http://www-igm.univ-mlv.fr/~berstel/.
6. M. Bird. The equivalence problem for deterministic two-tape automata. *Journal of Computer and System Sciences*, 7:218–236, 1973.
7. S.L. Bloom and Z. Ésik. *Iteration Theories.* Springer, Berlin, 1993.
8. A. Cardon and M. Crochemore. Détermination de la représentation standard d'une série reconnaissable. *Theoretical Informatics and Applications, RAIRO*, 14:371–379, 1980.
9. P. Caron and M. Flouret. Glushkov construction for multiplicities. In A. Paun and S. Yu, editors, *CIAA 2000*, volume 2088 of *Lecture Notes in Computer Science*, pages 67–79. Springer, Berlin, 2001.
10. J.-M. Champarnaud and D. Ziadi. Canonical derivatives, partial derivatives and finite automaton constructions. *Theoretical Computer Science*, 289:137–163, 2002.
11. P.M. Cohn. *Algebra.* Wiley, New York, 1974. 2nd edition: volume I, 1982; volume II, 1989; volume III, 1991.
12. J.H. Conway. *Regular Algebra and Finite Machines.* Chapman & Hall, London, 1971.
13. A. Ehrenfeucht, R. Parikh, and G. Rozenberg. Pumping lemmas for regular sets. *SIAM Journal on Computing*, 10:536–541, 1981.
14. S. Eilenberg. *Automata, Languages and Machines, volume A.* Academic Press, San Diego, 1974.
15. S. Eilenberg and M.P. Schützenberger. Rational sets in commutative monoids. *Journal of Algebra*, 13:173–191, 1969.
16. C.C. Elgot and J.E. Mezei. On relations defined by generalized finite automata. *IBM Journal of Research and Development*, 9:47–68, 1965.
17. M. Fliess. Formal languages and formal power series. In *Séminaire Logique et Automates, IRIA, 1971*, pages 77–85.
18. M. Fliess. Matrices de Hankel. *Journal de Mathématiques Pures et Appliquées*, 53:197–222, 1974. Erratum in: *Journal de Mathématiques Pures et Appliquées*, 54, 1975.
19. M. Fliess. Sur divers produit de séries formelles. *Bulletin de la Société Mathématique de France*, 102:181–191, 1974.
20. M. Flouret and E. Laugerotte. Noncommutative minimization algorithms. *Information Processing Letters*, 64:123–126, 1997.
21. W. Gröbner. *Matrizenrechnung.* Oldenburg, München, 1956.
22. T. Harju and J. Karhumäki. The equivalence problem of multitape finite automata. *Theoretical Computer Science*, 78:347–355, 1991.
23. G. Higman. Ordering by divisibility in abstract algebra. *Proceedings of the London Mathematical Society. Second Series*, 2:326–336, 1952.
24. G. Jacob. Représentations et substitutions matricielles dans la théorie algébrique des transductions. Thèse Sci. Math. Univ. Paris VII, 1975.
25. G. Jacob. Sur un théorème de Shamir. *Information and Control*, 27:218–261, 1975.

26. S.C. Kleene. Representation of events in nerve nets and finite automata. In C. Shannon and J. McCarthy, editors, *Automata Studies*, pages 3–41. Princeton University Press, Princeton, 1956.
27. D. Krob. Complete systems of B-rational identities. *Theoretical Computer Science*, 89:207–343, 1991.
28. D. Krob. The equality problem for rational series with multiplicities in the tropical semiring is undecidable. In W. Kuich, editor, *ICALP'92*, volume 623 of *Lecture Notes in Computer Science*, pages 101–112. Springer, Berlin, 1992.
29. W. Kuich and A. Salomaa. *Semirings, Automata, Languages.* Springer, Berlin, 1986.
30. F.W. Levi. On semigroups. *Bulletin of the Calcutta Mathematical Society*, 36:141–146, 1944 and 38:123–124, 1946.
31. D. Lind and B. Marcus. *An Introduction to Symbolic Dynamics and Coding.* Cambridge University Press, Cambridge, 1995.
32. S. Lombardy and J. Sakarovitch. Derivation of rational expressions with multiplicity. In *MFCS'02*, volume 2420 of *Lecture Notes in Computer Science*, pages 471–482. Springer, Berlin, 2002.
33. S. Lombardy and J. Sakarovitch. Derivation of rational expressions with multiplicity. *Theoretical Computer Science*, 332:141–177, 2005.
34. A.I. Malcev. On the embedding of group algebras in division algebras. *Doklady Akademii Nauk SSSR (N.S.)*, 60:1409–1501, 1948 (in Russian).
35. J. McKnight. Kleene's quotient theorems. *Pacific Journal of Mathematics*, 14:43–52, 1964.
36. B.H. Neumann. On ordered division ring. *Transactions of the American Mathematical Society*, 66:202–252, 1949.
37. M.O. Rabin and D. Scott. Finite automata and their decision problems. *IBM Journal of Research and Development*, 3:125–144, 1959. Reprinted in: E. Moore, editor, *Sequential Machines: Selected Papers*, Addison–Wesley, Reading, 1965.
38. A. Restivo and C. Reutenauer. On cancellation properties of languages which are support of rational series. *Journal of Computer and System Sciences*, 29:153–159, 1984.
39. A.R. Richardson. Simultaneous linear equations over a division algebra. *Proceedings of the London Mathematical Society*, 28:395–420, 1928.
40. J.M. Rutten. Automata, power series, and coinduction: Taking input derivatives seriously. In J. Wiedermann, P. van Emde Boas, and M. Nielsen, editors, *ICALP'99*, volume 1644 of *Lecture Notes in Computer Science*, pages 645–654. Springer, Berlin, 1999.
41. J.M. Rutten. Behavioural differential equations: A coinductive calculus of streams, automata, and power series. *Theoretical Computer Science*, 308:1–53, 2003.
42. J. Sakarovitch. Kleene's theorem revisited. In A. Kelemenova and K. Kelemen, editors, *Trends, Techniques and Problems in Theoretical Computer*

Science, volume 281 of *Lecture Notes in Computer Science*, pages 39–50. Springer, Berlin, 1987.
43. J. Sakarovitch. *Éléments de théorie des automates.* Vuibert, Paris, 2003. Corrected English edition: *Elements of Automata Theory*, Cambridge University Press, Cambridge, 2009.
44. J. Sakarovitch. The language, the expression and the (small) automaton. In *CIAA 2005*, volume 3845 of *Lecture Notes in Computer Science*, pages 15–30. Springer, Berlin, 2005.
45. A. Salomaa and M. Soittola. *Automata-Theoretic Aspects of Formal Power Series.* Springer, Berlin, 1977.
46. M.P. Schützenberger. On the definition of a family of automata. *Information and Control*, 4:245–270, 1961.
47. M.P. Schützenberger. Certain elementary families of automata. In *Symposium on Mathematical Theory of Automata, 1962*, pages 139–153.
48. M.P. Schützenberger. On a theorem of R. Jungen. *Proceedings of the American Mathematical Society*, 13:885–889, 1962.
49. I. Simon. Limited subsets of a free monoid. In *FOCS'78, 1978*, pages 143–150.
50. E.D. Sontag. On some questions of rationality and decidability. *Journal of Computer and System Sciences*, 11:375–385, 1975.
51. A. Tarski. *Cardinal Algebras.* Oxford University Press, London, 1949.
52. L.G. Valiant. The equivalence problem for deterministic finite-turn pushdown automata. *Information and Control*, 25:123–133, 1974.

Chapter 5: Weighted Automata and Weighted Logics

Manfred Droste[1] and Paul Gastin[2,*]

[1] Institut für Informatik, Universität Leipzig, 04009 Leipzig, Germany
droste@informatik.uni-leipzig.de
[2] LSV, ENS Cachan, INRIA, CNRS, 94230 Cachan, France
Paul.Gastin@lsv.ens-cachan.fr

1 Introduction

In automata theory, Büchi's and Elgot's fundamental theorems [6, 24, 7] established the coincidence of regular and ω-regular languages with languages definable in monadic second-order logic. At the same time, Schützenberger [56] investigated finite automata with weights and characterized their behaviors as rational formal power series. Both of these results have inspired a wealth of extensions and further research, cf. [4, 23, 41, 54, 59] for monographs and surveys as well as Chaps. 3 and 4 of this handbook [26, 55], and also led to recent practical applications, e.g., in verification of finite-state programs (model checking [3, 42, 45]), in digital image compression [11, 32, 34, 35], and

* The second author was partially supported by project ANR-06-SETIN-003 DOTS.

M. Droste, W. Kuich, H. Vogler (eds.), *Handbook of Weighted Automata*, Monographs in Theoretical Computer Science. An EATCS Series, DOI 10.1007/978-3-642-01492-5_5,

in speech-to-text processing [8, 48, 50]; cf. also Chaps. 10, 11, 6, and 14 of the present handbook [28, 1, 49, 37].

It is the goal of this chapter to introduce a logic with weights taken from an arbitrary semiring and to present conditions under which the behaviors of weighted finite automata are precisely the series definable in our weighted monadic second-order logic. We will deal with both finite and infinite words. In comparison to the essential predecessors [13, 14, 19], our logic will be defined in a purely syntactical way, and the results apply to arbitrary (also non-commutative) semirings.

Our motivation for this *weighted logic* is as follows. First, weighted automata and their behavior can be viewed as a quantitative extension of classical automata. The latter decide whether a given word is accepted or not, whereas weighted automata also compute, e.g., the resources, time, or cost used or the probability of its success when executing the word. We would like to have an extension of Büchi's and Elgot's theorems to this setting. Second, classical logic for automata describes whether a certain property (e.g., "there exist three consecutive a's") holds for a given word or not. One could be interested in knowing how often this property holds, i.e., again in extending the previous qualitative statement to a quantitative one.

Next, we describe the syntax of our weighted logics. Its definition incorporates weights taken as elements from a given abstract semiring S, just as done for weighted automata in order to model a variety of applications and situations. Also, our syntax should extend classical (unweighted) MSO logics. The semantics of a weighted logic formula φ should be a formal power series over an extended alphabet and with values in S. It is possible to assign a natural semantics to atomic formulas, to disjunction and conjunction, and to existential and universal quantifications, but a problem arises with negation. It would be natural to define the semantics of $\neg\varphi$ elementwise. But if S is not a Boolean algebra, S does not have a natural complement operation. Therefore, we restrict negation to atomic formulas whose semantics will take as values only 0 and 1 in S; then the negation of atomic formulas also has natural semantics. In comparison to classical MSO-logic, this is not an essential restriction, since the negation of a classical MSO-formula is equivalent (in the sense of defining the same language) to one in which negation is applied only to atomic formulas. This requires us to include conjunction and universal quantifications into our syntax (which we do). In this sense, our weighted MSO-logics then contains the classical MSO-logics which we obtain by letting $S = \mathbb{B}$, the 2-element Boolean algebra.

We define the semantics of sentences φ of our weighted MSO-logic by structural induction over φ. Thus, as usual, we also define the semantics of a formula φ with free variables, here as a formal power series over an extended alphabet. But even for the semiring of natural numbers or the tropical semiring, it turns out that neither universal first-order nor universal second-order quantification of formulas preserve recognizability, i.e., representability of their semantics as behavior of a weighted automaton, and for other (non-commutative) semirings, conjunction does not preserve recognizability. Therefore, we have to restrict conjunction and universal quantifications. We show that each formula in our logic which does not contain weights from the semiring (except 0 or 1) has a syntactic representation which is "unambiguous" and so its associated series takes on only 0 or 1 as values. We permit universal second-order quantification only for such syntactically unambiguous formulas, and universal first-order quantification for formulas in the disjunctive-conjunctive closure of arbitrary constants from the semiring and syntactically unambiguous formulas. With an additional restriction of conjunction, we obtain our class of syntactically restricted weighted MSO-formulas. Moreover, if we allow existential set quantifications only to occur at the beginning of a formula, we arrive at syntactically restricted existential MSO-logic.

Now, we give a summary of our results. First, we show for any semiring S that the behaviors of weighted automata with values in S are precisely the series definable by sentences of our syntactically restricted MSO-logic, or equivalently, of our syntactically restricted existential MSO-logic.

Second, if the semiring S is additively locally finite, we can apply universal first-order quantification even to the existential-disjunctive-conjunctive closure of the set of formulas described above and still obtain that the semantics of such sentences are representable by weighted automata. Third, if the semiring S is (additively and multiplicatively) locally finite, it suffices to just restrict universal second-order quantification, and we still obtain sentences with representable semantics. Locally finite resp. additively locally finite semirings were investigated in [12, 18]; they form large classes of semirings. Fourthly, we also deal with infinite words. As is well known and customary (see [10, 23] and Chap. 3 [26]), here one has to impose certain completeness properties on the semiring, i.e., infinite sums and products exist and interact nicely, in order to ensure that the behavior of weighted automata (and the semantics of weighted formulas) can be defined. Under such suitable completeness assumptions on the semiring, we again obtain that our syntactically restricted MSO-logic (syntactically defined in the same way, but now with semantics on infinite words) is expressively equivalent to a model of weighted Muller automata, and if the semiring is, furthermore, idempotent (like the max-plus- and min-plus-semirings), the same applies to our extension of syntactically restricted MSO-logic described above. We note that we obtain Büchi's and Elgot's theorems for languages of finite and infinite words as particular consequences. Moreover, if the semiring S is given in some effective way, then the constructions in our proofs yield effective conversions of sentences of our weighted logic to weighted automata, and vice versa. If, in addition, S is a field or locally finite, for the case of finite words, we also obtain decision procedures.

2 MSO-Logic and Weighted Automata

In this section, we summarize for the convenience of the reader our notation used for classical MSO-logic and basic background of weighted automata acting on finite words. We assume that the reader is familiar with the basics of monadic second-order logic and Büchi's theorem for languages of finite words; cf. [59, 36]. Let Σ be an alphabet. The syntax of formulas of MSO(Σ), the monadic second-order logic over Σ, is given by the grammar

$$\varphi ::= P_a(x) \mid x \leq y \mid x \in X \mid \varphi \vee \varphi \mid \neg\varphi \mid \exists x.\varphi \mid \exists X.\varphi$$

where a ranges over Σ, x, y are first-order variables and X is a set variable. We let Free(φ) be the set of all free variables of φ.

We let Σ^* be the free monoid of all finite words $w = w(1) \ldots w(n)$ ($n \geq 0$). If $w \in \Sigma^*$ has length n, we put $\mathrm{dom}(w) = \{1, \ldots, n\}$. The word $w \in \Sigma^*$ is usually represented by the structure $(\mathrm{dom}(w), \leq, (R_a)_{a\in\Sigma})$ where $R_a = \{i \in \mathrm{dom}(w) \mid w(i) = a\}$ for $a \in \Sigma$.

Let $\mathcal{V}$ be a finite set of first-order and second-order variables. A $(\mathcal{V}, w)$-assignment σ is a function mapping first-order variables in $\mathcal{V}$ to elements of $\mathrm{dom}(w)$ and second-order variables in $\mathcal{V}$ to subsets of $\mathrm{dom}(w)$. If x is a first-order variable and $i \in \mathrm{dom}(w)$ then $\sigma[x \to i]$ is the $(\mathcal{V} \cup \{x\}, w)$-assignment which assigns x to i and acts like σ on all other variables. Similarly, $\sigma[X \to I]$ is defined for $I \subseteq \mathrm{dom}(w)$. The definition that (w, σ) satisfies φ, denoted $(w, \sigma) \models \varphi$, is as usual assuming that the domain of σ contains Free(φ). Note that $(w, \sigma) \models \varphi$ only depends on the restriction $\sigma_{|\,\mathrm{Free}(\varphi)}$ of σ to Free(φ).

As usual, a pair (w, σ) where σ is a $(\mathcal{V}, w)$-assignment will be encoded using an extended alphabet $\Sigma_{\mathcal{V}} = \Sigma \times \{0,1\}^{\mathcal{V}}$. More precisely, we will write a word over $\Sigma_{\mathcal{V}}$ as a pair (w, σ) where w is the projection over Σ and σ is the projection over $\{0,1\}^{\mathcal{V}}$. Now, σ represents a *valid* assignment over $\mathcal{V}$ if for each first-order variable $x \in \mathcal{V}$, the x-row of σ contains exactly one 1. In this case, we identify σ with the $(\mathcal{V}, w)$-assignment such that for each first-order variable $x \in \mathcal{V}$, $\sigma(x)$ is the position of the 1 on the x-row, and for each second-order variable $X \in \mathcal{V}$, $\sigma(X)$ is the set of positions carrying a 1 on the X-row. Clearly, the language

$$N_{\mathcal{V}} = \{(w, \sigma) \in \Sigma_{\mathcal{V}}^* \mid \sigma \text{ is a valid } (\mathcal{V}, w)\text{-assignment}\}$$

is recognizable. We simply write $\Sigma_\varphi = \Sigma_{\mathrm{Free}(\varphi)}$ and $N_\varphi = N_{\mathrm{Free}(\varphi)}$. By Büchi's theorem, if $\mathrm{Free}(\varphi) \subseteq \mathcal{V}$ then the language

$$\mathcal{L}_{\mathcal{V}}(\varphi) = \{(w, \sigma) \in N_{\mathcal{V}} \mid (w, \sigma) \models \varphi\}$$

defined by φ over $\Sigma_{\mathcal{V}}$ is recognizable. Again, we simply write $\mathcal{L}(\varphi)$ for $\mathcal{L}_{\mathrm{Free}(\varphi)}(\varphi)$. Conversely, each recognizable language L in Σ^* is definable by an MSO-sentence φ, so $L = \mathcal{L}(\varphi)$.

Next, we turn to basic definitions and properties of semirings, formal power series, and weighted automata. For background, we refer the reader to [4, 23, 41, 54] and to Chaps. 1, 3, 4 in this handbook [15, 26, 55].

A *semiring* is a structure $(S, +, \cdot, 0, 1)$ where $(S, +, 0)$ is a commutative monoid, $(S, \cdot, 1)$ is a monoid, multiplication distributes over addition, and $0 \cdot s = s \cdot 0 = 0$ for each $s \in S$. If the multiplication is commutative, we say that S is *commutative*. If the addition is idempotent, then the semiring is called *idempotent*. Important examples include:

- The natural numbers $(\mathbb{N}, +, \cdot, 0, 1)$ with the usual addition and multiplication,
- the Boolean semiring $\mathbb{B} = (\{0, 1\}, \vee, \wedge, 0, 1)$,
- the tropical semiring Trop $= (\mathbb{N} \cup \{\infty\}, \mathsf{min}, +, \infty, 0)$ (also known as min-plus semiring), with min and $+$ extended to $\mathbb{N} \cup \{\infty\}$ in the natural way,
- the arctic semiring Arc $= (\mathbb{N} \cup \{-\infty\}, \mathsf{max}, +, -\infty, 0)$,
- the semiring $([0, 1], \mathsf{max}, \cdot, 0, 1)$ which can be used to compute probabilities,
- the semirings of languages $(\mathcal{P}(\Sigma^*), \cup, \cap, \emptyset, \Sigma^*)$ and $(\mathcal{P}(\Sigma^*), \cup, \cdot, \emptyset, \{\varepsilon\})$.

Given two subsets A, B of a semiring S, we say that A and B *commute elementwise*, if $a \cdot b = b \cdot a$ for all $a \in A$ and $b \in B$. We let S_A denote the subsemiring of S generated by A. Clearly, due to the distributivity law, the elements of S_A can be obtained by taking finite sums of finite products of elements of A. It follows that if $A, B \subseteq S$ and A and B commute elementwise, then S_A and S_B also commute elementwise. If S is a semiring and $n \in \mathbb{N}$, then $S^{n \times n}$ comprises all $(n \times n)$-matrices over S. With usual matrix multiplication and the unit matrix E, $(S^{n \times n}, \cdot, E)$ is a monoid.

A formal power series over a set $\mathcal{Z}$ is a mapping $r : \mathcal{Z} \to S$. In this paper, we will use for $\mathcal{Z}$ either the set Σ^* of finite words, or in Sects. 7 and 8 the set Σ^ω of infinite words. It is usual to write (r, w) for $r(w)$. The set $\text{supp}(r) := \{w \in \mathcal{Z} \mid (r, w) \neq 0\}$ is called the *support* of r. The set of all formal power series over S and $\mathcal{Z}$ is denoted by $S\langle\langle \mathcal{Z} \rangle\rangle$. Now let $r, r_1, r_2 \in S\langle\langle \mathcal{Z} \rangle\rangle$ and $s \in S$. The *sum* $r_1 + r_2$, the *Hadamard product* $r_1 \odot r_2$, and the *scalar products* $s \cdot r$ and $r \cdot s$ are each defined pointwise for $w \in \mathcal{Z}$:

$$\begin{aligned}
(r_1 + r_2, w) &:= (r_1, w) + (r_2, w), \\
(r_1 \odot r_2, w) &:= (r_1, w) \cdot (r_2, w), \\
(s \cdot r, w) &:= s \cdot (r, w), \\
(r \cdot s, w) &:= (r, w) \cdot s.
\end{aligned}$$

Then $(S\langle\langle \mathcal{Z} \rangle\rangle, +, \odot, 0, 1)$ where 0 and 1 denote the constant series with values 0 resp. 1, is again a semiring.

For $L \subseteq \mathcal{Z}$, we define the *characteristic series* $\mathbb{1}_L : \mathcal{Z} \to S$ by $(\mathbb{1}_L, w) = 1$ if $w \in L$, and $(\mathbb{1}_L, w) = 0$, otherwise. If $S = \mathbb{B}$, the correspondence $L \mapsto \mathbb{1}_L$ gives a useful and natural semiring isomorphism from $(\mathcal{P}(\mathcal{Z}), \cup, \cap, \emptyset, \mathcal{Z})$ onto $(\mathbb{B}\langle\langle \mathcal{Z} \rangle\rangle, +, \odot, 0, 1)$.

Now we turn to weighted automata over finite words. We fix a semiring S and an alphabet Σ. A *weighted finite automaton* over S and Σ is a quadruple $\mathcal{A} = (Q, \lambda, \mu, \gamma)$ where Q is a finite set of states, $\mu : \Sigma \to S^{Q \times Q}$ is the transition weight function and $\lambda, \gamma : Q \to S$ are weight functions for entering and leaving a state, respectively. Here, $\mu(a)$ is a $(Q \times Q)$-matrix whose (p,q)-entry $\mu(a)_{p,q} \in S$ indicates the weight (cost) of the transition $p \xrightarrow{a} q$. We also write $\mathrm{wt}(p,a,q) = \mu(a)_{p,q}$. Then μ extends uniquely to a monoid homomorphism (also denoted by μ) from Σ^* into $(S^{Q \times Q}, \cdot, E)$.

The *weight* of a path $P : q_0 \xrightarrow{a_1} q_1 \xrightarrow{a_2} \dots \xrightarrow{a_{n-1}} q_{n-1} \xrightarrow{a_n} q_n$ in $\mathcal{A}$ (where $n \geq 0$) is the product $\mathrm{weight}(P) := \lambda(q_0) \cdot \mu(a_1)_{q_0,q_1} \cdots \mu(a_n)_{q_{n-1},q_n} \cdot \gamma(q_n)$. This path has label $a_1 \dots a_n$. If $n = 0$ and $P = (q_0)$, we have $\mathrm{weight}(P) = \lambda(q_0) \cdot \gamma(q_0)$. The *weight* of a word $w = a_1 \dots a_n \in \Sigma^*$ in $\mathcal{A}$, denoted $(\| \mathcal{A} \|, w)$, is the sum of $\mathrm{weight}(P)$ over all paths P with label w. One can check that

$$(\| \mathcal{A} \|, w) = \sum_{p,q \in Q} \lambda(p) \cdot \mu(w)_{pq} \cdot \gamma(q) = \lambda \cdot \mu(w) \cdot \gamma$$

with usual matrix multiplication, considering λ as a row vector and γ as a column vector. If $w = \varepsilon$, we have $(\| \mathcal{A} \|, \varepsilon) = \lambda \cdot \gamma$. The formal power series $\| \mathcal{A} \| : \Sigma^* \to S$ is called the *behavior* of $\mathcal{A}$. A formal power series $r \in S\langle\langle \Sigma^* \rangle\rangle$ is called *recognizable*, if there exists a weighted finite automaton $\mathcal{A}$ such that $r = \| \mathcal{A} \|$. We let $\mathrm{Rec}(S, \Sigma^*)$ be the collection of all recognizable formal power series over S and Σ.

Lemma 2.1 ([23]; see also Chap. 4 [55]).

(a) For any recognizable language $L \subseteq \Sigma^*$, the series $\mathbb{1}_L$ is recognizable.
(b) Let $r, r_1, r_2 \in S\langle\langle \Sigma^* \rangle\rangle$ be recognizable, and let $s \in S$. Then $r_1 + r_2$, $s \cdot r$ and $r \cdot s$ are recognizable.
(c) Let $S_1, S_2 \subseteq S$ be two sub-semirings such that S_1 and S_2 commute elementwise. Let $r_1 \in \mathrm{Rec}(S_1, \Sigma^*)$ and $r_2 \in \mathrm{Rec}(S_2, \Sigma^*)$. Then $r_1 \odot r_2 \in \mathrm{Rec}(S, \Sigma^*)$.

As an immediate consequence of Lemma 2.1(c), for any recognizable series $r \in S\langle\langle \Sigma^* \rangle\rangle$ and recognizable language $L \subseteq \Sigma^*$, the series $r \odot \mathbb{1}_L$ is again recognizable.

Now let $h : \Sigma^* \to \Gamma^*$ be a homomorphism. For $r \in S\langle\langle \Gamma^* \rangle\rangle$ let $h^{-1}(r) = r \circ h \in S\langle\langle \Sigma^* \rangle\rangle$. That is, $(h^{-1}(r), w) = (r, h(w))$ for all $w \in \Sigma^*$. We call h *length-preserving*, if $|w| = |h(w)|$ for each $w \in \Sigma^*$. We say that h is *non-erasing*, if $h(a) \neq \varepsilon$ for each $a \in \Sigma$, or, equivalently, $|w| \leq |h(w)|$ for each $w \in \Sigma^*$. In this case, for $r \in S\langle\langle \Sigma^* \rangle\rangle$, define $h(r) : \Gamma^* \to S$ by $(h(r), v) := \sum_{w \in h^{-1}(v)} (r, w)$ $(v \in \Gamma^*)$, noting that the sum is finite.

Lemma 2.2 ([23]; see also Chap. 4 [55]). *Let $h : \Sigma^* \to \Gamma^*$ be a homomorphism.*

(a) $h^{-1} : S\langle\langle \Gamma^* \rangle\rangle \to S\langle\langle \Sigma^* \rangle\rangle$ preserves recognizability.
(b) Let h be non-erasing. Then $h : S\langle\langle \Sigma^* \rangle\rangle \to S\langle\langle \Gamma^* \rangle\rangle$ preserves recognizability.

We say $r : \Sigma^* \to S$ is a *recognizable step function*, if $r = \sum_{i=1}^{n} s_i \cdot \mathbb{1}_{L_i}$ for some $n \in \mathbb{N}$, $s_i \in S$ and recognizable languages $L_i \subseteq \Sigma^*$ $(i = 1, \dots, n)$. Then clearly r is a recognizable series by Lemma 2.1(a), (b). The following closure result is easy to see.

Lemma 2.3.

(a) (cf. [12]) The class of all recognizable step functions over Σ and S is closed under sum, scalar products, and Hadamard products.
(b) Let $h : \Sigma^* \to \Gamma^*$ be a homomorphism. Then $h^{-1} : S\langle\langle \Gamma^* \rangle\rangle \to S\langle\langle \Sigma^* \rangle\rangle$ preserves recognizable step functions.

Proof. (b) Let $r = \sum_{i=1}^{n} s_i \cdot \mathbb{1}_{L_i}$ be a recognizable step function with recognizable languages $L_i \subseteq \Gamma^*$. Then each language $h^{-1}(L_i) \subseteq \Sigma^*$ is also recognizable, hence $h^{-1}(r) = \sum_{i=1}^{n} s_i \cdot (\mathbb{1}_{L_i} \circ h) = \sum_{i=1}^{n} s_i \cdot \mathbb{1}_{h^{-1}(L_i)}$ is a recognizable step function. □

3 Weighted Logics

In this section, we introduce our weighted logic and study its first properties. We fix a semiring S and an alphabet Σ. For each $a \in \Sigma$, P_a denotes a unary predicate symbol.

Definition 3.1. *The syntax of formulas of the* weighted MSO-logic *is given by the grammar*

$$\begin{aligned} \varphi ::= {} & s \mid P_a(x) \mid \neg P_a(x) \mid x \leq y \mid \neg(x \leq y) \mid x \in X \mid \neg(x \in X) \\ & \mid \varphi \vee \varphi \mid \varphi \wedge \varphi \mid \exists x.\varphi \mid \exists X.\varphi \mid \forall x.\varphi \mid \forall X.\varphi \end{aligned}$$

where $s \in S$ and $a \in \Sigma$. We denote by $\mathrm{MSO}(S, \Sigma)$ *the collection of all such weighted* MSO*-formulas φ.*

Here, we do not permit negation of general formulas due to difficulties defining then their semantics: The semantics of a weighted logic formula φ should be a formal power series over an extended alphabet and with values in S. It would be natural to define the semantics of $\neg\varphi$ elementwise. In fact, this is possible if S is a bounded distributive lattice with complement function, like, e.g., any Boolean algebra or the semiring $S = ([0,1], \mathsf{max}, \mathsf{min}, 0, 1)$ with complement function $x \mapsto 1 - x$ ($x \in [0,1]$), cf. [16] and Chap. 12 [53]. But in general, arbitrary semirings as well as many important specific semirings do not have a natural complement function.

Therefore, as noted in the Introduction, we restrict negation to atomic formulas whose semantics will take as values only 0 and 1 in S; thus, the negation of atomic formulas takes as values 1 and 0. Since the negation of a classical MSO-formula is equivalent (in the sense of defining the same language) to one in which negation is applied only to atomic formulas, in this sense our weighted MSO-logic contains the classical MSO-logic which we obtain by letting $S = \mathbb{B}$. Note that in this case, the constant s in the logic is either 0 (false) or 1 (true).

Now we turn to the definition of the semantics of formulas $\varphi \in \mathrm{MSO}(S, \Sigma)$. As usual, a variable is said to be *free* in φ if there is an occurrence of it in

φ not in the scope of a quantifier. A pair (w, σ) where $w \in \Sigma^*$ and σ is a $(\mathcal{V}, w)$-assignment is represented by a word over the extended alphabet $\Sigma_\mathcal{V}$ as explained in Sect. 2. We will define the $\mathcal{V}$-semantics $[\![\varphi]\!]_\mathcal{V}$ of φ as a formal power series $[\![\varphi]\!]_\mathcal{V} : \Sigma_\mathcal{V}^* \to S$. This will enable us to investigate when $[\![\varphi]\!]_\mathcal{V}$ is a recognizable series. Also, by letting $S = \mathbb{B}$, the Boolean semiring, we can immediately compare our semantics with the classical one assigning languages to formulas.

Definition 3.2. *Let $\varphi \in \mathrm{MSO}(S, \Sigma)$ and $\mathcal{V}$ be a finite set of variables containing* $\mathrm{Free}(\varphi)$. *The $\mathcal{V}$-*semantics *of φ is a formal power series $[\![\varphi]\!]_\mathcal{V} \in S\langle\langle \Sigma_\mathcal{V}^* \rangle\rangle$. Let $(w, \sigma) \in \Sigma_\mathcal{V}^*$. If σ is not a valid $(\mathcal{V}, w)$-assignment, then we put $[\![\varphi]\!]_\mathcal{V}(w, \sigma) = 0$. Otherwise, we define $[\![\varphi]\!]_\mathcal{V}(w, \sigma) \in S$ inductively as follows:*

$$[\![s]\!]_\mathcal{V}(w,\sigma) = s$$

$$[\![P_a(x)]\!]_\mathcal{V}(w,\sigma) = \begin{cases} 1 & \text{if } w(\sigma(x)) = a \\ 0 & \text{otherwise} \end{cases}$$

$$[\![x \le y]\!]_\mathcal{V}(w,\sigma) = \begin{cases} 1 & \text{if } \sigma(x) \le \sigma(y) \\ 0 & \text{otherwise} \end{cases}$$

$$[\![x \in X]\!]_\mathcal{V}(w,\sigma) = \begin{cases} 1 & \text{if } \sigma(x) \in \sigma(X) \\ 0 & \text{otherwise} \end{cases}$$

$$[\![\neg\varphi]\!]_\mathcal{V}(w,\sigma) = \begin{cases} 1 & \text{if } [\![\varphi]\!]_\mathcal{V}(w,\sigma) = 0 \\ 0 & \text{if } [\![\varphi]\!]_\mathcal{V}(w,\sigma) = 1 \end{cases} \quad \begin{matrix} \text{if } \varphi \text{ is of the form } P_a(x), \\ (x \le y) \text{ or } (x \in X) \end{matrix}$$

$$[\![\varphi \vee \psi]\!]_\mathcal{V}(w,\sigma) = [\![\varphi]\!]_\mathcal{V}(w,\sigma) + [\![\psi]\!]_\mathcal{V}(w,\sigma)$$

$$[\![\varphi \wedge \psi]\!]_\mathcal{V}(w,\sigma) = [\![\varphi]\!]_\mathcal{V}(w,\sigma) \cdot [\![\psi]\!]_\mathcal{V}(w,\sigma)$$

$$[\![\exists x.\varphi]\!]_\mathcal{V}(w,\sigma) = \sum_{i \in \mathrm{dom}(w)} [\![\varphi]\!]_{\mathcal{V}\cup\{x\}}(w, \sigma[x \to i])$$

$$[\![\exists X.\varphi]\!]_\mathcal{V}(w,\sigma) = \sum_{I \subseteq \mathrm{dom}(w)} [\![\varphi]\!]_{\mathcal{V}\cup\{X\}}(w, \sigma[X \to I])$$

$$[\![\forall x.\varphi]\!]_\mathcal{V}(w,\sigma) = \prod_{i \in \mathrm{dom}(w)} [\![\varphi]\!]_{\mathcal{V}\cup\{x\}}(w, \sigma[x \to i])$$

$$[\![\forall X.\varphi]\!]_\mathcal{V}(w,\sigma) = \prod_{I \subseteq \mathrm{dom}(w)} [\![\varphi]\!]_{\mathcal{V}\cup\{X\}}(w, \sigma[X \to I])$$

where in the product over $\mathrm{dom}(w)$ *we follow the natural order, and we fix some order on the power set of $\{1, \ldots, |w|\}$ so that the last product is defined. We simply write $[\![\varphi]\!]$ for $[\![\varphi]\!]_{\mathrm{Free}(\varphi)}$.*

Note that if φ is a sentence, i.e., has no free variables, then $[\![\varphi]\!] \in S\langle\langle \Sigma^* \rangle\rangle$. We give several examples of possible interpretations for weighted formulas:

I. Let S be an arbitrary Boolean algebra $(B, \vee, \wedge, \bar{\ }, 0, 1)$. In this case, sums correspond to suprema, and products to infima. Here, we can define the semantics of $\neg\varphi$ for an arbitrary formula φ by $[\![\neg\varphi]\!](w, \sigma) :=$

$\overline{[\![\varphi]\!](w,\sigma)}$, the complement of $[\![\varphi]\!](w,\sigma)$ in B. Then clearly $[\![\varphi \wedge \psi]\!] = [\![\neg(\neg\varphi \vee \neg\psi)]\!]$, $[\![\forall x.\varphi]\!] = [\![\neg(\exists x.\neg\varphi)]\!]$ and $[\![\forall X.\varphi]\!] = [\![\neg(\exists X.\neg\varphi)]\!]$. This may be interpreted as a multi-valued logic. In particular, if $S = \mathbb{B}$, the 2-valued Boolean algebra, our semantics coincides with the usual semantics of unweighted MSO-formulas, identifying characteristic series with their supports. For the more general case where S is a bounded distributive lattice with complement function, we refer the reader to Chap. 12 [53].

II. Let $S = (\mathbb{N}, +, \cdot, 0, 1)$ and assume φ does not contain constants $s \in \mathbb{N}$. We may interpret $[\![\varphi]\!](w,\sigma)$ as the number of proofs or arguments we have that (w,σ) satisfies the formula φ. Here, the notion of "proof" should not be considered in an exact proof-theoretic, but in an intuitive sense. Indeed, for atomic formulas, the number of proofs should be 0 or 1, depending on whether φ holds for (w,σ) or not. Now if, e.g., $[\![\varphi]\!](w,\sigma) = m$ and $[\![\psi]\!](w,\sigma) = n$, the number of proofs that (w,σ) satisfies $\varphi \vee \psi$ should be $m+n$ (since any proof suffices), and for $\varphi \wedge \psi$ it should be $m \cdot n$ (since we may pair the proofs of φ and ψ arbitrarily). Similarly, the semantics of the existential and universal quantifiers can be interpreted.

III. The formula $\exists x.P_a(x)$ counts how often a occurs in the word. Here, *how often* depends on the semiring: e.g., natural numbers, Boolean semiring, integers modulo 2, ...

IV. Consider the probability semiring $S = ([0,1], \mathsf{max}, \cdot, 0, 1)$ and the alphabet $\Sigma = \{a_1, \ldots, a_n\}$. Assume that each letter a_i has a reliability p_i. Then the series assigning to a word its reliability can be given by the first-order formula $\forall x. \bigvee_{1 \le i \le n}(P_{a_i}(x) \wedge p_i)$.

V. Let $S = ([0,1], \mathsf{max}, \otimes, 0, 1)$ where $x \otimes y = \mathsf{max}(0, x+y-1)$, the semiring occurring in the MV-algebra used to define the semantics of Łukasiewicz multi-valued logic [30]. For this semiring, a restriction of Łukasiewicz logic coincides with our weighted MSO-logic [58].

Observe that if $\varphi \in \mathrm{MSO}(S,\Sigma)$, we have defined a semantics $[\![\varphi]\!]_{\mathcal{V}}$ for each finite set of variables $\mathcal{V}$ containing $\mathrm{Free}(\varphi)$. Now, we show that these semantics are consistent with each other.

Proposition 3.3. *Let* $\varphi \in \mathrm{MSO}(S,\Sigma)$ *and* $\mathcal{V}$ *a finite set of variables containing* $\mathrm{Free}(\varphi)$. *Then*

$$[\![\varphi]\!]_{\mathcal{V}}(w,\sigma) = [\![\varphi]\!](w,\sigma_{|\,\mathrm{Free}(\varphi)})$$

for each $(w,\sigma) \in \Sigma_{\mathcal{V}}^*$ *such that* σ *is a valid* $(\mathcal{V},w)$*-assignment. In particular,* $[\![\varphi]\!]$ *is recognizable iff* $[\![\varphi]\!]_{\mathcal{V}}$ *is recognizable, and* $[\![\varphi]\!]$ *is a recognizable step function iff* $[\![\varphi]\!]_{\mathcal{V}}$ *is a recognizable step function.*

Proof. The first claim can be shown by induction on the structure of φ.

For the final claim, consider the projection $\pi : \Sigma_{\mathcal{V}} \to \Sigma_{\varphi}$. For $(w,\sigma) \in \Sigma_{\mathcal{V}}^*$, we have $\pi(w,\sigma) = (w, \sigma_{|\,\mathrm{Free}(\varphi)})$. If $[\![\varphi]\!]$ is recognizable then $[\![\varphi]\!]_{\mathcal{V}} = \pi^{-1}([\![\varphi]\!]) \odot \mathbb{1}_{N_{\mathcal{V}}}$ is recognizable by Lemmas 2.1 and 2.2. This also shows that if $[\![\varphi]\!]$ is a recognizable step function, then so is $[\![\varphi]\!]_{\mathcal{V}}$ by Lemma 2.3.

Conversely, let F comprise the empty word and all $(w,\sigma) \in \Sigma_{\mathcal{V}}^{+}$ such that σ assigns to each variable x (resp. X) in $\mathcal{V} \setminus \mathrm{Free}(\varphi)$ position 1, i.e., $\sigma(x) = 1$ (resp. $\sigma(X) = \{1\}$). Then F is recognizable, and for each $(w,\sigma') \in \Sigma_{\varphi}^{*}$ there is a unique element $(w,\sigma) \in F$ such that $\pi(w,\sigma) = (w,\sigma')$. Thus, $[\![\varphi]\!] = \pi([\![\varphi]\!]_{\mathcal{V}} \odot \mathbb{1}_F)$, as is easy to check. Hence, if $[\![\varphi]\!]_{\mathcal{V}}$ is recognizable, then so is $[\![\varphi]\!]$ by Lemmas 2.1 and 2.2. Finally, note that $[\![\varphi]\!]$ assumes the same non-zero values as $[\![\varphi]\!]_{\mathcal{V}}$, and if $s \in S$, then $[\![\varphi]\!]^{-1}(s) = \pi([\![\varphi]\!]_{\mathcal{V}}^{-1}(s) \cap N_{\mathcal{V}})$ in case $s \neq 0$ or φ contains no free first order variable. If φ contains a free first order variable, then $[\![\varphi]\!]^{-1}(0) = \pi([\![\varphi]\!]_{\mathcal{V}}^{-1}(0) \cap N_{\mathcal{V}}) \cup (\Sigma_{\varphi}^{*} \setminus N_{\varphi})$. Hence, if $[\![\varphi]\!]_{\mathcal{V}}$ is a recognizable step function, so is $[\![\varphi]\!]$. □

Now, let $Z \subseteq \mathrm{MSO}(S,\Sigma)$. A series $r : \Sigma^{*} \to S$ is called *Z-definable*, if there is a sentence $\varphi \in Z$ such that $r = [\![\varphi]\!]$. The main goal of this paper is the comparison of Z-definable with recognizable series, for suitable fragments Z of $\mathrm{MSO}(S,\Sigma)$. Crucial for this will be closure properties of recognizable series under the constructs of our weighted logic. However, it is well known that $\mathrm{Rec}(S,\Sigma^{*})$ is in general not closed under the Hadamard product, and hence not under conjunction.

Example 3.4. Let $\Sigma = \{a,b\}$, $S = (\mathcal{P}(\Sigma^{*}), \cup, \cdot, \emptyset, \{\varepsilon\})$, and consider the formula $\varphi = \forall x.((P_a(x) \wedge \{a\}) \vee (P_b(x) \wedge \{b\}))$. Then $([\![\varphi]\!], w) = \{w\}$ for each $w \in \Sigma^{*}$. Clearly, $[\![\varphi]\!]$ is recognizable. However, $([\![\varphi \wedge \varphi]\!], w) = \{w\} \cdot \{w\} = \{w^2\}$ for each $w \in \Sigma^{*}$, and pumping arguments show that $[\![\varphi \wedge \varphi]\!]$ is not recognizable (cf. [22]).

Next we show that $\mathrm{Rec}(S,\Sigma^{*})$ is, in general, not closed under universal quantification.

Example 3.5 (cf. [14]). Let $S = (\mathbb{N}, +, \cdot, 0, 1)$. Then $[\![\forall x.2]\!](w) = 2^{|w|}$ and $[\![\forall y \forall x.2]\!](w) = (2^{|w|})^{|w|} = 2^{|w|^2}$. Clearly, the series $[\![\forall x.2]\!]$ is recognizable by the weighted automaton (Q,λ,μ,γ) with $Q = \{1\}$, $\lambda_1 = \gamma_1 = 1$ and $\mu(a)_{1,1} = 2$ for all $a \in \Sigma$. However, $[\![\forall y \forall x.2]\!]$ is not recognizable. Suppose there was an automaton $\mathcal{A}' = (Q',\lambda',\mu',\gamma')$ with behavior $[\![\forall y \forall x.2]\!]$. Let $M = \max\{|\lambda'_p|, |\gamma'_p|, |\mu'(a)_{p,q}| \mid p,q \in Q', a \in \Sigma\}$. Then for any $w \in \Sigma^{*}$ and for each path P labeled by w, we have $\mathrm{weight}(P) \leq M^{|w|+2}$ and since there are $|Q'|^{|w|+1}$ paths labeled w we obtain $(\|\mathcal{A}'\|, w) \leq |Q'|^{|w|+1} \cdot M^{|w|+2}$, a contradiction with $(\|\mathcal{A}'\|, w) = 2^{|w|^2}$.

A similar argument applies also for the tropical and the arctic semirings. Observe that in all these cases, $[\![\forall x.2]\!]$ has infinite image.

Example 3.6 (cf. [18]). Let $S = (\mathbb{N}, +, \cdot, 0, 1)$. Then $([\![\exists x.1]\!], w) = |w|$ and $[\![\forall y.\exists x.1]\!], w) = |w|^{|w|}$ for each $w \in \Sigma^{*}$. Hence $[\![\exists x.1]\!]$ is recognizable, but $[\![\forall y.\exists x.1]\!]$ is not, by the argument of the previous example. In contrast, if S is the tropical or arctic semiring (and 1 still the natural number 1), then $[\![\exists x.1]\!]$ takes on only two values, and $[\![\forall y.\exists x.1]\!]$ is recognizable.

Example 3.7. Let $S = (\mathbb{N}, +, \cdot, 0, 1)$. Then $[\![\forall X.2]\!](w) = 2^{2^{|w|}}$ for any $w \in \Sigma^*$, and as above $[\![\forall X.2]\!]$ is not recognizable due to its growth. Again, this counterexample also works for the tropical and the arctic semirings.

The examples show that unrestricted conjunction and universal quantification are in general too strong to preserve recognizability. Therefore we will consider fragments of $\mathrm{MSO}(S, \Sigma)$. Their syntactic definition needs a little preparation on unambiguous formulas.

4 Unambiguous Formulas

In all of this section, let S be a semiring and Σ an alphabet. Here we will define our concepts of unambiguous and of syntactically unambiguous MSO-formulas. The idea is that if φ, ψ are formulas whose semantics $[\![\varphi]\!]$, $[\![\psi]\!]$ each takes on only 0 and 1 as values, this is in general no longer true for $\varphi \vee \psi$, $\exists x.\varphi$ and $\exists X.\varphi$ (except if S is idempotent), but we can find "equivalent" constructs assuming only $0, 1$ as values and for these formulas, the Boolean semantics will coincide with the weighted semantics. The unambiguous formulas may be viewed as the logical counterpart of unambiguous rational expressions (and may therefore have independent interest). We let $\mathrm{MSO}^-(S, \Sigma)$ consist of all formulas of $\mathrm{MSO}(S, \Sigma)$ which do not contain constants $s \in S \setminus \{0, 1\}$.

Definition 4.1. *The class of* unambiguous *formulas in* $\mathrm{MSO}^-(S, \Sigma)$ *is defined inductively as follows:*

1. *All atomic formulas in* $\mathrm{MSO}^-(S, \Sigma)$ *and their negations are unambiguous.*
2. *If* φ, ψ *are unambiguous, then* $\varphi \wedge \psi$, $\forall x.\varphi$ *and* $\forall X.\varphi$ *are also unambiguous.*
3. *If* φ, ψ *are unambiguous and* $\mathrm{supp}([\![\varphi]\!]) \cap \mathrm{supp}([\![\psi]\!]) = \emptyset$, *then* $\varphi \vee \psi$ *is unambiguous.*
4. *Let* φ *be unambiguous and* $\mathcal{V} = \mathrm{Free}(\varphi)$. *If for any* $(w, \sigma) \in \Sigma_{\mathcal{V}}^*$, *there is at most one element* $i \in \mathrm{dom}(w)$ *such that* $[\![\varphi]\!]_{\mathcal{V} \cup \{x\}}(w, \sigma[x \to i]) \neq 0$, *then* $\exists x.\varphi$ *is unambiguous.*
5. *Let* φ *be unambiguous and* $\mathcal{V} = \mathrm{Free}(\varphi)$. *If for any* $(w, \sigma) \in \Sigma_{\mathcal{V}}^*$, *there is at most one subset* $I \subseteq \mathrm{dom}(w)$ *such that* $[\![\varphi]\!]_{\mathcal{V} \cup \{X\}}(w, \sigma[X \to I]) \neq 0$, *then* $\exists X.\varphi$ *is unambiguous.*

Note that, as for unambiguous rational expressions, this is not a purely syntactic definition since some restrictions are on the semantics of formulas. First we note the following proposition.

Proposition 4.2. *Let* $\varphi \in \mathrm{MSO}^-(S, \Sigma)$ *be unambiguous. We may also regard* φ *as a classical* MSO*-formula defining the language* $\mathcal{L}(\varphi) \subseteq \Sigma_\varphi^*$. *Then* $[\![\varphi]\!] = \mathbb{1}_{\mathcal{L}(\varphi)}$ *is a recognizable step function.*

Proof. Let $(w,\sigma) \in \Sigma_\varphi^*$. If $(w,\sigma) \notin N_\varphi$, then $[\![\varphi]\!](w,\sigma) = 0$ and $(w,\sigma) \notin \mathcal{L}(\varphi)$. Assume now that $(w,\sigma) \in N_\varphi$. We show by structural induction on φ that $[\![\varphi]\!](w,\sigma)$ equals 1 if $(w,\sigma) \models \varphi$ and equals 0, otherwise. This is clear for the atomic formulas and their negations. It is also trivial by induction for conjunction and universal quantifications. Using the unambiguity of the formulas, we also get the result by induction for disjunction and existential quantifications. Therefore, $[\![\varphi]\!] = \mathbb{1}_{\mathcal{L}(\varphi)}$ and since $\mathcal{L}(\varphi)$ is a recognizable language in Σ_φ^* we obtain that $[\![\varphi]\!]$ is a recognizable step function. □

Next we wish to give a purely syntactic definition of a class of unambiguous formulas and then show that any classical MSO-formula can be effectively transformed into an equivalent one which is syntactically unambiguous. We will proceed by structural induction on the given formula. Here (in contrast to [14]), we will include the case of formulas containing set quantifiers. When dealing with formulas of the form $\exists X.\varphi$ and $\forall X.\varphi$, we employ a linear order on the underlying structure (which is the power set of $\mathrm{dom}(w)$ where $w \in \Sigma^*$). For this, we recall that we identify (in assignments) subsets of $\mathrm{dom}(w)$ with their characteristic functions, and the set $\{0,1\}^{\mathrm{dom}(w)}$ carries the lexicographic order as a natural linear order. Let $y < x = \neg(x \leq y)$.

Definition 4.3. *For any $\varphi, \psi \in \mathrm{MSO}^-(S,\Sigma)$, we define inductively formulas φ^+, φ^-, $\varphi \xrightarrow{+} \psi$ and $\varphi \xleftrightarrow{+} \psi$ in $\mathrm{MSO}^-(S,\Sigma)$ by the following rules:*

1. *If φ is atomic or the negation of an atomic formula, we put $\varphi^+ = \varphi$ and $\varphi^- = \neg\varphi$ with the convention $\neg\neg\psi = \psi$, and $\neg 0 = 1$, $\neg 1 = 0$.*
2. $(\varphi \vee \psi)^+ = \varphi^+ \vee (\varphi^- \wedge \psi^+)$ *and* $(\varphi \vee \psi)^- = \varphi^- \wedge \psi^-$.
3. $(\varphi \wedge \psi)^- = \varphi^- \vee (\varphi^+ \wedge \psi^-)$ *and* $(\varphi \wedge \psi)^+ = \varphi^+ \wedge \psi^+$.
4. $(\exists x.\varphi)^+ = \exists x.(\varphi^+(x) \wedge \forall y.(y < x \wedge \varphi(y))^-)$ *and* $(\exists x.\varphi)^- = \forall x.\varphi^-$.
5. $(\forall x.\varphi)^- = \exists x.(\varphi^-(x) \wedge \forall y.(x \leq y \vee \varphi(y))^+)$ *and* $(\forall x.\varphi)^+ = \forall x.\varphi^+$.
6. $\varphi \xrightarrow{+} \psi = \varphi^- \vee (\varphi^+ \wedge \psi^+)$ *and* $\varphi \xleftrightarrow{+} \psi = (\varphi^+ \wedge \psi^+) \vee (\varphi^- \wedge \psi^-)$.
7. *For set variables X, Y, we define the following macros:*[3]

$$
\begin{aligned}
(X = Y) &= \forall z.(z \in X \xleftrightarrow{+} z \in Y) \\
(X < Y) &= \exists y.((y \in Y) \wedge \neg(y \in X) \wedge \forall z.(z < y \xrightarrow{+} (z \in X \xleftrightarrow{+} z \in Y))) \\
(X \leq Y) &= (X = Y) \vee (X < Y).
\end{aligned}
$$

8. $(\exists X.\varphi)^+ = \exists X.(\varphi^+(X) \wedge \forall Y.((Y < X) \wedge \varphi(Y))^-)$ *and* $(\exists X.\varphi)^- = \forall X.\varphi^-$.
9. $(\forall X.\varphi)^- = \exists X.(\varphi^-(X) \wedge \forall Y.((X \leq Y) \vee \varphi(Y))^+)$ *and* $(\forall X.\varphi)^+ = \forall X.\varphi^+$.

We define the class of *(unweighted) syntactically unambiguous formulas* as the smallest class of formulas containing all formulas of the form

- $\varphi^+, \varphi^-, \varphi \xrightarrow{+} \psi$ or $\varphi \xleftrightarrow{+} \psi$ for $\varphi, \psi \in \mathrm{MSO}^-(S,\Sigma)$, and

[3] The authors are thankful to Christian Mathissen for this formula $X < Y$ which simplifies an earlier more complicated formula of the authors.

- $\forall x.\varphi, \forall X.\varphi$ or $\varphi \wedge \psi$ if it contains φ and ψ.

By induction, it is easy to show the following lemma.

Lemma 4.4. *Let* $\varphi \in \mathrm{MSO}^-(S, \Sigma)$. *Then:*

- $\mathcal{L}(\varphi^+) = \mathcal{L}(\varphi)$ *and* $\mathcal{L}(\varphi^-) = \mathcal{L}(\neg\varphi)$.
- $[\![\varphi^+]\!] = \mathbb{1}_{\mathcal{L}(\varphi)}$ *and* $[\![\varphi^-]\!] = \mathbb{1}_{\mathcal{L}(\neg\varphi)}$.
- *Each syntactically unambiguous formula is unambiguous.*

The following result is a slight improvement of [14, Proposition 5.4].

Proposition 4.5. *For each classical* MSO*-sentence* φ*, we can effectively construct an unweighted syntactically unambiguous* $\mathrm{MSO}(S, \Sigma)$*-sentence* φ' *defining the same language, i.e.,* $[\![\varphi']\!] = \mathbb{1}_{\mathcal{L}(\varphi)}$.

Proof. Using also conjunctions and universal quantifications, transform φ into an equivalent MSO-sentence ψ in which negation is only applied to atomic formulas. Then put $\varphi' = \psi^+$. □

We define $\mathrm{aUMSO}(S, \Sigma)$, the collection of *almost unambiguous* formulas in $\mathrm{MSO}(S, \Sigma)$, to be the smallest subset of $\mathrm{MSO}(S, \Sigma)$ containing all constants s ($s \in S$) and all syntactically unambiguous formulas and which is closed under disjunction and conjunction.

We call two formulas $\varphi, \psi \in \mathrm{MSO}(S, \Sigma)$ *equivalent*, denoted $\varphi \equiv \psi$, if $[\![\varphi]\!] = [\![\psi]\!]$. Now we claim that each almost unambiguous formula ψ is equivalent to a formula ψ' of the form $\psi' = \bigvee_{j=1}^{n}(s_j \wedge \psi_j^+)$ for some $n \in \mathbb{N}$, $s_j \in S$ and $\psi_j \in \mathrm{MSO}^-(S, \Sigma)$ ($j = 1, \ldots, n$). Indeed, this follows from the following equivalences for any $\varphi, \xi, \zeta \in \mathrm{MSO}(S, \Sigma)$, $\pi, \rho \in \mathrm{MSO}^-(S, \Sigma)$ and $s, t \in S$:

$$\begin{aligned}
\varphi \wedge (\xi \vee \zeta) &\equiv (\varphi \wedge \xi) \vee (\varphi \wedge \zeta),\\
\pi^+ \wedge s &\equiv s \wedge \pi^+,\\
\pi^+ &\equiv 1 \wedge \pi^+,\\
s \wedge t &\equiv st,\\
\pi &\equiv \pi^+ \quad \text{if } \pi \text{ is unambiguous.}
\end{aligned}$$

Moreover, by forming suitable conjunctions of the formulas ψ_j^+, ψ_j^- in ψ' above, we can obtain that the languages $\mathcal{L}_{\mathrm{Free}(\psi')}(\psi_j)$ ($j = 1, \ldots, n$) are pairwise disjoint; then ψ' could be viewed as a "weighted unambiguous" formula similar to Definition 4.1 (we will not need this notion, but it also motivates the notion "almost unambiguous" for ψ).

As a consequence of this description (or Lemma 2.3) and Lemma 4.4, for each $\psi \in \mathrm{aUMSO}(S, \Sigma)$, $[\![\psi]\!]$ is a recognizable step function.

For an arbitrary formula $\varphi \in \mathrm{MSO}(S, \Sigma)$, let $\mathrm{val}(\varphi)$ denote the set containing all values of S occurring in φ.

Next, we turn to the definition of our (weighted) syntactically restricted $\mathrm{MSO}(S, \Sigma)$-formulas[4]:

[4] The authors would like to thank Dietrich Kuske for joint discussions which led to the development of this crucial concept.

Definition 4.6. *A formula $\varphi \in \mathrm{MSO}(S, \Sigma)$ is called* syntactically restricted, *if it satisfies the following conditions:*

1. *Whenever φ contains a conjunction $\psi \wedge \psi'$ as subformula, but not in the scope of a universal first order quantifier, then* $\mathrm{val}(\psi)$ *and* $\mathrm{val}(\psi')$ *commute elementwise.*
2. *Whenever φ contains $\forall X.\psi$ as a subformula, then ψ is an unweighted syntactically unambiguous formula.*
3. *Whenever φ contains $\forall x.\psi$ as a subformula, then ψ is almost unambiguous.*

We let $\mathrm{sRMSO}(S, \Sigma)$ *denote the set of all syntactically restricted formulas of* $\mathrm{MSO}(S, \Sigma)$.

Here, condition (1) requires us to be able to check for $s, s' \in S$ whether $s \cdot s' = s' \cdot s$. We assume this basic ability to be given in syntax checks of formulas from $\mathrm{MSO}(S, \Sigma)$. Note that for $\psi, \psi' \in \mathrm{MSO}(S, \Sigma)$, $\mathrm{val}(\psi)$ and $\mathrm{val}(\psi')$ trivially commute elementwise, if S is commutative (which was the general assumption of [14]) or if ψ or ψ' is in $\mathrm{MSO}^-(S, \Sigma)$, thus in particular, if ψ or ψ' is unambiguous. Hence, for each $\mathrm{MSO}(S, \Sigma)$-formula φ, it can be easily checked effectively whether φ is syntactically restricted or not.

A formula $\varphi \in \mathrm{MSO}(S, \Sigma)$ is *existential*, if it is of the form $\varphi = \exists X_1. \ldots \exists X_n.\psi$ where ψ does not contain any set quantifier. The set of all syntactically restricted and existential formulas of $\mathrm{MSO}(S, \Sigma)$ is denoted by $\mathrm{sREMSO}(S, \Sigma)$.

Our first main result which will be proved in Sect. 5 is the following theorem.

Theorem 4.7. *Let S be any semiring and Σ an alphabet. Let $r : \Sigma^* \to S$ be a series. The following are equivalent:*

1. *r is recognizable.*
2. *r is definable by some syntactically restricted sentence of* $\mathrm{MSO}(S, \Sigma)$.
3. *r is definable by some syntactically restricted existential sentence of* $\mathrm{MSO}(S, \Sigma)$.

We note that our proofs will be effective. That is, given a syntactically restricted sentence φ of $\mathrm{MSO}(S, \Sigma)$, we can construct a weighted automaton $\mathcal{A}$ with $\| \mathcal{A} \| = [\![\varphi]\!]$ (provided the operations of S are given effectively). For the converse, given $\mathcal{A}$, we will explicitly describe a sentence $\varphi \in \mathrm{sREMSO}(S, \Sigma)$ with $[\![\varphi]\!] = \| \mathcal{A} \|$.

Slightly extending [14], we call an $\mathrm{MSO}(S, \Sigma)$-formula φ *restricted*, if:

1. Whenever φ contains a conjunction $\psi \wedge \psi'$ as subformula, but not in the scope of a universal first order quantifier, then $\mathrm{val}(\psi)$ and $\mathrm{val}(\psi')$ commute elementwise.
2. Whenever φ contains $\forall X.\psi$ as a subformula, then ψ is an unambiguous formula.

3. Whenever φ contains $\forall x.\psi$ as a subformula, then $[\![\psi]\!]$ is a recognizable step function.

Note that in particular conditions (2) and (3) are not purely syntactic, but use the semantics of formulas. In [14], it was shown that if S is a field or a locally finite semiring (cf. Sect. 6), then it can be effectively checked whether an arbitrary $\mathrm{MSO}(S, \Sigma)$-sentence φ is restricted or not. For the general case, this remained open.

Since, as noted before, the semantics of almost unambiguous formulas are recognizable step functions, we have the following proposition.

Proposition 4.8. *Each syntactically restricted formula* $\varphi \in \mathrm{MSO}(S, \Sigma)$ *is restricted.*

5 Definability Equals Recognizability

In all of this section, let S be a semiring and Σ an alphabet. We wish to prove Theorem 4.7. For this, we first wish to show that whenever $\varphi \in \mathrm{MSO}(S, \Sigma)$ is restricted, then $[\![\varphi]\!]$ is recognizable. We proceed by induction over the structure of restricted MSO-formulas.

Lemma 5.1. *Let* $\varphi \in \mathrm{MSO}(S, \Sigma)$ *be atomic or the negation of an atomic formula. Then* $[\![\varphi]\!]$ *is a recognizable step function.*

Proof. If $\varphi = s$ with $s \in S$, we have $[\![\varphi]\!] = s \cdot \mathbb{1}_{\Sigma^*}$. If φ is one of the other atomic formulas or their negations, then $[\![\varphi]\!] = \mathbb{1}_{\mathcal{L}(\varphi)}$ is immediate from the definition. □

Lemma 5.2. *Let* $\varphi, \psi \in \mathrm{MSO}(S, \Sigma)$ *such that* $[\![\varphi]\!]$ *and* $[\![\psi]\!]$ *are recognizable. Then* $[\![\varphi \vee \psi]\!]$, $[\![\exists x.\varphi]\!]$ *and* $[\![\exists X.\varphi]\!]$ *are recognizable. Moreover, if* $[\![\varphi]\!]$ *and* $[\![\psi]\!]$ *are recognizable step functions, then* $[\![\varphi \vee \psi]\!]$ *is also a recognizable step function.*

Proof. For the disjunction, let $\mathcal{V} = \mathrm{Free}(\varphi) \cup \mathrm{Free}(\psi)$. By definition, we have $[\![\varphi \vee \psi]\!] = [\![\varphi]\!]_{\mathcal{V}} + [\![\psi]\!]_{\mathcal{V}}$. Hence the result follows from Proposition 3.3 and Lemma 2.1, resp. 2.3.

For the existential quantifiers, let $\mathcal{X}$ be the variable x or X. Let $\mathcal{V} = \mathrm{Free}(\exists \mathcal{X}.\varphi)$ and note that $\mathcal{X} \notin \mathcal{V}$ and $\mathrm{Free}(\varphi) \subseteq \mathcal{V} \cup \{\mathcal{X}\}$. Consider the projection $\pi : \Sigma^*_{\mathcal{V} \cup \{\mathcal{X}\}} \to \Sigma^*_{\mathcal{V}}$ which erases the $\mathcal{X}$-row. One can show that $[\![\exists \mathcal{X}.\varphi]\!] = \pi([\![\varphi]\!]_{\mathcal{V} \cup \{\mathcal{X}\}})$. Then Proposition 3.3 and Lemma 2.2(b) show that $[\![\exists \mathcal{X}.\varphi]\!]$ is recognizable. □

Next we deal with conjunction. For any formula $\varphi \in \mathrm{MSO}(S, \Sigma)$, we let $S_\varphi = S_{\mathrm{val}(\varphi)}$, the subsemiring of S generated by all constants occurring in φ.

Lemma 5.3. *Let* $\varphi, \psi \in \mathrm{MSO}(S, \Sigma)$.

(a) Assume that $\mathrm{val}(\varphi)$ *and* $\mathrm{val}(\psi)$ *commute elementwise, and that* $[\![\varphi]\!] \in \mathrm{Rec}(S_\varphi, \Sigma_\varphi^*)$ *and* $[\![\psi]\!] \in \mathrm{Rec}(S_\psi, \Sigma_\psi^*)$. *Then* $[\![\varphi \wedge \psi]\!]$ *is recognizable.*
(b) If $[\![\varphi]\!]$ *and* $[\![\psi]\!]$ *are recognizable step functions, so is* $[\![\varphi \wedge \psi]\!]$.

Proof. Let $\mathcal{V} = \mathrm{Free}(\varphi) \cup \mathrm{Free}(\psi)$. By definition, we have $[\![\varphi \wedge \psi]\!] = [\![\varphi]\!]_\mathcal{V} \odot [\![\psi]\!]_\mathcal{V}$.

(a) By Proposition 3.3, we get $[\![\varphi]\!]_\mathcal{V} \in \mathrm{Rec}(S_\varphi, \Sigma_\mathcal{V}^*)$ and $[\![\psi]\!]_\mathcal{V} \in \mathrm{Rec}(S_\psi, \Sigma_\mathcal{V}^*)$. As noted in Sect. 2, S_φ and S_ψ commute elementwise. Hence the result follows from Lemma 2.1(c).

(b) We apply Proposition 3.3 and Lemma 2.3. □

The most interesting case here arises from universal quantification. In [14], a corresponding result was proved under the assumption that S is commutative. The reason that this assumption can be avoided is due to the following. For a word (over an extended alphabet), the semantics of $\forall x.\varphi$ is evaluated along the sequence of positions, just as the weight of a path in a weighted automaton is computed following the sequence of transitions. This will be crucial in the proof.

Lemma 5.4. *Let* $\psi \in \mathrm{MSO}(S, \Sigma)$ *such that* $[\![\psi]\!]$ *is a recognizable step function. Then* $[\![\forall x.\psi]\!]$ *is recognizable.*

Proof. Let $\mathcal{W} = \mathrm{Free}(\psi) \cup \{x\}$ and $\mathcal{V} = \mathrm{Free}(\forall x.\psi) = \mathcal{W} \setminus \{x\}$. By Proposition 3.3 (in case $x \notin \mathrm{Free}(\psi)$), $[\![\psi]\!]_\mathcal{W}$ is a recognizable step function. We may write $[\![\psi]\!]_\mathcal{W} = \sum_{j=1}^n s_j \cdot \mathbb{1}_{L_j}$ with $n \in \mathbb{N}$, $s_j \in S$ and recognizable languages $L_1, \ldots, L_n \subseteq \Sigma_\mathcal{W}^*$ such that $(L_1, \ldots, L_n)$ is a partition of $N_\mathcal{W}$. Recall that if $(w, \sigma) \in (\Sigma_\mathcal{W})^* \setminus N_\mathcal{W}$ then $[\![\psi]\!](w, \sigma) = 0$.

Let $\widetilde{\Sigma} = \Sigma \times \{1, \ldots, n\}$. A word in $(\widetilde{\Sigma}_\mathcal{V})^*$ will be written (w, ν, σ) where $(w, \sigma) \in \Sigma_\mathcal{V}^*$ and $\nu \in \{1, \ldots, n\}^{|w|}$ is interpreted as a mapping from $\mathrm{dom}(w)$ to $\{1, \ldots, n\}$. Let $\widetilde{L}$ be the set of $(w, \nu, \sigma) \in (\widetilde{\Sigma}_\mathcal{V})^*$ such that $(w, \sigma) \in N_\mathcal{V}$ and for all $i \in \mathrm{dom}(w)$ and $j \in \{1, \ldots, n\}$ we have

$$\nu(i) = j \quad \text{implies} \quad (w, \sigma[x \to i]) \in L_j.$$

Observe that for each $(w, \sigma) \in N_\mathcal{V}$ there is a unique ν such that $(w, \nu, \sigma) \in \widetilde{L}$ since $(L_1, \ldots, L_n)$ is a partition of $N_\mathcal{W}$.

We claim that $\widetilde{L}$ is recognizable. In [14, proof of Lemma 4.4], we constructed directly an automaton recognizing $\widetilde{L}$. Here, we give an unpublished argument (already developed for [13, 14]) using Büchi's theorem.

First, let $\xi \in \mathrm{MSO}(\Sigma)$ be an arbitrary MSO formula. Define $\tilde{\xi}$ by replacing in ξ any occurrence of $P_a(y)$ by $\bigvee_{1 \le k \le n} P_{(a,k)}(y)$. Then assuming that $\mathrm{Free}(\xi) \subseteq \mathcal{U}$, it is easy to check by structural induction on ξ that for all $(w, \nu, \sigma) \in (\widetilde{\Sigma}_\mathcal{U})^*$ with $(w, \sigma) \in N_\mathcal{U}$ we have $(w, \nu, \sigma) \models \tilde{\xi}$ if and only if $(w, \sigma) \models \xi$.

By Büchi's theorem, there is an MSO formula ψ_j with $\mathrm{Free}(\psi_j) \subseteq \mathcal{W}$ such that for all $(w, \tau) \in N_\mathcal{W}$ we have $(w, \tau) \in L_j$ if and only if $(w, \tau) \models \psi_j$. Now, we define

$$\zeta = \forall x. \Big(\bigwedge_{1 \leq j \leq n} \bigvee_{a \in \Sigma} P_{(a,j)}(x) \to \widetilde{\psi_j} \Big).$$

Let $(w, \nu, \sigma) \in (\widetilde{\Sigma}_{\mathcal{V}})^*$ with $(w, \sigma) \in N_{\mathcal{V}}$. We have $(w, \nu, \sigma) \models \zeta$ if and only if for all $i \in \operatorname{dom}(w)$ and $j \in \{1, \ldots, n\}$ we have

$$\nu(i) = j \quad \text{implies} \quad (w, \nu, \sigma[x \mapsto i]) \models \widetilde{\psi_j}$$

and this last statement is equivalent with $(w, \sigma[x \mapsto i]) \models \psi_j$ which in turn is equivalent with $(w, \sigma[x \mapsto i]) \in L_j$. Therefore, the formula ζ defines the language $\widetilde{L}$ and our claim is proved.

Now, we proceed similar as in [14] with slight changes as in [18] since here S might not be commutative. There is a deterministic automaton $\widetilde{\mathcal{A}}$ over the alphabet $\widetilde{\Sigma}_{\mathcal{V}}$, recognizing $\widetilde{L}$. Now we obtain a weighted automaton $\mathcal{A}$ with the same state set by adding weights to the transitions of $\widetilde{\mathcal{A}}$ as follows: If $(p, (a, j, s), q)$ is a transition in $\widetilde{\mathcal{A}}$ with $(a, j, s) \in \widetilde{\Sigma}_{\mathcal{V}}$, we let this transition in $\mathcal{A}$ have weight s_j, i.e., $\mu_{\mathcal{A}}(a, j, s)_{p,q} = s_j$. All triples which are not transitions in $\widetilde{\mathcal{A}}$ get weight 0. Also, the initial state of $\widetilde{\mathcal{A}}$ gets initial weight 1 in $\mathcal{A}$, all non-initial states of $\widetilde{\mathcal{A}}$ get initial weight 0, and similarly for the final states and final weights.

Since $\widetilde{\mathcal{A}}$ is deterministic, for each $(w, \nu, \sigma) \in \widetilde{L}$ there is a unique path $P_w = (t_i)_{1 \leq i \leq |w|}$ in $\widetilde{\mathcal{A}}$ and we have in $\mathcal{A}$

$$\big(\| \mathcal{A} \|, (w, \nu, \sigma)\big) = \operatorname{weight}(P_w) = \prod_{i \in \operatorname{dom}(w)} \operatorname{wt}(t_i)$$

whereas $(\|\mathcal{A}\|, (w, \nu, \sigma)) = 0$ for each $(w, \nu, \sigma) \in \widetilde{\Sigma}^*_{\mathcal{V}} \setminus \widetilde{L}$. For each $i \in \operatorname{dom}(w)$, note that if $\nu(i) = j$, then $\operatorname{wt}(t_i) = s_j$ by construction of $\mathcal{A}$, and since $(w, \nu, \sigma) \in \widetilde{L}$ we get $(w, \sigma[x \to i]) \in L_j$ and $[\![\psi]\!]_{\mathcal{W}}(w, \sigma[x \to i]) = s_j$.

We consider now the strict alphabetic homomorphism $h : \widetilde{\Sigma}^*_{\mathcal{V}} \to \Sigma^*_{\mathcal{V}}$ defined by $h((a, k, s)) = (a, s)$ for each $(a, k, s) \in \widetilde{\Sigma}_{\mathcal{V}}$. Then for any $(w, \sigma) \in N_{\mathcal{V}}$ and the unique ν such that $(w, \nu, \sigma) \in \widetilde{L}$, we have

$$\begin{aligned}\big(h(\| \mathcal{A} \|), (w, \sigma)\big) &= \big(\| \mathcal{A} \|, (w, \nu, \sigma)\big) = \prod_{i \in \operatorname{dom}(w)} \operatorname{wt}(t_i) \\ &= \prod_{i \in \operatorname{dom}(w)} [\![\psi]\!]_{\mathcal{W}}(w, \sigma[x \to i]) = [\![\forall x.\psi]\!](w, \sigma).\end{aligned}$$

Therefore, $[\![\forall x.\psi]\!] = h(\| \mathcal{A} \|)$ which is recognizable by Lemma 2.2. □

Lemma 5.5. *Let $\psi \in \mathrm{MSO}(S, \Sigma)$ be unambiguous. Then $[\![\forall X.\psi]\!]$ is a recognizable step function.*

Proof. Since ψ is unambiguous, so is $\forall X.\psi$ and by Proposition 4.2 we deduce that $[\![\forall X.\psi]\!]$ is a recognizable step function.

The following result generalizes [14, Theorem 4.5] to non-commutative semirings.

Theorem 5.6. *Let S be any semiring, Σ be an alphabet and $\varphi \in \mathrm{MSO}(S,\Sigma)$ be restricted. Then $[\![\varphi]\!] \in \mathrm{Rec}(S,\Sigma_\varphi^*)$.*

Proof. Note that if $\varphi \in \mathrm{MSO}(S,\Sigma)$, then trivially $\varphi \in \mathrm{MSO}(S_\varphi,\Sigma)$. By induction over the structure of φ, we show that $[\![\varphi]\!] \in \mathrm{Rec}(S_\varphi,\Sigma_\varphi^*)$. But this is immediate by Lemmas 5.1–5.5. □

Next we aim at showing that, conversely, recognizable series are definable. First, for $s \in S$, we define

$$(x \in X) \xrightarrow{+} s = \neg(x \in X) \vee \big((x \in X) \wedge s\big).$$

This formula is almost unambiguous, and for any word w and valid assignment σ we have

$$[\![(x \in X) \xrightarrow{+} s]\!](w,\sigma) = \begin{cases} s & \text{if } \sigma(x) \in \sigma(X), \\ 1 & \text{otherwise.} \end{cases}$$

We introduce a few other abbreviations which are all unambiguous formulas. We let $\mathsf{min}(y) := \forall x. y \le x$, and $\mathsf{max}(z) := \forall x. x \le z$, and $(y = x+1) := (x \le y) \wedge \neg(y \le x) \wedge \forall z.(z \le x \vee y \le z)$. If $X_1,\dots,X_m$ are set variables, put

$$\mathrm{partition}(X_1,\dots,X_m) := \forall x. \bigvee_{i=1,\dots,m} \Big((x \in X_i) \wedge \bigwedge_{j \ne i} \neg(x \in X_j)\Big).$$

Now we show the following theorem.

Theorem 5.7. *Let S be any semiring, Σ be an alphabet and $r \in \mathrm{Rec}(S,\Sigma^*)$. Then r is* sREMSO*-definable.*

Proof. Let $\mathcal{A} = (Q,\lambda,\mu,\gamma)$ be a weighted automaton such that $r = \|\mathcal{A}\|$. For each triple $(p,a,q) \in Q \times \Sigma \times Q$ choose a set variable $X_{p,a,q}$, and let $\mathcal{V} = \{X_{p,a,q} \mid p,q \in Q, a \in \Sigma\}$. We choose an enumeration $\overline{X} = (X_1,\dots,X_m)$ of $\mathcal{V}$ with $m = |Q|^2 \cdot |\Sigma|$. Define the syntactically restricted formula

$$\begin{aligned} \psi(\overline{X}) := \mathrm{partition}(\overline{X}) \wedge \bigwedge_{p,a,q} \forall x.(x \in X_{p,a,q}) \xrightarrow{+} P_a(x) \\ \wedge\, \forall x \forall y.(y = x+1) \xrightarrow{+} \bigvee_{p,q,r \in Q, a,b \in \Sigma} (x \in X_{p,a,q}) \wedge (y \in X_{q,b,r}). \end{aligned}$$

Let $w = a_1 \dots a_n \in \Sigma^+$. If $P = (q_0 \xrightarrow{a_1} q_1 \xrightarrow{a_2} \cdots \xrightarrow{a_{n-1}} q_{n-1} \xrightarrow{a_n} q_n)$ is a path in $\mathcal{A}$ over w, we define the $(\mathcal{V},w)$-assignment σ_P by $\sigma_P(X_{p,a,q}) = \{i \mid (q_{i-1},a_i,q_i) = (p,a,q)\}$. Clearly, we have $[\![\psi]\!](w,\sigma_P) = 1$. Conversely, let σ be a $(\mathcal{V},w)$-assignment such that $[\![\psi]\!](w,\sigma) = 1$. For any $i \in \mathrm{dom}(w)$, there are uniquely determined $p_i, q_i \in Q$ such that $i \in \sigma(X_{p_i,a_i,q_i})$ and if $i < n$ then

$q_i = p_{i+1}$. Hence, with $q_0 = p_1$ we obtain a unique path $P = (q_0 \xrightarrow{a_1} q_1 \xrightarrow{a_2} \cdots \xrightarrow{a_{n-1}} q_{n-1} \xrightarrow{a_n} q_n)$ for w such that $\sigma_P = \sigma$. This gives a bijection between the set of paths in $\mathcal{A}$ over w and the set of $(\mathcal{V}, w)$-assignments σ satisfying ψ, i.e., such that $[\![\psi]\!](w, \sigma) = 1$.

Consider now the formula

$$\begin{aligned}\varphi(\overline{X}) :=& \psi(\overline{X}) \wedge \exists y. \Big(\mathsf{min}(y) \wedge \bigvee_{p,a,q} (y \in X_{p,a,q}) \wedge \lambda_p\Big) \\ & \wedge \forall x. \bigwedge_{p,a,q} (x \in X_{p,a,q}) \xrightarrow{+} \mu(a)_{p,q} \\ & \wedge \exists z. \Big(\mathsf{max}(z) \wedge \bigvee_{p,a,q} (z \in X_{p,a,q}) \wedge \gamma_q\Big).\end{aligned}$$

Let $P = (q_0 \xrightarrow{a_1} q_1 \xrightarrow{a_2} \cdots \xrightarrow{a_{n-1}} q_{n-1} \xrightarrow{a_n} q_n)$ be a path in $\mathcal{A}$ over w and let σ_P be the associated $(\mathcal{V}, w)$-assignment. We obtain

$$[\![\varphi]\!](w, \sigma_P) = \lambda_{q_0} \cdot \mu(a_1)_{q_0,q_1} \cdots \mu(a_n)_{q_{n-1},q_n} \cdot \gamma_{q_n} = \text{weight}(P) .$$

Note that $[\![\varphi(\overline{X})]\!](\varepsilon) = 0$ due to the subformula starting with $\exists y$ in φ. Hence, in order to deal with $w = \varepsilon$, let $\zeta = r(\varepsilon) \wedge \forall x.\neg(x \leq x)$. For $w \in \Sigma^+$, we have $[\![\forall x.\neg(x \leq x)]\!](w) = 0$. Now, $[\![\forall x.\neg(x \leq x)]\!](\varepsilon) = 1$ since an empty product is 1 by convention, hence we get $[\![\zeta]\!](\varepsilon) = r(\varepsilon)$.

Now let $\xi = \exists X_1 \cdots \exists X_m.(\varphi(X_1, \ldots, X_m) \vee \zeta)$. Then $\xi \in \text{MSO}(S, \Sigma)$ is existential, and $[\![\xi]\!](\varepsilon) = [\![\zeta]\!](\varepsilon) = r(\varepsilon)$. Using the bijection above, for $w \in \Sigma^+$ we get

$$\begin{aligned}[\![\xi]\!](w) &= \sum_{\sigma\ (\mathcal{V},w)\text{-assignment}} [\![\varphi]\!](w, \sigma) = \sum_{P \text{ path in } \mathcal{A} \text{ for } w} [\![\varphi]\!](w, \sigma_P) \\ &= \sum_{P \text{ path in } \mathcal{A} \text{ for } w} \text{weight}(P) = (\| \mathcal{A} \|, w).\end{aligned}$$

So $[\![\xi]\!] = \| \mathcal{A} \|$. In general, φ is not syntactically restricted due to the constants which may not commute. But it is known (cf. [23]) that we may choose $\mathcal{A}$ so that $\lambda(q), \gamma(q) \in \{0, 1\}$ for all $q \in Q$. In this case, φ is syntactically restricted and $\xi \in \text{sREMSO}(S, \Sigma)$. □

Now Theorem 4.7 is immediate by Proposition 4.8 and Theorems 5.6 and 5.7.

Next we consider the effectiveness of our proof of Theorem 4.7 implication (2) $\Rightarrow$ (1). Note that our proof of Theorem 5.6 in general was not effective, since in Lemma 5.4 we may not know the form of the step function $[\![\psi]\!]$. However, we have the following proposition.

Proposition 5.8. *Let S be an effectively given semiring and Σ an alphabet. Given $\varphi \in \text{sRMSO}(S, \Sigma)$, we can effectively compute a weighted automaton $\mathcal{A}$ for $[\![\varphi]\!]$.*

Proof. We follow the argument for Theorem 5.6 and proceed by induction on the structure of φ. Now, when dealing with a subformula $\forall x.\psi$ of φ, then we know the form of $\psi = \bigvee_{j=1}^{n}(s_j \wedge \psi_j^+)$ with $s_j \in S$ and $\psi_j \in \mathrm{MSO}^-(S, \Sigma)$ for $1 \leq j \leq n$, and we can use these constituents within the proof of Lemma 5.4.

All other lemmas employed are also constructive, meaning that if weighted automata are given for the *arguments*, then weighted automata can be effectively computed for the *results*. □

From this and decidability results for weighted automata, we immediately obtain decidability results for sRMSO-sentences. For instance, if S is an effectively given field (like $\mathbb{Q}$, the rational numbers), for any two sRMSO-sentences φ, ψ, we can decide whether $[\![\varphi]\!] = [\![\psi]\!]$: By Proposition 5.8, construct weighted automata $\mathcal{A}_\varphi$, $\mathcal{A}_\psi$ for φ, resp. ψ, and then decide whether $\|\mathcal{A}_\varphi\| = \|\mathcal{A}_\psi\|$ (cf. [4, 41]).

For the implication (1) ⇒ (3) of Theorem 4.7, given a weighted automaton $\mathcal{A}$, we can "write down" an sREMSO-sentence φ with $[\![\varphi]\!] = \|\mathcal{A}\|$. Using this, from the theory of formal power series (cf. [4, 41, 54]) we immediately obtain also undecidability results for the semantics of weighted MSO-sentences. For instance, it is undecidable whether a given sREMSO-sentence φ over $\mathbb{Q}$, the field of rational numbers, and an alphabet Σ, satisfies $\mathrm{supp}([\![\varphi]\!]) = \Sigma^*$. Also, by a result of Krob [38], the equality of given recognizable series over the tropical semiring is undecidable. Hence, the equality of two given sREMSO(Trop, Σ)-sentences is also undecidable.

6 Locally Finite Semirings

Here we will describe two larger classes of syntactically defined sentences which, for more particular semirings, are expressively equivalent to weighted automata.

First let us describe the semirings we will encounter. A monoid M is called *locally finite*, if each finitely generated submonoid of M is finite. Clearly, a commutative monoid M is locally finite iff each cyclic submonoid $\langle a \rangle$ of M is finite. Let us call a semiring S *additively locally finite* if its additive monoid $(S, +, 0)$ is locally finite. This holds iff the cyclic submonoid $\langle 1 \rangle$ of $(S, +, 0)$ is finite. Examples for additively locally finite semirings include:

- All idempotent semirings S (i.e., $x + x = x$ for each $x \in S$), in particular the arctic and the tropical semirings, the semiring $(\mathcal{P}(\Sigma^*), \cup, \cdot, \emptyset, \{\varepsilon\})$ of languages of Σ, and the semiring $([0,1], \mathsf{max}, \cdot, 0, 1)$ useful for describing probabilistic settings;
- all fields of characteristic p, for any prime p;
- all products $S_1 \times \cdots \times S_n$ (with operations defined pointwise) of additively locally finite semirings S_i $(1 \leq i \leq n)$;
- the semiring of polynomials $(S[X], +, \cdot, 0, 1)$ over a variable X and an additively locally finite semiring S;

- all locally finite semirings (see below).

Furthermore, a semiring $(S, +, \cdot, 0, 1)$ is *locally finite* [12], if each finitely generated subsemiring is finite. Clearly, equivalent to this is that both monoids $(S, +, 0)$ and $(S, \cdot, 1)$ are locally finite (cf. Chap. 1 [15]). Examples of such semirings include:

- Semirings S for which both addition and multiplication are idempotent and commutative; in particular, any bounded distributive lattice $(L, \vee, \wedge, 0, 1)$. Consequently, the chain $([0,1], \mathsf{max}, \mathsf{min}, 0, 1)$ and any Boolean algebra are locally finite;
- the Łukasiewicz semiring $([0,1], \mathsf{max}, \otimes, 0, 1)$ (cf. Sect. 3, example V);
- all matrix semirings $S^{n \times n}$ of $n \times n$-matrices over a locally finite semiring S for any $n \geq 2$, these semirings are non-commutative;
- the algebraic closures of the finite fields $\mathbb{Z}/p\mathbb{Z}$ (p prime) are (infinite) locally finite fields.

Next we turn to the formulas we will consider here. We define wUMSO(S, Σ), the collection of *weakly unambiguous* formulas in MSO(S, Σ), to be the smallest subset of MSO(S, Σ) containing all constants s ($s \in S$) and all syntactically unambiguous formulas φ^+, φ^- ($\varphi \in \mathrm{MSO}^-(S, \Sigma)$) which is closed under disjunction, conjunction and existential quantifications (both first and second order).

Definition 6.1. *A formula $\varphi \in \mathrm{MSO}(S, \Sigma)$ is called* syntactically weakly restricted, *if it satisfies the following conditions:*

1. *Whenever φ contains a conjunction $\psi \wedge \psi'$ as subformula but not in the scope of a universal first order quantifier, then* val(ψ) *and* val(ψ') *commute elementwise.*
2. *Whenever φ contains $\forall X.\psi$ as a subformula, then ψ is an unweighted syntactically unambiguous formula.*
3. *Whenever φ contains $\forall x.\psi$ as a subformula, then ψ is weakly unambiguous.*

We let swRMSO(S, Σ) denote the set of all syntactically weakly restricted formulas of MSO(S, Σ).

Our first goal will be to show that all syntactically weakly restricted formulas of MSO(S, Σ) have a recognizable semantics, provided S is additively locally finite.

Theorem 6.2. *Let S be any additively locally finite semiring, Σ be an alphabet, and $\varphi \in \mathrm{swRMSO}(S, \Sigma)$. Then $[\![\varphi]\!] \in \mathrm{Rec}(S, \Sigma_\varphi^*)$.*

As in Sect. 5, we will proceed by induction on the structure of φ. As preparation, first we aim to show that non-deleting homomorphisms preserve recognizable step functions provided S is additively locally finite.

Lemma 6.3 ([4, Corollaries III.2.4,2.5]). *Let $r : \Sigma^* \to \mathbb{N}$ be a recognizable series over the semiring $\mathbb{N}$. Then for any $a, b \in \mathbb{N}$ the languages $r^{-1}(a)$ and $r^{-1}(a + b\,\mathbb{N})$ are recognizable.*

Proposition 6.4. *Let S be additively locally finite. Let Σ, Γ be two alphabets and $h : \Sigma^* \to \Gamma^*$ be a non-erasing homomorphism.*

(a) Let $L \subseteq \Sigma^*$ be a recognizable language. Then $h(\mathbb{1}_L) : \Gamma^* \to S$ is a recognizable step function.
(b) $h : S\langle\langle\Sigma^*\rangle\rangle \to S\langle\langle\Gamma^*\rangle\rangle$ preserves recognizable step functions.

Proof. (a) We shall use the same technique as in the proof of [14, Lemma 7.8]. For any $s \in S$ and $n \geq 0$, we define $0 \otimes s = 0$ (of S) and $(n+1) \otimes s = s + (n \otimes s)$. Thus, $n \otimes s = s + \cdots + s$ with n summands s. For any $u \in \Gamma^*$, let $m(u) = |h^{-1}(u) \cap L|$. Then $(h(\mathbb{1}_L), u) = m(u) \otimes 1$. The additive monoid $\langle 1 \rangle$ generated by $\{1\}$ is finite. We choose a minimal element $a \in \mathbb{N}$ such that $a \otimes 1 = (a+x) \otimes 1$ for some $x > 0$ and we let b be the smallest such x. Then $\langle 1 \rangle = \{0, 1, \ldots, (a+b-1) \otimes 1\}$. Now for each $u \in \Gamma^*$ we have $m(u) \otimes 1 = d(u) \otimes 1$ for some uniquely determined $d(u) \in \mathbb{N}$ with $0 \leq d(u) \leq a+b-1$. Note that if $0 \leq d < a$, then $m(u) \otimes 1 = d \otimes 1$ iff $m(u) = d$, and if $a \leq d < a+b$, then $m(u) \otimes 1 = d \otimes 1$ iff $m(u) \in d + b\mathbb{N}$. For each $0 \leq d < a+b$, let $M_d = \{u \in \Gamma^* \mid d(u) = d\}$. Then $h(\mathbb{1}_L) = \sum_{d=0}^{a+b-1} d \cdot \mathbb{1}_{M_d}$.

Also, let $\mathbb{1}'_L \in \mathbb{N}\langle\langle\Sigma^*\rangle\rangle$ be the characteristic series of L over the semiring $\mathbb{N}$. Then by Lemma 2.2, the series $r = h(\mathbb{1}'_L) : \Gamma^* \to \mathbb{N}$ is recognizable, and $(r, u) = \sum_{w \in h^{-1}(u)} (\mathbb{1}'_L, w) = m(u)$ for each $u \in \Gamma^*$. Hence, $M_d = \{u \in \Gamma^* \mid m(u) = d\} = r^{-1}(d)$ if $0 \leq d < a$, and $M_d = \{u \in \Gamma^* \mid m(u) \in d + b\mathbb{N}\} = r^{-1}(d + b\mathbb{N})$ if $a \leq d < a+b$. In any case, M_d is recognizable by Lemma 6.3. Thus, $h(\mathbb{1}_L)$ is a recognizable step function.

(b) Let $r = \sum_{j=1}^{n} s_j \cdot \mathbb{1}_{L_j}$ be a recognizable step function in $S\langle\langle\Sigma^*\rangle\rangle$. Since $h : S\langle\langle\Sigma^*\rangle\rangle \to S\langle\langle\Gamma^*\rangle\rangle$ is a semiring homomorphism, we have $h(r) = \sum_{j=1}^{n} s_j \cdot h(\mathbb{1}_{L_j})$. Now, apply (a) and Lemma 2.3(a). □

Next we consider existential quantifications.

Lemma 6.5. *Let S be additively locally finite and $\varphi \in \mathrm{MSO}(S, \Sigma)$ such that $[\![\varphi]\!]$ is a recognizable step function. Then $[\![\exists x.\varphi]\!]$ and $[\![\exists X.\varphi]\!]$ are also recognizable step functions.*

Proof. Let $\mathcal{V} = \mathrm{Free}(\varphi)$ and let $\mathcal{X}$ be x or X. Following the proof of Lemma 5.2, we can write $[\![\exists\mathcal{X}.\varphi]\!]$ as the image under a length-preserving projection of $[\![\varphi]\!]_{\mathcal{V}\cup\{\mathcal{X}\}}$ which is a recognizable step function by assumption and Proposition 3.3. Now apply Proposition 6.4(b). □

Now we can prove Theorem 6.2.

Proof of Theorem 6.2. We proceed by induction over the structure of φ, aiming to show for each subformula ξ of φ that $[\![\xi]\!] \in \mathrm{Rec}(S_\varphi, \Sigma^*_\varphi)$. First, we claim that if ξ is weakly unambiguous, then $[\![\xi]\!] : \Sigma^*_\xi \to S_\xi$ is a recognizable step function. For constants and for syntactically unambiguous formulas, this is clear by Lemma 4.4. For disjunctions and conjunctions of such formulas, we apply Proposition 3.3 and Lemma 2.3(a), and for existential quantifications Lemma 6.5 to obtain our claim. Next we can proceed using Lemmas 5.2–5.5. □

Next we consider the case where the semiring S is locally finite. First we note the following proposition.

Proposition 6.6 ([12]). *Let S be locally finite. Then every recognizable series $r \in S\langle\langle \Sigma^* \rangle\rangle$ is a recognizable step function.*

We call a formula $\varphi \in \mathrm{MSO}(S, \Sigma)$ *weakly existential*, if whenever φ contains $\forall X.\psi$ as a subformula, then ψ is syntactically unambiguous. Now we show the following theorem.

Theorem 6.7. *Let S be any locally finite semiring, Σ be an alphabet, and $\varphi \in \mathrm{MSO}(S, \Sigma)$ be weakly existential. Then $[\![\varphi]\!]$ is recognizable.*

Proof. We claim that for each subformula ψ of φ, $[\![\psi]\!]$ is a recognizable step function. Due to Proposition 6.6, we only have to show that $[\![\psi]\!]$ is recognizable. Proceeding by induction, this follows from Lemmas 5.1–5.5. □

7 Weighted Automata on Infinite Words

In this section, we will consider weighted automata $\mathcal{A}$ acting on infinite words. As in the case of weighted automata on finite words, we will define the weight of an infinite path in $\mathcal{A}$ as the product of its—infinitely many—transitions, and the weight of a word w as the sum of all the weights of successful paths realizing w; in general, there might be infinitely (even uncountably) many such paths realizing w. Hence we need to be able to form infinite sums and products in the underlying semiring S. Such complete semirings have already been considered in Conway [10] and Eilenberg [23]; see also [31]. For weighted automata on infinite words and characterizations of their behaviors by rational series, the reader should consult Chap. 3 [26].

Assume that the semiring S is equipped with infinitary sum operations $\sum_I : S^I \to S$, for any index set I, such that for all I and all families $(s_i \mid i \in I)$ of elements of S the following hold:

$$\sum_{i\in\emptyset} s_i = 0, \qquad \sum_{i\in\{j\}} s_i = s_j, \qquad \sum_{i\in\{j,k\}} s_i = s_j + s_k \quad \text{for } j \neq k,$$

$$\sum_{j\in J}\Big(\sum_{i\in I_j} s_i\Big) = \sum_{i\in I} s_i, \quad \text{if } \bigcup_{j\in J} I_j = I \text{ and } I_j \cap I_{j'} = \emptyset \text{ for } j \neq j',$$

$$\sum_{i\in I}(c\cdot s_i) = c\cdot\Big(\sum_{i\in I} s_i\Big), \qquad \sum_{i\in I}(s_i\cdot c) = \Big(\sum_{i\in I} s_i\Big)\cdot c.$$

Then S together with the operations $\sum_I$ is called *complete* [23, 39].

A complete semiring is said to be *totally complete* [25] if it is endowed with a countably infinite product operation satisfying for all sequences $(s_i \mid i \geq 0)$ of elements of S the following conditions:

$$\prod_{i\geq 0} 1 = 1, \qquad s_0 \cdot \prod_{i\geq 0} s_{i+1} = \prod_{i\geq 0} s_i, \qquad \prod_{i\geq 0} s_i = \prod_{i\geq 0} (s_{n_i} \cdots s_{n_{i+1}-1}),$$

$$\prod_{j\geq 0} \sum_{i\in I_j} s_i = \sum_{(i_j)_{j\geq 0}\in \prod_{j\geq 0} I_j} \prod_{j\geq 0} s_{i_j},$$

where in the third equation $0 = n_0 < n_1 < n_2 < \cdots$ is a strictly increasing sequence and in the last equation $I_0, I_1, \ldots$ are arbitrary index sets and $\prod_{j\geq 0} I_j$ denotes the full Cartesian product of these sets.

Now we say that a totally complete semiring S is *conditionally completely commutative (ccc)*, if whenever $(s_i)_{i\geq 0}$ and $(s'_i)_{i\geq 0}$ are two sequences of elements of S such that $s_i \cdot s'_j = s'_j \cdot s_i$ for all $0 \leq j < i$, then

$$\left(\prod_{i\geq 0} s_i\right) \cdot \left(\prod_{i\geq 0} s'_i\right) = \prod_{i\geq 0} (s_i \cdot s'_i). \tag{1}$$

In [19], the authors considered totally complete semirings S satisfying (1) for *all* sequences $(s_i \mid i \geq 0)$ and $(s'_i \mid i \geq 0)$ in S. Such semirings are necessarily commutative.

Next we wish to show that there is an abundance of conditionally complete commutative semirings which are not commutative. For this, we recall the notions of ordered and continuous semirings (cf. Chap. 1 [15]).

A semiring $(S, +, \cdot, 0, 1)$ with a partial order $\leq$ is called *ordered*, if the partial order is preserved by addition and also by multiplication with elements $s \geq 0$. Now let S be an ordered semiring such that $s \geq 0$ for each $s \in S$. Then S is called *continuous*, if each directed subset D of S has a supremum (least upper bound) $\bigvee D$ in S, and addition and multiplication preserve suprema of directed subsets, i.e., $s + \bigvee D = \bigvee(s + D)$, $s \cdot \bigvee D = \bigvee(s \cdot D)$ and $(\bigvee D) \cdot s = \bigvee(D \cdot s)$ for each directed subset $D \subseteq S$ and each $s \in S$; here $s + D = \{s + d \mid d \in D\}$, $s \cdot D = \{s \cdot d \mid d \in D\}$ and analogously for $D \cdot s$. We may (and will) equip a continuous semiring with infinitary sum operations given by $\sum_{i\in I} s_i = \bigvee\{\sum_{i\in F} s_i \mid F \subseteq I \text{ finite}\}$ for any family $(s_i \mid i \in I)$ of elements of S; as is well known, then S is complete. We refer the reader to Chap. 1 [15] for many examples of (both commutative and non-commutative) continuous semirings. For instance, if S is continuous, the matrix semirings $S^{n\times n}$ and the power series semiring $S\langle\langle\Sigma^*\rangle\rangle$ (with addition and Cauchy product) are continuous and clearly non-commutative if $n \geq 2$, resp. $|\Sigma| \geq 2$. Now, we show the following proposition.

Proposition 7.1. *Let S be a continuous semiring and $S' = \{s \in S \mid s \geq 1\} \cup \{0\}$. We define an infinite product operation on S' by letting*

$$\prod_{i\geq 0} s_i = \begin{cases} \bigvee_{n\geq 0} \prod_{i=0}^{n} s_i & \text{if } s_i \neq 0 \text{ for all } i \geq 1, \\ 0 & \text{otherwise} \end{cases}$$

for each sequence $(s_i \mid i \geq 0)$ in S'. Then S' is a continuous ccc semiring.

Proof. Clearly S' is a continuous semiring. We claim that S' is totally complete. For this, it suffices to check the infinitary distributivity law. Let I_j ($j \geq 0$) be index sets and $s_i \in S'$ for $i \in I_j$. We may assume $I_j \neq \emptyset$ for each $j \geq 0$, and that $s_i \neq 0$, thus $s_i \geq 1$, for each $i \in I_j$ ($j \geq 0$).

By definitions of the infinite sum and product, we have

$$A := \prod_{j \geq 0} \sum_{i \in I_j} s_i = \bigvee_{n \geq 0} \prod_{j=0}^{n} \bigvee_{\substack{F_j \subseteq I_j \\ F_j \text{finite}}} \sum_{i \in F_j} s_i .$$

By continuity of multiplication and by distributivity, we obtain

$$A = \bigvee_{n \geq 0} \bigvee_{\substack{F_j \subseteq I_j \\ F_j \text{finite} \\ 0 \leq j \leq n}} \prod_{j=0}^{n} \sum_{i \in F_j} s_i = \bigvee_{n \geq 0} \bigvee_{\substack{F_j \subseteq I_j \\ F_j \text{finite} \\ 0 \leq j \leq n}} \sum_{(i_0, \ldots, i_n) \in F_0 \times \cdots \times F_n} \prod_{j=0}^{n} s_{i_j} .$$

We have to show that this quantity equals

$$B := \sum_{(i_j)_{j \geq 0} \in \prod_{j \geq 0} I_j} \prod_{j \geq 0} s_{i_j} = \bigvee_{\substack{F \text{finite} \\ F \subseteq \prod_{j \geq 0} I_j}} \sum_{(i_j)_{j \geq 0} \in F} \bigvee_{n \geq 0} \prod_{j=0}^{n} s_{i_j} .$$

By continuity of addition and using a diagonalization argument, we obtain

$$B = \bigvee_{\substack{F \text{finite} \\ F \subseteq \prod_{j \geq 0} I_j}} \bigvee_{n \geq 0} \sum_{(i_j)_{j \geq 0} \in F} \prod_{j=0}^{n} s_{i_j} .$$

We first show $A \leq B$. Fix $n \geq 0$ and for $0 \leq j \leq n$ let $F_j \subseteq I_j$ finite. For all $k > n$, choose $i_k \in I_k$ and let $F = F_0 \times \cdots \times F_n \times \prod_{j > n} \{i_k\}$ which is a finite subset of $\prod_{j \geq 0} I_j$. We have

$$\sum_{(i_0, \ldots, i_n) \in F_0 \times \cdots \times F_n} \prod_{j=0}^{n} s_{i_j} = \sum_{(i_j)_{j \geq 0} \in F} \prod_{j=0}^{n} s_{i_j}$$

and we deduce that $A \leq B$. Conversely, we show $B \leq A$. Fix a finite subset $F \subseteq \prod_{j \geq 0} I_j$ and some $n \geq 0$. Consider $m \geq n$ such that $|F'| = |F|$ where $F' = \{(i_0, \ldots, i_m) \mid (i_j)_{j \geq 0} \in F\}$. For $0 \leq j \leq m$, let F_j be the jth projection of F' so that $F' \subseteq F_0 \times \cdots \times F_m \subseteq I_0 \times \cdots \times I_m$. Then using $s_i \geq 1$ for each $i \in I_j$ and $j \geq 0$, we obtain

$$\begin{aligned} \sum_{(i_j)_{j \geq 0} \in F} \prod_{j=0}^{n} s_{i_j} = \sum_{(i_0, \ldots, i_m) \in F'} \prod_{j=0}^{n} s_{i_j} &\leq \sum_{(i_0, \ldots, i_m) \in F'} \prod_{j=0}^{m} s_{i_j} \\ &\leq \sum_{(i_0, \ldots, i_m) \in F_0 \times \cdots \times F_m} \prod_{j=0}^{m} s_{i_j} \end{aligned}$$

and we have shown $B \leq A$.

It remains to show that S' is ccc. Let $(s_i)_{i\geq 0}$ and $(s'_i)_{i\geq 0}$ be two sequences in S' such that $s_i \cdot s'_j = s'_j \cdot s_i$ for all $0 \leq j < i$. Then by continuity of the product, diagonalization, and our commutativity assumption we obtain

$$\left(\prod_{i\geq 0} s_i\right)\cdot\left(\prod_{i\geq 0} s'_i\right) = \bigvee_{m\geq 0}\bigvee_{n\geq 0}\left(\prod_{i=0}^{m} s_i\right)\cdot\left(\prod_{j=0}^{n} s'_j\right) = \bigvee_{n\geq 0}\left(\prod_{i=0}^{n} s_i\right)\cdot\left(\prod_{j=0}^{n} s'_j\right)$$
$$= \bigvee_{n\geq 0}\prod_{i=0}^{n}(s_i \cdot s'_i) = \prod_{i\geq 0}(s_i \cdot s'_i)$$

as required. □

Let S be a totally complete semiring. We call a subsemiring $S' \subseteq S$ a *totally complete subsemiring* of S if S' is closed in S under taking arbitrary sums and countably-infinite products. If $A \subseteq S$, the totally complete subsemiring generated by A is the smallest totally complete subsemiring of S containing A. Due to the infinitary distributivity law, it can be obtained by taking arbitrary sums of the closure A^{cl} of $A \cup \{0,1\}$ under countably-infinite products. To construct A^{cl}, in general it does not suffice to take all countably-infinite products of elements of $A \cup \{0,1\}$, since this set might not be closed under countably-infinite products; the process of taking countably-infinite products has to be iterated transfinitely (ω_1 steps suffice).

Lemma 7.2. *Let S be a ccc semiring and $A, B \subseteq S$ such that A and B commute elementwise. Let S_A^{tc} and S_B^{tc} be the totally complete subsemirings of S generated by A, resp. B. Then S_A^{tc} and S_B^{tc} commute elementwise.*

Proof. Choose any $s \in S_B^{\mathrm{tc}}$. First we show:

(1) If $(a_i)_{i\geq 1}$ is a sequence in S such that all a_i $(i \geq 1)$ commute with s, then $\prod_{i\geq 1} a_i$ commutes with s.
Indeed, put $a_0 = s_0 = s_i = 1$ for $i \geq 2$ and $s_1 = s$. Since S is ccc, we obtain:

$$\left(\prod_{i\geq 1} a_i\right)\cdot s = \left(\prod_{i\geq 0} a_i\right)\cdot\left(\prod_{i\geq 0} s_i\right) = \prod_{i\geq 0}(a_i \cdot s_i) = \prod_{i\geq 0}(s_i \cdot a_i)$$
$$= \left(\prod_{i\geq 0} s_i\right)\cdot\left(\prod_{i\geq 0} a_i\right) = s \cdot \prod_{i\geq 1} a_i.$$

(2) If $(a_i)_{i\in I}$ is a family in S such that all a_i $(i \in I)$ commute with s, then $\sum_I a_i$ commutes with s. Clearly, this holds in any complete semiring.

Now assume $s \in B$. Let A^{cl} be the closure of $A \cup \{0,1\}$ under countably infinite products. By the description of A^{cl} given above, by rule (1) and transfinite induction, we obtain that each element of A^{cl} commutes with s. Now S_A^{tc} consists of all sums of elements from A^{cl}. Hence rule (2) implies that each element of S_A^{tc} commutes with s.

So, S_A^{tc} and B commute elementwise. By a dual argument applied to S_B^{tc}, we obtain that S_A^{tc} and S_B^{tc} commute elementwise. □

We also note the following lemma.

Lemma 7.3. *Let S be a totally complete and idempotent semiring. Then $\Sigma_I 1 = 1$ for each set I of size at most continuum.*

Proof. By distributivity, we have $1 = \prod_{i\geq 0}(1+1) = \sum_{f\in 2^\omega} 1$. Now let $\leq$ be the natural partial order on the idempotent semiring S; i.e., for $x, y \in S$, we have $x \leq y$ iff $x + z = y$ for some $z \in S$. It follows that $1 \leq \Sigma_I 1 \leq \Sigma_{2^\omega} 1 = 1$ for each non-empty subset $I \subseteq 2^\omega$. Hence, $1 = \sum_I 1$. □

For the rest of this section, let S be a totally complete semiring. Now we present two weighted automata models acting on infinite words. We denote by Σ^ω the set of infinite words over Σ. Recall that a formal power series over infinite words is a mapping $r : \Sigma^\omega \to S$ and that we denote by $S\langle\langle\Sigma^\omega\rangle\rangle$ the set of formal power series over S and Σ^ω.

Definition 7.4.

(a) A weighted Muller automaton *(WMA for short) over S and Σ is a quadruple $\mathcal{A} = (Q, \lambda, \mu, \mathcal{F})$ where Q is a finite set of states, $\mu : \Sigma \to S^{Q\times Q}$ is the transition weight function, $\lambda : Q \to S$ is the weight function for entering a state, and $\mathcal{F} \subseteq \mathcal{P}(Q)$ is the family of final state sets.*
(b) A WMA $\mathcal{A}$ is a weighted Büchi automaton *(WBA for short) if there is a set $F \subseteq Q$ such that $\mathcal{F} = \{S \subseteq Q \mid S \cap F \neq \emptyset\}$.*

As for weighted finite automata, the value $\mu(a)_{p,q} \in S$ indicates the weight of the transition $p \xrightarrow{a} q$. We also write $\mathrm{wt}(p, a, q) = \mu(a)_{p,q}$.

The *weight* of an infinite path $P : q_0 \xrightarrow{a_0} q_1 \xrightarrow{a_1} q_2 \to \cdots$ in $\mathcal{A}$ is the product $\mathrm{weight}(P) := \lambda(q_0) \cdot \prod_{i\geq 0} \mathrm{wt}(q_i, a_i, q_{i+1})$. This path has label $a_0a_1\ldots$ and it is *successful*, if $\{q \in Q \mid q = q_i$ for infinitely many $i\} \in \mathcal{F}$. The *weight* of a word $w = a_0a_1\ldots \in \Sigma^\omega$ in $\mathcal{A}$, denoted $(\|\mathcal{A}\|, w)$, is the sum of $\mathrm{weight}(P)$ over all successful paths P with label w. The formal power series $\|\mathcal{A}\| : \Sigma^\omega \to S$ is called the *ω-behavior* of $\mathcal{A}$.

A series $r : \Sigma^\omega \to S$ is called *Muller recognizable* (resp. *Büchi recognizable* or *ω-recognizable*) if there is a WMA (resp. WBA) $\mathcal{A}$ such that $S = \|\mathcal{A}\|$. The class of all Muller recognizable (resp. ω-recognizable) series over S and Σ is denoted by $\text{M-Rec}(S, \Sigma^\omega)$ (resp. $\omega\text{-Rec}(S, \Sigma^\omega)$).

The following result was proved in [19].

Theorem 7.5. $\text{M-Rec}(S, \Sigma^\omega) = \omega\text{-Rec}(S, \Sigma^\omega)$.

In the sequel, we wish to provide a logical characterization of the class of ω-recognizable series in our weighted MSO logics interpreted over infinite words. For this goal, we shall need closure properties of ω-recognizable series which we recall in the following.

Lemma 7.6.

(a) For any ω-recognizable language $L \subseteq \Sigma^\omega$, the series $\mathbb{1}_L$ is ω-recognizable.
(b) Let $r, r_1, r_2 \in S\langle\langle \Sigma^\omega \rangle\rangle$ be ω-recognizable, and let $s \in S$. Then $r_1 + r_2$ and $s \cdot r$ are ω-recognizable.

Next we show the following lemma.

Lemma 7.7. *Let S be a ccc semiring. Let $S_1, S_2 \subseteq S$ be two totally complete subsemirings such that S_1 and S_2 commute elementwise. Let $r_1 \in \omega\text{-Rec}(S_1, \Sigma^\omega)$ and $r_2 \in \omega\text{-Rec}(S_2, \Sigma^\omega)$. Then $r_1 \odot r_2 \in \omega\text{-Rec}(S, \Sigma^\omega)$.*

Proof. We show that a classical construction of a weighted Muller automaton for $r_1 \odot r_2$ (cf. [19, 51]) works under the present assumptions on S.

Let $\mathcal{A}_1 = (Q_1, \lambda_1, \mu_1, \mathcal{F}_1)$ and $\mathcal{A}_2 = (Q_2, \lambda_2, \mu_2, \mathcal{F}_2)$ be two WMA. We construct the WMA $\mathcal{A} = (Q, \lambda, \mu, \mathcal{F})$ in the following way. Its state set is $Q = Q_1 \times Q_2$, and the initial distribution is given by $\lambda(q, q') = \lambda_1(q)\lambda_2(q')$ for all $(q, q') \in Q$. Its weight transition mapping is specified by $\mathrm{wt}((q, q'), a, (p, p')) = \mathrm{wt}_1(q, a, p)\,\mathrm{wt}_2(q', a, p')$ for all $(q, q'), (p, p') \in Q$, $a \in A$. Finally, the family $\mathcal{F}$ is constructed as follows: $\mathcal{F} = \{F \mid \pi_1(F) \in \mathcal{F}_1,\ \pi_2(F) \in \mathcal{F}_2\}$ where $\pi_i : Q \to Q_i$ is the projection of Q on Q_i $(i = 1, 2)$. Now let $w = a_0 a_1 \ldots \in \Sigma^\omega$, and let $P_i = (q_0^i \xrightarrow{a_0} q_1^i \xrightarrow{a_1} q_2^i \to \cdots)$ be a path for w in $\mathcal{A}_i$ $(i = 1, 2)$. Then $P = ((q_0^1, q_0^2) \xrightarrow{a_0} (q_1^1, q_1^2) \xrightarrow{a_1} (q_2^1, q_2^2) \to \cdots)$ is a path for w in $\mathcal{A}$. Clearly, P is successful in $\mathcal{A}$ iff both P_1 and P_2 are successful in $\mathcal{A}_1$ resp. $\mathcal{A}_2$. Moreover, since S is ccc and S_1 and S_2 commute elementwise, we obtain

$$\begin{aligned}
\mathrm{weight}(P) &= \lambda_1(q_0^1)\lambda_2(q_0^2)\prod_{i\ge 0}\left(\mathrm{wt}_1(q_i^1, a_i, q_{i+1}^1)\cdot \mathrm{wt}_2(q_i^2, a_i, q_{i+1}^2)\right) \\
&= \left(\lambda_1(q_0^1)\prod_{i\ge 0}\mathrm{wt}_1(q_i^1, a_i, q_{i+1}^1)\right)\cdot\left(\lambda_2(q_0^2)\prod_{i\ge 0}\mathrm{wt}_2(q_i^2, a_i, q_{i+1}^2)\right) \\
&= \mathrm{weight}(P_1)\cdot \mathrm{weight}(P_2).
\end{aligned}$$

From this, it easily follows that $(\| \mathcal{A} \|, w) = (\| \mathcal{A}_1 \|, w) \cdot (\| \mathcal{A}_2 \|, w)$. Hence $\| \mathcal{A} \| = \| \mathcal{A}_1 \| \odot \| \mathcal{A}_2 \| = r_1 \odot r_2$. □

Now let $h : \Sigma^* \to \Gamma^*$ be a length-preserving homomorphism. Then h can be extended to a mapping $h : \Sigma^\omega \to \Gamma^\omega$ by letting $h(w) = h(w(0))h(w(1))\ldots$.

For $r \in S\langle\langle \Gamma^\omega \rangle\rangle$, let $h^{-1}(r) = r \circ h \in S\langle\langle \Sigma^\omega \rangle\rangle$. For $r \in S\langle\langle \Sigma^\omega \rangle\rangle$, define $h(r) : \Gamma^\omega \to S$ by $(h(r), v) := \sum_{w \in h^{-1}(v)} (r, w)$ for $v \in \Gamma^\omega$.

Lemma 7.8 ([19]). *Let $h : \Sigma^* \to \Gamma^*$ be a length-preserving homomorphism. Then $h^{-1} : S\langle\langle \Gamma^\omega \rangle\rangle \to S\langle\langle \Sigma^\omega \rangle\rangle$ and $h : S\langle\langle \Sigma^\omega \rangle\rangle \to S\langle\langle \Gamma^\omega \rangle\rangle$ preserve ω-recognizability.*

We say that $r \in S\langle\langle \Sigma^\omega \rangle\rangle$ is an *ω-recognizable step function*, if $r = \sum_{i=1}^n s_i \cdot \mathbb{1}_{L_i}$ for some $n \in \mathbb{N}$, $s_i \in S$ and ω-recognizable languages $L_i \subseteq \Sigma^\omega$ $(i = 1, \ldots, n)$. Then clearly r is an ω-recognizable series by Lemma 7.6. The following closure result is easy to see.

Lemma 7.9.

(a) The class of all ω-recognizable step functions over Σ and S is closed under sum, scalar products, and Hadamard products.
(b) Let $h : \Sigma^* \to \Gamma^*$ be a length-preserving homomorphism. Then $h^{-1} : S\langle\langle \Gamma^\omega \rangle\rangle \to S\langle\langle \Sigma^\omega \rangle\rangle$ preserves ω-recognizable step functions.
(c) Let $h : \Sigma^\omega \to \Gamma^\omega$ be a length-preserving homomorphism and assume that S is idempotent. Then $h : S\langle\langle \Sigma^\omega \rangle\rangle \to S\langle\langle \Gamma^\omega \rangle\rangle$ preserves ω-recognizable step functions.

Proof. (a) Straightforward.

(b) We follow the proof of Lemma 2.3(b) and note that the class of ω-recognizable languages is closed under inverses of length-preserving homomorphisms (cf. [51]).

(c) For any language $L \subseteq \Sigma^\omega$, we have $h(\mathbb{1}_L) = \mathbb{1}_{h(L)}$ by Lemma 7.3. Now, follow the argument for Proposition 6.4(b). □

8 Weighted Logics on Infinite Words

In this section, we wish to develop weighted logics for infinite words which are expressively equivalent to weighted Büchi automata. In particular, we will derive analogues of Theorems 4.7 and 6.2 for infinite words.

MSO-logic over infinite words is defined as in Sect. 2. The only difference is that the domain of an infinite word is now $\mathbb{N}$. Again, the language

$$N_{\mathcal{V}}^\omega = \{(w,\sigma) \in \Sigma_{\mathcal{V}}^\omega \mid \sigma \text{ is a valid } (\mathcal{V},w)\text{-assignment}\}$$

is recognizable and by Büchi's theorem, if $\mathrm{Free}(\varphi) \subseteq \mathcal{V}$, the language

$$\mathcal{L}_{\mathcal{V}}^\omega(\varphi) = \{(w,\sigma) \in N_{\mathcal{V}}^\omega \mid (w,\sigma) \models \varphi\}$$

defined by φ over $\Sigma_{\mathcal{V}}$ is recognizable. We simply write $\mathcal{L}^\omega(\varphi)$ for $\mathcal{L}^\omega_{\mathrm{Free}(\varphi)}(\varphi)$.

In all of this section, let S be a totally complete semiring and Σ an alphabet. Given weighted MSO-formulas as in Definition 3.1, we first have to define their semantics for infinite words.

Definition 8.1. *Let $\varphi \in \mathrm{MSO}(S,\Sigma)$ and $\mathcal{V}$ be a finite set of variables containing $\mathrm{Free}(\varphi)$. The ω-$\mathcal{V}$-semantics of φ is a formal power series $[\![\varphi]\!]_{\mathcal{V}}^\omega \in S\langle\langle \Sigma_{\mathcal{V}}^\omega \rangle\rangle$. For short, in this section, we write $[\![\varphi]\!]_{\mathcal{V}}$ for $[\![\varphi]\!]_{\mathcal{V}}^\omega$. Let $(w,\sigma) \in \Sigma_{\mathcal{V}}^\omega$. If σ is not a valid $(\mathcal{V},w)$-assignment, then we put $[\![\varphi]\!]_{\mathcal{V}}(w,\sigma) = 0$. Otherwise, we define $[\![\varphi]\!]_{\mathcal{V}}(w,\sigma) \in S$ inductively just as in Definition 3.2.*

To define the semantics of $\forall X.\varphi$, we assume that in S products over index sets of size continuum exist. Then we put

$$[\![\forall X.\varphi]\!]_{\mathcal{V}}(w,\sigma) = \prod_{I \subseteq \mathrm{dom}(w)} [\![\varphi]\!]_{\mathcal{V}\cup\{X\}}(w,\sigma[X \to I]).$$

We simply write $[\![\varphi]\!]$ for $[\![\varphi]\!]_{\mathrm{Free}(\varphi)}$.

We note that the additional assumption here on products in S can be lifted again in a moment, since we will only consider formulas $\varphi \in \mathrm{MSO}(S,\Sigma)$ in which universal set quantification is only applied to syntactically unambiguous formulas, and we define uncountable products of the elements 0, 1 in the obvious way.

Indeed, from now on, we will consider syntactically unambiguous, almost unambiguous, syntactically restricted, and weakly unambiguous formulas in $\mathrm{MSO}(S,\Sigma)$, precisely as defined before. Our two main results will be the following.

Theorem 8.2. *Let S be a totally complete semiring which is ccc, let Σ be an alphabet, and let $r : \Sigma^\omega \to S$ be a series. The following are equivalent:*

1. *r is ω-recognizable.*
2. *r is definable by some syntactically restricted sentence of* $\mathrm{MSO}(S,\Sigma)$.
3. *r is definable by some syntactically restricted existential sentence of* $\mathrm{MSO}(S,\Sigma)$.

Theorem 8.3. *Let S be a totally complete semiring which is ccc and idempotent, and let Σ be an alphabet. Let $\varphi \in \mathrm{swRMSO}(S,\Sigma)$. Then $[\![\varphi]\!] \in \omega\text{-}\mathrm{Rec}(S,\Sigma_\varphi^\omega)$.*

For the proof of these results, we proceed almost exactly as before. For the convenience of the reader, we just indicate the main steps below where we assume that S is a totally complete semiring which is ccc.

As in the finitary case, the definition of the ω-semantics of a weighted MSO-formula $\varphi \in \mathrm{MSO}(S,\Sigma)$ depends on the set $\mathcal{V}$. In the following, we show that $[\![\varphi]\!]_\mathcal{V}$ in fact depends only on $\mathrm{Free}(\varphi)$.

Proposition 8.4. *Let $\varphi \in \mathrm{MSO}(S,\Sigma)$ and $\mathcal{V}$ a finite set of variables containing* $\mathrm{Free}(\varphi)$*. Then*

$$[\![\varphi]\!]_\mathcal{V}(w,\sigma) = [\![\varphi]\!](w,\sigma_{|\,\mathrm{Free}(\varphi)})$$

for each $(w,\sigma) \in \Sigma_\mathcal{V}^\omega$ such that σ is a valid $(\mathcal{V},w)$-assignment. In particular, $[\![\varphi]\!]$ is ω-recognizable iff $[\![\varphi]\!]_\mathcal{V}$ is ω-recognizable, and $[\![\varphi]\!]$ is an ω-recognizable step function iff $[\![\varphi]\!]_\mathcal{V}$ is an ω-recognizable step function.

Proof. We can follow the proof of Proposition 3.3 taking into account Lemmas 7.6(a), 7.7, 7.8, and 7.9(a), (b). □

We define the notion of *unambiguous* formulas (but now with respect to infinite words) as in Definition 4.1. Then we have the following proposition.

Proposition 8.5. *Let $\varphi \in \mathrm{MSO}(S,\Sigma)$ be unambiguous. We may also regard φ as a classical* MSO*-formula defining the language $\mathcal{L}^\omega(\varphi) \subseteq \Sigma_\varphi^\omega$. Then $[\![\varphi]\!] = \mathbb{1}_{\mathcal{L}^\omega(\varphi)}$ is an ω-recognizable step function.*

Now we obtain the following lemma.

Lemma 8.6. *Let* $\varphi \in \mathrm{MSO}^-(S, \Sigma)$. *Then:*

- $\mathcal{L}^\omega(\varphi^+) = \mathcal{L}^\omega(\varphi)$ *and* $\mathcal{L}^\omega(\varphi^-) = \mathcal{L}^\omega(\neg\varphi)$.
- $[\![\varphi^+]\!] = \mathbb{1}_{\mathcal{L}^\omega(\varphi)}$ *and* $[\![\varphi^-]\!] = \mathbb{1}_{\mathcal{L}^\omega(\neg\varphi)}$.
- φ^+ *and* φ^- *are unambiguous.*

As a by-product, we have the following proposition.

Proposition 8.7. *For each classical* MSO*-sentence* φ*, we can effectively construct an unweighted syntactically unambiguous* $\mathrm{MSO}(S, \Sigma)$*-sentence* φ' *defining the same language, i.e.,* $[\![\varphi']\!] = \mathbb{1}_{\mathcal{L}^\omega(\varphi)}$.

The definition of ω-restricted formulas is precisely as for restricted formulas, just replacing recognizable step functions by ω-recognizable step functions.

Now we proceed as in Sect. 5.

Lemma 8.8. *Let* $\varphi \in \mathrm{MSO}(S, \Sigma)$ *be atomic or the negation of an atomic formula. Then* $[\![\varphi]\!]$ *is an* ω*-recognizable step function.*

Lemma 8.9. *Let* $\varphi, \psi \in \mathrm{MSO}(S, \Sigma)$ *such that* $[\![\varphi]\!]$ *and* $[\![\psi]\!]$ *are* ω*-recognizable. Then* $[\![\varphi \vee \psi]\!]$, $[\![\exists x.\varphi]\!]$ *and* $[\![\exists X.\varphi]\!]$ *are* ω*-recognizable. Moreover, if* $[\![\varphi]\!]$ *and* $[\![\psi]\!]$ *are* ω*-recognizable step functions, then* $[\![\varphi \vee \psi]\!]$ *is also an* ω*-recognizable step function.*

Proof. We proceed analogously to Lemma 5.2, now using Proposition 8.4 and Lemmas 7.6, 7.8, and 7.9(a). □

Next we deal with conjunction. If $\varphi \in \mathrm{MSO}(S, \Sigma)$, we let S_φ^{tc} be the totally complete subsemiring of S generated by $\mathrm{val}(\varphi)$.

Lemma 8.10. *Let* $\varphi, \psi \in \mathrm{MSO}(S, \Sigma)$.

(a) Assume that $\mathrm{val}(\varphi)$ *and* $\mathrm{val}(\psi)$ *commute elementwise, and that* $[\![\varphi]\!] \in \omega\text{-Rec}(S_\varphi^{\mathrm{tc}}, \Sigma_\varphi^\omega)$ *and* $[\![\psi]\!] \in \omega\text{-Rec}(S_\psi^{\mathrm{tc}}, \Sigma_\psi^\omega)$*. Then* $[\![\varphi \wedge \psi]\!]$ *is* ω*-recognizable.*
(b) If $[\![\varphi]\!]$ *and* $[\![\psi]\!]$ *are* ω*-recognizable step functions, so is* $[\![\varphi \wedge \psi]\!]$.

Proof. (a) As shown in Lemma 7.2, S_φ^{tc} and S_ψ^{tc} commute elementwise. Now apply Proposition 8.4 and Lemma 7.7.

(b) We apply Proposition 8.4 and Lemma 7.9(a). □

Next we turn to universal quantification.

Lemma 8.11. *Let* $\psi \in \mathrm{MSO}(S, \Sigma)$ *such that* $[\![\psi]\!]$ *is an* ω*-recognizable step function. Then* $[\![\forall x.\psi]\!]$ *is* ω*-recognizable.*

Proof. We proceed as for Lemma 5.4, utilizing that the class of ω-recognizable languages is closed under Boolean operations. Then the corresponding ω-recognizable language $\widetilde{L}$ can be accepted by a deterministic Muller automaton $\widetilde{\mathcal{A}}$. We can transform $\widetilde{\mathcal{A}}$ into a weighted Muller automaton $\mathcal{A}$ by keeping its state set and the set of final states and defining initial weights and weights of transitions as before. Proceeding as before, we obtain $[\![\forall x.\varphi]\!] = h(\|\mathcal{A}\|)$ which is ω-recognizable by Lemma 7.8. □

Now we can prove Theorem 8.2.

Proof of Theorem 8.2. (3) $\Rightarrow$ (2): Trivial.

(2) $\Rightarrow$ (1): Combine Proposition 8.5 and Lemmas 8.8–8.11.

(1) $\Rightarrow$ (3): (Here, we only need that S is totally complete.) Let $\mathcal{A} = (Q, \lambda, \mu, F)$ be a weighted Büchi automaton with $r = \| \mathcal{A} \|$. By possibly adding a new initial state, we may assume that $\lambda(q) \in \{0, 1\}$ for each $q \in Q$. We define the formula $\psi(\overline{X})$ as in the proof of Theorem 5.7. Consider now the formula

$$\varphi(\overline{X}) := \psi(\overline{X}) \wedge \exists y. \Big(\mathsf{min}(y) \wedge \bigvee_{p,a,q} (y \in X_{p,a,q}) \wedge \lambda_p \Big)$$
$$\wedge \forall x. \bigwedge_{p,a,q} (x \in X_{p,a,q}) \xrightarrow{+} \mu(a)_{p,q}$$
$$\wedge \Bigg(\bigvee_{(p,a,q) \in F \times \Sigma \times Q} \forall x. \exists y. \big(x < y \wedge (y \in X_{p,a,q}) \big) \Bigg)^{+} .$$

Intuitively, the last conjunct ensures that the considered paths are accepting. The proof is now similar to the finitary case (Theorem 5.7). □

Next, we turn to the proof of Theorem 8.3. We will need the following lemma.

Lemma 8.12. *Let S be idempotent and $\varphi \in \mathrm{MSO}(S, \Sigma)$ such that $[\![\varphi]\!]$ is an ω-recognizable step function. Then $[\![\exists x.\varphi]\!]$ and $[\![\exists X.\varphi]\!]$ are also ω-recognizable step functions.*

Proof. We follow the proof of Lemma 6.5, applying Proposition 8.4 and Lemma 7.9(c). □

Now we can prove Theorem 8.3.

Proof of Theorem 8.3. Following the proof of Theorem 6.2, we proceed by induction on the structure of φ. Here, we apply Lemmas 8.6, 7.9(a), 8.8–8.12, and Propositions 8.4 and 8.5. □

Finally, we note that all constructions for the proofs of Theorems 8.2 and 8.3 are again effective (if S is given effectively).

9 Conclusions and Open Problems

In this chapter, we have presented a weighted logic which is expressively equivalent to weighted automata, both if interpreted over finite or infinite words, respectively. In the case of finite words, together with Schützenberger's theorem (see [56] and Chaps. 3, 4 [26, 55]), we thus obtain for arbitrary semirings an equivalence between weighted automata, rational expressions for formal power series, and our logical formalism by syntactically restricted MSO-logic. In the case of infinite words, we needed completeness assumptions on the semiring. Further equivalences were obtained in case the semiring is additively locally finite or locally finite or (for infinite words) idempotent.

In [14], we also investigated weighted first-order logic and could show an equivalence result to a concept of aperiodic series (cf. [12]), thus also extending the classical equivalence result [46, 57] between aperiodic, starfree, and first-order definable languages into a weighted setting. This needed the semiring to be bi-aperiodic and commutative (in which case it is also locally finite, but not conversely). We refer the reader to [14] for these results.

Weighted automata with discounting have been investigated in [17]. Discounting is a well-known concept in mathematical economics as well as systems theory in which later events get less value than earlier ones, cf., e.g., [2]. In [17], the possible behaviors of such weighted automata with discounting were characterized by rational resp. ω-rational expressions; also see [40] for further results on this. In [18], they were further characterized by a discounted restricted weighted logic. Somewhat surprisingly, the discounting only had to be reflected in the semantics of the universal first-order quantifier.

In [20], also cf. Chap. 9 [29], an equivalence result for weighted automata and a weighted logic over ranked trees was obtained for all commutative semirings. In recent work [21], our present approach has been applied to unranked trees, a syntactically defined weighted logic and arbitrary semirings.

Our approach has also been extended to pictures [27], traces [47], distributed processes [5], also cf. Chap. 10 [28] in this handbook, and very recently to texts, sp-biposets and nested words with an application to algebraic formal power series; see [43, 44]. In each case, crucial differences occur when dealing with the universal first-order quantifier. In [52], weighted automata and weighted logics for infinite trees were investigated. In [16], weighted logics with values in bounded distributive lattices were considered; cf. also Chap. 12 [53].

These results show the robustness of our approach. One could also try to define weighted temporal logics and study not only expressiveness but also decidability and complexity of natural problems such as quantitative model checking.

Open problems:

1. Given any signature $\mathcal{S}$ of predicate calculus and a semiring S, we might define the syntax of a weighted logic as in Definition 3.1, employing the new atomic formulas and their negations. The semantics can then be defined similarly as in Definition 3.2 for arbitrary finite $\mathcal{S}$-structures, and for arbitrary $\mathcal{S}$-structures assuming S is totally complete. Which results of model theory [9, 33] can be developed for such a general weighted logic?
2. Find a model of weighted automata which is expressively equivalent to the full logic $\mathrm{MSO}(S, \Sigma)$.

3. Find a weighted temporal logic which is expressively equivalent to suitable fragments of MSO(S, Σ).
4. Find applications.

References

1. J. Albert, and J. Kari. Digital image compression. In this *Handbook*. Chapter 11. Springer, Berlin, 2009.
2. L. de Alfaro, T.A. Henzinger, and R. Majumda. Discounting the future in systems theory. In *Proc. of Automata, Languages and Programming, 30th ICALP, Eindhoven*, volume 2719 of *Lecture Notes in Computer Science*, pages 1022–1037. Springer, Berlin, 2003.
3. A. Arnold. *Finite Transition Systems, International Series in Computer Science.* Prentice Hall, Englewood Cliffs, 1994.
4. J. Berstel and Ch. Reutenauer. *Rational Series and Their Languages*, volume 12 of *Monographs in Theoretical Computer Science. An EATCS Series.* Springer, Berlin, 1988.
5. B. Bollig and I. Meinecke. Weighted distributed systems and their logics. In *Proc. of Logical Foundations of Computer Science, LFCS, New York*, volume 4514 of *Lecture Notes in Computer Science*, pages 54–68. Springer, Berlin, 2007.
6. J.R. Büchi. Weak second-order arithmetic and finite automata. *Zeitschrift für Mathematische Logik und Grundlagen der Mathematik*, 6:66–92, 1960.
7. J.R. Büchi. On a decision method in restricted second order arithmetic. In *Proc. 1960 Int. Congress for Logic, Methodology and Philosophy of Science*, pages 1–11, 1962.
8. A.L. Buchsbaum, R. Giancarlo, and J.R. Westbrook. On the determinization of weighted finite automata. *SIAM Journal on Computing*, 30:1502–1531, 2000.
9. C.C. Chang and H.J. Keisler. *Model Theory.* North-Holland, Amsterdam, 1990.
10. J.H. Conway. *Regular Algebra and Finite Machines.* Chapman & Hall, London, 1971.
11. K. Culik and J. Kari. Image compression using weighted finite automata. *Computer and Graphics*, 17:305–313, 1993.
12. M. Droste and P. Gastin. On aperiodic and star-free formal power series in partially commuting variables. *Theory of Computing Systems*, 42:608–631, 2008.
13. M. Droste and P. Gastin. Weighted automata and weighted logics. In *Proc. of Automata, Languages and Programming, 32nd ICALP, Lisbon*, volume 3580 of *Lecture Notes in Computer Science*, pages 513–525. Springer, Berlin, 2005.
14. M. Droste and P. Gastin. Weighted automata and weighted logics. *Theoretical Computer Science*, 380:69–86, 2007.

15. M. Droste, and W. Kuich. Semirings and formal power series. In this *Handbook*. Chapter 1. Springer, Berlin, 2009.
16. M. Droste, W. Kuich, and G. Rahonis. Multi-valued MSO logics over words and trees. *Fundamenta Informaticae*, 84:305–327, 2008.
17. M. Droste and D. Kuske. Skew and infinitary formal power series. *Theoretical Computer Science*, 366:189–227, 2006. Extended abstract in: *Proc. of Automata, Languages and Programming, 30th ICALP, Eindhoven*, volume 2719 of *Lecture Notes in Computer Science*, pages 426–438. Springer, Berlin, 2003.
18. M. Droste and G. Rahonis. Weighted automata and weighted logics with discounting. In *Proc. of Implementation and Application of Automata, 12th CIAA, Prague*, volume 4783 of *Lecture Notes in Computer Science*, pages 73–84. Springer, Berlin, 2007.
19. M. Droste and G. Rahonis. Weighted automata and weighted logics on infinite words. In *Izvestiya VUZ. Matematika*, 2009, in press.
20. M. Droste and H. Vogler. Weighted tree automata and weighted logics. *Theoretical Computer Science*, 366:228–247, 2006.
21. M. Droste and H. Vogler. Weighted logic for unranked tree automata, *Theory of Computing Systems*, 2009, in press.
22. M. Droste and G.-Q. Zhang. On transformations of formal power series. *Information and Computation*, 184:369–383, 2003.
23. S. Eilenberg. *Automata, Languages and Machines, volume A*. Academic Press, New York, 1974.
24. C.C. Elgot. Decision problems of finite automata design and related arithmetics. *Transactions of the American Mathematical Society*, 98:21–52, 1961.
25. Z. Ésik and W. Kuich. On iteration semiring-semimodule pairs. *Semigroup Forum*, 75:129–159, 2007.
26. Z. Ésik and W. Kuich. Finite automata. In this *Handbook*. Chapter 3. Springer, Berlin, 2009.
27. I. Fichtner. Weighted picture automata and weighted logics. *Theory of Computing Systems*, in press.
28. I. Fichtner, D. Kuske, and I. Meinecke. Traces, series-parallel posets, and pictures: A weighted study. In this *Handbook*. Chapter 10. Springer, Berlin, 2009.
29. Z. Fülöp and H. Vogler. Weighted tree automata and tree transducers. In this *Handbook*. Chapter 9. Springer, Berlin, 2009.
30. B. Gerla. Automata over MV-algebras. In *Proc. of IEEE International Symposium on Multiple-Valued Logic, 34th ISMVL, Toronto*, pages 49–54. IEEE Computer Society Press, Los Alamitos, 2004.
31. S. Golan. *Semirings and Their Applications*. Kluwer Academic, Dordrecht, 1999.
32. U. Hafner. Low bit-rate image and video coding with weighted finite automata. PhD thesis, Universität Würzburg, Germany, 1999.

33. W. Hodges. *Model Theory.* Cambridge University Press, Cambridge, 1993.
34. Z. Jiang, B. Litow, and O. de Vel. Similarity enrichment in image compression through weighted finite automata. In *Proc. of Computing and Combinatorics, 6th COCOON, Sydney*, volume 1858 of *Lecture Notes in Computer Science*, pages 447–456. Springer, Berlin, 2000.
35. F. Katritzke. Refinements of data compression using weighted finite automata. PhD thesis, Universität Siegen, Germany, 2001
36. B. Khoussainov and A. Nerode. *Automata Theory and Its Applications.* Birkhäuser, Boston, 2001.
37. K. Knight and J. May. Applications of weighted automata in natural language processing. In this *Handbook.* Chapter 14. Springer, Berlin, 2009.
38. D. Krob. The equality problem for rational series with multiplicities in the tropical semiring is undecidable. *International Journal of Algebra and Computation*, 4:405–425, 1994.
39. W. Kuich. Semirings and formal power series. In G. Rozenberg and A. Salomaa, editors, *Handbook of Formal Languages, volume 1*, Chapter 9, pages 609–677. Springer, Berlin, 1997.
40. W. Kuich. On skew formal power series. In: S. Bozapalidis, A. Kalampakas, and G. Rahonis, editors, *Proceedings of the Conference on Algebraic Informatics, Thessaloniki*, pages 7–30, 2005.
41. W. Kuich and A. Salomaa. *Semirings, Automata, Languages*, volume 6 of *Monographs in Theoretical Computer Science. An EATCS Series.* Springer, Berlin, 1986.
42. R.P. Kurshan. *Computer-Aided Verification of Coordinating Processes. Princeton Series in Computer Science.* Princeton University Press, Princeton, 1994.
43. Ch. Mathissen. Definable transductions and weighted logics for texts. In *Proc. of Developments in Language Theory, 11th DLT, Turku*, volume 4588 of *Lecture Notes in Computer Science*, pages 324–336. Springer, Berlin, 2007.
44. Ch. Mathissen. Weighted logics for nested words and algebraic formal power series. In *Proc. of Automata, Languages and Programming, 35th ICALP, Part II, Reykjavik*, volume 5126 of *Lecture Notes in Computer Science*, pages 221–232. Springer, Berlin, 2008.
45. K. McMillan *Symbolic Model Checking.* Kluwer Academic, Dordrecht, 1993.
46. R. McNaughton and S. Papert. *Counter-Free Automata.* MIT Press, Cambridge, 1971.
47. I. Meinecke. Weighted logics for traces. In *Proc. of Computer Science—Theory and Applications, 1st CSR, St. Petersburg*, volume 3967 of *Lecture Notes in Computer Science*, pages 235–246. Springer, Berlin, 2006.
48. M. Mohri. Finite-state transducers in language and speech processing. *Computational Linguistics*, 23:269–311, 1997.

49. M. Mohri. Weighted automata algorithms. In this *Handbook*. Chapter 6. Springer, Berlin, 2009.
50. M. Mohri, F. Pereira, and M. Riley. The design principles of a weighted finite-state transducer library. *Theoretical Computer Science*, 231:17–32, 2000.
51. D. Perrin and J.-É. Pin. *Infinite Words*. Elsevier, Amsterdam, 2004.
52. G. Rahonis. Weighted Muller tree automata and weighted logics. *Journal of Automata, Languages and Combinatorics*, 12:455–483, 2007.
53. G. Rahonis. Fuzzy languages. In this *Handbook*. Chapter 12. Springer, Berlin, 2009.
54. A. Salomaa and M. Soittola. *Automata-Theoretic Aspects of Formal Power Series. Texts and Monographs in Computer Science*. Springer, Berlin, 1978.
55. J. Sakarovitch. Rational and recognisable power series. In this *Handbook*. Chapter 4. Springer, Berlin, 2009.
56. M.P. Schützenberger. On the definition of a family of automata. *Information and Control*, 4:245–270, 1961.
57. M.P. Schützenberger. On finite monoids having only trivial subgroups. *Information and Control*, 8:190–194, 1965.
58. S. Schwarz. Łukasiewicz logics and weighted logics over MV-semirings. *Journal of Automata, Languages and Combinatorics*, 12:485–499, 2007.
59. W. Thomas. Languages, automata and logic. In G. Rozenberg and A. Salomaa, editors, *Handbook of Formal Languages, volume 3*, Chapter 7, pages 389–485. Springer, Berlin, 1997.

Chapter 6: Weighted Automata Algorithms

Mehryar Mohri[1,2]

[1] Courant Institute of Mathematical Sciences,
251 Mercer Street, New York, NY 10012, USA
mohri@cims.nyu.edu

[2] Google Research,
76 Ninth Avenue, New York, NY 10011, USA
mohri@google.com

M. Droste, W. Kuich, H. Vogler (eds.), *Handbook of Weighted Automata*,
Monographs in Theoretical Computer Science. An EATCS Series,
DOI 10.1007/978-3-642-01492-5_6,

1 Introduction

This chapter presents several fundamental algorithms for weighted automata and transducers. While the mathematical counterparts of weighted transducers, *rational power series*, have been extensively studied in the past [22, 54, 13, 36], several essential weighted transducer algorithms, e.g., composition, determinization, and minimization, have been devised only in the last decade [38, 43], in part motivated by novel applications in speech recognition, speech synthesis, machine translation, other areas of natural language processing, image processing, optical character recognition, and more recently machine learning.

These algorithms can be viewed as the generalization to the weighted transducer case of the standard algorithms for un-weighted acceptors. However, this generalization is often not straightforward and has required a number of specific studies either because the old schema could not be applied in the presence of weights and a novel technique was required, as in the case of composition [50, 46], or because of the analysis of the conditions of application of an algorithm as in the case of determinization [38, 3].

The chapter favors a presentation of weighted automata and transducers in terms of graphs, the natural concepts for an algorithmic description and complexity analysis. Also, while power series lead to more concise and rigorous proofs in most cases [36], proofs related to questions of ambiguity naturally require the introduction of paths and reasoning on graph concepts.

2 Preliminaries

This section introduces the definitions and notation related to *weighted finite-state transducers*, weighted transducers for short, and *weighted automata*.

2.1 Semirings

For various operations to be well defined, the weight set associated to a weighted transducer must have the structure of a *semiring* (see [20]). A system $(S, \oplus, \otimes, \overline{0}, \overline{1})$ is a *semiring* if $(S, \oplus, \overline{0})$ is a commutative monoid with identity element $\overline{0}$, $(S, \otimes, \overline{1})$ is a monoid with identity element $\overline{1}$, $\otimes$ distributes over $\oplus$, and $\overline{0}$ is an annihilator for $\otimes$: for all $a \in S, a \otimes \overline{0} = \overline{0} \otimes a = \overline{0}$. Thus, a semiring is a ring that may lack negation.

Table 1 lists several semirings. In addition to the Boolean semiring, and the probability semiring used to combine probabilities, two semirings often used in applications are the *log semiring*, which is isomorphic to the probability semiring via the negative-log morphism, and the *tropical semiring*, which is derived from the log semiring using the *Viterbi approximation*. In the following definitions, S will be used to denote a semiring.

Table 1. Semiring examples. $\oplus_{\log}$ is defined by: $x \oplus_{\log} y = -\log(e^{-x} + e^{-y})$

Semiring	Set	$\oplus$	$\otimes$	$\overline{0}$	$\overline{1}$
Boolean	$\{0,1\}$	$\vee$	$\wedge$	0	1
Probability	$\mathbb{R}_+ \cup \{+\infty\}$	$+$	$\times$	0	1
Log	$\mathbb{R} \cup \{-\infty, +\infty\}$	$\oplus_{\log}$	$+$	$+\infty$	0
Tropical	$\mathbb{R} \cup \{-\infty, +\infty\}$	min	$+$	$+\infty$	0

A semiring is said to be *commutative* when the multiplicative operation $\otimes$ is commutative. The semirings listed in Table 1 are all commutative. It is said to be *idempotent* if $x \oplus x = x$ for all $x \in S$. The Boolean semiring and the tropical semiring are idempotent.

2.2 Weighted Transducers and Automata

Given an alphabet Σ, we will denote by $|x|$ the length of a string $x \in \Sigma^*$ and by ϵ the *empty string* for which $|\epsilon| = 0$. The *mirror image* of a string $x = x_1 \ldots x_n$ is the string $x^R = x_n x_{n-1} \ldots x_1$.

Finite-state transducers are finite automata in which each transition is augmented with an output label in addition to the familiar input label [12, 22, 54, 36]. Output labels are concatenated along a path to form an output sequence and similarly with input labels. *Weighted transducers* are finite-state transducers in which each transition carries some weight in addition to the input and output labels [54, 36, 52]. The weights are elements of a semiring $(S, \oplus, \otimes, \overline{0}, \overline{1})$.

The $\otimes$-operation is used to compute the weight of a path by $\otimes$-multiplying the weights of the transitions along that path. The $\oplus$-operation computes the weight of a pair of input and output strings (x, y) by $\oplus$-summing the weights of the paths labeled with (x, y). The following gives a formal definition of weighted transducers.

Definition 2.1. *A* weighted transducer T *over a semiring* $(S, \oplus, \otimes, \overline{0}, \overline{1})$ *is an 8-tuple* $T = (\Sigma, \Delta, Q, I, F, E, \lambda, \rho)$ *where* Σ *is a finite input alphabet,* Δ *a finite output alphabet,* Q *is a finite set of states,* $I \subseteq Q$ *the set of initial states,* $F \subseteq Q$ *the set of final states,* E *a finite multi-set*[3] *of transitions, which are elements of* $Q \times (\Sigma \cup \{\epsilon\}) \times (\Delta \cup \{\epsilon\}) \times S \times Q$*,* $\lambda : I \to S$ *an initial weight function, and* $\rho : F \to S$ *a final weight function mapping* F *to* S*.*

For a state $q \in Q$, we will denote by $E[q]$ the outgoing transitions of q and more generally by $E[Q']$, the outgoing transitions of all states q in a subset

[3] Thus, there can be two transitions from state p to state q with the same input and output label, and even the same weight. In practice, this is avoided by keeping only one such transition whose weight is the $\oplus$-sum of the weights of the original redundant transitions. We will denote by $\uplus$ the standard join operation of multi-sets as in $\{1,2\} \uplus \{1,3\} = \{1,1,2,3\}$.

of states $Q' \subseteq Q$. An ϵ-transition is a transition with both input and output label equal to ϵ.

A path π of a transducer is an element of E^* with consecutive transitions. We denote by $p[\pi]$ its origin or previous state and by $n[\pi]$ its destination or next state. A cycle π is a path with $p[\pi] = n[\pi]$. An ϵ-cycle is a cycle with both input and output label equal to ϵ. We also denote by:

- $P(Q_1, Q_2)$ the set of all paths from a subset $Q_1 \subseteq Q$ to a subset $Q_2 \subseteq Q$
- $P(Q_1, x, Q_2)$ the subset of all paths of $P(Q_1, Q_2)$ with input label x
- $P(Q_1, x, y, Q_2)$ the subset of all paths of $P(Q_1, x, Q_2)$ with output label y

A path in $P(I, F)$ is said to be *accepting* or *successful*. The weight of a path π obtained by $\otimes$-multiplying the weights of its constituent transitions is denoted by $w[\pi]$. For any transducer T, we denote by T^{-1} its *inverse*, that is the transducer obtained from T by swapping the input and output label of each transition.

A transducer T is said to be *regulated* if the output weight associated by T to any pair of strings $(x, y) \in \Sigma^* \times \Delta^*$ defined as:

$$T(x, y) = \bigoplus_{\pi \in P(I,x,y,F)} \lambda(p[\pi]) \otimes w[\pi] \otimes \rho(n[\pi]) \quad (1)$$

is an element of S and its definition does not depend on the order of the terms in the $\oplus$-sum. $T(x, y)$ is defined to be $\overline{0}$ when $P(I, x, y, F) = \emptyset$.[4] Note that in the absence of ϵ-cycles, the set of accepting paths $P(I, x, y, F)$ is finite for any $(x, y) \in \Sigma^* \times \Delta^*$ and thus T is regulated. Also, as we shall see later, in some semirings, such as the four semirings of Table 1, all weighted transducers are regulated. The weighted transducers we will be considering in this chapter will be regulated. Figure 1(a) shows an example of a weighted transducer.

While our definition allows for multiple initial states with initial weights, in all our examples there will be a unique initial state with initial weight $\overline{1}$, and thus that weight is not indicated in figures. Since any weighted transducer can be represented by an equivalent one with this property, this does not represent a real limitation.

A state $q \in Q$ is said to be *non-accessible* (*non-coaccessible*) when there is no path from I to q (resp. from q to F). Non-accessible and non-coaccessible states are called *useless states*. They can be removed using a connection (or trimming) algorithm in linear time without affecting the weight T associates to any pair. A transducer with no useless state is said to be *trim*.

A transducer is said to be *unambiguous* if for any string $x \in \Sigma^*$ it admits at most one accepting path with input label x. It is said to be *deterministic* or

[4] Our definition of *regulated transducers* is more general that the standard one which assumes that transducers do not have cycles with input or output ϵ [54, 36, 52]. The usual definition leads to a simpler presentation but it rules out weighted transducers that are crucial in applications or that can be obtained as a result of application of various algorithms.

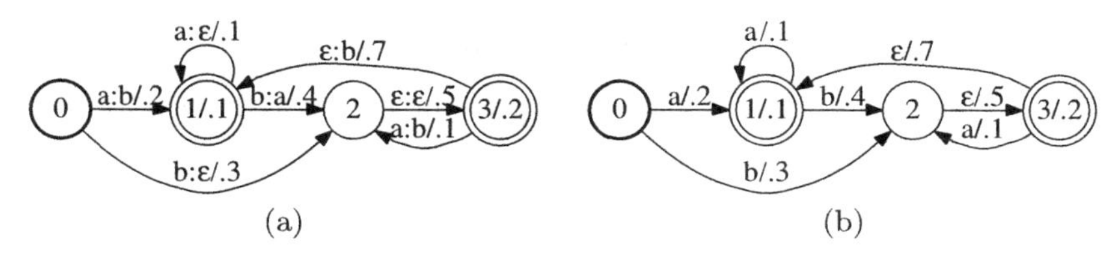

Fig. 1. (**a**) Example of a weighted transducer T over the probability semiring. (**b**) Example of a weighted automaton A over the probability semiring. A can be obtained from T by removing output labels. A *bold circle* indicates an initial state with initial weight $\overline{1}$ and a *double-circle* a final state. A final state q's weight $\rho(q)$ is indicated after the *slash* symbol representing the state number

sequential if it has at most one initial state and at any state no two outgoing transitions share the same input label.

A *weighted automaton* A can be defined as a weighted transducer with identical input and output labels, for any transition. Thus, only string pairs of the form (x, x) can have a non-zero weight by A, which is why the weight associated by A to (x, x) is abusively denoted by $A(x)$ and identified with the *weight associated by* A *to* x. Similarly, in the graph representation of weighted automata, the output (or input) label is omitted. Figure 1(b) shows an example of a weighted automaton. The language accepted by A is the one accepted by the un-weighted automaton obtained by ignoring its weights and is denoted by $L(A)$.

Note that (un-weighted) finite automata [51] can be viewed as weighted automata over the Boolean semiring and, similarly, (un-weighted) finite-state transducers [22, 54, 12, 36] as weighted transducers defined over the Boolean semiring.

3 Shortest-Distance Algorithms

Shortest-paths problems are familiar problems in computer science and mathematics. In these problems, edge weights may represent distances, costs, or any other real-valued quantity that can be added along a path, and that one may wish to minimize. Thus, edge weights are real numbers and the specific operations used are addition to compute the weight of a path and minimum to select the best path weight.

This section introduces a generalization of this problem to the case where the operations are those of a semiring. These problems turn out to be crucial in the design of several algorithms such as ϵ-removal or pushing and in many other contexts. Different algorithmic solutions will be presented depending on the semiring properties.

We will consider directed graphs $G = (Q, E, w)$ over a semiring S, where Q is a set of vertices, E a set of edges, and $w : E \to S$ the edge weight function which we can extend to any path $\pi = e_1 \dots e_k$ by $w[\pi] = \bigotimes_{i=1}^{k} w[e_i]$.

3.1 All-Pairs Shortest-Distance Problems

The general all-pairs shortest-distance algorithm described in this section is defined for any *complete semiring*.[5]

3.1.1 Complete Semirings

A semiring $(S, \oplus, \otimes, \overline{0}, \overline{1})$ is said to be *complete* if for any index set I and any family $(a_i)_{i \in I}$ of elements of S, $\bigoplus_{i \in I} a_i$ is an element of S whose definition does not depend on the order of the terms in the $\oplus$-sum and that has the following properties [22, 20]:

$$\bigoplus_{i \in I} a_i = \overline{0} \qquad \text{if } \operatorname{card}(I) = 0, \tag{2}$$

$$\bigoplus_{i \in I} a_i = a_i \qquad \text{if } \operatorname{card}(I) = 1, \tag{3}$$

$$\bigoplus_{i \in I} a_i = \bigoplus_{j \in J} \Big(\bigoplus_{i \in I_j} a_i \Big) \qquad \text{for any disjoint partition } I = \bigcup_{j \in J} I_j, \tag{4}$$

$$a \otimes \Big(\bigoplus_{i \in I} a_i \Big) = \bigoplus_{i \in I} (a \otimes a_i) \qquad \text{for any } a \in S, \tag{5}$$

$$\Big(\bigoplus_{i \in I} a_i \Big) \otimes a = \bigoplus_{i \in I} (a_i \otimes a) \qquad \text{for any } a \in S. \tag{6}$$

A straightforward consequence of these axioms is that in a complete semiring the identity $(\bigoplus_{i \in I} a_i)(\bigoplus_{j \in J} b_j) = \bigoplus_{(i,j) \in I \times J} (a_i \otimes b_j)$ holds for any two families $(a_i)_{i \in I}$ and $(b_j)_{j \in J}$ of elements of S. Note that in a complete semiring all weighted transducers are regulated since all infinite sums are elements of S.

A complete semiring S is a *starsemiring* [20], that is a semiring that can be augmented with an internal unary *closure* operation * defined by $a^* = \bigoplus_{n=0}^{\infty} a^n$ for any $a \in S$.[6] Furthermore, associativity, commutativity, and distributivity apply to these infinite sums.

The *Boolean semiring* $(\{0,1\}, \vee, \wedge, 0, 1)$ with $a^* = 1$ for $a \in \{0,1\}$, and the *tropical semiring* $(\mathbb{R}_+ \cup \{+\infty\}, \mathsf{min}, +, +\infty, 0)$, with $a^* = 0$ for all $a \in \mathbb{R}_+ \cup \{+\infty\}$, implicitly used in shortest-paths problems, are familiar examples of complete semirings. The more general tropical semiring $(\mathbb{R} \cup \{-\infty, +\infty\}$, $\mathsf{min}, +, +\infty, 0)$ with

[5] The algorithm applies in fact more generally to any *closed semiring* as defined in [41], which, unlike the definition given by [17], does not require idempotence. Note that the earlier definition of closed semirings given by Aho et al. [1] is not axiomatically correct (see [37, 26]). Any complete semiring is a closed semiring.

[6] Thus, with the terminology of [20], it is a *complete starsemiring*. All complete starsemirings are Conway semirings [20].

$$a^* = \begin{cases} 0 & \text{if } a \in \mathbb{R}_+; \\ -\infty & \text{otherwise,} \end{cases} \tag{7}$$

and $(+\infty) + (-\infty) = (-\infty) + (+\infty) = +\infty$, also defines a complete semiring. Note that the family of complete semirings includes non-idempotent semirings such as the probability semiring $(\mathbb{R}_+ \cup \{+\infty\}, +, \times, 0, 1)$ with the closure operation defined by

$$a^* = \begin{cases} \frac{1}{1-a} & \text{if } 0 \le a < 1; \\ +\infty & \text{otherwise.} \end{cases} \tag{8}$$

The log semiring $(\mathbb{R}_+ \cup \{-\infty, +\infty\}, \oplus_{\log}, +, +\infty, 0)$ which is isomorphic to the probability semiring is also a non-idempotent complete semiring with

$$a^* = \begin{cases} \log(1-a) & \text{if } 0 \le a < 1; \\ -\infty & \text{otherwise.} \end{cases} \tag{9}$$

The *lattice semiring* $(\mathbb{L}, \vee, \wedge, \perp, \top)$ where $\mathbb{L}$ is a complete and distributive lattice with infimum $\perp$ and supremum $\top$ is a complete semiring with $a^* = \top$ for all $a \in \mathbb{L}$, when it verifies properties (5) and (6) [20]. Thus, all weighted transducers are regulated in the semirings just examined.

3.1.2 All-Pairs Shortest-Distance Algorithm

For a complete semiring, we can define the *distance* or *shortest-distance* from vertex p to vertex q in $G = (Q, E, w)$ by

$$d[p, q] = \bigoplus_{\pi \in P(p,q)} w[\pi], \tag{10}$$

where the $\oplus$-sum runs over the set of all paths from p to q. This definition coincides with the classical definition of shortest-distance where the weights are summed along the path and where the shortest path is sought for the tropical semiring $(\mathbb{R}_+ \cup \{+\infty\}, \min, +, +\infty, 0)$. The general *all-pair shortest-distance problem* is that of computing the shortest distances $d[p, q]$ for all pairs (p, q) with $p, q \in Q$.

This problem can be solved by computing the closure of the matrix $\mathbf{M} = (\mathbf{M}_{pq}) \in S^{|Q| \times |Q|}$ defined by $\mathbf{M}_{pq} = \bigoplus_{e \in E \cap P(p,q)} w[e]$ for all $p, q \in Q$. Indeed, using the semiring operations in matrix multiplication [20], for $n \in \mathbb{N}$, the coefficient $\mathbf{M}^n_{pq}$ of $\mathbf{M}^n$ gives the $\oplus$-sum of the weights of all paths of length at most n from p to q. For idempotent semirings such as the tropical semiring for which $\overline{1} \oplus x = \overline{1}$ for all $x \in S$, only simple paths (paths with no cycle) need to be considered in the computation of the shortest distances and thus $\mathbf{M}^* = \mathbf{M}^{|Q|-1}$. Using the standard repeated squaring technique [17], $\mathbf{M}^{|Q|-1}$ can be computed in time $\mathbf{\Theta}(|Q|^3(T_\oplus + T_\otimes) \log |Q|)$, where $T_\oplus$ denotes the computational cost of the $\oplus$ operation, and $T_\otimes$ that of the $\otimes$ operation.

```
GEN-ALL-PAIRS(G)
 1  for i ← 1 to |Q| do
 2      for j ← 1 to |Q| do
 3          d[i, j] ← ⊕_{e∈E∩P(i,j)} w[e]
 4  for k ← 1 to |Q| do
 5      for i ← 1 to |Q|, i ≠ k do
 6          for j ← 1 to |Q|, j ≠ k do
 7              d[i, j] ← d[i, j] ⊕ (d[i, k] ⊗ d[k, k]* ⊗ d[k, j])
 8      for i ← 1 to |Q|, i ≠ k do
 9          d[k, i] ← d[k, k]* ⊗ d[k, i]
10          d[i, k] ← d[i, k] ⊗ d[k, k]*
11      d[k, k] ← d[k, k]*
```

Fig. 2. Generic all-pairs shortest-distance algorithm

There exists however a more efficient method for computing all-pairs shortest-distances for all complete semirings based on a generalization of the Floyd–Warshall algorithm.[7]

The Floyd–Warshall algorithm [25, 57] originally designed for the Boolean semiring can be generalized to compute all-pair shortest-distances in all complete semirings. Figure 2 gives the pseudo-code of an in-place implementation of the algorithm where $d[i,j]$ corresponds to the tentative shortest distance from vertex i to vertex j. Lines 1–3 initialize each distance $d[i,j]$ to the sum of the weights of the transitions between i and j. By convention, the $\oplus$-sum is $\overline{0}$ if i and j are not adjacent. The loops of lines 4–11 update the tentative shortest-distances in a way that is similar to the steps of the standard Floyd–Warshall algorithm but using operations of an arbitrary complete semiring.

Let T_* denote the cost of the closure operation.

Theorem 3.1. *Let $G = (Q, E, w)$ be a weighted directed graph over a complete semiring S. Then the algorithm* GEN-ALL-PAIRS *computes the shortest-distances $d[i,j]$ between all pairs of vertices (i,j) of G in time $\Theta(|Q|^3(T_\oplus + T_\otimes + T_*))$ and space $\Theta(|Q|^2)$.*

Proof. Let $P^k(i,j)$ denote the set of paths from i to j with all intermediate vertices within $\{1,\ldots,k\}$. For any $i,j \in Q$, $k \in \{0\} \cup Q$, let d_{ij}^k be the sum of all paths from i to j with all intermediate vertices within $\{1,\ldots,k\}$: $d_{ij}^k = \bigoplus_{w \in P^k(i,j)} w[\pi]$. Since the semiring is complete, d_{ij}^k is well defined and in S.

Let π be a path in $P^k(i,j)$. It is either a path from i to j with all intermediate vertices within $\{1,\ldots,k-1\}$ or it can be decomposed into a path from i to k with all intermediate vertices within $\{1,\ldots,k-1\}$, followed by any

[7] Ésik and Kuich also gave a cubic-time algorithm for computing $\mathbf{M}^*$ for all *Conway semirings* (see [20] for the definition), which include complete semirings [23].

number of cycles at k with all intermediate vertices in $\{1, \ldots, k-1\}$, followed by a path from k to j with all intermediate vertices within $\{1, \ldots, k-1\}$. Thus, for all $i, j, k \in Q$,

$$P^k(i,j) = P^{k-1}(i,j) \cup \big(P^{k-1}(i,k)\big(P^{k-1}(k,k)\big)^* P^{k-1}(k,j)\big). \tag{11}$$

By definition, a path in $P^{k-1}(i,j)$ does not go through k, thus

$$P^{k-1}(i,j) \cap \big(P^{k-1}(i,k)\big(P^{k-1}(k,k)\big)^* P^{k-1}(k,j)\big) = \emptyset. \tag{12}$$

Thus, even if S is not idempotent, d_{ij}^k can be decomposed, for all $i, j, k \in Q$, as

$$d_{ij}^k = d_{i,j}^{k-1} \oplus \big(d_{ik}^{k-1} \otimes \big(d_{kk}^{k-1}\big)^* \otimes d_{kj}^{k-1}\big). \tag{13}$$

This identity leads directly to an algorithm for computing all-pairs shortest distances using a triple-indexed array. An in-place implementation of the algorithm limits the space used to that of a single $|Q| \times |Q|$-matrix (Fig. 2), and thus the space complexity of the algorithm to $O(|Q|^2)$. The cubic-time complexity follows directly the definition of the algorithm. □

The efficiency of the algorithm can be improved for graphs G with relatively small strongly connected components (SCCs) by decomposing G into its SCCs, which can be done in linear time, then running GEN-ALL-PAIRS on each SCC.

The GEN-ALL-PAIRS algorithm is useful in a variety of applications. With the Boolean semiring, it can be used to compute the transitive closure of any vertex of a graph and then coincides with the classical Floyd–Warshall algorithm [25, 57]. With the tropical semiring, the algorithm can compute the all-pairs shortest distances in the classical case including for graphs with negative cycles using the general topical semiring $(\mathbb{R} \cup \{-\infty, +\infty\}, \mathsf{min}, +, +\infty, 0)$. The algorithm of [28] based on Dijkstra's algorithm and that of Bellman–Ford has a better time complexity for graphs with real-valued weights, $O(|Q|^2 \log |Q| + |Q||E|)$, but it cannot be used with graphs that have a negative cycle. GEN-ALL-PAIRS can also be used to compute the minimum spanning tree of a directed graph using the complete semiring $(\mathbb{R} \cup \{-\infty, \infty\}, \mathsf{min}, \mathsf{max}, \infty, -\infty)$ [17]. Finally, it is also useful for computing the epsilon-removal of a weighted automaton in the general case of complete semirings [40] where Johnson's algorithm does not apply, which is the main motivation for our presentation of the algorithm.

GEN-ALL-PAIRS can be used of course to compute single-source shortest distances in graphs G weighted over a complete semiring. The complexity of the GEN-ALL-PAIRS algorithm in this case, $(|Q|^3)$, makes it impractical for large graphs. The next section describes a single-source shortest-distance algorithm which can be significantly more efficient in many cases.

3.2 Single-Source Shortest-Distance Problems

The general single-source shortest-distance algorithm described in this section is defined for any *k-closed semiring* [41, 20].[8]

3.2.1 *k*-Closed Semirings

Let $k \geq 0$ be an integer. A semiring $(S, \oplus, \otimes, \overline{0}, \overline{1})$ is said to be *k-closed* if

$$\forall a \in S, \quad \bigoplus_{n=0}^{k+1} a^n = \bigoplus_{n=0}^{k} a^n. \tag{14}$$

A k-closed semiring is thus a starsemiring with $a^* = \bigoplus_{n=0}^{k} a^n$ for all $a \in S$ (as defined in [20]). The Boolean semiring, the tropical semiring $(\mathbb{R}_+ \cup \{+\infty\}, \mathsf{min}, +, +\infty, 0)$, or $(\mathbb{R} \cup \{-\infty, \infty\}, \mathsf{min}, \mathsf{max}, \infty, -\infty)$ are examples of k-closed semirings with $k = 0$.

3.2.2 General Single-Source Shortest-Distance Algorithm

The shortest-distance $d[i, j]$ from any vertex i to any vertex j is well defined in a k-closed semiring S. Given a source vertex $s \in Q$, the general single-source shortest-distance problem consists of computing all distances $d[s, q]$, $q \in Q$.

Figure 3 gives the pseudo-code of an algorithm computing the single-source shortest-distances for any k-closed semiring [41]. The algorithm is based on a generalization of the relaxation technique to the k-closed semirings.

The algorithm maintains two arrays $d[q]$ and $r[q]$ indexed with vertices. $d[q]$ denotes the tentative shortest distance from the source s to q. $r[q]$ keeps track of the sum of the weights $\oplus$-added to $d[q]$ since the last queue extraction of q. The attribute r is needed for the shortest-distance algorithm to work in non-idempotent cases. The algorithm uses a queue $\mathcal{Q}$ to store the set of states to consider for the relaxation steps of lines 11–15 [41]. Any queue discipline, e.g., FIFO, shortest-first, topological (in the acyclic case), can be used.

Different queue disciplines yield different running times for our algorithm. The choice of the best queue discipline to use depends on the semiring and the graph structure.

If the graph is acyclic, then using the topological order queue discipline gives a linear-time algorithm: $O(|Q| + (T_\oplus + T_\otimes)|E|)$. For the tropical semiring $(\mathbb{R}_+ \cup \{+\infty\}, \mathsf{min}, +, +\infty, 0)$ and the best-first queue discipline, the algorithm coincides with Dijkstra's algorithm and its complexity is $O(|E| + |Q| \log |Q|)$ using Fibonacci heaps. In the presence of negative weights but no negative cycles, using a FIFO queue discipline, the algorithm coincides with the Bellman–Ford algorithm.

[8] See also [24, 20] for the related definition of *locally closed semirings*.

```
GEN-SINGLE-SOURCE(G, s)
 1  for i ← 1 to |Q| do
 2        d[i] ← r[i] ← 0̄
 3  d[s] ← r[s] ← 1̄
 4  𝒬 ← {s}
 5  while 𝒬 ≠ ∅ do
 6        q ← HEAD(𝒬)
 7        DEQUEUE(𝒬)
 8        r' ← r[q]
 9        r[q] ← 0̄
10        for each e ∈ E[q] do
11              if d[n[e]] ≠ d[n[e]] ⊕ (r' ⊗ w[e]) then
12                    d[n[e]] ← d[n[e]] ⊕ (r' ⊗ w[e])
13                    r[n[e]] ← r[n[e]] ⊕ (r' ⊗ w[e])
14                    if n[e] ∉ 𝒬 then
15                          ENQUEUE(𝒬, n[e])
```

Fig. 3. Generic single-source shortest-distance algorithm

The initialization step of the algorithm (lines 1–3) takes $O(|Q|)$ time, each relaxation (lines 11–13) takes $O(T_\oplus + T_\otimes + C(A))$ time. There are exactly $N(q)|E[q]|$ relaxations at q. The total cost of the relaxations is thus: $O((T_\oplus + T_\otimes + C(A))|E|\max_{q\in Q}N(q))$. Since each vertex q is inserted in $\mathcal{Q}$ $N(q)$ times (line 15), it is also extracted from $\mathcal{Q}$ $N(q)$ times (lines 6–7), and the general expression of the complexity is

$$O\Big(|Q| + (T_\oplus + T_\otimes + C(A))|E|\max_{q\in Q}N(q) + (C(I) + C(E))\sum_{q\in Q}N(q)\Big), \quad (15)$$

where $C(E)$ is the worst cost of removing a vertex q from the queue $\mathcal{Q}$, $C(I)$ that of inserting q in $\mathcal{Q}$, and $C(A)$ that of an assignment, including the possible necessary cost of reorganizing the queue.

Theorem 3.2. *Let $G = (Q, E, w)$ be a weighted directed graph over a k-closed commutative semiring S and let $s \in Q$ be a distinguished source vertex. Then, the algorithm* GEN-SINGLE-SOURCE *computes the single-source shortest-distances $d[s, q]$ to all vertices $q \in Q$ regardless of the queue discipline used for $\mathcal{Q}$.*

The proof of theorem is given in [41].

4 Rational Operations

Regulated weighted transducers are closed under the following three standard operations called *rational operations*:

- The *sum* (or *union*) of two weighted transducers T_1 and T_2 is defined by

$$\forall(x,y) \in \Sigma^* \times \Delta^*, \quad (T_1 \oplus T_2)(x,y) = T_1(x,y) \oplus T_2(x,y). \tag{16}$$

- The *product* (or *concatenation*) of two weighted transducers T_1 and T_2 is defined by

$$\forall(x,y) \in \Sigma^* \times \Delta^*, \quad (T_1 \otimes T_2)(x,y) = \bigoplus_{\substack{x=x_1x_2\\ y=y_1y_2}} T_1(x_1,y_1) \otimes T_2(x_2,y_2). \tag{17}$$

The sum runs over all possible ways of decomposing x into a prefix $x_1 \in \Sigma^*$ and a suffix $x_2 \in \Sigma^*$ and similarly $y \in \Delta^*$ into a prefix $y_1 \in \Delta^*$ and a suffix y_2. The product of $n > 0$ instances of T, $\overbrace{T \otimes \cdots \otimes T}^{n}$, is denoted by T^n, and by convention $T^0 = \mathcal{E}$, where $\mathcal{E}$ is the transducer defined by

$$\begin{cases} \mathcal{E}(x,y) = \overline{1} & \text{if } (x,y) = (\epsilon,\epsilon); \\ \overline{0} & \text{otherwise.} \end{cases} \tag{18}$$

- The *closure* (or *Kleene-closure*) of a weighted transducer T is defined by

$$\forall(x,y) \in \Sigma^* \times \Delta^*, \quad T^*(x,y) = \bigoplus_{n=0}^{+\infty} T^n(x,y), \tag{19}$$

when $\bigoplus_{n=0}^{+\infty} T^n(x,y)$ is an element of S for all $(x,y) \in \Sigma^* \times \Delta^*$. Note that in the absence of accepting ϵ-paths, that is when $P(I,\epsilon,\epsilon,F) = \emptyset$, $T^n(x,y) = \overline{0}$ for $n > |x| + |y|$, thus $T^*(x,y)$ is defined by a finite sum and is always an element of S. In complete semirings, the closure operation is defined for all weighted transducers.

Rational operations can be used to create complex weighted transducers from simpler ones as in the standard case of un-weighted acceptors. They admit simple and efficient algorithms. Figures 4(c)–(e) illustrate these algorithms for the particular cases of the transducers T_1 and T_2 of Figs. 4(a)–(b).

The transducer sum of two transducers T_1 and T_2 can be constructed from T_1 and T_2 by introducing a new state, made the unique initial state, with ϵ-transitions to the initial states of T_1 and T_2 carrying the weight $\overline{1}$. By construction, the sum of the weights of the paths with input label x and output label y in the resulting transducer is exactly the sum of the weights of the paths with these labels in T_1 and those with these labels in T_2, which matches precisely the definition of $T_1 \oplus T_2$ (Fig. 4(c)). The time and space complexity of the algorithm is thus linear, $O(|T_1| + |T_2|)$. Furthermore, the algorithm admits a natural *on-demand* or *on-the-fly construction*: states and transitions of the transducer sum can be created only as required by the algorithm using $T_1 \oplus T_2$. This is because the outgoing transition of a state of $T_1 \oplus T_2$ can be constructed only by using that state and T_1 and T_2 without

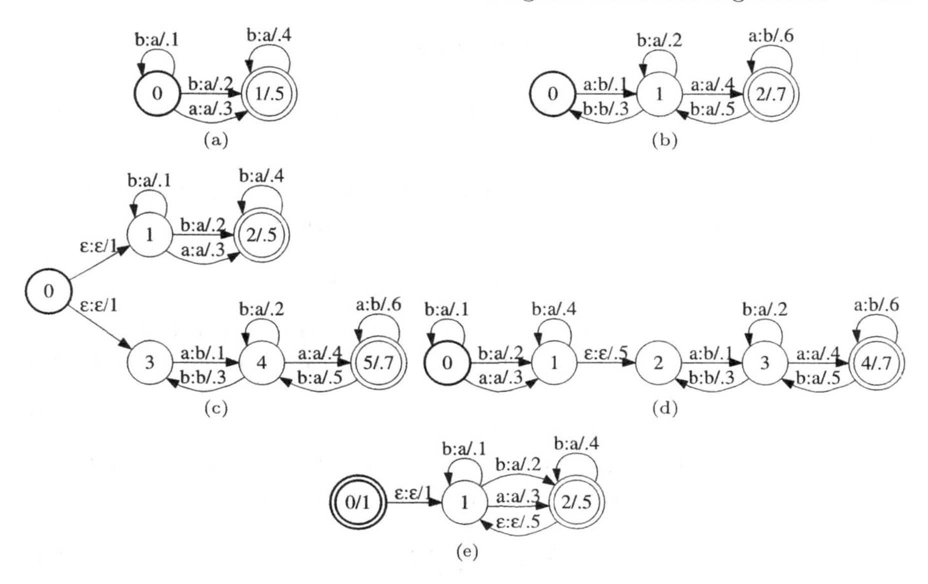

Fig. 4. (**a**) Weighted transducer T_1 and (**b**) weighted transducer T_2 over the probability semiring. (**c**) Sum of T_1 and T_2, $T_1 \oplus T_2$. (**d**) Product of T_1 and T_2, $T_1 \otimes T_2$. (**e**) Closure of T_1, T_1^*

inspecting other states of $T_1 \oplus T_2$. This *local* availability of the information needed to construct the output is what characterizes algorithms admitting natural on-the-fly constructions.

Similarly, the product (or concatenation) of two transducers T_1 and T_2 can be constructed from these transducers by making the final states of T_1 non-final and by creating an ϵ-transition from each final state p of T_1 to each initial state q of T_2 carrying the final weight of p (Fig. 4(d)). It is straightforward to verify the correctness of this construction. The time and space complexity of the algorithm is $O(|T_1| + |T_2| + |F_1||I_2|)$. The complexity of the product computation is linear for transducers with a single initial state, which is the typical situation in practice. As with the sum, the product algorithm admits a natural on-demand implementation.

The closure of a transducer T_1 can be constructed as in the standard case of un-weighted acceptors. A new initial state is created that is also final with final weight $\overline{1}$. An ϵ-transition with weight $\overline{1}$ is created from this state to the previously initial state of the transducer. Finally, an ϵ-transition is added from each final state p to the previously initial state carrying the final weight of p (Fig. 4(e)). The correctness of the construction follows the definition of the closure. The complexity of the algorithm is linear $O(|T_1|)$ and the algorithm admits a natural on-demand implementation as in the case of the other rational operations.

5 Elementary Unary Operations

This section briefly describes three elementary unary operations that are often useful in application.

- The *reversal* of a weighted transducers T produces a transducer T^R that assigns to each pair of strings (x, y) what T assigns to their mirror images (x^R, y^R):
$$T^R(x,y) = T(x^R, y^R). \tag{20}$$
- The *inversion* (or *transposition*) of a weighted transducer T produces a new weighted transducer by swapping the input and output label of each transition
$$T^{-1}(x,y) = T(y,x). \tag{21}$$
- The *projection* of a weighted transducer T on the *input side* (or *left projection*) yields an acceptor $\downarrow T$ by omitting output labels:
$$\downarrow T(x) = \bigoplus_y T(x,y). \tag{22}$$
Projection on the output side (or *right projection*), $T\downarrow$, is defined in a similar way.

These operations admit straightforward linear-time algorithms that are illustrated by Fig. 5. Inversion and projection are trivial and clearly admit a linear-time algorithm. When the semiring S is commutative, reversal can be obtained by reverting the direction of each transition and making initial states final and final states initial. It can also be obtained as in Fig. 5(a) by reverting the direction of all transitions, creating a new state p made the unique initial state, with ϵ-transitions to each previously final state q carrying the final weight of q, and making previously initial states final with the same weights. In all cases, reversal does not admit a natural on-demand computation since the computation of the outgoing transitions of a state of the output transducer requires creating or inspecting other output states.

6 Fundamental Binary Operations

In this section, the semiring S is assumed to be commutative.

6.1 Composition

Composition is a general operation for combining two or more weighted transducers [22, 54, 36, 35]. It is a powerful tool used in a variety of applications to create a complex weighted transducer from simpler ones representing statistical models or discriminative models.

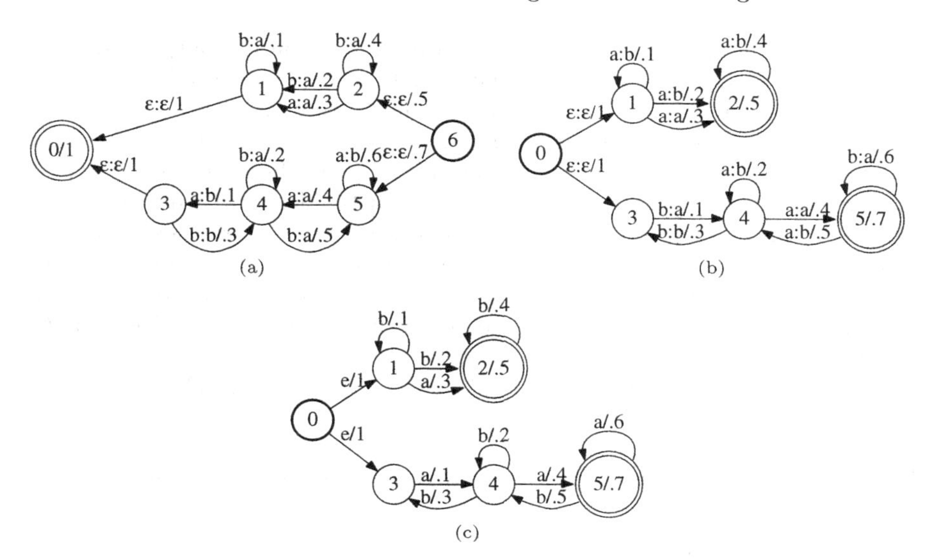

Fig. 5. Elementary operations applies to the transducer $T = T_1 \oplus T_2$ of Fig. 4(c). (**a**) Reversed transducer T^R. (**b**) Inverted transducer T^{-1}. (**c**) Projected transducer $\downarrow T$

The algorithm for the composition of weighted transducers is a generalization of the standard composition algorithm for un-weighted finite-state transducers. However, as we shall see later, the weighted case requires a more subtle technique to deal with ϵ-path multiplicity issues [50, 46]. The algorithm takes as input two weighted transducers

$$T_1 = (\Sigma^*, \Delta^*, Q_1, I_1, F_1, E_1, \lambda_1, \rho_1) \quad \text{and}$$
$$T_2 = (\Delta^*, \Omega^*, Q_2, I_2, F_2, E_2, \lambda_2, \rho_2)$$

such that the input alphabet of T_2, Δ, coincides with the output alphabet of T_1, outputs a weighted transducer $T = (\Sigma^*, \Omega^*, Q, I, F, E, \lambda, \rho)$ realizing the composition of T_1 and T_2.

Let T_1 and T_2 be two weighted transducers defined over S such that the input alphabet of T_2 coincides with the output alphabet of T_1. Assume that the infinite sum $\bigoplus_{z \in \Delta^*} T_1(x, z) \otimes T_2(z, y)$ is defined and in S for all $(x, y) \in \Sigma^* \times \Omega^*$. This condition holds for all transducers defined over a complete semiring such as the Boolean semiring, the tropical semiring, the probability semiring and the log semiring, and for all acyclic transducers defined over an arbitrary semiring. Then the result of the composition of T_1 and T_2 is a weighted transducer denoted by $T_1 \circ T_2$ and defined for all x, y by

$$(T_1 \circ T_2)(x, y) = \bigoplus_{z \in \Delta^*} T_1(x, z) \otimes T_2(z, y). \tag{23}$$

The sum runs over all strings z labeling a path of T_1 on the output side and a path of T_2 on input label z. The matrix notation we have used emphasizes the connection of composition with matrix multiplication.[9]

There exists a general and efficient algorithm to compute the composition of two weighted transducers. In the absence of ϵs on the input side of T_1 or the output side of T_2, the states of $T_1 \circ T_2$ can be identified with pairs of a state of T_1 and a state of T_2, $Q \subseteq Q_1 \times Q_2$. Initial states are those obtained by pairing initial states of the original transducers, $I = I_1 \times I_2$, and similarly final states are defined by $F = Q \cap (F_1 \times F_2)$. Transitions are obtained by matching a transition of T_1 with one of T_2 from appropriate transitions of T_1 and T_2:

$$E = \biguplus_{\substack{(q_1,a,b,w_1,q_2)\in E_1 \\ (q_1',b,c,w_2,q_2')\in E_2}} \left\{\left((q_1,q_1'),a,c,w_1 \otimes w_2,(q_2,q_2')\right)\right\}.$$

The following is the pseudo-code of the algorithm in the ϵ-free case.

```
Weighted-Composition(T1, T2)
 1  Q ← I1 × I2
 2  𝒬 ← I1 × I2
 3  while 𝒬 ≠ ∅ do
 4        q = (q1, q2) ← Head(𝒬)
 5        Dequeue(𝒬)
 6        if q ∈ I1 × I2 then
 7              I ← I ∪ {q}
 8              λ(q) ← λ1(q1) ⊗ λ2(q2)
 9        if q ∈ F1 × F2 then
10              F ← F ∪ {q}
11              ρ(q) ← ρ1(q1) ⊗ ρ2(q2)
12        for each (e1, e2) ∈ E[q1] × E[q2] such that o[e1] = i[e2] do
13              if (q' = (n[e1], n[e2]) ∉ Q) then
14                    Q ← Q ∪ {q'}
15                    Enqueue(𝒬, q')
16              E ← E ⊎ {(q, i[e1], o[e2], w[e1] ⊗ w[e2], q')}
17  return T
```

E, I, and F are all assumed to be initialized to the empty set. The algorithm uses a queue $\mathcal{Q}$ containing the set of pairs of states yet to be examined. The queue discipline of $\mathcal{Q}$ can be arbitrarily chosen and does not affect the termination of the algorithm. The set of states Q is originally reduced to the set of pairs of the initial states of the original transducers and $\mathcal{Q}$ is initialized to the same (lines 1–2). At each execution of the loop of lines 3–16, a new pair of states (q_1, q_2) is extracted from $\mathcal{Q}$ (lines 4–5). The initial weight of

[9] Our choice of a matrix notation as opposed to a functional notation is motivated by its convenience in applications.

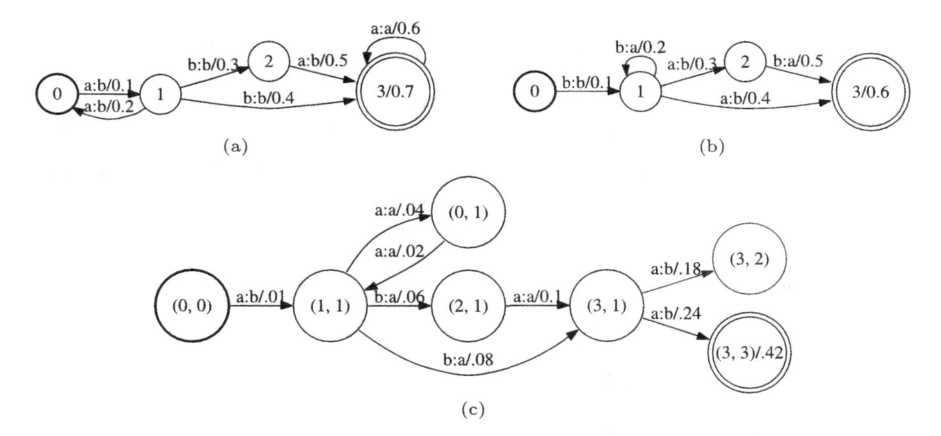

Fig. 6. Weighted transducers (**a**) T_1 and (**b**) T_2 over the probability semiring. (**c**) Illustration of composition of T_1 and T_2, $T_1 \circ T_2$. Some states might be constructed during the execution of the algorithm that are not co-accessible, e.g., *(3, 2)*. Such states and the related transitions can be removed by a trimming (or connection) algorithm in linear-time

(q_1, q_2) is computed by $\otimes$-multiplying the initial weights of q_1 and q_2 when they are both initial states (lines 6–8). Similar steps are followed for final states (lines 9–11). Then for each pair of matching transitions (e_1, e_2), a new transition is created according to the rules specified earlier (line 16). If the destination state $(n[e_1], n[e_2])$ has not been found earlier on, it is added to Q and inserted in $\mathcal{Q}$ (lines 14–15).

In the worst case, all transitions of T_1 leaving a state q_1 match all those of T_2 leaving state q_1', thus the space and time complexity of composition is quadratic: $O(|T_1||T_2|)$. However, an important feature of composition is that it admits a natural on-demand computation which can be used to construct only the part of the composed transducer that is needed. Figures 6(a)–(c) illustrate the algorithm when applied to the transducers of Figs. 6(a)–(b) defined over the probability semiring.

More care is needed when T_1 admits output ϵ labels or T_2 input ϵ labels. Indeed, as illustrated by Fig. 7, a straightforward generalization of the ϵ-free case would generate redundant ϵ-paths and, in the case of non-idempotent semirings, would lead to an incorrect result. The weight of the matching paths of the original transducers would be counted p times, where p is the number of redundant paths in the result of composition.

To cope with this problem, all but one ϵ-path must be filtered out of the composite transducer. Figure 7 indicates in boldface one possible choice for that path, which in this case is the shortest. Remarkably, that filtering mechanism itself can be encoded as a finite-state transducer F (Fig. 7(b)).

To apply that filter, we need to first augment T_1 and T_2 with auxiliary symbols that make the semantics of ϵ explicit. Thus, let $\tilde{T}_1$ ($\tilde{T}_2$) be the weighted

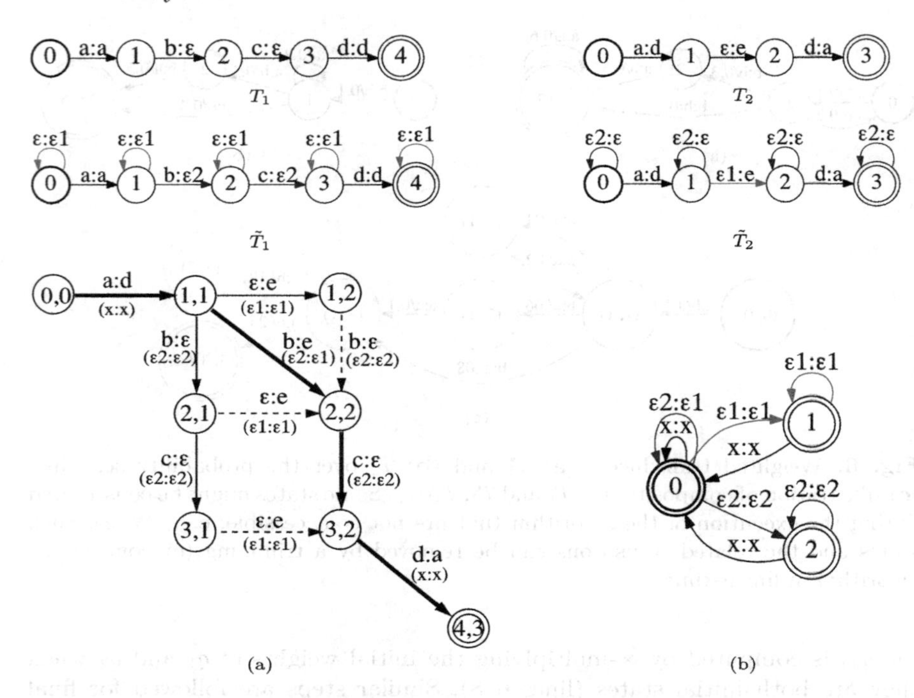

Fig. 7. Redundant ϵ-paths in composition. All transition and final weights are equal to $\overline{1}$. (**a**) A straightforward generalization of the ϵ-free case would generate all the paths from *(1, 1)* to *(3, 2)* when composing T_1 and T_2 and produce an incorrect result in non-idempotent semirings. (**b**) Filter transducer F [46]. The shorthand x is used to represent an element of Σ

transducer obtained from T_1 (resp. T_2) by replacing the output (resp. input) ϵ labels with ϵ_2 (resp. ϵ_1) as illustrated by Fig. 7. Thus, matching with the symbol e_1 corresponds to remaining at the same state of T_1 and taking a transition of T_2 with input ϵ. e_2 can be described in a symmetric way. The filter transducer F disallows a matching (ϵ_2, ϵ_2) immediately after (ϵ_1, ϵ_1) since this can be done instead via (ϵ_2, ϵ_1). By symmetry, it also disallows a matching (ϵ_1, ϵ_1) immediately after (ϵ_2, ϵ_2). In the same way, a matching (ϵ_1, ϵ_1) immediately followed by (ϵ_2, ϵ_1) is not permitted by the filter F since a shorter path via the matchings $(\epsilon_2, \epsilon_1)(\epsilon_1, \epsilon_1)$ is possible. Similarly, $(\epsilon_2, \epsilon_2)(\epsilon_2, \epsilon_1)$ is ruled out. It is not hard to verify that the filter transducer F is precisely a finite automaton over pairs accepting the complement of the language

$$L = \sigma^*\big((\epsilon_1, \epsilon_1)(\epsilon_2, \epsilon_2) + (\epsilon_2, \epsilon_2)(\epsilon_1, \epsilon_1) + (\epsilon_1, \epsilon_1)(\epsilon_2, \epsilon_1) + (\epsilon_2, \epsilon_2)(\epsilon_2, \epsilon_1)\big)\sigma^*,$$

where $\sigma = \{(\epsilon_1, \epsilon_1), (\epsilon_2, \epsilon_2), (\epsilon_2, \epsilon_1), x\}$ [4]. Thus, the filter F guarantees that exactly one ϵ-path is allowed in the composition of each ϵ sequences. To obtain the correct result of composition, it suffices then to use the ϵ-free composition algorithm already described and compute

$$\tilde{T}_1 \circ F \circ T_2. \tag{24}$$

Indeed, the two compositions in $\tilde{T}_1 \circ F \circ \tilde{T}_2$ no more involve ϵs. Since the size of the filter transducer F is constant, the complexity of general composition is the same as that of ϵ-free composition, that is $O(|T_1||T_2|)$. In practice, the augmented transducers $\tilde{T}_1$ and $\tilde{T}_2$ are not explicitly constructed, instead the presence of the auxiliary symbols is simulated. Further filter optimizations help limit the number of non-coaccessible states created, for example, by examining more carefully the case of states with only outgoing non-ϵ-transitions or only outgoing ϵ-transitions [46].

Composition of weighted transducers can be further generalized to the *N-way composition* of weighted transducers [4]. Furthermore, N-way composition of three or more transducers can be substantially faster than the use of the standard composition [5].

6.2 Intersection

The intersection (or Hadamard product) of two weighted automata A_1 and A_2 is defined by [22, 54, 36]:

$$(A_1 \cap A_2)(x) = A_1(x) \otimes A_2(x). \tag{25}$$

It coincides with the special case of composition of weighted transducers where the input label of each transition matches its output label. Thus, the same algorithm can be used to compute intersection with the same complexity. Figure 8 illustrates the application of the algorithm to two weighted automata extracted from the weighted transducers of Fig. 6.

6.3 Difference

Negation is not defined for all semirings, but a *difference* operation can be defined for a weighted automata A_1 and an un-weighted deterministic automaton A_2 as follows:[10]

$$\forall x \in \Sigma^*, \quad (A_1 - A_2)(x) = \begin{cases} A_1(x) & \text{if } x \notin L(A_2); \\ \overline{0} & \text{otherwise.} \end{cases} \tag{26}$$

Thus, $(A_1 - A_2)$ is the weighted automaton A_1 from which all accepting paths labeled with a string accepted by A_2 are removed, which leads to the following equivalent formulation:

$$\forall x \in \Sigma^*, \quad (A_1 - A_2)(x) = (A_1 \cap \overline{A}_2)(x), \tag{27}$$

[10] Of course, when negation is defined, $A_1 \oplus (\ominus A_2)$ defined by $\forall x \in \Sigma^*, A_1 \oplus (\ominus A_2) = A_1(x) \ominus A_2(x)$ can be computed by applying the sum algorithm to A_1 and $\ominus A_2$. The semantics of the difference operation considered here is different.

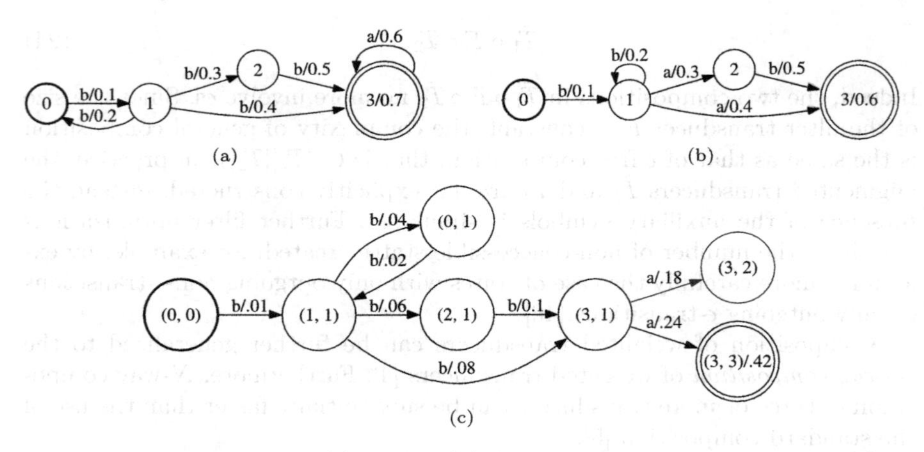

Fig. 8. Weighted automata (**a**) A_1 and (**b**) A_2 over the probability semiring. (**c**) Illustration of intersection of A_1 and A_2, $A_1 \cap A_2$. Some states might be constructed during the execution of the algorithm that are not co-accessible, e.g., *(3, 2)*. Such states and the related transitions can be removed by a trimming (or connection) algorithm in linear-time

where $\overline{A}_2$ is a weighted automaton over the semiring S accepting exactly the complement of $L(A_2)$ and assigning weight $\overline{1}$ to each string accepted. Since A_2 is deterministic, its complement $\overline{A}_2$ can be computed from A_2 in linear time, with the following two steps:

- *Completion*: first making A_2 *complete*, that is creating an equivalent automaton to A_2 such that all alphabet symbols can be read from any state. This can be done by augmenting A_2 with a new state p with self-loops labeled with all alphabet symbols, and by adding a transition labeled with $a \in \Sigma$ from state q to p when no transition labeled with a is available at q in A_2.
- *Complementation*: then making all final states of the modified automaton A_2 non-final and vice versa. Finally, all weights of the automaton are set to $\overline{1}$ to make it an automaton over S.

Both of these steps can be executed in linear time $O(|A_2| + |\Sigma|)$ and admit a natural on-demand implementation. Note that the complementation of arbitrary finite automata is PSPACE-complete [1]; this is the reason why A_2 was assumed to be deterministic here. The difference can then be obtained by computing the intersection of A_1 and $\overline{A}_2$. Since intersection or composition also admit a natural on-the-demand computation, the same is true of the difference algorithm. Note that using that property, the alphabet symbol actually used in complementation can be limited to the symbols appearing in A_1 and A_2. Thus, the overall complexity of difference is $O(|A_1|(|A_2| + |\Sigma'|)$.

Figure 9 illustrates the difference algorithm.

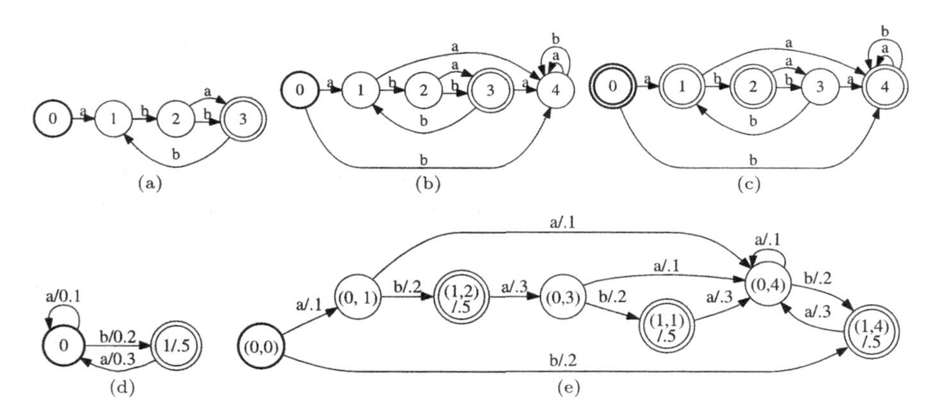

Fig. 9. (**a**) Un-weighted automaton A_2. (**b**) Complete automaton equivalent to A_2. (**c**) Complement of A_2, $\overline{A}_2$, all weights are equal to $\overline{1}$ and thus not indicated. (**d**) Weighted automaton A_1 defined over the probability semiring. (**e**) Difference of A_1 and A_2, $A_1 - A_2$, obtained by intersection of A_1 and $\overline{A}_2$

7 Optimization Algorithms

7.1 Epsilon-Removal

The use of various automata or transducer operations such as rational operations generate ϵ-transitions. These transitions cause some delay in the use of the resulting transducers since the search for an alphabet symbol to match in composition or other similar operations requires reading some sequences of ϵs first. To make these weighted transducers more efficient to use, it may be preferable to *remove* all ϵ-transitions, that is, to create an equivalent weighted transducer with no ϵ-transition. This section describes a general ϵ-removal algorithm that precisely achieves this task.

Simply removing ϵ-transitions from the input transducer clearly does not result in an equivalent one. Instead, for a given state p, the non-ϵ-transitions of all states q reachable from p via ϵ-transitions should be added to those of p. In the weighted case, this does not result in an equivalent transducer since the weights of the ϵ-transitions from p to q would be ignored. Thus, before adding an outgoing transition of state q to p, the weight it carries must be pre-$\otimes$-multiplied by the sum of the weights of the ϵ-paths from p to q. To ensure that this weight is an element of the semiring, we will assume that S is complete.[11]

This leads to a two-step algorithm [40]. Given a transducer T, let T_ϵ denote the transducer derived from T by keeping only ϵ-transitions and let $d_\epsilon[p, q] = \bigoplus_{\pi \in P(p,\epsilon,q)} w[\pi]$ denote the distance from state p to state q in T_ϵ. Then the following are the two main steps of the algorithms:

[11] The results presented also hold in the case of closed semirings.

- ϵ-Closure computation: at each state p, the *weighted ϵ-closure* defined by

$$C(p) = \{(q, w) : P(p, \epsilon, q) \neq \emptyset,\ w = d_\epsilon[p, q]\} \tag{28}$$

  is computed.
- Actual removal of ϵs: all ϵ-transitions are removed and for each p and each $(q, w) \in C(p)$, the transition set of p is augmented with the following transitions

$$\{(p, a, b, d_\epsilon[p, q] \otimes w, r) : (q, a, b, w, r) \in E,\ (a, b) \neq (\epsilon, \epsilon)\}. \tag{29}$$

  If there exists $(q, w) \in C(p)$ with $q \in F$, then $d_\epsilon[p, q] \otimes \rho(q)$ must be $\oplus$-added to the final weight of p.

The following is the pseudo-code of the algorithm which follows the main steps just discussed.

Epsilon-Removal(T)

```
 1  for each p ∈ Q do
 2      Compute-Closure(C(p))
 3  E' ← Ē_ε ← {(p, a, b, w, q) ∈ E : (a, b) ≠ (ε, ε)}
 4  F' ← F
 5  ρ' ← ρ
 6  for each p ∈ Q do
 7      for each (q, w') ∈ C[p] do
 8          E'[p] ← E'[p] ⊎ {(p, a, b, w' ⊗ w, r) : (q, a, b, w, r) ∈ Ē_ε}
 9          if q ∈ F then
10              if p ∉ F then
11                  F' ← F ∪ {p}
12                  ρ'[p] ← 0̄
13              ρ'[p] ← ρ'[p] ⊕ (w' ⊗ ρ(q))
14  return T' = (Σ, Δ, Q, I, F', E', λ, ρ')
```

Theorem 7.1 ([40]). *Let T be a weighted transducer over a complete semiring S. Assume that the closures $C(p)$ can be computed for any state p of T. Then the weighted transducer T' returned by the epsilon-removal algorithm just described is equivalent to T.*

The proof is simple and is given in [40].

The ϵ-closures $C(p)$ can be computed using an all-pair shortest-distance algorithm over T_ϵ when the semiring S is complete, or by applying the single-shortest distance algorithm from each source p when the semiring is k-closed, as described in Sect. 3. The complexity of the second stage of the algorithm (lines 6–13) is in $O(|Q|^2 + |Q||E|)$ since in the worst case each $C(p)$ contains all states of the transducer. Thus, the overall complexity of the algorithm is

$$O\big(|\text{Compute-Closure}| + |Q|^2 + |Q||E|(T_\oplus + T_\otimes)\big), \tag{30}$$

where $O(|\text{COMPUTE-CLOSURE}|)$ denotes the total cost of the closure computation. We are now examining several special cases of practical interest:

- T_ϵ is acyclic, that is T admits no ϵ-cycle, a rather frequent case in practice. In that case, the single-source shortest distance algorithm can be used for any semiring S and has linear time complexity. The total complexity of applying the algorithm at each state is $O(|Q|^2 + |Q||E|(T_\oplus + T_\otimes))$ and matches that of the second stage. Thus, the overall complexity of epsilon-removal is then
$$O\big(|Q|^2 + |Q||E|(T_\oplus + T_\otimes)\big). \tag{31}$$
The algorithm can in fact be improved in that special case. The complexity of the computation of the all-pairs shortest distances can be substantially improved if the states of T_ϵ are visited in reverse topological order and if the single-source shortest-distance algorithm is interleaved with the actual removal of ϵs as follows: for each state p of T_ϵ visited in reverse topological order,
  - Run a single-source shortest-distance algorithm with source p to compute the distance from p to each state q in T_ϵ
  - Then remove ϵ-transitions leaving q and update the final weight as already described

  The reverse topological order guarantees that the ϵ-paths leaving p are reduced to the ϵ-transitions leaving p. Thus, the cost of the shortest-distance algorithm run from p only depends on the number of ϵ-transitions leaving p and the total cost of the computation of the shortest-distances is linear: $O(|Q| + (T_\oplus + T_\otimes)|E|)$.
- S is the tropical semiring. In that case, the complexity of the first stage of the algorithm is that of a standard shortest-path algorithm from each state of T_ϵ. Using Fibonacci heaps, the complexity of the first stage of the algorithm is thus $O(|Q||E| + |Q|^2 \log |Q|)$. Thus, the overall complexity of epsilon-removal is again
$$O\big(|Q|^2 + |Q||E|(T_\oplus + T_\otimes)\big). \tag{32}$$
- S is a complete semiring. In that case, when the all-pairs shortest-distance algorithm of Sect. 3 is the only algorithm available, the complexity of the first stage of the algorithm is $\Theta(|Q|^3(T_\oplus + T_\otimes + T_*))$ and the overall complexity of epsilon-removal is also
$$O\big(|Q|^3(T_\oplus + T_\otimes + T_*) + |Q||E|(T_\oplus + T_\otimes)\big). \tag{33}$$

Epsilon-removal does not create any new state. However, not all states of the original transducer may be necessary. States with only incoming (or outgoing) ϵ-transitions become non-accessible (resp. non-coaccessible) after removal of these transitions, which causes other states not to be accessible or co-accessible. All of these states and corresponding transitions can be removed in linear time using a standard trimming or connection algorithm.

During the epsilon-removal construction, it may happen quite often in practice that several transitions from the same state p to the same state q, with the same input and output label, need to be constructed in the resulting transducer T'. To avoid this redundancy, the weights of these transitions are $\oplus$-summed to maintain at any time a single transition instead.

Note that epsilon-removal admits a natural on-demand computation since the outgoing transitions of state q of the output automaton can be computed directly using the ϵ-closure of q. However, the reverse topological order described in the case of an acyclic T_ϵ requires examining all states of T_ϵ and thus that version of the algorithm cannot be viewed as a natural on-demand construction. In practice, both versions of the algorithm can be useful.

When epsilon-removal is to be followed immediately by the application of determinization, the integration of these two operations often results in much more efficient overall computation. This is because the ϵ-closure of a subset of states created by determinization can be computed by a single shortest-distance algorithm with all states of the subset serving as sources, rather than a distinct one for each state of the subset. Since some states can be reached by several elements of the subset, the first method provides more sharing.

This integration of determinization and epsilon-removal can be extended to the weighted case where the weighted determinization [38] presented in the next section is used.[12]

Epsilon-removal can be straightforwardly modified to remove transitions with input label a and output label b, with $(a, b) \neq (\epsilon, \epsilon)$. This can be done for example by relabeling ϵ-transitions with a new label and replacing (a, b) by (ϵ, ϵ), applying epsilon-removal, and then restoring original ϵs. The resulting transducer is equivalent to the original if (a, b) is assigned the semantics of (ϵ, ϵ).

Figure 10 illustrates the epsilon-removal algorithm. Figure 10(c) shows the transducer T' resulting from the transducer T of Fig. 10(a) by application of epsilon-removal. Note that only three of the original states remain in T'. As already discussed, since state 2 and 3 admit only incoming ϵ-transitions (and only outgoing ϵ-transitions in the case of state 3), after removal of ϵ-transitions they become inaccessible and can thus be removed. Figure 10(b) indicates all non-$\overline{0}$ shortest-distances between states in T_ϵ, which summarizes the closure information. These distances are used to determine the weight of the new transitions added.

Instead of removing an ϵ-transition from p to q by adding to state p all non-ϵ-transitions leaving q, one can equivalently proceed by adding to q all non-ϵ-transitions entering p. This is equivalent to applying epsilon-removal in the same way as before but to the reverse of T. We will thus refer to *reverse epsilon-removal* as the algorithm that consists of the following sequence of op-

[12] It is, however, limited to the cases where the result of epsilon-removal is *determinizable*, that is cases where the determinization algorithm terminates, which as we shall see later, does not always hold in the weighted case.

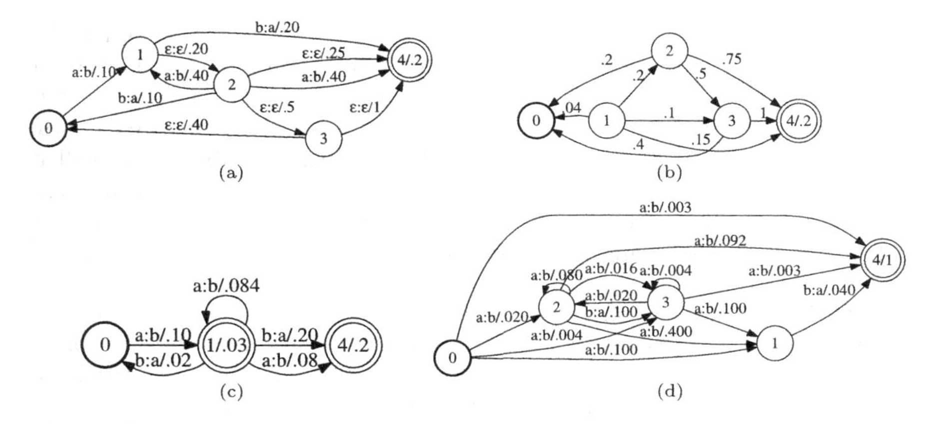

Fig. 10. (**a**) Weighted transducer T defined over the probability semiring. (**b**) Weighted graph showing non-$\overline{0}$ all-pair distance $d_\epsilon[p, q]$. From each state p, the outgoing weighted edges give the closure $C(p)$. (**c**) Weighted transducer T' resulting from T by epsilon-removal. (**d**) Weighted transducer T'' resulting from T by reverse epsilon-removal

erations: reversal, epsilon-removal, reversal. Figure 10(d) shows T'' the result of the application of reverse epsilon-removal to T. T'' is equivalent to T and T'' and has the same number of states as T, but in this case has more transitions than T'. Which algorithm epsilon-removal or reverse epsilon-removal, produces the smallest transducer depends on the number of outgoing transitions of the states q reached by an ϵ-path in the epsilon-removal case, or the number of incoming transitions of the states p with incoming ϵs in the reverse epsilon-removal case. The decision of the direction of epsilon-removal can be made in fact for each pair of states (p, q) based upon these quantities.

7.2 Determinization

This section describes a general determinization algorithm for weighted automata and transducers [38] which generalizes the standard powerset construction for un-weighted finite automata. The presentation will focus on the case of weighted automata, the weighted transducer case can be treated in a similar way or as a special case of the general algorithm we present [38].

A weighted automaton is said to be *deterministic* or *subsequential* if it has a unique initial state and if no two transitions leaving any state share the same input label. There exists a natural extension of the classical subset construction to the case of weighted automata called *determinization.* Weighted determinization requires some technical conditions on the semiring or the weighted automaton which we will first introduce. These conditions hold in most cases in practice.

Weighted determinization is a generic algorithm: it works with any *weakly divisible semiring*. A semiring is said to be *divisible* if all non-$\overline{0}$ elements admit an inverse, that is if $S - \{\overline{0}\}$ is a group. $(S, \oplus, \otimes, \overline{0}, \overline{1})$ is said to be *weakly divisible* if for any x and y in S such that $x \oplus y \neq \overline{0}$, there exists at least one z such that $x = (x \oplus y) \otimes z$. The $\otimes$-operation is *cancellative* if z is unique and we can write: $z = (x \oplus y)^{-1}x$. When z is not unique, we can still assume that we have an algorithm to find one of the possible z and call it $(x \oplus y)^{-1}x$. Furthermore, we will assume that z can be found in a consistent way, that is: $((u \otimes x) \oplus (u \otimes y))^{-1}(u \otimes x) = (x \oplus y)^{-1}x$ for any $x, y, u \in S$ such that $u \neq \overline{0}$. A semiring is *zero-sum-free* if for any x and y in S, $x \oplus y = \overline{0}$ implies $x = y = \overline{0}$.

Additionally, we assume that for any string $x \in \Sigma^*$, the sum of the weights of the paths labeled with x and starting at an initial state is non-$\overline{0}$: $w[P(I, x, Q)] \neq \overline{0}$. This condition is always satisfied with trim weighted automata over the tropical semiring or any zero-sum-free semiring.

The pseudo-code of the algorithm is given below with Q', I', F', and E' all initialized to the empty set.

Weighted-Determinization(A)

1 $i' \leftarrow \{(i, \lambda(i)) : i \in I\}$
2 $\lambda'(i') \leftarrow \overline{1}$
3 $\mathcal{Q} \leftarrow \{i'\}$
4 **while** $\mathcal{Q} \neq \emptyset$ **do**
5 $\quad p' \leftarrow \text{Head}(\mathcal{Q})$
6 $\quad \text{Dequeue}(\mathcal{Q})$
7 $\quad$ **for each** $x \in i[E[Q[p']]]$ **do**
8 $\quad\quad w' \leftarrow \bigoplus\{v \otimes w : (p, v) \in p', (p, x, w, q) \in E\}$
9 $\quad\quad q' \leftarrow \{(q, \bigoplus\{w'^{-1} \otimes (v \otimes w) : (p, v) \in p', (p, x, w, q) \in E\}) :$
$\quad\quad\quad q = n[e], i[e] = x, e \in E[Q[p']]\}$
10 $\quad\quad E' \leftarrow E' \cup \{(p', x, w', q')\}$
11 $\quad\quad$ **if** $q' \notin Q'$ **then**
12 $\quad\quad\quad Q' \leftarrow Q' \cup \{q'\}$
13 $\quad\quad\quad$ **if** $Q[q'] \cap F \neq \emptyset$ **then**
14 $\quad\quad\quad\quad F' \leftarrow F' \cup \{q'\}$
15 $\quad\quad\quad\quad \rho'(q') \leftarrow \bigoplus\{v \otimes \rho(q) : (q, v) \in q', q \in F\}$
16 $\quad\quad\quad \text{Enqueue}(\mathcal{Q}, q')$
17 **return** T'

A *weighted subset* p' of Q is a set of pairs $(q, x) \in Q \times S$. We will denote by $Q[p']$ the set of states q of the weighted subset p'. $E[Q[p']]$ represents the set of transitions leaving these states, and $i[E[Q[p']]]$ the set of input labels of these transitions.

The states of the output automaton can be identified with (weighted) subsets of the states of the original automaton. A state r of the output automaton that can be reached from the start state by a path π is identified with the set of pairs $(q, x) \in Q \times S$ such that q can be reached from an initial state of the origi-

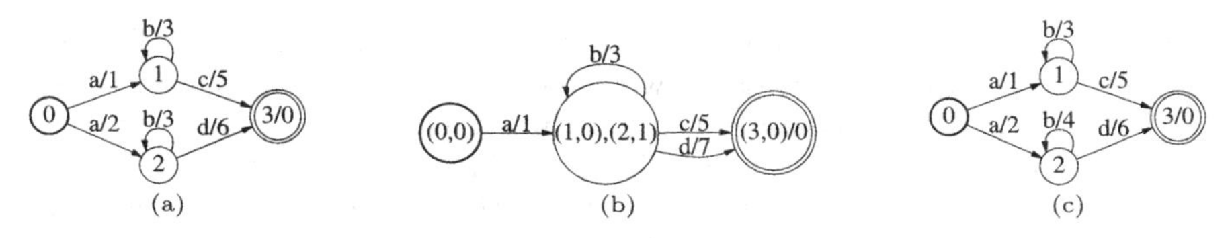

Fig. 11. Determinization of weighted automata. (**a**) Weighted automaton over the tropical semiring A. (**b**) Equivalent weighted automaton B obtained by determinization of A. (**c**) Non-determinizable weighted automaton over the tropical semiring, states *1* and *2* are non-twin siblings

nal machine by a path σ with $i[\sigma] = i[\pi]$ and $\lambda(p[\sigma]) \otimes w[\sigma] = \lambda(p[\pi]) \otimes w[\pi] \otimes x$. Thus, x can be viewed as the *residual* weight at state q.

Determinization does not terminate for all weighted automata. As we shall see, not all weighted automata are *determinizable* by the algorithm just described. When it terminates, the algorithm returns a subsequential weighted automaton $A' = (\Sigma, Q', I', F', E', \lambda', \rho')$, equivalent to the input $A = (\Sigma, Q, I, F, E, \lambda, \rho)$.

The algorithm uses a queue $\mathcal{Q}$ containing the set of states of the resulting automaton A', yet to be examined. The queue discipline of $\mathcal{Q}$ can be arbitrarily chosen and does not affect the termination of the algorithm. A' admits a unique initial state, i', defined as the set of initial states of A augmented with their respective initial weights. Its input weight is $\overline{1}$ (lines 1–2). $\mathcal{Q}$ originally contains only the subset i' (line 3). At each execution of the loop of lines 4–16, a new subset p' is extracted from $\mathcal{Q}$ (lines 5–6). For each x labeling at least one of the transitions leaving a state p of the subset p', a new transition with input label x is constructed. The weight w' associated to that transition is the sum of the weights of all transitions in $E[Q[p']]$ labeled with x pre-$\otimes$-multiplied by the residual weight v at each state p (line 8). The destination state of the transition is the subset containing all the states q reached by transitions in $E[Q[p']]$ labeled with x. The weight of each state q of the subset is obtained by taking the $\oplus$-sum of the residual weights of the states p $\otimes$-times the weight of the transition from p leading to q and by *dividing* that by w'. The new subset q' is inserted in the queue $\mathcal{Q}$ when it is a new state (line 15). If any of the states in the subset q' is final, q' is made a final state and its final weight is obtained by summing the final weights of all the final states in q', pre-$\otimes$-multiplied by their residual weight v (line 14).

Figure 11 illustrates the determinization of a weighted automaton over the tropical semiring. The worst case complexity of determinization is exponential even in the un-weighted case. However, in many practical cases such as for weighted automata used in large-vocabulary speech recognition, this blow-up does not occur. It is also important to notice that just like composition, determinization admits a natural lazy implementation which can be useful for saving space.

Unlike the unweighted case, determinization does not halt on all input weighted automata. In fact, some weighted automata, non-*subsequentiable* automata, do not even admit equivalent subsequential machines. But even for some subsequentiable automata, the algorithm does not halt. We say that a weighted automaton A is *determinizable* if the determinization algorithm halts for the input A. With a determinizable input, the algorithm outputs an equivalent subsequential weighted automaton.

There exists a general *twins property* for weighted automata that provides a characterization of determinizable weighted automata under some general conditions. Let A be a weighted automaton over a weakly divisible semiring S. Two states q and q' of A are said to be *siblings* if there exist two strings x and y in A^* such that both q and q' can be reached from I by paths labeled with x and there is a cycle at q and a cycle at q' both labeled with y. When S is a commutative and cancellative semiring, two sibling states are said to be *twins* iff for any string y:

$$w[P(q,y,q)] = w[P(q',y,q')]. \qquad (34)$$

A has *the twins property* if any two sibling states of A are twins.[13] Figure 11(c) shows an unambiguous weighted automaton over the tropical semiring that does not have the twins property: states 1 and 2 can be reached by paths labeled with a from the initial state and admit cycles with the same label b, but the weights of these cycles (3 and 4) are different.

The following theorem is proven in [38].

Theorem 7.2 ([38]). *Let A be a weighted automaton over the tropical semiring. If A has the twins property, then A is determinizable.*

With trim unambiguous weighted automata, the condition is also necessary [38, 3].

Theorem 7.3 ([38, 3]). *Let A be a trim unambiguous weighted automaton over the tropical semiring. Then the three following properties are equivalent:*

1. *A is determinizable.*
2. *A has the twins property.*
3. *A is subsequentiable.*

There exists an efficient algorithm for testing the twins property for trim unambiguous and even *cycle-unambiguous* weighted automata in time

[13] The notion of twins property was originally introduced for un-weighted finite-state transducers by [15, 16] and was shown to be decidable. Polynomial-time algorithms were later given to test this property for functional transducers in time $O(|Q|^4(|Q|^2+|E|^2)|\Delta|)$ by [58], $O(|Q|^4(|Q|^2+|E|^2))$ by [11], and $O(|Q|^2(|Q|^2+|E|^2))$ by [3], where Q is the set of states of the input transducer, E the set of its transitions and Δ the output alphabet.

$O(|Q|^2 + |E|^2)$ [3].[14] Note that any acyclic weighted automaton over a zero-sum-free semiring has the twins property and is determinizable.

The existence of an equivalent sequential weighted automaton for a finitely ambiguous weighted automaton over the tropical semiring was shown to be decidable [32]. The twins property has also been shown more recently to be a necessary and sufficient condition for the determinizability of finitely ambiguous trim weighted automata, that is, trim automata for which at most a fixed finite number of accepting paths are labeled by any string that are defined over the tropical semiring of integers $(\mathbb{Z} \cup \{+\infty\}, \min, +, +\infty, 0)$ [31]. A more general notion of *clones property* was introduced by the same author and shown to be a decidable necessary and sufficient condition characterizing determinizability for *polynomially ambiguous* automata over the tropical semiring, that is weighted automata over the tropical semiring for which the number of accepting paths of any string x is bounded by a fixed polynomial defined over the length of x.

7.3 Weight Pushing

The choice of the distribution of the total weight along each successful path of a weighted automaton does not affect the definition of the function realized by that automaton, but it may have a critical impact on efficiency in many applications, e.g., information extraction or natural language processing, where a heuristic pruning can often be used to visit only a subpart of the automaton. There exists an algorithm, *weight pushing*, for normalizing the distribution of the weights along the paths of a weighted automaton or more generally a weighted directed graph [38, 43].

Let A be a weighted automaton over a semiring S. Assume that S is zero-sum-free and weakly divisible. For any state $q \in Q$, assume that the following sum is defined and in S:

$$d[q] = \bigoplus_{\pi \in P(q,F)} \big(w[\pi] \otimes \rho(n[\pi])\big). \tag{35}$$

$d[q]$ is the *shortest-distance* from q to F including the final weight. $d[q]$ is well defined for all $q \in Q$ when S is a k-closed semiring. The weight pushing algorithm consists of computing each shortest-distance $d[q]$ and of *re-weighting* the transition weights, initial weights, and final weights in the following way:

$$\forall e \in E \text{ s.t. } d\big[p[e]\big] \neq \overline{0}, \quad w[e] \leftarrow d\big[p[e]\big]^{-1} \otimes w[e] \otimes d\big[n[e]\big], \tag{36}$$

$$\forall q \in I, \quad \lambda(q) \leftarrow \lambda(q) \otimes d[q], \tag{37}$$

$$\forall q \in F \text{ s.t. } d[q] \neq \overline{0}, \quad \rho(q) \leftarrow d[q]^{-1} \otimes \rho(q). \tag{38}$$

[14] An automaton is cycle-unambiguous if for any state q and any string x there exists at most one cycle at q labeled with x.

Roughly speaking, the algorithm *pushes the weights* of each path as much as possible toward the initial states. Figures 12(a)–(c) illustrate the application of the algorithm in a special case both for the tropical and probability semirings.

Each of the operations described can be assumed to be done in constant time, thus re-weighting can be done in linear time $O(T_{\otimes}|A|)$ where $T_{\otimes}$ denotes the worst cost of an $\otimes$-operation. The complexity of the computation of the shortest-distances depends on the semiring and the algorithm used (see Sect. 3). In the case of k-closed semirings such as the tropical semiring, $d[q]$ can be computed using a single-source shortest-path algorithm. The complexity of the algorithm is linear in the case of an acyclic automaton: $O(|Q| + (T_{\oplus} + T_{\otimes})|E|)$, where $T_{\oplus}$ denotes the worst cost of an $\oplus$-operation. In the case of a general weighted automaton over the tropical semiring, the complexity of the algorithm is $O(|E| + |Q| \log |Q|)$.

In the case of complete semirings such as $(\mathbb{R}_+, +, \times, 0, 1)$, a generalization of the Floyd–Warshall algorithm for computing all-pairs shortest-distances can be used. The complexity of the algorithm is $\mathbf{\Theta}(|Q|^3(T_{\oplus} + T_{\otimes} + T_*))$ where T_* denotes the worst cost of the closure operation. The space complexity of these algorithms is $\mathbf{\Theta}(|Q|^2)$. These complexities make it impractical to use the Floyd–Warshall algorithm for computing $d[q]$, $q \in Q$, for relatively large graphs or automata of several hundred million states or transitions. An approximate version of a generic shortest-distance algorithm can be used instead to compute $d[q]$ efficiently.

Note that if $d[q] = \overline{0}$, then since S is zero-sum-free, the weight of all paths from q to F is $\overline{0}$. Let A be a weighted automaton over the semiring S. Assume that S is complete or k-closed and that the shortest-distances $d[q]$ are all well defined and in $S - \{\overline{0}\}$. Note that in both cases we can use the distributivity over the infinite sums defining shortest distances. Let e' (π') denote the transition e (path π) after application of the weight pushing algorithm. e' (π') differs from e (resp. π) only by its weight. Let λ' denote the new initial weight function, and ρ' the new final weight function.

The following proposition is proven in [38, 43].

Proposition 7.4 ([38, 43]). *Let $B = (A, Q, I, F, E', \lambda', \rho')$ be the result of the weight pushing algorithm applied to the weighted automaton A, then:*

1. *The weight of a successful path π is unchanged after application of weight pushing:*

$$\lambda'[p[\pi']] \otimes w[\pi'] \otimes \rho'[n[\pi']] = \lambda(p[\pi]) \otimes w[\pi] \otimes \rho(n[\pi]). \tag{39}$$

2. *The weighted automaton B is* stochastic, *i.e.,*

$$\forall q \in Q, \quad \bigoplus_{e' \in E'[q]} w[e'] = \overline{1}. \tag{40}$$

These two properties of weight pushing are illustrated by Figs. 12(a)–(c): the total weight of a successful path is unchanged after pushing; at each state

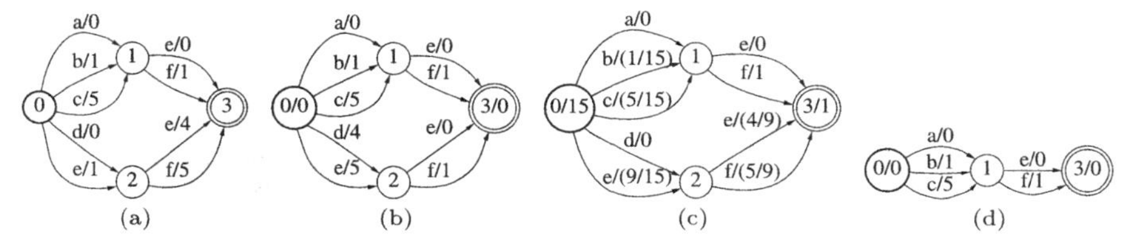

Fig. 12. Weight pushing algorithm. (**a**) Weighted automaton A. (**b**) Equivalent weighted automaton B obtained by weight pushing in the tropical semiring. (**c**) Weighted automaton C obtained from A by weight pushing in the probability semiring. (**d**) Minimal weighted automaton over the tropical semiring equivalent to A

of the weighted automaton of Fig. 12(b), the minimum weight of the outgoing transitions is 0, and at each state of the weighted automaton of Fig. 12(c), the weights of outgoing transitions sum to 1.

Weight pushing can also be used to test the equivalence of two subsequential weighted automata [38, 43]. Let A and B be two subsequential weighted automata to which weight pushing can be applied and let A' and B' be the resulting automata after weight pushing. Then the equivalence of A and B can be tested by applying the standard equivalence algorithm for un-weighted automata [1] to A' and B' after considering each pair of (transition label, transition weight) as a single label. The equivalence of two arbitrary weighted automata A and B over the field of real numbers with alphabet Σ can be tested in time $O(|\Sigma|(|Q_A|+|Q_B|)^3)$, where $|Q_A|$ denotes the number of states of A and $|Q_B|$ the number of states of B, using an algorithm [19] based on the standardization technique of Schützenberger [55]. The equivalence of arbitrary weighted automata over the tropical semiring is known to be undecidable [34].

7.4 Minimization

A deterministic weighted automaton is said to be *minimal* if there exists no other deterministic weighted automaton with a smaller number of states and realizing the same function. Two states of a deterministic weighted automaton are said to be *equivalent* if exactly the same set of strings with the same weights label paths from these states to a final state; the final weights being included. Thus, two equivalent states of a deterministic weighted automaton can be merged without affecting the function realized by that automaton. A weighted automaton is minimal when it admits no two distinct equivalent states after any redistribution of the weights along its paths.

There exists a general algorithm for computing a minimal deterministic automaton equivalent to a given weighted automaton [38]. It is thus a generalization of the minimization algorithms for un-weighted finite automata. In fact, minimization of both un-weighted [39] and weighted finite-state transducers can be viewed as special instances of this algorithm.

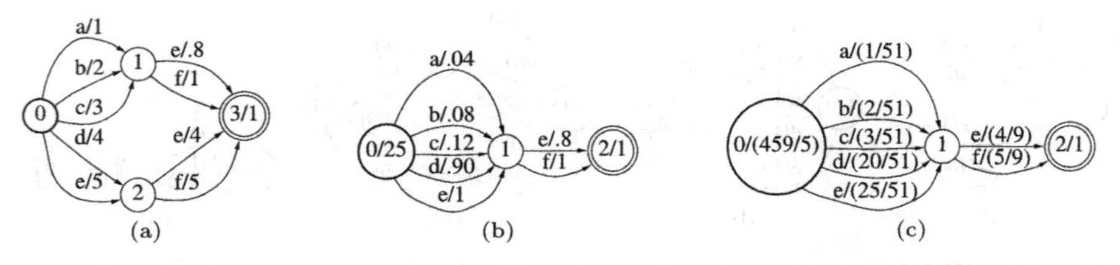

Fig. 13. Minimization of weighted automata. (**a**) Weighted automaton A' over the probability semiring. (**b**) Minimal weighted automaton B' equivalent to A'. (**c**) Minimal weighted automaton C' equivalent to A'

The algorithm consists of first applying weight pushing to normalize the distribution of the weights along the paths of the input automaton, and then applying the classical un-weighted automata minimization while treating each pair (label, weight) as a single label.

Theorem 7.5 ([38]). *Let A be a deterministic weighted automaton over a semiring S. Assume that the conditions of application of the weight pushing algorithm hold, then the execution of the following steps:*

1. *Weight pushing*
2. *(Un-weighted) automata minimization, treating each pair (label, weight) as a single label*

yield a minimal weighted automaton equivalent to A.

The complexity of automata minimization is linear in the case of acyclic automata $O(|Q| + |E|)$ [53] and in $O(|E| \log |Q|)$ in the general case [1]. Thus, in view of the complexity results given in the previous section, in the case of the tropical semiring, the total complexity of the weighted minimization algorithm is linear in the acyclic case $O(|Q| + |E|)$ and in $O(|E| \log |Q|)$ in the general case.

Figures 12(a), (b), and (d) illustrate the application of the algorithm in the tropical semiring. The automaton of Fig. 12(a) cannot be further minimized using the classical un-weighted automata minimization since no two states are equivalent in that machine. After weight pushing, the automaton (Fig. 12(b)) has two states (1 and 2) that can be merged by the classical un-weighted automata minimization.

Figures 13(a)–(c) illustrate the minimization of an automaton defined over the probability semiring. Unlike the un-weighted case, a minimal weighted automaton is not unique, but all minimal weighted automata have the same graph topology, they only differ by the way the weights are distributed along each path. The weighted automata B' and C' are both minimal and equivalent to A'. B' is obtained from A' using the algorithm described above in the probability semiring and it is thus a stochastic weighted automaton in the probability semiring.

For a deterministic weighted automaton, the first operation of the semiring can be arbitrarily chosen without affecting the definition of the function it realizes. This is because by definition a deterministic weighted automaton admits at most one path labeled with any given string. Thus, in the algorithm described in Theorem 7.5, the weight pushing step can be executed in any semiring S' whose multiplicative operation matches that of S. The minimal weighted automaton obtained by pushing the weights in S' is also minimal in S since it can be interpreted as a (deterministic) weighted automaton over S.

In particular, A' can be interpreted as a weighted automaton over the semiring $(\mathbb{R}_+, \mathsf{max}, \times, 0, 1)$. The application of the weighted minimization algorithm to A' in this semiring leads to the minimal weighted automaton C' of Fig. 13(c). C' is also a *stochastic* weighted automaton in the sense that at any state the maximum weight of all outgoing transitions is one.

This fact leads to several interesting observations. One is related to the complexity of the algorithms. Indeed, we can choose a semiring S' in which the complexity of weight pushing is better than in S. The resulting automaton is still minimal in S and has the additional property of being stochastic in S'. It only differs from the weighted automaton obtained by pushing weights in S in the way weights are distributed along the paths. They can be obtained from each other by application of weight pushing in the appropriate semiring. In the particular case of a weighted automaton over the probability semiring, it may be preferable to use weight pushing in the $(\mathsf{max}, \times)$-semiring since the complexity of the algorithm is then equivalent to that of classical single-source shortest-paths algorithms. The corresponding algorithm is a special instance of the generic shortest-distance algorithm for k-closed semirings presented earlier in the chapter.

Another important point is that the weight pushing algorithm may not be defined in S because the machine is not zero-sum-free or for other reasons. But an alternative semiring S' can sometimes be used to minimize the input weighted automaton.

The results just presented were all related to the minimization of the number of states of a deterministic weighted automaton. The following simple proposition shows that minimizing the number of states coincides with minimizing the number of transitions.

Proposition 7.6. *Let A be a minimal deterministic weighted automaton, then A has the minimal number of transitions.*

Proof. Let A be a deterministic weighted automaton with the minimal number of transitions. If two distinct states of A were equivalent, they could be merged, thereby strictly reducing the number of its transitions. Thus, A must be a minimal deterministic automaton. Since, minimal deterministic automata have the same topology, in particular the same number of states and transitions, this proves the proposition. □

7.5 Synchronization

The weight pushing algorithm normalizes the way the weights are distributed along the paths. The algorithm presented in this section, *synchronization of weighted transducers*, normalizes instead the way the input and output labels are shifted with respect to each other along the paths. Roughly speaking, the objective of the algorithm is to synchronize the consumption of non-ϵ symbols by the input and output tapes of a transducer, to the extent that is possible.

The following concept helps analyze and describe the domain of application of the algorithm.

Definition 7.7. *The* delay *of a path π is defined as the difference of length between its output and input labels:*

$$d[\pi] = |o[\pi]| - |i[\pi]|. \tag{41}$$

The delay of a path is thus simply the sum of the delays of its constituent transitions. A trim transducer T is said to have *bounded delays* if the delay along all paths of T is bounded. We then denote by $d[T] \geq 0$ the maximum delay in absolute value of a path in T. The following lemma gives a straightforward characterization of transducers with bounded delays.

Lemma 7.8. *A transducer T has bounded delays iff the delay of any cycle in T is zero.*

Proof. If T admits a cycle π with non-zero delay, then $d[T] \geq |d[\pi^n]| = n|d[\pi]|$ is not bounded. Conversely, if all cycles have zero delay, then the maximum delay in T is that of the simple paths which are of finite number. □

We define the *string delay* of a path π as the string $\sigma[\pi]$ defined by

$$\sigma[\pi] = \begin{cases} \text{suffix of } o[\pi] \text{ of length } |d[\pi]| & \text{if } d[\pi] \geq 0; \\ \text{suffix of } i[\pi] \text{ of length } |d[\pi]| & \text{otherwise.} \end{cases} \tag{42}$$

For any state $q \in Q$, the *string delay at state* q, $s[q]$, is defined by the set of string delays of the paths from an initial state to q:

$$s[q] = \{\sigma[\pi] : \pi \in P(I, q)\}. \tag{43}$$

Lemma 7.9. *If T has bounded delays then the set $s[q]$ is finite for any $q \in Q$.*

Proof. The lemma follows immediately the fact that the elements of $s[q]$ are all of length less than $d[T]$. □

A weighted transducer T is said to be *synchronized* if along any successful path of T the delay is zero or varies strictly monotonically. An algorithm that takes as input a transducer T and computes an equivalent synchronized transducer T' is called a *synchronization* algorithm.The synchronization algorithm described here [42] applies to all weighted transducers with bounded delays. The following is the pseudo-code of the algorithm.

```
SYNCHRONIZATION(T)
 1  F' ← Q' ← E' ← ∅
 2  𝒬 ← i' ← {(i, ε, ε) : i ∈ I}
 3  while 𝒬 ≠ ∅ do
 4       p' = (q, x, y) ← HEAD(𝒬)
 5       DEQUEUE(𝒬)
 6       if (q ∈ F and |x| + |y| = 0) then
 7            F' ← F' ∪ {p'}; ρ'(p') ← ρ(q)
 8       elseif (q ∈ F and |x| + |y| > 0) then
 9            q' ← (f, cdr(x), cdr(y))
10            E' ← E' ⊎ (p', car(x), car(y), ρ(q), q')
11            if (q' ∉ Q') then
12                 Q' ← Q' ∪ {q'}; ENQUEUE(𝒬, q')
13       for each e ∈ E[q] do
14            if (|x i[e]| > 0 and |y o[e]| > 0) then
15                 q' ← (n[e], cdr(x i[e]), cdr(y o[e]))
16                 E' ← E' ⊎ {(p', car(x i[e]), car(y o[e]), w[e], q')}
17            else q' ← (n[e], x i[e], y o[e])
18                 E' ← E' ⊎ {(p', ε, ε, w[e], q')}
19            if (q' ∉ Q') then
20                 Q' ← Q' ∪ {q'}; ENQUEUE(𝒬, q')
21  return T'
```

To simplify the presentation of the algorithm, we augment Q and F with a new state f and set: $\rho(f) = \overline{1}$ and $E[f] = \emptyset$. We denote by $\mathrm{car}(x)$ the first symbol of a string x if x is not empty, ϵ otherwise, and denote by $\mathrm{cdr}(x)$ the suffix of x such that $x = \mathrm{car}(x)\,\mathrm{cdr}(x)$.

Each state of the resulting transducer T' corresponds to a triplet (q, x, y) where $q \in Q$ is a state of the original machine T and where $x \in \Sigma^*$ and $y \in \Delta^*$ are strings over the input and output alphabet of T.

The algorithm maintains a queue $\mathcal{Q}$ that contains at any time the set of states of T' to examine. At each execution of the loop of lines 3–19, a new state $p' = (q, x, y)$ is extracted from $\mathcal{Q}$ (line 4) and its outgoing transitions are computed and added to E'. The state p' is final iff q is final and $x = y = \epsilon$ and in that case the final weight at p' is simply the final weight at the original state q (lines 5–6). If q is final but the string x and y are not both empty, then the algorithm constructs a sequence of transitions from p' to (f, ϵ, ϵ) to consume the remaining input and output strings x and y (lines 7–11).

For each transition e of q, an outgoing transition e' is created for p' with weight $w[e]$. The input and output labels of e' are both ϵ if $x\,i[e]$ or $y\,o[e]$ is the empty string, the first symbol of these strings otherwise. The remaining suffixes of these strings are stored in the destination state q' (lines 12–19). Note that in all cases, the transitions created by the steps of the algorithm

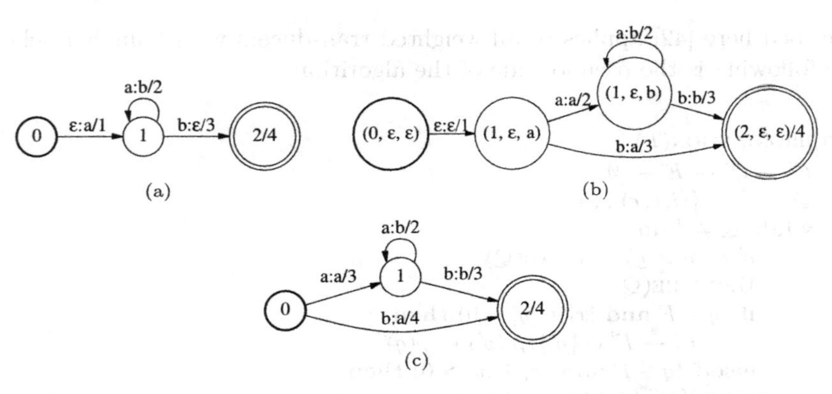

Fig. 14. (**a**) Weighted transducer T_1 over the tropical semiring. (**b**) Equivalent synchronized transducer T_2. (**c**) Synchronized weighted transducer T_3 equivalent to T_1 and T_2 obtained by ϵ-removal from T_2

described in lines 14–17 have zero delay. The state q' is inserted in $\mathcal{Q}$ if it has never been found before (lines 18–19). Figures 14(a)–(b) illustrate the synchronization algorithm just presented.

Theorem 7.10 ([42]). *Let T be a weighted transducer with bounded delays. Then if we run the synchronization algorithm just described with input T, the algorithm terminates and returns a synchronized transducer T' equivalent to T.*

The algorithm creates a distinct state (q, x, ϵ) or (q, ϵ, y) for each string delay $x, y \in s[q]$ at state $q \neq f$. The paths from a state (q, x, ϵ) or (q, ϵ, y), $q \in F$, to (f, ϵ, ϵ) are of length $|x|$ or $|y|$. The length of a string delay is bounded by $d[T]$. Thus, there are at most $|\Sigma|^{\leq d[T]} + |\Delta|^{\leq d[T]} = O(|\Sigma|^{d[T]} + |\Delta|^{d[T]})$ distinct string delays at each state. Thus, in the worst case, the size of the resulting transducer T' is

$$O\big((|Q| + |E|)\big(|\Sigma|^{d[T]} + |\Delta|^{d[T]}\big)\big). \tag{44}$$

The string delays can be represented in a compact and efficient way using a suffix tree. Indeed, let U be a tree representing all the input and output labels of the paths in T found in a depth-first search of T. The size of U is linear in that of T and a suffix tree V of U can be built in time proportional to the number of nodes of U times the size of the alphabet [27], that is, in $O((|\Sigma| + |\Delta|)(|Q| + |E|))$. Since each string delay x is a suffix of a string represented by U, it can be represented by two nodes n_1 and n_2 of V and a position in the string labeling the edge from n_1 to n_2. The operations performed by the algorithm to construct a new transition require either computing xa or $a^{-1}x$ where a is a symbol of the input or output alphabet. Clearly, these operations can be performed in constant time: xa is obtained by going down one position in the suffix tree, and $a^{-1}x$ by using the suffix link at node n_1. Thus, using this representation, the operations performed for the construction of each new transition can be done in constant time. This includes the cost of comparison of a newly created state (q', x', ϵ) with an existing state (q, x, ϵ), since the comparison of the string delays x and x' can be done in constant time. Thus, the worst case space and time complexity of the algorithm is

$$O\big((|Q| + |E|)\big(|\Sigma|^{d[T]} + |\Delta|^{d[T]}\big)\big). \tag{45}$$

This is not a tight evaluation of the complexity since it is not clear if the worst case previously described can ever occur, but the algorithm can indeed produce an exponentially larger transducer in some cases.

Note that the algorithm does not depend on the queue discipline used for $\mathcal{Q}$ and that the construction of the transitions leaving a state $p' = (q, x, y)$ of T' only depends on p' and not on the states and transitions previously constructed. Thus, the transitions of T' can be naturally computed on demand. Note also that the additive and multiplicative operations of the semiring are not used in the definition of the algorithm. Only $\overline{1}$, the identity element of $\otimes$, was used for the definition of the final weight of f. Thus, to a large extent, the algorithm is independent of the semiring S. In particular, the behavior of the algorithm is identical for two semirings having the same identity elements, such as, for example, the tropical and log semirings.

The result of the synchronization algorithm may contain ϵ-transitions even if the input contains none. An equivalent weighted transducer with no ϵ-transitions can be computed from T' using the general epsilon-removal algorithm described in a previous section [40]. Figure 14(c) illustrates the result of that algorithm when applied to the synchronized transducer of Fig. 14(b). Since epsilon-removal does not shift input and output labels with respect to each other, the result of its application to T' is also a synchronized transducer.

Note that the synchronization algorithm does not produce any ϵ-cycle if the original machine T does not contain any. Thus, in that case, the computation of the ϵ-closures in T can be done in linear time [40] and the total time complexity of epsilon-removal is $O(|Q'|^2 + (T_\oplus + T_\otimes)|Q'||E'|)$. Also, on-demand synchronization can be combined with on-demand epsilon-removal to *directly* create synchronized transducers with no ϵ-transition on-the-fly.

A transducer T is said to be *double-tape unambiguous*, if when the input and output labels of a transition are treated as single pair label (input label, output label), no two accepting paths have the same label. A by-product of synchronization followed by epsilon-removal is that the resulting transducer is double-tape unambiguous. Note that the definition does not entail any requirement on the weights.

Proposition 7.11 ([42]). *Let T be a synchronized transducer and assume that T has no ϵ-transition. Then T is double-tape unambiguous.*

Table 2. Properties of several transducer algorithms. The second column indicates the time or space complexity of each algorithm, the third whether or not it admits a natural on-demand computation

Algorithm	Complexity	On-demand
Sum	$O(\|T_1\| + \|T_2\|)$	+
Product	$O(\|T_1\| + \|T_2\| + \|F_1\|\|I_2\|)$	+
Closure	$O(\|T\|)$	+
Reversal	$O(\|T\|)$	−
Inversion	$O(\|T\|)$	+
Projection	$O(\|T\|)$	+
Composition	$O(\|T_1\|\|T_2\|)$	+
Intersection	$O(\|A_1\|\|A_2\|)$	+
Completion	$O(\|A\| + \|\Sigma\|)$	+
Complementation[a]	$O(\|A\|)$	+
Difference	$O(\|A_1\|\|A_2\|)$	+
Epsilon-removal[b]	$O(\|Q\|^2 + \|Q\|\|E\|)$	+
Determinization	exponential	+
Minimization[b]	$O(\|E\| \log \|Q\|)$	−
Synchronization[c]	$O((\|Q\| + \|E\|)(\|\Sigma\|^{d[T]} + \|\Delta\|^{d[T]}))$	+

[a] For A deterministic
[b] For the tropical semiring
[c] $d(T)$ denotes the maximum delay in an accepting path of T

8 Conclusion

Table 2 summarizes some of the essential properties of the algorithms described in this chapter. There are of course many other algorithms related to weighted automata and transducers. But those presented here constitute some of the core algorithms. Many other algorithms related to weighted automata can be derived either directly from these algorithms or as a combination.

These algorithms are useful in a variety of applications including statistical language modeling [7], parsing [45, 44, 8], phonological rule compilation [29, 30, 49], speech recognition [38, 43, 48], speech synthesis [56, 6], image processing [2], bio-informatics [21, 9], sequence modeling and prediction [18], optical character recognition [14], and more generally any problem related to sequences and probabilistic models defined over sequences [43, 33]. An efficient implementation of these algorithms and several others, including an on-demand implementation when possible, is available from the FSM library (executables only) [47] and the OpenFst library (source and executables) [10].

References

1. A.V. Aho, J.E. Hopcroft, and J.D. Ullman. *The Design and Analysis of Computer Algorithms.* Addison–Wesley, Reading, 1974.

2. J. Albert and J. Kari. Digital image compression. In this *Handbook*. Chapter 11. Springer, Berlin, 2009.
3. C. Allauzen and M. Mohri. Efficient algorithms for testing the twins property. *Journal of Automata, Languages and Combinatorics*, 8(2):117–144, 2003.
4. C. Allauzen and M. Mohri. N-way composition of weighted finite-state transducers. Technical Report TR2007-902, Courant Institute of Mathematical Sciences, New York University, August 2007.
5. C. Allauzen and M. Mohri. 3-way composition of weighted finite-state transducers. In *Proceedings of the 13th International Conference on Implementation and Application of Automata (CIAA 2008), San Francisco, California, July 2008*, volume 5148 of *Lecture Notes in Computer Science*. Springer, Heidelberg, 2008.
6. C. Allauzen, M. Mohri, and M. Riley. Statistical modeling for unit selection in speech synthesis. In *42nd Meeting of the Association for Computational Linguistics (ACL 2004), Proceedings of the Conference, Barcelona, Spain, July 2004.*
7. C. Allauzen, M. Mohri, and B. Roark. Generalized algorithms for constructing statistical language models. In *41st Meeting of the Association for Computational Linguistics (ACL 2003), Proceedings of the Conference, Sapporo, Japan, July 2003.*
8. C. Allauzen, M. Mohri, and B. Roark. The design principles and algorithms of a weighted grammar library. *International Journal of Foundations of Computer Science*, 16(3):403–421, 2005.
9. C. Allauzen, M. Mohri, and A. Talwalkar. Sequence kernels for predicting protein essentiality. In *Proceedings of the Twenty-Fifth International Conference on Machine Learning (ICML 2008), Helsinki, Finland, July 2008.*
10. C. Allauzen, M. Riley, J. Schalkwyk, W. Skut, and M. Mohri. OpenFst: A general and efficient weighted finite-state transducer library. In *Proceedings of the 12th International Conference on Implementation and Application of Automata (CIAA 2007), Prague, Czech Republic, July 2007*, volume 4783 of *Lecture Notes in Computer Science*, pages 11–23. Springer, Heidelberg, 2007.
11. M.-P. Béal, O. Carton, C. Prieur, and J. Sakarovitch. Squaring transducers: An efficient procedure for deciding functionality and sequentiality. In *Proceedings of LATIN'2000*, volume 1776 of *Lecture Notes in Computer Science*. Springer, Heidelberg, 2000.
12. J. Berstel. *Transductions and Context-Free Languages*. Teubner Studienbucher, Stuttgart, 1979.
13. J. Berstel and C. Reutenauer. *Rational Series and Their Languages*. Springer, Berlin, 1988.
14. T.M. Breuel. The OCRopus open source OCR system. In *Proceedings of IS&T/SPIE 20th Annual Symposium, 2008.*

15. C. Choffrut. Une caractérisation des fonctions séquentielles et des fonctions sous-séquentielles en tant que relations rationnelles. *Theoretical Computer Science*, 5:325–338, 1977.
16. C. Choffrut. Contributions à l'étude de quelques familles remarquables de fonctions rationnelles. PhD thesis (thèse de doctorat d'Etat), Université Paris 7, LITP, Paris, 1978.
17. T.H. Cormen, C.E. Leiserson, and R.E. Rivest. *Introduction to Algorithms.* MIT Press, Cambridge, 1992.
18. C. Cortes, P. Haffner, and M. Mohri. A machine learning framework for spoken-dialog classification. In L. Rabiner and F. Juang, editors, *Handbook on Speech Processing and Speech Communication, Part E: Speech Recognition.* Springer, Heidelberg, 2008.
19. C. Cortes, M. Mohri, and A. Rastogi. L_p distance and equivalence of probabilistic automata. *International Journal of Foundations of Computer Science*, 18(4):761–780, 2007.
20. M. Droste and W. Kuich. Semirings and formal power series. In this *Handbook.* Chapter 1. Springer, Berlin, 2009.
21. R. Durbin, S.R. Eddy, A. Krogh, and G.J. Mitchison. *Biological Sequence Analysis: Probabilistic Models of Proteins and Nucleic Acids.* Cambridge University Press, Cambridge, 1998.
22. S. Eilenberg. *Automata, Languages and Machines, volume A.* Academic Press, San Diego, 1974.
23. Z. Ésik and W. Kuich. Modern automata theory. www.dmg.tuwien.ac.at/kuich.
24. Z. Esik and W. Kuich. Rationally additive semirings. *Journal of Universal Computer Science*, 8:173–183, 2002.
25. R.W. Floyd. Algorithm 97 (SHORTEST PATH). *Communications of the ACM*, 18, 1968.
26. M. Goldstern. Vervollständigung von Halbringen. Master's thesis, Technische Universität Wien, 1985.
27. S. Inenaga, H. Hoshino, A. Shinohara, M. Takeda, and S. Arikawa. Construction of the CDAWG for a Trie. In *Proceedings of the Prague Stringology Conference (PSC'01).* Czech Technical University, Prague, 2001.
28. D.B. Johnson. Efficient algorithms for shortest paths in sparse networks. *Journal of the ACM*, 24(1):1–13, 1977.
29. R.M. Kaplan and M. Kay. Regular models of phonological rule systems. *Computational Linguistics*, 20(3), 1994.
30. L. Karttunen. The replace operator. In *33rd Annual Meeting of the Association for Computational Linguistics*, pages 16–23. Association for Computational Linguistics, Stroudsburg, 1995. Distributed by Morgan Kaufmann, San Francisco.
31. D. Kirsten. A Burnside approach to the termination of Mohri's algorithm for polynomially ambiguous min-plus-automata. *Informatique Théorique et Applications, RAIRO, Special Issue on Journées Montoises d'Informatique Théorique 2006 (JM'06)*, 42(3):553–581, 2008.

32. I. Klimann, S. Lombardy, J. Mairesse, and C. Prieur. Deciding unambiguity and sequentiality from a finitely ambiguous max-plus automaton. *Theoretical Computer Science*, 327(3):349–373, 2004.
33. K. Knight and J. May. Applications of weighted automata in natural language processing. In this *Handbook*. Chapter 14. Springer, Berlin, 2009.
34. D. Krob. The equality problem for rational series with multiplicities in the tropical semiring is undecidable. *Journal of Algebra and Computation*, 4, 1994.
35. W. Kuich. Semirings and formal languages: Their relevance to formal languages and automata. In G. Rozenberg and A. Salomaa, editors, *Handbook of Formal Languages, volume 1: Word, Language, Grammar*, pages 609–677. Springer, New York, 1997.
36. W. Kuich and A. Salomaa. *Semirings, Automata, Languages*, volume 5 of *Monographs in Theoretical Computer Science. An EATCS Series.* Springer, Berlin, 1986.
37. D.J. Lehmann. Algebraic structures for transitive closures. *Theoretical Computer Science*, 4:59–76, 1977.
38. M. Mohri. Finite-state transducers in language and speech processing. *Computational Linguistics*, 23:2, 1997.
39. M. Mohri. Minimization algorithms for sequential transducers. *Theoretical Computer Science*, 234:177–201, 2000.
40. M. Mohri. Generic epsilon-removal and input epsilon-normalization algorithms for weighted transducers. *International Journal of Foundations of Computer Science*, 13(1):129–143, 2002.
41. M. Mohri. Semiring frameworks and algorithms for shortest-distance problems. *Journal of Automata, Languages and Combinatorics*, 7(3):321–350, 2002.
42. M. Mohri. Edit-distance of weighted automata: General definitions and algorithms. *International Journal of Foundations of Computer Science*, 14(6):957–982, 2003.
43. M. Mohri. Statistical natural language processing. In M. Lothaire, editor, *Applied Combinatorics on Words*. Cambridge University Press, Cambridge, 2005.
44. M. Mohri and M.-J. Nederhof. Regular approximation of context-free grammars through transformation. In *Robustness in Language and Speech Technology*, pages 153–163. Kluwer Academic, Dordrecht, 2001.
45. M. Mohri and F.C.N. Pereira. Dynamic compilation of weighted context-free grammars. In *36th Meeting of the Association for Computational Linguistics (ACL'98), Proceedings of the Conference, Montréal, Québec, Canada, 1998*, pages 891–897.
46. M. Mohri, F.C.N. Pereira, and M. Riley. Weighted automata in text and speech processing. In *Proceedings of the 12th Biennial European Conference on Artificial Intelligence (ECAI-96), Workshop on Extended Finite State Models of Language, Budapest, Hungary, 1996*. Wiley, Chichester, 1996.

47. M. Mohri, F.C.N. Pereira, and M. Riley. The design principles of a weighted finite-state transducer library. *Theoretical Computer Science*, 231:17–32, 2000.
48. M. Mohri, F.C.N. Pereira, and M. Riley. Speech recognition with weighted finite-state transducers. In L. Rabiner, F. Juang, editors, *Handbook on Speech Processing and Speech Communication, Part E: Speech Recognition.* Springer, Heidelberg, 2008.
49. M. Mohri and R. Sproat. An efficient compiler for weighted rewrite rules. In *34th Meeting of the Association for Computational Linguistics (ACL'96), Proceedings of the Conference, Santa Cruz, California, 1996.*
50. F.C.N. Pereira and M.D. Riley. Speech recognition by composition of weighted finite automata. In *Finite-State Language Processing*, pages 431–453. MIT Press, Cambridge, 1997.
51. D. Perrin. Finite automata. In J. Van Leuwen, editor, *Handbook of Theoretical Computer Science, volume B: Formal Models and Semantics*, pages 1–57. Elsevier, Amsterdam, 1990.
52. I. Petre and A. Salomaa. Algebraic systems and pushdown automata. In this *Handbook.* Chapter 7. Springer, Berlin, 2009.
53. D. Revuz. Minimisation of acyclic deterministic automata in linear time. *Theoretical Computer Science*, 92(1):181–189, 1992.
54. A. Salomaa and M. Soittola. *Automata-Theoretic Aspects of Formal Power Series.* Springer, New York, 1978.
55. M.-P. Schützenberger. On the definition of a family of automata. *Information and Control*, 4, 1961.
56. R. Sproat. A finite-state architecture for tokenization and grapheme-to-phoneme conversion in multilingual text analysis. In *Proceedings of the ACL SIGDAT Workshop, Dublin, Ireland.* Association for Computational Linguistics, Stroudsburg, 1995.
57. S. Warshall. A theorem on Boolean matrices. *Journal of the ACM*, 9(1):11–12, 1962.
58. A. Weber and R. Klemm. Economy of description for single-valued transducers. *Information and Computation*, 118(2):327–340, 1995.

Part III

Weighted Discrete Structures

Chapter 7: Algebraic Systems and Pushdown Automata

Ion Petre[1] and Arto Salomaa[2]

[1] Academy of Finland and Åbo Akademi University, Turku 20520, Finland
ion.petre@abo.fi
[2] Turku Centre for Computer Science, Turku 20520, Finland
arto.salomaa@utu.fi

1 Introduction

The theory of algebraic power series in non-commuting variables, as we understand it today, was initiated in [2] and developed in its early stages by the French school. The main motivation was the interconnection with context-free grammars: the defining equations were made to correspond to context-free productions. Then the coefficient of a word w in the series equals the degree of ambiguity of w according to the grammar.

We concentrate in this chapter on the core aspects of algebraic series, pushdown automata, and their relation to formal languages. We choose to follow

M. Droste, W. Kuich, H. Vogler (eds.), *Handbook of Weighted Automata*, Monographs in Theoretical Computer Science. An EATCS Series,
DOI 10.1007/978-3-642-01492-5_7,

here a presentation of their theory based on the concept of *properness*. Alternatively, one may present the theory in terms of complete semirings, as done for example in [11] and [3]. The main difference between the two presentations is in the handling of infinite sums, especially in connection with products and stars of infinite matrices. While such sums are well defined in the case of complete semirings, special care must be taken for arbitrary semirings. In the case of proper algebraic systems and proper pushdown automata, the infinite matrices have locally-finite stars. Consequently, if one considers the ambiguity of context-free grammars, one option is to assume that the ambiguity is in the semiring $\mathbb{N}^\infty$ and follow the results based on complete semirings, while the other option is to consider proper grammars and take the ambiguity in the semiring $\mathbb{N}$.

We introduce in Sect. 2 some auxiliary notions and results needed throughout the chapter, in particular the notions of discrete convergence in semirings and C-cycle free infinite matrices. In Sect. 3, we introduce the algebraic power series in terms of algebraic systems of equations. We focus on interconnections with context-free grammars and on normal forms. We then conclude the section with a presentation of the theorems of Shamir and Chomsky–Schützenberger. We discuss in Sect. 4 the algebraic and the regulated rational transductions, as well as some representation results related to them. Section 5 is dedicated to pushdown automata and focuses on the interconnections with classical (non-weighted) pushdown automata and on the interconnections with algebraic systems. We then conclude the chapter with a brief discussion of some of the other topics related to algebraic systems and pushdown automata.

2 Auxiliary Notions and Results

We introduce in this section the notion of discrete convergence in arbitrary semirings and in connection to it, a notion of convergence for column finite (infinite) matrices. This allows us then to define the notion of C-cycle free (infinite) matrices, needed in this chapter in connection with proper pushdown transition matrices. We give here only the elements that are essential for the purpose of this chapter, referring to [12] for more details, including other notions of convergence.

Definition 2.1. *A* sequence *in the semiring S is a mapping $\alpha : \mathbb{N} \to S$. We denote $\alpha_n = \alpha(n)$, for all $n \in \mathbb{N}$ and $\alpha = (\alpha_n)_{n\in\mathbb{N}}$. We denote the set of all sequences over S by $S^{\mathbb{N}}$. We say that α is* convergent *in S if there exist $a \in S$ and $n_0 \in \mathbb{N}$ such that $\alpha_n = a$, for all $n \geq n_0$. In this case, a is called the* limit *of α, denoted as $\lim_{n\to\infty} \alpha_n = a$.*

The notion of convergence defined above is often called the *discrete convergence* in the semiring S.

We note that the notion of cycle-free formal power series defined in Chap. 1 is equivalent with saying that, for $r \in S\langle\langle\Sigma^*\rangle\rangle$, the sequence $(r^n, \epsilon)_{n\in\mathbb{N}}$ is convergent in S and its limit is 0. It has been observed already in Chap. 1 that the star of any cycle-free series exists. We give here a short proof of this result that will be mirrored by a similar result for matrices.

Lemma 2.2. *For any cycle-free formal power series $r \in S\langle\langle\Sigma^*\rangle\rangle$ and any $w \in \Sigma^*$, there exists $n_w \in \mathbb{N}$ such that $(r^n, w) = 0$, for all $n \geq n_w$. Consequently, r^* exists and is locally finite.*

Proof. We prove the claim by induction on $|w|$. For $|w| = 0$, it follows from the definition of the limit in the discrete convergence that there exists $n_0 \in \mathbb{N}$ such that $(r^n, \epsilon) = 0$, for all $n \geq n_0$. Consider now an arbitrary $w \in \Sigma^+$ and assume that the claim holds for all words shorter than w. Then for any $n \geq n_0$,

$$\begin{aligned}(r^n, w) &= \sum_{w=uv} (r^{n-n_0}, u)(r^{n_0}, v) \\ &= (r^{n-n_0}, w)(r^{n_0}, \epsilon) + \sum_{\substack{w=uv \\ |u|<|w|}} (r^{n-n_0}, u)(r^{n_0}, v) \\ &= \sum_{\substack{w=uv \\ |u|<|w|}} (r^{n-n_0}, u)(r^{n_0}, v).\end{aligned}$$

If we choose $n_w \geq n_0 + n_u$, for all $|u| < |w|$, then the claim follows for $n \geq n_w$ based on the induction hypothesis. Indeed, in this case, $(r^{n-n_0}, u) = 0$. In particular, one may choose $n_w = n_0(|w| + 1)$, for all $w \in \Sigma^*$.

The second part of the lemma follows by observing that

$$(r^*, w) = \sum_{n\geq 0} (r^n, w) = \sum_{n=0}^{n_w - 1} (r^n, w). \quad \square$$

Definition 2.3. *Let $(M_n)_{n\in\mathbb{N}} \in (S^{I\times I})^{\mathbb{N}}$ be a sequence of column finite (infinite) matrices. We say that $(M_n)_{n\geq 0}$ is convergent if the following two conditions are satisfied:*

(i) For all $j \in I$, there exists a finite set $I(j) \subseteq I$ such that $(M_n)_{i,j} = 0$, for all $n \in \mathbb{N}$ and all $i \in I \setminus I(j)$.
(ii) For all $i, j \in I$, the sequence $((M_n)_{i,j})_{n\in\mathbb{N}} \in S^{\mathbb{N}}$ is convergent in S.

For $m_{i,j} = \mathrm{lim}_{n\to\infty}((M_n)_{i,j})$, we say that the matrix $M = (m_{i,j})_{i,j\in I}$ is the limit *of the sequence $(M_n)_{n\in\mathbb{N}}$, denoted $M = \mathrm{lim}_{n\to\infty} M_n$.*

Note that condition (i) in Definition 2.3 is not equivalent with the matrices M_n being column finite for all $n \in \mathbb{N}$. Note also that a different notion of convergence can be defined for row finite matrices; see [12].

Definition 2.4. *Let $M \in (S\langle\langle \Sigma^* \rangle\rangle)^{I \times I}$ be a column finite matrix. We say that M is* C-cycle free *if the sequence $(M^n, \epsilon)_{n \in \mathbb{N}} \in S^{I \times I}$ is convergent and its limit is the zero matrix.*

The notion of C-cycle free matrix is very similar with the notion of cycle-free power series. We indicate explicitly the letter C to stress that our notion is applied to column finite matrices only and also, to preserve the terminology in [12]. Note that a related notion of R-cycle free matrices may also be defined for row finite matrices. Note also that, based on the definition of discrete convergence and that of convergent matrices, a matrix $M \in (S\langle\langle \Sigma^* \rangle\rangle)^{I \times I}$ is C-cycle free if and only if the following two conditions are satisfied:

(i) For all $j \in I$, there exists a finite set $I(j) \subseteq I$ such that $(M^n, \epsilon)_{i,j} = 0$, for all $i \notin I(j)$ and all $n \geq 0$.
(ii) For all $j \in I$, there exists $n(j) \in \mathbb{N}$ such that $(M^n, \epsilon)_{i,j} = 0$, for all $n \geq n(j)$ and all $i \in I$.

The following result will be needed in connection with proper pushdown transition matrices and their stars.

Lemma 2.5. *The star of any C-cycle free matrix is locally finite.*

Proof. Let $M \in (S\langle\langle \Sigma^* \rangle\rangle)^{I \times I}$ be a C-cycle free matrix. We claim that for any $j \in I$ and any $w \in \Sigma^*$, there exists a non negative integer $n(j, w)$ such that $((M^n)_{i,j}, w) = 0$, for all $i \in I$ and all $n \geq n(j, w)$. Then we obtain that

$$\sum_{n \geq 0} ((M^n)_{i,j}, w) = \sum_{n=0}^{n(j,w)} ((M^n)_{i,j}, w),$$

showing that M^* exists and is locally finite.

We prove the claim by induction on $|w|$. For $|w| = 0$, the claim follows from the definition of C-cycle free matrices. Indeed, since $\lim_{n \to \infty}(M^n, \epsilon) = 0$, there exists $n(j) \in \mathbb{N}$ for all $j \in I$ such that $(M^n, \epsilon)_{i,j} = 0$, for all $i \in I$ and all $n \geq n(j)$.

Consider now $w \in \Sigma^*$ with $|w| \geq 1$ and assume inductively that the claim holds for all words shorter than w. Then for any $i, j \in I$, $n \geq n(j)$, $w \in \Sigma^*$, we have that

$$\begin{aligned} ((M^n)_{i,j}, w) &= \sum_{\substack{w=uv \\ k \in I}} ((M^{n-n(j)})_{i,k}, u)((M^{n(j)})_{k,j}, v) \\ &= \sum_{\substack{w=uv \\ |u|<|w| \\ k \in I}} ((M^{n-n(j)})_{i,k}, u)((M^{n(j)})_{k,j}, v) \\ &\quad + \sum_{k \in I} ((M^{n-n(j)})_{i,k}, w)((M^{n(j)})_{k,j}, \epsilon). \end{aligned}$$

Note now that $((M^{n(j)})_{k,j}, \epsilon) = 0$. Also, if $n - n(j) \geq n(k, u)$, then it follows by the induction hypothesis that $((M^{n-n(j)})_{i,k}, u) = 0$. Since there are only finitely many $k \in I$ with $(M^{n(j)})_{k,j} \neq 0$, to obtain that $((M^n)_{i,j}, w) = 0$, it is enough to define $n(j, w) \in \mathbb{N}$ as follows:

$$n(j,w) = \begin{cases} n(j), & \text{if } w = \epsilon, \\ \max\{n(j) + n(k,u) \mid (M^{n(j)})_{k,j} \neq 0, |u| < |w|\}, & \text{otherwise.} \end{cases} \qquad \square$$

3 Algebraic Power Series

This section introduces algebraic power series in terms of algebraic systems of equations and discusses various reduction, normal form, and characterization results. Special emphasis will be in the interconnection with context-free grammars and languages. The defining equations are algebraic, that is, polynomial equations. Moreover, they are of a somewhat special form. The form makes the interconnection with context-free grammars very direct.

The first comprehensive treatment about algebraic power series in non-commuting variables is in [15] where also references to earlier work, mainly by M.P. Schützenberger, are given.

3.1 Definition and Basic Reductions

Consider an alphabet $\Sigma = \{x_1, \ldots, x_k\}$, $k \geq 1$, and a commutative semiring S. Let $Y = \{y_1, \ldots, y_n\}$, $n \geq 1$, be another alphabet, the alphabet of *variables.*

Definition 3.1. *An S-*algebraic system *is a set of equations of the form*

$$y_i = p_i, \quad i = 1, \ldots, n,$$

where $p_i \in S\langle(\Sigma \cup Y)^*\rangle$. The system is termed proper *if, for all i and j, $(p_i, \epsilon) = 0$ and $(p_i, y_j) = 0$.*

A *solution* to the algebraic system consists of n power series $r_1, \ldots, r_n$ in $S\langle\langle\Sigma^*\rangle\rangle$ "satisfying" the system in the sense that if each variable y_i is replaced by the series r_i, then n valid equations result. This can be formalized as follows. Consider a column vector

$$R = \begin{pmatrix} r_1 \\ \vdots \\ r_n \end{pmatrix} \in (S\langle\langle(\Sigma \cup Y)^*\rangle\rangle)^{n \times 1}$$

consisting of n power series, and define the morphism

$$h_R : (\Sigma \cup Y)^* \to S\langle\langle(\Sigma \cup Y)^*\rangle\rangle$$

by $h_R(y_i) = r_i$, $1 \le i \le n$, and $h_R(x) = x$, for $x \in \Sigma$. Defining

$$h_R(p) = \sum_{w \in (\Sigma \cup Y)^*} (p, w) h_R(w),$$

where $p \in S\langle(\Sigma \cup Y)^*\rangle$, we now term R a *solution* of the original algebraic system if $r_i = h_R(p_i)$, for all i, $1 \le i \le n$.

So far, we have not used the assumption of the semiring S being commutative. Indeed, some parts of the theory such as the interconnection with pushdown automata remain valid without this assumption. We make this assumption because it is needed in important parts of the theory and, moreover, our main interest is in the semirings $\mathbb{Z}$ and $\mathbb{N}$ of integers and non-negative integers, as well as in the Boolean semiring $\mathbb{B}$.

An S-algebraic system does not always possess a solution because, for instance, an equation may be contradictory, such as the equation $y_1 = y_1 + x_1$ in the semiring $\mathbb{N}$. However, every proper S-algebraic system possesses a solution.

Theorem 3.2. *Every proper S-algebraic system possesses exactly one solution where each component is quasi-regular. In addition, it may have other solutions.*

Proof. The theorem is established by considering an *"approximation sequence"* R^i, $i = 0, 1, \dots$, of n-tuples (or column vectors) of power series. By definition, R^0 consists of 0's, and R^{i+1} is obtained by applying h_{R^i} to each component of R^i, for $i \ge 0$. For $j \ge 0$, we consider also the truncation operator T_j defined for power series $r \in S\langle\langle\Sigma^*\rangle\rangle$ by

$$T_j(r) = \sum_{|w| \le j} (r, w) w.$$

The operator T_j is applied to n-tuples componentwise. An obvious induction on j shows that $T_j(R^j) = T_j(R^{j+t})$, for all j and t. This shows that the approximation sequence R^i converges (with respect to discrete convergence) to a specific

$$R = \begin{pmatrix} r_1 \\ \vdots \\ r_n \end{pmatrix} \in (S\langle\langle\Sigma^*\rangle\rangle)^{n \times 1},$$

where each r_i is quasi-regular. Denoting

$$P = \begin{pmatrix} p_1 \\ \vdots \\ p_n \end{pmatrix},$$

we see again inductively that $T_j(R) = T_j(h_R(P))$ holds for all $j \ge 0$, and consequently, R is a solution. Finally, if R' is another solution with quasi-regular components, we have $T_0(R) = T_0(R')$. Assuming inductively that $T_j(R) = T_j(R')$, we deduce

$$T_{j+1}(R) = T_{j+1}\big(h_R(P)\big) = T_{j+1}\big(h_{T_j(R)}(P)\big) = T_{j+1}\big(h_{T_j(R')}(P)\big)$$
$$= T_{j+1}\big(h_{R'}(P)\big) = T_{j+1}(R'),$$

which completes the induction and shows that $R = R'$. □

A proper S-algebraic system may have other solutions where the components are not quasi-regular. For instance, the $\mathbb{N}$-algebraic system consisting of the single equation $y_1 = y_1 y_1$ has both of the power series $r_1 = 0$ and $r_2 = \epsilon$ as solutions. The solution constructed above as *the limit of the approximation sequence* is in the sequel referred to as the *strong solution.*

We are now ready for the basic definition.

Definition 3.3. *A formal power series $r \in S\langle\langle \Sigma^* \rangle\rangle$ is* S-algebraic, *in symbols $r \in S^{\mathrm{alg}}\langle\langle \Sigma^* \rangle\rangle$, if $r = (r, \epsilon)\epsilon + r'$, where r' is some component of the strong solution of a proper S-algebraic system.*

We have stated and established Theorem 3.2 for *proper* S-algebraic systems. Then the approximation sequence converges with respect to the discrete convergence, and the resulting solution was called strong. However, a more general result is valid for continuous semirings S. By the *fixpoint theorem*, the *least* solution of an S-algebraic system exists and is obtained by computing the least upper bound of the approximation sequence associated to it. The least solution is not necessarily strong. More details of this approach can be found in Chap. 2 and [11]. More specific results can be obtained if we are dealing with the Boolean semiring. The proofs of the following result can be found in [12].

Theorem 3.4. *Every $\mathbb{B}$-algebraic system possesses a strong solution. If r is a component of the strong solution of a $\mathbb{B}$-algebraic system, then the quasi-regular part of r is a component of the strong solution of a proper $\mathbb{B}$-algebraic system.*

We use in the next example and in several other places throughout the chapter the so-called *Dyck language* and *Dyck mapping.* The Dyck language L_D over the alphabet $\{x_1, x_2\}$ is the language of all correctly nested parentheses when x_1 and x_2 are viewed as the left and right parenthesis, respectively. More generally, for $X = X_1 \cup \overline{X}_1$, $\overline{X}_1 = \{\overline{x} \mid x \in X_1\}$, the Dyck language $L_D(X)$ consists of all words w such that $D(w) = \epsilon$, where D is the Dyck mapping,

$$D : X^* \to X^*$$

defined as follows. Intuitively, we view X_1 as a set of left parenthesis and $\overline{X}_1$ as the set of corresponding right parenthesis. Then D removes from a word over X all pairs of adjacent matching parenthesis, until no further removals are possible. Thus, $D(wx) = D(w)x$ for $x \in X_1$, and

$$D(w\overline{x}) = \begin{cases} w_1 & \text{for } D(w) = w_1 x, \\ D(w)\overline{x} & \text{for } D(w) \notin X^* x, \end{cases}$$

for $x \in X_1$.

Example 3.5. We consider supports of some $\mathbb{N}$-algebraic series. The considerations are also preparatory for the next subsection. For the proper $\mathbb{N}$-algebraic system

$$y = yy + x_1 y x_2 + x_1 x_2,$$

the approximation sequence R^i, $i = 0, 1, \ldots$, consists of singletons of power series because there is only one variable y. We obtain

$$R^0 = 0, \qquad R^1 = x_1x_2, \qquad R^2 = (x_1x_2)^2 + x_1^2x_2^2 + x_1x_2,$$
$$R^3 = (x_1x_2)^4 + (x_1x_2)^2x_1^2x_2^2 + x_1^2x_2^2(x_1x_2)^2 + (x_1^2x_2^2)^2 + 2(x_1x_2)^3 + x_1^2x_2^2x_1x_2$$
$$+ x_1x_2x_1^2x_2^2 + x_1^2x_2x_1x_2^2 + x_1^3x_2^3 + x_1^2x_2^2 + (x_1x_2)^2 + x_1x_2.$$

The support of the resulting power series r equals the Dyck language L_D over the alphabet $\{x_1, x_2\}$ (without the empty word). However, r is not the characteristic series r_D of L_D because, as seen already from R^3, some coefficients in r are greater than 1. The characteristic series r_D is the first component of the solution of the $\mathbb{N}$-algebraic system

$$y = x_1yx_2 + x_1yx_2y + x_1x_2y + x_1x_2.$$

Also, the $\mathbb{N}$-algebraic system

$$y_1 = y_2 + y_1y_2,$$
$$y_2 = x_1y_1x_2 + x_1x_2$$

(which is not proper) can be used for the same purpose. Then the first components of the vectors R^i, $0 \le i \le 4$, in the approximation sequence are

$$0, \quad 0, \quad x_1x_2, \quad (x_1x_2)^2 + x_1x_2,$$
$$(x_1x_2)^2x_1^2x_2^2 + (x_1x_2)^3 + x_1x_2x_1^2x_2^2 + (x_1x_2)^2 + x_1^2x_2^2 + x_1x_2.$$

Also, now the approximation sequence converges, and r_D is the first component of the (strong) solution of the $\mathbb{N}$-algebraic system. Finally, also the $\mathbb{N}$-algebraic system

$$y = yx_1yx_2y + \epsilon$$

(which is also not proper) possesses a (strong) solution whose support equals L_D. The same result is obtained from the asymmetric systems, where either the first, or the last y on the right-hand side has been erased.

The next theorem deals with S-algebraic systems, such as the last one in our example, where the polynomials are not necessarily quasi-regular. The proof, similar to that of Theorem 3.2, is given in [12].

Theorem 3.6. *Assume that in an S-algebraic system $y_i = p_i$, $i = 1, \ldots, n$, the support of each polynomial p_i is contained in the language*

$$(\Sigma \cup Y)^* \Sigma (\Sigma \cup Y)^* \cup \{\epsilon\}.$$

Then the system possesses a unique solution, which moreover is strong.

3.2 Interconnections with Context-Free Grammars

Every context-free grammar and semiring S give rise to an S-algebraic system. Conversely, every S-algebraic system gives rise to a context-free grammar. Explicitly, this correspondence can be described as follows.

Given a context-free grammar G with Σ and $Y = \{y_1, \ldots, y_n\}$ as the terminal and non-terminal alphabets, y_1 as the initial symbol and R as the set of production rules, the *corresponding* S-algebraic system consists of the equations $y_i = p_i$, $i = 1, \ldots, n$, where $(p_i, w) = 1$ if $y_i \to w$ is a production in R, and $(p_i, w) = 0$, otherwise. (Since R is finite, each p_i is a polynomial.) Conversely, given an S-algebraic system $y_i = p_i$, $i = 1, \ldots, n$, the *corresponding* context-free grammar $\mathcal{G} = (\Sigma, Y, y_1, R)$ is defined by the condition: $y_i \to w$ is in R if and only if $(p_i, w) \neq 0$.

If we begin with an S-algebraic system, form the corresponding context-free grammar and then again the corresponding S-algebraic system, then the latter system does not necessarily coincide with the original one.

The most natural semiring for considerations dealing with formal languages is $\mathbb{N}$. A word generated by a context-free grammar appears in the support of the corresponding $\mathbb{N}$-algebraic power series. Moreover, its degree of ambiguity according to the grammar equals its coefficient in the series. If we want to deal with arbitrary context-free grammars, we should consider the semiring $\mathbb{N}^\infty$. We prefer dealing with $\mathbb{N}$ because every context-free grammar can be transformed to an equivalent one where no word has infinitely many leftmost derivations.

Definition 3.7. *A language $L \subseteq \Sigma^*$ is S-algebraic if it equals the support of a power series in $S^{\mathrm{alg}}\langle\langle \Sigma^* \rangle\rangle$.*

Theorem 3.8. *A language is context-free if and only if it is $\mathbb{N}$-algebraic.*

Proof. It suffices to establish the result for ϵ-free languages and quasi-regular $\mathbb{N}$-algebraic series. Let $\mathcal{G}_r$ be the grammar corresponding to a given proper $\mathbb{N}$-algebraic system $y_j = p_j$, $1 \leq j \leq n$, where r is the first component of its strong solution. To show that $\mathrm{supp}(r) = L(\mathcal{G}_r)$, we first establish the inclusion $L(\mathcal{G}_r) \subseteq \mathrm{supp}(r)$ inductively. We consider the approximation sequence R^i as in the proof of Theorem 3.2, and denote by r_j^i, $1 \leq j \leq n$, the jth component of R^i. It is now straightforward to establish inductively on t the following claim. Whenever a word $w \in \Sigma^*$ possesses a derivation of length at most t from y_j, $1 \leq j \leq n$, then $w \in \mathrm{supp}(r_j^t)$. Indeed, the claim holds for $t = 1$ by the definition of the corresponding grammar. The inductive step is proven by dividing a $(t+1)$-step derivation into a 1-step and t-step derivation, and applying the induction hypothesis to the latter.

The inclusion $L(\mathcal{G}_r) \subseteq \mathrm{supp}(r)$ follows. To prove the reverse inclusion, it suffices to establish inductively on t the following claim. Whenever $w \in \mathrm{supp}(r_j^t)$, $1 \leq j \leq n$, then there is a derivation of w from y_j according to $\mathcal{G}_r$. (Observe that we do not specify the length of the derivation.) For $t = 0$, the claim holds vacuously. Assume the claim holds for a fixed value t and consider a word $w \in \mathrm{supp}(r_j^{t+1})$. Consequently, for some

$$w' = u_1 y_{j_1} u_2 \ldots u_m y_{j_m} u_{m+1} \in \mathrm{supp}(p_j), \quad u_k \in \Sigma^*,$$

we have

$$w = u_1 w_{j_1} u_2 \ldots u_m w_{j_m} u_{m+1}, \quad w_{j_k} \in \mathrm{supp}(r_{j_k}^t),\ 1 \leq k \leq m.$$

We now use the induction hypothesis and the fact that, according to the definition of $\mathcal{G}_r$, y_j directly derives w', and conclude that the claim holds for the value $t+1$. Thus, we have shown that $L(\mathcal{G}_r) = \mathrm{supp}(r)$.

The argument above shows that every $\mathbb{N}$-algebraic language is context-free. Given an ϵ-free context-free grammar, we first eliminate from it all chain productions, where a non-terminal goes to a non-terminal. Then the corresponding $\mathbb{N}$-algebraic system will be proper, and we can show the equality of the two languages exactly as above. □

The following generalization of Theorem 3.8 is easily obtained, [18]. Recall that a semiring S is *positive* if the mapping h of S into $\mathbb{B}$ defined by

$$h(0) = 0, \qquad h(s) = 1 \quad \text{for } s \neq 0,$$

is a morphism.

Theorem 3.9. *All of the following five statements are equivalent for a language L:*

(i) L is a context-free language.
(ii) L is $\mathbb{N}$-algebraic.
(iii) L is $\mathbb{B}$-algebraic.
(iv) L is S-algebraic for all positive semirings S.
(v) L is S-algebraic for some positive semiring S.

There are $\mathbb{Z}$-algebraic languages that are not context-free. An obvious way to obtain such languages is to consider the difference between the characteristic series of Σ^* and L, where $L \subseteq \Sigma^*$ is an unambiguous context-free language whose complement is not context-free.

There are several open language-theoretic problems in this area. For instance, no characterization is known for $\mathbb{Z}$-algebraic languages, in terms of some of the well-known language hierarchies. $\mathbb{N}$-algebraic languages over a one-letter alphabet coincide with regular languages, but it is not known whether this holds for $\mathbb{Z}$-algebraic languages as well.

Apart from the language generated by a context-free grammar $\mathcal{G}$, the corresponding $\mathbb{N}$-algebraic power series $r_{\mathcal{G}}$ tells the degree of ambiguity of each word in the language. In the following theorem, we assume that $\mathcal{G}$ is a context-free grammar without ϵ-rules and chain rules. Then the corresponding $\mathbb{N}$-algebraic system is proper.

Theorem 3.10. *The coefficient of each word w in the $\mathbb{N}$-algebraic power series $r_{\mathcal{G}}$ equals the degree of ambiguity of w according to $\mathcal{G}$. Consequently, $\mathcal{G}$ is unambiguous (resp. of bounded ambiguity) if and only if the coefficients in $r_{\mathcal{G}}$ are at most* 1 *(resp. bounded).*

Proof (outline). As before, we assume that the non-terminals of the given grammar are $y_1, \ldots, y_n$. We let each of them be the initial letter, obtaining the grammars $\mathcal{G}_i$, $1 \le i \le n$. We consider also the proper $\mathbb{N}$-algebraic system $y_i = p_i$, $1 \le i \le n$, corresponding to the grammar $\mathcal{G} = \mathcal{G}_1$. If the n-tuple $(r_1, \ldots, r_n)$ is the solution of the $\mathbb{N}$-algebraic system, it follows by Theorem 3.8 that $L(\mathcal{G}_i) = \mathrm{supp}(r_i)$, $1 \le i \le n$. For a word w, we denote by $\mathrm{amb}(\mathcal{G}_i, w)$ the ambiguity of w according to the grammar $\mathcal{G}_i$. It can now be shown by induction on $|w|$ that $\mathrm{amb}(\mathcal{G}_i, w) = (r_i, w)$, $1 \le i \le n$, whence the theorem follows. Indeed, it suffices to consider the approximation sequence for the solution, and separate the first step in a derivation according to $\mathcal{G}_i$. The details are given in [18]. □

The following generalization is again immediate. Observe that now the conditions corresponding to points (iii) and (v) in Theorem 3.9 are not applicable.

Theorem 3.11. *The following three statements are equivalent for a language L:*

(i) L is an unambiguous context-free language.
(ii) The characteristic series of L is $\mathbb{N}$-algebraic.
(iii) The characteristic series of L is S-algebraic for all positive semirings S.

We mention finally that Theorem 3.10 can be stated also without restrictions on the productions of $\mathcal{G}$. Then we have to deal with $\mathbb{N}^\infty$-algebraic series because the ambiguity of w may be ∞. The first component of the least solution of the corresponding $\mathbb{N}^\infty$-algebraic system indicates the ambiguity of each word in the language, [11]. However, the approximation sequence does not necessarily converge with respect to the discrete convergence.

3.3 Normal Forms

We already pointed out that the algebraic systems under consideration are of a special form, resembling the productions in a context-free grammar. (In fact, very little is known about more general algebraic systems.) We now take a step further by considering several "normal forms": we may assume that the polynomials p_i appearing on the right-hand sides of the equations satisfy certain additional conditions, without losing any power series as solutions. Such normal forms are customary in language theory, and indeed the ones considered below resemble those introduced for context-free grammars.

Definition 3.12. *An S-algebraic system*

$$y_i = p_i, \quad 1 \leq i \leq n, \qquad Y = \{y_1, \ldots, y_n\},$$

over the alphabet Σ is in the Chomsky *(resp.* operator, Greibach*) normal form if, for each i, $1 \leq i \leq n$,*

$$\mathrm{supp}(p_i) \subseteq \Sigma \cup Y^2$$

(resp. $\mathrm{supp}(p_i) \subseteq \{\epsilon\} \cup Y\Sigma \cup Y\Sigma Y$, $\mathrm{supp}(p_i) \subseteq \Sigma \cup \Sigma Y \cup \Sigma Y^2$*).*

Theorem 3.13. *The components of the strong solution of a proper S-algebraic system appear also as components of the strong solution of such a system in the Chomsky normal form. Moreover, the latter system can be effectively constructed from the former.*

Proof (outline). We first transform the given system into one, where the supports of the right-hand sides are contained in $\Sigma \cup YY^+$, by replacing letters $x \in \Sigma$ with new variables y and introducing the equations $y = x$. A similar introduction of new variables is then applied to catenations larger than 2. For instance, the equation $y_1 = y_1y_2y_3$ becomes

$$y_1 = y_1y_4, \qquad y_4 = y_2y_3. \quad \square$$

In the constructions in the next theorem, we need the operators w^{-1}, $w \in \Sigma^*$, customary in language theory. By definition, for $u \in \Sigma^*$, $w^{-1}u = v$ if $u = wv$, and $w^{-1}u = 0$, otherwise. The operator w^{-1} is defined similarly from the right, and extended additively to concern power series. The application of this operator explains the presence of ϵ in the supports defining the operator normal form.

Theorem 3.14. *The first component of the strong solution of a proper S-algebraic system appears also as the first component of the strong solution of such a system in the operator normal form, effectively obtainable from the given system.*

Proof (outline). By the preceding theorem, we may assume that the given S-algebraic system is in the Chomsky normal form. We separate in the equations the Σ-parts and Y^2-parts, obtaining the system

$$y_i = \sum_{x \in \Sigma} (p_i, x)x + \sum_{k,m=1}^{n} (p_i, y_ky_m)y_ky_m, \quad 1 \leq i \leq n.$$

We now construct a new S-algebraic system, with the set of variables

$$Y_1 = y_0 \cup \{y_{i,x} \mid 1 \leq i \leq n, \ x \in \Sigma\}.$$

The equations in the new system are

$$y_0 = \sum_{x \in \Sigma} y_{1,x} x,$$

$$y_{i,x} = (p_i, x)\epsilon + \sum_{x' \in \Sigma} \sum_{k,m=1}^{n} (p_i, y_k y_m) y_{k,x'} x' y_{m,x} = q_{i,x},$$

where x ranges over Σ and $1 \le i \le n$. Clearly, the new system is in the operator normal form. We now claim that the $y_{i,x}$-component of the strong solution of the new system is obtained by applying the operator x^{-1} from the right to the y_i-component of the strong solution of the original system. The theorem follows from this claim, by the equation for y_0.

Let R^j (resp. Q^j) be the approximation sequence associated to the system $y_i = p_i$ (resp. $y_{i,x} = q_{i,x}$), with the components r_i^j (resp. $r_{i,x}^j$). It can be shown, by an induction on j that $r_{i,x}^j = r_i^j x^{-1}$, whence the claim follows. This holds for $j = 0$, by the definition of the new system. The details of the inductive step are presented in [12]. Thereby the equations

$$\sum_{x \in \Sigma} (rx^{-1})x = r \quad \text{and} \quad r'(rx^{-1}) = (r'r)x^{-1}$$

are needed. The equations are valid only for quasi-regular power series r and r'. □

Theorem 3.15. *The first component of the strong solution of a proper S-algebraic system appears also as the first component of the strong solution of such a system in the Greibach normal form, effectively obtainable from the given system.*

Proof (outline). The proof consists of eliminating the left recursion from the equations. The elimination can be based either on the Chomsky normal form, [18], or on the operator normal form, [12]. Suppose we are dealing with the Chomsky normal form. If the given system is $y_i = p_i$, $1 \le i \le n$, we separate on the right-hand sides the Σ-parts and the Y^2-parts as in the preceding proof. The result can be written in the matrix form

$$Y = YM + P$$

where the ith entry of the row vector P equals $\sum_{x \in \Sigma} (p_i, x)x$ and M is an $n \times n$ matrix whose (j,k)th entry equals the polynomial

$$\sum_{i=1}^{n} (p_k, y_j y_i) y_i,$$

for $1 \le j, k \le n$. We now introduce a new variable y_{jk} for each entry of the matrix M. The resulting equations have the required form. For details, we refer to [18]. The theorem now follows by observing that the matrix M^+ exists. □

The arguments applied above show also many well-known facts about context-free languages. For instance, the argument in Theorem 3.14 shows that the family of context-free languages is closed under left and right derivatives. The following theorem summarizes some of the results obtainable in this fashion.

Theorem 3.16. *The following five statements are equivalent for a language L:*

(i) L is context-free.
(ii) $L - \epsilon$ is generated by a context-free grammar without chain rules and ϵ-rules.
(iii) $L - \epsilon$ is generated by a context-free grammar where the right-hand side of every production is in $Y^2 \cup \Sigma$.
(iv) L is generated by a context-free grammar where the right-hand side of every production is in $Y\Sigma Y \cup Y\Sigma \cup \epsilon$.
(v) $L - \epsilon$ is generated by a context-free grammar where the right-hand side of every production is in $\Sigma Y^2 \cup \Sigma Y \cup \Sigma$.

We now take a step further by considering "meta" normal forms, that is, classes of normal forms with parameters such that each of the (infinitely many) values of the parameter gives rise to a normal form. The approach has turned out to be very useful in language theory: in some cases a characterization of all possible normal forms has been obtained. The results below are stated in a form producing only quasi-regular series. The transition to general series is straightforward. If r is the first component of the strong solution of the system $y_i = p_i$, $1 \le i \le n$, then for any $s \in S$. $s\epsilon + r$ is the first component of the strong solution of the system

$$y_0 = s\epsilon + p_1, \qquad y_i = p_i, \quad 1 \le i \le n.$$

Theorem 3.17. *Assume that m_1, m_2, m_3 are non-negative integers. Then every power series $r \in S^{\mathrm{alg}}\langle\langle \Sigma^* \rangle\rangle$ can be effectively obtained from the strong solution of a proper S-algebraic system where the supports of the right-hand sides of the equations are included in the set*

$$\Sigma^+ \cup \Sigma^{m_1} Y \Sigma^{m_2} Y \Sigma^{m_3},$$

with Y being the alphabet of variables.

The proof of Theorem 3.17 is given in [12], the original ideas being due to [1, 13, 17]. Observe that the Chomsky, operator, and Greibach normal forms are essentially obtained from the triples $(0,0,0)$, $(0,1,0)$, and $(1,0,0)$. Theorem 3.17 can be generalized. Instead of the triple (m_1, m_2, m_3), one can consider an arbitrary t-tuple $(m_1, \ldots, m_t)$, $t \ge 3$, and show that the supports can be included in the set

$$\Sigma^+ \cup \Sigma^{m_1} Y \ldots Y \Sigma^{m_t}.$$

A further strengthening of Theorem 3.17 consists in restricting the powers of Σ in Σ^+ to lengths belonging to the length set of the support of the original series. If only such powers are used, the system is said to be *terminally balanced.* For instance, the system

$$y_1 = y_2y_1 + x, \qquad y_2 = x^2$$

is in $(0,0,0)$-form, but not terminally balanced. The series

$$r_1 = \sum_{j=0}^{\infty} x^{2j+1}$$

is, however, obtained also from the following terminally balanced system in $(0,0,0)$-form:

$$y_1 = y_2y_1 + x, \qquad y_2 = y_3y_3, \qquad y_3 = x.$$

In general, such a simple construction does not work. The general construction of terminal balancing, due to [14], is presented in [12]. The construction works only for the Boolean semiring $\mathbb{B}$. It is an open problem to what extent the result can be extended to other semirings.

In conclusion for this subsection, some remarks about *closure properties* are in order. The general closure theory, the theory of abstract families of power series, [12, 7–9], is beyond the scope of this contribution. Some basic results are rather easily obtainable [18]. The family of S-algebraic power series generated by proper systems is closed under sum, product, and quasi-inverse. It is closed under semiring morphisms and non-erasing monoid morphisms, but not under arbitrary monoid morphisms. The Hadamard (or pointwise) product of an S-algebraic and S-rational series is S-algebraic [19]. This result corresponds to the well-known result about the intersection of context-free and regular languages. Every $\mathbb{Z}$-algebraic power series can be represented as the difference of two $\mathbb{N}$-algebraic series.

3.4 Theorems of Shamir and Chomsky–Schützenberger

We now discuss two famous theorems concerning algebraic power series. Both deal with the computation of the coefficients and consequently, also with degrees of ambiguity in derivations according to a context-free grammar. The methods for computing the coefficients, obtained by these theorems, are more direct than the iterative method of the approximation sequence, discussed above. The theorems of Shamir and Chomsky–Schützenberger were originally presented in [20] and [2], respectively. Our discussion uses also ideas from [15] and [18].

We consider first Shamir's theorem. An important auxiliary concept is that of an *involutive monoid.* Let X_1 be an alphabet, and denote $\overline{X}_1 = \{\overline{x} \mid x \in X_1\}$

and $X = X_1 \cup \overline{X}_1$. Then the *involutive monoid* $M(X)$ is the monoid generated by X, with the defining relations

$$x\overline{x} = \epsilon, \quad \text{for all } x \in X_1.$$

The monoid $M(X)$ can also be defined in terms of the Dyck mapping $D : X^* \to X^*$. The relation E_D defined by

$$wE_Dw' \iff D(w) = D(w')$$

is a congruence, and $M(X)$ can be defined as the factor monoid X^*/E_D. Observe that power series and polynomials can be defined for arbitrary monoids in the same way as for the free monoid Σ^*. Thus, $S\langle M(X)\rangle$ stands for polynomials over $M(X)$, with coefficients in S.

Theorem 3.18. *Let $r \in S\langle\langle \Sigma^* \rangle\rangle$ be a component in the strong solution of a proper S-algebraic system. Then there exist an alphabet $X = X_1 \cup \overline{X}_1$, $x_1 \in X_1$, and a morphism $h : \Sigma^+ \to S\langle M(X)\rangle$ such that the condition*

$$(r, w) = \big(h(w), \overline{x}_1\big)$$

is satisfied for all $w \in \Sigma^+$.

Proof. By Theorem 3.15, we assume that $r = r_1$, where $(r_1, \ldots, r_n)$ is the strong solution of the S-algebraic system $y_i = p_i$, $1 \le i \le n$, with

$$\text{supp}(p_i) \subseteq \Sigma \cup \Sigma Y \cup \Sigma Y^2, \quad 1 \le i \le n, \quad Y = \{y_1, \ldots, y_n\}.$$

Define the alphabets $X_1 = \{x_1, \ldots, x_n\}$ and $X = X_1 \cup \overline{X}_1$, as well as the morphism $h : \Sigma^+ \to S\langle M(X)\rangle$ by the condition

$$h(a) = \sum_{i,j,k} (p_i, ay_jy_k)\overline{x}_i x_k x_j + \sum_{i,j} (p_i, ay_j)\overline{x}_i x_j + \sum_i (p_i, a)\overline{x}_i,$$

for all $a \in \Sigma$. (Observe that the support of each $h(a)$ is contained in the set $\overline{X}_1 X_1^2 \cup \overline{X}_1 X_1 \cup \overline{X}_1$.) We have to prove that, for all $w \in \Sigma^+$,

$$(r_1, w) = \big(h(w), \overline{x}_1\big).$$

We do this by establishing the stronger *claim*

$$(r_i, w) = \big(h(w), \overline{x}_i\big), \quad \text{for all } i = 1, \ldots, n.$$

The proof is by induction on the length, $|w|$. The basis $|w| = 1$ is clear. Then $w = a \in \Sigma$, and we have

$$(r_i, w) = (p_i, a) = \big(h(w), \overline{x}_i\big).$$

Assume that the claim holds for all words of length at most t, and consider a word $w = aw_1$, where $a \in \Sigma$ and $|w_1| = t$. Considering the form of the supports $\mathrm{supp}(p_i)$, we obtain first

$$(r_i, w) = (r_i, aw_1) = \sum_{\substack{j,k \\ w_1=u_1u_2}} (p_i, ay_jy_k)(r_j, u_1)(r_k, u_2) + \sum_j (p_i, ay_j)(r_j, w_1).$$

This implies, by the inductive hypothesis and a slight modification of the first sum,

$$(r_i, w) = \sum_{j,k} (p_i, ay_jy_k) \sum_{w_1=u_1u_2} \big(h(u_1), \overline{x}_j\big)\big(h(u_2), \overline{x}_k\big) + \sum_j (p_i, ay_j)\big(h(w_1), \overline{x}_j\big).$$

Hence, because h is a morphism, we have to establish the equation

$$\begin{aligned}\big(h(a)h(w_1), \overline{x}_i\big) = &\sum_{j,k} (p_i, ay_jy_k) \sum_{w=u_1u_2} \big(h(u_1), \overline{x}_j\big)\big(h(u_2), \overline{x}_k\big) \\ &+ \sum_j (p_i, ay_j)\big(h(w_1), \overline{x}_j\big)\end{aligned}$$

to complete the induction. Denote the two sums on the right-hand side of the equation by A and B, respectively. We are interested in those terms of $h(w_1)$ only which together with $h(a)$ cancel in the Dyck mapping, to yield $\overline{x}_i$. This means that if $|w_1| = 1$ (resp. $|w_1| > 1$), we have to consider only B (resp. A) on the right side.

Assume that $w_1 = 1$, that is, w_1 is a letter. Then $\big(h(a)h(w_1), \overline{x}_i\big) = B$. This follows because the only terms in $h(a)h(w_1)$ canceling to $\overline{x}_i$ are obtained by multiplying a term with support $\overline{x}_i x_j$ in $h(a)$ and a term with support $\overline{x}_j$ in $h(w_1)$. The sum of such products equals B.

If $|w_1| > 1$, we have $\big(h(a)h(w_1), \overline{x}_i\big) = A$. Considering possible cancelations, the validity of this equation is first reduced to the validity of the equation

$$\big(h(w_1), \overline{x}_j\overline{x}_k\big) = \sum_{w_1=u_1u_2} \big(h(u_1), \overline{x}_j\big)\big(h(u_2), \overline{x}_k\big).$$

This equation holds because

$$\begin{aligned}\sum_{w_1=u_1u_2} \big(h(u_1), \overline{x}_j\big)\big(h(u_2), \overline{x}_k\big) &= \sum_{w_1=u_1u_2} (r_j, u_1)(r_k, u_2) \\ &= (r_jr_k, w_1) = \big(h(w_1), \overline{x}_j\overline{x}_k\big).\end{aligned}$$

This completes the induction, and we obtain Shamir's theorem. □

Example 3.19. Consider the $\mathbb{N}$-algebraic system over $\Sigma = \{a_0, a_1, a_2\}$, consisting of the single equation

$$y_1 = a_0 + a_1y_1 + a_2y_1^2.$$

The system is proper and in Greibach normal form. Following the notation in Shamir's theorem, we obtain

$$h(a_0) = \overline{x}_1, \qquad h(a_1) = \overline{x}_1 x_1, \qquad h(a_2) = \overline{x}_1 x_1^2.$$

Hence, all values $h(w)$ are monomials, with the coefficient 1. This implies that the resulting power series is the characteristic series of its language. The language is customarily referred to as the *Łukasiewicz language* and can be characterized as follows. Consider the morphism g of Σ^+ into the additive monoid of integers, defined by $g(a_i) = i - 1$, $0 \leq i \leq 2$. Then the Łukasiewicz language consists of all words w such that $g(w) = -1$ and $g(w') \geq 0$ for all proper prefixes w' of w.

Example 3.20. Consider the alphabet $\Sigma = \{a, b\}$ and the proper $\mathbb{N}$-algebraic (actually right linear) system

$$y_1 = ay_1 + 2by_2,$$
$$y_2 = 3ay_1 + by_2 + b.$$

We obtain now

$$h(a) = \overline{x}_1 x_1 + 3\overline{x}_2 x_1, \qquad h(b) = 2\overline{x}_1 x_2 + \overline{x}_2 x_2 + \overline{x}_2.$$

In this case, it is easy to analyze cancelations to $\overline{x}_1$. Corresponding to words of the form $a^i b^j$, we have the polynomial

$$(\overline{x}_1 x_1 + 3\overline{x}_2 x_1)^i (2\overline{x}_1 x_2 + \overline{x}_2 x_2 + \overline{x}_2)^j.$$

For any $i \geq 0$ and $j = 1$, we obtain the coefficient 2 and still have to cancel x_2. This requires arbitrarily many multiplications with $\overline{x}_2 x_2$ and the final multiplication with $\overline{x}_2$. Thus, every word in a^*bb^* has the coefficient 2, whereas all other words in a^*b^* have the coefficient 0. A similar analysis shows that every word in $b^+a^+bb^+$ has the coefficient 12. In general, the possibility of the cancelation to $\overline{x}_1$ shows that a change between the two letters in a word introduces a factor 2 or 3 to its coefficient. Observe that this example can be viewed also as a weighted grammar or a weighted finite automaton.

The converse of Shamir's theorem can be stated as follows. For a proof, see [18].

Theorem 3.21. *Assume that* $r \in S\langle\langle \Sigma^* \rangle\rangle$ *is quasi-regular,* $h : \Sigma^+ \to S\langle M(X) \rangle$, $X = X_1 \cup \overline{X}_1$, *is a morphism with the property that* $h(a)$ *is quasi-regular and non-zero for every* $a \in \Sigma$, *and* $\gamma \in M(X)$, *such that*

$$(r, w) = \big(h(w), \gamma\big)$$

holds for all $w \in \Sigma^+$. *Then* r *is* S*-algebraic.*

Instead of the involutive monoid $M(X)$, Shamir's theorem can be stated for the free group $\mathcal{G}(X_1)$ generated by X_1. (Thus, elements of $\overline{X}_1$ are inverses, not only right inverses as for $M(X)$.) Then the morphism h will be more complicated and more general normal forms for algebraic power series will be needed. The details are given in [18].

Finally, we present the *Chomsky–Schützenberger theorem*. It gives a method, similarly as Shamir's theorem, for computing the coefficients of an algebraic power series. While Shamir's theorem uses a morphism of the free monoid into a multiplicative monoid of polynomials, the Chomsky–Schützenberger theorem produces the coefficients by a morphism from the characteristic series of the intersection between a Dyck language and a regular language.

We omit the proof, [18], of the following Chomsky–Schützenberger theorem. The proof runs along the same lines as the corresponding result for context-free languages.

Theorem 3.22. *Let $r \in S\langle\langle \Sigma^* \rangle\rangle$ be a component in the strong solution of a proper S-algebraic system. Then there exist an alphabet $X = X_1 \cup \overline{X}_1$ and a regular language R over X such that r is a morphic image of the characteristic series of the intersection $L_D(X) \cap R$.*

4 Transductions

The theory of transductions originates from considerations about finite automata with outputs, generalized sequential machines, and pushdown transducers. Transductions can be viewed as mappings from $\mathbb{B}\langle\langle \Sigma_1^* \rangle\rangle$ into $\mathbb{B}\langle\langle \Sigma_2^* \rangle\rangle$ if only languages without multiplicities are considered. In general transductions between families of power series, the Boolean semiring is replaced by an arbitrary commutative semiring S. Direct generalizations of customary transductions between languages lead into difficulties because infinite sums over S may occur. Either one has to make strong summability assumptions about S, or else restrict the attention to cases not leading to infinite sums over S. The notion of a *regulated representation* is a convenient tool in the latter approach. We say that a morphism

$$h : \Sigma_1^* \to (S\langle\langle \Sigma_2^* \rangle\rangle)^{m \times m}$$

is a *regulated representation* if, for some positive integer t, all entries in all matrices $h(w)$ with $|w| \geq t$ are quasi-regular.

Let r be some component of the strong solution of a proper S-algebraic system. For brevity, we refer to such series r as *proper S-algebraic*. Hence, all proper S-algebraic series are quasi-regular.

Definition 4.1. *A mapping $\tau : S\langle\langle \Sigma_1^* \rangle\rangle \to S\langle\langle \Sigma_2^* \rangle\rangle$ is termed a* regulated semi-algebraic transduction *if, for $r \in S\langle\langle \Sigma_1^* \rangle\rangle$,*

$$\tau(r) = (r,\epsilon)r_0 + \sum_{w\in\Sigma_1^+} (r,w)\big(h(w)\big)_{1m},$$

where $r_0 \in S^{\mathrm{alg}}\langle\langle\Sigma_2^*\rangle\rangle$ *and*

$$h : \Sigma_1^* \to \big(S^{\mathrm{alg}}\langle\langle\Sigma_2^*\rangle\rangle\big)^{m\times m}$$

is a regulated representation. (As usual, M_{ij} *denotes the* (i,j)*th entry of a matrix* M*.) If in addition* r_0 *and all entries in every matrix* $h(a)$*,* $a \in \Sigma_1$*, are proper* S*-algebraic series, then* τ *is termed a* regulated algebraic transduction.

We are now ready for the fundamental result concerning regulated algebraic transductions.

Theorem 4.2. *A regulated semi-algebraic (resp. regulated algebraic) transduction maps every algebraic (resp. proper algebraic) series into an algebraic (resp. a proper algebraic) series.*

Proof. We use the notation in the definition above. We establish first the claim in parentheses, concerning regulated algebraic transductions. Consider a proper S-algebraic series r, and assume that all entries in the matrices $h(a)$, $a \in \Sigma_1$, are proper S-algebraic series. Hence, they are quasi-regular. Let $y_i = p_i$, $i = 1,\ldots,n$, be the proper S-algebraic system defining r. For each of the variables y_i, we associate the $m \times m$ matrix of variables

$$\begin{pmatrix} y^i_{11} & \cdots & y^i_{1m} \\ \vdots & & \vdots \\ y^i_{m1} & \cdots & y^i_{mm} \end{pmatrix}.$$

(Observe that each $a \in \Sigma_1$ is replaced by $h(a)$. No terms of S appear additively in any p_i, since the system is proper. Such terms would have to be multiplied by the identity matrix.) When the variables y_i in the original S-algebraic system are replaced by the associated matrices and the resulting equations are written out entry-wise, we obtain a proper S-algebraic system for the entries in the matrices, in particular, for the $(1,m)$th entry. However, in this new system, the coefficients on right-hand sides of the equations are power series in $S^{\mathrm{alg}}\langle\langle\Sigma^*\rangle\rangle$. It is shown in [11] that Theorem 3.2 holds for such systems as well. This establishes the claim concerning regulated algebraic transductions. (Observe that the commutativity of S is needed.)

Consider next the claim concerning regulated semi-algebraic transductions. It is no loss of generality to assume that the given series r is proper S-algebraic. For if $r = (r,\epsilon)\epsilon + r'$ and the claim holds for the proper S-algebraic r', then it clearly holds for r as well. Thus, we assume that the entries in all matrices $h(w)$, $|w| = t \geq 1$, are quasi-regular. The proof is now carried out by considering the words in Σ_1^t as new letters and reducing the argument to the (already established) case of regulated algebraic transductions.

Thus, consider the alphabet $\Sigma_3 = \{z_1, \ldots, z_l\}$ where $z_1, \ldots, z_l$ are all the words in Σ_1^t. Let

$$g : \Sigma_3^* \to \Sigma_1^*$$

be the natural morphism, mapping each z_i to the appropriate product of letters of Σ_1. Let $w \in \Sigma_1^+$ and define the series

$$r_w = \sum_{u \in \Sigma_3^*} (r, wg(u))u.$$

It is easy to see that r_w is proper S-algebraic. (In fact, proper S-algebraic series are closed under inverse monoid morphisms.) By the first part of the proof, it follows that the entries of

$$\sum_{u \in \Sigma_3^*} (r, wg(u))h(g(u))$$

are proper S-algebraic. Because we can write

$$\tau(r) = \sum_{|w|<t} \tau(w)\tau(r_w),$$

we conclude that $\tau(r)$ is S-algebraic, which completes the proof. □

We now relax the requirement of the representation being regulated. In the following definition, we assume that our commutative semiring S is also *complete.*

Definition 4.3. *A mapping* $\tau : S\langle\langle \Sigma_1^* \rangle\rangle \to S\langle\langle \Sigma_2^* \rangle\rangle$ *is termed an* algebraic transduction *if, for* $r \in S\langle\langle \Sigma_1^* \rangle\rangle$,

$$\tau(r) = (r, \epsilon)r_0 + \sum_{w \in \Sigma_1^+} (r, w)\big(h(w)\big)_{1m},$$

where $r_0 \in S^{\mathrm{alg}}\langle\langle \Sigma_2^* \rangle\rangle$ *and*

$$h : \Sigma_1^* \to \big(S^{\mathrm{alg}}\langle\langle \Sigma_2^* \rangle\rangle\big)^{m \times m}$$

is a semiring morphism.

It is not known whether an algebraic transduction maps an algebraic series into an algebraic series. The problem goes essentially back to applying erasing morphisms to algebraic series. However, if S is continuous, then an algebraic transduction always maps an algebraic series to an algebraic series, [11]. The following result can be obtained in the general case.

Theorem 4.4. *Every algebraic transduction can be represented as the composition of a projection and a regulated semi-algebraic transduction.*

Proof (outline). The argument is commonly used in language theory: introduce a new letter x' to the alphabet Σ_2. Multiply then in each entry of the matrices of the algebraic transduction the coefficient of ϵ by x'. A regulated semi-algebraic transduction results (with $\Sigma_2 \cup \{x'\}$ instead of Σ_2 as the target alphabet.) After applying this regulated transduction, apply the projection erasing x' and keeping the letters of Σ_2 fixed. □

A *regulated rational transduction* is defined exactly as a regulated semi-algebraic transduction (Definition 4.1) except that now $r_0 \in S^{\text{rat}}\langle\langle \Sigma_2^* \rangle\rangle$ and the target semiring of h is $(S^{\text{rat}}\langle\langle \Sigma_2^* \rangle\rangle)^{m \times m}$. A regulated rational transduction maps a series in $S^{\text{rat}}\langle\langle \Sigma_1^* \rangle\rangle$ into a series in $S^{\text{rat}}\langle\langle \Sigma_2^* \rangle\rangle$. The reader is referred to [18] for further details, as well as for the proof of the following result which can be viewed as another formulation of the Chomsky–Schützenberger Theorem.

Theorem 4.5. *For every proper S-algebraic series r, there is an alphabet $X = X_1 \cup \overline{X}_1$ and a regulated rational transduction τ such that*

$$r = \tau\big(\text{char}\big(L_D(X)\big)\big).$$

An alternative way of presenting the theory of transductions is to consider power series in the product monoid $\Sigma_1^* \times \Sigma_2^*$. We now define (general) rational transductions using this approach.

Definition 4.6. *Assume that S is complete and $\tau : S\langle\langle \Sigma_1^* \rangle\rangle \to S\langle\langle \Sigma_2^* \rangle\rangle$ is a mapping such that $\tau(r) = \sum (r, w)\tau(w)$. If*

$$\sum w \times \tau(w) \in S^{\text{rat}}\langle\langle \Sigma_1^* \times \Sigma_2^* \rangle\rangle,$$

then τ is said to be a rational transduction.

Both rational transductions and regulated rational transductions are closed under composition. The following result is a restatement of relations concerning rational power series in product monoids. (See [18].)

Theorem 4.7. *A mapping $\tau : S\langle\langle \Sigma_1^* \rangle\rangle \to S\langle\langle \Sigma_2^* \rangle\rangle$ is a rational transduction if and only if there are a series $r_0 \in S^{\text{rat}}\langle\langle \Sigma_2^* \rangle\rangle$ and a representation $h : \Sigma_1^* \to (S^{\text{rat}}\langle\langle \Sigma_2^* \rangle\rangle)^{m \times m}$ such that*

$$\tau(r) = (r, \epsilon)r_0 + \sum_{w \neq \epsilon} (r, w)\big(h(w)\big)_{1m}.$$

Observe that this theorem shows that every regulated rational transduction is a rational transduction. The following theorem tells explicitly the interconnection between rational transductions and regulated rational transductions. The result may be proved with arguments similar as those used for Theorem 4.4.

Theorem 4.8. *Every rational transduction can be expressed as the composition of a projection and a regulated rational transduction.*

We conclude this section with the following *Nivat's theorem.* The Hadamard product is denoted by $\odot$.

Theorem 4.9. *A mapping $\tau : S\langle\langle \Sigma_1^* \rangle\rangle \to S\langle\langle \Sigma_2^* \rangle\rangle$ is a rational transduction if and only if, for some alphabet Σ_3, projections $g : \Sigma_3^* \to \Sigma_1^*$ and $h : \Sigma_3^* \to \Sigma_2^*$, and for some series $r_0 \in S^{\mathrm{rat}}\langle\langle \Sigma_3^* \rangle\rangle$, we have*

$$\tau(r) = h\big(g^{-1}(r) \odot r_0\big).$$

Much of the fundamental work concerning algebraic transductions is due to [15] and [6]. We have also used the above arguments from [18].

5 Pushdown Automata

This section introduces $S\langle\langle \Sigma^* \rangle\rangle$-pushdown automata and discusses on one hand the interconnection with classical pushdown automata over finite alphabets (without weights) and on the other hand, the interconnection with algebraic systems.

5.1 Pushdown Transition Matrices

Throughout this section, $\Sigma = \{x_1, \ldots, x_k\}$ will denote a finite alphabet and S a commutative semiring. Also, Q will denote a finite non-empty set (of states) and Γ a finite alphabet (of pushdown symbols), not necessarily distinct from Σ.

Definition 5.1. *A matrix $M \in ((S\langle\langle \Sigma^* \rangle\rangle)^{Q\times Q})^{\Gamma^* \times \Gamma^*}$ is called an $S\langle\langle \Sigma^* \rangle\rangle$-*pushdown transition matrix *if the following two conditions are satisfied:*

(i) For any $p \in \Gamma$, there exist only finitely many $\pi \in \Gamma^*$ such that $M_{p,\pi} \neq 0$.
(ii) For any $\pi_1, \pi_2 \in \Gamma^*$,

$$M_{\pi_1,\pi_2} = \begin{cases} M_{p,\pi}, & \text{if } \pi_1 = p\pi', \pi_2 = \pi\pi', \text{ for some } \pi' \in \Gamma^*, \\ 0, & \text{otherwise.} \end{cases}$$

If all entries of M are in $S\langle \Sigma \cup \{\epsilon\} \rangle$, then we call M an $S\langle \Sigma \cup \{\epsilon\} \rangle$-pushdown transition matrix.

It follows directly from the definition that any pushdown transition matrix is finitely specified by the blocks $M_{p,\pi}$, with $p \in \Gamma$ and $\pi \in \Gamma^*$. In particular, any such matrix is both row and column finite. Consequently, the product of pushdown transition matrices and their arbitrary powers are well defined. However, without special assumptions about either the semiring or the matrix itself, the star may not exist because infinite sums may arise. For instance, if the semiring is complete, the infinite sums are well defined and the star always exists. For a presentation of pushdown transition matrices and pushdown automata in the case of complete semirings, we refer to [11] and [3]. We give here a different presentation where the semiring is not assumed to be complete, but rather the matrices are assumed to satisfy such properties as to obtain a locally finite star matrix, thus avoiding infinite sums. This allows us, e.g., to consider pushdown automata with multiplicities in $\mathbb{N}$ rather than $\mathbb{N}^\infty$, which is a desirable feature from the point of view of weighted automata and formal languages. In many essentials, we follow here the presentation in [12].

It is important to note that based on the semiring isomorphisms described in Chap. 1 of this handbook a pushdown transition matrix may be considered in $((S\langle\langle\Sigma^*\rangle\rangle)^{Q\times Q})^{\Gamma^*\times\Gamma^*}$, but also in $((S^{Q\times Q})\langle\langle\Sigma^*\rangle\rangle)^{\Gamma^*\times\Gamma^*}$, or in $(S^{Q\times Q})^{\Gamma^*\times\Gamma^*}\langle\langle\Sigma^*\rangle\rangle$. We will use also both of the latter semirings in our considerations without risk of confusion. For example, when discussing the star of a pushdown matrix, we will base our discussion on the semiring $(S^{Q\times Q})^{\Gamma^*\times\Gamma^*}\langle\langle\Sigma^*\rangle\rangle$, but the definition below of a proper pushdown transition matrix is based on $((S^{Q\times Q})\langle\langle\Sigma^*\rangle\rangle)^{\Gamma^*\times\Gamma^*}$.

Definition 5.2. *An $S\langle\langle\Sigma^*\rangle\rangle$-pushdown transition matrix is called proper if for all $p\in\Gamma$ and $\pi\in\Gamma^*$, $(M_{p,\pi},\epsilon)\neq 0$ implies that $|\pi|\geq 2$.*

The next result shows that a proper pushdown transition matrix is C-cycle free, and so based on Lemma 2.5, its star exists.

Theorem 5.3. *Let M be an $S\langle\langle\Sigma^*\rangle\rangle$-pushdown transition matrix. If M is proper, then it is C-cycle free. Moreover, $(M^*)_{p,\epsilon}$ is quasi-regular for all $p\in\Gamma$.*

Proof. We prove first that $(M^n,\epsilon)_{\pi_1,\pi_2}=0$, for all $\pi_1,\pi_2\in\Gamma^*$ and all $n\geq 0$, with $|\pi_2|\leq|\pi_1|+n-1$. We prove the claim by induction on n.

For $n=0$, the claim holds vacuously. Also, in case $|\pi_1|=0$, the claim follows directly from the definition of a pushdown transition matrix. Let $n\geq 1$ and $\pi_1,\pi_2\in\Gamma^*$ such that $|\pi_2|\leq|\pi_1|+n-1$. We may assume without loss of generality that $|\pi_1|\geq 1$, i.e., $\pi_1=p\pi_1'$, for some $p\in\Gamma$, $\pi_1'\in\Gamma^*$. Then

$$\begin{aligned}(M^n,\epsilon)_{p\pi_1',\pi_2} &= \sum_{\pi\in\Gamma^*,|\pi|\geq 2}(M,\epsilon)_{p\pi_1',\pi\pi_1'}(M^{n-1},\epsilon)_{\pi\pi_1',\pi_2}\\ &= \sum_{\pi\in\Gamma^*,|\pi|\geq 2}(M,\epsilon)_{p,\pi}(M^{n-1},\epsilon)_{\pi\pi_1',\pi_2}.\end{aligned}$$

Note now that $|\pi\pi_1'|+(n-1)-1\geq|\pi_1'|+n=|\pi_1|+n-1\geq|\pi_2|$, and so by the induction hypothesis, $(M^{n-1},\epsilon)_{\pi\pi_1',\pi_2}=0$, proving our claim.

To prove that M is C-cycle free, we have to show, by definition that $\lim^C_{n\to\infty}(M^n,\epsilon)=0$. This is equivalent with the following two conditions:

(i) For all $\pi_2\in\Gamma^*$, there exists a finite set $I(\pi_2)\subseteq\Gamma^*$ such that $(M^n,\epsilon)_{\pi_1,\pi_2}=0$, for all $\pi_1\notin I(\pi_2)$ and all $n\geq 0$.

(ii) For all $\pi_2 \in \Gamma^*$, there exists a non-negative integer $n(\pi_2)$ such that $(M^n, \epsilon)_{\pi_1,\pi_2} = 0$, for all $n \geq n(\pi_2)$ and all $\pi_1 \in \Gamma^*$.

Part (i) follows from our claim for $I(\pi_2) = \{\pi_1 \in \Gamma^* \mid |\pi_1| \leq |\pi_2|\}$. Part (ii) follows from our claim for $n(\pi_2) = |\pi_2| + 1$.

Applying again our claim, this time for $\pi_1 = p \in \Gamma$, $\pi_2 = \epsilon$, it follows that $(M^n, \epsilon)_{p,\epsilon} = 0$, for all $n \geq 0$, i.e., $(M^*)_{p,\epsilon}$ is quasi-regular. □

The following two results will be useful in the next section when proving the equivalence of algebraic systems and pushdown automata. For proofs, we refer to [12], where the results are stated also for C-cycle free (and other types of) pushdown transition matrices.

Theorem 5.4. *Let M be a proper $S\langle\langle\Sigma^*\rangle\rangle$-pushdown transition matrix. Then $(M^*)_{\pi_1\pi_2,\epsilon} = (M^*)_{\pi_1,\epsilon}(M^*)_{\pi_2,\epsilon}$, for all $\pi_1, \pi_2 \in \Gamma^*$.*

Theorem 5.5. *Let M be a proper $S\langle\langle\Sigma^*\rangle\rangle$-pushdown transition matrix. For any $p \in \Gamma$, let $S_p \in (S\langle\langle\Sigma^*\rangle\rangle)^{Q\times Q}$ be quasi-regular. Also, let S_ϵ be the $Q \times Q$ unity matrix and $S_{p\pi} = S_p S_\pi$, for all $p \in \Gamma$, $\pi \in \Gamma^*$. If*

$$S_p = \sum_{\pi\in\Gamma^*} M_{p,\pi} S_\pi,$$

then $S_\pi = (M^*)_{\pi,\epsilon}$, for all $\pi \in \Gamma^*$.

5.2 $S\langle\langle\Sigma^*\rangle\rangle$-Pushdown Automata

We define in this section the notion of $S\langle\langle\Sigma^*\rangle\rangle$-pushdown automata and their behavior.

Definition 5.6. *An* $S\langle\langle\Sigma^*\rangle\rangle$-pushdown automaton $\mathcal{P}$ *is a structure*

$$\mathcal{P} = (Q, \Gamma, M, q_0, p_0, P),$$

where:

(i) Q *is a finite set of* states.
(ii) Γ *is a finite alphabet of* pushdown symbols.
(iii) M *is an* $S\langle\langle\Sigma^*\rangle\rangle$*-pushdown transition matrix.*
(iv) $q_0 \in Q$ *is an* initial state.
(v) $p_0 \in \Gamma$ *is an* initial pushdown symbol.
(vi) $P \in (S\langle\{\epsilon\}\rangle)^{Q\times 1}$ *is a* final state vector.

We say that $\mathcal{P}$ is an $S\langle\Sigma \cup \{\epsilon\}\rangle$-pushdown automaton if M is an $S\langle\Sigma \cup \{\epsilon\}\rangle$-pushdown transition matrix. We also say that $\mathcal{P}$ is proper *if M is a proper pushdown transition matrix.*

The behavior $\|\mathcal{P}\| \in S\langle\langle\Sigma^*\rangle\rangle$ of $\mathcal{P}$ is defined by

$$\|\mathcal{P}\| = e_{q_0}(M^*)_{p_0,\epsilon}P = \left((M^*)_{p_0,\epsilon}P\right)_{q_0},$$

provided that M^* *exists, where* $e_{q_0} \in (S\langle\{\epsilon\}\rangle)^Q$*, with* $(e_{q_0})_{q_0} = \epsilon$ *and* $(e_{q_0})_q = 0$*, for all* $q \in Q \setminus \{q_0\}$*. We say that two pushdown automata* $\mathcal{P}_1$ *and* $\mathcal{P}_2$ *are* equivalent *if* $\|\mathcal{P}_1\| = \|\mathcal{P}_2\|$.

Note that the behavior of a pushdown automaton is well defined if its pushdown transition matrix is proper. Note also that, using standard terminology of formal language theory, the mode of acceptance of a pushdown automaton is defined here through reaching a final state, while emptying the pushdown stack. Two other (equivalent) modes of acceptance are often considered: through emptying the pushdown stack (regardless of the state), or by reaching a final state (regardless of the pushdown stack).

Example 5.7. Let $\Sigma = \{a, b\}$ and consider the $\mathbb{N}\langle\Sigma \cup \{\epsilon\}\rangle$-pushdown automaton $\mathcal{P} = (Q, \Gamma, M, q_1, p_0, P)$, where $Q = \{q_1, q_2\}$, $\Gamma = \{p_0, a, b\}$, $P_{q_1} = 0$, $P_{q_2} = \epsilon$ and $M \in ((\mathbb{N}\langle\langle\Sigma^*\rangle\rangle)^{Q\times Q})^{\Gamma^*\times\Gamma^*}$ is defined as follows:

$$M_{p_0,a} = \begin{pmatrix} a & 0 \\ 0 & 0 \end{pmatrix}, \qquad M_{a,aa} = \begin{pmatrix} a & 0 \\ 0 & 0 \end{pmatrix}, \qquad M_{a,\epsilon} = \begin{pmatrix} 0 & b \\ 0 & b \end{pmatrix}.$$

Clearly, M and by consequence $\mathcal{P}$, are proper. Then M^* exists and $\|\mathcal{P}\| = ((M^*)_{p_0,\epsilon})_{q_1,q_2}$.

Based on the definition of M, we obtain that

$$(M^*)_{p_0,\epsilon} = M_{p_0,a}(M^*)_{a,\epsilon} = \begin{pmatrix} a & 0 \\ 0 & 0 \end{pmatrix}(M^*)_{a,\epsilon},$$

$$(M^*)_{a,\epsilon} = M_{a,\epsilon} + M_{a,aa}(M^*)_{aa,\epsilon} = \begin{pmatrix} 0 & b \\ 0 & b \end{pmatrix} + \begin{pmatrix} a & 0 \\ 0 & 0 \end{pmatrix}\left((M^*)_{a,\epsilon}\right)^2.$$

If

$$(M^*)_{a,\epsilon} = \begin{pmatrix} p_1 & p_2 \\ p_3 & p_4 \end{pmatrix},$$

with $p_i \in \mathbb{N}\langle\langle\Sigma^*\rangle\rangle$, $1 \le i \le 4$, then it follows that $p_3 = 0$, $p_4 = b$, $p_1 = ap_1^2$, and $p_2 = ap_1p_2 + ap_2b + b$. Consequently, $p_1 = 0$ and $p_2 = ap_2b + b$. Then $p_2 = \sum_{n\ge 0} a^n b^{n+1}$, and so,

$$\|\mathcal{P}\| = \sum_{n\ge 0} a^{n+1}b^{n+1}.$$

Definition 5.8. *An* $S\langle\langle\Sigma^*\rangle\rangle$*-pushdown automaton* $\mathcal{P} = (Q, \Gamma, M, q_0, p_0, P)$ *is called* normalized *if:*

(i) $(M_{\pi_1,\pi_2})_{q,q_0} = 0$*, for all* $\pi_1, \pi_2 \in \Gamma^*$*,* $q \in Q$*.*
(ii) There is $t \in Q \setminus \{q_0\}$ *such that* $P_t = \epsilon$*,* $P_q = 0$*, for all* $q \in Q \setminus \{t\}$*, and* $(M_{\pi_1,\pi_2})_{t,q} = 0$*, for all* $\pi_1, \pi_2 \in \Gamma^*$*,* $q \in Q$*.*

It is not difficult to see that for any $S\langle\Sigma \cup \{\epsilon\}\rangle$-pushdown automaton, an equivalent normalized one can be constructed. The construction is the one often encountered in automata theory: one adds a new initial state and a new final state, and extends the pushdown matrix in a suitable way so that no transitions into the initial state and no transitions from the final state exist. Moreover, the new pushdown automaton remains proper if the initial one was so. We state the following result without proof, referring to [12] for a detailed construction and proof.

Theorem 5.9. *For any proper $S\langle\Sigma\cup\{\epsilon\}\rangle$-pushdown automaton, an equivalent normalized proper one exists.*

We recall that a (classical) pushdown automaton over an alphabet is a structure $\mathcal{A} = (Q, \Gamma, M, q_0, p_0, F)$, where $F \subseteq Q$ is a set of final states, $M \in (\mathcal{P}(\Sigma\cup\{\epsilon\})^{Q\times Q})^{\Gamma^*\times\Gamma^*}$ is a row and column finite pushdown transition matrix and the significance of the other components is the same as in the case of an $S\langle\langle\Sigma^*\rangle\rangle$-pushdown automaton. In particular, note that the matrix M may be seen as an $\mathbb{B}\langle\Sigma \cup \{\epsilon\}\rangle$-pushdown transition matrix, and F may be seen as a final state vector in $(\mathbb{B}\langle\{\epsilon\}\rangle)^{Q\times 1}$. Consequently, we may consider any (classical) pushdown automaton $\mathcal{A}$ as an $\mathbb{B}\langle\Sigma \cup \{\epsilon\}\rangle$-pushdown automaton. In this case M^* always exists over $\mathbb{B}$ and $\|\mathcal{A}\| \in \mathbb{B}\langle\langle\Sigma^*\rangle\rangle$ is the characteristic series of the language accepted by $\mathcal{A}$. Similarly, if M is proper, $\mathcal{A}$ may also be seen as an $\mathbb{N}\langle\Sigma\cup\{\epsilon\}\rangle$-pushdown automaton. The correspondence between the language $\mathcal{L}(\mathcal{A})$ accepted by $\mathcal{A}$ and its behavior $\|\mathcal{A}\| \in \mathbb{N}\langle\langle\Sigma^*\rangle\rangle$ is given in the next result.

Theorem 5.10. *For any (classical) proper pushdown automaton $\mathcal{A}$ over Σ and any $w \in \Sigma^*$, $(\|\mathcal{A}\|, w)$ is the number of distinct successful computations of $\mathcal{A}$ on input w, where the acceptance mode of $\mathcal{A}$ is with empty pushdown stack and final state.*

Proof. We prove the more general claim that for any $\pi_1, \pi_2 \in \Gamma^*$, $q_1, q_2 \in Q$, $w \in \Sigma^*$, $n \geq 0$, $(((M^n)_{\pi_1,\pi_2})_{q_1,q_2}, w)$ is equal to the number of distinct n-step computations in $\mathcal{A}$ changing the state from q_1 to q_2 and the stack content from π_1 to π_2 while reading the input w. Then the theorem follows with $\pi_1 = p_0$, $\pi_2 = \epsilon$, $q_1 = q_0$, and $q_2 \in F$.

For $n = 0$ and $n = 1$ the claim is trivial. For $n > 0$, assume the claim holds up to n. Then any $n+1$-step computation with input w, changing the state from q_1 to q_2 and the stack from π_1 to π_2, can be decomposed into:

(i) An n-step computation with input u, changing the state from q_1 to q and the stack from π_1 to π, and
(ii) A 1-step computation with input v, changing the state from q to q_2 and the stack from π to π_2

where $w = uv$, $q \in Q$, and $\pi \in \Gamma^*$. Thus, based on the induction hypothesis, the number of such distinct $(n+1)$-step computations is the following:

$$\sum_{\pi\in\Gamma^*}\sum_{q\in Q}\sum_{\substack{u,v\in\Sigma^*\\ w=uv}}\Big(\big((M^n)_{\pi_1,\pi_2}\big)_{q_1,q},u\Big)\Big(\big(M_{\pi,\pi_2}\big)_{q,q_2},v\Big)$$
$$=\Big(\big((M^{n+1})_{\pi_1,\pi_2}\big)_{q_1,q_2},w\Big).\quad\square$$

5.3 Equivalence with Algebraic Systems

We prove in this section that the algebraic systems and the pushdown automata are equivalent, in the sense that the set of behaviors of proper $S\langle\Sigma\cup\{\epsilon\}\rangle$-pushdown automata is exactly $S^{\mathrm{alg}}\langle\langle\Sigma^*\rangle\rangle$. We prove first that any algebraic series is the behavior of a pushdown automaton.

Theorem 5.11. *Let r be the first component of the strong solution of a proper algebraic system. Then there exists a proper $S\langle\Sigma\cup\{\epsilon\}\rangle$-pushdown automaton $\mathcal{P}$ such that $\|\mathcal{P}\|=r$.*

Proof. Let
$$y_i=p_i,\quad i=1,\dots,n, \tag{1}$$
be a proper algebraic system with r as the first component of its strong solution, where $p_i\in S\langle(\Sigma\cup Y)^*\rangle$, for all $1\le i\le n$. We consider the $S\langle\Sigma\cup\{\epsilon\}\rangle$-pushdown automaton
$$\mathcal{P}=(\{q\},\Sigma\cup Y,M,q,y_1,(\epsilon)),$$
where M is defined as follows:

$$M_{y_i,y_j\gamma}=(p_i,y_j\gamma)\epsilon+\sum_{x\in\Sigma}(p_i,xy_j\gamma)x,\quad \text{for } \gamma\in(\Sigma\cup Y)^*,\ 1\le i,j,\le n,$$
$$M_{y_i,x\gamma}=\sum_{x'\in\Sigma}(p_i,x'x\gamma)x',\quad \text{for } \gamma\in(\Sigma\cup Y)^*,\ x\in\Sigma,\ 1\le i\le n,$$
$$M_{y_i,\epsilon}=\sum_{x\in\Sigma}(p_i,x)x,\quad \text{for } 1\le i\le n,$$
$$M_{x,\epsilon}=x,\ \text{for } x\in\Sigma,$$
$$M_{\pi_1,\pi_2}=0,\quad \text{in all other cases.}$$

Note that M is a proper pushdown matrix. Indeed, if $(M_{q,\pi},\epsilon)\neq 0$, for some $q\in\Gamma$, $\pi\in\Gamma^*$, it implies that $q=p_i$ and $(p_i,\pi)\neq 0$, for some $1\le i\le n$. However, since the algebraic system is proper, it follows that $|\pi|\ge 2$. Thus, it follows by Theorem 5.3 that M^* exists and $(M^*)_{y_i,\epsilon}$ is quasi-regular for all $1\le i\le n$.

We write the algebraic system (1) as follows:

$$\begin{aligned}
y_i =& \sum_{j=1}^{n} \sum_{\gamma\in(\Sigma\cup Y)^*} (p_i, y_j\gamma)y_j\gamma + \sum_{j=1}^{n} \sum_{\gamma\in(\Sigma\cup Y)^*} \sum_{x\in\Sigma} (p_i, xy_j\gamma)xy_j\gamma \\
&+ \sum_{\gamma\in(\Sigma\cup Y)^*} \sum_{x\in\Sigma} \sum_{x'\in\Sigma} (p_i, x'x\gamma)x'x\gamma + \sum_{x\in\Sigma}(p_i, x)x \qquad (2)\\
=& \sum_{j=1}^{n} \sum_{\gamma\in(\Sigma\cup Y)^*} M_{y_i,y_j\gamma}y_j\gamma + \sum_{x\in\Sigma} \sum_{\gamma\in(\Sigma\cup Y)^*} M_{y_i,x\gamma}x\gamma + M_{y_i,\epsilon}.
\end{aligned}$$

We claim now that the system (1) is satisfied when substituting $(M^*)_{y_i,\epsilon}$ for y_i, for all $1 \le i \le n$. Based on Theorem 5.4, that means that when checking the equalities in (1), we will substitute $(M^*)_{\pi,\epsilon}$ for all $\pi \in (\Sigma \cup Y)^*$.

It is easy to see that $M^n_{x,\epsilon} = 0$, for all $x \in \Sigma$ and $n \ge 2$ and so, $(M^*)_{x,\epsilon} = x$. For $(M^*)_{y_i,\epsilon}$, based on the definition of M, we obtain that

$$\begin{aligned}
(M^*)_{y_i,\epsilon} =& \sum_{\pi\in(\Sigma\cup Y)^*} M_{y_i,\pi}(M^*)_{\pi,\epsilon} \\
=& \sum_{j=1}^{n} \sum_{\gamma\in(\Sigma\cup Y)^*} M_{y_i,y_j\gamma}(M^*)_{y_j\gamma,\epsilon} \\
&+ \sum_{x\in\Sigma} \sum_{\gamma\in(\Sigma\cup Y)^*} M_{y_i,x\gamma}(M^*)_{x\gamma,\epsilon} + M_{y_i,\epsilon},
\end{aligned}$$

for all $1 \le i \le n$, i.e., the refined version (2) of system (1) is verified, proving the claim.

Note now that based on Theorem 5.3, $(M^*)_{y_i,\epsilon}$ is a quasi-regular series. Since a proper algebraic system has only one solution with all components quasi-regular, see Theorem 3.2. It follows now that $r = (M^*)_{y_1,\epsilon} = \|\mathcal{A}\|$, concluding our proof. □

Example 5.12. Consider the proper $\mathbb{N}$-algebraic system

$$y = yy + x_1yx_2 + x_1x_2$$

of Example 3.5. Based on Theorem 5.11, we construct a pushdown automaton $\mathcal{P}$ such that $\|\mathcal{P}\|$ is the strong solution of the system. We consider $\mathcal{P} = (\{q\}, \{x_1, x_2, y\}, M, q, y, (\epsilon))$, where the pushdown transition matrix M is defined as follows: $M_{x_1,\epsilon} = x_1$, $M_{x_2,\epsilon} = x_2$, $M_{y,x_2} = x_1$, $M_{y,x_2y} = x_1$, $M_{y,yx_2} = x_1$, $M_{y,yx_2y} = x_1$. It follows then by Theorem 5.11 that $\|\mathcal{P}\| = (M^*)_{y_1,\epsilon}$ is the strong solution of the algebraic system above.

We prove now the reverse transition, from a pushdown automaton to an algebraic system.

Theorem 5.13. *Let $\mathcal{P}$ be a proper $S\langle\Sigma \cup \{\epsilon\}\rangle$-pushdown automaton. Then there exists a proper $S\langle\langle\Sigma^*\rangle\rangle$-algebraic system with $\|\mathcal{P}\|$ as the first component of its strong solution.*

Proof. By Theorem 5.9, we may assume without loss of generality that $\mathcal{P} = (Q, \Gamma, M, q_0, p_0, P)$ is a normalized proper $S\langle \Sigma \cup \{\epsilon\}\rangle$-pushdown automaton. Thus, $\|\mathcal{P}\| = ((M^*)_{p_0,\epsilon})_{q_0,t}$, where $t \in Q$, $P_t = \epsilon$ (and it is the only non-zero component of P).

Consider the alphabet

$$Y = \{y^p_{q_1,q_2} \mid p \in \Gamma,\ q_1, q_2 \in Q\}.$$

We consider the matrices $Y_p \in (S\langle Y \rangle)^{Q\times Q}$, defined by $(Y_p)_{q_1,q_2} = y^p_{q_1,q_2}$, for all $q_1, q_2 \in Q$. We then extend our definition to $Y_\pi \in (S\langle Y\rangle)^{Q\times Q}$, for all $\pi \in \Gamma^*$ in the following way:

$$Y_\epsilon = E, \qquad Y_{p\pi} = Y_p Y_\pi,$$

for all $\pi \in \Gamma^*$, where we denote by E the unity matrix (in this case a $Q \times Q$ matrix).

Consider now the algebraic system written in the following matrix notation:

$$Y_p = \sum_{\pi \in \Gamma^*} M_{p,\pi} Y_\pi, \quad \text{for all } p \in \Gamma. \tag{3}$$

Clearly, since M is proper, so is our algebraic system. Consequently, it follows by Theorem 5.5 that the strong solution of (3) is given by $(M^*)_{p,\epsilon}$, $p \in \Gamma$ (substituted for Y_p in the system (3)). It follows in particular that the component of the strong solution of (3) corresponding to $y^{p_0}_{q_0,t}$ is $((M^*)_{p_0,\epsilon})_{q_0,t} = \|\mathcal{P}\|$. □

One should observe that the variables $y^p_{q_1,q_2}$ in the proof of Theorem 5.13 correspond to the well-known triple construction $[q_1, p, q_2]$, used in the transition from (classical) pushdown automata to context-free grammars. The construction and the transition are originally due to Evey [4].

Example 5.14. Let $\Sigma = \{a, b\}$ and consider the proper $\mathbb{N}\langle\langle \Sigma \cup \{\epsilon\}\rangle\rangle$-pushdown automaton in Example 5.7. We construct a proper algebraic system with $\|\mathcal{P}\|$ as a component of its strong solution as follows. Let

$$Y = \{y^p_{q,q'} \mid p \in \{p_0, a, b\},\ q, q' \in \{q_1, q_2\}\}.$$

Let also

$$Y_{p_0} = \begin{pmatrix} y^{p_0}_{q_1,q_1} & y^{p_0}_{q_1,q_2} \\ y^{p_0}_{q_2,q_1} & y^{p_0}_{q_2,q_2} \end{pmatrix}, \qquad Y_a = \begin{pmatrix} y^a_{q_1,q_1} & y^a_{q_1,q_2} \\ y^a_{q_2,q_1} & y^a_{q_2,q_2} \end{pmatrix},$$

and consider the following algebraic system:

$$\begin{cases} Y_{p_0} = \left(\begin{smallmatrix} a & 0 \\ 0 & 0 \end{smallmatrix}\right) Y_a, \\ Y_a = \left(\begin{smallmatrix} a & 0 \\ 0 & 0 \end{smallmatrix}\right) Y_a^2 + \left(\begin{smallmatrix} 0 & b \\ 0 & b \end{smallmatrix}\right). \end{cases} \tag{4}$$

A simple calculation shows that (4) implies that $y^a_{q_2,q_1} = 0$, $y^a_{q_2,q_2} = b$, $y^a_{q_1,q_1} = a(y^a_{q_1,q_1})^2$, $y^a_{q_1,q_2} = a y^a_{q_1,q_1} y^a_{q_1,q_2} + a y^a_{q_1,q_2} b + b$. However, this implies that any solution of (4) will have 0 on the component corresponding to $y^a_{q_1,q_1}$ and so, we obtain that in (4), we may replace the equation corresponding to $y^a_{q_1,q_2}$ with the equation $y^a_{q_1,q_2} = ay^a_{q_1,q_2}b + b$. A suitable change of notation leads to the following proper algebraic system:

$$\begin{cases} z_1 = az_2, \\ z_2 = az_2b + b, \end{cases}$$

where $\|\mathcal{P}\|$ is the first component of its strong solution.

As noted already earlier in this chapter, any (classical) proper pushdown automaton may be seen as an $\mathbb{B}\langle\Sigma \cup \{\epsilon\}\rangle$- and as an $\mathbb{N}\langle\Sigma \cup \{\epsilon\}\rangle$-pushdown automaton. Based on this analogy and on Theorems 5.11 and 5.13, the following result may be proved.

Theorem 5.15.

(i) An ϵ-free language is context-free if and only if it is the behavior of a proper pushdown automaton.
(ii) For any epsilon-free context-free grammar G without chain rules, there exists a proper pushdown automaton $\mathcal{A}_G$ such that the ambiguity of any word $w \in \Sigma^*$ in $L(G)$ is $(\|\mathcal{A}_G\|, w)$.
(iii) For any proper pushdown automaton $\mathcal{A}$, there exists a context-free grammar $G_{\mathcal{A}}$ such that the ambiguity of any word $w \in \Sigma^*$ in $L(G_{\mathcal{A}})$ is $(\|\mathcal{A}\|, w)$.

6 Other Topics

Several other topics may be considered in connection with algebraic systems and pushdown automata. We mention here briefly two such topics. A result of Gruska [5] on a characterization of context-free languages may be generalized to a Kleene theorem for algebraic power series. One may prove (see [10] for a presentation in terms of complete semirings) that the algebraic power series coincide with the least equationally closed semiring containing all monomials. One may also consider the algebraic power series over the free commutative monoid $\Sigma^\oplus$ rather than Σ^*: this corresponds to the case where all variables are commuting. As it is well known from the theory of formal languages, the commuting case yields very different behavior; one example in this respect is the theorem of Parikh [16]. In the case of formal power series, several interesting decidability results may be given in the commutative case, based on tools from mathematical analysis and algebraic geometry. We refer to [12] for more details on the topic.

Acknowledgement. We thank the editors, in particular Werner Kuich, for a careful reading of our original version of the chapter and for many useful suggestions. The choice toward a more combinatorial, rather than a purely algebraic approach is ours.

References

1. M. Blattner and S. Ginsburg. Position-restricted grammar forms and grammars. *Theoretical Computer Science*, 17:1–27, 1982.
2. N. Chomsky and M.P. Schützenberger. The algebraic theory of context-free languages. In P. Braffort and D. Hirschberg, editors, *Computer Programming and Formal Systems*, pages 118–161. North-Holland, Amsterdam, 1963.
3. Z. Ésik and W. Kuich. Modern automata theory. http://www.gmd.tuwien.ac.at/kuich/, 2007.
4. R.J. Evey. The theory and application of pushdown store machines. Mathematical Linguistics and Automatic Translation, Harvard Univ. Comput. Lab. Rept. NSF-IO, 1963.
5. J. Gruska. A characterization of context-free languages. *Journal of Computer and System Sciences*, 5:353–364, 1971.
6. G. Jacob. Représentations et substitutions matricielles dans la théorie algébrique des transductions. Thèses de doctorat d'état, Université Paris, VII, 1975.
7. W. Kuich. Formal power series and one-way stack automata. *Bulletin of the European Association for Theoretical Computer Science*, 15:50–54, 1981.
8. W. Kuich. An algebraic characterization of some principal regulated rational cones. *Journal of Computer and System Sciences*, 25:377–401, 1982.
9. W. Kuich. Formal power series, cycle-free automata and algebraic systems. Bericht F 103, Institute für Informationsverarbeitung Graz, 1982.
10. W. Kuich. Gaussian elimination and a characterization of algebraic power series. In *Proc. of the 23rd International Symposium on Mathematical Foundations of Computer Science*, volume 1450 of *Lecture Notes in Computer Science*, pages 512–521. Springer, Berlin, 1998.
11. W. Kuich. Semirings and formal power series: their relevance to formal languages and automata. In G. Rozenberg and A. Salomaa, editors, *Handbook of Formal Languages*, pages 609–678. Springer, Berlin, 1997.
12. W. Kuich and A. Salomaa. *Semirings, Automata, Languages.* Springer, Berlin, 1986.
13. H.A. Maurer, A. Salomaa, and D. Wood. Completeness of context-free grammar forms. *Journal of Computer and System Sciences*, 23:1–10, 1981.
14. H.A. Maurer, A. Salomaa, and D. Wood. A supernormalform theorem for context-free grammar forms. *Journal of the ACM*, 30:95–102, 1983.
15. M. Nivat. Transduction des langages de Chomsky. *Université de Grenoble. Annales de l'Institut Fourier*, 18:339–455, 1968.
16. R.J. Parikh. Language generating devices. M.I.T. Res. Lab. Electron. Quart. Prog. Rep. 60, 1961.
17. D.J. Rosenkrantz. Matrix equation and normal forms for context-free grammars. *Journal of the ACM*, 14:501–507, 1967.

18. A. Salomaa and M. Soittola. *Automata-Theoretic Aspects of Formal Power Series*. Springer, Berlin, 1978.
19. M.P. Schützenberger. On the definition of a family of automata. *Information and Control*, 4:245–270, 1961.
20. E. Shamir. A representation theorem for algebraic and context-free power series in noncommuting variables. *Information and Control*, 11:239–254, 1967.

Chapter 8: Lindenmayer Systems

Juha Honkala

Department of Mathematics, University of Turku, 20014 Turku, Finland
juha.honkala@utu.fi

1 Introduction

The theory of Lindenmayer systems studies free monoid morphisms, free monoid substitutions and their iterations. In this chapter, we discuss similar ideas in a more general framework. Instead of a free monoid, we consider the free semi-algebra $S\langle \Sigma^* \rangle$ consisting of polynomials with non-commuting variables in Σ and coefficients in a semiring S and we study the iteration of endomorphisms of $S\langle \Sigma^* \rangle$. We allow various modes of iteration and we consider various classes of morphisms. Classical L systems are obtained as special cases by taking S to be the Boolean semiring. Our approach also generalizes the theory of algebraic series in non-commuting variables.

A brief outline of the contents of the chapter follows. In Sect. 2, we discuss the connections between classical L systems and rational power series. This topic is discussed in detail in [37, 30, 31]. Our discussion is brief and we will mostly not repeat the material covered in these references. However, we will recall the Berstel–Nielsen theorem stating that it is decidable whether or not two given D0L systems are growth range equivalent and discuss the applications of this result discovered by Ruohonen. In Sect. 3, we consider L algebraic systems and series. This theory assumes its simplest form if the basic semiring S is continuous. Here, in order to include also the cases where

M. Droste, W. Kuich, H. Vogler (eds.), *Handbook of Weighted Automata*,
Monographs in Theoretical Computer Science. An EATCS Series,
DOI 10.1007/978-3-642-01492-5_8,

the basic semiring is a ring, we will not assume that S is continuous. In Sects. 4 and 5, we discuss restricted classes of L algebraic systems which give the power series versions of D0L systems, DT0L systems and other L systems studied in the classical theory. Various aspects of L algebraic series are discussed in detail in [3–20, 22–24, 26, 27].

Below we will use the customary notation concerning L systems, which makes it very easy to say precisely what kind of iteration and what kind of morphisms are intended. This notation was originally inspired by biological applications. For us, the main motivation to study L systems comes from the very basic mathematical ideas involved.

2 Iterated Morphisms and Rational Series

Let X and Y be finite alphabets. A mapping $h : X^* \to Y^*$ is called a *morphism* if

$$h(uv) = h(u)h(v)$$

whenever $u, v \in X^*$. If $h : X^* \to Y^*$ is a morphism and ε is the empty word, then $h(\varepsilon) = \varepsilon$.

An *HDT0L system* is a construct $G = (X, Y, h_1, \ldots, h_n, g, w)$, where X and Y are finite alphabets, n is a positive integer, $h_i : X^* \to X^*$ $(i = 1, \ldots, n)$ and $g : X^* \to Y^*$ are morphisms and $w \in X^*$ is a word. An HDT0L system $G = (X, Y, h_1, \ldots, h_n, g, w)$ is called a *DT0L system* if $X = Y$ and g is the identity morphism. If G is a DT0L system, we write $G = (X, h_1, \ldots, h_n, w)$. An HDT0L system $G = (X, Y, h_1, \ldots, h_n, g, w)$ is called an *HD0L system* if $n = 1$. Finally, an HD0L system $G = (X, Y, h, g, w)$ is called a *D0L system* if $X = Y$ and g is the identity morphism. In other words, a D0L system is a triple $G = (X, h, w)$, where X is a finite alphabet, $h : X^* \to X^*$ is a morphism and $w \in X^*$ is a word.

Assume that $G = (X, Y, h_1, \ldots, h_n, g, w)$ is an HDT0L system. Let $\mathbf{n} = \{1, \ldots, n\}$ be an alphabet with n letters. Then the *sequence* $S(G)$ of G is the mapping $S(G) : \mathbf{n}^* \to Y^*$ defined by

$$S(G)(i_1 \ldots i_t) = gh_{i_t} \ldots h_{i_1}(w)$$

for $t \geq 0$ and $i_1, \ldots, i_t \in \mathbf{n}$. The *language* $L(G)$ of G is the image of $S(G)$. In other words,

$$L(G) = \{gh_{i_t} \ldots h_{i_1}(w) \mid t \geq 0,\ i_1, \ldots, i_t \in \mathbf{n}\}.$$

The HDT0L systems G and H are *sequence equivalent* (resp. *language equivalent*) if $S(G) = S(H)$ (resp. $L(G) = L(H)$). Clearly, sequence equivalence implies language equivalence but not vice versa.

In the special case of a D0L system $G = (X, h, w)$, the sequence $S(G)$ of G consists of the words

$$w, h(w), h^2(w), h^3(w), \ldots .$$

Two D0L systems $G = (X, g, w)$ and $H = (X, h, v)$ are sequence equivalent if and only if

$$g^i(w) = h^i(v) \quad \text{for all } i \geq 0.$$

We next establish the basic connection between iterated morphisms and $\mathbb{N}$-rational series.

Assume that $G = (X, Y, h_1, \ldots, h_n, g, w)$ is an HDT0L system. Then the *length series* (or *growth series*) $r(G)$ of G is defined by

$$r(G) = \sum_{u \in \mathbf{n}^*} |S(G)(u)|u.$$

If $G = (X, Y, h, g, w)$ is an HD0L system, we consider the *length sequence* $(a(n))_{n \geq 0}$ of G defined by

$$a(n) = \left|gh^n(w)\right|$$

for $n \geq 0$.

To prove that length series of HDT0L systems are $\mathbb{N}$-rational, let $X = \{x_1, \ldots, x_m\}$, $Y = \{y_1, \ldots, y_s\}$ and let $h : X^* \to Y^*$ be a morphism. If w is a word and z is a letter, then $|w|_z$ is the number of occurrences of z in w. The *Parikh mapping* $\psi_X : X^* \to \mathbb{N}^m$ associated to X is defined by

$$\psi_X(w) = (|w|_{x_1}, \ldots, |w|_{x_m}) \quad \text{for } w \in X^*.$$

The *growth matrix* of h is the $m \times s$-matrix A defined by

$$A_{ij} = |h(x_i)|_{y_j}, \quad i = 1, \ldots, m, \ j = 1, \ldots, s.$$

Next, assume that $w \in X^*$. Then we have

$$\psi_Y\big(h(w)\big) = \psi_X(w)A. \tag{1}$$

To prove (1), it suffices to observe that

$$|h(w)|_{y_j} = \sum_{i=1}^{m} |w|_{x_i} |h(x_i)|_{y_j} = \sum_{i=1}^{m} \big(\psi_X(w)\big)_i A_{ij} = \big(\psi_X(w)A\big)_j$$

for $j = 1, \ldots, s$. (Here, if v is a vector, then v_j is the jth component of v.)

Let now $G = (X, Y, h_1, \ldots, h_n, g, w)$ be an HDT0L system. Let $A_1, \ldots, A_n$ and A be the growth matrices of $h_1, \ldots, h_n$ and g, respectively, and let $\eta = (1, \ldots, 1)$ be the vector which has card(Y) entries all equal to 1. Then it follows by (1) that

$$|gh_{i_t} \ldots h_{i_1}(w)| = \psi_X(w) A_{i_1} \ldots A_{i_t} A \eta^T$$

for $t \geq 0$, $i_1, \ldots, i_t \in \mathbf{n}$. This implies the first claim of the following result. The second claim is a consequence of the definitions.

Theorem 1. *Let G be an HDT0L system. Then the length series $r(G)$ of G is $\mathbb{N}$-rational. Conversely, if r is an $\mathbb{N}$-rational series, then there is an HDT0L system G such that r is the length series of G.*

For the HD0L length sequences, we obtain a similar result. Recall that a sequence $(a(n))_{n\geq 0}$ is called $\mathbb{N}$-rational if the series $\sum_{n\geq 0} a(n)z^n \in A\langle\langle z^* \rangle\rangle$ is $\mathbb{N}$-rational.

Theorem 2. *Let G be an HD0L system. Then the length sequence of G is $\mathbb{N}$-rational. Conversely, if $(a(n))_{n\geq 0}$ is an $\mathbb{N}$-rational sequence, then there exists an HD0L system G such that $(a(n))_{n\geq 0}$ is the length sequence of G.*

The following theorem characterizes D0L length sequences among $\mathbb{N}$-rational sequences.

Theorem 3. *Let $(a(n))_{n\geq 0}$ be an $\mathbb{N}$-rational sequence. Then $(a(n))_{n\geq 0}$ is a D0L length sequence if and only if there exists a positive integer C such that*

$$a(n+1) \leq Ca(n)$$

holds for all $n \geq 0$.

We refer to [37, 30] for the proof of Theorem 3 and for other characterizations of D0L length sequences. For connections between DT0L systems and rational series, see also [29].

We discuss next a remarkable result due to Berstel and Nielsen concerning D0L length sequences [1]. We state the result in two ways.

Theorem 4. *It is decidable, given D0L length sequences $(s(n))_{n\geq 0}$ and $(t(n))_{n\geq 0}$ whether or not*

$$\{s(n) \mid n \geq 0\} = \{t(n) \mid n \geq 0\}.$$

For the second version of the Berstel–Nielsen theorem, call a mapping $\varphi : \mathbb{Z} \to \mathbb{Z}$ *piecewise affine* if there exist integers $a \geq 1$ and $u_j \geq 1$, v_j for $0 \leq j < a$ such that

$$\varphi(an+j) = u_j n + v_j$$

whenever $n \in \mathbb{Z}$ and $0 \leq j < a$.

Theorem 5. *Suppose $(s(n))_{n\geq 0}$ and $(t(n))_{n\geq 0}$ are D0L length sequences such that $\{s(n) \mid n \geq 0\}$ and $\{t(n) \mid n \geq 0\}$ are infinite sets. If*

$$\{s(n) \mid n \geq 0\} = \{t(n) \mid n \geq 0\},$$

then there is a piecewise affine mapping $\varphi : \mathbb{Z} \to \mathbb{Z}$ such that

$$s(n) = t(\varphi(n))$$

for almost all $n \geq 0$.

For generalizations of Theorem 5, see [36, 25].

Theorem 5 has been used in a very effective way by Ruohonen to obtain deep decidability results concerning D0L systems and their generalizations. As an example of Ruohonen's method, we will consider the equivalence problem between DF0L and D0L languages. By definition, a *DF0L system* is a construct $G = (\Sigma, h, F)$, where Σ is an alphabet, $h : \Sigma^* \to \Sigma^*$ is a morphism and $F \subseteq \Sigma^*$ is a finite nonempty set. The language $L(G)$ of G is defined by

$$L(G) = \{h^i(w) \mid i \geq 0,\ w \in F\}.$$

Let now $G = (X, g, \{v_0, \ldots, v_{p-1}\})$ be a DF0L system and let $H = (X, h, v)$ be a D0L system. We will assume that $L(G)$ and $L(H)$ are infinite and each letter of X occurs in every word of $L(G)$ and $L(H)$. First, construct a polynomial P of $\text{card}(X)$ variables with non-negative integer coefficients such that P induces an injective mapping $P : \mathbb{N}^{\text{card}(X)} \to \mathbb{N}$. If $\text{card}(X) = 2$ we may choose the polynomial

$$P_2(x, y) = (x + y)^2 + 3x + y,$$

where $\frac{1}{2}P_2$ is the Cantor pairing polynomial. If $P_n : \mathbb{N}^n \to \mathbb{N}$ is injective, then the polynomial

$$P_{n+1}(x_1, \ldots, x_{n+1}) = P_2\big(P_n(x_1, \ldots, x_n), x_{n+1}\big)$$

induces an injective mapping from $\mathbb{N}^{n+1}$ into $\mathbb{N}$.

Second, define the sequence $(s(n))_{n\geq 0}$ by

$$s(pi + j) = P\big(\psi(g^i(v_j))\big)$$

for $i \geq 0$, $j = 0, 1, \ldots, p - 1$, and define the sequence $(t(n))_{n\geq 0}$ by

$$t(n) = P\big(\psi(h^n(v))\big)$$

for $n \geq 0$.

Now $(s(n))_{n\geq 0}$ and $(t(n))_{n\geq 0}$ are D0L length sequences (for details, see [34]).

Next, decide whether or not $\{s(n) \mid n \geq 0\} = \{t(n) \mid n \geq 0\}$. If not, the injectivity of P implies that

$$\psi\big(L(G)\big) \neq \psi\big(L(H)\big)$$

and hence also $L(G) \neq L(H)$. Suppose that $\{s(n) \mid n \geq 0\} = \{t(n) \mid n \geq 0\}$. Then Theorem 5 implies that there exists a piecewise affine mapping $\varphi : \mathbb{Z} \to \mathbb{Z}$ and a non-negative integer n_0 such that

$$s(n) = t\big(\varphi(n)\big)$$

for $n \geq n_0$. It is clear that φ and n_0 can be computed effectively. Again, the injectivity of P implies that

$$\psi(g^i(v_j)) = \psi(h^{\varphi(pi+j)}(v))$$

for all $i \geq 0$, $j = 0, 1, \ldots, p-1$ such that $pi + j \geq n_0$. Because ψ is injective on $L(H)$, the equality $L(G) = L(H)$ holds only if

$$g^i(v_j) = h^{\varphi(pi+j)}(v)$$

for all $i \geq 0$, $j = 0, 1, \ldots, p-1$ such that $pi+j \geq n_0$. Hence, using Theorem 5, we have reduced the DF0L–D0L language equivalence problem to a finite number of instances of the D0L sequence equivalence problem.

Ruohonen has actually proved the following stronger result (see [34]).

Theorem 6. *The equivalence problem between F0L and D0L languages is decidable.*

Informally, F0L systems are obtained from DF0L systems by replacing the underlying morphism by a finite substitution.

For various other important decidability results obtained by Ruohonen, see [32, 33, 35]. Berstel–Nielsen theorem together with methods due to Ruohonen have also been used to prove Theorem 20 below and the following result from [21].

Theorem 7. *It is decidable whether or not*

$$L(H_1) = L(H_2),$$

if H_1 and H_2 are HD0L systems such that the length sequences of H_1 and H_2 are D0L length sequences.

3 Lindenmayerian Algebraic Series

In what follows S will always be a commutative semiring. Assume that Σ and Δ are finite alphabets. A mapping $h : S\langle\Sigma^*\rangle \to S\langle\Delta^*\rangle$ is called a *semi-algebra morphism* if $h(1) = 1$ and

$$\begin{aligned} h(p_1 + p_2) &= h(p_1) + h(p_2), \\ h(p_1 p_2) &= h(p_1)h(p_2), \\ h(ap_1) &= ah(p_1), \end{aligned}$$

for all $p_1, p_2 \in S\langle\Sigma^*\rangle$ and $a \in S$. A semi-algebra morphism $h : S\langle\Sigma^*\rangle \to S\langle\Delta^*\rangle$ is called *propagating* if we have

$$\varepsilon \notin \mathrm{supp}(h(\sigma))$$

for all $\sigma \in \Sigma$. A semi-algebra morphism $h : S\langle\Sigma^*\rangle \to S\langle\Delta^*\rangle$ is called a *monomial morphism* if for each $\sigma \in \Sigma$ there exist a non-zero $a \in S$ and a word $w \in \Delta^*$ such that $h(\sigma) = aw$.

If $h : S\langle \Sigma^* \rangle \to S\langle \Delta^* \rangle$ is a propagating semi-algebra morphism, h can be extended in a natural way to a mapping from $S\langle\langle \Sigma^* \rangle\rangle$ into $S\langle\langle \Delta^* \rangle\rangle$.

Suppose Σ and $Z = \{z_{11}, \ldots, z_{1s}, \ldots, z_{n1}, \ldots, z_{ns}\}$ are alphabets and $Y = \{y_1, \ldots, y_n\}$ is an alphabet of variables. It is assumed that the sets Σ, Z and Y are pairwise disjoint.

A *Lindenmayerian algebraic system* with variables in Y (briefly, an L algebraic system) is a system of equations

$$y_i = p_i\bigl(y_1, \ldots, y_n, h_{11}(y_1), \ldots, h_{1s}(y_1), \ldots, h_{n1}(y_n), \ldots, h_{ns}(y_n)\bigr),\quad 1 \le i \le n, \tag{2}$$

where $p_i(y_1, \ldots, y_n, z_{11}, \ldots, z_{1s}, \ldots, z_{n1}, \ldots, z_{ns})$ is a polynomial in $S\langle(\Sigma \cup Y \cup Z)^*\rangle$ and $h_{\alpha\beta} : S\langle \Sigma^* \rangle \to S\langle \Sigma^* \rangle$ is a semi-algebra morphism for $1 \le i, \alpha \le n$, $1 \le \beta \le s$. Here, we do not assume that each $z_{\alpha\beta}$ actually has an occurrence in p_i, $1 \le i, \alpha \le n$, $1 \le \beta \le s$.

If there is no danger of confusion, we use a vectorial notation. We write y for $y_1, \ldots, y_n$, p for $p_1, \ldots, p_n$, h for $h_{11}, \ldots, h_{1s}, \ldots, h_{n1}, \ldots, h_{ns}$, and z for $z_{11}, \ldots, z_{1s}, \ldots, z_{n1}, \ldots, z_{ns}$. Moreover, we write $h(y)$ for $h_{11}(y_1), \ldots, h_{1s}(y_1)$, $\ldots, h_{n1}(y_n), \ldots, h_{ns}(y_n)$. By this vectorial notation, an L algebraic system as defined above is now written as

$$y = p\bigl(y, h(y)\bigr). \tag{3}$$

Consider the L algebraic system $y = p(y, h(y))$ given by (3). The system is called a *propagating L algebraic system* (briefly, a PL algebraic system) if $h_{\alpha\beta}$ is propagating for all $1 \le \alpha \le n$, $1 \le \beta \le s$. Similarly, the system is called a *deterministic L algebraic system* (briefly, a DL algebraic system) if $h_{\alpha\beta}$ is a monomial morphism for all $1 \le \alpha \le n$, $1 \le \beta \le s$. L algebraic systems with only one equation play an important role. We will call such systems LS algebraic systems. Hence, the system given by (2) is an *LS algebraic system* (or an LS system) if $n = 1$. PLS and DLS systems are now defined in the natural way.

We also discuss L rational systems. By definition, the L algebraic system $y = p(y, h(y))$ given by (3) is an *L rational system* if each p_i is linear in $Y \cup Z$, $1 \le i \le n$. PL rational, DL rational, LS rational, PLS rational and DLS rational systems are defined in the natural way. Compared with the definition of a rational series our definition appears to be too general. We will justify the definition later.

Next, fix a convergence $\mathcal{D}$ in S and transfer $\mathcal{D}$ to $S\langle\langle \Sigma^* \rangle\rangle$ as explained in [28]. Unless stated otherwise, we assume that $\mathcal{D}$ is the discrete convergence.

Consider the L algebraic system (3). The *approximation sequence* $(r^j)_{j \ge 0}$ associated to (3) is defined by

$$r^0 = 0, \qquad r^{j+1} = p\bigl(r^j, h(r^j)\bigr), \quad j \ge 0.$$

Then if (r^j) converges with respect to $\mathcal{D}$,

$$\lim r^j$$

is called the *vector of series* generated by (3).

Now, a series $r \in S\langle\langle \Sigma^* \rangle\rangle$ is called *L algebraic* if there exists an L algebraic system $y = p(y, h(y))$ such that the associated approximation sequence (r^j) converges and r equals the first component of $\lim r^j$. The set of L algebraic series as defined above is denoted by $S^{\mathrm{Lalg}}\langle\langle \Sigma^* \rangle\rangle$. In what follows, Σ_∞ is a fixed countably infinite alphabet. We denote

$$S^{\mathrm{Lalg}}\{\{\Sigma_\infty^*\}\} = \bigcup_{\Sigma \subset \Sigma_\infty,\ \Sigma \text{ finite}} S^{\mathrm{Lalg}}\langle\langle \Sigma^* \rangle\rangle.$$

Next, the classes $S^{\mathrm{PLalg}}\langle\langle \Sigma^* \rangle\rangle$, $S^{\mathrm{DLalg}}\langle\langle \Sigma^* \rangle\rangle$, $S^{\mathrm{LSalg}}\langle\langle \Sigma^* \rangle\rangle$, $S^{\mathrm{Lrat}}\langle\langle \Sigma^* \rangle\rangle$, $S^{\mathrm{LSrat}}\langle\langle \Sigma^* \rangle\rangle$ and other similar classes are defined in the natural way. The power series in these classes are called *PL algebraic*, *DL algebraic*, *LS algebraic*, *L rational* and *LS rational*, respectively. LS algebraic series are also called LS series.

In the equations defining an L rational system, we can freely use words of Σ^*. This differs from the situation with rational series. The next result from [13] shows that our definition of L rational series is not too general.

Theorem 8. *Suppose $r \in S\langle\langle \Sigma^* \rangle\rangle$ is an L rational series. Then there exists an L rational system $y_i = p_i(y, h(y))$, $1 \le i \le n$, generating r such that*

$$\mathrm{supp}(p_i) \subseteq \Sigma_\infty^* \cup Z$$

for $1 \le i \le n$.

In the classical theory of Lindenmayer systems, the letter E, prefixed to the name of a family of languages, denotes the set of all languages that can be obtained by taking the languages of the family and intersecting them with Δ^*, for some alphabet Δ. Here, we define analogously

$$S^{\mathrm{ELalg}}\{\{\Sigma_\infty^*\}\} = \{r \mid r = s \odot \mathrm{char}(\Delta^*),\ s \in S^{\mathrm{Lalg}}\{\{\Sigma_\infty^*\}\},\ \Delta \subset \Sigma_\infty\}.$$

The classes $S^{\mathrm{ELSalg}}\{\{\Sigma_\infty^*\}\}$, $S^{\mathrm{ELrat}}\{\{\Sigma_\infty^*\}\}$ and $S^{\mathrm{ELSrat}}\{\{\Sigma_\infty^*\}\}$ are defined similarly. Power series in the classes $S^{\mathrm{ELalg}}\{\{\Sigma_\infty^*\}\}$, $S^{\mathrm{ELSalg}}\{\{\Sigma_\infty^*\}\}$, $S^{\mathrm{ELrat}}\{\{\Sigma_\infty^*\}\}$ and $S^{\mathrm{ELSrat}}\{\{\Sigma_\infty^*\}\}$ are called *EL algebraic*, *ELS algebraic*, *EL rational* and *ELS rational* power series, respectively. ELS algebraic series are also called ELS series.

Suppose $r \in S\langle\langle \Sigma^* \rangle\rangle$ and $\Delta \subseteq \Sigma$. Define the semi-algebra morphism $h : S\langle \Sigma^* \rangle \to S\langle \Sigma^* \rangle$ by $h(\sigma) = \sigma$ if $\sigma \in \Delta$ and $h(\sigma) = 0$ if $\sigma \notin \Delta$. Then $r \odot \mathrm{char}(\Delta^*) = h(r)$. This observation implies the following theorem.

Theorem 9. $S^{\mathrm{ELalg}}\{\{\Sigma_\infty^*\}\} = S^{\mathrm{Lalg}}\{\{\Sigma_\infty^*\}\}$.

On the other hand, if we use the E-mechanism, then we can restrict attention to L algebraic systems with only one equation.

Theorem 10. *Let $r \in S\langle\langle \Sigma^* \rangle\rangle$ be a proper L algebraic series. Then r is an ELS series.*

Proof. Let

$$y_i = p_i\big(y, h(y)\big), \quad 1 \le i \le n, \tag{4}$$

be an L algebraic system such that the associated approximation sequence $(r^j)_{j\ge 0}$ converges and

$$\lim r_1^j = r.$$

We will give the proof with the assumption that r_i^j is proper for all $j \ge 0$ and $1 \le i \le n$.

Let now $Z = \{z_{\alpha\beta\gamma} \mid 1 \le \alpha, \beta \le n, 1 \le \gamma \le s\}$ be an alphabet. Suppose that

$$p_i = p_i(z_{i11}, \ldots, z_{i1s}, \ldots, z_{in1}, \ldots, z_{ins}),$$

for $1 \le i \le n$ and suppose that (4) is given by

$$y_i = p_i\big(h_{i11}(y_1), \ldots, h_{i1s}(y_1), \ldots, h_{in1}(y_n), \ldots, h_{ins}(y_n)\big), \quad 1 \le i \le n.$$

It is clear that the assumption that (4) has this form involves no loss of generality.

Next, we assume that there exist disjoint alphabets $\Sigma_1, \Sigma_2, \ldots, \Sigma_n$ such that

$$\operatorname{supp}(r_i^j) \subseteq \Sigma_i^*$$

for $1 \le i \le n$. Furthermore, each $h_{\alpha\beta\gamma}$ maps each letter of $\bigcup_{i \ne \beta} \Sigma_i$ to 0. If these conditions do not hold initially, we proceed as follows. Let $\Sigma_1 = \Sigma$ and let $\Sigma_2, \ldots, \Sigma_n$ be new disjoint alphabets of the same cardinality as Σ and let $\mathrm{copy}_i : \Sigma^* \to \Sigma_i^*$ be an isomorphism for $1 \le i \le n$. (We take copy_1 to be the identity mapping.) Then each p_i, $2 \le i \le n$, is replaced by $\mathrm{copy}_i(p_i)$ and each $h_{\alpha\beta\gamma}$ by the extension of $\mathrm{copy}_\alpha \circ h_{\alpha\beta\gamma} \circ \mathrm{copy}_\beta^{-1}$ mapping each letter of $\bigcup_{i \ne \beta} \Sigma_i$ to 0.

Consider now the LS system

$$y = \sum_{i=1}^{n} p_i\big(h_{i11}(y), \ldots, h_{i1s}(y), \ldots, h_{in1}(y), \ldots, h_{ins}(y)\big). \tag{5}$$

Let $(s^j)_{j\ge 0}$ be the approximation sequence associated to (5). We claim that

$$s^j = r_1^j + \cdots + r_n^j \tag{6}$$

for $j \ge 0$. Clearly, (6) holds if $j = 0$. If (6) holds for $j \ge 0$, we have

$$\begin{aligned} s^{j+1} &= \sum_{i=1}^{n} p_i\big(h_{i11}(s^j), \ldots, h_{i1s}(s^j), \ldots, h_{in1}(s^j), \ldots, h_{ins}(s^j)\big) \\ &= \sum_{i=1}^{n} p_i\big(h_{i11}(r_1^j + \cdots + r_n^j), \ldots, h_{i1s}(r_1^j + \cdots + r_n^j), \end{aligned}$$

$$\ldots, h_{in1}(r_1^j + \cdots + r_n^j), \ldots, h_{ins}(r_1^j + \cdots + r_n^j))$$
$$= \sum_{i=1}^{n} p_i(h_{i11}(r_1^j), \ldots, h_{i1s}(r_1^j), \ldots, h_{in1}(r_n^j), \ldots, h_{ins}(r_n^j))$$
$$= \sum_{i=1}^{n} r_i^{j+1}.$$

Here, the third equation follows because the series r_i^j are proper. Hence, (6) is true for all $j \geq 0$. This implies that $\lim s^j$ exists and

$$\lim s^j = \sum_{i=1}^{n} \lim r_i^j.$$

Therefore,

$$r = \lim r_1^j = (\lim s^j) \odot \mathrm{char}(\Sigma_1^*)$$

is an ELS series. □

It can be shown in a similar way that proper L rational series are ELS rational series.

Next, we discuss briefly the fixed point properties of L algebraic series. (For more details, see [24, 13].) Suppose that the L algebraic system (2) generates the vector $r = (r_1, \ldots, r_n)$. Then r is called a *fixed point* of (2) if $h_{\alpha\beta}(r_\alpha)$ exists for all $1 \leq \alpha \leq n$, $1 \leq \beta \leq s$, and

$$r_i = p_i(r_1, \ldots, r_n, h_{11}(r_1), \ldots, h_{ns}(r_n))$$

for $1 \leq i \leq n$.

Theorem 11. *Let S be a continuous semiring and use the natural convergence in S. Let r be the vector generated by the L algebraic system (2). Then r is a fixed point of (2).*

Proof. The fixed point theorem given as Corollary 2.6 in Chap. 2 is applicable. □

Theorem 12. *Let (2) be a PL algebraic system generating the vector r. Then r is a fixed point of (2).*

We next consider the preservation of L algebraicness under various operations. Many results of this kind are obtained as corollaries of the next theorem.

Suppose $\Sigma = \{\sigma_1, \ldots, \sigma_m\}$ and consider an L algebraic system

$$y_i = p_i(\sigma_1, \ldots, \sigma_m, y_1, \ldots, y_n, h_{i11}(y_1), \ldots, h_{i1s}(y_1), \ldots, h_{in1}(y_n), \ldots, h_{ins}(y_n)), \quad 1 \leq i \leq n \tag{7}$$

where

$$p_i = p_i(\sigma_1, \ldots, \sigma_m, y_1, \ldots, y_n, z_{i11}, \ldots, z_{i1s}, \ldots, z_{in1}, \ldots, z_{ins})$$
$$\in S\langle(\Sigma \cup Y \cup Z)^*\rangle.$$

We say that (7) is a *strong PL system* if (7) is a PL system and for no i, $1 \leq i \leq n$, $\mathrm{supp}(p_i)$ contains a word in $\varepsilon \cup Y \cup Z$. A series $r \in A\langle\langle \Sigma^* \rangle\rangle$ is called a *strong PL series* if there is a strong PL system (7) such that r equals the first component of the vector generated by (7).

Theorem 13. *Suppose $\alpha : \Sigma^* \to S^{\mathrm{Lalg}}\langle\langle \Sigma^* \rangle\rangle$ is a mapping such that $\alpha(\sigma)$ is proper for all $\sigma \in \Sigma$. If (7) is a strong PL system then there exist L algebraic series $q_1, \ldots, q_n \in S\langle\langle \Sigma^* \rangle\rangle$ such that*

$$q_i = p_i\big(\alpha(\sigma_1), \ldots, \alpha(\sigma_m), q_1, \ldots, q_n, h_{i11}(q_1), \ldots, h_{i1s}(q_1),$$
$$\ldots, h_{in1}(q_n), \ldots, h_{ins}(q_n)\big), \quad 1 \leq i \leq n$$

holds true.

Proof. Let $(r^t)_{t\geq 0}$ be the approximation sequence associated to (7). Let $\alpha(\sigma_j)$ be the first component of the vector generated by the L algebraic system

$$y_{jk} = p_{jk}, \quad 1 \leq k \leq n_j, \tag{8}$$

$1 \leq j \leq m$. Denote by $(s^{jt})_{t\geq 0}$ the approximation sequence associated to (8). We again assume that s_1^{jt} is proper for $1 \leq j \leq m$, $t \geq 0$.

We now construct a new L algebraic system

$$y_i = p_i\big(y_{11}, \ldots, y_{m1}, y_1, \ldots, y_n, h_{i11}(y_1), \ldots, h_{i1s}(y_1),$$
$$\ldots, h_{in1}(y_n), \ldots, h_{ins}(y_n)\big), \quad 1 \leq i \leq n,$$
$$y_{1k} = p_{1k}, \quad 1 \leq k \leq n_1,$$
$$\vdots$$
$$y_{mk} = p_{mk}, \quad 1 \leq k \leq n_m,$$

with variables in $\{y_i \mid 1 \leq i \leq n\} \cup \{y_{ij} \mid 1 \leq i \leq m, 1 \leq j \leq n_i\}$. Let the approximation sequence associated to this L algebraic system be $(q^t)_{t\geq 0}$ where q^t has components

$$q_1^t, \ldots, q_n^t, q_{11}^t, \ldots, q_{1n_1}^t, \ldots, q_{m1}^t, \ldots, q_{mn_m}^t.$$

By the definition of (q^t), the sequences (q_{ij}^t), $1 \leq i \leq m$, $1 \leq j \leq n_i$, converge. We claim that also (q_i^t) converges if $1 \leq i \leq n$. To prove this, it is enough to show that there exists a sequence of integers $(\beta_t)_{t\geq 0}$ such that

$$\big(q_i^{\beta}, w\big) = \big(q_i^{\beta_t}, w\big) \quad \text{for } 1 \leq i \leq n, \tag{9}$$

whenever $|w| \le t$ and $\beta \ge \beta_t$ for $t \ge 0$. For $t = 0$, we may choose $\beta_0 = 0$ because q_i^t is proper for $1 \le i \le n$, $t \ge 0$. Assume then that (9) holds for $t \ge 0$ whenever $|w| \le t$ and $\beta \ge \beta_t$. Because the sequences (q_{ij}^t) converge, there is a positive integer γ_0 such that

$$(q_{ij}^{\gamma}, w) = (q_{ij}^{\gamma_0}, w) \quad \text{for } 1 \le i \le m,\ 1 \le j \le n_i$$

whenever $|w| \le t + 1$ and $\gamma \ge \gamma_0$. Because the system (7) is a strong PL system, it follows from

$$\begin{aligned} q_i^{\gamma+1} = p_i\big(&q_{11}^{\gamma}, \ldots, q_{m1}^{\gamma}, q_1^{\gamma}, \ldots, q_n^{\gamma}, h_{i11}(q_1^{\gamma}), \ldots, h_{i1s}(q_1^{\gamma}), \\ &\ldots, h_{in1}(q_n^{\gamma}), \ldots, h_{ins}(q_n^{\gamma})\big), \end{aligned} \tag{10}$$

that we may take

$$\beta_{t+1} = \max\{\beta_t, \gamma_0\} + 1.$$

This proves the existence of β_t for any $t \ge 0$ and shows that (q_i^t) converges for $1 \le i \le n$. Therefore, also (q^t) converges. Denote

$$\lim q^t = (q_1, \ldots, q_n, q_{11}, \ldots, q_{1n_1}, \ldots, q_{m1}, \ldots, q_{mn_m}).$$

Now, because we are working with the discrete convergence, (10) implies that

$$\begin{aligned} q_i &= p_i\big(q_{11}, \ldots, q_{m1}, q_1, \ldots, q_n, h_{i11}(q_1), \ldots, h_{i1s}(q_1), \\ &\qquad \ldots, h_{in1}(q_n), \ldots, h_{ins}(q_n)\big) \\ &= p_i\big(\alpha(\sigma_1), \ldots, \alpha(\sigma_m), q_1, \ldots, q_n, h_{i11}(q_1), \ldots, h_{i1s}(q_1), \\ &\qquad \ldots, h_{in1}(q_n), \ldots, h_{ins}(q_n)\big), \quad 1 \le i \le n, \end{aligned}$$

where $q_1, \ldots, q_n$ are L algebraic series. This concludes the proof. □

The following theorems are consequences of Theorem 13; see [13] for details.

Theorem 14. *The classes $S^{\mathrm{Lalg}}\langle\langle \Sigma^* \rangle\rangle$ and $S^{\mathrm{PLalg}}\langle\langle \Sigma^* \rangle\rangle$ are rationally closed (i.e., closed under sum, product and quasi-inverse of a proper series).*

Theorem 15. *Suppose $\alpha : \Sigma^* \to S^{\mathrm{Lalg}}\langle\langle \Sigma^* \rangle\rangle$ is a morphism such that $\alpha(\sigma)$ is proper for all $\sigma \in \Sigma$. If $r \in S^{\mathrm{alg}}\langle\langle \Sigma^* \rangle\rangle$ then $\alpha(r) \in S^{\mathrm{Lalg}}\langle\langle \Sigma^* \rangle\rangle$.*

4 D0L Power Series

In this section, S will again be a commutative semiring.

The simplest class of L algebraic series consists of D0L power series. By definition, a power series $r \in S\langle\langle X^* \rangle\rangle$ is a *D0L power series* if r is generated by a deterministic LS system

$$y = aw + h(y),$$

where $a \in S$ is non-zero, $w \in X^*$ and $h : S\langle X^* \rangle \to S\langle X^* \rangle$ is a monomial morphism. Hence, if $r \in S\langle\langle X^* \rangle\rangle$ is a D0L *power series*, there exist a non-zero $a \in S$, a word $w \in X^*$ and a monomial morphism $h : S\langle X^* \rangle \to S\langle X^* \rangle$ such that

$$r = \sum_{n=0}^{\infty} ah^n(w). \tag{11}$$

To exclude trivial cases, we will require that in (11)

$$\text{supp}(ah^i(w)) \neq \text{supp}(ah^j(w))$$

whenever $0 \leq i < j$.

Consider the series r given by (11) and denote

$$ah^n(w) = c_n w_n,$$

where $c_n \in S$ and $w_n \in X^*$ for $n \geq 0$. Then we have

$$r = \sum_{n=0}^{\infty} c_n w_n. \tag{12}$$

In what follows, the right-hand side of (12) is called the *normal form* of r. A sequence $(c_n)_{n \geq 0}$ of elements of S is called a *D0L multiplicity sequence* over S if there exists a D0L power series r such that (12) is the normal form of r.

It is easy to see that the characteristic series of an infinite D0L language is a D0L power series. Conversely, the support of a D0L power series is a D0L language.

Next, we characterize D0L multiplicity sequences over commutative semirings.

By definition, a sequence $(a_n)_{n \geq 0}$ of non-negative integers is a *modified PD0L length sequence* if there exists a non-negative integer t such that $a_0 = a_1 = \cdots = a_{t-1} = 0$ and $(a_{n+t})_{n \geq 0}$ is a PD0L length sequence. A sequence $(a_n)_{n \geq 0}$ of non-negative integers is a modified PD0L length sequence if and only if the sequence $(a_{n+1} - a_n)_{n \geq 0}$ is $\mathbb{N}$-rational (see [30]).

If $h : S\langle X^* \rangle \to S\langle Y^* \rangle$ is a monomial morphism, the *underlying monoid morphism* $g : X^* \to Y^*$ is defined by $g(x) = \text{supp}(h(x))$ for $x \in X$.

Theorem 16. *A sequence $(c_n)_{n \geq 0}$ of non-zero elements of S is a D0L multiplicity sequence over S if and only if there exist a positive integer k, non-zero $a_1, \ldots, a_k \in S$ and modified PD0L length sequences $(s_{in})_{n \geq 0}$ for $1 \leq i \leq k$ such that*

$$c_n = \prod_{i=1}^{k} a_i^{s_{in}} \tag{13}$$

for all $n \geq 0$.

Proof. Suppose first that $r = \sum_{n=0}^{\infty} ah^n(w)$ is a D0L power series over S with the normal form

$$r = \sum_{n=0}^{\infty} c_n w_n.$$

Without loss of generality, we assume that $a = c_0 = 1$. Let $g : X^* \to X^*$ be the underlying monoid morphism of the monomial morphism $h : S\langle X^* \rangle \to S\langle X^* \rangle$. Then we have $g^n(w_0) = w_n$ for all $n \geq 0$. Let $\overline{X} = \{\overline{x} \mid x \in X\}$ be a new alphabet with the same cardinality as X. Define the monoid morphism $g_1 : (X \cup \overline{X})^* \to (X \cup \overline{X})^*$ by

$$g_1(x) = \overline{x}g(x), \quad g_1(\overline{x}) = \varepsilon, \quad x \in X.$$

For each $x \in X$, let $a_x \in S$ be such that $h(x) = a_x g(x)$. Define the semi-algebra morphism $\alpha : S\langle (X \cup \overline{X})^* \rangle \to S$ by

$$\alpha(x) = 1, \quad \alpha(\overline{x}) = a_x, \quad x \in X.$$

Then we have

$$h(u) = \alpha\big(g_1(u)\big)g(u) \tag{14}$$

and

$$g_1\big(g(u)\big) = g_1^2(u) \tag{15}$$

for any word $u \in X^*$. Equation (15) implies inductively that

$$g_1\big(g^n(u)\big) = g_1^{n+1}(u) \tag{16}$$

for any $n \geq 1$ and $u \in X^*$. We claim that

$$c_n = \alpha\big(w_0 g_1(w_0) g_1^2(w_0) \ldots g_1^n(w_0)\big) \tag{17}$$

for all $n \geq 0$.

The claim is trivially true for $n = 0$. If the claim holds for $n = k$, we have by (14), (16), and (17)

$$\begin{aligned} h^{k+1}(w_0) &= h(c_k w_k) = c_k h(w_k) = c_k \alpha\big(g_1(w_k)\big)g(w_k) \\ &= c_k \alpha\big(g_1\big(g^k(w_0)\big)\big)w_{k+1} = c_k \alpha\big(g_1^{k+1}(w_0)\big)w_{k+1} \\ &= \alpha\big(w_0 g_1(w_0) g_1^2(w_0) \ldots g_1^{k+1}(w_0)\big)w_{k+1}, \end{aligned}$$

which implies the claim for $n = k + 1$, and hence for all $n \geq 0$.

Next, for each $x \in X$, define the sequence $(s(x)_n)_{n \geq 0}$ by

$$s(x)_n = \big|w_0 g_1(w_0) g_1^2(w_0) \ldots g_1^n(w_0)\big|_{\overline{x}}.$$

Because

$$s(x)_{n+1} - s(x)_n = \big|g_1^{n+1}(w_0)\big|_{\overline{x}},$$

Theorem 2 implies that the sequence $(s(x)_{n+1} - s(x)_n)_{n\geq 0}$ is $\mathbb{N}$-rational for all $x \in X$. Hence, the sequences $(s(x)_n)_{n\geq 0}$ are modified PD0L length sequences. By (17), we have

$$c_n = \prod_{x\in X} a_x^{s(x)_n}$$

for all $n \geq 0$. This concludes the proof in one direction.

Suppose then that there exist a positive integer k, non-zero $a_1, \ldots, a_k \in S$ and modified PD0L length sequences $(s_{in})_{n\geq 0}$ for $1 \leq i \leq k$ such that (13) holds for all $n \geq 0$. We have to show that $(c_n)_{n\geq 0}$ is a D0L multiplicity sequence over S. Because D0L multiplicity sequences over S are closed under finite product provided that no term of the product sequence is zero, it suffices to consider the case $k = 1$. Denote $a = a_1$ and $s_n = s_{1n}$ for $n \geq 0$. Without restriction, we suppose that (s_n) is a PD0L length sequence. If the set $\{s_n \mid n \geq 0\}$ is finite, $(c_n)_{n\geq 0}$ is clearly a D0L multiplicity sequence over S. Suppose therefore that $\{s_n \mid n \geq 0\}$ is an infinite set and let $G = (\Sigma, f, w_0)$ be a PD0L system generating the sequence $S(G) = (w_n)_{n\geq 0}$ with $|w_n| = s_n$ for $n \geq 0$. Define the monomial morphism $h : S\langle\Sigma^*\rangle \to S\langle\Sigma^*\rangle$ by

$$h(\sigma) = a^{|f(\sigma)|-1} f(\sigma)$$

for $\sigma \in \Sigma$. It follows inductively that

$$a^{s_0} h^n(w_0) = a^{s_n} w_n$$

for $n \geq 0$. Hence, the series r defined by

$$r = \sum_{n=0}^{\infty} a^{s_0} h^n(w_0)$$

is a D0L power series over S and the sequence $(c_n)_{n\geq 0} = (a^{s_n})_{n\geq 0}$ is a D0L multiplicity sequence over S. □

The following two results are consequences of Theorem 16. For details, see [12].

Theorem 17. *Suppose S is a field. A sequence $(c_n)_{n\geq 0}$ of elements of S is a D0L multiplicity sequence over S if and only if there exist a positive integer k, nonzero $a_1, \ldots, a_k \in S$, and $\mathbb{Z}$-rational sequences $(s_{in})_{n\geq 0}$ for $1 \leq i \leq k$ such that*

$$c_n = \prod_{i=1}^{k} a_i^{s_{in}}$$

for all $n \geq 0$.

Theorem 18. *Suppose S is a field. A sequence $(c_n)_{n\geq 0}$ of non-zero elements of S is a D0L multiplicity sequence over S if and only if there exist a positive integer t and integers $\beta_1, \ldots, \beta_t$ such that*

$$c_{n+t} = c_{n+t-1}^{\beta_1} c_{n+t-2}^{\beta_2} \cdots c_n^{\beta_t} \tag{18}$$

for $n \geq 0$.

Theorem 18 is used in [12] to prove the following result.

Theorem 19. *Suppose S is a computable field. It is decidable whether or not two given D0L power series over S are equal.*

Theorem 19 has been extended for DF0L power series in [17]. By definition, a power series $r \in S\langle\langle X^* \rangle\rangle$ is a *DF0L power series* over S if there exists a polynomial $a_1 v_1 + \cdots + a_k v_k \in S\langle X^* \rangle$ where $a_j \in S$, $v_j \in X^*$, $1 \leq j \leq k$, and a monomial morphism $h : S\langle X^* \rangle \to S\langle X^* \rangle$ such that

$$r = \sum_{n=0}^{\infty} h^n(a_1 v_1 + \cdots + a_k v_k)$$

and, furthermore, the series

$$r_j = \sum_{n=0}^{\infty} a_j h^n(v_j)$$

are D0L power series over S for $1 \leq j \leq k$.

Theorem 20. *Suppose S is a computable field. It is decidable whether or not two given DF0L power series over S are equal.*

The proof of Theorem 20 uses the Berstel–Nielsen theorem, methods developed by Ruohonen, elementary morphisms, and somewhat lengthy arguments concerning periodicity properties of free monoid morphisms. Theorem 20 implies that language equivalence is decidable for DF0L systems.

The following results are from [14, 10].

Theorem 21. *It is decidable whether or not a given D0L power series over $\mathbb{Q}$ is $\mathbb{Q}$-rational.*

Theorem 22. *It is decidable whether or not a given D0L power series over $\mathbb{Q}_+$ is $\mathbb{Q}_+$-algebraic.*

No algorithm is known for deciding whether or not a given D0L power series over $\mathbb{Q}$ is $\mathbb{Q}$-algebraic. However, the problem is known to be decidable. An algorithm is obtained if it is known for which values of k the series

$$P_k = \sum_{n=1}^{\infty} x_1^n x_2^n \dots x_k^n$$

is $\mathbb{Q}$-algebraic. (Here, $x_1, \dots, x_k$ are distinct letters.)

5 Other Power Series Generalizations of L Systems

In the previous section, we defined D0L power series. The framework of L algebraic series can be used in a similar way to define power series generalizations of other L systems.

The definition of 0L power series results if we do not require determinism in the definition of D0L power series. More specifically, a power series $r \in S\langle\langle \Sigma^* \rangle\rangle$ is a *0L power series* if r is generated by an LS system

$$y = aw + h(y), \tag{19}$$

where now $a \in S$, $w \in \Sigma^*$ and $h : S\langle \Sigma^* \rangle \to S\langle \Sigma^* \rangle$ is a semi-algebra morphism.

Proceeding in another direction, a series $r \in S\langle\langle \Sigma^* \rangle\rangle$ is a *DT0L power series* if there exists a deterministic LS system

$$y = aw + h_1(y) + \cdots + h_n(y) \tag{20}$$

such that r is generated by (20). In (20), $a \in S$ is non-zero, $w \in \Sigma^*$ and $h_1, \ldots, h_n : S\langle \Sigma^* \rangle \to S\langle \Sigma^* \rangle$ are monomial morphisms.

Similarly, a power series $r \in S\langle\langle \Sigma^* \rangle\rangle$ is a *T0L power series* if r is generated by an LS system having the form (20), where now $a \in S$, $w \in \Sigma^*$ and $h_1, \ldots, h_n : S\langle \Sigma^* \rangle \to S\langle \Sigma^* \rangle$ are semi-algebra morphisms.

In the theory of L systems the letters C, E, F, H, and P are used in a standard way denoting codings, intersection with terminal alphabets, finite sets of axioms, homomorphisms, and propagating systems, respectively. These definitions extend to power series in a natural way. The use of the letter P simply means that the considered morphisms are assumed to be propagating. If $X \in \{\varepsilon, D, T, DT\}$, a series $r \in S\langle\langle \Sigma^* \rangle\rangle$ is a *CX0L power series* (resp. an *HX0L power series*) if there exist a X0L power series $s \in S\langle\langle \Delta^* \rangle\rangle$ and a coding (resp. monomial morphism) $\alpha : S\langle \Delta^* \rangle \to S\langle \Sigma^* \rangle$ such that $r = \alpha(s)$. Here, a semi-algebra morphism $\alpha : S\langle \Delta^* \rangle \to S\langle \Sigma^* \rangle$ is called a *coding* if for each $x \in \Delta$ there exist a non-zero $a \in S$ and a letter $\sigma \in \Sigma$ such that

$$\alpha(x) = a\sigma.$$

Similarly, a series $r \in S\langle\langle \Sigma^* \rangle\rangle$ is an *EX0L power series* if there exist a X0L power series $s \in S\langle\langle \Delta^* \rangle\rangle$ and a subset $\Delta_1 \subseteq \Delta$ such that

$$r = s \odot \mathrm{char}(\Delta_1^*).$$

We next discuss some results concerning DT0L power series.

Suppose that the DT0L power series r is generated by (20). Then

$$r = \sum_{u \in \mathbf{n}^*} S(u)$$

where, if $u = i_1 i_2 \ldots i_k$ ($i_\alpha \in \mathbf{n}$),

$$S(u) = h_{i_1} h_{i_2} \dots h_{i_k}(aw).$$

For $u \in \mathbf{n}^*$, write

$$S(u) = c(u)s(u)$$

where $c(u) \in S$ and $s(u) \in \Sigma^*$. The mapping $c : \mathbf{n}^* \to S$ is called the *DT0L multiplicity sequence* of r. A mapping $c : \mathbf{n}^* \to S$ is called a DT0L multiplicity sequence over S if there is a DT0L power series r over S such that c equals the multiplicity sequence of r.

Theorem 16 has an analogue for DT0L multiplicity sequences.

Theorem 23. *Suppose $n \geq 2$. A mapping $c : \mathbf{n}^* \to \mathbb{Q}$ is a DT0L multiplicity sequence over $\mathbb{Q}$ if and only if there exist a positive integer k, positive primes $p_1, \dots, p_k$ and $\mathbb{Z}$-rational series $s_0, s_1, \dots, s_k \in \mathbb{Z}\langle\langle \mathbf{n}^* \rangle\rangle$ such that*

$$c(u) = (-1)^{s_0(u)} \prod_{i=1}^{k} p_i^{s_i(u)}$$

for all $u \in \mathbf{n}^*$.

The next result can be proved by using Theorem 23 (see [9]).

Theorem 24. *Suppose $B \subseteq \mathbb{N}$ is a recursively enumerable set. Then there exists a DT0L power series r over $\mathbb{Q}$ such that*

$$\mathrm{Im}(r) \cap \{x \mid x \geq 1\} = \{2^a \mid a \in B\}.$$

The following result is a direct consequence of Theorem 24.

Theorem 25. *It is undecidable whether or not a given DT0L power series over $\mathbb{Q}$ has coefficient* 1.

Next we discuss a power series generalization of the classical *E0L = C0L* theorem (see [2, 30]). Suppose

$$r = \sum_{n=0}^{\infty} ag^n(w) \odot \mathrm{char}(\Delta^*)$$

is an E0L power series, where $g : S\langle \Sigma^* \rangle \to S\langle \Sigma^* \rangle$ is a semi-algebra morphism, $a \in S$, $w \in \Sigma^*$ and $\Delta \subseteq \Sigma$. Then we say that r satisfies the *ε-condition* if

$$(g(c), \varepsilon) = (g^n(c), \varepsilon)$$

for all $n \geq 1$, $c \in \Sigma$.

Theorem 26. *If $r \in S\langle\langle \Delta^* \rangle\rangle$ is a proper E0L power series which satisfies the ε-condition, then r is a C0L power series.*

The necessity of the ε-condition in the above theorem is an open problem. The remaining part of the $E0L = C0L$ theorem generalizes without additional assumptions.

Theorem 27. *If $r \in S\langle\langle \Delta^* \rangle\rangle$ is a proper C0L power series, then r is an E0L power series.*

For the proofs of Theorems 26 and 27, we refer to [16].

Our final theorem shows that ET0L power series are very closely related to L rational series.

Theorem 28. *Suppose $r \in S\langle\langle \Sigma^* \rangle\rangle$ and choose a new letter $\# \notin \Sigma$. Then $\#r$ is an ET0L power series if and only if $\#r$ is L rational.*

For this result and other ways to define ET0L power series, see [13, 26, 23, 24].

Above we have indicated some open problems concerning L algebraic series and L systems. A very important open problem is the HD0L language equivalence problem. A special case of the problem is solved by Theorem 7, but the general case remains open.

References

1. J. Berstel and M. Nielsen. The growth range equivalence problem for D0L systems is decidable. In A. Lindenmayer and G. Rozenberg, editors, *Automata, Languages, Development*, pages 161–178. North-Holland, Amsterdam, 1976.
2. A. Ehrenfeucht and G. Rozenberg. The equality of E0L languages and codings of 0L languages. *International Journal of Computer Mathematics*, 4:95–104, 1974.
3. Z. Ésik and W. Kuich. A Kleene theorem for Lindenmayerian algebraic power series. *Journal of Automata, Languages and Combinatorics*, 5:109–122, 2000.
4. J. Honkala. On Lindenmayerian series in complete semirings. In G. Rozenberg and A. Salomaa, editors, *Developments in Language Theory*, pages 179–192. World Scientific, Singapore, 1994.
5. J. Honkala. An iteration property of Lindenmayerian power series. In J. Karhumäki, H. Maurer, and G. Rozenberg, editors, *Results and Trends in Theoretical Computer Science*, pages 159–168. Springer, Berlin, 1994.
6. J. Honkala. On morphically generated formal power series. *Theoretical Informatics and Applications, RAIRO*, 29:105–127, 1995.
7. J. Honkala. On the decidability of some equivalence problems for L algebraic series. *International Journal of Algebra and Computation*, 7:339–351, 1997.

8. J. Honkala. On Lindenmayerian algebraic sequences. *Theoretical Computer Science*, 183:143–154, 1997.
9. J. Honkala. Decision problems concerning a power series generalization of DT0L systems. *Fundamenta Informaticae*, 32:341–348, 1997.
10. J. Honkala. On algebraicness of D0L power series. *Journal of Universal Computer Science*, 5:11–19, 1999.
11. J. Honkala. The equivalence problem of D0L and DF0L power series. *Fundamenta Informaticae*, 38:201–208, 1999.
12. J. Honkala. On sequences defined by D0L power series. *Theoretical Informatics and Applications, RAIRO*, 33:125–132, 1999.
13. J. Honkala. On formal power series generated by Lindenmayer systems. *Journal of Automata, Languages and Combinatorics*, 5:123–144, 2000.
14. J. Honkala. On D0L power series. *Theoretical Computer Science*, 244:117–134, 2000.
15. J. Honkala. A Kleene–Schützenberger theorem for Lindenmayerian rational power series. *Theoretical Informatics and Applications, RAIRO*, 34:297–305, 2000.
16. J. Honkala. Results concerning E0L and C0L power series. *Acta Cybernetica*, 14:597–605, 2000.
17. J. Honkala. The equivalence problem for DF0L languages and power series. *Journal of Computer and System Sciences*, 65:377–392, 2002.
18. J. Honkala. On the simplification of HD0L power series. *Journal of Universal Computer Science*, 8:1040–1046, 2002.
19. J. Honkala. On images of D0L and DT0L power series. *Theoretical Computer Science*, 290:1869–1882, 2003.
20. J. Honkala. On D0L power series over various semirings. In C. Martin-Vide and V. Mitrana, editors, *Grammars and Automata for String Processing: From Mathematics and Computer Science to Biology, and Back*, pages 263–273. Taylor & Francis, London, 2003.
21. J. Honkala. The language equivalence problem for HD0L systems having D0L growths. *Theoretical Computer Science*, 330:123–133, 2005.
22. J. Honkala and W. Kuich. On four classes of Lindenmayerian power series. *Journal of Universal Computer Science*, 1:131–135, 1995.
23. J. Honkala and W. Kuich. On a power series generalization of ET0L languages. *Fundamenta Informaticae*, 25:257–270, 1996.
24. J. Honkala and W. Kuich. On Lindenmayerian algebraic power series. *Theoretical Computer Science*, 183:113–142, 1997.
25. J. Honkala and K. Ruohonen. On the images of ℕ-rational sequences counting multiplicities. *International Journal of Algebra and Computation*, 13:303–321, 2003.
26. W. Kuich. Lindenmayer systems generalized to formal power series and their growth functions. In G. Rozenberg and A. Salomaa, editors, *Developments in Language Theory*, pages 171–178. World Scientific, Singapore, 1994.

27. W. Kuich. Generalized Lindenmayerian algebraic systems. In G. Păun and A. Salomaa, editors, *New Trends in Formal Languages*, pages 412–421. Springer, Berlin, 1997.
28. W. Kuich and A. Salomaa. *Semirings, Automata, Languages.* Springer, Berlin, 1986.
29. C. Reutenauer. Sur les séries associées à certains systèmes de Lindenmayer. *Theoretical Computer Science*, 9:363–375, 1979.
30. G. Rozenberg and A. Salomaa. *The Mathematical Theory of L Systems.* Academic Press, New York, 1980
31. G. Rozenberg and A. Salomaa, editors. *Handbook of Formal Languages, volumes 1–3.* Springer, Berlin, 1997.
32. K. Ruohonen. On the decidability of the 0L–D0L equivalence problem. *Information and Control*, 40:301–318, 1979.
33. K. Ruohonen. The inclusion problem for D0L languages. *Elektronische Informationsverarbeitung und Kybernetik*, 15:535–548, 1979.
34. K. Ruohonen. The decidability of the F0L–D0L equivalence problem. *Information Processing Letters*, 8:257–260, 1979.
35. K. Ruohonen. The decidability of the D0L–DT0L equivalence problem. *Journal of Computer and System Sciences*, 22:42–52, 1981.
36. K. Ruohonen. On a variant of a method of Berstel's and Nielsen's. *Fundamenta Informaticae*, 4:369–400, 1981.
37. A. Salomaa and M. Soittola. *Automata-Theoretic Aspects of Formal Power Series.* Springer, Berlin, 1978.

Chapter 9: Weighted Tree Automata and Tree Transducers

Zoltán Fülöp[1,*] and Heiko Vogler[2]

[1] Department of Computer Science, University of Szeged,
Árpád tér 2., 6720 Szeged, Hungary
fulop@inf.u-szeged.hu
[2] Faculty of Computer Science, Technische Universität Dresden,
01062 Dresden, Germany
vogler@tcs.inf.tu-dresden.de

Dedicated to Symeon Bozapalidis

* Research of the first author was supported by the Hungarian Scientific Research Fund, grant T 46686, and by the Austrian–Hungarian Action Fund, grant 68öu2.

M. Droste, W. Kuich, H. Vogler (eds.), *Handbook of Weighted Automata*, Monographs in Theoretical Computer Science. An EATCS Series,
DOI 10.1007/978-3-642-01492-5_9, © Springer-Verlag Berlin Heidelberg 2009

1 Introduction

Over the past four decades, the theory of finite state tree automata and tree transducers has been developed intensively (cf. [68, 69, 29, 64] for a survey). This classical theory deals with (formal) tree languages and with relations over trees; it contains, e.g., characterizations of the class of recognizable tree languages and composition results for certain classes of tree transformations. Tree automata and tree transducers have also proved useful as formal models for analyzing and transforming trees in applications like natural language processing [85, 114, 72, 83, 84], syntax-directed semantics [79, 51, 64, 34], picture generation [35], or the processing of semi-structured documents [136, 117, 126, 54, 111].

Now it is natural to generalize tree automata and tree transducers by changing from the qualitative point of view to a quantitative one. For instance, besides knowing that a pattern occurs in a tree, one might want to know also the number of such occurrences. Another example is that we would like to know the probability of the event that an output tree is the translation of an input tree. Then a tree language becomes a mapping, called a tree series, from the set of trees to a set S of quantities. Similarly, a tree transformation is turned into a mapping, called a tree series transformation, from trees into tree series over S. This extension results in the formal models of *weighted tree automaton* and *weighted tree transducer*, respectively. In order to be able to calculate with quantities, an algebraic structure is needed; and it has turned out that semirings are the most appropriate ones for this purpose. Then for an arbitrary run on an input tree, the weights of the involved transitions are combined by using the semiring multiplication and, if there exist several runs on a tree (which, in the case of transducers, lead to the same translation), then the semiring addition is applied to the weights of all these runs. The classical *unweighted case* is reobtained by considering the Boolean semiring $\mathbb{B}$ with disjunction and conjunction as addition and multiplication, respectively. In fact, for string automata the quantitative point of view has been investigated since the 1960s of the previous century and it led to the rich theory of formal power series [45, 125, 96, 8, 87].

Weighted tree automata have been studied quite intensively by now. The approaches in the studies differ in the class of semirings they employ, e.g., completely distributive lattices [78, 59], fields [7], commutative semirings [1], and continuous semirings [88, 58]. Every such class has its own benefits: Using lattices, a bridge to fuzzy sets and concepts is built; taking fields, the tools and results of linear algebra are available; using commutative semirings, more general results can be proved; continuous semirings allow for the solution of systems of linear equations which is a fundamental concept. Using the semiring of real numbers (with the usual addition and multiplication), probabilistic tree automata [100, 47] can be defined; such automata associate with every transition a weight in the interval $[0, 1]$, and the weights of all possible transitions in a state on a symbol sum up to 1. The investigation into weighted tree transducers was started in [91] and continued in [53, 65, 61, 66, 101, 104] and others. Results regarding composition, decomposition, and hierarchies were lifted from the unweighted to the weighted case. As for weighted tree automata, this lifting had to be done with much care because properties of $\mathbb{B}$ (like idempotency, finiteness, commutativity), which are used quite often in the unweighted case, are now gone. In fact, this makes the weighted case interesting. For a survey on some results on weighted tree automata and weighted tree transducers, we refer to [88, 58].

In this chapter of the *Handbook of Weighted Automata*, we have collected some important results for weighted tree automata and weighted tree transducers. We restrict ourselves to finite trees and we consider only ranked trees (in contrast to unranked trees such as those used to model fully structured XML-documents). In particular, we address closure properties of the class of recognizable tree series, results on the support of such tree series, the determinization of weighted tree automata, pumping lemmata and decidability results, and finite algebraic characterizations of recognizable tree series. We discuss the equivalence between recognizable tree series and equational, rational, and MSO-definable tree series, and we present a comparison of several other models of recognizability. The part on weighted tree automata ends with a list of further results which we will not discuss in detail. For weighted tree transducers, we show composition and decomposition results, an inclusion diagram of some fundamental classes of tree series transformations, and hierarchies obtained by composing weighted tree transducers. We briefly discuss other models of weighted tree transducers. Finally, we give a short list of further results on weighted tree transducers that are not addressed in our main sections.

We have tried to produce a self-contained chapter; thus, the reader who has some background in automata theory and formal languages can easily follow the development. For many theorems, we have included sketches of their proofs, and we have always indicated the original source where the reader can sometimes find more details.

The different topics which we address require different additional properties for the used semiring, e.g., commutativity, zero-divisor freeness, or that the semiring is a semifield. In order to avoid repetitions of the respective list of additional properties during the development of a topic, we adopt the following convention: We will place, if appropriate, at the beginning of a section or subsection a general statement about the additional properties which we assume to hold throughout that section or subsection, and we do *not* explicitly mention these assumptions in the individual statements.

2 Preliminaries

2.1 General Notation

Let $\mathbb{N}$ denote the set $\{0, 1, 2, \dots\}$ of natural numbers. For a set A, we denote its set of subsets by $\mathcal{P}(A)$ and the set of strings over A by A^*. The empty string is denoted by ε and the length of a string w by $|w|$. We denote the cardinality of a finite set A by $|A|$.

Let H, I, and J be sets. An *$I \times J$ matrix over H* is a mapping $\mathcal{M} : I \times J \to H$; the set of all $I \times J$ matrices over H is denoted by $H^{I \times J}$. We write an entry $\mathcal{M}(i, j) \in H$ as $\mathcal{M}_{i,j}$. An *I-vector v over H* is defined analogously; the set of all I-vectors over H is denoted by H^I and an element $v(i) \in H$ of v is denoted by v_i.

For two functions $f : A \to B$ and $g : B \to C$, we denote their composition by $g \circ f$ where $(g \circ f)(a) = g(f(a))$ for every $a \in A$.

2.2 Trees

A *ranked alphabet* is a tuple (Σ, rk) where Σ is a finite set and $rk : \Sigma \to \mathbb{N}$ is a mapping called rank mapping. For every $k \geq 0$, we define $\Sigma^{(k)} = \{\sigma \in \Sigma \mid rk(\sigma) = k\}$. Sometimes, we write $\sigma^{(k)}$ to mean that $\sigma \in \Sigma^{(k)}$. Moreover, let H be a set disjoint with Σ. The set of *Σ-terms over H*, denoted by $T_\Sigma(H)$, is the smallest set T such that (i) $\Sigma^{(0)} \cup H \subseteq T$ and (ii) if $k \geq 1$, $\sigma \in \Sigma^{(k)}$, and $\xi_1, \dots, \xi_k \in T$, then $\sigma(\xi_1, \dots, \xi_k) \in T$. We denote $T_\Sigma(\emptyset)$ by T_Σ; obviously $T_\Sigma \neq \emptyset$ iff $\Sigma^{(0)} \neq \emptyset$. If H is finite, then we will also view $T_\Sigma(H)$ as $T_{\Sigma \cup H}$ where $(\Sigma \cup H)^{(0)} = \Sigma^{(0)} \cup H$ and $(\Sigma \cup H)^{(k)} = \Sigma^{(k)}$ for every $k \geq 1$. Since terms can be depicted in a very illustrative way as trees, i.e., particular graphs, it has become a custom to call Σ-terms also *Σ-trees*. In this chapter, we follow this custom. Every subset $L \subseteq T_\Sigma$ is called a *Σ-tree language*. Frequently, we will consider a tree $\xi \in T_\Sigma$ which has the form $\xi = \sigma(\xi_1, \dots, \xi_k)$ for some $k \geq 0$, $\sigma \in \Sigma^{(k)}$, and $\xi_1, \dots, \xi_k \in T_\Sigma$. Whenever we use this notation, for $k = 0$, the string $\sigma(\xi_1, \dots, \xi_k)$ stands for σ, rather than $\sigma()$. In order to avoid repetition of the quantifications of k, σ, and $\xi_1, \dots, \xi_k$, we henceforth only write that we consider a $\xi \in T_\Sigma$ of the form $\xi = \sigma(\xi_1, \dots, \xi_k)$.

In the rest of this chapter, Σ and Δ will denote arbitrary ranked alphabets if not specified otherwise. Moreover, we assume that $\Sigma^{(0)} \neq \emptyset$ and $\Delta^{(0)} \neq \emptyset$.

We define the *height*, *size*, and *set of positions* of trees as the functions height : $T_\Sigma \to \mathbb{N}$, size : $T_\Sigma \to \mathbb{N}$, and pos : $T_\Sigma \to \mathcal{P}(\mathbb{N}^*)$, respectively, as follows: (i) for every $\alpha \in \Sigma^{(0)}$, we define $\text{height}(\alpha) = 0$, $\text{size}(\alpha) = 1$, and $\text{pos}(\alpha) = \{\varepsilon\}$, and (ii) for every $\xi = \sigma(\xi_1, \ldots, \xi_k)$, where $k \geq 1$, we define $\text{height}(\xi) = 1 + \max\{\text{height}(\xi_i) \mid 1 \leq i \leq k\}$, $\text{size}(\xi) = 1 + \sum_{1 \leq i \leq k} \text{size}(\xi_i)$, and $\text{pos}(\xi) = \{\varepsilon\} \cup \{iv \mid 1 \leq i \leq k, v \in \text{pos}(\xi_i)\}$.

Now let $\xi, \zeta \in T_\Sigma$ and $w \in \text{pos}(\xi)$. The *label of ξ at w*, denoted by $\xi(w)$, the *subtree of ξ at w*, denoted by $\xi|_w$, and the *replacement of the subtree of ξ at w by ζ*, denoted by $\xi[\zeta]_w$ are defined as follows: (i) for every $\alpha \in \Sigma^{(0)}$, we define $\alpha(\varepsilon) = \alpha$, $\alpha|_\varepsilon = \alpha$, and $\alpha[\zeta]_\varepsilon = \zeta$, and (ii) for every $\xi = \sigma(\xi_1, \ldots, \xi_k)$ with $k \geq 1$, we define $\xi(\varepsilon) = \sigma$, $\xi|_\varepsilon = \xi$, and $\xi[\zeta]_\varepsilon = \zeta$, and for every $1 \leq i \leq k$ and $v \in \text{pos}(\xi_i)$, we define $\xi(iv) = \xi_i(v)$, $\xi|_{iv} = \xi_i|_v$, and $\xi[\zeta]_{iv} = \sigma(\xi_1, \ldots, \xi_{i-1}, \xi_i[\zeta]_v, \xi_{i+1}, \ldots, \xi_k)$. For a subset $Q \subseteq \Sigma$, we define $\text{pos}_Q : T_\Sigma \to \mathcal{P}(\mathbb{N}^*)$ by $\text{pos}_Q(\xi) = \{w \in \text{pos}(\xi) \mid \xi(w) \in Q\}$.

We will often use the notion of variable. Let $Z = \{z_1, z_2, \ldots\}$ be a set of variables, disjoint with Σ, and $Z_k = \{z_1, \ldots, z_k\}$ for every $k \geq 0$.

Next, we define *tree substitution*. Let H be a set disjoint with Σ. For $\xi \in T_\Sigma(Z \cup H)$, a finite set $I \subseteq \mathbb{N}$, and a family $(\xi_i \mid i \in I)$ with $\xi_i \in T_\Sigma(H)$, the expression $\xi(\xi_i \mid i \in I)$ denotes the result of substituting in ξ every occurrence of z_i by ξ_i for every $i \in I$. In case $I = \{1, \ldots, n\}$, we write $\xi(\xi_1, \ldots, \xi_n)$. Moreover, if $I = \{1\}$ and $z = z_1$, then we write $\xi \cdot_z \xi_1$ instead of $\xi(\xi_1)$. The operation $\cdot_z$ is associative in the sense that for every $\xi' \in T_\Sigma(Z \cup H)$ we have $(\xi \cdot_z \xi') \cdot_z \xi_1 = \xi \cdot_z (\xi' \cdot_z \xi_1)$.

2.3 Algebraic Concepts

In this chapter, we will often denote an algebraic structure just by its carrier set, if its operations are clear from the context.

Let $(S, +, 0)$ be a commutative monoid. Then S is *naturally ordered* if the binary relation $\sqsubseteq$ on S is a partial order on S, where $\sqsubseteq$ is defined by $a \sqsubseteq b$ iff there is a $c \in S$ such that $a + c = b$. A monoid S is *locally finite* if, for every finite $S' \subseteq S$, the sub-monoid of S generated by S' is finite.

An *infinitary sum operation* $\sum$ associates with every countable index set I and family $(a_i \mid i \in I)$ of elements $a_i \in S$ an element $\sum_{i \in I} a_i$. If $\sum$ is commutative, associative, and extends $+$, then S is a *$\sum$-complete monoid* (cf., e.g., [76, 88, 40]); in particular, $\sum_{i \in \emptyset} a_i = 0$. A $\sum$-complete and naturally ordered monoid S is *$\sum$-continuous* if, for every I, family $(a_i \mid i \in I)$, and $b \in S$, the following implication holds: if $\sum_{i \in E} a_i \sqsubseteq b$ for every finite subset E of I, then $\sum_{i \in I} a_i \sqsubseteq b$. We call a monoid *complete* (resp., *continuous*) if there is an infinitary sum operation $\sum$ such that S is $\sum$-complete (resp., $\sum$-continuous).

A *semiring* $(S, +, \cdot, 0, 1)$ is an algebra which consists of a commutative monoid $(S, +, 0)$, called the additive monoid of S, and a monoid $(S, \cdot, 1)$, called the multiplicative monoid of S, such that multiplication distributes (from left and right) over addition, and moreover, $0 \neq 1$ and 0 is absorbing with respect to $\cdot$ (also from left and right). We call S *idempotent* if $a + a = a$; *zero-sum free* if $a + b = 0$ implies $a = b = 0$; *commutative* if its multiplicative monoid is commutative; *zero-divisor free* if $a \cdot b = 0$ implies $a = 0$ or $b = 0$ for every $a, b \in S$; *positive* if it is zero-sum free and zero-divisor free. Finally, S is *locally finite* if, for every finite $S' \subseteq S$, the subsemiring of S generated by S' is finite.

Let S be commutative and $(a_i \mid i \in I)$ be a finite family of elements $a_i \in S$. Then we denote the product of all the elements of the family by $\prod_{i \in I} a_i$; in particular, we have that $\prod_{i \in \emptyset} a_i = 1$.

Let Q be a finite set and $u, v \in S^Q$ two Q-vectors over S. Then we define the *inner product* of u and v as $u \cdot v = \sum_{q \in Q} u_q \cdot v_q$.

In the rest of this chapter, S will denote an arbitrary semiring $(S, +, \cdot, 0, 1)$ if not specified otherwise.

Among other semirings, we consider the following particular ones: the Boolean semiring $(\mathbb{B}, \vee, \wedge, 0, 1)$ where $\mathbb{B} = \{0, 1\}$, the semiring $\mathsf{Nat} = (\mathbb{N}, +, \cdot, 0, 1)$ of natural numbers, the arctic semiring $\mathsf{Arct} = (\mathbb{N} \cup \{-\infty\}, \mathsf{max}, +, -\infty, 0)$, the tropical semiring $\mathsf{Trop} = (\mathbb{N} \cup \{\infty\}, \mathsf{min}, +, \infty, 0)$, and the semiring of formal languages $\mathsf{Lang}_A = (\mathcal{P}(A^*), \cup, \cdot, \emptyset, \{\varepsilon\})$ over any set A (where $L \cdot L' = \{uv \mid u \in L, v \in L'\}$ for languages L, L').

Again, let $\sum$ be an infinitary sum operation. The semiring S is *$\sum$-complete* if its additive monoid is $\sum$-complete and the following distributive laws hold: $c \cdot (\sum_{i \in I} a_i) = \sum_{i \in I} (c \cdot a_i)$ and $(\sum_{i \in I} a_i) \cdot c = \sum_{i \in I} (a_i \cdot c)$ for every $c \in S$ and family $(a_i \mid i \in I)$ with $a_i \in S$. Moreover, S is *naturally ordered* if its additive monoid is naturally ordered, and S is *$\sum$-continuous* if it is $\sum$-complete and its additive monoid is $\sum$-continuous.

We call S a *ring (with unit)* if $(S, +, 0)$ is a group; the additive inverse of $a \in S$ is denoted by $-a$. A semiring which is not a ring is called *proper*. Thus, every positive semiring is proper. We call S a *semifield* if $(S \setminus \{0\}, \cdot, 1)$ is a commutative group, i.e., every element $a \in S \setminus \{0\}$ has an inverse, which we denote by a^{-1}. Moreover, S is a *field* if it is a ring and a semifield.

For a semiring S, an *S-semimodule* is a commutative monoid $(V, +, 0)$ equipped with a scalar multiplication $\circ : S \times V \to V$ satisfying the following laws:

$$
\begin{aligned}
(a \cdot a') \circ v &= a \circ (a' \circ v), \\
a \circ (v + v') &= (a \circ v) + (a \circ v'), \\
(a + a') \circ v &= (a \circ v) + (a' \circ v), \\
1 \circ v &= v, \\
a \circ 0 &= 0 \circ v = 0
\end{aligned}
$$

for every $a, a' \in S$ and $v, v' \in V$ (cf. [71], page 149). Note that the symbols $+$ and 0 are overloaded because they denote operations over both S and V. Also, at other places in this chapter, such overloading of symbols may occur. However, it will always be clear from the context which operation is meant. As usual, we drop $\circ$ from $a \circ v$ and just write av.

An S-semimodule $(V, +, 0)$ with scalar multiplication $\circ$ is *complete* if the monoid $(V, +, 0)$ is complete, say, with the infinitary sum $\sum$, and $a \circ \sum_{i \in I} v_i = \sum_{i \in I}(a \circ v_i)$ holds for every $a \in S$ and family $(v_i \mid i \in I)$ with $v_i \in V$.

Let V and V' be S-semimodules. A mapping $f : V^k \to V'$ is *multilinear* if $f(v_1, \ldots, v_{i-1}, au+bv, v_{i+1}, \ldots, v_k) = af(v_1, \ldots, v_{i-1}, u, v_{i+1}, \ldots, v_k)+bf(v_1, \ldots, v_{i-1}, v, v_{i+1}, \ldots, v_k)$ for every $1 \le i \le k$, $u, v, v_1, \ldots, v_k \in V$, and $a, b \in S$. A multilinear unary mapping $f : V \to V'$ is called *linear*.

If S is a field and $(V, +, 0)$ is a commutative group, then the S-semimodule $(V, +, 0)$ is an *S-vector space*. Later, we will consider the particular S-vector space $(S^Q, +, 0)$ where Q is a finite set. The zero element 0 is the vector with $0_q = 0$ for every $q \in Q$. Moreover, for every $u, v \in S^Q$, $q \in Q$, and $a \in S$, we have that $(u + v)_q = u_q + v_q$ and $(au)_q = a \cdot u_q$. We call a linear mapping $\gamma : V \to S$ (where S is viewed as S-vector space) a *linear form*.

Let V be an S-vector space. The vectors $v_1, \ldots, v_m \in V$ are *linearly independent* if, for every $a_1, \ldots, a_m \in S$, the equality $a_1v_1 + \cdots + a_mv_m = 0$ implies that $a_1 = \cdots = a_m = 0$. A subset V' of V is linearly independent if the vectors in every finite subset of V' are linearly independent. Moreover, V' *generates* V if, for every $v \in V$, there are $m \ge 1$, $v_i \in V'$ and $a_i \in S$ for $1 \le i \le m$ such that $v = a_1v_1 + \cdots + a_mv_m$. Finally, V' is a *basis of* V if it is linearly independent and generates V. If V admits a basis consisting of $\kappa \in \mathbb{N}$ elements, then it is called *κ-dimensional*; V is *finite-dimensional* if it is κ-dimensional for some $\kappa \in \mathbb{N}$. In a finite-dimensional vector space each basis has the same number of elements.

A *Σ-algebra* (V, θ) consists of a nonempty set V (*carrier set*) and an arity preserving interpretation θ of symbols from Σ as operations over V, i.e., $\theta(\sigma) : V^k \to V$ for every $k \ge 0$ and $\sigma \in \Sigma^{(k)}$. The *Σ-term algebra* (T_Σ, top), defined by $\mathrm{top}(\sigma)(\xi_1, \ldots, \xi_k) = \sigma(\xi_1, \ldots, \xi_k)$, is *initial* in the class of all Σ-algebras, i.e., for every Σ-algebra (V, θ), there is a unique Σ-algebra homomorphism from T_Σ to V, which we denote by h_V (if not specified otherwise). That means that for every $\sigma(\xi_1, \ldots, \xi_k) \in T_\Sigma$ we have $h_V(\sigma(\xi_1, \ldots, \xi_k)) = \theta(\sigma)(h_V(\xi_1), \ldots, h_V(\xi_k))$. Now let (V, θ) be a Σ-algebra and z a nullary symbol such that $z \notin \Sigma$. For every $v \in V$, we define the *v-extension of* (V, θ) to be the $\Sigma \cup \{z\}$-algebra (V, θ^v) where $\theta^v(z) = v$ and $\theta^v(\sigma) = \theta(\sigma)$ for every $k \ge 0$ and $\sigma \in \Sigma^{(k)}$. We denote the unique $\Sigma \cup \{z\}$-algebra homomorphism from $T_{\Sigma \cup \{z\}}$ to V by h_V^v. For more details, we refer to [73, 70, 139].

An *S-Σ-semimodule* $(V, +, 0, \theta)$ consists of an S-semimodule $(V, +, 0)$ and a Σ-algebra (V, θ) where $\theta(\sigma)$ is multilinear for every $\sigma \in \Sigma$. If S is a field and $(V, +, 0)$ is an S-vector space, then we call $(V, +, 0, \theta)$ an *S-Σ-vector space*. An S-Σ-semimodule $(V, +, 0, \theta)$ is *complete* if the S-semimodule $(V, +, 0)$ is complete and for every $\sigma \in \Sigma$, the operation $\theta(\sigma)$ preserves infinite sums in each of its arguments, i.e.,

$$\theta(\sigma)\Big(\dots, \sum_{i\in I} v_i, \dots\Big) = \sum_{i\in I} \theta(\sigma)(\dots, v_i, \dots).$$

Moreover, a complete S-Σ-semimodule $(V, +, 0, \theta)$ is *continuous* if $(V, +, 0)$ is continuous.

2.4 Tree Series

Let H be a set with $\Sigma \cap H = \emptyset$. A *tree series over* Σ, H, *and* S (or for short: tree series) is a mapping $r : T_\Sigma(H) \to S$. For every $\xi \in T_\Sigma(H)$, the element $r(\xi) \in S$ is called the *coefficient* of ξ and it is denoted by (r, ξ). Moreover, the tree series r is written as the formal sum $\sum_{\xi\in T_\Sigma(H)} (r, \xi).\xi$.

The *support of the tree series* r is defined as the set $\mathrm{supp}(r) = \{\xi \in T_\Sigma(H) \mid (r, \xi) \neq 0\}$. Moreover, r is *polynomial* (resp., a *monomial*) if $\mathrm{supp}(r)$ is finite (resp., a singleton). We will denote a polynomial r by $a_1.\xi_1+\dots+a_k.\xi_k$, where $\mathrm{supp}(r) = \{\xi_1, \dots, \xi_k\}$ and $a_i = (r, \xi_i)$ for $1 \leq i \leq k$. The set of all (resp., polynomial) tree series is denoted by $S\langle\langle T_\Sigma(H)\rangle\rangle$ (resp., $S\langle T_\Sigma(H)\rangle$).

Let $r \in S\langle\langle T_\Sigma(H)\rangle\rangle$ be a tree series. We call r *Boolean* if $(r, \xi) \in \{0, 1\}$ holds for every $\xi \in T_\Sigma(H)$. If there is an $a \in S$ such that for every $\xi \in T_\Sigma(H)$, we have $(r, \xi) = a$, then r is a *constant* and also denoted by $\widetilde{a}$. Note that the constants $\widetilde{0}$ and $\widetilde{1}$ are Boolean.

For a set A and $B \subseteq A$, the *characteristic function of* B *with respect to* S is the mapping $1_{(S,B)} : A \to S$ such that $1_{(S,B)}(a) = 1$ if $a \in B$, and $1_{(S,B)}(a) = 0$ otherwise for every $a \in A$. For a tree language $L \subseteq T_\Sigma$, we call the tree series $1_{(S,L)}$ the *characteristic tree series of* L *with respect to* S. Certainly, we have $\mathrm{supp}(1_{(S,L)}) = L$.

3 Weighted Tree Automata

3.1 Bottom-up Tree Automata

The theory of finite-state string automata and of recognizable string languages has been successfully generalized to trees. For instance, the class of recognizable tree languages is characterized by solutions of linear equations [116], rational expressions [135], monadic-second order logic [135, 33], congruences of finite index [26, 100], and finitely generated congruences [63, 86]. An excellent, detailed survey on recognizable tree languages and finite-state tree automata can be found in [68, 69] (also cf. [29]). Let us recall here the concept of a finite-state bottom-up tree automaton.

A *(finite-state) bottom-up tree automaton* is a tuple $\mathcal{A} = (Q, \Sigma, \delta, F)$, where Q is a finite nonempty set (*states*), δ is a Σ-indexed family $(\delta_\sigma \mid \sigma \in \Sigma)$

where $\delta_\sigma \subseteq Q^k \times Q$ for $\sigma \in \Sigma^{(k)}$ (*set of transitions at* σ), and $F \subseteq Q$ (*final states*). We call Σ the *input ranked alphabet*, and an element $\xi \in T_\Sigma$ an *input tree*. Here and in the rest of the chapter, we view Q^k as the set of strings over Q of length k. To define the semantics of $\mathcal{A}$, we consider the Σ-algebra $(\mathcal{P}(Q), \delta_\mathcal{A})$ with $\delta_\mathcal{A}(\sigma)(P_1, \ldots, P_k) = \{q \in Q \mid \exists (q_1 \ldots q_k, q) \in \delta_\sigma : q_i \in P_i$ for every $1 \leq i \leq k\}$ for every $P_1, \ldots, P_k \in \mathcal{P}(Q)$. *The tree language recognized by* $\mathcal{A}$ is $L_\mathcal{A} = \{\xi \in T_\Sigma \mid h_{\mathcal{P}(Q)}(\xi) \cap F \neq \emptyset\}$. The class of all *recognizable* Σ*-tree languages* is denoted by $\mathrm{Rec}(\Sigma)$.

It is well known that every bottom-up tree automaton can be transformed into an equivalent deterministic one. A bottom-up tree automaton is *(total) deterministic* if the relation δ_σ is a total function for every $\sigma \in \Sigma$. Then we view $h_{\mathcal{P}(Q)}$ as a mapping of type $T_\Sigma \to Q$ and write h_Q rather than $h_{\mathcal{P}(Q)}$.

Example 3.1. Consider the ranked alphabet $\Sigma = \{\sigma^{(2)}, \gamma^{(1)}, \alpha^{(0)}\}$ and the pattern $\sigma(z, \alpha)$. We say that the pattern $\sigma(z, \alpha)$ occurs in a tree $\xi \in T_\Sigma$ if there is a position $w \in \mathrm{pos}(\xi)$ and a tree $\xi' \in T_\Sigma$ such that $\xi = \xi[\sigma(\xi', \alpha)]_w$. We construct a bottom-up tree automaton $\mathcal{A}$ such that $L_\mathcal{A}$ is the set of all Σ-trees in which $\sigma(z, \alpha)$ occurs at least once. The automaton performs a nondeterministic guess-and-verify strategy; it selects nondeterministically an occurrence of α and verifies whether it is a right child of a σ. For this, let $\mathcal{A} = (Q, \Sigma, \delta, F)$ be defined by $Q = \{q, \overline{\alpha}, f\}$, $F = \{f\}$, and $\delta_\alpha = \{(\varepsilon, q), (\varepsilon, \overline{\alpha})\}$, $\delta_\gamma = \{(q, q), (f, f)\}$, and $\delta_\sigma = \{(qq, q), (qf, f), (fq, f), (q\overline{\alpha}, f)\}$. Then for every $\xi \in T_\Sigma$, we have that $f \in h_{\mathcal{P}(Q)}(\xi)$ iff the pattern $\sigma(z, \alpha)$ occurs in ξ.

3.2 Recognizable Tree Series

Bottom-up tree automata can be reformulated such that the reformulation easily leads to the concept of weighted tree automata. The idea behind this is to represent a tree language $L \subseteq T_\Sigma$ as a characteristic tree series $1_{(\mathbb{B}, L)} : T_\Sigma \to \mathbb{B}$ and then, in a second step, to replace the Boolean semiring $\mathbb{B}$ by S.

Consider now the system $\mathcal{A} = (Q, \Sigma, \mathbb{B}, \mu, \nu)$, called a *weighted tree automaton over* $\mathbb{B}$, where Q is as in Sect. 3.1, while μ is a family $(\mu_k : \Sigma^{(k)} \to \mathbb{B}^{Q^k \times Q} \mid k \geq 0)$ of mappings and $\nu \in \mathbb{B}^Q$ is a Q-vector over $\mathbb{B}$. We define the semantics of $\mathcal{A}$ as a mapping $r_\mathcal{A} : T_\Sigma \to \mathbb{B}$ in the following way. Let us introduce the Σ-algebra $(\mathbb{B}^Q, \mu_\mathcal{A})$, where

$$\mu_\mathcal{A}(\sigma)(v_1, \ldots, v_k)_q = \bigvee_{q_1, \ldots, q_k \in Q} (v_1)_{q_1} \wedge \cdots \wedge (v_k)_{q_k} \wedge \mu_k(\sigma)_{q_1 \cdots q_k, q},$$

for every $k \geq 0$, $\sigma \in \Sigma^{(k)}$, and $v_1, \ldots, v_k \in \mathbb{B}^Q$. Now let

$$r_\mathcal{A}(\xi) = h_\mu(\xi) \wedge \nu$$

for every $\xi \in T_\Sigma$, where h_μ is the unique Σ-homomorphism from the Σ-term algebra T_Σ to $(\mathbb{B}^Q, \mu_\mathcal{A})$. (Recall that, according to the notion of inner product of Q-vectors over $\mathbb{B}$ from Sect. 2.3, $h_\mu(\xi) \wedge \nu = \bigvee_{q \in Q} h_\mu(\xi)_q \wedge \nu_q$.)

It should be clear that bottom-up tree automata and weighted tree automata over $\mathbb{B}$ are semantically equivalent: for every bottom-up tree automaton $\mathcal{A} = (Q, \Sigma, \delta, F)$ one can construct the weighted tree automaton $\mathcal{B} = (Q, \Sigma, \mathbb{B}, \mu, \nu)$, where $\mu_k(\sigma) = 1_{(\mathbb{B},\delta_\sigma)}$ and $\nu = 1_{(\mathbb{B},F)}$. Then $r_\mathcal{B} = 1_{(\mathbb{B},L_\mathcal{A})}$. Clearly, the construction can be reversed.

Now we observe that weighted tree automata over $\mathbb{B}$ can be generalized to weighted tree automata over S in an obvious way: then S, $+$, and $\cdot$ take over the role of $\mathbb{B}$, $\vee$, and $\wedge$, respectively. The semantics of a weighted tree automaton $\mathcal{A}$ over S will be a tree series $r_\mathcal{A} : T_\Sigma \to S$. We expect that weighted tree automata can compute tree series like:

- height : $T_\Sigma \to \mathbb{N}$ over Arct,
- $\text{size}_\delta : T_\Sigma \to \mathbb{N}$ and size : $T_\Sigma \to \mathbb{N}$ over Nat and also over Trop, where $\text{size}_\delta(\xi) = |\text{pos}_{\{\delta\}}(\xi)|$,
- $\#_{\sigma(z,\alpha)} : T_\Sigma \to \mathbb{N}$ over Nat, where $\#_{\sigma(z,\alpha)}(\xi)$ is the number of occurrences of the pattern $\sigma(z, \alpha)$ in ξ,
- $\text{shortest}_\alpha : T_\Sigma \to \mathbb{N}$ over Trop, where $\text{shortest}_\alpha(\xi)$ is the length of a shortest path in ξ from its root to one of its leaves with label α,
- yield : $T_\Sigma \to \mathcal{P}(\Sigma^*)$ over Lang_Σ, where yield(ξ) is the concatenation of the nullary symbols occurring in ξ from left to right,
- revpos : $T_\Sigma \to \mathcal{P}(\mathbb{N}^*)$ over $\text{Lang}_\mathbb{N}$, where revpos(ξ) is the set of reversals of elements in pos(ξ),
- $\text{revpos}_{\sigma(z,\alpha)} : T_\Sigma \to \mathcal{P}(\mathbb{N}^*)$ over $\text{Lang}_\mathbb{N}$, where $\text{revpos}_{\sigma(z,\alpha)}(\xi)$ is the set of reversals of positions of ξ at which the pattern $\sigma(z, \alpha)$ occurs; note that $\#_{\sigma(z,\alpha)}(\xi)$ is the cardinality of $\text{revpos}_{\sigma(z,\alpha)}(\xi)$.

Now let us start with the formal definition of weighted tree automata. We follow the approach of [1], where this model was called an S-Σ-tree automaton.

Definition 3.2. *A* weighted tree automaton (over S) *(for short: wta) is a tuple* $\mathcal{A} = (Q, \Sigma, S, \mu, \nu)$ *where:*

- Q *is a finite nonempty set, the* set of states.
- Σ *is the* ranked input alphabet.
- $\mu = (\mu_k \mid k \in \mathbb{N})$ *is a* family of transition mappings[3] $\mu_k : \Sigma^{(k)} \to S^{Q^k \times Q}$.
- $\nu \in S^Q$ *is a Q-vector over S, the* root weight vector.

For every transition $(w, q) \in Q^k \times Q$, the element $\mu_k(\sigma)_{w,q} \in S$ is the weight of (w, q). We denote the set $\{\mu_k(\sigma)_{w,q} \mid k \geq 0, \sigma \in \Sigma^{(k)}, w \in Q^k, q \in Q\} \cup \{\nu_q \mid q \in Q\}$ of all weights which occur in $\mathcal{A}$, by wts$(\mathcal{A})$. Note that wts$(\mathcal{A}) \subseteq S$.

For a wta $\mathcal{A}$, we consider the Σ-algebra $(S^Q, \mu_\mathcal{A})$ where, for every $k \geq 0$ and $\sigma \in \Sigma^{(k)}$, the k-ary operation $\mu_\mathcal{A}(\sigma) : S^Q \times \cdots \times S^Q \to S^Q$ is defined by

$$\mu_\mathcal{A}(\sigma)(v_1, \ldots, v_k)_q = \sum_{q_1,\ldots,q_k \in Q} (v_1)_{q_1} \cdot \cdots \cdot (v_k)_{q_k} \cdot \mu_k(\sigma)_{q_1 \cdots q_k, q}$$

[3] In the literature, μ is also called a *tree representation*.

for every $q \in Q$ and $v_1, \ldots, v_k \in S^Q$. Let us denote here the unique Σ-algebra homomorphism from T_Σ to S^Q by h_μ. The tree series $r_\mathcal{A} \in S\langle\langle T_\Sigma \rangle\rangle$ *recognized* by $\mathcal{A}$ is defined by

$$(r_\mathcal{A}, \xi) = h_\mu(\xi) \cdot \nu$$

for every $\xi \in T_\Sigma$. (Again recall that, according to the notion of inner product of Q-vectors over S from Sect. 2.3, $h_\mu(\xi) \cdot \nu = \sum_{q \in Q} h_\mu(\xi)_q \cdot \nu_q$.) A tree series $r \in S\langle\langle T_\Sigma \rangle\rangle$ is *recognizable* if there is a wta $\mathcal{A}$ such that $r = r_\mathcal{A}$. The class of all tree series over Σ and S which are recognizable is denoted by $\mathrm{Rec}(\Sigma, S)$.

Due to the definitions of $\mu_\mathcal{A}$ and h_μ, we can observe that

$$h_\mu\big(\sigma(\xi_1, \ldots, \xi_k)\big)_q = \sum_{q_1, \ldots, q_k \in Q} h_\mu(\xi_1)_{q_1} \cdot \cdots \cdot h_\mu(\xi_k)_{q_k} \cdot \mu_k(\sigma)_{q_1 \ldots q_k, q}$$

for every $\sigma(\xi_1, \ldots, \xi_k) \in T_\Sigma$ and $q \in Q$.

Example 3.3. We construct a wta $\mathcal{A} = (Q, \Sigma, \mathsf{Arct}, \mu, \nu)$ which recognizes the tree series height. Let $Q = \{p_1, p_2\}$, $\Sigma = \{\sigma^{(2)}, \alpha^{(0)}\}$, and $\nu_{p_1} = 0$ and $\nu_{p_2} = -\infty$. Moreover, let

$$\begin{aligned}
\mu_0(\alpha)_{\varepsilon, p_1} &= \mu_0(\alpha)_{\varepsilon, p_2} = 0, \\
\mu_2(\sigma)_{p_1 p_2, p_1} &= \mu_2(\sigma)_{p_2 p_1, p_1} = 1, \\
\mu_2(\sigma)_{p_2 p_2, p_2} &= 0,
\end{aligned}$$

and for every other transition $(q_1 q_2, q)$ we have $\mu_2(\sigma)_{q_1 q_2, q} = -\infty$. We consider the tree $\xi = \sigma(\alpha, \alpha)$ and compute $h_\mu(\xi)_{p_1}$ and $h_\mu(\xi)_{p_2}$. Clearly, $h_\mu(\alpha)_{p_1} = \mu_0(\alpha)_{\varepsilon, p_1} = 0$ and $h_\mu(\alpha)_{p_2} = 0$. Then

$$h_\mu(\sigma(\alpha, \alpha))_{p_1} = \max_{q_1, q_2 \in Q} \{h_\mu(\alpha)_{q_1} + h_\mu(\alpha)_{q_2} + \mu_2(\sigma)_{q_1 q_2, p_1}\} = 1$$

(note that $\mu_2(\sigma)_{p_1 p_1, p_1} = \mu_2(\sigma)_{p_2 p_2, p_1} = -\infty$ and $-\infty$ is unit for max) and similarly, $h_\mu(\sigma(\alpha, \alpha))_{p_2} = 0$. In general, we can prove by structural induction on ξ that $h_\mu(\xi)_{p_1} = \mathrm{height}(\xi)$ and $h_\mu(\xi)_{p_2} = 0$ for every $\xi \in T_\Sigma$. Thus, $r_\mathcal{A} = \mathrm{height}$, and hence height $\in \mathrm{Rec}(\Sigma, \mathsf{Arct})$.

We have defined recognizable tree series in an initial algebra semantics style [70]. An alternative way is to define the semantics of a wta by means of its runs. A *run of $\mathcal{A}$ on $\xi \in T_\Sigma$* is a mapping $\kappa : \mathrm{pos}(\xi) \to Q$; the *set of all runs of $\mathcal{A}$ on ξ* is denoted by $R_\mathcal{A}(\xi)$. For every $\kappa \in R_\mathcal{A}(\xi)$ and $w \in \mathrm{pos}(\xi)$, the *run induced by κ at position w* is the run $\kappa|_w \in R_\mathcal{A}(\xi|_w)$ and defined for every $w' \in \mathrm{pos}(\xi|_w)$ by $\kappa|_w(w') = \kappa(ww')$. For every $\xi = \sigma(\xi_1, \ldots, \xi_k) \in T_\Sigma$, the *weight* $\mathrm{wt}(\kappa)$ *of* κ is $\mathrm{wt}(\kappa) = \mathrm{wt}(\kappa|_1) \cdot \cdots \cdot \mathrm{wt}(\kappa|_k) \cdot \mu_k(\sigma)_{\kappa(1) \ldots \kappa(k), \kappa(\varepsilon)}$. The *run semantics of $\mathcal{A}$* is the tree series $r_\mathcal{A}^{\mathrm{run}} \in S\langle\langle T_\Sigma \rangle\rangle$ such that

$$(r_\mathcal{A}^{\mathrm{run}}, \xi) = \sum_{\kappa \in R_\mathcal{A}(\xi)} \mathrm{wt}(\kappa) \cdot \nu_{\kappa(\varepsilon)}$$

for every $\xi \in T_\Sigma$. In fact, for every wta $\mathcal{A}$, the run semantics of $\mathcal{A}$ and the tree series recognized by $\mathcal{A}$ are the same. More precisely, the equation $h_\mu(\xi)_q = \sum_{\kappa \in R_\mathcal{A}(\xi), \kappa(\varepsilon)=q} \mathrm{wt}(\kappa)$ holds for every $\xi \in T_\Sigma$ and $q \in Q$, which easily implies $r_\mathcal{A} = r_\mathcal{A}^{\mathrm{run}}$.

Example 3.4. We construct a wta $\mathcal{A}'$ which recognizes the tree series $\#_{\sigma(z,\alpha)} : T_\Sigma \to \mathbb{N}$. This generalizes Example 3.1 in the sense that we not only consider whether the pattern $\sigma(z, \alpha)$ occurs in ξ, but also compute the number of those occurrences. For this, recall the bottom-up tree automaton $\mathcal{A} = (Q, \Sigma, \delta, F)$ of Example 3.1 and construct $\mathcal{A}' = (Q, \Sigma, \mathsf{Nat}, \mu, \nu)$ such that $\mu_k(\theta) = 1_{(\mathsf{Nat}, \delta_\theta)}$ for every $k \geq 0$, $\theta \in \Sigma^{(k)}$; moreover, let $\nu = 1_{(\mathsf{Nat}, F)}$.

Then for every $\xi \in T_\Sigma$ and run $\kappa \in R_\mathcal{A}(\xi)$ with $\kappa(\varepsilon) = f$, the weight $\mathrm{wt}(\kappa)$ is 1 iff at exactly one occurrence of σ the transition $(q\overline{\alpha}, f)$ was applied (and $\mathrm{wt}(\kappa) = 0$ otherwise). Since the application of this transition indicates an occurrence of the pattern $\sigma(z, \alpha)$, we have that $(r_\mathcal{A}^{\mathrm{run}}, \xi) = \#_{\sigma(z,\alpha)}(\xi)$. Thus, $\#_{\sigma(z,\alpha)} \in \mathrm{Rec}(\Sigma, \mathsf{Nat})$.

In fact, wta generalize in a natural way weighted finite automata as they are presented, e.g., in Part I of this handbook. Here, we follow the formal approach of [39], where a weighted finite automaton over a semiring S and an (unranked) alphabet Γ is a tuple $\mathcal{A} = (Q, \lambda, \mu, \gamma)$ and Q is a finite set of states, $\mu : \Gamma \to S^{Q \times Q}$ is the transition weight function, and $\lambda, \gamma \in S^Q$ are weight functions for entering and leaving a state. The behavior $\|\mathcal{A}\| : \Gamma^* \to S$ of $\mathcal{A}$ associates with every word $w = a_1 a_2 \dots a_n \in \Gamma^*$ the value $(\|\mathcal{A}\|, w) = \sum_{p \in P(w)} \mathrm{wt}(p)$ where $P(w)$ is the set of all paths with label w, and $\mathrm{wt}(p)$ is the weight of p defined to be $\lambda_{q_0} \cdot \mu(a_1)_{q_0,q_1} \cdot \dots \cdot \mu(a_n)_{q_{n-1},q_n} \cdot \gamma_{q_n}$ assuming that p has the form $q_0 \xrightarrow{a_1} q_1 \xrightarrow{a_2} \cdots \xrightarrow{a_{n-1}} q_{n-1} \xrightarrow{a_n} q_n$. Let $\mathrm{Rec}_\mathrm{w}(\Gamma, S)$ denote the class of all behaviors of weighted finite automata, i.e., of all recognizable formal power series over Γ and S.

By rotating a word w counterclockwise by 90°, we obtain the tree $\mathrm{tree}(w)$. Formally, let Γ_t be the ranked alphabet $\{a^{(1)} \mid a \in \Gamma\} \cup \{e^{0)}\}$, and $\mathrm{tree} : \Gamma^* \to T_{\Gamma_t}$ be defined by $\mathrm{tree}(\varepsilon) = e$ and $\mathrm{tree}(wa) = a(\mathrm{tree}(w))$ for every $a \in \Gamma$ and $w \in \Gamma^*$. Clearly, tree is a bijection; moreover, $L \subseteq \Gamma^*$ is a recognizable language iff the tree language $\mathrm{tree}(L)$ is recognizable. Then, given a weighted finite automaton $\mathcal{A} = (Q, \lambda, \mu, \gamma)$, we can construct the wta $\mathcal{A}_t = (Q, \Gamma_t, S, \theta, \gamma)$ over S and Γ_t where $\theta_0(e)_{\varepsilon,q} = \lambda_q$ and $\theta_1(a)_{q,p} = \mu(a)_{q,p}$ for every $a \in \Gamma$ and $q, p \in Q$. It is obvious that $\|\mathcal{A}\| = r_{\mathcal{A}_t}^{\mathrm{run}} \circ \mathrm{tree}$. Vice versa, given a wta $\mathcal{B}$ over some ranked alphabet Γ_t which results from an (unranked) alphabet Γ, the wta $\mathcal{B}$ can be viewed in an obvious way as a weighted finite automaton $\mathcal{B}'$ such that $r_\mathcal{B}^{\mathrm{run}} = \|\mathcal{B}'\| \circ \mathrm{tree}^{-1}$. By extending the mapping tree in the usual way to languages and classes of languages, we obtain that $\mathrm{tree}(\mathrm{Rec}_\mathrm{w}(\Gamma, S)) = \mathrm{Rec}(\Gamma_t, S)$ for every alphabet Γ.

As first type of restriction on wta, we define deterministic wta. A wta $\mathcal{A} = (Q, \Sigma, S, \mu, \nu)$ is:

- *bottom-up deterministic* (for short: bu-deterministic) if for every $k \geq 0$, $\sigma \in \Sigma^{(k)}$, and $w \in Q^k$ there is at most one $q \in Q$ such that $\mu_k(\sigma)_{w,q} \neq 0$,
- *total bu-deterministic* if for every $k \geq 0$, $\sigma \in \Sigma^{(k)}$, and $w \in Q^k$, there is exactly one state q such that $\mu_k(\sigma)_{w,q} \neq 0$,
- *top-down deterministic* (for short: td-deterministic) if for every $k \geq 0$, $\sigma \in \Sigma^{(k)}$, and $q \in Q$ there is at most one $w \in Q^k$ such that $\mu_k(\sigma)_{w,q} \neq 0$; moreover, $\nu_q \neq 0$ for at most one state q.

Note that, if the wta $\mathcal{A}$ is bu-deterministic, then for every input tree $\xi \in T_\Sigma$, there is at most one $q \in Q$ such that $h_\mu(\xi)_q \neq 0$. In this case, the operation $+$ of S is not used for the computation of $r_\mathcal{A}$. Also in the td-deterministic case $+$ is not used to compute $r_\mathcal{A}$.

We also note that, for every bu-deterministic wta $\mathcal{A}$, there exists a total bu-deterministic wta $\mathcal{A}'$ such that $r_\mathcal{A} = r_{\mathcal{A}'}$. This normal form can always be achieved in a standard way by using an additional dummy state for which the root weight vector ν yields 0.

Let $g \in \{\text{bu}, \text{td}\}$. Then a tree series $r \in S\langle\langle T_\Sigma\rangle\rangle$ is *g-deterministically recognizable* if there is a g-deterministic wta $\mathcal{A}$ such that $r = r_\mathcal{A}$. The corresponding classes of recognizable tree series are denoted by bud-Rec(Σ, S) and tdd-Rec(Σ, S). In Sect. 3.5, we will deal with the question under which conditions a wta can be determinized.

As second type of restriction on wta, we consider their root weights. A wta $\mathcal{A}$ has *Boolean root weights* if $\{\nu_q \mid q \in Q\} \subseteq \{0, 1\}$; in this case, we replace ν by the set $F = \{q \in Q \mid \nu_q = 1\}$. Then $(r_\mathcal{A}, \xi) = \sum_{q \in F} h_\mu(\xi)_q$. In fact, the wta of Examples 3.3 and 3.4 have Boolean root weights.

A tree series $r \in S\langle\langle T_\Sigma\rangle\rangle$ is *recognizable with Boolean root weights* if there is a wta $\mathcal{A}$ with Boolean root weights such that $r = r_\mathcal{A}$. The corresponding classes of recognizable tree series are denoted by indexing the original class with a capital B, e.g., bud-Rec$_\text{B}(\Sigma, S)$ is the class of all tree series which are bu-deterministically recognizable with Boolean root weights.

Example 3.5. The remaining tree series shown in the list on page 322 are also recognizable as follows: size$_\delta$ and size are in Rec$_\text{B}(\Sigma, \mathsf{Nat}) \cap$bud-Rec$_\text{B}(\Sigma, \mathsf{Trop})$; shortest$_\alpha \in$ Rec$_\text{B}(\Sigma, \mathsf{Trop})$; yield $\in$ bud-Rec$_\text{B}(\Sigma, \mathsf{Lang}_\Sigma)$; revpos $\in$ bud-Rec$_\text{B}$ $(\Sigma, \mathsf{Lang}_\mathbb{N})$; and revpos$_{\sigma(z,\alpha)} \in$ Rec$_\text{B}(\Sigma, \mathsf{Lang}_\mathbb{N})$, cf. [13].

In general, wta and wta with Boolean root weights are equally powerful.

Theorem 3.6 ([14], Theorems 6.1.6 and 6.2.2). Rec$(\Sigma, S) =$ Rec$_\text{B}(\Sigma, S)$ *and* tdd-Rec$(\Sigma, S) =$ tdd-Rec$_\text{B}(\Sigma, S)$.

Proof. For a given wta $\mathcal{A} = (Q, \Sigma, S, \mu, \nu)$, we construct the wta $\mathcal{A}' = (Q', \Sigma, S, \mu', \{q_f\})$ with Boolean root weights by defining $Q' = Q \cup \{q_f\}$ for a new state q_f. Moreover, for every $\sigma \in \Sigma^{(k)}$, $w \in Q^k$, and $q \in Q'$, we define $\mu'_k(\sigma)_{w,q} = \mu_k(\sigma)_{w,q}$ if $q \in Q$, and $\mu'_k(\sigma)_{w,q_f} = \sum_{q \in Q} \mu_k(\sigma)_{w,q} \cdot \nu_q$; and for every $w \in (Q')^k \setminus Q^k$ and $q \in Q'$, we define $\mu'_k(\sigma)_{w,q} = 0$. We note that this construction preserves td-determinism but not bu-determinism. It is obvious that $h_\mu(\xi) \cdot \nu = h_{\mu'}(\xi)_{q_f}$ and thus $r_\mathcal{A} = r_{\mathcal{A}'}$. □

In fact, the classes bud-Rec(Σ, S) and bud-Rec$_\mathrm{B}$(Σ, S) may indeed differ. More precisely, if Σ contains at least one non-nullary symbol σ, and there is an element $a \in S \setminus \{0\}$ which has no multiplicative right inverse, then bud-Rec(Σ, S) $\setminus$ bud-Rec$_\mathrm{B}$(Σ, S) $\neq \emptyset$. A witness of this set is the polynomial tree series $r = a.\alpha + 1.\sigma(\alpha, \ldots, \alpha)$ (cf. [14], Lemma 6.1.3). However, for semifields bu-deterministic wta and bu-deterministic wta with Boolean root weights are equally powerful.

Theorem 3.7 ([14], Lemma 6.1.4). *Let S be a semifield. Then* bud-Rec(Σ, S) = bud-Rec$_\mathrm{B}$(Σ, S).

Proof. Let $\mathcal{A} = (Q, \Sigma, S, \mu, \nu)$ be a bu-deterministic wta. We construct the bu-deterministic wta $\mathcal{A}' = (Q, \Sigma, S, \mu', F)$ with $F = \{q \in Q \mid \nu_q \neq 0\}$ and $\mu'_k(\sigma)_{w,q} = \nu'(q_k)^{-1} \cdot \ldots \cdot \nu'(q_1)^{-1} \cdot \mu_k(\sigma)_{w,q} \cdot \nu'(q)$ for every $\sigma \in \Sigma^{(k)}$, $w = q_1 \ldots q_k \in Q^k$, and $q \in Q$; the auxiliary function $\nu' : Q \to S$ is defined by $\nu'(q) = \nu_q$ if $q \in F$ and $\nu'(q) = 1$ otherwise. Then $h_{\mu'}(\xi)_q = h_\mu(\xi)_q \cdot \nu'(q)$ for every $\xi \in T_\Sigma$ and $q \in Q$. Finally, we have $(r_\mathcal{A}, \xi) = h_\mu(\xi) \cdot \nu = \sum_{q \in F} h_\mu(\xi)_q \cdot \nu_q = \sum_{q \in F} h_\mu(\xi)_q \cdot \nu'(q) = \sum_{q \in F} h_{\mu'}(\xi)_q = (r_{\mathcal{A}'}, \xi)$. □

3.3 Closure Properties

As for tree languages, one can define operations on tree series in $S\langle\langle T_\Sigma \rangle\rangle$, e.g., the multiplication of a tree series with a semiring element, sum, Hadamard-product, top-concatenation, OI-substitution, α-concatenation (i.e., Cauchy-product), where α is a nullary symbol in Σ, α-Kleene star, and relabeling. Let us define these operations and show the corresponding closure properties of Rec(Σ, S).

Let $a \in S$ and $r \in S\langle\langle T_\Sigma \rangle\rangle$. Then the *scalar multiplication of a and r* is the tree series $ar \in S\langle\langle T_\Sigma \rangle\rangle$ defined by $(ar, \xi) = a \cdot (r, \xi)$ for every $\xi \in T_\Sigma$.

Let $r_1, r_2 \in S\langle\langle T_\Sigma \rangle\rangle$. The *sum of r_1 and r_2* and the *Hadamard product of r_1 and r_2* are the tree series $r_1 + r_2 \in S\langle\langle T_\Sigma \rangle\rangle$ and $r_1 \odot r_2 \in S\langle\langle T_\Sigma \rangle\rangle$, respectively, defined by $(r_1 + r_2, \xi) = (r_1, \xi) + (r_2, \xi)$ and $(r_1 \odot r_2, \xi) = (r_1, \xi) \cdot (r_2, \xi)$ for every $\xi \in T_\Sigma$. We can also sum up over an infinite family of tree series assuming that this family is locally finite. A family $(r_i \mid i \in I)$ of tree series is *locally finite* if for every $\xi \in T_\Sigma$, the set $I_\xi = \{i \in I \mid (r_i, \xi) \neq 0\}$ is finite. Then we define the sum $\sum_{i \in I} r_i \in S\langle\langle T_\Sigma \rangle\rangle$ by $(\sum_{i \in I} r_i, \xi) = \sum_{i \in I_\xi} (r_i, \xi)$ for every $\xi \in T_\Sigma$.

For every $\sigma \in \Sigma^{(k)}$, the *top-concatenation (with σ)* $\mathrm{top}_\sigma : S\langle\langle T_\Sigma \rangle\rangle^k \to S\langle\langle T_\Sigma \rangle\rangle$ is defined, for every $r_1, \ldots, r_k \in S\langle\langle T_\Sigma \rangle\rangle$ and $\xi \in T_\Sigma$ as follows: if $\xi = \sigma(\xi_1, \ldots, \xi_k)$, then $(\mathrm{top}_\sigma(r_1, \ldots, r_k), \xi) = (r_1, \xi_1) \cdot \ldots \cdot (r_k, \xi_k)$, otherwise $(\mathrm{top}_\sigma(r_1, \ldots, r_k), \xi) = 0$.

Next, we define the OI-substitution of tree series, which generalizes the OI-substitution of tree languages [55, 56]. Let $n \geq 0$, $\bar{\alpha} = (\alpha_1, \ldots, \alpha_n) \in (\Sigma^{(0)})^n$, and $\bar{r} = (r_1, \ldots, r_n) \in S\langle\langle T_\Sigma \rangle\rangle^n$. For every $\xi \in T_\Sigma$, the tree series $\xi \leftarrow_{\mathrm{OI},\bar{\alpha}} \bar{r}$ (abbreviated by s in this definition) is defined inductively on the structure of ξ: (i) if $\xi = \alpha_i$, then $s = r_i$, and (ii) if $\xi = \sigma(\xi_1, \ldots, \xi_k)$ and $k \geq 1$ or $k = 0$ and $\sigma \notin \{\alpha_1, \ldots, \alpha_n\}$, then $s = \mathrm{top}_\sigma(\xi_1 \leftarrow_{\mathrm{OI},\bar{\alpha}} \bar{r}, \ldots, \xi_k \leftarrow_{\mathrm{OI},\bar{\alpha}} \bar{r})$. We note that $(s, \xi') = 0$ unless ξ' can be obtained from ξ by substituting the α_i's with suitable trees. Then the *OI-substitution of $\bar{r}$ into $r \in S\langle\langle T_\Sigma \rangle\rangle$ at $\bar{\alpha}$* is the tree series $r \leftarrow_{\mathrm{OI},\bar{\alpha}} \bar{r}$ in $S\langle\langle T_\Sigma \rangle\rangle$ which is defined to be $\sum_{\xi \in T_\Sigma} (r, \xi)(\xi \leftarrow_{\mathrm{OI},\bar{\alpha}} \bar{r})$. Note that the family $((r, \xi)(\xi \leftarrow_{\mathrm{OI},\bar{\alpha}} \bar{r}) \mid \xi \in T_\Sigma)$ of tree series is locally finite, and thus the summation is well defined.

Let $r_1, r_2 \in S\langle\langle T_\Sigma \rangle\rangle$ and $\alpha \in \Sigma^{(0)}$. The *α-concatenation of r_1 and r_2* is the tree series $r_1 \leftarrow_{\mathrm{OI},(\alpha)} (r_2)$, abbreviated by $r_1 \circ_\alpha r_2$.

Let $r \in S\langle\langle T_\Sigma \rangle\rangle$ and $\alpha \in \Sigma^{(0)}$. The *nth α-iteration of r* is the tree series $r^n_\alpha \in S\langle\langle T_\Sigma \rangle\rangle$ defined inductively as follows: $r^0_\alpha = \widetilde{0}$ and for every $n \geq 0$, $r^{n+1}_\alpha = r \circ_\alpha r^n_\alpha + 1.\alpha$. A tree series $r \in S\langle\langle T_\Sigma \rangle\rangle$ is *α-proper* if $(r, \alpha) = 0$. For every α-proper $r \in S\langle\langle T_\Sigma \rangle\rangle$, $\xi \in T_\Sigma$, and $n \geq \mathrm{height}(\xi) + 1$, we have that $(r^{n+1}_\alpha, \xi) = (r^n_\alpha, \xi)$ (cf. [41], Lemma 3.10). Then, for every $r \in S\langle\langle T_\Sigma \rangle\rangle$, the *$\alpha$-Kleene star of r* is the tree series $r^*_\alpha \in S\langle\langle T_\Sigma \rangle\rangle$ defined as follows. If r is α-proper, then $(r^*_\alpha, \xi) = (r^{\mathrm{height}(\xi)+1}_\alpha, \xi)$ for every $\xi \in T_\Sigma$; otherwise, $r^*_\alpha = \widetilde{0}$.

Next, let $\tau : \Sigma \to \mathcal{P}(\Delta)$ be a *relabeling (from Σ to Δ)*, i.e., a mapping such that $\tau(\sigma) \subseteq \Delta^{(k)}$ for every $k \geq 0$ and $\sigma \in \Sigma^{(k)}$. This mapping is extended in a canonical way to a mapping $\tau : T_\Sigma \to \mathcal{P}(T_\Delta)$, and then to a mapping $\tau : S\langle\langle T_\Sigma \rangle\rangle \to S\langle\langle T_\Delta \rangle\rangle$ by defining $(\tau(r), \xi) = \sum_{\zeta \in T_\Sigma, \xi \in \tau(\zeta)} (r, \zeta)$ for every $r \in S\langle\langle T_\Sigma \rangle\rangle$ and $\xi \in T_\Delta$ (note that the summation is finite).

In fact, recognizability of tree series is preserved under the aforementioned operations. The proofs of these closure results are sometimes folklore and sometimes straightforward generalizations of the corresponding results for formal power series over strings or for recognizable tree languages (cf., e.g., [45, 125, 7, 96, 69, 21, 91]). In particular, in view of Theorem 3.6, we can use results of [41] in which wta is defined with Boolean root weights.

Theorem 3.8. *Let S be commutative. Then* $\mathrm{Rec}(\Sigma, S)$ *is closed under:*

- *scalar multiplication ([41], Lemma 6.3),*
- *sum ([41], Lemma 6.4),*
- *Hadamard-product ([12], Corollary 3.9, also cf. [7], Proposition 5.1 for a field S),*
- *top-concatenation ([41], Lemmas 6.1 and 6.2),*
- *α-concatenation ([41], Lemma 6.5),*
- *α-Kleene star ([41], an easy adaptation of Lemma 6.7), and*
- *OI-substitution.*

Moreover, if $r \in$ $\mathrm{Rec}(\Sigma, S)$ *and τ is a relabeling from Σ to Δ, then $\tau(r) \in$* $\mathrm{Rec}(\Delta, S)$ *([43], Lemma 3.4, also cf. [87], Corollary 14 for continuous semirings).*

Proof. We prove the closure under OI-substitution. It has been proved for so-called well ω-additive semirings in [21], Lemma 24. Now we show that it also holds for commutative semirings. For this, let $n \geq 0$, $\bar{\alpha} = (\alpha_1, \ldots, \alpha_n) \in (\Sigma^{(0)})^n$, $\bar{r} = (r_1, \ldots, r_n) \in S\langle\langle T_\Sigma \rangle\rangle^n$, and $r \in S\langle\langle T_\Sigma \rangle\rangle$. Then

$$r \leftarrow_{\mathrm{OI},\bar{\alpha}} \bar{r} = \tau_1'\big((r \circ_{\alpha_1} \tau_1(r_1)) \leftarrow_{\mathrm{OI},(\alpha_2,\ldots,\alpha_n)} (r_2, \ldots, r_n)\big)$$

where $\tau_1 : \Sigma \to \mathcal{P}(\Sigma_1)$ is the relabeling with $\Sigma_1 = \{\sigma_1 \mid \sigma \in \Sigma\}$ and $\tau_1(\sigma) = \{\sigma_1\}$; moreover, $\tau_1' : \Sigma \cup \Sigma_1 \to \mathcal{P}(\Sigma)$ is the relabeling with $\tau_1'(\sigma) = \tau_1'(\sigma_1) = \{\sigma\}$. This process can be repeated $n-1$ times where in the ith step we use τ_i, τ_i', and Σ_i instead of τ_1, τ_1', and Σ_1, respectively. Since recognizability is preserved under α-concatenation and also under relabeling, we obtain that $\mathrm{Rec}(\Sigma, S)$ is closed under OI-substitution for commutative semirings. □

Recognizability is also preserved under semiring homomorphisms. Formally, let S' be another semiring and $h : S \to S'$ a mapping; h extends to the mapping $h : S\langle\langle T_\Sigma \rangle\rangle \to S'\langle\langle T_\Sigma \rangle\rangle$ by defining $h(r) = h \circ r$ for every $r \in S\langle\langle T_\Sigma \rangle\rangle$. If h is a semiring homomorphism, then the preservation of recognizability is obtained by replacing every transition weight $\mu_k(\sigma)_{w,q}$ and every root weight ν_q of the wta over S by $h(\mu_k(\sigma)_{w,q})$ and $h(\nu_q)$, respectively.

Theorem 3.9 ([16], Lemma 3). *Recognizability is preserved under semiring homomorphisms, i.e., for every semiring homomorphism $h : S \to S'$ and $r \in S\langle\langle T_\Sigma \rangle\rangle$, if $r \in \mathrm{Rec}(\Sigma, S)$, then $h(r) \in \mathrm{Rec}(\Sigma, S')$.*

3.4 Support of Recognizable Tree Series

It is straightforward to embed the class $\mathrm{Rec}(\Sigma)$ of recognizable tree languages into the class $\text{bud-Rec}(\Sigma, S)$ by using the support function. First, however, we prove a useful technical lemma.

Lemma 3.10. *Let (Q, θ) be a finite Σ-algebra and $f : Q \to S$ a mapping. Then $f \circ h_Q \in \text{bud-Rec}(\Sigma, S)$.*

Proof. We construct the bu-deterministic wta $\mathcal{A} = (Q, \Sigma, S, \mu, f)$ by defining $\mu_k(\sigma) = 1_{(S,\theta(\sigma))}$; note that f is an element of S^Q. Clearly, $h_\mu(\xi)_q = 1$ if $h_Q(\xi) = q$, and 0, otherwise. Then $(r_\mathcal{A}, \xi) = h_\mu(\xi) \cdot f = h_\mu(\xi)_{h_Q(\xi)} \cdot f(h_Q(\xi)) = f(h_Q(\xi))$ for every $\xi \in T_\Sigma$. Hence, $f \circ h_Q \in \text{bud-Rec}(\Sigma, S)$. □

Lemma 3.11. *If $L \in \mathrm{Rec}(\Sigma)$, then $1_{(S,L)} \in \text{bud-Rec}(\Sigma, S)$. In particular, $\mathrm{Rec}(\Sigma) \subseteq \mathrm{supp}(\text{bud-Rec}(\Sigma, S))$.*

Proof. Let $L \in \mathrm{Rec}(\Sigma)$. Hence, there is a deterministic bottom-up tree automaton $\mathcal{A} = (Q, \Sigma, \delta, F)$ with $L = L_\mathcal{A}$. Then $1_{(S,L)} = 1_{(S,F)} \circ h_Q$, and by Lemma 3.10 it follows that $1_{(S,L)} \in \text{bud-Rec}(\Sigma, S)$. □

In the Boolean case, wta computes exactly the class of recognizable tree languages, i.e., $\mathrm{supp}(\mathrm{Rec}(\Sigma, \mathbb{B})) = \mathrm{Rec}(\Sigma)$. However, this equality also holds for a larger class of semirings (also cf. [125], Corollary II.5.3 for formal power series).

Theorem 3.12. *Let S be positive. Then* $\mathrm{supp}(\mathrm{Rec}(\Sigma, S)) = \mathrm{Rec}(\Sigma)$.

Proof. Consider the particular mapping $h : S \to \mathbb{B}$ defined by $h(a) = 1$ for $a \neq 0$ and $h(0) = 0$. Since S is positive, h is a semiring homomorphism from S to $\mathbb{B}$. Moreover, $\mathrm{supp}(h(r)) = \mathrm{supp}(r)$ for every $r \in S\langle\langle T_\Sigma \rangle\rangle$. By Theorem 3.9, we have $h(r) \in \mathrm{Rec}(\Sigma, \mathbb{B})$ for every $r \in \mathrm{Rec}(\Sigma, S)$, and thus $\mathrm{supp}(r) = \mathrm{supp}(h(r)) \in \mathrm{supp}(\mathrm{Rec}(\Sigma, \mathbb{B})) = \mathrm{Rec}(\Sigma)$. The other inclusion holds by Lemma 3.11. □

On the other hand, there is a ranked alphabet Σ such that $\mathrm{supp}(\mathrm{Rec_B}(\Sigma, \mathbb{Z})) \setminus \mathrm{Rec}(\Sigma) \neq \emptyset$, where $\mathbb{Z}$ is the semiring of integers. This follows directly from the fact that there is a weighted finite automaton $\mathcal{A}$ such that $\mathrm{supp}(\|\mathcal{A}\|)$ is not recognizable (cf. [124], Example 6.2 and [8], Example III.3.1) and that weighted finite automata can be simulated by our wta in the way which we described on page 324.

We note that Theorem 3.12 holds for formal power series over commutative and (so-called) quasi-positive semirings, cf. [138], Corollary 5.2. Obviously, Theorem 3.12 shows that also the inverse of the implication of Lemma 3.11 holds for positive semirings. This is even true for the larger class of proper semirings.

Theorem 3.13. *Let S be a commutative and proper semiring. Then $L \in \mathrm{Rec}(\Sigma)$ iff $1_{(S,L)} \in \mathrm{Rec}(\Sigma, S)$ for every Σ-tree language L.*

Proof. Let $1_{(S,L)} \in \mathrm{Rec}(\Sigma, S)$. By [137], Theorem 2.1, the fact that S is not a ring implies that there is a semiring homomorphism $h : S \to \mathbb{B}$. By Theorem 3.9, we have $h(1_{(S,L)}) = 1_{(\mathbb{B},L)} \in \mathrm{Rec}(\Sigma, \mathbb{B})$. Thus $L = \mathrm{supp}(1_{(\mathbb{B},L)}) \in \mathrm{supp}(\mathrm{Rec}(\Sigma, \mathbb{B})) = \mathrm{Rec}(\Sigma)$. The inverse is proved in Lemma 3.11. □

In the following we show that also the inverse of Lemma 3.10 holds provided that the semiring is locally finite. Then a tree series $r \in \mathrm{Rec}(\Sigma, S)$ can be computed by a finite Σ-algebra. The construction in the next lemma is the one of [12], Sect. 4 (also cf. [5], Theorem 2.1; [80], Sect. 3.1; [16], Theorem 9; and [43], Lemma 6.1).

Lemma 3.14. *Let S be locally finite and $r \in \mathrm{Rec}(\Sigma, S)$. Then there is a finite Σ-algebra Q and a mapping $f : Q \to S$ such that $r = f \circ h_Q$.*

Proof. Let $\mathcal{A} = (P, \Sigma, S, \mu, \nu)$ be a wta such that $r_\mathcal{A} = r$. Let S' be the smallest subsemiring containing $\mathrm{wts}(\mathcal{A})$. Since S is locally finite, S' is finite. Now we consider the Σ-algebra $(Q, \mu'_\mathcal{A})$ where $Q = (S')^P$ and $\mu'_\mathcal{A}(\sigma)$ is the restriction of $\mu_\mathcal{A}(\sigma)$ to Q^k for every $k \geq 0$. Moreover, we define the mapping f by $f(v) = v \cdot \nu$. Clearly, $h_\mu(\xi) \in Q$ and $h_\mu(\xi) = h_Q(\xi)$ for every $\xi \in T_\Sigma$, and thus we have $(r, \xi) = h_\mu(\xi) \cdot \nu = h_Q(\xi) \cdot \nu = f(h_Q(\xi))$. □

Using Lemma 3.14, we can prove the following inverse image theorem.

Theorem 3.15 ([19], Theorem 4). *Let S be locally finite, $E \subseteq S$, and $r \in \mathrm{Rec}(\Sigma, S)$. Then $r^{-1}(E) \in \mathrm{Rec}(\Sigma)$.*

Proof. By Lemma 3.14, there is a finite Σ-algebra (Q, θ) and a mapping $f : Q \to S$ such that $r = f \circ h_Q$. Now we construct the deterministic bottom-up tree automaton $\mathcal{A} = (Q, \Sigma, \delta, F)$, where $\delta_\sigma = \theta(\sigma)$ and $F = \{q \in Q \mid f(q) \in E\}$. Then $(r, \xi) \in E$ iff $f(h_Q(\xi)) \in E$ iff $h_Q(\xi) \in F$ iff $\xi \in L_{\mathcal{A}}$. □

From the inverse image theorem, we can derive the following results for a recognizable tree series r over a locally finite S. We can show that the support of r is recognizable. Also, assuming that $(S, \leq)$ is a partially ordered set, all cut sets of r are recognizable; for every $a \in S$, the *a-cut* of r is the set $r_a = \{\xi \in T_\Sigma \mid (r, \xi) \geq a\}$ (cf. [129]). Moreover, we can prove that r is a *recognizable step function*, i.e., there are $n \geq 0$, recognizable tree languages $L_1, \ldots, L_n \subseteq T_\Sigma$, and $a_1, \ldots, a_n \in S$ such that $r = \sum_{i=1}^{n} a_i 1_{(S,L_i)}$. (Clearly, a recognizable step function is a recognizable tree series provided S is commutative, by Lemma 3.11 and Theorem 3.8.)

Corollary 3.16. *Let S be locally finite and $r \in \mathrm{Rec}(\Sigma, S)$. Then*

(A) $\mathrm{supp}(r) \in \mathrm{Rec}(\Sigma)$.

(B) If $(S, \leq)$ is a partially ordered set, then $r_a \in \mathrm{Rec}(\Sigma)$ for every $a \in S$ ([16], Theorem 9).

(C) r is a recognizable step function ([43], Lemma 6.1).

Proof. Since $\mathrm{supp}(r) = r^{-1}(S \setminus \{0\})$ and $r_a = r^{-1}(\{b \in S \mid b \geq a\})$, (A) and (B) follow from Theorem 3.15. To prove (C), we observe that $r = \sum_{a \in R} a 1_{(S, r^{-1}(a))}$ where $R = \{(r, \xi) \mid \xi \in T_\Sigma\}$ is the range of r. Since R is finite by Lemma 3.14 (because $R \subseteq f(Q)$) and $r^{-1}(a) \in \mathrm{Rec}(\Sigma)$ by Theorem 3.15, r is a recognizable step function. □

3.5 Determinization of Weighted Tree Automata

It is well known that the usual power set construction for finite-state string automata can be extended in a straightforward way to bottom-up tree automata [68, 69]. This means that bottom-up tree automata and deterministic bottom-up tree automata accept the same class of tree languages. On the other hand, there are recognizable tree languages which are not recognizable by deterministic top-down tree automata.

If the power set construction is extended to a wta (by identifying $\mathcal{P}(Q)$ with $\mathbb{B}^Q$ and then turning $\mathbb{B}$ into S), this might lead to a deterministic wta with infinitely many states because its state set is the set of all reachable Q-vectors over S. However, infinity can be avoided if the semiring S is locally finite. The following theorem generalizes [14], Theorem 6.3.3 and [12], Theorem 4.8.

Theorem 3.17. *Let S be locally finite, $r \in \mathrm{Rec}(\Sigma, S)$, and $g : S \to S$. Then $g(r) \in \text{bud-Rec}(\Sigma, S)$. In particular, $\mathrm{Rec}(\Sigma, S) = \text{bud-Rec}(\Sigma, S)$.*

Proof. By Lemma 3.14, we have that $r = f \circ h_Q$ for some finite Σ-algebra Q and mapping $f : Q \to S$. Then $g(r) = g \circ (f \circ h_Q) = (g \circ f) \circ h_Q$, and thus by Lemma 3.10, $g(r) \in \text{bud-Rec}(\Sigma, S)$. □

In fact, for $S = \mathbb{B}$, the construction of Lemma 3.14 is exactly the usual power set construction for bottom-up tree automata. Also we note that, in general, the condition that S is locally finite cannot be dropped from Theorem 3.17. For this, we consider the tree series r over the field $(\mathbb{Q}, +, \cdot, 0, 1)$ of rational numbers such that $(r, \gamma^n(\alpha)) = 1 + 2^n$, where $\Sigma = \{\gamma^{(1)}, \alpha^{(0)}\}$. Then $r \in \text{Rec}(\Sigma, \mathbb{Q})$ by constructing a wta with two states q_1 and q_2 such that $h_\mu(\xi)_{q_1} = 1$ and $h_\mu(\xi)_{q_2} = 2^n$. Now assume that there is a bu-deterministic wta $\mathcal{A}$ such that $r = r_\mathcal{A}$. Then we can derive a contradiction by using the observation that, for every $\xi \in T_\Sigma$, the coefficient $(r_\mathcal{A}, \xi)$ is an element of the carrier set of the smallest sub-monoid of $(\mathbb{Q}, \cdot, 1)$ containing $\text{wts}(\mathcal{A})$; thus, $r \notin \text{bud-Rec}(\Sigma, \mathbb{Q})$ (cf. [17], Lemma 6.3).

Finally we mention that, similarly to the unweighted case, the easy tree series $r = 1.\sigma(\alpha, \ldots, \alpha, \beta) + 1.\sigma(\beta, \alpha, \ldots, \alpha)$ separates the classes $\text{Rec}(\Sigma, S)$ and $\text{tdd-Rec}(\Sigma, S)$.

Theorem 3.18 ([14], Theorem 6.3.5). *Let S be commutative or zero-divisor free. Moreover, let Σ contain at least two nullary symbols and at least one symbol with rank ≥ 2. Then* $\text{Rec}(\Sigma, S) \setminus \text{tdd-Rec}(\Sigma, S) \neq \emptyset$.

3.6 Pumping Lemmata and Decidability

For the class of recognizable tree languages, there is a well-known pumping lemma (cf. [69], Proposition 5.2). Here, we will present pumping lemmata and decidability results for recognizable tree series.

As a technical concept, we need contexts. The set C_Σ of *contexts* is the set of Σ-trees over $\{z\}$ in which z occurs exactly once. In fact, C_Σ is the free monoid freely generated by the set C'_Σ with operation $\cdot_z$ and $z \in C_\Sigma$ as neutral element, cf. [7], Proposition 9.1; $C'_\Sigma \subseteq C_\Sigma$ is the set of those contexts in which the z occurs at a child of the root. For every $\zeta \in C_\Sigma$, we define $\zeta^0 = z$ and $\zeta^{n+1} = \zeta^n \cdot_z \zeta$.

As additional technical preparation, we extend the homomorphism induced by a wta to contexts. Formally, let $\mathcal{A} = (Q, \Sigma, S, \mu, \nu)$ be a wta and $v \in S^Q$. Now we consider the v-extension $(S^Q, \mu^v_\mathcal{A})$ of the Σ-algebra $(S^Q, \mu_\mathcal{A})$ (as defined in Sect. 2.3) and denote the unique $\Sigma \cup \{z\}$-algebra homomorphism from $T_{\Sigma \cup \{z\}}$ to S^Q by h^v_μ. Then $h_\mu(\zeta \cdot_z \xi) = h^{h_\mu(\xi)}_\mu(\zeta)$ for every $\zeta \in C_\Sigma$ and $\xi \in T_\Sigma$. If v has the particular form that there is a $q \in Q$ with $v_q = 1$ and $v_p = 0$ for every $p \neq q$, then we abbreviate h^v_μ by h^q_μ.

The pumping lemma that we present first deals with recognizable tree series over fields.

Theorem 3.19 ([7], Theorem 9.2). *Let S be a field and $r \in \text{Rec}(\Sigma, S)$. There is an $m \geq 1$ such that, for every $\xi \in \text{supp}(r)$ with $\text{height}(\xi) \geq m$,* *there are* $\zeta, \zeta' \in C_\Sigma$, *and* $\xi' \in T_\Sigma$ *such that* $\xi = \zeta' \cdot_z \zeta \cdot_z \xi'$ *and we have that* $\{\zeta' \cdot_z \zeta^n \cdot_z \xi' \mid n \geq 0\} \cap \mathrm{supp}(r)$ *is an infinite set.*

The second pumping lemma concerns bu-deterministically recognizable tree series over an arbitrary semiring.

Theorem 3.20 ([12], Theorem 5.6). *Let* $r \in \text{bud-Rec}(\Sigma, S)$. *There is an* $m \geq 1$ *such that, for every* $\xi \in \mathrm{supp}(r)$ *and position* $w = i_1 \ldots i_l \in \mathrm{pos}(\xi)$ *with* $i_1, \ldots, i_l \in \mathbb{N}$ *and* $l \geq m$, *there are indices* j, k *with* $0 \leq j < k \leq l$ *and* $a, a', b, b', c \in S$ *such that:*

- $l - j \leq m$ *and*
- $(r, \zeta' \cdot_z \zeta^n \cdot_z \xi') = a' \cdot a^n \cdot c \cdot b^n \cdot b'$ *for every* $n \geq 0$ *where*
  - $\zeta' = \xi[z]_u$ *with* $u = i_1 \ldots i_j$,
  - $\zeta = (\xi|_u)[z]_v$ *with* $v = i_{j+1} \ldots i_k$, *and*
  - $\xi' = \xi|_{uv}$.

If S *is zero-divisor free, then* $\zeta' \cdot_z \zeta^n \cdot_z \xi' \in \mathrm{supp}(r)$ *for every* $n \geq 0$.

We give a sketch of the proof. Assume that $\mathcal{A} = (Q, \Sigma, S, \mu, \nu)$ is a total bu-deterministic wta which recognizes r and let $m = |Q|$. Now let $\xi \in T_\Sigma$ be an input tree. Then there is a unique run κ on ξ for which all transitions have nonzero weight, i.e., for every $w \in \mathrm{pos}(\xi)$, the condition $\mu_k(\sigma)_{q_1 \ldots q_k, q} \neq 0$ holds with $\sigma^{(k)} = \xi(w)$, $q = \kappa(w)$, and $q_i = \kappa(wi)$ for $1 \leq i \leq k$. Let us denote the state $\kappa(\varepsilon)$ by $\widetilde{\mu}(\xi)$. Now assume there is a $w \in \mathrm{pos}(\xi)$ with $|w| \geq m$, which implies $\mathrm{height}(\xi) \geq m$. Then the standard pumping can be done because there is a repetition of states along w, i.e., there are contexts ζ and ζ' and a tree ξ' such that $\xi = \zeta' \cdot_z \zeta \cdot_z \xi'$ and $\widetilde{\mu}(\zeta \cdot_z \xi') = \widetilde{\mu}(\xi')$. Let u (and v) be the position of z in ζ' (and ζ, resp.). Then the element a of S is the product of the weights of all the transitions which are performed (in κ) at positions v' of ζ such that v' is lexicographically smaller than v and v' is not a prefix of v; b is the product of the weights of all the transitions (in κ) at the other positions of ζ except v; in both cases the order of the factors is determined by the left-to-right traversal over ζ. The elements a' and b' are defined similarly for u and ζ' instead of v and ζ, except that b' contains the root weight $\nu(\widetilde{\mu}(\xi))$ as an additional factor. Finally, c is the product of the weights of all the transitions which are performed at positions of ξ'. Since S may contain zero-divisors, even the unique run κ on ξ can have weight 0. However, if S is zero-divisor free and $\xi \in \mathrm{supp}(r)$, then $\mathrm{wt}(\kappa) \neq 0$, and thus also $\zeta' \cdot_z \zeta^n \cdot_z \xi' \in \mathrm{supp}(r)$.

Using Theorem 3.20, it can, e.g., be proved that the tree series height is not in bud-Rec(Σ, Arct) (cf. [12], Example 5.9). Also, this pumping lemma can be used to prove decidability results which we discuss here for the question whether a tree series is constant. We assume that the semiring S is effectively given and also the considered tree series r is effectively given by a total bu-deterministic wta $\mathcal{A} = (Q, \Sigma, S, \mu, \nu)$. Let P be the set of all those $\widetilde{\mu}(\xi) \in Q$, where $\xi \in T_\Sigma$ with $\mathrm{height}(\xi) \leq |Q| - 1$ and there is $\zeta \in C_\Sigma$ such that $\mathrm{height}(\zeta) \leq 2 \cdot |Q| - 2$, and $\zeta \cdot_z \xi \in \mathrm{supp}(r)$. Intuitively, P is the set of states which are reachable by small trees ξ and there is a small context ζ such that $\zeta \cdot_z \xi$ is in the support of r. As the final preparation, we define the sets B_1 and B_2. The set B_1 is the set of weights of small contexts that can be pumped, i.e.,

$$B_1 = \{h_\mu^{\widetilde{\mu}(\xi)}(\zeta)_{\widetilde{\mu}(\xi)} \mid \xi \in T_\Sigma,\ \zeta \in C_\Sigma,\ \text{height}(\zeta \cdot_z \xi) \leq |Q|,$$
$$\widetilde{\mu}(\xi) = \widetilde{\mu}(\zeta \cdot_z \xi) \in P\}.$$

The set B_2 is the set of weights of small trees, i.e.,

$$B_2 = \{(r, \xi) \mid \xi \in T_\Sigma \text{ and } \text{height}(\xi) \leq |Q| - 1\}.$$

The sets P, B_1, and B_2 are finite sets which can be constructed effectively. Then the following key lemma can be shown, making essential use of Theorem 3.20.

Lemma 3.21 ([12], Lemma 6.3). *Let S be commutative and $d \in S$. Moreover, let $r \in$ bud-Rec(Σ, S). Then $(r, \xi) = d$ for every $\xi \in$ supp(r) iff (i) $b \cdot d \in \{0, d\}$ for every $b \in B_1$ and (ii) $B_2 \subseteq \{0, d\}$.*

Theorem 3.22 ([12], Sect. 6). *Let S be commutative and $r \in$ bud-Rec(Σ, S). Then the following problems are decidable:*

(A) Constant-on-its-support problem, i.e., is there an $a \in S$ such that $(r, \xi) = a$ for every $\xi \in$ supp(r)?
(B) Constant tree series problem, i.e., is there an $a \in S$ such that $(r, \xi) = a$ for every $\xi \in T_\Sigma$?
(C) Emptiness problem, i.e., is $r = \widetilde{0}$ (or equivalently, is supp$(r) = \emptyset$)?
(D) Boolean tree series problem, i.e., is $(r, \xi) \in \{0, 1\}$ for every $\xi \in T_\Sigma$?

Proof. First, we prove (A). By Lemma 3.21, we know that r is constant on its support iff there is a $d \in S$ such that conditions (i) and (ii) hold for this d. Now the decision procedure computes B_2. If $|B_2| > 2$ or $|B_2| = 2$ and $0 \notin B_2$, then r is not constant on its support. If $B_2 = \{0\}$, then (i) holds with $d = 0$, hence r is constant on its support, in fact, $r = \widetilde{0}$. If $B_2 = \{d\}$ or $B_2 = \{0, d\}$ for some $d \neq 0$, then check whether $b \cdot d \in \{0, d\}$ for every $b \in B_1$. If yes, then r is constant on its support with value d, if no, then it is not constant.

Proof of (B): This can be proved in a similar way to the first statement by first proving the modification of Lemma 3.21 in which $\xi \in$ supp(r) and $\{0, d\}$ are replaced by $\xi \in T_\Sigma$ and $\{d\}$, respectively. Proof of (C) and (D): These statements follow directly from Lemma 3.21 with $d = 0$ and $d = 1$, respectively. □

For the decision of the finiteness of supp(r) we additionally require that S is zero-divisor free. Then, by Theorem 3.20, if $(r, \zeta' \cdot_z \zeta \cdot_z \xi') \neq 0$, then also $(r, \zeta' \cdot_z \zeta^n \cdot_z \xi') \neq 0$ for every $n \geq 0$. Hence, supp(r) is finite iff height$(\xi) \leq$

$|Q|-1$ for every $\xi \in \mathrm{supp}(r)$. Thus, $\mathrm{supp}(r)$ is finite iff $\mathrm{supp}(r') = \emptyset$, where r' is the tree series defined by $(r', \xi) = (r, \xi)$ if $\mathrm{height}(\xi) \geq |Q|$, and $(r', \xi) = 0$ otherwise. It is not difficult to show that r' is in bud-Rec(Σ, S), effectively (see [12], Lemma 6.10). Thus, finiteness of r can be decided by Theorem 3.22(C): the emptiness problem.

Theorem 3.23 ([12], Theorem 6.11). *Let S be commutative and zero-divisor free, and $r \in$ bud-Rec(Σ, S) a recognizable tree series. Then the finiteness problem is decidable, i.e., it is decidable whether $\mathrm{supp}(r)$ is a finite set.*

In [103], the emptiness problem has been considered for arbitrary tree series in $r \in \mathrm{Rec}(\Sigma, S)$. This result, which is reported in the next theorem, is based on (i) a pumping lemma for deterministic wta over distributive Ω-algebras (cf. [103], Theorem 4), (ii) a decidable property which characterizes the emptiness of the tree series recognized by deterministic wta over zero-sum free distributive Ω-algebras (cf. [103], Proposition 4), and (iii) the simulation of a wta by a wta over a particular distributive Ω-algebra (cf. [103], Proposition 2).

Theorem 3.24 ([103], Corollary 3). *Let S be commutative and zero-sum free and $r \in \mathrm{Rec}(\Sigma, S)$. Then it is decidable whether $r = \widetilde{0}$.*

The emptiness problem is also decidable if S is a field. The proof, given in [127] on the base of [45], exploits some methods of linear algebra in an elegant way. This is possible because now the Σ-algebra $(S^Q, \mu_{\mathcal{A}})$ which is associated to a wta $\mathcal{A}$, is a finite-dimensional S-Σ-vector space; recall that this means that $(S^Q, +, 0)$ is an S-vector space and the mappings $\mu_{\mathcal{A}}(\sigma)$ are multilinear. As preparation, we recall a well-known statement from linear algebra.

Lemma 3.25. *Let V be a finite-dimensional S-vector space and let $V' \subseteq V$ be a subspace of V. Then $\dim(V') \leq \dim(V)$; moreover, if $\dim(V') = \dim(V)$, then $V' = V$.*

Theorem 3.26 ([127], Theorem 4.2; [18], Lemma 2). *Let S be a field and $r \in \mathrm{Rec}(\Sigma, S)$. Then it is decidable whether $r = \widetilde{0}$.*

Proof. Let $\mathcal{A} = (Q, \Sigma, S, \mu, \nu)$ be a wta with $r_{\mathcal{A}} = r$. Assume that $|Q| = n$. It suffices to prove that

$$r_{\mathcal{A}} = \widetilde{0} \quad \text{iff} \quad (r_{\mathcal{A}}, \xi) = 0 \text{ for every } \xi \in T_\Sigma \text{ with } \mathrm{height}(\xi) \leq n \tag{1}$$

because the latter property is decidable.

Note that S^Q is an n-dimensional S-vector space. For every $m \geq 0$, we define the subspace $V_m = \langle \{h_\mu(\xi) \mid \xi \in T_\Sigma, \mathrm{height}(\xi) \leq m\} \rangle$ generated by the vectors $h_\mu(\xi)$ for trees ξ of height at most m. This forms the chain $V_0 \subseteq V_1 \subseteq \cdots \subseteq V_m \subseteq V_{m+1} \subseteq \cdots \subseteq S^Q$ of subspaces. By Lemma 3.25, we have $\dim(V_m) \leq \dim(V_{m+1}) \leq n$ for every $m \geq 0$. Moreover, it is easy to see that $V_{m+1} = \langle\{\mu_{\mathcal{A}}(\sigma)(v_1,\ldots,v_k) \mid k \geq 0, \sigma \in \Sigma^{(k)}, v_1,\ldots,v_k \in V_m\}\rangle$. This characterization of V_{m+1} in terms of V_m proves (by a straightforward induction on l) that if $V_m = V_{m+1}$ for some $m \geq 0$, then $V_m = V_{m+l}$ for every $l \geq 1$.

Moreover, there must be an m_0 such that $0 \leq m_0 \leq n$ and $\dim(V_{m_0}) = \dim(V_{m_0+1})$. By Lemma 3.25, we obtain that $V_{m_0} = V_{m_0+1}$, and thus $V_{m_0} = V_{m_0+l}$ for every $l \geq 1$. Hence, $V_n = \bigcup_{m \geq 0} V_m = \langle\{h_\mu(\xi) \mid \xi \in T_\Sigma\}\rangle$.

Now we can verify equivalence (1) as follows: $r_{\mathcal{A}} = \widetilde{0}$ iff $(r_{\mathcal{A}}, \xi) = 0$ for every $\xi \in T_\Sigma$ iff $h_\mu(\xi) \cdot \nu = 0$ for every $\xi \in T_\Sigma$ iff $h_\mu(\xi) \cdot \nu = 0$ for every $\xi \in T_\Sigma$ with $\text{height}(\xi) \leq n$ iff $(r_{\mathcal{A}}, \xi) = 0$ for every $\xi \in T_\Sigma$ with $\text{height}(\xi) \leq n$, where we prove the "if" part of the last, but one equivalence in the following way. Let $\xi \in T_\Sigma$. Since $h_\mu(\xi) \in V_n$, it can be written as a linear combination $h_\mu(\xi) = \sum_{i=1}^{l} a_i h_\mu(\xi_i)$ for some $l \geq 1$, $a_1,\ldots,a_l \in S$, and trees $\xi_1,\ldots,\xi_l \in T_\Sigma$ of height at most n. Then by an easy calculation in S^Q, we obtain $h_\mu(\xi) \cdot \nu = \sum_{i=1}^{l} a_i \cdot (h_\mu(\xi_i) \cdot \nu)$. This proves equivalence (1), and thus the theorem. □

As a corollary, we obtain that the equivalence problem of recognizable tree series over a field is decidable.

Corollary 3.27 ([127], Theorem 4.2; [18], Lemma 2). *Let S be a field and $r_1, r_2 \in \text{Rec}(\Sigma, S)$. Then it is decidable whether $r_1 = r_2$.*

Proof. Let $r_1, r_2 \in \text{Rec}(\Sigma, S)$ be effectively given. Then by Theorem 3.8, which is effective, also $r = r_1 + (-1) \cdot r_2$ is in $\text{Rec}(\Sigma, S)$. Clearly, $r_1 = r_2$ iff $r = \widetilde{0}$, which is decidable by Theorem 3.26. □

For the particular semirings $(\mathbb{R}_+, +, \cdot, 0, 1)$ (of non-negative reals) and Nat, the decidability of the equivalence of recognizable tree series has been proved in [20].

3.7 Finite Algebraic Characterizations of Recognizable Tree Series

The Myhill–Nerode theorem for recognizable string languages has been extended to recognizable tree languages [100, 130, 86, 69]. That is, a Σ-tree language is recognizable if and only if its syntactic Σ-algebra is finite. In this section, we discuss three similar characterizations of recognizable tree series where the characterizations are based on fields (cf. Theorem 3.31), semifields (cf. Theorem 3.35), and commutative and zero-divisor free semirings (cf. Theorem 3.36), respectively.

In this section, we will again use the notations C_Σ, h_μ^v, and h_μ^q introduced for contexts in the beginning of Sect. 3.6.

For the development of these characterizations, we will use both the right and the left quotient of a tree series. For every $r \in S\langle\langle T_\Sigma\rangle\rangle$ and $\zeta \in C_\Sigma$, the *right quotient of r with respect to ζ* is the tree series $r\zeta^{-1} \in S\langle\langle T_\Sigma\rangle\rangle$, where

$(r\zeta^{-1}, \xi) = (r, \zeta \cdot_z \xi)$ for every $\xi \in T_\Sigma$. For the definition of the left quotient, we need mappings of the type $C_\Sigma \to S$. Since they are very similar to tree series, we can adapt the notions and operations from tree series to this setting. We call a mapping of this type a *context series* and denote the class of all context series by $S\langle\langle C_\Sigma\rangle\rangle$. Then for every $r \in S\langle\langle T_\Sigma\rangle\rangle$ and $\xi \in T_\Sigma$, the *left quotient of r with respect to* ξ is the context series $\xi^{-1}r \in S\langle\langle C_\Sigma\rangle\rangle$ defined by $(\xi^{-1}r, \zeta) = (r, \zeta \cdot_z \xi)$ for every $\zeta \in C_\Sigma$.

Characterizations for Fields

In this subsection, we assume that S is a field.

Since S is a field, both $(S\langle\langle T_\Sigma\rangle\rangle, +, \widetilde{0})$ and $(S\langle\langle C_\Sigma\rangle\rangle, +, \widetilde{0})$ are S-vector spaces. For every $r \in S\langle\langle T_\Sigma\rangle\rangle$, we denote by RQ_r the subspace of $S\langle\langle T_\Sigma\rangle\rangle$ generated by all the right quotients $r\zeta^{-1}$ for $\zeta \in C_\Sigma$, and by LQ_r the subspace of $S\langle\langle C_\Sigma\rangle\rangle$ generated by all the left quotients $\xi^{-1}r$ for $\xi \in T_\Sigma$. Then we can prove the following relation.

Lemma 3.28 ([24], Theorem 3.1). *Let $r \in S\langle\langle T_\Sigma\rangle\rangle$. Then RQ_r is finite-dimensional iff LQ_r is finite-dimensional, and in this case* $\dim(RQ_r) = \dim(LQ_r)$.

Proof. Assume LQ_r is n-dimensional and let $\xi_1^{-1}r, \ldots, \xi_n^{-1}r$ be a basis of LQ_r. Consider the mapping $\psi : RQ_r \to S^n$, where $\psi(r\zeta^{-1}) = ((r, \zeta \cdot_z \xi_1), \ldots, (r, \zeta \cdot_z \xi_n))$ for every context $\zeta \in C_\Sigma$, and then ψ is linearly extended to RQ_r. We can prove that ψ is injective, which implies $\dim(RQ_r) \leq n$. To prove the injectivity of ψ, we take an arbitrary element $s = a_1(r\zeta_1^{-1}) + \cdots + a_m(r\zeta_m^{-1})$ of RQ_r and show that $\psi(s) = 0^n$ implies $s = \widetilde{0}$. In fact, $\psi(s) = 0^n$ means $\sum_{i=1}^m a_i \cdot (r, \zeta_i \cdot_z \xi_j) = 0$ for $j = 1, \ldots, n$. Now for every $\xi \in T_\Sigma$, we have $(s, \xi) = \sum_{i=1}^m a_i \cdot (r\zeta_i^{-1}, \xi) = \sum_{i=1}^m a_i \cdot (\xi^{-1}r, \zeta_i)$. By letting $\xi^{-1}r = \sum_{j=1}^n b_j(\xi_j^{-1}r)$ and reordering the members of the sum appropriately, we obtain $(s, \xi) = 0$. Analogously, by assuming that RQ_r is finite-dimensional, we can prove that $\dim(LQ_r) \leq \dim(RQ_r)$. □

For every recognizable tree series r, the S-vector space LQ_r is finite-dimensional. In [24], Theorem 2.1 even the equivalence was proved, i.e., $r \in \mathrm{Rec}(\Sigma, S)$ iff LQ_r is finite-dimensional. However, since we will have a slightly different proof of this equivalence (cf. Theorem 3.31), we now cite only the mentioned implication.

Lemma 3.29 ([24], Theorem 2.1). *Let $r \in S\langle\langle T_\Sigma\rangle\rangle$. If $r \in \mathrm{Rec}(\Sigma, S)$, then LQ_r is finite-dimensional.*

Proof. Let $\mathcal{A} = (Q, \Sigma, S, \mu, \nu)$ be a wta such that $r_\mathcal{A} = r$. We define the mapping $\varphi : S^Q \to S\langle\langle C_\Sigma\rangle\rangle$ by $(\varphi(v), \zeta) = h_\mu^v(\zeta) \cdot \nu$ for every $v \in S^Q$ and $\zeta \in C_\Sigma$. The fact that all the mappings $\mu_\mathcal{A}^v$ are multilinear implies that φ

is a linear mapping between the two S-vector spaces S^Q and $S\langle\langle C_\Sigma\rangle\rangle$. Then $(\varphi(h_\mu(\xi)),\zeta) = h_\mu^{h_\mu(\xi)}(\zeta)\cdot\nu = h_\mu(\zeta\cdot_z\xi)\cdot\nu = (r,\zeta\cdot_z\xi) = (\xi^{-1}r,\zeta)$, and thus $\varphi(h_\mu(\xi)) = \xi^{-1}r$. Hence LQ_r is a subspace of the range $\mathrm{ran}(\varphi)$ of φ, and thus $\dim(LQ_r) \leq \dim(\mathrm{ran}(\varphi))$. Since, in general, for a linear mapping $\varphi : V \to V'$ between two S-vector spaces V and V' the dimension of $\mathrm{ran}(\varphi)$ cannot be larger than $\dim(V)$, we obtain that in our case $\dim(\mathrm{ran}(\varphi)) \leq \dim(S^Q) = |Q|$. Thus, LQ_r is finite-dimensional. □

Next, we recall from [23, 18] the facts that the S-vector space LQ_r can be enriched to an S-Σ-vector space and that LQ_r and the so-called syntactic S-Σ-vector space of r are isomorphic.

The enrichment of the S-vector space $(LQ_r,+,\widetilde{0})$ with a Σ-algebraic structure is done in the following way. Since every S-vector space has a basis (assuming Zorn's lemma), also LQ_r has a basis, say B. Now for every $\sigma \in \Sigma^{(k)}$, we define the mapping $\theta_r(\sigma) : B^k \to LQ_r$ by $\theta_r(\sigma)(\eta_1^{-1}r,\ldots,\eta_k^{-1}r) = \sigma(\eta_1,\ldots,\eta_k)^{-1}r$ for all base vectors $\eta_1^{-1}r,\ldots,\eta_k^{-1}r \in B$. Then we extend $\theta_r(\sigma)$ to a k-ary multilinear mapping on LQ_r which we also denote by $\theta_r(\sigma)$. Thus, we obtain the S-Σ-vector space $(LQ_r,+,\widetilde{0},\theta_r)$.

For the definition of the syntactic S-Σ-vector space of r, we consider the S-Σ-vector space $(S\langle T_\Sigma\rangle,+,\widetilde{0},\mathrm{top})$ of polynomials with $\mathrm{top}(\sigma) = \mathrm{top}_\sigma$ for every $\sigma \in \Sigma$. This is the initial algebra in the class of all S-Σ-vector spaces. Now let $r \in S\langle\langle T_\Sigma\rangle\rangle$ be a tree series. For a polynomial $s = a_1.\xi_1 + \cdots + a_k.\xi_k$ in $S\langle T_\Sigma\rangle$, we define the *left quotient of* r *with respect to* s by letting $s^{-1}r = a_1(\xi_1^{-1}r) + \cdots + a_k(\xi_k^{-1}r)$. Then we define the equivalence relation $\sim_r$ over $S\langle T_\Sigma\rangle$ such that for every $s_1,s_2 \in S\langle T_\Sigma\rangle$ we have $s_1 \sim_r s_2$ if and only if $s_1^{-1}r = s_2^{-1}r$. It is not difficult to prove that $\sim_r$ is a congruence relation over the S-Σ-vector space $S\langle T_\Sigma\rangle$. We call $\sim_r$ the *syntactic congruence* of r, and we call the quotient space $(S\langle T_\Sigma\rangle/\sim_r,+_{\sim_r},[\widetilde{0}]_{\sim_r},\mathrm{top}_{\sim_r})$ the *syntactic* S-Σ-*vector space of* r.

Next, we relate the two S-Σ-vector spaces $S\langle T_\Sigma\rangle/\sim_r$ and LQ_r. For the initial homomorphism $\Phi_r : S\langle T_\Sigma\rangle \to LQ_r$, it is easy to prove that $\Phi_r(s) = s^{-1}r$ for every $s \in S\langle T_\Sigma\rangle$ and that Φ_r is surjective. Since the kernel of Φ_r is $\sim_r$, we immediately obtain the following result by applying the homomorphism theorem of universal algebra ([73], Theorem 11.1).

Lemma 3.30 ([23], Proposition 3). *For every $r \in S\langle\langle T_\Sigma\rangle\rangle$, the S-Σ-vector spaces $S\langle T_\Sigma\rangle/\sim_r$ and LQ_r are isomorphic.*

Now we can prove the first Myhill–Nerode-like theorem for recognizable tree series. For this, let $r \in S\langle\langle T_\Sigma\rangle\rangle$ and $\sim$ be a congruence on the S-Σ-vector space $S\langle T_\Sigma\rangle$. We say that $\sim$ *respects* r if there is a linear form $\gamma : S\langle T_\Sigma\rangle/\sim \to S$ such that $(r,\xi) = \gamma([\xi]_\sim)$ for every $\xi \in T_\Sigma$.

Theorem 3.31 ([24], Theorems 2.1 and 3.1; [23], Propositions 2 and 3). *Let $r \in S\langle\langle T_\Sigma\rangle\rangle$. Then the following five statements are equivalent:*

(A) $r \in \mathrm{Rec}(\Sigma,S)$.

(B) There is a congruence $\sim$ on $S\langle T_\Sigma\rangle$ such that $S\langle T_\Sigma\rangle/\sim$ is finite-dimensional and $\sim$ respects r.
(C) The S-vector space $S\langle T_\Sigma\rangle/\sim_r$ is finite-dimensional.
(D) The S-vector space RQ_r is finite-dimensional.
(E) The S-vector space LQ_r is finite-dimensional.

Proof. Statement (A) implies statement (E) by Lemma 3.29. Statements (C), (D), and (E) are equivalent due to Lemmata 3.28 and 3.30. Next, we prove that statement (C) implies statement (B). For this, let $\sim = \sim_r$ and define the linear form $\gamma : S\langle T_\Sigma\rangle/\sim \to S$ such that $\gamma([s]_\sim) = (s^{-1}r, z)$. Since $(s^{-1}r, z) = a_1 \cdot (r, \xi_1) + \cdots + a_k \cdot (r, \xi_k)$ for every polynomial $s = a_1.\xi_1 + \cdots + a_k.\xi_k$, γ is a linear form. Moreover, we have $\gamma([\xi]_\sim) = (r, \xi)$.

Finally, we prove that statement (B) implies statement (A), where we abbreviate an equivalence class $[\xi]_\sim$ by $[\xi]$. Let $Q = \{[\xi_1], \ldots, [\xi_n]\}$, where $\xi_1, \ldots, \xi_n \in T_\Sigma$, be a basis of $S\langle T_\Sigma\rangle/\sim$ and construct the wta $\mathcal{A} = (Q, \Sigma, S, \mu, \nu)$ in the following way. For every $k \geq 0$, $\sigma \in \Sigma^{(k)}$, and $1 \leq i, i_1, \ldots, i_k \leq n$, let $\mu_k(\sigma)_{[\xi_{i_1}]\ldots[\xi_{i_k}],[\xi_i]} = [\sigma(\xi_{i_1}, \ldots, \xi_{i_k})]_{[\xi_i]}$, where the latter denotes the coefficient of $[\xi_i]$ in the representation of $[\sigma(\xi_{i_1}, \ldots, \xi_{i_k})]$ as a linear combination of the base vectors. Moreover, for every $1 \leq i \leq n$, let $\nu_{[\xi_i]} = \gamma([\xi_i])$. Note that, in general, $\mathcal{A}$ is nondeterministic. We can prove easily that $h_\mu(\xi) = h_\sim(\xi)$ for every $\xi \in T_\Sigma$, where $h_\sim : S\langle T_\Sigma\rangle \to S\langle T_\Sigma\rangle/\sim$ is the canonical S-Σ-vector space homomorphism and we identify the isomorphic vector spaces S^Q and $S\langle T_\Sigma\rangle/\sim$. Now let $\xi \in T_\Sigma$ and assume that $[\xi] = a_1.[\xi_1] + \cdots + a_n.[\xi_n]$ for some $a_1, \ldots, a_n \in S$. Then $(r_\mathcal{A}, \xi) = h_\mu(\xi) \cdot \nu = \sum_{i=1}^n [\xi]_{[\xi_i]} \cdot \gamma([\xi_i]) = \gamma(\sum_{i=1}^n a_i \cdot [\xi_i]) = \gamma([\xi]) = (r, \xi)$, which proves that $\mathcal{A}$ recognizes r. □

Characterizations for Semifields

Now we show a second Myhill–Nerode-like theorem, which characterizes recognizable tree series in terms of congruences of finite index over the term algebra T_Σ. However, this characterization holds only for bu-deterministically recognizable tree series, while we can relax from fields to semifields (cf. Theorem 3.35). Since some of the auxiliary results which we need and which are interesting on their own hold even for arbitrary commutative and zero-divisor free semirings (and every semifield is zero-divisor free), we do not immediately require that S is a semifield, but we make the following assumption.

In this subsection, we assume that S is commutative and zero-divisor free.

Now we define the Myhill–Nerode relation $\equiv_r \subseteq T_\Sigma \times T_\Sigma$ by $\xi_1 \equiv_r \xi_2$ iff there are $a, b \in S \setminus \{0\}$ such that $a(\xi_1^{-1}r) = b(\xi_2^{-1}r)$. In fact, the factors a and b are needed, because if they were dropped then, e.g., $\equiv_{\mathsf{size}}$ would have infinite index for $S = \mathsf{Trop}_{\mathsf{sf}}$, where $\mathsf{Trop}_{\mathsf{sf}}$ is the tropical semifield of reals. However, size $\in$ bud-Rec$(\Sigma, \mathsf{Trop}_{\mathsf{sf}})$ (cf. [11], Example 3), and thus recognizability would not imply a finite index of the Myhill–Nerode relation. Note that for different semirings with the same carrier set S, a tree series $r \in S\langle\langle T_\Sigma \rangle\rangle$ may yield different $\equiv_r$ relations with respect to those semirings.

It is straightforward to prove that $\equiv_r$ is an equivalence relation (the zero-divisor freeness guarantees transitivity) and is invariant under contexts, i.e., $\equiv_r$ is a congruence with respect to the Σ-term algebra (cf. [109], Lemma 2).

Next, let us prove that $\equiv_r$ has finite index if $r \in \text{bud-Rec}(\Sigma, S)$. As technical preparation, for a total bu-deterministic wta $\mathcal{A} = (Q, \Sigma, S, \mu, \nu)$, we define the *underlying deterministic bottom-up tree automaton* $B(\mathcal{A}) = (Q, \Sigma, \delta, F)$ such that $F = \{q \in Q \mid \nu_q \neq 0\}$, $\delta = (\delta_\sigma | \sigma \in \Sigma)$ and $\delta_\sigma(q_1, \ldots, q_k) = q$, where q is the unique state such that $\mu_k(\sigma)_{q_1 \ldots q_k, q} \neq 0$. Let $\equiv_{B(\mathcal{A})}$ be the kernel of the homomorphism $h_Q : T_\Sigma \to Q$. By standard arguments, it follows that $\equiv_{B(\mathcal{A})}$ is a congruence on T_Σ which has finite index at most $|Q|$.

We note that, for a total bu-deterministic wta $\mathcal{A}$ and every $\zeta \in C_\Sigma$, $\xi \in T_\Sigma$, and $q \in Q$, we have that $h_\mu(\zeta \cdot_z \xi)_q = h^p_\mu(\zeta)_q \cdot h_\mu(\xi)_p$ where $p = h_Q(\xi)$. This property will be used in the proof of the next lemma.

Lemma 3.32 ([109], Theorem 4). *For every total bu-deterministic wta $\mathcal{A}$, the index of $\equiv_{r_\mathcal{A}}$ is at most the number of states of $\mathcal{A}$. In particular, $\equiv_r$ has finite index for every $r \in \text{bud-Rec}(\Sigma, S)$.*

Proof. Let $\mathcal{A} = (Q, \Sigma, S, \mu, \nu)$ be a total bu-deterministic wta. Since the index of $\equiv_{B(\mathcal{A})}$ is at most $|Q|$, it suffices to show that $\equiv_{B(\mathcal{A})} \subseteq \equiv_{r_\mathcal{A}}$. To prove this inclusion, let $\xi_1, \xi_2 \in T_\Sigma$ such that $\xi_1 \equiv_{B(\mathcal{A})} \xi_2$. Hence, $h_Q(\xi_1) = h_Q(\xi_2)$, and thus also $h_Q(\zeta \cdot_z \xi_1) = h_Q(\zeta \cdot_z \xi_2)$ for every $\zeta \in C_\Sigma$. Let p abbreviate $h_Q(\xi_1)$. Now consider $\zeta \in C_\Sigma$ and let q abbreviate $h_Q(\zeta \cdot_z \xi_1)$. Then $h_\mu(\xi_2)_p \cdot (r, \zeta \cdot_z \xi_1)$ $= h_\mu(\xi_2)_p \cdot h_\mu(\zeta \cdot_z \xi_1)_q \cdot \nu_q = h_\mu(\xi_2)_p \cdot h^p_\mu(\zeta)_q \cdot h_\mu(\xi_1)_p \cdot \nu_q = h_\mu(\xi_1)_p \cdot h_\mu(\zeta \cdot_z \xi_2)_q \cdot \nu_q$ $= h_\mu(\xi_1)_p \cdot (r, \zeta \cdot_z \xi_2)$. Hence, $\xi_1 \equiv_{r_\mathcal{A}} \xi_2$.

For the proof of the second claim, recall that for every $r \in \text{bud-Rec}(\Sigma, S)$ there is a total bu-deterministic wta $\mathcal{A}$ such that $r = r_\mathcal{A}$. Then the second claim follows from the first one. □

Next, we prove that $\equiv_r$ has a particular property, called (MN). In the definition of (MN), we will have to discard those trees $\xi \in T_\Sigma$ that cannot be completed to a tree in $\text{supp}(r)$. Formally, we define $L_r = \{\xi \in T_\Sigma \mid \xi^{-1}r = \widetilde{0}\}$. Now let $\equiv$ be an equivalence relation on T_Σ. Then we say that $\equiv$ *satisfies (MN) for* r if there is a *representation mapping* φ for $\equiv$, i.e., a mapping $\varphi : T_\Sigma/\!\equiv \;\to T_\Sigma$ such that $\varphi([\xi]_\equiv) \in [\xi]_\equiv$ for every $\xi \in T_\Sigma$, and there is a mapping $a_\varphi : T_\Sigma \to S \setminus \{0\}$ such that:

(MN1) For every $\xi \in T_\Sigma$, we have that $(r, \xi) = a_\varphi(\xi) \cdot (r, \varphi([\xi]_\equiv))$.

(MN2) For every $\xi = \sigma(\xi_1, \ldots, \xi_k) \in T_\Sigma \setminus L_r$ and $\xi' = \sigma(\xi'_1, \ldots, \xi'_k) \in T_\Sigma$ with $\xi_i \equiv \xi'_i$ for every $1 \leq i \leq k$, we have that

$$a_\varphi(\xi'_k) \cdot \cdots \cdot a_\varphi(\xi'_1) \cdot a_\varphi(\xi) = a_\varphi(\xi_k) \cdot \cdots \cdot a_\varphi(\xi_1) \cdot a_\varphi(\xi').$$

We note that, if S is a semifield, Condition (MN2) amounts to say that for every $\xi = \sigma(\xi_1, \ldots, \xi_k) \in T_\Sigma \setminus L_r$ there is a $b \in S \setminus \{0\}$ such that $a_\varphi(\xi) =$

$b \cdot \prod_{i=1}^{k} a_\varphi(\xi_i)$. In the sequel, we will abbreviate $[\xi]_\equiv$ by $[\xi]$ and $\varphi([\xi]_\equiv)$ by $\overline{\xi}$, because $\equiv$ and φ will be clear from the context.

Lemma 3.33. *Let $r \in S\langle\langle T_\Sigma \rangle\rangle$, where S is a semifield. Then $\equiv_r$ satisfies (MN) for r.*

Proof. Take any representation mapping $\varphi : T_\Sigma/{\equiv_r} \to T_\Sigma$. Since $\xi \equiv_r \overline{\xi}$ for every $\xi \in T_\Sigma$, there are $a, b \in S \setminus \{0\}$ such that $a(\xi^{-1}r) = b(\overline{\xi}^{-1}r)$. Let us fix some arbitrary such a and b, and call them a_ξ and b_ξ henceforth. Then for every context $\zeta \in C_\Sigma$, we have $a_\xi \cdot (r, \zeta \cdot_z \xi) = b_\xi \cdot (r, \zeta \cdot_z \overline{\xi})$. Now we define the mapping $a_\varphi : T_\Sigma \to S \setminus \{0\}$ by $a_\varphi(\xi) = a_\xi^{-1} \cdot b_\xi$ and get

$$(r, \zeta \cdot_z \xi) = a_\varphi(\xi) \cdot (r, \zeta \cdot_z \overline{\xi}). \tag{2}$$

In particular, with $\zeta = z$, we obtain $(r, \xi) = a_\varphi(\xi) \cdot (r, \overline{\xi})$, which proves that (MN1) holds.

Now let $\xi_1, \ldots, \xi_k \in T_\Sigma$ such that $\xi = \sigma(\xi_1, \ldots, \xi_k) \notin L_r$. Thus, there is a context $\zeta_0 \in C_\Sigma$ such that $(r, \zeta_0 \cdot_z \xi) \neq 0$, hence by (2) also $(r, \zeta_0 \cdot_z \overline{\xi}) \neq 0$.

Let us compute as follows:

$$\begin{aligned}(r, \zeta_0 \cdot_z \xi) &= \big(r, (\zeta_0 \cdot_z \sigma(z, \xi_2, \ldots, \xi_k)) \cdot_z \xi_1\big)\\ &= a_\varphi(\xi_1) \cdot \big(r, (\zeta_0 \cdot_z \sigma(z, \xi_2, \ldots, \xi_k)) \cdot_z \overline{\xi_1}\big)\\ &= a_\varphi(\xi_1) \cdot \big(r, \zeta_0 \cdot_z \sigma(\overline{\xi_1}, \xi_2, \ldots, \xi_k)\big),\end{aligned}$$

where at the second equation we applied (2) with $\xi = \xi_1$. Clearly, this process can be applied to all the ξ_i's, thus we obtain

$$(r, \zeta_0 \cdot_z \xi) = \prod_{i=1}^{k} a_\varphi(\xi_i) \cdot \big(r, \zeta_0 \cdot_z \sigma(\overline{\xi_1}, \ldots, \overline{\xi_k})\big). \tag{3}$$

Using (2) with $\zeta = \zeta_0$, the fact that the inverse of $(r, \zeta_0 \cdot_z \overline{\xi})$ exists, and (3), we obtain that $\prod_{i=1}^{k} a_\varphi(\xi_i)^{-1} \cdot a_\varphi(\xi) = (r, \zeta_0 \cdot_z \sigma(\overline{\xi_1}, \ldots, \overline{\xi_k})) \cdot (r, \zeta_0 \cdot_z \overline{\xi})^{-1}$.

Now we consider $\xi_1', \ldots, \xi_k' \in T_\Sigma$ such that $\xi_i' \in [\xi_i]$ for every $1 \leq i \leq k$, and we denote $\sigma(\xi_1', \ldots, \xi_k')$ by ξ'. Thus, $\xi' \equiv_r \xi$, because $\equiv_r$ is a congruence. Since $(r, \zeta_0 \cdot_z \xi) \neq 0$, also $(r, \zeta_0 \cdot_z \xi') \neq 0$, and hence $\xi' \notin L_r$. Thus, we can prove (3) in the same way as above for ξ' instead of ξ, and obtain $\prod_{i=1}^{k} a_\varphi(\xi_i')^{-1} \cdot a_\varphi(\xi') = (r, \zeta_0 \cdot_z \sigma(\overline{\xi_1'}, \ldots, \overline{\xi_k'})) \cdot (r, \zeta_0 \cdot_z \overline{\xi'})^{-1}$. Since $\overline{\xi_i} = \overline{\xi_i'}$ and $\overline{\xi} = \overline{\xi'}$, we obtain eventually $\prod_{i=1}^{k} a_\varphi(\xi_i)^{-1} \cdot a_\varphi(\xi) = \prod_{i=1}^{k} a_\varphi(\xi_i')^{-1} \cdot a_\varphi(\xi')$ which is the same as (MN2) after multiplying with $\prod_{i=1}^{k} a_\varphi(\xi_i)$ and $\prod_{i=1}^{k} a_\varphi(\xi_i')$. □

We will need the following auxiliary result.

Lemma 3.34. *Let $r \in S\langle\langle T_\Sigma \rangle\rangle$ and let $\equiv$ be a congruence on T_Σ which satisfies (MN) for r. Then $\equiv$ saturates L_r, i.e., L_r is the union of some equivalence classes.*

Proof. Let φ and a_φ be the mappings such that (MN) is satisfied for r. Moreover, let $\xi, \xi' \in T_\Sigma$ such that $\xi \equiv \xi'$ and $\xi \in L_r$. Since $\equiv$ is a congruence, we have that $[\zeta \cdot_z \xi] = [\zeta \cdot_z \xi']$ for every context $\zeta \in C_\Sigma$, and thus $\overline{\zeta \cdot_z \xi} = \overline{\zeta \cdot_z \xi'}$. By (MN1), we have that $(r, \zeta \cdot_z \xi) = a_\varphi(\zeta \cdot_z \xi) \cdot (r, \overline{\zeta \cdot_z \xi})$ for every context $\zeta \in C_\Sigma$. Since $a_\varphi(\zeta \cdot_z \xi) \neq 0$ and S is zero-divisor free, it follows that $(r, \overline{\zeta \cdot_z \xi}) = 0$, and thus $(r, \overline{\zeta \cdot_z \xi'}) = 0$. Hence, $(r, \zeta \cdot_z \xi') = a_\varphi(\zeta \cdot_z \xi') \cdot (r, \overline{\zeta \cdot_z \xi'}) = 0$ again by (MN1). Since this implication holds for every context ζ, we obtain that $\xi' \in L_r$. □

Now we can prove a Myhill–Nerode-like theorem for bu-deterministically recognizable tree series over semifields.

Theorem 3.35 ([14], Theorem 7.3.1). *Let S be a semifield and $r \in S\langle\langle T_\Sigma \rangle\rangle$. Then the following three statements are equivalent:*

(A) $r \in \text{bud-Rec}(\Sigma, S)$.
(B) There is a congruence $\equiv$ on T_Σ which has finite index and satisfies (MN) for r.
(C) $\equiv_r$ has finite index.

Proof. Statement (A) implies statement (C) by Lemma 3.32. By Lemma 3.33, we have that statement (C) implies statement (B).

For the proof that statement (B) implies statement (A), let $\varphi : T_\Sigma/\equiv \to T_\Sigma$ and $a_\varphi : T_\Sigma \to S \setminus \{0\}$ such that (MN1) and (MN2) hold. We construct the bu-deterministic wta $\mathcal{A} = (Q, \Sigma, S, \mu, \nu)$ where $Q = T_\Sigma/\equiv$, $\nu_{[\xi]} = (r, \varphi([\xi]))$ for every $\xi \in T_\Sigma$, and for every $k \geq 0$, $\sigma \in \Sigma^{(k)}$, and $[\xi_1], \ldots, [\xi_k], [\xi] \in Q$:

$$\mu_k(\sigma)_{[\xi_1]\ldots[\xi_k],[\xi]} = \begin{cases} \prod_{i=1}^k a_\varphi(\xi_i)^{-1} \cdot a_\varphi(\xi) & \text{if } [\xi] = [\sigma(\xi_1, \ldots, \xi_k)] \text{ and } \xi \notin L_r, \\ 0 & \text{otherwise.} \end{cases}$$

We note that $\mu_k(\sigma)$ is well defined. To see this, let $[\xi'_1] = [\xi_1], \ldots, [\xi'_k] = [\xi_k]$ and $[\xi'] = [\xi]$. Then $[\sigma(\xi'_1, \ldots, \xi'_k)] = [\sigma(\xi_1, \ldots, \xi_k)]$, hence $[\xi] = [\sigma(\xi_1, \ldots, \xi_k)]$ iff $[\xi'] = [\sigma(\xi'_1, \ldots, \xi'_k)]$. Moreover, $\xi \notin L_r$ iff $\xi' \notin L_r$ by Lemma 3.34. Finally, the property (MN2) of $\equiv$ assures that $\mu_k(\sigma)_{[\xi'_1]\ldots[\xi'_k],[\xi']}$ has the same value.

Next, it is straightforward to prove by induction on $\xi \in T_\Sigma$ that for every $[\xi'] \in Q$ we have

$$h_\mu(\xi)_{[\xi']} = \begin{cases} a_\varphi(\xi) & \text{if } [\xi] = [\xi'] \text{ and } \xi \notin L_r, \\ 0 & \text{otherwise.} \end{cases}$$

In the proof, we have to use the fact that for every tree $\xi = \sigma(\xi_1, \ldots, \xi_k)$, if $\xi \notin L_r$, then $\xi_i \notin L_r$.

Finally, for every $\xi \in T_\Sigma$, we obtain $(r_{\mathcal{A}}, \xi) = h_\mu(\xi) \cdot \nu = h_\mu(\xi)_{[\xi]} \cdot \nu_{[\xi]}$ because $[\xi] \neq [\xi']$ implies $h_\mu(\xi)_{[\xi']} = 0$. If in addition $\xi \notin L_r$, then $h_\mu(\xi)_{[\xi]} \cdot \nu_{[\xi]} = a_\varphi(\xi) \cdot (r, \varphi([\xi])) = (r, \xi)$ by (MN1). If $\xi \in L_r$, and thus, in particular, $(r, \xi) = 0$, then $h_\mu(\xi)_{[\xi]} \cdot \nu_{[\xi]} = 0 \cdot (r, \varphi([\xi])) = 0 = (r, \xi)$. □

To show the use of Theorem 3.35, let us consider the tree series size. By an obvious automaton construction, we have that size $\in$ bud-Rec$(\Sigma, \mathsf{Trop}_{\mathrm{sf}})$, and hence $\equiv_{\mathsf{size}}$ over $\mathsf{Trop}_{\mathrm{sf}}$ is of finite index, in fact, the index is 1. On the other hand, we can prove that size $\notin$ bud-Rec$(\Sigma, \mathbb{Q})$ if Σ contains at least a binary symbol σ and a nullary symbol α and $(\mathbb{Q}, +, \cdot, 0, 1)$ is the field of rational numbers (cf. [14], Example 7.3.2). For this, we prove that $\xi_1 \equiv_{\mathsf{size}} \xi_2$ iff $(\mathrm{size}, \xi_1) = (\mathrm{size}, \xi_2)$ which shows that $\equiv_{\mathsf{size}}$ has infinite index over $\mathbb{Q}$. Assume that $(\mathrm{size}, \xi_1) = (\mathrm{size}, \xi_2)$. Then for every context $\zeta \in C_\Sigma$, we have $(\mathrm{size}, \zeta \cdot_z \xi_1) = (\mathrm{size}, \zeta \cdot_z \xi_2)$ and hence $\xi_1 \equiv_{\mathsf{size}} \xi_2$. Now assume that $\xi_1 \equiv_{\mathsf{size}} \xi_2$. Hence, there is an $a \in \mathbb{Q} \backslash \{0\}$ such that for every $\zeta \in C_\Sigma$ we have $(\mathrm{size}, \zeta \cdot_z \xi_1) = a \cdot (\mathrm{size}, \zeta \cdot_z \xi_2)$. Instantiating this equation twice, with $\zeta = z$ and $\zeta = \sigma(\alpha, z)$, we obtain: $(\mathrm{size}, \xi_1) = a \cdot (\mathrm{size}, \xi_2)$ and $2 + (\mathrm{size}, \xi_1) = a \cdot (2 + (\mathrm{size}, \xi_2))$, respectively; this implies that $a = 1$, and hence $(\mathrm{size}, \xi_1) = (\mathrm{size}, \xi_2)$.

Characterizations for Commutative, Zero-Divisor Free Semirings

Now let us recall the third Myhill–Nerode-like characterization which is due to [109]. It shows, for the class of commutative and zero-divisor free semirings, a characterization of bud-Rec(Σ, S) in terms of a slightly different property. We say that a congruence $\equiv$ on T_Σ *respects a tree series* $r \in S\langle\langle T_\Sigma \rangle\rangle$ if there exists a mapping $f : T_\Sigma/\equiv \to S$ and a mapping $c : T_\Sigma \to S \setminus \{0\}$ such that:

- $(r, \xi) = c(\xi) \cdot f([\xi]_\equiv)$ for every $\xi \in T_\Sigma$.
- For every $k \geq 0$, $\sigma \in \Sigma^{(k)}$, there is a mapping $b_\sigma : (T_\Sigma/\equiv)^k \to S$ such that for every $\xi_1, \ldots, \xi_k \in T_\Sigma$ we have that $c(\sigma(\xi_1, \ldots, \xi_k)) = c(\xi_1) \cdot \cdots \cdot c(\xi_k) \cdot b_\sigma([\xi_1]_\equiv, \ldots, [\xi_k]_\equiv)$.

Theorem 3.36 ([109], Theorem 19). *Let S be a commutative and zero-divisor free semiring. Moreover, let $r \in S\langle\langle T_\Sigma \rangle\rangle$. Then the following two statements are equivalent:*

(A) $r \in$ bud-Rec(Σ, S).
(B) There is a congruence $\equiv$ on T_Σ which has finite index and respects r.

We note that, for every semifield S, the second statements of Theorems 3.36 and 3.35 are equivalent.

3.8 Equational Tree Series

By definition, an equational subset of a Σ-algebra $\mathcal{A}$ is a component of the least solution of some system of linear equations [116, 32, 69]. It was shown in [116] that every equational subset of $\mathcal{A}$ is the homomorphic image of a recognizable tree language and vice versa. In particular, the class of equational subsets of T_Σ is the class of recognizable Σ-tree languages. Here, we show how these results are generalized to recognizable tree series [7, 88, 21, 58]. Since the solutions of systems of linear equations are obtained by fixpoints, we refer the reader to [57] for an introduction to that part of the theory of fixpoints that has applications to weighted automata.

A *system of linear equations* (for short: system) is a finite family E of equations $z_i = s_i$ where $1 \le i \le n$ for some $n \ge 1$, and $Z_n = \{z_1, \ldots, z_n\}$ is a set of variables, and $s_i \in S\langle T_\Sigma(Z_n)\rangle$. The system E is *proper* if for every $1 \le i \le n$, the tree series s_i is z_j-proper, i.e., $(s_i, z_j) = 0$, for every $1 \le j \le n$.

We solve such systems in S-Σ-semimodules. For this, we generalize the concept of OI-substitution as introduced in Sect. 3.3 as follows. Let $(V, +, 0, \theta)$ be an S-Σ-semimodule and $\bar{v} = (v_1, \ldots, v_n) \in V^n$. Moreover, let $s \in S\langle T_\Sigma(Z_n)\rangle$. The *OI-substitution of* $\bar{v}$ *into* s is the element $s \leftarrow_{\mathrm{OI}} \bar{v}$ in V which is defined to be $\sum_{\zeta \in \mathrm{supp}(s)} (s, \zeta)(\zeta \leftarrow_{\mathrm{OI}} \bar{v})$; the element $\zeta \leftarrow_{\mathrm{OI}} \bar{v} \in V$ is defined in exactly the same way as $\zeta \leftarrow_{OI,\bar{z}} \bar{v}$ (with $\bar{z} = (z_1, \ldots, z_n)$ and $\bar{v} \in S\langle\langle T_\Sigma\rangle\rangle^n$) except that in case (ii) the symbol σ is not interpreted by top_σ but by $\theta(\sigma)$. Note that in the expression $(s, \zeta)(\zeta \leftarrow_{\mathrm{OI}} \bar{v})$ the subexpressions (s, ζ) and $\zeta \leftarrow_{\mathrm{OI}} \bar{v}$ are combined by means of the scalar multiplication of the S-semimodule $(V, +, 0)$.

Now let E be the system $z_i = s_i$ with $1 \le i \le n$ and $(V, +, 0, \theta)$ an S-Σ-semimodule. A *solution of* E in V is a vector $\bar{v} = (v_1, \ldots, v_n) \in V^n$ such that $v_i = (s_i \leftarrow_{\mathrm{OI}} \bar{v})$ for every i. In other words, $\bar{v}$ is a fixpoint of the mapping $\Phi_E : V^n \to V^n$ defined by $\Phi_E(\bar{u}) = (s_1 \leftarrow_{\mathrm{OI}} \bar{u}, \ldots, s_n \leftarrow_{\mathrm{OI}} \bar{u})$ for every $\bar{u} \in V^n$. Let additionally the monoid $(V, +, 0)$ be naturally ordered by $\sqsubseteq$; this relation is extended componentwise to V^n. An element $v \in V$ is *equational* (*p-equational*) if it is a component of the least solution of a system (resp., proper system), if it exists. The *class of all equational elements (p-equational elements)* in the S-Σ-semimodule V is denoted by $\mathrm{Eq}(V)$ (resp., $\mathrm{Eq_p}(V)$).

Solutions over the S-Σ-Semimodule of Tree Series

Before dealing with equational elements in general, we first solve proper systems in the particular S-Σ-semimodule $(S\langle\langle T_\Sigma\rangle\rangle, +, \widetilde{0}, \mathrm{top})$ where $\mathrm{top}(\sigma) = \mathrm{top}_\sigma$ for every $\sigma \in \Sigma$. (Note that for the concept of solution we do not need a partial order.)

Lemma 3.37 ([7], Proposition 6.1). *Every proper system has a unique solution in* $(S\langle\langle T_\Sigma\rangle\rangle, +, \widetilde{0}, \mathrm{top})$.

Proof. Let E be a proper system $z_i = s_i$ with $1 \le i \le n$. Moreover, let $\bar{r} = (r_1, \ldots, r_n)$ be a vector of tree series in $S\langle\langle T_\Sigma\rangle\rangle$, and assume that $\bar{r}$ is a solution of E. Hence, $r_i = (s_i \leftarrow_{\mathrm{OI}} \bar{r}) = \sum_{\zeta \in \mathrm{supp}(s_i)} (s_i, \zeta)(\zeta \leftarrow_{\mathrm{OI}} \bar{r})$. We prove that this solution is the only solution.

Since E is proper, every ζ has the form $\zeta = \delta(\zeta_1, \ldots, \zeta_k)$ for some $k \ge 0$, $\delta \in \Sigma^{(k)}$, and $\zeta_1, \ldots, \zeta_k \in T_\Sigma(Z_n)$, and either $\zeta \in T_\Sigma$ or $\zeta \in T_\Sigma(Z_n) \setminus T_\Sigma$. Hence, we can continue with

$$r_i = \sum_{\zeta \in T_\Sigma} (s_i, \zeta).\zeta + \sum_{\delta(\zeta_1, \ldots, \zeta_k) \in T_\Sigma(Z_n) \setminus T_\Sigma} (s_i, \delta(\zeta_1, \ldots, \zeta_k))\, \mathrm{top}_\delta(\zeta_1', \ldots, \zeta_k')$$

where $\zeta_j' = (\zeta_j \leftarrow_{\mathrm{OI}} \bar{r})$. Then for every $\alpha \in \Sigma^{(0)}$, we have that

$$(r_i, \alpha) = (s_i, \alpha), \tag{4}$$

and for every $\xi = \sigma(\xi_1, \ldots, \xi_k) \in T_\Sigma$ with $k \geq 1$, we have that

$$(r_i, \xi) = (s_i, \xi) + \sum_{\sigma(\zeta_1,\ldots,\zeta_k) \in T_\Sigma(Z_n) \setminus T_\Sigma} \big(s_i, \sigma(\zeta_1, \ldots, \zeta_k)\big) \cdot \prod_{i=1}^{k} (\zeta_i', \xi_i). \tag{5}$$

In the summation, we can restrict to those $\sigma(\zeta_1, \ldots, \zeta_k) \in T_\Sigma(Z_n) \setminus T_\Sigma$ such that $\xi_j \in \mathrm{supp}(\zeta_j \leftarrow_{\mathrm{OI}} \bar{r})$ for every $1 \leq j \leq k$. Hence, $(\zeta_j \leftarrow_{\mathrm{OI}} \bar{r}, \xi_j)$ is the product of coefficients of the form $(r_j, \widehat{\xi})$, where $1 \leq j \leq n$ and $\widehat{\xi}$ is a subtree of ξ_j (equal to ξ_j if $\zeta_j \in Z_n$), and hence a strict subtree of ξ. All in all, the value of (r_i, ξ) is uniquely determined by s_i and by the values of the r_j's on strict subtrees of ξ. Hence, $\bar{r}$ is uniquely determined.

On the other hand, (4) and (5) can be used as defining equations. Thus, $\bar{r}$ exists. □

We note that the solution $\bar{r}$ of E in Lemma 3.37 can be explicitly given by $(r_i, \xi) = (\Phi_E^{m+1}((\widetilde{0}, \ldots, \widetilde{0}))_i, \xi)$ where $m = \mathrm{height}(\xi)$ and $\Phi_E^{m+1}((\widetilde{0}, \ldots, \widetilde{0}))_i$ denotes the ith component of $\Phi_E^{m+1}((\widetilde{0}, \ldots, \widetilde{0}))$.

Example 3.38. Consider $\Sigma = \{\sigma^{(2)}, \gamma^{(1)}, \alpha^{(0)}\}$ and the tree series $\#_{\sigma(z,\alpha)} : T_\Sigma \to \mathbb{N}$ which maps every tree ξ to the number of occurrences of the pattern $\sigma(z, \alpha)$ in ξ (cf. page 322). We consider the proper system E

$$\begin{aligned} z_1 &= \gamma(z_1) + \sigma(z_1, z_2) + \sigma(z_2, z_1) + \sigma(z_2, \alpha), \\ z_2 &= \alpha + \gamma(z_2) + \sigma(z_2, z_2). \end{aligned}$$

It is easy to see that (r_1, r_2) is a solution of E in the S-Σ-semimodule $S\langle\langle T_\Sigma \rangle\rangle$ with $r_1 = \#_{\sigma(z,\alpha)}$ and $r_2 = 1_{(\mathsf{Nat}, T_\Sigma)}$.

There is a close relationship between wta with Boolean root weights and particularly simple proper systems; it is based on the idea of identifying states with variables. A proper system E is *simple* if its equations have the form $z_i = s_i$ with $\mathrm{supp}(s_i) \subseteq \{\sigma(z_{i_1}, \ldots, z_{i_k}) \mid k \geq 0, \sigma \in \Sigma^{(k)}, z_{i_1}, \ldots, z_{i_k} \in Z_n\}$. Then let $\mathcal{A} = (Q, \Sigma, S, \mu, F)$ be a wta with Boolean root weights such that F is a singleton, and E a simple system $z_i = s_i$ with $1 \leq i \leq n$. We call $\mathcal{A}$ and E *related* if $Q = Z_n$ and

$$s_i = \sum_{\substack{k \geq 0,\ \sigma \in \Sigma^{(k)} \\ q_1, \ldots, q_k \in Q}} \mu_k(\sigma)_{q_1 \ldots q_k, z_i} . \sigma(q_1, \ldots, q_k).$$

If $\mathcal{A}$ and E are related, then $\bar{r} = (r_{z_1}, \ldots, r_{z_n})$ is a solution of E, where $r_{z_i} \in S\langle\langle T_\Sigma \rangle\rangle$ is defined by $(r_{z_i}, \xi) = h_\mu(\xi)_{z_i}$ for every $\xi \in T_\Sigma$. This can be proved by a straightforward induction on ξ.

Obviously, if $\mathcal{A}$ and E are related, then E has to be simple. But we can extend this relationship to arbitrary proper systems, because every proper system E can be simulated by a simple system $\widetilde{E}$. For instance, the system E of Example 3.38 is not simple, because the term $\sigma(z_2, \alpha)$ on the right-hand side of the z_1-equation does not have the appropriate form. However, if we define a new system $\widetilde{E}$ which is the same as E except that the disturbing α is replaced by z_3 and the equation $z_3 = \alpha$ is added, then $\widetilde{E}$ is simple; moreover, $\widetilde{E}$ is related to the wta of Example 3.4 (renaming $f, q, \overline{\alpha}$ into z_1, z_2, z_3, resp.).

In general, for two proper systems E_1 and E_2, we say that E_1 is *simulated* by E_2 if every component of a solution of E_1 is also a component of a solution of E_2. The construction of the simple system $\widetilde{E}$ is easy and it proceeds in two steps: in the first step, a system E' is constructed in which the height of every tree $\zeta \in \text{supp}(s_i)$ where $z_i = s_i$ is an equation of E', is not greater than 1; this can be achieved by introducing appropriate auxiliary equations which break down too high trees. In the second step, equations of the form $z_i = s_i$ of E' where $\text{supp}(s_i)$ contains a tree of height 1 with nullary symbols (e.g., $z_i = \sigma(\alpha, z_j)$) are split up into appropriate equations (like $z_i = \sigma(y_1, z_j)$ and $y_1 = \alpha$ with a new variable y_1).

Lemma 3.39 ([7], Lemma 6.3; [58], Corollary 3.6). *For every proper system E, there is a simple system $\widetilde{E}$ which simulates E.*

Now we can show that the recognizable tree series in $S\langle\langle T_\Sigma\rangle\rangle$ are exactly the p-equational elements in $(S\langle\langle T_\Sigma\rangle\rangle, +, \widetilde{0}, \theta)$ where we assume that S is naturally ordered, i.e., $\sqsubseteq$ is a partial order; this partial order extends to the set $S\langle\langle T_\Sigma\rangle\rangle$ by defining $r \sqsubseteq s$ iff $(r, \xi) \sqsubseteq (s, \xi)$ for every $\xi \in T_\Sigma$, and to $(S\langle\langle T_\Sigma\rangle\rangle)^n$ by componentwise comparison. Hence, we can speak about equational elements in $(S\langle\langle T_\Sigma\rangle\rangle, +, \widetilde{0}, \theta)$.

Theorem 3.40 ([58], Corollary 3.6). *Let S be naturally ordered. Then the following two statements hold:*

(A) $\text{Rec}(\Sigma, S) = \text{Eq}_\text{p}(S\langle\langle T_\Sigma\rangle\rangle)$.
(B) If S is commutative and continuous, then $\text{Rec}(\Sigma, S) = \text{Eq}(S\langle\langle T_\Sigma\rangle\rangle)$.

Proof. For the proof of statement (A), let $\mathcal{A}$ be a wta. By the construction in the proof of Theorem 3.6, we can assume that $\mathcal{A}$ is a wta (Q, Σ, S, μ, F) with Boolean root weights and $F = \{q_f\}$; note that $(r_\mathcal{A}, \xi) = h_\mu(\xi)_{q_f}$ for every $\xi \in T_\Sigma$. Now construct the proper system E with equations $z_q = s_q$ for $q \in Q$ that is related to $\mathcal{A}$. Then $\bar{r} = (r_q \mid q \in Q)$ is a solution of E in $S\langle\langle T_\Sigma\rangle\rangle$, where $(r_q, \xi) = h_\mu(\xi)_q$. Since E is proper, it follows from Lemma 3.37 that this solution is the unique solution in the S-Σ-semimodule $S\langle\langle T_\Sigma\rangle\rangle$, which then is also its least solution. Thus, $r_\mathcal{A}$ is equational.

Conversely, let $r \in \text{Eq}_\text{p}(S\langle\langle T_\Sigma\rangle\rangle)$. Then there is a proper system E of equations $z_i = s_i$ with $1 \le i \le n$ such that r is the, say, first component of the least solution $\bar{r} = (r_1, \ldots, r_n)$ of E. By Lemma 3.39, we can assume that E is simple. Clearly, we can construct a wta $\mathcal{A} = (Z_n, \Sigma, S, \mu, \{z_1\})$ that is related to E. Note that $(r_{\mathcal{A}}, \xi) = h_\mu(\xi)_{z_1}$ for every $\xi \in T_\Sigma$. Then $r_{\mathcal{A}} = r_1$ and $r \in \mathrm{Rec}(\Sigma, S)$.

For statement (B), it remains to show that every system E can be simulated by a proper system E' (here simulation means that every component of the least solution of E, if it exists, is a component of the solution of E'). This has been proved in [88], Theorem 3.2 (also cf. [58], Theorem 3.2) for commutative and continuous semirings. □

Solutions over Arbitrary S-Σ-Semimodules

Now let us turn to equational elements in an arbitrary S-Σ-semimodule $(V, +, 0, \theta)$. In order to guarantee the existence of least solutions, we require that $(V, +, 0)$ is a continuous monoid for the natural order $\sqsubseteq$. Thus, by [88], Theorem 2.3, $(V, \sqsubseteq)$ is a complete partially ordered set (for short: cpo). Since $\sqsubseteq$ can be extended to V^n (by componentwise comparison) and Φ_E is continuous on the cpo $(V^n, \sqsubseteq)$ by [21], Theorem 8 (also cf. [58], Proposition 2.6), i.e., preserves least upper bounds of ω-chains, it follows from Tarski's fixpoint theorem [139] that the least fixpoint of Φ_E exists and is the least upper bound of the ω-chain $\Phi_E^n(\perp)$ where $\perp = (0, \ldots, 0)$ is the least element of V^n.

Least solutions are preserved under homomorphisms. To see this, let $(V', \oplus, 0, \theta')$ be another S-Σ-semimodule. A mapping $h : V \to V'$ is an S-Σ-semimodule homomorphism if h is both a monoid homomorphism from $(V, +, 0)$ to $(V', \oplus, 0)$ and a Σ-algebra homomorphism from (V, θ) to (V', θ'), and moreover it satisfies the law $h(av) = ah(v)$ for every $a \in S$ and $v \in V$.

Lemma 3.41 ([21], Theorem 16). *Let $(V, +, 0, \theta)$ and $(V', \oplus, 0, \theta')$ be two continuous S-Σ-semimodules. Moreover, let $h : V \to V'$ be an S-Σ-semimodule homomorphism. If E is a system $z_i = s_i$ with $1 \le i \le n$ and $\bar{v} \in V^n$ is the least solution of E in V, then $h(\bar{v})$ is the least solution of E in V' (where h is extended to V^n componentwise).*

The following Mezei–Wright-like theorem is based on the idea that one can compute equational elements of V by first calculating "symbolically" in $S\langle\langle T_\Sigma\rangle\rangle$ and then evaluating the resulting tree series by a homomorphism from $S\langle\langle T_\Sigma\rangle\rangle$ to V. Let S be continuous. Then the S-Σ-semimodule $(S\langle\langle T_\Sigma\rangle\rangle, +, \widetilde{0}, \mathrm{top})$ is initial in the class of all continuous S-Σ-semimodules $(V, +, 0, \theta)$ (cf. [21], Theorem 4). Hence, there is exactly one S-Σ-semimodule homomorphism from $S\langle\langle T_\Sigma\rangle\rangle$ to V; this homomorphism we denote by h_V.

Theorem 3.42 (Mezei–Wright-like Theorem, cf. [21]). *Let S be commutative and continuous. Moreover, let $(V, +, 0, \theta)$ be a continuous S-Σ-semimodule. Then* $\mathrm{Eq}(V) = h_V(\mathrm{Rec}(\Sigma, S))$.

Proof. By Theorem 3.40(B), we only have to prove that $\mathrm{Eq}(V) = h_V(\mathrm{Eq}(S\langle\langle T_\Sigma\rangle\rangle))$. By Lemma 3.41, the least solution of a system E in V is the image under h_V of its least solution in $S\langle\langle T_\Sigma\rangle\rangle$. Hence, the same holds for the components of these solutions. □

There is another way of defining equational tree series which is based on so-called [IO]-substitution [25]; we note that in Sect. 4 we will define another variant called IO-substitution. For $s \in S\langle T_\Sigma(Z_n)\rangle$ and $\bar{r} = (r_1, \ldots, r_n) \in S\langle\langle T_\Sigma\rangle\rangle^n$, the *[IO]-substitution of* $\bar{r}$ *into* s is the tree series $s \leftarrow_{[\mathrm{IO}]} \bar{r}$ defined by $s \leftarrow_{[\mathrm{IO}]} \bar{r} = \sum_{\zeta \in \mathrm{supp}(s)} (s, \zeta)(\zeta \leftarrow_{[\mathrm{IO}]} \bar{r})$ and $\zeta \leftarrow_{[\mathrm{IO}]} \bar{r} = \sum_{\zeta_1, \ldots, \zeta_n \in T_\Sigma} (r_{i_1}, \zeta_{i_1}) \cdot \cdots \cdot (r_{i_l}, \zeta_{i_l}).\zeta(\zeta_1, \ldots, \zeta_n)$ for every $\zeta \in T_\Sigma(Z_n)$ where $z_{i_1}, \ldots, z_{i_l}$, $1 \leq i_1 < \cdots < i_l \leq n$, are all the variables which occur in ζ. An *[IO]-solution* of a system and the *class* $[\mathrm{IO}]\text{-}\mathrm{Eq}(S\langle\langle T_\Sigma\rangle\rangle)$ *of [IO]-equational tree series* are defined in the same way as above except that OI-substitution is replaced by [IO]-substitution.

Then [IO]-equational tree series can be characterized as the image of recognizable tree series under nondeleting tree homomorphisms. For this, we consider a family $h_{\Sigma,\Delta} = (h_\sigma \mid \sigma \in \Sigma)$ such that $h_\sigma \in T_\Delta(Z_k)$ if σ has rank k. The *tree homomorphism induced by* $h_{\Sigma,\Delta}$ is the mapping $h : T_\Sigma \to T_\Delta$ defined inductively by $h(\sigma(\xi_1, \ldots, \xi_k)) = h_\sigma(h(\xi_1), \ldots, h(\xi_k))$. A tree homomorphism is *nondeleting* if z_i occurs in h_σ for every $\sigma \in \Sigma^{(k)}$ and $1 \leq i \leq k$. Let $H^{\mathrm{nd}}_{\Sigma,\Delta}$ denote the class of all nondeleting tree homomorphisms induced by some family $\mathrm{h}_{\Sigma,\Delta}$.

We generalize a tree homomorphism $h : T_\Sigma \to T_\Delta$ to a mapping $h : S\langle\langle T_\Sigma\rangle\rangle \to S\langle\langle T_\Delta\rangle\rangle$ by defining $(h(r), \xi) = \sum_{\xi' \in T_\Sigma, h(\xi') = \xi} (r, \xi')$. In order to get the sum (which might have an infinite index set) well defined, we assume that S is complete.

Theorem 3.43 ([25], Theorem 16). *Let* S *be commutative and complete. Then* $[\mathrm{IO}]\text{-}\mathrm{Eq}(S\langle\langle T_\Delta\rangle\rangle) = H^{\mathrm{nd}}_{\Sigma,\Delta}(\mathrm{Rec}(\Sigma, S))$.

3.9 Rational Tree Series

Kleene's fundamental theorem on the equivalence between recognizable and rational string languages [81] has been extended to trees [135] and to tree series over commutative, complete, and continuous semirings [21, 88, 58], and over commutative semirings [41, 119, 120]. In [1], a characterization of weighted regular tree grammars in terms of rational tree series is sketched. We refer to [124] for a survey on rational formal power series over strings.

In order to keep the technical overhead of this chapter small, we will only show here some characteristic details of the equivalence proof along the approach of [41].

Throughout Sect. 3.9, we assume that S *is commutative.*

The set of *rational tree series expressions over* Σ *and* S, denoted by $\mathrm{RatExp}(\Sigma, S)$, is the smallest set R which satisfies conditions (1)–(5). For every $\eta \in \mathrm{RatExp}(\Sigma, S)$, we define $[\![\eta]\!] \in S\langle\langle T_\Sigma\rangle\rangle$ simultaneously.

1. For every $k \geq 0$, $\sigma \in \Sigma^{(k)}$, and $\eta_1, \ldots, \eta_k \in R$, the expression $\sigma(\eta_1, \ldots, \eta_k) \in R$ and $[\![\sigma(\eta_1, \ldots, \eta_k)]\!] = \mathrm{top}_\sigma([\![\eta_1]\!], \ldots, [\![\eta_k]\!])$.

2. For every $\eta \in R$ and $a \in S$, the expression $(a\eta) \in R$ and $[\![(a\eta)]\!] = a[\![\eta]\!]$.
3. For every $\eta_1, \eta_2 \in R$, the expression $(\eta_1 + \eta_2) \in R$ and $[\![(\eta_1 + \eta_2)]\!] = [\![\eta_1]\!] + [\![\eta_2]\!]$.
4. For every $\eta_1, \eta_2 \in R$ and $\alpha \in \Sigma^{(0)}$, the expression $(\eta_1 \circ_\alpha \eta_2) \in R$ and $[\![(\eta_1 \circ_\alpha \eta_2)]\!] = [\![\eta_1]\!] \circ_\alpha [\![\eta_2]\!]$.
5. For every $\eta \in R$ and $\alpha \in \Sigma^{(0)}$, the expression $(\eta_\alpha^*) \in R$ and $[\![(\eta_\alpha^*)]\!] = [\![\eta]\!]_\alpha^*$.

A tree series $r \in S\langle\!\langle T_\Sigma \rangle\!\rangle$ is a *rational tree series over* Σ *and* S if there is an $\eta \in \mathrm{RatExp}(\Sigma, S)$ such that $r = [\![\eta]\!]$. The *class of all rational tree series over* Σ *and* S is denoted by $\mathrm{Rat}(\Sigma, S)$. We say that a class $\mathcal{C} \subseteq S\langle\!\langle T_\Sigma \rangle\!\rangle$ is *closed under the rational operations* if it is closed under top-concatenation top_σ for every $\sigma \in \Sigma$, multiplication with coefficients in S, sum, α-concatenation and α-Kleene star for every $\alpha \in \Sigma^{(0)}$.

Obviously, every polynomial is a rational tree series (note that $\widetilde{0} = [\![0\alpha]\!]$ for any $\alpha \in \Sigma^{(0)}$; recall that we required in general that $\Sigma^{(0)} \neq \emptyset$). Thus, $\mathrm{Rat}(\Sigma, S)$ is the smallest subclass of $S\langle\!\langle T_\Sigma \rangle\!\rangle$ that contains $S\langle T_\Sigma \rangle$, and is closed under the rational operations.

Example 3.44. Consider $\Sigma = \{\sigma^{(2)}, \gamma^{(1)}, \alpha^{(0)}\}$ and again the tree series $\#_{\sigma(z,\alpha)}$ from page 322. We fix a $z \notin \Sigma$ and define the ranked alphabet $\Delta = \Sigma \cup \{z^{(0)}\}$. Then we define the rational expressions $\eta, \eta_1, \eta_2 \in \mathrm{RatExp}(\Delta, \mathsf{Nat})$ by

$$\begin{aligned} \eta &= \eta_1 \circ_z \sigma(z, \alpha) \circ_z \eta_2, \\ \eta_1 &= \big(\gamma(z) + \sigma(\eta_2, z) + \sigma(z, \eta_2)\big)_z^*, \\ \eta_2 &= \big(\gamma(z) + \sigma(z, z)\big)_z^* \circ_z \alpha. \end{aligned}$$

It is obvious that $[\![\eta_1]\!], [\![\eta_2]\!] \in \mathsf{Nat}\langle\!\langle T_\Delta \rangle\!\rangle$ with $[\![\eta_1]\!] = 1_{(\mathsf{Nat}, C_\Sigma)}$ and $[\![\eta_2]\!] = 1_{(\mathsf{Nat}, T_\Sigma)}$. Then $[\![\eta]\!] = \#_{\sigma(z,\alpha)}$.

In Example 3.44, we did not distinguish between the two tree series $[\![\eta]\!]$ and $[\![\eta]\!]|_{T_\Sigma}$. In general, we will not distinguish between a tree series $r \in S\langle\!\langle T_\Sigma \rangle\!\rangle$ and a tree series $r' \in S\langle\!\langle T_\Sigma(Q) \rangle\!\rangle$ for which $r = r'|_{T_\Sigma}$ and $\mathrm{supp}(r') \subseteq T_\Sigma$.

First, let us show why every recognizable tree series is rational. More precisely, we consider the wta $\mathcal{A} = (Q, \Sigma, S, \mu, F)$ with Boolean root weights and we will show that $r_\mathcal{A} \in \mathrm{Rat}(\Sigma \cup Q, S)$ where the states are assumed to be nullary symbols. (In fact, similar to the case of tree languages, we will need the states as substitution symbols in q-concatenations and q-Kleene stars of tree series for $q \in Q$.) The basic idea of the proof is the same as in Kleene's proof [81]: starting with the empty set, the set of permitted intermediate states is enlarged until it reaches the set of all states; at every level of this process, a tree series is defined as rational expression over the tree series of the previous level.

Formally, for every $P \subseteq Q$ and $q \in Q$, we define the tree series $r_\mathcal{A}(P, q) \in S\langle\!\langle T_\Sigma(Q) \rangle\!\rangle$ such that for every $\xi \in T_\Sigma(Q)$,

$$(r_\mathcal{A}(P, q), \xi) = \begin{cases} \sum_{\kappa \in R_\mathcal{A}^P(\xi, q)} \mathrm{wt}(\kappa) & \text{if } \xi \in T_\Sigma(Q) \setminus Q, \\ 0 & \text{if } \xi \in Q \end{cases}$$

where $R^P_{\mathcal{A}}(\xi, q)$ is the set of all those runs $\kappa \in R_{\mathcal{A}}(\xi)$ such that (i) $\kappa(\varepsilon) = q$, (ii) $\kappa(w) \in P$ for every $w \in \mathrm{pos}(\xi) \setminus (\mathrm{pos}_Q(\xi) \cup \{\varepsilon\})$, and (iii) $\kappa(w) = \xi(w)$ for every $w \in \mathrm{pos}_Q(\xi)$. The next lemma shows what happens when one state is added to the set P of inner states.

Lemma 3.45 ([41], Lemma 5.1). *Let $M = (Q, \Sigma, S, \mu, F)$ be a wta with Boolean root weights. Let $P \subseteq Q$ and $q \in Q$, and let $p \in Q \setminus P$. Then $r_{\mathcal{A}}(P \cup \{p\}, q) = r_{\mathcal{A}}(P, q) \circ_p r_{\mathcal{A}}(P, p)^*_p$.*

In the sequel, we will denote $\bigcup_{Q \text{ finite set}} \mathrm{Rat}(\Sigma \cup Q, S)$ by $\mathrm{Rat}(\Sigma \cup Q_\infty, S)$; similarly $\mathrm{Rec}(\Sigma \cup Q_\infty, S)$ is defined.

Theorem 3.46 ([41], Theorem 5.2). $\mathrm{Rec}(\Sigma, S) \subseteq \mathrm{Rat}(\Sigma \cup Q_\infty, S)$.

Proof. Let $r \in \mathrm{Rec}(\Sigma, S)$. By Theorem 3.6, we can assume that there is a wta $\mathcal{A} = (Q, \Sigma, S, \mu, F)$ with Boolean root weights such that $r_{\mathcal{A}} = r$. Let $Q = \{q_1, \ldots, q_n\}$. We prove that $r_{\mathcal{A}} \in \mathrm{Rat}(\Sigma \cup Q, S)$.

Since $\mathrm{supp}(r_{\mathcal{A}}(Q, q))$ may contain trees in which states occur, whereas this is not true for $\mathrm{supp}(r_{\mathcal{A}})$, we filter out from the tree series $r_{\mathcal{A}}(Q, q)$ trees in $T_\Sigma(Q) \setminus T_\Sigma$. Obviously, $r_{\mathcal{A}} = \sum_{q \in F}(\ldots(r_{\mathcal{A}}(Q, q) \circ_{q_1} \widetilde{0}) \circ_{q_2} \widetilde{0} \ldots) \circ_{q_n} \widetilde{0}$. Thus, it remains to show that $r_{\mathcal{A}}(Q, q) \in \mathrm{Rat}(\Sigma \cup Q, S)$.

For this, we prove the following statement by induction on the number of elements in P: for every $P \subseteq Q$ and $q \in Q$, the tree series $r_{\mathcal{A}}(P, q)$ is in $\mathrm{Rat}(\Sigma \cup Q, S)$. For the induction base, i.e., $P = \emptyset$, we can easily observe that

$$r_{\mathcal{A}}(\emptyset, q) = \sum_{\substack{k \geq 0,\ \sigma \in \Sigma^{(k)} \\ p_1, \ldots, p_k \in Q}} \mu_k(\sigma)_{p_1 \ldots p_k, q} . \sigma(p_1, \ldots, p_k),$$

which is a polynomial, and hence $r_{\mathcal{A}}(\emptyset, q)$ is rational. For the induction step, we assume that $r_{\mathcal{A}}(P, q)$ is rational for every $q \in Q$. Now let $p \in Q \setminus P$. Then it follows from Lemma 3.45 that also $r_{\mathcal{A}}(P \cup \{p\}, q)$ is rational because it is built up from rational tree series by rational operations. □

Second, the inclusion $\mathrm{Rat}(\Sigma, S) \subseteq \mathrm{Rec}(\Sigma, S)$ follows from the fact that $\mathrm{Rec}(\Sigma, S)$ contains every polynomial in $S\langle\langle T_\Sigma \rangle\rangle$ and that $\mathrm{Rec}(\Sigma, S)$ is closed under the rational operations (cf. Theorem 3.8). Thus, we obtain the following Kleene theorem for recognizable tree series and commutative semirings.

Theorem 3.47 ([41], Theorem 7.1). $\mathrm{Rec}(\Sigma \cup Q_\infty, S) = \mathrm{Rat}(\Sigma \cup Q_\infty, S)$.

3.10 MSO-Definable Tree Series

Büchi's and Elgot's fundamental theorem [28, 46] shows the equivalence of recognizability and definability by means of formulas of monadic second order logic (MSO-logic) for the class of string languages. This result was extended to various other structures, including trees [135, 33] and unranked trees [117,

98, 99]. Then weighted MSO-logic was introduced in [37, 38], see also [39], and the equivalence between recognizability and definability of power series was proved. Most recently, this equivalence on the quantitative level has been extended to finite and infinite strings with discounting [42], trees [43], unranked trees [44], infinite trees [122], trace languages [115], picture languages [60], and texts and nested words [112, 113]. Here, we will report on Büchi–Elgot's theorem for recognizable tree series [43, 44] and we follow the approach of [37, 38], see also [39].

The main idea of [37, 38] for defining weighted MSO-logic is to consider formulas of MSO-logic in their negation normal form (i.e., all negation operators are moved down to the atoms) and then to allow elements of S to occur additionally as atomic formulas. Formally, the *set of all formulas of weighted MSO-logic over Σ and S on trees*, denoted by $\mathrm{MSO}(\Sigma, S)$, is defined to be the smallest set G such that:

(i) G contains all the *atomic formulas* a, $\mathrm{label}_\sigma(x)$, $\mathrm{edge}_i(x,y)$, $(x \sqsubseteq y)$, and $(x \in X)$, and the negations $\neg\mathrm{label}_\sigma(x)$, $\neg\mathrm{edge}_i(x,y)$, $\neg(x \sqsubseteq y)$, and $\neg(x \in X)$, and
(ii) if $\varphi, \psi \in G$, then also $\varphi \vee \psi$, $\varphi \wedge \psi$, $\exists x.\varphi$, $\forall x.\varphi$, $\exists X.\varphi$, $\forall X.\varphi \in G$,

where $a \in S$, x, y are first order variables, $\sigma \in \Sigma$, $1 \leq i \leq \max\{rk(\sigma) \mid \sigma \in \Sigma\}$ and X is a second order variable.

Next, we define the semantics of a formula $\varphi \in \mathrm{MSO}(\Sigma, S)$. We denote the set of free variables of φ by $\mathrm{Free}(\varphi)$. Let $\mathcal{V}$ be a finite set of variables containing $\mathrm{Free}(\varphi)$ and $\xi \in T_\Sigma$. A $(\mathcal{V}, \xi)$-assignment is a function ρ that maps the first order variables in $\mathcal{V}$ to elements of $\mathrm{pos}(\xi)$ and the second order variables in $\mathcal{V}$ to subsets of $\mathrm{pos}(\xi)$. We call a $(\mathrm{Free}(\varphi), \xi)$-assignment also simply an assignment for (φ, ξ), or a (φ, ξ)-assignment. We let the $\mathcal{V}$-semantics $[\![\varphi]\!]_\mathcal{V}$ of φ be the function which maps each pair $\zeta = (\xi, \rho)$ with $\xi \in T_\Sigma$ and $(\mathcal{V}, \xi)$-assignment ρ to the value $([\![\varphi]\!]_\mathcal{V}, \zeta) \in S$. We define this value inductively (over the structure of φ) as follows, where $\leq_\xi$ denotes the linear order on $\mathrm{pos}(\xi)$ induced by the postorder tree walk on ξ:

$$
\begin{aligned}
([\![a]\!]_\mathcal{V}, \zeta) &= a, \\
([\![\mathrm{label}_\sigma(x)]\!]_\mathcal{V}, \zeta) &= 1 \text{ if } \xi(\rho(x)) = \sigma, \text{ and } 0 \text{ otherwise}, \\
([\![\mathrm{edge}_i(x,y)]\!]_\mathcal{V}, \zeta) &= 1 \text{ if } \rho(y) = \rho(x)\, i, \text{ and } 0 \text{ otherwise}, \\
([\![x \sqsubseteq y]\!]_\mathcal{V}, \zeta) &= 1 \text{ if } \rho(x) \leq_\xi \rho(y), \text{ and } 0 \text{ otherwise}, \\
([\![x \in X]\!]_\mathcal{V}, \zeta) &= 1 \text{ if } \rho(x) \in \rho(X), \text{ and } 0 \text{ otherwise}, \\
([\![\neg\varphi]\!]_\mathcal{V}, \zeta) &= 1 \text{ if } ([\![\varphi]\!]_\mathcal{V}, \zeta) = 0, \text{ and } 0 \text{ if } ([\![\varphi]\!]_\mathcal{V}, \zeta) = 1 \\
&\quad \big(\text{where } \varphi \text{ is of the form } \mathrm{label}_\sigma(x),\ \mathrm{edge}_i(x,y),\ x \sqsubseteq y, \text{ or } x \in X\big), \\
([\![\varphi \vee \psi]\!]_\mathcal{V}, \zeta) &= ([\![\varphi]\!]_\mathcal{V}, \zeta) + ([\![\psi]\!]_\mathcal{V}, \zeta), \\
([\![\varphi \wedge \psi]\!]_\mathcal{V}, \zeta) &= ([\![\varphi]\!]_\mathcal{V}, \zeta) \cdot ([\![\psi]\!]_\mathcal{V}, \zeta), \\
([\![\exists x.\varphi]\!]_\mathcal{V}, \zeta) &= \sum_{w \in \mathrm{pos}(\zeta)} ([\![\varphi]\!]_{\mathcal{V} \cup \{x\}}, \zeta[x \to w]),
\end{aligned}
$$

$$([\![\forall x.\varphi]\!]_{\mathcal{V}}, \zeta) = \prod_{w \in \mathrm{pos}(\zeta)} ([\![\varphi]\!]_{\mathcal{V} \cup \{x\}}, \zeta[x \to w]),$$

$$([\![\exists X.\varphi]\!]_{\mathcal{V}}, \zeta) = \sum_{I \subseteq \mathrm{pos}(\zeta)} ([\![\varphi]\!]_{\mathcal{V} \cup \{X\}}, \zeta[X \to I]),$$

$$([\![\forall X.\varphi]\!]_{\mathcal{V}}, \zeta) = \prod_{I \subseteq \mathrm{pos}(\zeta)} ([\![\varphi]\!]_{\mathcal{V} \cup \{X\}}, \zeta[X \to I]).$$

The factors in the product over $\mathrm{pos}(\zeta)$ are ordered according to $\leq_\zeta$; moreover, for the product over subsets I of $\mathrm{pos}(\zeta)$, we employ the lexicographical linear order on the set $\{0,1\}^{\mathrm{pos}(\zeta)}$, where the elements of $\mathrm{pos}(\zeta)$ are ordered by $\leq_\zeta$.

We write $[\![\varphi]\!]$ for $[\![\varphi]\!]_{\mathrm{Free}(\varphi)}$. Now let $Y \subseteq \mathrm{MSO}(\Sigma, S)$. A tree series $r \in S\langle\langle T_\Sigma \rangle\rangle$ is *Y-definable* if there is a sentence $\varphi \in Y$ such that $S = [\![\varphi]\!]$. By [135, 33], $\mathrm{Rec}(\Sigma)$ is the class of all $\mathrm{MSO}(\Sigma, \mathbb{B})$-definable tree languages.

Example 3.48. We consider the ranked alphabet $\Sigma = \{\sigma^{(2)}, \gamma^{(1)}, \alpha^{(0)}\}$ and the formula $\varphi = \exists x.\mathrm{label}_\sigma(x) \wedge (\exists y.\mathrm{edge}_2(x,y) \wedge \mathrm{label}_\alpha(y))$ in $\mathrm{MSO}(\Sigma, S)$ for an arbitrary S. Clearly, for $S = \mathbb{B}$, the tree series $[\![\varphi]\!]$ maps $\xi \in T_\Sigma$ to 1 iff the pattern $\sigma(z, \alpha)$ occurs at least once in ξ. For $S = \mathsf{Nat}$, $[\![\varphi]\!] = \#_{\sigma(z,\alpha)}$, i.e., it computes the number of occurrences of $\sigma(z, \alpha)$ in a given tree ξ.

In order to obtain an equivalence between recognizability and definability, we have to restrict either the $\mathrm{MSO}(\Sigma, S)$-logic or the class of involved semirings because, e.g., the tree series $[\![\forall x.\forall y.2]\!]$ over the semiring of natural numbers is not recognizable. For the restriction of the logic, we first recall the concept of unambiguous formulas (cf. [38], Definition 5.1). We denote by $\mathrm{MSO}^-(\Sigma, S)$ the fragment of $\mathrm{MSO}(\Sigma, S)$ obtained by not permitting atomic subformulas of the form a, where $a \in S$. Such formulas can be viewed as classical (unweighted) MSO-formulas over trees. Then roughly speaking, an unambiguous formula is a formula of $\mathrm{MSO}^-(\Sigma, S)$ in which the disjunction $\varphi \vee \psi$ is allowed only if, for every $\xi \in T_\Sigma$, there is no assignment ρ for $(\varphi \vee \psi, \xi)$ such that (ξ, ρ) satisfies both φ and ψ, and finally, for every first order and second order existential quantification $\exists x.\varphi$ and $\exists X.\varphi$, respectively, and every $\xi \in T_\Sigma$, there is at most one variable assignment to x and X, respectively, which fulfills φ. In fact, for every unambiguous formula φ, $\xi \in T_\Sigma$, and (φ, ξ)-assignment ρ, we have that $([\![\varphi]\!], (\xi, \rho)) = 1 \in S$ if (ξ, ρ) satisfies φ, and $([\![\varphi]\!], (\xi, \rho)) = 0$ otherwise (cf. [38], Proposition 5.2).

Note that the definition of unambiguity is based on the semantics of the formula. However, due to the algorithm in [39], Definition 4.3, for every $\varphi \in \mathrm{MSO}^-(\Sigma, S)$, there is a purely syntactic definition of formulas φ^+ and φ^- in $\mathrm{MSO}^-(\Sigma, S)$ such that:

- The formulas φ^+ and φ^- are unambiguous and $\mathrm{Free}(\varphi^+) = \mathrm{Free}(\varphi^-) = \mathrm{Free}(\varphi)$.
- For every $\xi \in T_\Sigma$ and (φ, ξ)-assignment ρ, (ξ, ρ) satisfies φ iff (ξ, ρ) satisfies φ^+ iff (ξ, ρ) does not satisfy φ^-.

We note that the atomic formula $(x \sqsubseteq y)$ is needed for this disambiguation. Moreover, for any $\varphi, \psi \in \mathrm{MSO}^-(\Sigma, S)$, we define the formulas $\varphi \stackrel{+}{\rightarrow} \psi$ and $\varphi \stackrel{+}{\leftrightarrow} \psi$ in $\mathrm{MSO}^-(\Sigma, S)$ as follows: $\varphi \stackrel{+}{\rightarrow} \psi = \varphi^- \vee (\varphi^+ \wedge \psi^+)$ and $\varphi \stackrel{+}{\leftrightarrow} \psi = (\varphi^+ \wedge \psi^+) \vee (\varphi^- \wedge \psi^-)$. Using this, we can define a formula to be *syntactically unambiguous* if it is of the form φ^+, φ^-, $\varphi \stackrel{+}{\rightarrow} \psi$, or $\varphi \stackrel{+}{\leftrightarrow} \psi$ for $\varphi, \psi \in \mathrm{MSO}^-(\Sigma, S)$. Clearly, each syntactically unambiguous formula is unambiguous.

The collection of *almost unambiguous* formulas is the smallest subset of $\mathrm{MSO}(\Sigma, S)$ which contains a for every $a \in S$ and all syntactically unambiguous formulas, and which is closed under disjunction and conjunction. In fact, for every almost unambiguous sentence φ, the tree series $[\![\varphi]\!]$ is a recognizable step function and vice versa, each recognizable step function is definable by some almost unambiguous sentence (cf. [44], Proposition 5.5).

Now we can define the fragment of $\mathrm{MSO}(\Sigma, S)$-logic which characterizes $\mathrm{Rec}(\Sigma, S)$. A formula $\varphi \in \mathrm{MSO}(\Sigma, S)$ is called *syntactically restricted* [39], if it satisfies the following conditions:

1. Whenever φ contains a conjunction $\psi \wedge \psi'$ as subformula but not in the scope of a universal first order quantifier, then each value of S occurring in ψ commutes with each value of S occurring in ψ'.
2. Whenever φ contains $\forall X.\psi$ as a subformula, then ψ is a syntactically unambiguous formula.
3. Whenever φ contains $\forall x.\psi$ as a subformula, then ψ is almost unambiguous.

We let $\mathrm{srMSO}(\Sigma, S)$ denote the set of all syntactically restricted formulas of $\mathrm{MSO}(\Sigma, S)$.

Theorem 3.49. *Let $r \in S\langle\langle T_\Sigma \rangle\rangle$. Then the following statements hold:*

(A) $r \in \mathrm{Rec}(\Sigma, S)$ if and only if r is $\mathrm{srMSO}(\Sigma, S)$-definable ([44], Theorem 7.2).

(B) Let S be commutative and locally finite. Then $r \in \mathrm{Rec}(\Sigma, S)$ if and only if r is $\mathrm{MSO}(\Sigma, S)$-definable ([43], Theorem 6.5).

We only indicate a sketch of the proof. For (A), given a wta $\mathcal{A}$ with $r = r_\mathcal{A}$, we can use the structure of $\mathcal{A}$ to explicitly write down an $\mathrm{srMSO}(\Sigma, S)$-sentence φ with $[\![\varphi]\!] = r_\mathcal{A}$. Conversely, given an $\mathrm{srMSO}(\Sigma, S)$-sentence ψ, we proceed by induction over the structure of ψ to construct a wta $\mathcal{A}$ such that $r_\mathcal{A} = [\![\psi]\!]$. For this, we encode pairs (ξ, ρ), where $\xi \in T_\Sigma$ and ρ is a $(\mathcal{V}, \xi)$-assignment, as trees over an extended alphabet $\Sigma_\mathcal{V} = \Sigma \times \{0,1\}^\mathcal{V}$ (as it is done also in the unweighted case). This enables us to view the function $[\![\varphi]\!]_\mathcal{V}$, where φ is an arbitrary $\mathrm{srMSO}(\Sigma, S)$-formula, as a formal tree series over S and $\Sigma_\mathcal{V}$ and we show that it is recognizable. When dealing with conjunctions and universal quantifications, we need the assumptions (1) and (2)–(3), respectively, on φ given above. For (B), there are alternative arguments exploiting that S is locally finite.

3.11 Other Models Related to Recognizable Tree Series

In the literature, several other concepts of recognizability of tree series were investigated: recognizability (a) by S-Σ-tree automata [21] where S is a commutative semiring, (b) by finite, polynomial tree automata over S [88, 58] where S is a commutative and continuous semiring, (c) by multilinear representations over fields [7], (d) by S-Σ-representations [19] where S is a field, (e) by polynomially-weighted tree automata [128], and (f) by wta over distributive multioperator monoids [90, 103, 62, 132] (which were already mentioned in Sect. 3.6). Here, we recall the models (a)–(d) and sketch how they are related to wta (for more details cf. [67]; for an informal comparison of model (a) with wta cf. [12]). Finally, we briefly discuss the concepts (e) and (f).

S-Σ-Tree Automata

In this subsection, we assume that S is commutative.

In [21], the recognizability of tree series by S-Σ-tree automata was defined. For the finite nonempty set $Q = \{q_1, \ldots, q_\kappa\}$, we consider the S-semimodule S^Q. Now let $k \geq 0$ and consider a mapping $\mu : Q^k \to S^Q$. A *multilinear extension of* μ is a mapping $\overline{\mu} : (S^Q)^k \to S^Q$ such that $\overline{\mu}$ is multilinear and for every $p_1, \ldots, p_k \in Q$ we have $\overline{\mu}(1_{p_1}, \ldots, 1_{p_k}) = \mu(p_1, \ldots, p_k)$ where 1_{p_i} is the p_i-unit vector in S^Q. It can easily be seen that such a multilinear extension of μ exists and is unique. In fact, it has the form

$$\overline{\mu}(v_1, \ldots, v_k)_q = \sum_{p_1, \ldots, p_k \in Q} (v_1)_{p_1} \cdot \cdots \cdot (v_k)_{p_k} \cdot \mu(p_1, \ldots, p_k)_q.$$

Thus, we speak about *the* multilinear extension of μ. Since scalar factors can be pulled out from arguments of $\overline{\mu}$ in different order, the definition of multilinear extension only makes sense if S is commutative.

The S-Σ-tree automaton [21] is the same as the wta of Definition 3.2 except that for every $\sigma^{(k)}$, $\mu_k(\sigma)$ is a function from Q^k to S^Q and $\mu_{\mathcal{A}}(\sigma)$ is the multilinear extension of $\mu_k(\sigma)$. Obviously, this leads to the same Σ-algebra $(S^Q, \mu_{\mathcal{A}})$. Thus, the S-Σ-tree automaton is just a reformulation of the wta.

Finite, Polynomial Tree Automata

In this subsection, we assume that S is commutative and continuous.

In [88] and [58], the following tree automaton model was defined. A *finite tree automaton (over S and Σ)* (for short: fta) is a tuple $\mathcal{A} = (Q, \mathcal{M}, I, F)$ where Q is a finite nonempty set (of *states*), $\mathcal{M} = (\mathcal{M}_k \mid k \geq 1)$ is a family of *transition matrices* $\mathcal{M}_k$ such that $\mathcal{M}_k \in S\langle\langle T_\Sigma(Z_k)\rangle\rangle^{Q \times Q^k}$ and for almost every k it holds that every entry of $\mathcal{M}_k$ is $\widetilde{0} \in S\langle\langle T_\Sigma(Z_k)\rangle\rangle$ (recall that $Z_k =$

$\{z_1, \ldots, z_k\})$, $I \in S\langle\langle T_\Sigma(Z_1)\rangle\rangle^Q$ is the *initial state vector*, and $F \in S\langle\langle T_\Sigma\rangle\rangle^Q$ is the *final state vector*.

Intuitively, an fta $\mathcal{A}$ produces a tree series in a top-down fashion. It starts with the tree series I_q for every $q \in Q$ and then repeatedly "unfolds" transition matrices; finally, it replaces the remaining occurrences of variables in Z by elements of F_p for appropriate $p \in Q$.

The semantics of an fta $\mathcal{A}$ is defined by means of the fixpoint of a mapping $\Phi_\mathcal{A}$, i.e., in a bottom-up fashion. As preparation, we define the substitution of matrices over tree series. Let $\mathcal{M}_k \in S\langle\langle T_\Sigma(Z_k)\rangle\rangle^{Q \times Q^k}$ be a transition matrix and $v_1, \ldots, v_k \in S\langle\langle T_\Sigma\rangle\rangle^Q$. Then we define the matrix $\mathcal{M}_k(v_1, \ldots, v_k) \in S\langle\langle T_\Sigma\rangle\rangle^Q$ of tree series for every $q \in Q$ by

$$\mathcal{M}_k(v_1, \ldots, v_k)_q = \sum_{q_1, \ldots, q_k \in Q} (\mathcal{M}_k)_{q, q_1 \ldots q_k} \leftarrow_{\text{OI}} \big((v_1)_{q_1}, \ldots, (v_k)_{q_k}\big).$$

Since S is continuous with $\sqsubseteq$ as partial order, $(S, \sqsubseteq)$ is a cpo. By extending $(S, \sqsubseteq)$ to $(S\langle\langle T_\Sigma\rangle\rangle, \sqsubseteq)$, and in its turn, extending $(S\langle\langle T_\Sigma\rangle\rangle, \sqsubseteq)$ to $(S\langle\langle T_\Sigma\rangle\rangle^Q, \sqsubseteq)$ componentwise, also $(S\langle\langle T_\Sigma\rangle\rangle^Q, \sqsubseteq)$ is a cpo. Moreover, the mapping $\Phi_\mathcal{A} : S\langle\langle T_\Sigma\rangle\rangle^Q \to S\langle\langle T_\Sigma\rangle\rangle^Q$ defined for every $v \in S\langle\langle T_\Sigma\rangle\rangle^Q$ by $\Phi_\mathcal{A}(v) = \sum_{k \geq 1} \mathcal{M}_k(v, \ldots, v) + F$ is continuous (cf. [58], page 228). Thus, by Tarski's fixpoint theorem, $\Phi_\mathcal{A}$ has a least fixpoint $\text{fix}\,\Phi_\mathcal{A}$ and this is the least upper bound of the *approximation sequence* $(\Phi^n_\mathcal{A}(\bot) \mid n \geq 0)$ *of* M with $\bot_q = \widetilde{0}$ for every $q \in Q$, i.e., $\text{fix}\,\Phi_\mathcal{A} = \mathsf{sup}\{\Phi^n_\mathcal{A}(\bot) \mid n \geq 0\}$. Then the *tree series recognized by* M is

$$r_\mathcal{A} = \sum_{q \in Q} \big(I_q \leftarrow_{\text{OI}} ((\text{fix}\,\Phi_\mathcal{A})_q)\big).$$

In order to relate this notion of recognizability with the one induced by our wta, we have to restrict the fta. We call an fta $\mathcal{A} = (Q, \mathcal{M}, I, F)$ *polynomial* if for every $k \geq 1$ all the entries of $\mathcal{M}_k$ are in $S\langle T_\Sigma(Z_k)\rangle$; moreover, for every $q \in Q$, there is an $a \in S$ such that $I_q = a.z_1$; and finally, for every $q \in Q$, the entry F_q is in $S\langle T_\Sigma\rangle$. A tree series $r \in S\langle\langle T_\Sigma\rangle\rangle$ is *recognizable by a polynomial fta over* S if there is a polynomial fta $\mathcal{A} = (Q, \mathcal{M}, I, F)$ over S and Σ such that $r = r_\mathcal{A}$. Note that $r_\mathcal{A} = \sum_{q \in Q} (I_q, z_1)(\text{fix}\,\Phi_\mathcal{A})_q$. The class of all tree series recognizable by some polynomial fta over S is denoted by $\text{Rec-FPTA}(\Sigma, S)$.

Now we will compare the concept of recognizability by polynomial fta with that of wta. Since polynomial fta are close to systems of linear equations (as discussed in Sect. 3.8), we will use the equivalence of recognizable and equational (cf. Theorem 3.40). For every polynomial fta $\mathcal{A} = (Q, \mathcal{M}, I, F)$ over Σ and S, one easily associates the system E of equations $z_q = s_q$ for every $q \in Q$, with $s_q = F_q + \sum_{k \geq 1, q_1, \ldots, q_k \in Q} (\mathcal{M}_k)_{q, q_1 \ldots q_k} \leftarrow_{\text{OI}} (z_{q_1}, \ldots, z_{q_k})$. Obviously, $\Phi_E = \Phi_\mathcal{A}$. Thus, $(\text{fix}\,\Phi_\mathcal{A})_q$ is a recognizable tree series for every $q \in Q$ (by Theorem 3.40). Since recognizability is preserved under scalar multiplication and summation (by Theorem 3.8), also $r_\mathcal{A}$ is recognizable. For the other direction, let $\mathcal{A} = (Q, \Sigma, S, \mu, \nu)$ be a wta. One can construct the polynomial fta $\mathcal{B} = (Q, \mathcal{M}, I, F)$ by defining $(\mathcal{M}_k)_{q, q_1 \ldots q_k} = \sum_{\sigma \in \Sigma^{(k)}} \mu_k(\sigma)_{q_1 \ldots q_k, q} . \sigma(z_1, \ldots, z_k)$,

$F_q = \sum_{\alpha \in \Sigma^{(0)}} \mu_0(\alpha)_{\varepsilon,q}.\alpha$, and $I_q = \nu_q.z_1$ for every $k \geq 1$ and $q, q_1, \ldots, q_k \in Q$. Then for every $\xi \in T_\Sigma$, $q \in Q$, $n \geq 0$, if $n \geq \text{height}(\xi)$, then $((\Phi_{\mathcal{B}}^{n+1}(\bot))_q, \xi) = h_\mu(\xi)_q$; this can be proved by induction on n. Then, using the fact that $((\Phi_{\mathcal{B}}^{n+1}(\bot))_q, \xi) = (\text{fix}\,(\Phi_{\mathcal{B}})_q, \xi)$ for every $n \geq \text{height}(\xi)$, it follows that $r_{\mathcal{A}} = r_{\mathcal{B}}$.

Theorem 3.50 ([58], Corollary 3.6). $\text{Rec}(\Sigma, S) = \text{Rec-FPTA}(\Sigma, S)$.

Multilinear Representations

In this subsection, we assume that S is a field.

In [7], the recognizability of tree series was defined in terms of multilinear mappings over finite-dimensional S-vector spaces in the following way. A *multilinear representation* of T_Σ is a tuple (V, μ, γ) where $(V, +, 0, \mu)$ is a non-trivial S-Σ-vector space and $\gamma : V \to S$ is a linear form. Then (V, μ, γ) defines the tree series $r \in S\langle\langle T_\Sigma \rangle\rangle$, where $(r, \xi) = \gamma(h_V(\xi))$ for every $\xi \in T_\Sigma$. A tree series $r \in S\langle\langle T_\Sigma \rangle\rangle$ is *recognizable by multilinear mappings over an S-vector space* if there is a multilinear representation (V, μ, γ) which defines r such that V is finite-dimensional. We denote the class of all tree series over Σ and S recognizable by multilinear mappings over an S-vector space by $\text{Rec-ML}(\Sigma, S)$.

Example 3.51 ([7], Example 4.1). We consider the tree series size_δ as defined on page 322. Then size_δ is recognizable by multilinear mappings over the $\mathbb{Q}$-vector space $(\mathbb{Q}^2, +, (0,0))$, where $(\mathbb{Q}, +, \cdot, 0, 1)$ is the field of rational numbers. Let $\{e_1, e_2\}$ be the basis of $\mathbb{Q}^2$ with $e_1 = (1,0)$ and $e_2 = (0,1)$. We define the multilinear representation $(\mathbb{Q}^2, \mu, \gamma)$ as follows. For every $k \geq 0$, $\sigma \in \Sigma^{(k)}$, and $i_1, \ldots, i_k \in \{1,2\}$ let

$$\mu(\sigma)(e_{i_1}, \ldots, e_{i_k}) = \begin{cases} e_1 + e_2 & \text{if } i_1 = \cdots = i_k = 1 \text{ and } \sigma = \delta, \\ e_1 & \text{if } i_1 = \cdots = i_k = 1 \text{ and } \sigma \neq \delta, \\ e_2 & \text{if } i_j = 2 \text{ for exactly one } 1 \leq j \leq k, \\ 0_2 & \text{otherwise.} \end{cases}$$

Since $\mu(\sigma)$ is multilinear, it suffices to define it on the base vectors. Finally, let $\gamma(e_1) = 0$ and $\gamma(e_2) = 1$ (since γ is a linear form, this definition extends to arbitrary vectors). Then we can prove by a straightforward induction that $h_{\mathbb{Q}^2}(\xi) = e_1 + (\text{size}_\delta, \xi)e_2$ for every $\xi \in T_\Sigma$. Thus, we obtain $\gamma(h_{\mathbb{Q}^2}(\xi)) = (\text{size}_\delta, \xi)$.

In fact, the concepts of recognizability by multilinear mappings over a finite-dimensional S-vector space V and recognizability by a wta over the field S coincide. This is based on the idea of viewing the state set Q of the wta as a basis of V; then V and S^Q are isomorphic S-vector spaces. Then roughly speaking, a multilinear representation (V, μ', γ) of T_Σ and a wta $\mathcal{A} = (Q, \Sigma, S, \mu, \nu)$ are related if μ is the restriction of μ' to base vectors (or, equivalently: μ' is the multilinear extension of μ) and $\nu(q) = \gamma(1_q)$ for every $q \in Q$ where 1_q is the q-unit vector. Obviously, $h_V = h_\mu$, and thus (V, μ', γ) defines the tree series $r_{\mathcal{A}}$. Since the relatedness implicitly contains a construction for both directions, we obtain the following characterization.

Theorem 3.52 ([12]; [67], Theorem 4.6). $\mathrm{Rec}(\Sigma, S) = \mathrm{Rec\text{-}ML}(\Sigma, S)$.

The question arises whether the concept of recognizability by multilinear mappings generalizes recognizability by bottom-up tree automata, i.e., whether $\mathrm{Rec}(\Sigma) \subseteq \mathrm{supp}(\mathrm{Rec\text{-}ML}(\Sigma, S))$ for every field S. The answer is yes, and this follows from Lemma 3.11, the fact that $\text{bud-Rec}(\Sigma, S) \subseteq \mathrm{Rec}(\Sigma, S)$, and Theorem 3.52. If we combine the involved two constructions, then we realize that every deterministic bottom-up tree automaton can be straightforwardly transformed into a multilinear representation. We note that the application of this transformation to a *non*deterministic bottom-up tree automaton would, in general, lead to a semantically different multilinear representation. This can be easily seen for the field $\mathbb{Z}_2 = \{0, 1\}$ where two accepting runs cancel each other because $1 + 1 = 0$.

S-*Σ*-Representations

Also, in this subsection, we assume that S is a field.

Next, we compare recognizability by wta over fields with the concept of representability [19]; this comparison is due to [23]. For this, let $n \geq 1$ and consider the *dual monoid* $(C_\Sigma, \diamond_z, z)$ of $(C_\Sigma, \cdot_z, z)$, where $\zeta \diamond_z \zeta' = \zeta' \cdot_z \zeta$ for every $\zeta, \zeta' \in C_\Sigma$. Then an *S-Σ-representation of dimension n* is a triple $R = (\varphi, \psi, \lambda)$ where $\varphi : C_\Sigma \to S^{n \times n}$ is a morphism from the monoid $(C_\Sigma, \diamond_z, z)$ to the monoid $(S^{n\times n}, \cdot, \mathcal{I}_n)$ of $n \times n$-matrices over S with the usual matrix multiplication and the unit matrix $\mathcal{I}_n$; moreover, $\psi : \Sigma^{(0)} \to S^{1\times n}$, and $\lambda \in S^{n\times 1}$ such that the following consistency condition is true: $\psi(\alpha) \cdot \varphi(\zeta) = \psi(\alpha') \cdot \varphi(\zeta')$ for every $\alpha, \alpha' \in \Sigma^{(0)}$ and $\zeta, \zeta' \in C_\Sigma$ such that $\zeta \cdot_z \alpha = \zeta' \cdot_z \alpha'$. The *tree series r_R represented by R* is defined by $(r_R, \xi) = \psi(\alpha) \cdot \varphi(\zeta) \cdot \lambda$ for every $\xi \in T_\Sigma$, $\zeta \in C_\Sigma$, and $\alpha \in \Sigma^{(0)}$ such that $\xi = \zeta \cdot_z \alpha$. A tree series r is *S-Σ-representable* if there is an S-Σ-representation R of some dimension n such that $r = r_R$. We denote the class of all S-Σ-representable tree series by $\mathrm{Rep}(\Sigma, S)$.

Example 3.53. Let us again consider the tree series $\#_{\sigma(z,\alpha)}$ as defined in the list of tree series on page 322, but now we use the field $\mathbb{Q}$ of rational numbers as the underlying semiring rather than Nat, i.e., $\#_{\sigma(z,\alpha)} \in \mathbb{Q}\langle\langle T_\Sigma \rangle\rangle$.

Now we consider the $\mathbb{Q}$-Σ-representation $R = (\varphi, \psi, \lambda)$ of dimension 3 where

$$\varphi(\zeta) = \begin{pmatrix} 1 & 0 & \#_{\sigma(z,\alpha)}(\zeta) \\ 0 & a(\zeta) & b(\zeta) \\ 0 & 0 & 1 \end{pmatrix}, \qquad \psi(\alpha) = (1\,1\,0), \qquad \lambda = \begin{pmatrix} 0 \\ 0 \\ 1 \end{pmatrix},$$

and $a(\zeta) = 1$ if $\zeta = z$ and 0 otherwise, and $b(\zeta) = 1$ if z occurs at the second child of a position of ζ and 0 otherwise. Here, we have generalized $\#_{\sigma(z,\alpha)}$ in the obvious way such that it also works on contexts, e.g., $\#_{\sigma(z,\alpha)}(\sigma(z,\alpha)) = 1$. Since for every $\xi \in T_\Sigma$ and $\zeta \in C_\Sigma$ such that $\xi = \zeta \cdot_z \alpha$, we have that $(r_R, \xi) = \#_{\sigma(z,\alpha)}(\zeta) + b(\zeta)$, we obtain that $r_R = \#_{\sigma(z,\alpha)}$.

Theorem 3.54 ([23], Theorems 1 and 2). $\mathrm{Rec}(\Sigma, S) = \mathrm{Rep}(\Sigma, S)$.

Proof. Let $r \in \mathrm{Rec}(\Sigma, S)$. By Lemma 3.29, the S-vector space LQ_r (as defined on page 336) is finite-dimensional. Let $\{\eta_1^{-1}r, \ldots, \eta_n^{-1}r\}$ be a basis of LQ_r for some $\eta_1, \ldots, \eta_n \in T_\Sigma$. Then we define the S-Σ-representation $R = (\varphi, \psi, \lambda)$ of dimension n, where $\varphi : C_\Sigma \to S^{n\times n}$ is defined for every $\zeta \in C_\Sigma$ and $1 \le p, q \le n$ by $\varphi(\zeta)_{p,q} = ((\zeta \cdot_z \eta_p)^{-1}r)_q$ (note that the vector $(\zeta \cdot_z \eta_p)^{-1}r$ of LQ_r can be written as a linear combination of the base vectors; then $((\zeta \cdot_z \eta_p)^{-1}r)_q$ is the coefficient of $\eta_q^{-1}r$ in this representation). Moreover, for every $\alpha \in \Sigma^{(0)}$ and $1 \le p \le n$ we define $\psi(\alpha)_{1,p} = (\alpha^{-1}r)_p$ and $\lambda_{p,1} = (r, \eta_p)$.

For the proof of the consistency condition, we consider an arbitrary $\xi \in T_\Sigma$ such that $\xi = \zeta \cdot_z \alpha$ for some $\zeta \in C_\Sigma$ and $\alpha \in \Sigma^{(0)}$. Then using the concept of left quotient $\zeta^{-1}s$ also for $\zeta \in C_\Sigma$ and $s \in S\langle\langle C_\Sigma\rangle\rangle$, it is easy to compute that $\xi^{-1}r = \sum_{q=1}^{n}(\psi(\alpha) \cdot \varphi(\zeta))_q(\eta_q^{-1}r)$ (this follows from the facts (1) $(\zeta \cdot_z \xi)^{-1}r = \zeta^{-1}(\xi^{-1}r)$, and (2) the mapping $s \mapsto \zeta^{-1}s$ is linear). Since $\eta_1^{-1}r, \ldots, \eta_n^{-1}r$ is a basis of LQ_r, the coefficients of $\eta_q^{-1}r$ are uniquely determined; thus, if $\xi = \zeta' \cdot_z \alpha'$ is another decomposition of ξ, then we obtain that $\psi(\alpha) \cdot \varphi(\zeta) = \psi(\alpha') \cdot \varphi(\zeta')$. This proves the consistency condition. Using the above equation for $\xi^{-1}r$, it is straightforward to prove that $r = r_R$. Thus, r is S-Σ-representable.

Next, let $R = (\varphi, \psi, \lambda)$ be an S-Σ-representation of dimension n. We extend the mapping $R_{\varphi\psi} : T_\Sigma \to S^n$ defined by $R_{\varphi\psi}(\xi) = \psi(\alpha) \cdot \varphi(\zeta)$ for every $\xi \in T_\Sigma$ with $\xi = \zeta \cdot_z \alpha$, linearly to the mapping $R_{\varphi\psi} : S\langle T_\Sigma\rangle \to S^n$; thus $R_{\varphi\psi}$ is an S-Σ-semimodule homomorphism. Moreover, we extend the mapping $\varphi : C_\Sigma \to S^{n\times n}$ linearly to the mapping $\varphi : S\langle C_\Sigma\rangle \to S^{n\times n}$.

In the sequel, we will use that $R_{\varphi\psi}(r' \circ_z r) = R_{\varphi\psi}(r) \cdot \varphi(r')$ for every $r \in S\langle T_\Sigma\rangle$ and $r' \in S\langle C_\Sigma\rangle$, which can be seen as follows. First, for every $\zeta \in C_\Sigma$ and $\xi \in T_\Sigma$, we can compute $R_{\varphi\psi}(\zeta \cdot_z \xi) = R_{\varphi\psi}(\zeta \cdot_z \zeta' \cdot_z \alpha)$, where $\xi = \zeta' \cdot_z \alpha$ is an arbitrary decomposition. This is equal to $\psi(\alpha) \cdot \varphi(\zeta \cdot_z \zeta') = \psi(\alpha) \cdot \varphi(\zeta') \cdot \varphi(\zeta)$ because φ is a monoid homomorphism. Finally, this is equal to $R_{\varphi\psi}(\zeta' \cdot_z \alpha) \cdot \varphi(\zeta) = R_{\varphi\psi}(\xi) \cdot \varphi(\zeta)$. Secondly, we can generalize the equation proved in the first step by using the fact that $R_{\varphi\psi}$ and φ are linear.

Now we define the multilinear representation (V, μ, γ) where $V = \mathrm{im}(R_{\varphi\psi})$ is the image of $S\langle T_\Sigma\rangle$ under $R_{\varphi\psi}$. For every $k \ge 0$, $\sigma \in \Sigma^{(k)}$, we define the mapping $\mu(\sigma)$ by $\mu(\sigma)(v_1, \ldots, v_k) = R_{\varphi\psi}(\mathrm{top}_\sigma(s_1, \ldots, s_k))$ for every $v_1, \ldots, v_k \in \mathrm{im}(R_{\varphi\psi})$, where $s_i \in S\langle T_\Sigma\rangle$ is a preimage of v_i, i.e., $R_{\varphi\psi}(s_i) = v_i$; in fact, one can prove the independence of this definition from the chosen preimages by using the equation $R_{\varphi\psi}(r' \circ_z r) = R_{\varphi\psi}(r) \cdot \varphi(r')$. Note that μ is in fact multilinear because $R_{\varphi\psi}$ is a homomorphism. Moreover, we define $\gamma : \mathrm{im}(R_{\varphi\psi}) \to S$ by $\gamma(v) = v \cdot \lambda$.

We can see that $R_{\varphi\psi} : S\langle T_\Sigma\rangle \to \mathrm{im}(R_{\varphi\psi})$ is a surjective S-Σ-semimodule homomorphism. Since $(S\langle T_\Sigma\rangle, +, \widetilde{0}, \mathrm{top})$ is initial, we obtain that $R_{\varphi\psi} = h_V$. Then $(r_R, \xi) = R_{\varphi\psi}(\xi) \cdot \lambda = \gamma(h_V(\xi))$ for every $\xi \in T_\Sigma$, and hence r_R can be recognized by multilinear mappings over an S-vector space. By Theorem 3.52, it follows that $r \in \mathrm{Rec}(\Sigma, S)$. □

We note that there is also a direct way to construct from a given wta over a field S an S-Σ-representation. For instance, the $\mathbb{Q}$-Σ-representation R which we have considered in Example 3.53, can be obtained from the wta $\mathcal{A} = (Q, \Sigma, \mathbb{Q}, \mu, \nu)$ of Example 3.4 (with Nat already replaced by $\mathbb{Q}$) by identifying the states q, $\overline{\alpha}$, and f with the indices 1, 2, and 3, respectively, and defining $\varphi(\zeta)_{p,q} = h^p_\mu(\zeta)_q$, $\psi(\alpha)_q = h_\mu(\alpha)_q$, and $\lambda_q = \nu(q)$ for every $\zeta \in C_\Sigma$ and $p, q \in Q$.

Theorem 3.54 provides a characterization of recognizable S-Σ-tree series in terms of S-Σ-representations of finite dimension. This characterization can now be used to prove that if Σ contains at least one binary symbol and $\mathbb{N} \subseteq S$, then height $\notin \mathrm{Rec}(\Sigma, S)$ (at the same time, height $\in \mathrm{Rec}(\Sigma, \mathsf{Arct})$, cf. Example 3.3). The proof of this non-membership is done by contradiction, and it exploits the well-known Cayley–Hamilton theorem which states a property of the *characteristic polynomial* $\chi_\mathcal{M}$ *of a matrix* $\mathcal{M}$. For every $\mathcal{M} \in S^{n\times n}$, this polynomial is defined by $\chi_\mathcal{M}(x) = \det(\mathcal{M} - x\mathcal{I}_n)$, where $\det(\mathcal{N})$ is the determinant of a matrix $\mathcal{N}$. Clearly, the degree of $\chi_M(x)$ is n, and the coefficient of x^n is $(-1)^n$.

Theorem 3.55 ([97], Chap. XIV, Theorem 3.1). *Let* $n \geq 1$ *and* $\mathcal{M} \in S^{n\times n}$. *Then* $\chi_\mathcal{M}(\mathcal{M}) = 0$.

Now assuming that height $\in \mathrm{Rec}(\Sigma, S)$, we know that there is an S-Σ-representation $R = (\varphi, \psi, \lambda)$ such that height $= r_R$. In the next lemma, we apply Theorem 3.55 to matrices of the form $\varphi(\zeta)$ where $\zeta \in C_\Sigma$. We abbreviate $\zeta_1 \cdot_z \zeta_2^n$ (and $\zeta_1 \cdot_z \zeta_2^n \cdot_z \alpha$) by $\zeta_1\zeta_2^n$ ($\zeta_1\zeta_2^n\alpha$, respectively) for all contexts $\zeta_1, \zeta_2 \in C_\Sigma$ (for ζ^n see the beginning of Sect. 3.6).

Lemma 3.56 ([7], Proposition 9.3; [131]). *Let* $R = (\varphi, \psi, \lambda)$ *be an* S-Σ-*representation of dimension* n *and* $\zeta_2 \in C_\Sigma$. *Then there are* $a_1, \ldots, a_n \in S$ *such that for every* $\zeta_1 \in C_\Sigma$, $\alpha \in \Sigma^{(0)}$ *and* $i \geq 0$ *we have that*

$$(-1)^n \cdot (r_R, \zeta_1\zeta_2^{i+n}\alpha) + a_1 \cdot (r_R, \zeta_1\zeta_2^{i+n-1}\alpha) + \cdots + a_n \cdot (r_R, \zeta_1\zeta_2^{i}\alpha) = 0.$$

Proof. Let $\chi_{\varphi(\zeta_2)}(x) = (-1)^n x^n + a_1 x^{n-1} + \cdots + a_n$ for some $a_1, \ldots, a_n \in S$. By Theorem 3.55, we have $\chi_{\varphi(\zeta_2)}(\varphi(\zeta_2)) = 0$.

Now let $\zeta_1 \in C_\Sigma$, $\alpha \in \Sigma^{(0)}$, and $i \geq 0$. Then, by multiplying both sides of the equation $\chi_{\varphi(\zeta_2)}(\varphi(\zeta_2)) = 0$ with the vector $\psi(\alpha) \cdot \varphi(\zeta_2)^i$ from the left, and with the matrix $\varphi(\zeta_1)$ from the right, applying algebraic laws, and the fact that φ is a monoid morphism, we obtain the equation

$$\sum_{l=0}^{n} a_{n-l} \cdot \big(\psi(\alpha) \cdot \varphi(\zeta_1 \zeta_2^{i+l})\big) = 0$$

where $a_0 = (-1)^n$. By multiplying the above equation with λ from the right and taking into account that $\psi(\alpha) \cdot \varphi(\zeta_1 \zeta_2^{i+l}) \cdot \lambda = (r_R, \zeta_1 \zeta_2^{i+l} \alpha)$, we obtain the statement of this lemma. □

Before continuing with the proof that height $\notin \mathrm{Rec}(\Sigma, S)$, let us give an example of a characteristic polynomial and the application of Lemma 3.56.

Example 3.57. Recall the $\mathbb{Q}$-Σ-representation $R = (\varphi, \psi, \lambda)$ of dimension 3 from Example 3.53 and the matrix

$$\varphi(\zeta) = \begin{pmatrix} 1 & 0 & \#_{\sigma(z,\alpha)}(\zeta) \\ 0 & a(\zeta) & b(\zeta) \\ 0 & 0 & 1 \end{pmatrix}$$

for any $\zeta \in C_\Sigma$. Since $\chi_{\varphi(\zeta)}$ does not depend on $\varphi(\zeta)_{q,f}$ and $\varphi(\zeta)_{\overline{\alpha},f}$, we obtain that for every $\zeta \in C_\Sigma$:

- If $\zeta \neq z$, then $a(\zeta) = 0$ and $\chi_{\varphi(\zeta)}(x) = -x^3 + 2x^2 - x$.
- If $\zeta = z$, then $a(\zeta) = 1$ and $\chi_{\varphi(\zeta)}(x) = -x^3 + 3x^2 - 3x + 1$.

Thus, in particular, for every $\zeta \neq z$, $\zeta' \in C_\Sigma$, and $i \geq 0$, we obtain the following recurrence equation by applying Lemma 3.56:

$$\#_{\sigma(z,\alpha)}(\zeta' \zeta^{i+3} \alpha) = 2 \cdot \#_{\sigma(z,\alpha)}(\zeta' \zeta^{i+2} \alpha) - \#_{\sigma(z,\alpha)}(\zeta' \zeta^{i+1} \alpha).$$

Theorem 3.58 ([7], Example 9.2; [131]). *If Σ contains at least one binary symbol and $\mathbb{N} \subseteq S$, then* height $\notin \mathrm{Rec}(\Sigma, S)$.

Proof. Assume that height $\in \mathrm{Rec}(\Sigma, S)$. Then by Theorem 3.54, there is an S-Σ-representation $R = (\varphi, \psi, \lambda)$ of some dimension n with height $= r_R$.

Consider the tree $\zeta_2 = \sigma(z, \alpha)$ for some $\sigma \in \Sigma^{(2)}$. By Lemma 3.56, there are $a_1, \ldots, a_n \in S$ such that for every $\zeta_1 \in C_\Sigma$ and $\alpha \in \Sigma^{(0)}$ and $i \geq 0$ we have that

$$(-1)^n \cdot \big(\mathrm{height}, \zeta(i+n)\big) + a_1 \cdot \big(\mathrm{height}, \zeta(i+n-1)\big) + \cdots + a_n \cdot \big(\mathrm{height}, \zeta(i)\big) = 0 \tag{6}$$

where $\zeta(l)$ abbreviates $\zeta_1 \zeta_2^l \alpha$. Now let $\xi \in T_\Sigma$ be an arbitrary tree such that $(\mathrm{height}, \xi) = n$. Moreover, let $\zeta_1 = \sigma(z, \xi)$. Then

$$\big(\mathrm{height}, \zeta(j)\big) = 1 + \max(j, n) \tag{7}$$

for every $j \geq 0$. Using (6) with $i = 0$, and (7) with $j = n, \ldots, 0$ we obtain that $(-1)^n \cdot (1 + n) + a_1 \cdot (1 + n) + \cdots + a_n \cdot (1 + n) = 0$. On the other hand, using (6) with $i = 1$, and (7) with $j = n + 1, \ldots, 1$, we obtain that $(-1)^n \cdot (1 + (n + 1)) + a_1 \cdot (1 + n) + \cdots + a_n \cdot (1 + n) = 0$. We obtain $(-1)^n = 0$, which is a contradiction. □

Polynomially-Weighted Tree Automata

In this subsection, we assume that S is commutative.

Let us now briefly discuss the concept (e) of recognizability (cf. the beginning of this subsection). This is the recognizability by polynomially-weighted tree automata, which were defined in [128]. Such a tree automaton uses a polynomial over S to compute the weight of a transition at a node by applying it to the weights of the subtrees of that node. In fact, polynomially-weighted tree automata are strictly more powerful than wta, cf. [67], Theorem 7.2 and Theorem 7.5. Moreover, it is decidable whether a polynomially-weighted tree automaton $\mathcal{A} = (Q, \Sigma, S, \mu, \nu)$ is bounded, where S can be any of the arctic semiring, the tropical semiring, and the semiring of finite subsets of $\mathbb{N}$, i.e., whether there is an $a \in S$ such that for every $\xi \in T_\Sigma$ and $q \in Q$ we have that $h_\mu(\xi)_q \sqsubseteq a$, cf. [128]. This result is generalized in [15] to finitely factorizing, monotonic, and naturally ordered semirings.

Wta over Distributive Multioperator Monoids

Finally, we consider the concept (f) of recognizability. In [90], the semiring was generalized to the distributive multioperator monoid (for short: DM-monoid) and wta over DM-monoids were introduced. A DM-monoid $(S, +, 0, \Omega)$ consists of a monoid $(S, +, 0)$ which is equipped with an Ω-algebraic structure, where the operations of Ω distribute over $+$. Such wta generalize polynomially-weighted tree automata in the sense that the weight of a transition on $\sigma \in \Sigma^{(k)}$ is a finite sum of Ω-polynomials; an Ω-polynomial is inductively built up from variables $z_1, \ldots, z_k$, elements of S, and operations of Ω. In other words, the weight is taken from $S\langle T_\Omega(Z_k)\rangle$. In [103, 62, 132], simple wta over DM-monoids have been investigated; there every transition is weighted by a single k-ary operation taken from Ω. In fact, simple (and hence, also arbitrary) wta over DM-monoids are strictly more powerful than polynomially-weighted tree automata, cf. [67], Theorems 8.6 and 8.9.

3.12 Further Results

There are further results on recognizable tree series which we did not address in the previous sections. Here, we list (some of) them in a very rough form.

In [62], Kleene's theorem has been proved for simple wta over distributive multioperator monoids and, as a consequence, for wta over arbitrary (i.e., not necessarily commutative) semirings. The latter result generalizes Theorem 3.47. Moreover, in [92, 95], Kleene's theorem has been proved for sorted algebras. In [10], a general Kleene-type theorem has been proved which is applicable to all grove theories that are Conway theories.

In [89], the concept of full abstract family of tree series, for short: AFT, has been defined. This is a family of tree series which contains $\widetilde{0}$ and is closed under sum, top-concatenation, and least solution of equational systems, and additionally under linear nondeleting recognizable tree transductions. In fact, $\text{Rec}(\Sigma, S)$ is an AFT if S is a commutative and continuous semiring, cf. [89], Theorem 3.5.

In [36], it has been shown how aa-deterministically recognizable tree series over a semifield can be learned by a minimal adequate teacher. The aa stands for *all accepting* and means that the root weight vector of the used wta maps every state to 1. A minimal adequate teacher [2] answers coefficient queries and equivalence queries faithfully. In [108], this result has been extended to arbitrary tree series in bud-$\text{Rec}(\Sigma, S)$ where S is a commutative semifield. We refer to [75] for an investigation about learning nondeterministic recognizable tree series over a field.

In [77], forward and backward bisimulation minimization algorithms for weighted tree automata have been investigated. For the origin of forward and backward bisimulation on weighted string automata, we refer to [27].

In [59], fuzzy tree automata are investigated. Such automata can be considered as weighted tree automata over a completely distributive lattice. In that paper, Kleene's theorem and the equivalence between equational and rational fuzzy sets over an arbitrary algebra is proved by using the theory of fixpoints and μ-clones of monotonic functions over a complete lattice.

In [9], recognizable tree series are studied in a coalgebraic way and several representation theorems are proved.

There also exist weighted pushdown tree automata (for the string case cf. [121]). In [94], algebraic tree series have been characterized by (weighted) pushdown tree automata; this generalizes the result of [74] from the unweighted to the weighted case. In [22], algebraic (or: context-free) tree series have been characterized as the closure of polynomials under second-order substitution of tree series and iteration; moreover, they have been compared with recursive program schemes [30, 31]. In [93] Corollary 3.6, it is proved that the class of algebraic tree series is closed under linear and nondeleting algebraic tree transductions.

4 IO-Substitution and Tree Series Transformations

As a preparation for Sect. 5 on weighted tree transducers, we recall the concept and the most important properties of the IO-substitution of tree series as well as the concept of tree series transformation and their composition.

Besides Z, we use a further set $X = \{x_1, x_2, \ldots\}$ of variables and we define $X_k = \{x_1, \ldots, x_k\}$ for every $k \geq 0$. For a finite set Q and a set $U \subseteq X$, we define $Q(U) = \{q(x) \mid q \in Q, x \in U\}$. We write just $Q(U)^*$ for $(Q(U))^*$. For every $w \in Q(X)^*$ and $x \in X$, we denote by $|w|_x$ the number of occurrences of x in w; thus $|w| = \sum_{x \in X} |w|_x$; e.g., $|q(x_1)p(x_2)p(x_1)|_{x_1} = 2$ and $|q(x_1)p(x_2)p(x_1)| = 3$. Hence, $|w|$ coincides with the usual definition of the length of a string w provided we consider $Q(U)$ as an alphabet. We say that

$w \in Q(U)^*$ is *linear in* U (resp., *nondeleting in* U), if every $x \in U$ occurs at most once (resp., at least once) in w. Moreover, we use the notation $|\xi|_z$ for $\xi \in T_\Delta(Z)$ and $z \in Z$, and the notions of linearity and nondeletion in $U \subseteq Z$ also for $\xi \in T_\Delta(U)$, accordingly.

For $r \in S\langle\langle T_\Delta(Z)\rangle\rangle$, we define $\mathrm{var}(r) = \bigcup_{\xi \in \mathrm{supp}(r)} \mathrm{var}(\xi)$, with $\mathrm{var}(\xi) = \{z \in Z \mid |\xi|_z > 0\}$. A tree series $r \in S\langle\langle T_\Delta(U)\rangle\rangle$, with $U \subseteq Z$, is *linear* (resp., *nondeleting*) in U, if every $\xi \in \mathrm{supp}(r)$ is linear (resp., nondeleting) in U.

Next, we extend tree substitution (cf. the end of Sect. 2.2) to IO-substitution of polynomial tree series. Certain statements that we need concerning this IO-substitution, e.g., property P5 below, only hold for commutative S. Since we do not want to monitor all the time whether we need commutativity or not and for the sake of succinctness, we make the following convention.

In the rest of Sect. 4, we assume that S is commutative.

Let $r \in S\langle T_\Delta(Z)\rangle$, $I \subseteq \mathbb{N}$ finite, and $(s_i \mid i \in I)$ a family of tree series $s_i \in S\langle T_\Delta(H)\rangle$ for some set H. The *IO-substitution* of tree series $(s_i \mid i \in I)$ into r, denoted by $r \leftarrow_{\mathrm{IO}} (s_i \mid i \in I)$, is defined by

$$r \leftarrow_{\mathrm{IO}} (s_i \mid i \in I) = \sum_{\substack{\xi \in T_\Delta(Z), \\ (\forall i \in I): \zeta_i \in T_\Delta(H)}} \left((r,\xi) \cdot \prod_{i \in I} (s_i, \zeta_i)\right) . \xi(\zeta_i \mid i \in I).$$

Note that the above sum is (locally) finite because there are finitely many choices of ξ and ζ_i, $i \in I$, such that all coefficients (r,ξ) and (s_i,ζ_i), $i \in I$, are not 0. In case $I = \{1,\ldots,n\}$, we write $r \leftarrow_{\mathrm{IO}} (s_1,\ldots,s_n)$.

In the following, we summarize some properties of IO-substitution. The notations r, I, and $(s_i \mid i \in I)$ stand for the same as above.

P0 *Empty substitution*: If $I = \emptyset$, then we have $r \leftarrow_{\mathrm{IO}} (s_i \mid i \in I) = r$.

P1 *Missing variables* ([53], Observation 2.6a): If a variable z_j with $j \in I$ does not occur in a tree $\xi \in T_\Delta(Z)$, then the tree obtained by the substitution $\xi(\zeta_i \mid i \in I)$ does not depend on ζ_j. Still, the value (s_j, ζ_j) contributes to the coefficient of $\xi(\zeta_i \mid i \in I)$ in the tree series $r \leftarrow_{\mathrm{IO}} (s_i \mid i \in I)$ for every choice of ζ_j, hence that coefficient is the result of an infinite summation in S.

P2 *Dropping an index* ([104], Observation 3): For every $j \in I$ such that $z_j \notin \mathrm{var}(r)$ and $s_j = 1.\zeta$ for some $\zeta \in T_\Delta(H)$, we have $r \leftarrow_{\mathrm{IO}} (s_i \mid i \in I) = r \leftarrow_{\mathrm{IO}} (s_i \mid i \in I \setminus \{j\})$.

P3 *Zero propagation* ([53], Observation 2.6b): If $r = \widetilde{0}$ or $s_i = \widetilde{0}$ for some $i \in I$, then $r \leftarrow_{\mathrm{IO}} (s_i \mid i \in I) = \widetilde{0}$.

P4 *Preserving polynomials* ([53], Proposition 2.7): The tree series $r \leftarrow_{\mathrm{IO}} (s_i \mid i \in I)$ is polynomial.

P5 *Linearity in coefficients* ([53], Proposition 2.8): For every $a \in S$ and family $(a_i \mid i \in I)$ of elements of S, we have $ar \leftarrow_{\mathrm{IO}} (a_i s_i \mid i \in I) = (a \cdot \prod_{i \in I} a_i)(r \leftarrow_{\mathrm{IO}} (s_i \mid i \in I))$.

P6 *Linearity in variables*: Let $r \in S\langle T_\Delta(Z_n)\rangle$ for some $n \geq 0$ and $r_1, \ldots, r_n \in S\langle T_\Delta(Z)\rangle$ such that $\mathrm{var}(r_i) \cap \mathrm{var}(r_j) = \emptyset$ for $1 \leq i \neq j \leq n$. If r is linear (and nondeleting) in Z_n and every r_i is linear (and nondeleting) in $\mathrm{var}(r_i)$, then the tree series $r \leftarrow_{\mathrm{IO}} (r_1, \ldots, r_n)$ is linear (and nondeleting) in $\bigcup_{i=1}^n \mathrm{var}(r_i)$.

P7 *Distributivity* ([53], Proposition 2.9): Let $J \subseteq \mathbb{N}$ and $J_i \subseteq \mathbb{N}$ for every $i \in I$ be further finite sets, and furthermore $(r_j \mid j \in J)$ and $(s_{j_i} \mid j_i \in J_i)$ families of tree series in $S\langle T_\Delta(Z)\rangle$ and $S\langle T_\Delta(H)\rangle$, respectively. Then

$$\Big(\sum_{j\in J} r_j\Big) \leftarrow_{\mathrm{IO}} \Big(\sum_{j_i\in J_i} s_{j_i} \mid i \in I\Big) = \sum_{\substack{j\in J,\\ \forall i\in I: j_i \in J_i}} r_j \leftarrow_{\mathrm{IO}} (s_{j_i} \mid i \in I).$$

P8 *Weak associativity* ([104], Corollary 7): Let $J \subseteq \mathbb{N}$ be finite and assume that $\mathrm{var}(r) \subseteq \{z_j \mid j \in J\}$. Moreover, let $(r_j \mid j \in J)$ be a family with $r_j \in S\langle T_\Delta(Z)\rangle$ and $(I_j \mid j \in J)$ be a partition of I such that $\mathrm{var}(r_j) \subseteq \{z_{i_j} \mid i_j \in I_j\}$ for every $j \in J$. Then

$$\begin{aligned}&\big(r \leftarrow_{\mathrm{IO}} (r_j \mid j \in J)\big) \leftarrow_{\mathrm{IO}} (s_i \mid i \in I)\\ &= r \leftarrow_{\mathrm{IO}} \big(r_j \leftarrow_{\mathrm{IO}} (s_i \mid i \in I_j) \mid j \in J\big).\end{aligned}$$

A mapping $\tau : T_\Sigma \to \mathcal{P}_{\mathrm{fin}}(T_\Delta)$ is called a *tree transformation*, where $\mathcal{P}_{\mathrm{fin}}(T_\Delta)$ denotes the set of finite subsets of T_Δ. The *composition of the tree transformations* τ *and* $\tau' : T_\Delta \to \mathcal{P}_{\mathrm{fin}}(T_\Gamma)$ is the tree transformation $\tau;\tau' : T_\Sigma \to \mathcal{P}_{\mathrm{fin}}(T_\Gamma)$ defined by $(\tau;\tau')(\xi) = \bigcup_{\eta\in\tau(\xi)} \tau'(\eta)$. A mapping $\tau : T_\Sigma \to S\langle T_\Delta\rangle$ is a *(tree to) tree series transformation (over S)*. Then τ extends to a mapping of type $S\langle T_\Sigma\rangle \to S\langle T_\Delta\rangle$ by letting $\tau(r) = \sum_{\xi\in T_\Sigma}(r,\xi)\tau(\xi)$ for every $r \in S\langle T_\Sigma\rangle$. Also, for a finite set Q, a mapping $\tau : T_\Sigma \to S\langle T_\Delta\rangle^Q$ extends to a mapping of type $S\langle T_\Sigma\rangle \to S\langle T_\Delta\rangle^Q$ by letting $\tau(r)_q = \sum_{\xi\in T_\Sigma}(r,\xi)\tau(\xi)_q$ for every $r \in S\langle T_\Sigma\rangle$ and $q \in Q$. The *composition of the tree series transformations* $\tau : T_\Sigma \to S\langle T_\Delta\rangle$ *and* $\tau' : T_\Delta \to S\langle T_\Gamma\rangle$ is the tree series transformation $\tau;\tau' : T_\Sigma \to S\langle T_\Gamma\rangle$ defined by $(\tau;\tau')(\xi) = \tau'(\tau(\xi))$ for every $\xi \in T_\Sigma$. Then we extend composition to classes of tree series transformations in the following way. Let $\mathrm{C}(S)$ and $\mathrm{D}(S)$ be classes of polynomial tree series transformations over S. The composition of $\mathrm{C}(S)$ and $\mathrm{D}(S)$ is the class $\mathrm{C}(S);\mathrm{D}(S) = \{\tau;\tau' \mid$ there are ranked alphabets Σ, Δ, and Γ such that $\tau : T_\Sigma \to S\langle T_\Delta\rangle, \tau' : T_\Delta \to S\langle T_\Gamma\rangle, \tau \in \mathrm{C}(S)$, and $\tau' \in \mathrm{D}(S)\}$.

A tree series transformation $\tau : T_\Sigma \to S\langle T_\Sigma\rangle$ is called a *weighted identity* if, for every $\xi \in T_\Sigma$, we have $\tau(\xi) = a.\xi$ for some $a \in S$. The particular weighted identity $\iota : T_\Sigma \to S\langle T_\Sigma\rangle$, defined by $\iota(\xi) = 1.\xi$ for every $\xi \in T_\Sigma$, is called *identity (tree series transformation)*. It should be clear that $\iota;\tau = \tau$ for every $\tau : T_\Sigma \to S\langle T_\Delta\rangle$ and $\tau;\iota = \tau$ for every $\tau : T_\Gamma \to S\langle T_\Sigma\rangle$.

5 Weighted Tree Transducers

5.1 Tree Transducers

Tree transducers generalize finite-state tree automata in the way that, besides processing an input tree, they produce output trees. The classical generalizations are the *bottom-up tree transducers* [134, 48], the *top-down tree transducers* [123, 133, 48], and the *top-down tree transducers with regular look-ahead* [49]; their names come from the direction in which they process the input tree, where we assume that trees grow downward. Here, we recall their definition, however, we start with the definition of *generalized finite-state tree transducers* (for short: gfst) [48] which generalize all of the three classical tree transducers.

A gfst is a system $\mathcal{M} = (Q, \Sigma, \Delta, R, F)$, where Q is a finite set (of *states*) with $Q \cap (\Sigma \cup \Delta) = \emptyset$, R is a finite set of *rules*, and $F \subseteq Q$ is a set of *distinguished states*. We call Σ and Δ the *input* and the *output ranked alphabet*, respectively. Each rule in R has the form $(q, k, l, \sigma \to \zeta, \varphi)$, where $q \in Q$, $k \geq 0$, $l \geq 0$, $\sigma \in \Sigma^{(k)}$, $\zeta \in T_\Delta(Z_l)$, and $\varphi : Z_l \to Q(X_k)$ such that if $k = 0$, then $l = 0$. Such a rule can be visualized as

$$q\big(\sigma(x_1, \ldots, x_k)\big) \to \zeta, \langle q_1(x_{i_1}) \ldots q_l(x_{i_l})\rangle, \tag{8}$$

where $\varphi(z_j) = q_j(x_{i_j})$ for every $1 \leq j \leq l$, and it can be interpreted as follows. A q-translation of a Σ-tree rooted by the k-ary symbol σ is a Δ-tree which is computed in the way that a q_j-translation of the i_jth descendant of σ is substituted for every $1 \leq j \leq l$ and for every occurrence of z_j in ζ (if any). Of course, there may be several q-translations of such a σ-rooted tree. For a gfst $\mathcal{M}$, we consider the Σ-algebra $(\mathcal{P}_{\text{fin}}(T_\Delta)^Q, \mu_\mathcal{M})$ where, for every $k \geq 0$ and $\sigma \in \Sigma^{(k)}$, the k-ary operation $\mu_\mathcal{M}(\sigma) : \mathcal{P}_{\text{fin}}(T_\Delta)^Q \times \cdots \times \mathcal{P}_{\text{fin}}(T_\Delta)^Q \to \mathcal{P}_{\text{fin}}(T_\Delta)^Q$ is defined by

$$\mu_\mathcal{M}(\sigma)(v_1, \ldots, v_k)_q = \bigcup_{\substack{\text{for every rule} \\ \text{of the form (8)}}} \{\zeta(\zeta_1, \ldots, \zeta_l) \mid \zeta_j \in (v_{i_j})_{q_j},\ 1 \leq j \leq l\}.$$

Let us denote the unique Σ-algebra homomorphism from T_Σ to $\mathcal{P}_{\text{fin}}(T_\Delta)^Q$ by $h_\mathcal{M}$. Then the *tree transformation* $\tau_\mathcal{M} : T_\Sigma \to \mathcal{P}_{\text{fin}}(T_\Delta)$ *computed by* $\mathcal{M}$ is defined as

$$\tau_\mathcal{M}(\xi) = \bigcup_{q \in F} h_\mathcal{M}(\xi)_q$$

for every $\xi \in T_\Sigma$. Let GFST denote the class of all tree transformations computed by gfst.

Top-down tree transducers, top-down tree transducers with regular look-ahead, and bottom-up tree transducers can be derived from gfst as follows, cf. [48, 49]. The gfst $\mathcal{M}$ is a top-down tree transducer (resp., with regular look-ahead) if, for every rule (8), the tree ζ is linear and nondeleting (resp., linear) in Z_l. For the sake of simplicity, we consider only top-down tree transducers. Then the rule (8) is written as $q(\sigma(x_1,\ldots,x_k)) \to \overline{\zeta}$, where $\overline{\zeta} = \zeta(q_1(x_{i_1}),\ldots,q_l(x_{i_l}))$. Moreover, the gfst $\mathcal{M}$ is a bottom-up tree transducer if, for every rule (8), we have $k = l$ and $i_1 = 1,\ldots,i_k = k$. Then the rule (8) is written as $\sigma(q_1(x_1),\ldots,q_k(x_k)) \to q(\zeta(x_1,\ldots,x_k))$. This allows to consider both top-down and bottom-up tree transducers as special term rewrite systems [82, 3] and to define their semantics, i.e., the computed tree transformation, in terms of term rewriting in the standard way, cf. [123, 48, 49]. See [48] for the exact definition of the term rewrite semantics of top-down and bottom-up tree transducers, and cf. [48], Lemmas 5.5 and 5.6 for the equivalence of the initial algebra semantics (as defined above) and the term rewrite semantics. We denote the classes of all tree transformations computed by top-down tree transducers and bottom-up tree transducers by TOP and BOT, respectively.

5.2 The Basic Model

Tree transducers can be generalized to weighted tree transducers over a semiring in a similar way as bottom-up tree automata were generalized to wta (cf. Sect. 3.2). The idea behind this is to represent a tree transformation $\tau : T_\Sigma \to \mathcal{P}_{\text{fin}}(T_\Delta)$ as a tree series transformation $\tau : T_\Sigma \to \mathbb{B}\langle T_\Delta\rangle$ over the Boolean semiring $\mathbb{B}$ and then to generalize from $\mathbb{B}$ to an arbitrary semiring S.

For the first step, consider the system $\mathcal{M} = (Q,\Sigma,\Delta,\mathbb{B},\mu,F)$, called a *weighted tree transducer over* $\mathbb{B}$, where Q, Σ, Δ, and F are as for a gfst, while $\mu = (\mu_k \mid k \geq 0)$ is a family of mappings

$$\mu_k : \Sigma^{(k)} \to \mathbb{B}\langle T_\Delta(Z)\rangle^{Q(X_k)^* \times Q}$$

such that $\mu_k(\sigma)_{w,q} \neq \widetilde{0}$ only for finitely many $(w,q) \in Q(X_k)^* \times Q$. Moreover, $\mu_k(\sigma)_{w,q} \in \mathbb{B}\langle T_\Delta(Z_l)\rangle$ with $l = |w|$, for every $(w,q) \in Q(X_k)^* \times Q$. Then we consider the Σ-algebra $(\mathbb{B}\langle T_\Delta\rangle^Q, \mu_\mathcal{M})$ where for every $k \geq 0$ and $\sigma \in \Sigma^{(k)}$, the k-ary operation $\mu_\mathcal{M}(\sigma) : \mathbb{B}\langle T_\Delta\rangle^Q \times \cdots \times \mathbb{B}\langle T_\Delta\rangle^Q \to \mathbb{B}\langle T_\Delta\rangle^Q$ is defined as follows. For every $q \in Q$ and $v_1,\ldots,v_k \in \mathbb{B}\langle T_\Delta\rangle^Q$, we have

$$\mu_\mathcal{M}(\sigma)(v_1,\ldots,v_k)_q = \bigvee_{\substack{w \in Q(X_k)^*,\\ w = q_1(x_{i_1})\ldots q_l(x_{i_l})}} \mu_k(\sigma)_{w,q} \leftarrow_{\text{IO}} \big((v_{i_1})_{q_1},\ldots,(v_{i_l})_{q_l}\big).$$

Let us denote the unique Σ-algebra homomorphism from T_Σ to $\mathbb{B}\langle T_\Delta\rangle^Q$ by h_μ. Now the tree series transformation $\tau_\mathcal{M} : T_\Sigma \to \mathbb{B}\langle T_\Delta\rangle$ *computed by* $\mathcal{M}$ is defined by

$$\tau_\mathcal{M}(\xi) = \bigvee_{q \in F} h_\mu(\xi)_q$$

for every $\xi \in T_\Sigma$.

It should be clear that gfst and weighted tree transducers over $\mathbb{B}$ are semantically equivalent in the sense that, for every gfst $\mathcal{M}$, we can construct a weighted tree transducer $\mathcal{N}$ over $\mathbb{B}$ such that $\tau_{\mathcal{M}}(\xi) = \text{supp}(\tau_{\mathcal{N}}(\xi))$ for every $\xi \in T_\Sigma$, and vice versa. In fact, if $\mathcal{M} = (Q, \Sigma, \Delta, R, F)$ and $\mathcal{N} = (Q, \Sigma, \Delta, \mathbb{B}, \mu, F)$, then the connection between them is the following: for every $k \geq 0$, $\sigma \in \Sigma^{(k)}$, $w \in Q(X_k)^*$, and $q \in Q$, we have $\mu_k(\sigma)_{w,q} = \bigvee(1.\zeta \mid q(\sigma(x_1, \ldots, x_k)) \to \zeta, \langle w \rangle \text{ is in } R)$.

For the second step, we observe that, as in the case of wta, weighted tree transducers over $\mathbb{B}$ can easily be generalized to weighted tree transducers over an arbitrary semiring S. For the same reason as in Sect. 4, we make the following conventions.

In the rest of Sect. 5, we assume that S is commutative. Moreover, $\leftarrow_{\text{IO}}$ will be abbreviated by $\leftarrow$.

Definition 5.1. *A* weighted tree transducer (over S)[4] *(for short: wtt) is a tuple $\mathcal{M} = (Q, \Sigma, \Delta, S, \mu, F)$, where:*

- *Q is a finite nonempty set, the* set of states, *with $Q \cap (\Sigma \cup \Delta) = \emptyset$.*
- *Σ and Δ are the* input and the output ranked alphabets, *respectively.*
- *$\mu = (\mu_k \mid k \in \mathbb{N})$ is a* family of rule mappings
$$\mu_k : \Sigma^{(k)} \to S\langle T_\Delta(Z)\rangle^{Q(X_k)^* \times Q}$$
such that $\mu_k(\sigma)_{w,q} \neq \widetilde{0}$ only for finitely many $(w, q) \in Q(X_k)^* \times Q$, and $\mu_k(\sigma)_{w,q} \in S\langle T_\Delta(Z_l)\rangle$ with $l = |w|$, for every $(w, q) \in Q(X_k)^* \times Q$.
- *$F \subseteq Q$ is the* set of designated states.

For such a wtt $\mathcal{M}$ we consider the Σ-algebra $(S\langle T_\Delta\rangle^Q, \mu_{\mathcal{M}})$ where, for every $k \geq 0$ and $\sigma \in \Sigma^{(k)}$, the k-ary operation $\mu_{\mathcal{M}}(\sigma) : S\langle T_\Delta\rangle^Q \times \cdots \times S\langle T_\Delta\rangle^Q \to S\langle T_\Delta\rangle^Q$ is defined as follows. For every $q \in Q$ and $v_1, \ldots, v_k \in S\langle T_\Delta\rangle^Q$, we have

$$\mu_{\mathcal{M}}(\sigma)(v_1, \ldots, v_k)_q = \sum_{\substack{w \in Q(X_k)^*,\\ w = q_1(x_{i_1})\ldots q_l(x_{i_l})}} \mu_k(\sigma)_{w,q} \leftarrow \big((v_{i_1})_{q_1}, \ldots, (v_{i_l})_{q_l}\big).$$

Let us denote the unique Σ-algebra homomorphism from T_Σ to $S\langle T_\Delta\rangle^Q$ by h_μ. Then the tree series transformation $\tau_{\mathcal{M}} : T_\Sigma \to S\langle T_\Delta\rangle$ *computed by* $\mathcal{M}$ is defined as

$$\tau_{\mathcal{M}}(\xi) = \sum_{q \in F} h_\mu(\xi)_q$$

for every $\xi \in T_\Sigma$. We denote by $\text{WTT}(S)$ the class of all tree series transformations over S that are computable by a wtt.

An equivalent definition of $\tau_{\mathcal{M}}$ can be given as follows. For every $q \in Q$, we define the tree series transformation $\tau_{\mathcal{M},q} : T_\Sigma \to S\langle T_\Delta\rangle$ by induction: for every $\xi = \sigma(\xi_1, \ldots, \xi_k) \in T_\Sigma$, let

[4] In the literature, a weighted tree transducer is also known as a (polynomial) *tree series transducer*. Moreover, μ is also called a *tree representation*.

$$\tau_{\mathcal{M},q}(\xi) = \sum_{\substack{w \in Q(X_k)^*,\\ w=q_1(x_{i_1})\dots q_l(x_{i_l})}} \mu_k(\sigma)_{w,q} \leftarrow \big(\tau_{\mathcal{M},q_1}(\xi_{i_1}), \dots, \tau_{\mathcal{M},q_l}(\xi_{i_l})\big).$$

Then we can easily show that $\tau_{\mathcal{M},q}(\xi) = h_\mu(\xi)_q$, hence $\tau_{\mathcal{M}}(\xi) = \sum_{q \in F} \tau_{\mathcal{M},q}(\xi)$.

Example 5.2. As an example, we consider the wtt $\mathcal{M} = (Q, \Sigma, \Delta, \mathsf{Trop}, \mu, F)$ with $Q = \{q, q_0, q_\alpha, q_\beta\}$, $\Sigma = \{\delta^{(2)}, \gamma^{(1)}, \alpha^{(0)}, \beta^{(0)}\}$, $\Delta = \{\sigma^{(2)}, \gamma_1^{(1)}, \gamma_2^{(1)}, \alpha^{(0)}, \beta^{(0)}\}$, and $F = \{q_0\}$. Moreover, we specify the rule mappings such that, for every $\xi \in T_\Sigma$ and $\theta \in \{\alpha, \beta\}$, the following statements hold (where we have dropped the parentheses in subtrees of the form $\gamma(\zeta)$):

$$\tau_{\mathcal{M},q_\theta}(\xi) = \begin{cases} k.\theta & \text{if } \xi = \gamma^k\theta \text{ for } k \geq 0, \\ \widetilde{\infty} & \text{otherwise,} \end{cases}$$

$$\tau_{\mathcal{M},q}(\xi) = \begin{cases} \mathsf{min}_{\zeta \in L(n,\theta)} |\zeta|_{\gamma_2}.\zeta & \text{if } \xi = \gamma^n\theta \text{ for } n \geq 0, \\ \widetilde{\infty} & \text{otherwise,} \end{cases}$$

where $L(n, \theta) = \{\gamma_{i_1} \dots \gamma_{i_n}\theta \mid i_1, \dots, i_n \in \{1, 2\}\}$, and

$$\tau_{\mathcal{M},q_0}(\xi) = \begin{cases} \mathsf{min}_{\zeta \in L(n,\alpha)}(|\zeta|_{\gamma_2} + k) \,.\, \sigma(\zeta, \zeta) \\ \qquad \text{if } \xi = \delta(\gamma^n\theta, \gamma^k\alpha) \text{ for } n, k \geq 0, \\ \mathsf{min}_{\zeta_1,\zeta_2 \in L(n,\beta)}(|\sigma(\zeta_1, \zeta_2)|_{\gamma_2} + k) \,.\, \sigma(\zeta_1, \zeta_2) \\ \qquad \text{if } \xi = \delta(\gamma^n\theta, \gamma^k\beta) \text{ for } n, k \geq 0, \\ \widetilde{\infty} \quad \text{otherwise.} \end{cases}$$

For this purpose, we define μ such that, for every $\theta \in \{\alpha, \beta\}$,

$$\begin{aligned} \mu_1(\gamma)_{q_\theta(x_1),q_\theta} &= 1.z_1, \\ \mu_0(\theta)_{\varepsilon,q_\theta} &= 0.\theta, \\ \mu_1(\gamma)_{q(x_1),q} &= 0.\gamma_1(z_1) + 1.\gamma_2(z_1), \\ \mu_0(\theta)_{\varepsilon,q} &= 0.\theta, \\ \mu_2(\delta)_{q(x_1)q_\alpha(x_2),q_0} &= 0.\sigma(z_1, z_1), \\ \mu_2(\delta)_{q(x_1)q(x_1)q_\beta(x_2),q_0} &= 0.\sigma(z_1, z_2), \end{aligned}$$

and all the other not mentioned tuples $(w, p) \in Q(X_k)^* \times Q$ lead to $\widetilde{\infty}$.

Now we consider the input tree $\xi = \delta(\gamma^n\theta, \gamma^k\beta)$. Assuming that $\tau_{\mathcal{M},q}(\gamma^n\theta)$, $\tau_{\mathcal{M},q_\alpha}(\gamma^k\beta)$, and $\tau_{\mathcal{M},q_\beta}(\gamma^k\beta)$ have the desired form (as reported above), the tree series transformation $\tau_{\mathcal{M},q_0}$ can be evaluated on ξ as follows (where we abbreviate $\tau_{\mathcal{M},p}$ by τ_p for every $p \in Q$):

$$\begin{aligned} &\tau_{q_0}\big(\delta(\gamma^n\theta, \gamma^k\beta)\big) \\ &= \mathsf{min}\,\{\mu_2(\delta)_{q(x_1)q_\alpha(x_2),q_0} \leftarrow \big(\tau_q(\gamma^n\theta), \tau_{q_\alpha}(\gamma^k\beta)\big), \\ &\qquad \mu_2(\delta)_{q(x_1)q(x_1)q_\beta(x_2),q_0} \leftarrow \big(\tau_q(\gamma^n\theta), \tau_q(\gamma^n\theta), \tau_{q_\beta}(\gamma^k\beta)\big)\} \end{aligned}$$

$$
\begin{aligned}
&= \min\{0.\sigma(z_1,z_1) \leftarrow (\tau_q(\gamma^n\theta), \widetilde{\infty}),\\
&\qquad 0.\sigma(z_1,z_2) \leftarrow (\tau_q(\gamma^n\theta), \tau_q(\gamma^n\theta), \tau_{q_\beta}(\gamma^k\beta))\}\\
&\overset{(P3)}{=} \min\{\widetilde{\infty}, 0.\sigma(z_1,z_2) \leftarrow (\tau_q(\gamma^n\theta), \tau_q(\gamma^n\theta), \tau_{q_\beta}(\gamma^k\beta))\}\\
&= 0.\sigma(z_1,z_2) \leftarrow (\tau_q(\gamma^n\theta), \tau_q(\gamma^n\theta), \tau_{q_\beta}(\gamma^k\beta))\\
&= \min_{\substack{\xi' \in T_\Delta(Z_3)\\ \zeta_1,\zeta_2,\zeta_3 \in T_\Delta}} a.\ \xi'(\zeta_1,\zeta_2,\zeta_3)\\
&\quad (\text{where } a = (0.\sigma(z_1,z_2),\xi') + (\tau_q(\gamma^n\theta),\zeta_1) + (\tau_q(\gamma^n\theta),\zeta_2)\\
&\qquad + (\tau_{q_\beta}(\gamma^k\beta),\zeta_3))\\
&= \min_{\zeta_1,\zeta_2 \in T_\Delta} ((\tau_q(\gamma^n\theta),\zeta_1) + (\tau_q(\gamma^n\theta),\zeta_2) + k).\sigma(\zeta_1,\zeta_2)\\
&\quad (\text{choosing } \xi' = \sigma(z_1,z_2) \text{ and } \zeta_3 = \beta)\\
&= \min_{\zeta_1,\zeta_2 \in L(n,\theta)} (|\sigma(\zeta_1,\zeta_2)|_{\gamma_2} + k).\sigma(\zeta_1,\zeta_2).
\end{aligned}
$$

We shortly discuss the copy and the deletion capabilities of weighted tree transducers where ξ stands for an input tree and ξ' is a subtree of ξ.

(T) A wtt can process several copies of ξ'; the weight of each processing of ξ' is included into that of ξ (even if some of the copies are processed in the same way). For instance, in Example 5.2 for $\xi = \delta(\xi', \gamma^k\beta)$ and $\xi' = \gamma^n\theta$, we have that

$$\tau_{\mathcal{M},q_0}(\xi) = 0.\sigma(z_1,z_2) \leftarrow (\tau_{\mathcal{M},q}(\xi'), \tau_{\mathcal{M},q}(\xi'), \dots)$$

due to the equation $\mu_2(\delta)_{q(x_1)q(x_1)q_\beta(x_2),q_0} = 0.\sigma(z_1,z_2)$. Then by the definition of IO-substitution, two trees ζ_1, ζ_2 are chosen from the support of $\tau_{\mathcal{M},q}(\xi')$, and z_i is replaced by ζ_i in $\sigma(z_1,z_2)$; moreover, both values $(\tau_{\mathcal{M},q}(\xi'), \zeta_1)$ and $(\tau_{\mathcal{M},q}(\xi'), \zeta_2)$ are included in $\tau_{\mathcal{M},q_0}(\xi)$:

$$\tau_{\mathcal{M},q_0}(\xi) = \min_{\zeta_1,\zeta_2 \in T_\Delta} ((\tau_{\mathcal{M},q}(\xi'),\zeta_1) + (\tau_{\mathcal{M},q}(\xi'),\zeta_2) + \cdots).\sigma(\zeta_1,\zeta_2).$$

We note that, for top-down tree transducers, this corresponds to the property "(T) copying an input subtree followed by processing these copies nondeterministically (and independently)" [48].

(B1) A wtt can process ξ' and then copy the result of the processing; the weight of the processing of ξ' is included into that of ξ only once, no matter how many times the result is copied. This happens, e.g., in Example 5.2 for $\xi = \delta(\xi', \gamma^k\alpha)$ and $\xi' = \gamma^n\theta$: due to the equation $\mu_2(\delta)_{q(x_1)q_\alpha(x_2),q_0} = 0.\sigma(z_1,z_1)$, we have that

$$\tau_{\mathcal{M},q_0}(\xi) = 0.\sigma(z_1,z_1) \leftarrow (\tau_{\mathcal{M},q}(\xi'), \dots)$$

and, by the definition of IO-substitution, a tree ζ in the support of $\tau_{\mathcal{M},q}(\xi')$ is chosen and then copied to both occurrences of z_1 in $\sigma(z_1,z_1)$, whereas its weight $(\tau_{\mathcal{M},q}(\xi'), \zeta)$ is included in $\tau_{\mathcal{M},q}(\xi)$ only once:

$$\tau_{\mathcal{M},q_0}(\xi) = \min_{\zeta \in T_\Delta} \big((\tau_{\mathcal{M},q}(\xi'), \zeta) + \cdots\big).\sigma(\zeta, \zeta).$$

For bottom-up tree transducers, this corresponds to the property "(B1) non-deterministically processing an input subtree followed by copying the result of this process" [48].

(B2) A wtt can process ξ' (and thereby check a property of ξ') and then it can delete the result of this processing. In this case, the weight of the processing of ξ' is included into that of ξ. This situation occurs, e.g., in Example 5.2 for $\xi_1 = \delta(\xi', \gamma^k\alpha)$ and $\xi_2 = \delta(\xi', \gamma^k\beta)$ where $\xi' = \gamma^n\theta$. Then

$$\tau_{\mathcal{M},q_0}(\xi_1) = 0.\sigma(z_1, z_1) \leftarrow \big(\ldots,\ \tau_{\mathcal{M},q_\alpha}(\gamma^k\alpha)\big),$$
$$\tau_{\mathcal{M},q_0}(\xi_2) = 0.\sigma(z_1, z_2) \leftarrow \big(\ldots,\ \ldots,\ \tau_{\mathcal{M},q_\beta}(\gamma^k\beta)\big)$$

due to the equations

$$\mu_2(\delta)_{q(x_1)q_\alpha(x_2),q_0} = 0.\sigma(z_1, z_1),$$
$$\mu_2(\delta)_{q(x_1)q(x_1)q_\beta(x_2),q_0} = 0.\sigma(z_1, z_2),$$

resp., and by the definition of IO-substitution the value k is included in $\tau_{\mathcal{M},q_0}(\xi_1)$ and $\tau_{\mathcal{M},q_0}(\xi_2)$:

$$\tau_{\mathcal{M},q_0}(\xi_2) = \min_{\zeta \in T_\Delta} (\cdots + k).\sigma(\zeta, \zeta),$$
$$\tau_{\mathcal{M},q_0}(\xi_1) = \min_{\zeta_1, \zeta_2 \in T_\Delta} (\cdots + \cdots + k).\sigma(\zeta_1, \zeta_2).$$

For bottom-up tree transducers, this phenomenon is known as "(B2) checking a property of an input subtree followed by deletion" [48].

In the literature (e.g., in [91, 53, 65, 102, 104]), wtt are often defined such that a rule mapping has the type

$$\mu_k : \Sigma^{(k)} \to S\langle\langle T_\Delta(Z)\rangle\rangle^{Q(X_k)^* \times Q},$$

i.e., the tree series $\mu_k(\sigma)_{w,q}$ is not necessarily polynomial, and there the wtt of our Definition 5.1 is called polynomial. We would like to ask the reader to keep this is mind, when we later on refer to statements proved in that literature for the more general type of wtt. Clearly, we will only refer to statements which hold for the polynomial restriction of that general type.

In some works, a wtt $\mathcal{M}$ has, instead of designated states, a so-called *root output*. This root output is specified by a mapping $\nu : Q \to S\langle C_\Delta\rangle$ [102, 104] or more generally by $\nu : Q \to S\langle T_\Delta(Z_1)\rangle$ [91] and the tree series transformation computed by $\mathcal{M}$ is defined as $\tau_{\mathcal{M}}(\xi) = \sum_{q \in Q} \nu(q) \leftarrow h_\mu(\xi)_q$ for every $\xi \in T_\Sigma$. Certainly, the wtt with root output generalizes our wtt because the set F of designated states can be simulated by the particular root output defined by $\nu(q) = 1.z_1$ if $q \in F$ and $\nu(q) = \widetilde{0}$, otherwise for $q \in Q$. On the other hand, our wtt and the wtt with root output are semantically equivalent, because for every wtt $\mathcal{M}$ with root output one can construct a wtt $\mathcal{M}'$ with $\tau_{\mathcal{M}} = \tau_{\mathcal{M}'}$, see [104], Lemma 10.

5.3 Restricted Models

We define three types of restrictions of wtt. The first of them is made concerning the possible directions in which weighted tree transducers process the input trees. We say that a wtt $\mathcal{M} = (Q, \Sigma, \Delta, S, \mu, F)$ is a:

- *top-down wtt with regular look-ahead* (resp., *top-down wtt*) if, for every $k \geq 0$, $\sigma \in \Sigma^{(k)}$, $w \in Q(X_k)^*$, and $q \in Q$, the tree series $\mu_k(\sigma)_{w,q}$ is linear (resp., linear and nondeleting) in $Z_{|w|}$,
- *bottom-up wtt* if, for every $k \geq 0$, $\sigma \in \Sigma^{(k)}$, $w \in Q(X_k)^*$, and $q \in Q$, such that $\mu_k(\sigma)_{w,q} \neq \widetilde{0}$ we have that $w = q_1(x_1) \ldots q_k(x_k)$ for some $q_1, \ldots, q_k \in Q$.

The corresponding class of tree series transformations over S is denoted by $\mathrm{TOP}^{\mathrm{R}}(S)$ (resp., $\mathrm{TOP}(S)$, $\mathrm{BOT}(S)$). If $\mathcal{M}$ is a top-down wtt with regular look-ahead or a top-down wtt (resp., bottom-up wtt), then F is called the set of *initial* (resp., *final* states).

Example 5.3. If we drop from the wtt $\mathcal{M}$ of Example 5.2 the state q_α (and all the equations that involve q_α), then we obtain a top-down wtt $\mathcal{M}'$ with regular look-ahead such that for every $\xi \in T_\Sigma$ and $\theta \in \{\alpha, \beta\}$

$$\tau_{\mathcal{M}',q_0}(\xi) = \begin{cases} \min_{\zeta_1,\zeta_2 \in L(n,\theta)}(|\sigma(\zeta_1,\zeta_2)|_{\gamma_2} + k) \,.\, \sigma(\zeta_1,\zeta_2) \\ \qquad \text{if } \xi = \delta(\gamma^n\theta, \gamma^k\beta) \text{ for } n, k \geq 0, \\ \widetilde{\infty} \quad \text{otherwise.} \end{cases}$$

In fact, the state q_β can be called a *look-ahead state.*

If we replace in the rule mapping μ_2 of $\mathcal{M}'$ the equation

$$\mu_2(\delta)_{q(x_1)q(x_1)q_\beta(x_2),q_0} = 0.\sigma(z_1, z_2)$$

by

$$\mu_2(\delta)_{q(x_1)q(x_1),q_0} = 0.\sigma(z_1, z_2),$$

and we drop all the equations that involve q_β, then we obtain a top-down wtt $\mathcal{M}''$ such that for every $\xi \in T_\Sigma$

$$\tau_{\mathcal{M}'',q_0}(\xi) = \begin{cases} \min_{\zeta_1,\zeta_2 \in L(n,\theta)} |\sigma(\zeta_1,\zeta_2)|_{\gamma_2} \,.\, \sigma(\zeta_1,\zeta_2) \\ \qquad \text{if } \xi = \delta(\gamma^n\theta, \zeta) \text{ for } n \geq 0 \text{ and } \zeta \in T_\Sigma, \\ \widetilde{\infty} \quad \text{otherwise.} \end{cases}$$

If we drop from the wtt $\mathcal{M}$ of Example 5.2 the state q_β, then we obtain a bottom-up wtt $\mathcal{M}'''$ such that for every $\xi \in T_\Sigma$

$$\tau_{\mathcal{M}''',q_0}(\xi) = \begin{cases} \min_{\zeta \in L(n,\theta)}(|\zeta|_{\gamma_2} + k) \,.\, \sigma(\zeta,\zeta) \\ \qquad \text{if } \xi = \delta(\gamma^n\theta, \gamma^k\alpha) \text{ for } n, k \geq 0, \\ \widetilde{\infty} \quad \text{otherwise.} \end{cases}$$

(a)	(b)
$\gamma(\sigma(\sigma(\alpha,\alpha),\alpha))$	$\gamma(\sigma(\sigma(\alpha,\alpha),\alpha))$
$\Rightarrow \gamma(\sigma(\sigma(q(\alpha),\alpha),\alpha))$	$\Rightarrow \gamma(\sigma(\sigma(q(\alpha),\alpha),\alpha))$
$\Rightarrow \gamma(\sigma(\sigma(q(\alpha),\overline{\alpha}(\alpha)),\alpha))$	$\Rightarrow \gamma(\sigma(\sigma(q(\alpha),q(\alpha)),\alpha))$
$\Rightarrow \gamma(\sigma(f(\alpha),\alpha))$	$\Rightarrow \gamma(\sigma(q(\sigma(\alpha,\alpha)),\alpha))$
$\Rightarrow \gamma(\sigma(f(\alpha),q(\alpha)))$	$\Rightarrow \gamma(\sigma(q(\sigma(\alpha,\alpha)),\overline{\alpha}(\alpha)))$
$\Rightarrow \gamma(f(\sigma(\alpha,\alpha)))$	$\Rightarrow \gamma(f(\sigma(\alpha,\alpha)))$
$\Rightarrow f(\gamma(\sigma(\alpha,\alpha)))$	$\Rightarrow f(\gamma(\sigma(\alpha,\alpha)))$

Fig. 1. Two leftmost derivations of $f(\gamma(\sigma(\alpha,\alpha)))$ from $\gamma(\sigma(\sigma(\alpha,\alpha),\alpha))$

Example 5.4. Let us consider again $\Sigma = \{\sigma^{(2)}, \gamma^{(1)}, \alpha^{(0)}\}$ and the pattern $\sigma(z,\alpha)$. We construct a bottom-up tree transducer $\mathcal{N}$ whose domain is the set of all Σ-trees in which $\sigma(z,\alpha)$ occurs at least once, i.e., the tree language recognized by the bottom-up tree automaton $\mathcal{A}$ of Example 3.1. Given an input tree $\xi \in T_\Sigma$, the bottom-up tree transducer $\mathcal{N}$ nondeterministically selects an occurrence of $\sigma(z,\alpha)$ in ξ, if any, then either leaves it unchanged or deletes σ and α from it. If $\sigma(z,\alpha)$ does not occur in ξ, then $\mathcal{N}$ will not compute any output tree for ξ. For this, let $\mathcal{N} = (Q,\Sigma,\Sigma,R,F)$, where $Q = \{q,\overline{\alpha},f\}$, $F = \{f\}$, and R is the set of the following rules:

$$\begin{aligned} \alpha &\to q(\alpha), & \alpha &\to \overline{\alpha}(\alpha),\\ \gamma(q(x_1)) &\to q(\gamma(x_1)), & \gamma(f(x_1)) &\to f(\gamma(x_1)),\\ \sigma(q(x_1),q(x_2)) &\to q(\sigma(x_1,x_2)), & \sigma(q(x_1),\overline{\alpha}(x_2)) &\to f(\sigma(x_1,x_2)) \mid f(x_1),\\ \sigma(f(x_1),q(x_2)) &\to f(\sigma(x_1,x_2)), & \sigma(q(x_1),f(x_2)) &\to f(\sigma(x_1,x_2)). \end{aligned}$$

Let $\Rightarrow_\mathcal{N}$ be the term rewrite relation induced by R. We consider the input tree $\xi = \gamma(\sigma(\sigma(\alpha,\alpha),\alpha))$ and show a derivation starting from ξ in Fig. 1(a) (where we have dropped $\mathcal{N}$ from $\Rightarrow_\mathcal{N}$). Hence, we have $\gamma(\sigma(\alpha,\alpha)) \in \tau_\mathcal{N}(\gamma(\sigma(\sigma(\alpha,\alpha),\alpha)))$. In fact, this derivation is a *leftmost derivation* [66] meaning that in every step we applied a rule to the leftmost redex. Note that there is another leftmost derivation starting also from ξ with the same result; see Fig. 1(b).

Now we view $\mathcal{N}$ as a bottom-up wtt over the semiring Nat. More exactly, let $\mathcal{M} = (Q,\Sigma,\Sigma,\mathsf{Nat},\mu,F)$ be the bottom-up wtt, where μ is defined as follows. For every $k \in \{0,1,2\}$, $\delta \in \Sigma^{(k)}$, and $q_1,\ldots,q_k,p \in Q$, we have

$$\mu_k(\delta)_{q_1(x_1)\ldots q_k(x_k),p} = \sum_{\delta(q_1(x_1),\ldots,q_k(x_k))\to p(\zeta(x_1,\ldots,x_k))\in R} 1.\zeta.$$

For instance, $\mu_2(\sigma)_{q(x_1)\overline{\alpha}(x_2),f} = 1.\sigma(z_1,z_2) + 1.z_1$. One can show that $(\tau_\mathcal{M}(\xi),\eta)$ is the number of leftmost derivations of $f(\eta)$ from ξ using the rewrite relation $\Rightarrow_\mathcal{N}$, for every $\xi \in T_\Sigma$ and $\eta \in T_\Delta$, cf. [66], Example 5.6.

The second type of restriction concerns the state behavior: it can be total and/or deterministic. Here, we will only impose these restrictions to top-down wtt and bottom-up wtt. Let $\mathcal{M} = (Q, \Sigma, \Delta, S, \mu, F)$ be a top-down wtt or a bottom-up wtt.

- The top-down wtt $\mathcal{M}$ is *total* (resp., *deterministic*) if, for every $k \geq 0$, $\sigma \in \Sigma^{(k)}$, and $q \in Q$, there is at least one (resp., at most one) $w \in Q(X_k)^*$ and $\zeta \in T_\Delta(Z)$, such that $\zeta \in \text{supp}(\mu_k(\sigma)_{w,q})$. Moreover, in a deterministic top-down wtt, we additionally require that the set F is a singleton.
- The bottom-up wtt $\mathcal{M}$ is *total* (resp., *deterministic*) if, for every $k \geq 0$, $\sigma \in \Sigma^{(k)}$, and $w \in Q(X_k)^*$, there is at least one (resp., at most one) $q \in Q$ and $\zeta \in T_\Delta(Z)$, such that $\zeta \in \text{supp}(\mu_k(\sigma)_{w,q})$.

We note that if $\mathcal{M}$ is a deterministic top-down or a deterministic bottom-up wtt, then the operation $+$ of S is not used for the computation of $\tau_\mathcal{M}$. Moreover, in the case that $\mathcal{M}$ is total we do not require $F = Q$.

The restrictions total and deterministic are abbreviated by t and d, respectively. Any combination x over $\{\text{t}, \text{d}\}$ can be applied to a top-down (resp., bottom-up) wtt, hence we obtain an x top-down (resp., bottom-up) wtt. The class of tree series transformations computed by all so-obtained x top-down wtt or x bottom-up wtt is denoted by prefixing with x the notation of the class of tree series transformations computed by that kind of wtt. For example, the class of tree series transformations over S computed by total and deterministic top-down wtt is denoted by td-TOP(S). Moreover, a total and deterministic top-down (resp., bottom-up) wtt $\mathcal{M}$ is called a *homomorphism top-down wtt* (resp., *homomorphism bottom-up wtt*), provided $Q = F = \{q\}$. The class of tree series transformations over S which are computable by homomorphism top-down wtt (resp., homomorphism bottom-up wtt) is denoted by h-TOP(S) (resp., h-BOT(S)).

Finally, the third type of restriction concerns the form of the polynomials that may occur in the rule mappings. A wtt $\mathcal{M} = (Q, \Sigma, \Delta, S, \mu, F)$ is:

- *Boolean*, if for every $k \geq 0$, $\sigma \in \Sigma^{(k)}$, $w \in Q(X_k)^*$, and $q \in Q$, the tree series $\mu_k(\sigma)_{w,q}$ is Boolean,
- *linear* (resp., *nondeleting*), if for every $k \geq 0$, $\sigma \in \Sigma^{(k)}$, $w \in Q(X_k)^*$, and $q \in Q$ such that $\mu_k(\sigma)_{w,q} \neq \widetilde{0}$, both $\mu_k(\sigma)_{w,q}$ is linear (resp., nondeleting) in $Z_{|w|}$ and w is linear (resp., nondeleting) in X_k.

Moreover, if $\mathcal{M}$ is a bottom-up wtt, then $\mathcal{M}$ is:

- *wta*, if $\Sigma = \Delta$ and for every $k \geq 0$, $\sigma \in \Sigma^{(k)}$, $w \in Q(X_k)^*$, and $q \in Q$, we have that $\mu_k(\sigma)_{w,q} = a.\sigma(z_1, \ldots, z_k)$ for some $a \in S$.

The first three restrictions are abbreviated by b, l, and n, respectively. Any combination of $\{\text{b}, \text{l}, \text{n}\}$ can be applied to any of the wtt classes defined above. For example, the class of tree series transformations over S computed by linear top-down wtt with regular look-ahead is denoted by l-TOP$^{\text{R}}(S)$, while the class of tree series transformations over S computed by deterministic, linear, and nondeleting bottom-up wtt is denoted by dln-BOT(S).

Each wta bottom-up wtt computes a weighted identity. The class of all weighted identities over S computed by wta bottom-up wtt is denoted by WTA(S).

By two easy constructions, we can show that linear and nondeleting top-down wtt have the same computation power as linear and nondeleting bottom-up wtt and hereby generalize the corresponding result concerning tree transducers, cf. [48], Theorem 2.9.

Theorem 5.5 ([53], Theorem 5.24). ln-TOP(S) = ln-BOT(S) *and* bln-TOP(S) = bln-BOT(S).

Although bottom-up homomorphism tree transducers compute the same class of tree transformations as top-down homomorphism tree transducers (cf. [48]), this equivalence does not hold for the corresponding classes of tree series transformations, cf. [53], Proposition 3.14. However, bh-TOP(S) = bh-BOT(S), therefore, we do not use different notations but denote this class by b-HOM(S), cf. [53], Corollary 4.15. We note that the equation was proved in [53] only for idempotent S. However, it is easy to see that idempotency is not necessary because, as mentioned, $+$ is not used for the computation of the tree series transformation computed by any homomorphism wtt.

We also note that the identity tree series transformation ι can be computed by any of the above kind of wtt.

Finally, we show that wtt over positive semirings have the same computational power as gfst. This relation has been discussed in [53], Sect. 4.1 for Boolean wtt and gfst; and it has been proved in detail in [53], Sects. 4.2 and 4.3 for the case of bottom-up wtt and top-down wtt, respectively, over idempotent semirings.

The heart of such a comparison is the concept of the relatedness of wtt and gfst. We define a wtt $\mathcal{M} = (Q, \Sigma, \Delta, S, \mu, F)$ and a gfst $\mathcal{N} = (Q, \Sigma, \Delta, R, F)$ as being *related* if the following holds:

$$q(\sigma(x_1, \ldots, x_k)) \to \zeta, \langle w \rangle \text{ is in } R \quad \text{iff} \quad \zeta \in \mathrm{supp}(\mu_k(\sigma)_{w,q})$$

for every $q \in Q$, $k \geq 0$, $\sigma \in \Sigma^{(k)}$, $w \in Q(X_k)^*$, and $\zeta \in T_\Delta(Z_{|w|})$.

The same concept of relatedness can be defined for top-down tree transducers and top-down wtt, and for bottom-up tree transducers and bottom-up wtt. Assuming that S is positive, it is easy to prove by induction on ξ that $h_{\mathcal{N}}(\xi)_q = \mathrm{supp}(h_\mu(\xi)_q)$ for every $\xi \in T_\Sigma$ and $q \in Q$. The key equation in this proof is that for $\xi = \sigma(\xi_1, \ldots, \xi_k)$ and $q \in Q$,

$$\bigcup_{\substack{\text{for every rule} \\ \text{of the form (8)}}} \{\zeta(\zeta_1, \ldots, \zeta_l) \mid \zeta_j \in h_{\mathcal{N}}(\xi_{i_j})_{q_j},\ 1 \leq j \leq l\}$$

$$= \mathrm{supp}\Bigg(\sum_{w = q_1(x_{i_1}) \ldots q_l(x_{i_l}) \in Q(X_k)^*} \mu_k(\sigma)_{w,q} \leftarrow \big(h_\mu(\xi_{i_1})_{q_1}, \ldots, h_\mu(\xi_{i_l})_{q_l}\big)\Bigg).$$

To prove this, we need the following facts. If S is positive, then for all tree series $r_1, r_2 \in S\langle T_\Delta \rangle$, we have $\mathrm{supp}(r_1 + r_2) = \mathrm{supp}(r_1) \cup \mathrm{supp}(r_2)$, and for $r \in S\langle T_\Delta(Z) \rangle$, finite $I \subseteq \mathbb{N}$, and family $(s_i \mid i \in I)$ with $s_i \in S\langle T_\Delta \rangle$ we have $\mathrm{supp}(r \leftarrow (s_i \mid i \in I)) = \{\zeta(\zeta_i \mid i \in I) \mid \zeta \in \mathrm{supp}(r), \zeta_i \in \mathrm{supp}(s_i)\}$. Then the equation follows from the induction hypothesis, the relatedness of $\mathcal{M}$ and $\mathcal{N}$, and the fact that S is positive.

In order to relate the semantics of a wtt $\mathcal{M}$ with that of a gfst $\mathcal{N}$ we will use a generalization of the mapping supp. For this, define the mapping

$$\mathrm{supp}_{S,\Sigma,\Delta} : (T_\Sigma \to S\langle T_\Delta \rangle) \to \big(T_\Sigma \to \mathcal{P}_{\mathrm{fin}}(T_\Delta)\big)$$

by $(\mathrm{supp}_{S,\Sigma,\Delta}(\tau))(\xi) = \mathrm{supp}(\tau(\xi))$ for every $\tau : T_\Sigma \to S\langle T_\Delta \rangle$ and $\xi \in T_\Sigma$. Let supp_S denote the class of all mappings $\mathrm{supp}_{S,\Sigma,\Delta}$.

Now it easily follows that $\tau_\mathcal{N} = \mathrm{supp}_{S,\Sigma,\Delta}(\tau_\mathcal{M})$ if $\mathcal{M}$ and $\mathcal{N}$ are related and S is positive. Since, for every given wtt $\mathcal{M}$ we can construct a gfst $\mathcal{N}$ which is related to $\mathcal{M}$, and vice versa, we obtain the following result.

Theorem 5.6 ([53], Theorem 4.13 and Theorem 4.6). *Let S be positive. Then:*

(A) $\mathrm{supp}_S(\mathrm{WTT}(S)) = \mathrm{GFST}$.
(B) $\mathrm{supp}_S(\mathrm{TOP}(S)) = \mathrm{TOP}$.
(C) $\mathrm{supp}_S(\mathrm{BOT}(S)) = \mathrm{BOT}$.

The next theorem shows that the range of a wta bottom-up wtt is a recognizable tree series and vice versa. For an arbitrary weighted identity $\tau : T_\Sigma \to S\langle T_\Sigma \rangle$, we define the tree series $\mathrm{range}(\tau) \in S\langle T_\Sigma \rangle$ by $(\mathrm{range}(\tau), \xi) = (\tau(\xi), \xi)$ for every $\xi \in T_\Sigma$. We let $\mathrm{WTA}(\Sigma, S)$ denote the class of all those weighted identities in $\mathrm{WTA}(S)$ which have the type $T_\Sigma \to S\langle T_\Sigma \rangle$.

Theorem 5.7. $\mathrm{Rec}(\Sigma, S) = \mathrm{range}(\mathrm{WTA}(\Sigma, S))$.

Proof. A wta $\mathcal{A} = (Q, \Sigma, S, \mu, F)$ with Boolean root weights and a wta bottom-up wtt $\mathcal{M} = (Q, \Sigma, \Sigma, S, \nu, F)$ are related if for every $k \geq 0$, $\sigma \in \Sigma^{(k)}$, $q_1, \ldots, q_k, q \in Q$, and $a \in S$: $\mu_k(\sigma)_{q_1 \ldots q_k, q} = a$ iff $\nu_k(\sigma)_{q_1(x_1) \ldots q_k(x_k), q} = a.\sigma(z_1, \ldots, z_k)$. Then it is straightforward to prove by induction that $h_\mu(\xi)_q.\xi = h_\nu(\xi)_q$ for every $\xi \in T_\Sigma$ and $q \in Q$; this implies that $(r_\mathcal{A}, \xi).\xi = \tau_\mathcal{M}(\xi)$, and finally that $r_\mathcal{A} = \mathrm{range}(\tau_\mathcal{M})$. Then the statement of the theorem is implied by Theorem 3.6. □

5.4 Composition and Decomposition

In this subsection, we will investigate composition and decomposition results for $\mathrm{WTT}(S)$. A composition result has the form $\mathrm{C}(S);\mathrm{D}(S) \subseteq \mathrm{E}(S)$, where $\mathrm{C}(S)$, $\mathrm{D}(S)$, and $\mathrm{E}(S)$ are subclasses of $\mathrm{WTT}(S)$. Clearly, the smaller the class $\mathrm{E}(S)$ is (for fixed $\mathrm{C}(S)$ and $\mathrm{D}(S)$) the stronger the composition result is. If $\mathrm{C}(S) = \mathrm{E}(S)$, then we say that $\mathrm{E}(S)$ is closed under right-composition with D(S), and if D(S) = E(S), then we say that E(S) is closed under left-composition with C(S). If C(S) = D(S) = E(S), then C(S) is closed under composition. The first composition results were obtained for gfst in [48] and for top-down and bottom-up tree transducers in [48], [49], and [4].

Decomposition results have the form E(S) $\subseteq$ C(S);D(S) where again C(S), D(S), and E(S) are subclasses of WTT(S). Of course, a decomposition E(S) $\subseteq$ C(S);D(S) makes sense only if C(S) and D(S) are subclasses of E(S). Such decomposition results were first obtained for gfst, top-down, and bottom-up tree transducers in [48] and [49]. In the rest of this section, we will generalize some of the composition and decomposition results obtained in the above cited papers.

5.4.1 Results Concerning wtt

First, we consider composition results of the form C(S);D(S) $\subseteq$ E(S), where C(S) $\subseteq$ TOP(S) and D(S) $\subseteq$ BOT(S). Thus, for every top-down wtt $\mathcal{M}_1$ of type c and bottom-up wtt $\mathcal{M}_2$ of type d, we have to construct a wtt $\mathcal{M}$ of type e such that $\tau_{\mathcal{M}_1}; \tau_{\mathcal{M}_2} = \tau_{\mathcal{M}}$. There is a general approach to achieve this goal: $\mathcal{M}$ is the *syntactic composition of* $\mathcal{M}_1$ *and* $\mathcal{M}_2$, denoted by $\mathcal{M}_1 \circ \mathcal{M}_2$ (cf. [4], pages 195 and 199). The wtt $\mathcal{M}_1 \circ \mathcal{M}_2$ is obtained, roughly speaking, by letting $\mathcal{M}_2$ work on the pieces of output produced by $\mathcal{M}_1$. However, since these pieces of output may contain variables from Z, which $\mathcal{M}_2$ cannot process, we first extend $\mathcal{M}_2$ appropriately, cf. [104], Definition 14.

To define this extension, let $\mathcal{M} = (Q, \Sigma, \Delta, S, \mu, F)$ be a bottom-up wtt. Let $l \geq 0$ and $\overline{q} \in Q^J$ for some $J \subseteq Z_l$. We define the $(\Sigma \cup Z_l)$-algebra $(S\langle T_\Delta(Z_l)\rangle^Q, \mu^{\overline{q}}_{\mathcal{M}})$ where every $z \in Z_l$ is a nullary symbol and we define $\mu^{\overline{q}}_{\mathcal{M}}(z)()_q = 1.z$ if $z \in J$ and $q = \overline{q}_z$, and $\mu^{\overline{q}}_{\mathcal{M}}(z)()_q = \widetilde{0}$ otherwise; and for every $k \geq 0$ and $\sigma \in \Sigma^{(k)}$, the k-ary operation $\mu^{\overline{q}}_{\mathcal{M}}(\sigma) : S\langle T_\Delta(Z_l)\rangle^Q \times \cdots \times S\langle T_\Delta(Z_l)\rangle^Q \to S\langle T_\Delta(Z_l)\rangle^Q$ is defined in the same way as $\mu_{\mathcal{M}}(\sigma)$. Let us denote the unique $(\Sigma \cup Z_l)$-algebra homomorphism from $T_{\Sigma \cup Z_l}$ to $S\langle T_\Delta(Z_l)\rangle^Q$ by $h^{\overline{q}}_\mu$, and let us denote its extension to a mapping of type $S\langle T_{\Sigma \cup Z_l}\rangle \to S\langle T_\Delta(Z_l)\rangle^Q$ (cf. page 363) also by $h^{\overline{q}}_\mu$.

Now we define the concept of syntactic composition and, in fact, we compose an *arbitrary* wtt with a bottom-up wtt; this will be useful for the results concerning bottom-up wtt. The *syntactic composition* (called simple composition in [104]) of a wtt $\mathcal{M}_1 = (Q_1, \Sigma, \Delta, S, \mu^1, F_1)$ and a bottom-up wtt $\mathcal{M}_2 = (Q_2, \Delta, \Gamma, S, \mu^2, F_2)$, see [104], Definition 22, is the wtt $\mathcal{M}_1 \circ \mathcal{M}_2 = (Q_1 \times Q_2, \Sigma, \Gamma, S, \mu, F_1 \times F_2)$, such that for every $k, l \geq 0$, $\sigma \in \Sigma^{(k)}$, $p, p_1, \ldots, p_l \in Q_1$, $q, q_1, \ldots, q_l \in Q_2$, and $1 \leq i_1, \ldots, i_l \leq k$, we have

$$\mu_k(\sigma)_{(p_1,q_1)(x_{i_1})\ldots(p_l,q_l)(x_{i_l}),(p,q)} = h^{\overline{q}}_{\mu^2}\left(\mu^1_k(\sigma)_{p_1(x_{i_1})\ldots p_l(x_{i_l}),p}\right)_q$$

where $\overline{q} \in Q_2^{Z_l}$ with $\overline{q}_{z_i} = q_i$ for every $z_i \in Z_l$. We note that this composition generalizes the syntactic composition of bottom-up tree transducers introduced in [4], page 199.

Remark 5.8. If both $\mathcal{M}_1$ and $\mathcal{M}_2$ are (deterministic) bottom-up wtt, then $\mathcal{M}_1 \circ \mathcal{M}_2$ is a (deterministic) bottom-up wtt. Moreover, if both $\mathcal{M}_1$ and $\mathcal{M}_2$ are homomorphism bottom-up wtt and $\mathcal{M}_2$ is Boolean, then $\mathcal{M}_1 \circ \mathcal{M}_2$ is a homomorphism bottom-up wtt. If, in addition, $\mathcal{M}_1$ is Boolean too, then $\mathcal{M}_1 \circ \mathcal{M}_2$ is a Boolean homomorphism bottom-up wtt.

If $\mathcal{M}_1$ is a (linear) top-down wtt and $\mathcal{M}_2$ is a linear bottom-up wtt, then $\mathcal{M}_1 \circ \mathcal{M}_2$ is a (linear) top-down wtt with regular look-ahead. If $\mathcal{M}_1$ is a (linear, nondeleting, or linear and nondeleting) top-down wtt and $\mathcal{M}_2$ is a linear and nondeleting bottom-up wtt, then $\mathcal{M}_1 \circ \mathcal{M}_2$ is a (linear, nondeleting, or linear and nondeleting) top-down wtt. The proof needs property P6 of IO-substitution.

Now we will show that, under certain conditions, $\mathcal{M}_1 \circ \mathcal{M}_2$ computes $\tau_{\mathcal{M}_1} ; \tau_{\mathcal{M}_2}$. For this, we formulate an important property of (certain restricted) bottom-up wtt, namely, that h_μ distributes over substitutions $\xi(\xi_i \mid z_i \in \mathrm{var}(\xi))$ for $\xi \in T_\Sigma(Z)$ and $\xi_i \in T_\Sigma$.

Lemma 5.9 ([104], Proposition 18). *Let $\mathcal{M} = (Q, \Sigma, \Delta, S, \mu, F)$ be a bottom-up wtt, $q \in Q$, $l \geq 0$, $\xi \in T_\Sigma(Z_l)$, and $\xi_i \in T_\Sigma$ for every $z_i \in \mathrm{var}(\xi)$. If (a) $\mathcal{M}$ is Boolean and deterministic or (b) ξ is linear in Z_l, then*

$$h_\mu\big(\xi(\xi_i \mid z_i \in \mathrm{var}(\xi))\big)_q = \sum_{\overline{q} \in Q^{\mathrm{var}(\xi)}} h_\mu^{\overline{q}}(\xi)_q \leftarrow \big(h_\mu(\xi_i)_{\overline{q}_{z_i}} \mid z_i \in \mathrm{var}(\xi)\big).$$

Proof. The proof is performed by induction on ξ. In the proof, properties P7 and P8 of IO-substitution are used. Moreover, for item (a), a version of P8 is used which also assures associativity of tree series substitution, cf. [104], Lemma 8. □

Next we show sufficient conditions which guarantee that $\mathcal{M}_1 \circ \mathcal{M}_2$ computes $\tau_{\mathcal{M}_1} ; \tau_{\mathcal{M}_2}$. Note that the following lemma generalizes [4], Theorem 6.

Lemma 5.10 ([104], Lemma 23). *Let $\mathcal{M}_1$ be a wtt and $\mathcal{M}_2$ a bottom-up wtt. If (a) $\mathcal{M}_1$ is a bottom-up wtt and $\mathcal{M}_2$ is total, deterministic, and Boolean or (b) $\mathcal{M}_1$ is a top-down wtt, then for every $\xi \in T_\Sigma$, $p \in Q_1$, and $q \in Q_2$, we have $h_{\mu^2}(h_{\mu^1}(\xi)_p)_q = h_\mu(\xi)_{(p,q)}$ and $\tau_{\mathcal{M}_1 \circ \mathcal{M}_2} = \tau_{\mathcal{M}_1} ; \tau_{\mathcal{M}_2}$.*

Proof. The first equation is proved by induction on ξ. The proof needs items (a) and (b) of Lemma 5.9 in the cases (a) and (b), respectively. Moreover, it needs properties P2, P5, and P7 of IO-substitution of the tree series. Then the second equation follows straightforwardly. □

We are ready to prove the following composition results.

Lemma 5.11 ([102], Lemma 2). *For every combination* x *over* $\{\mathrm{l}, \mathrm{n}\}$*:*

(A) x-TOP(S); x-BOT$(S) \subseteq$ x-WTT(S).
(B) x-TOP(S); xl-BOT$(S) \subseteq$ x-TOP$^{\mathrm{R}}(S)$.

(C) x-TOP(S); ln-BOT$(S) \subseteq$ x-TOP(S).

Proof. For the proof of (A), let $\mathcal{M}_1$ be a top-down wtt and $\mathcal{M}_2$ a bottom-up wtt. By Lemma 5.10(b), we have $\tau_{\mathcal{M}_1}; \tau_{\mathcal{M}_2} = \tau_{\mathcal{M}_1 \circ \mathcal{M}_2}$. The preservation of properties x follows by Remark 5.8. Statements (B) and (C) are proved in the same way as (A) by using once again Remark 5.8. □

Now we turn to a decomposition result of WTT(S) that generalizes [48], Lemma 5.8.

Lemma 5.12 ([102], Lemma 1). *For every combination* x *over* {b, l}*:*

(A) x-WTT$(S) \subseteq$ xbn-HOM(S); x-BOT(S).
(B) x-TOP$^{\mathrm{R}}(S) \subseteq$ xbn-HOM(S); xl-BOT(S).

Proof. First, we prove (A). Let $\mathcal{M} = (Q, \Sigma, \Delta, S, \mu, F)$ be a wtt. We construct a homomorphism top-down wtt $\mathcal{M}_1$ and a bottom-up wtt $\mathcal{M}_2$ such that $\tau_{\mathcal{M}} = \tau_{\mathcal{M}_1}; \tau_{\mathcal{M}_2}$. Note that $\mathcal{M}_2$, being a bottom-up wtt, is allowed to make exactly one computation on every subtree. The idea behind the decomposition is that $\mathcal{M}_1$ copies the input subtrees so that $\mathcal{M}_2$ can simulate different computations of $\mathcal{M}$ on a subtree using different copies of that subtree. More exactly, let $mx = \max(\{1\} \cup \{|w|_{x_i} \mid 1 \leq i \leq k, \sigma \in \Sigma^{(k)}, (w, q) \in Q(X_k)^* \times Q, \mu_k(\sigma)_{w,q} \neq \widetilde{0}\})$, i.e., the maximal number of copies of a subtree taken by a rule, and consider the ranked alphabet $\Gamma = \{\sigma^{(k \cdot mx)} \mid k \geq 0, \sigma \in \Sigma^{(k)}\}$. Note that $mx = 1$ if $\mathcal{M}$ is linear. Now we construct $\mathcal{M}_1 = (\{\star\}, \Sigma, \Gamma, S, \mu^1, \{\star\})$ with

$$\mu^1_k(\sigma)\underbrace{\star(x_1)\ldots\star(x_1)}_{mx \text{ times}}\ldots\underbrace{\star(x_k)\ldots\star(x_k)}_{mx \text{ times}},\star = 1.\sigma(z_1,\ldots,z_{k\cdot mx})$$

for every $k \geq 0$ and $\sigma \in \Sigma^{(k)}$. Clearly, $\mathcal{M}_1$ is a Boolean and nondeleting homomorphism top-down wtt; moreover, $\mathcal{M}_1$ is linear and computes the identity if $\mathcal{M}$ is linear.

Then let $d \notin Q$ be a new state and $Q' = Q \cup \{d\}$. For every $k \geq 0$, $\sigma \in \Sigma^{(k)}$, and $w \in Q(X_k)^*$ such that $|w|_{x_i} \leq mx$ for $1 \leq i \leq k$, construct the string $w' \in Q'(X_{k \cdot mx})^*$ in two steps as follows. First, construct $\widetilde{w} \in Q(X_{k \cdot mx})^*$ by replacing, for every $1 \leq i \leq k$, the jth occurrence of x_i in w by $x_{(i-1)\cdot mx + j}$. Note that $\widetilde{w}$ is linear in $X_{k \cdot mx}$. Then for every $1 \leq j \leq k \cdot mx$ such that $|\widetilde{w}|_{x_j} = 0$, append $d(x_j)$ to $\widetilde{w}$. Certainly, the string w' obtained in this way is linear and nondeleting in $X_{k \cdot mx}$. Then construct the wtt $\mathcal{M}' = (Q', \Gamma, \Delta, S, \mu', F)$, where μ' is defined as follows. For every $k \geq 0$, $\sigma \in \Sigma^{(k)}$, let $\mu'_{k \cdot mx}(\sigma)_{d(x_1)\ldots d(x_{k \cdot mx}),d} = 1.\alpha$, where $\alpha \in \Delta^{(0)}$ is arbitrary; and, for every $(w, q) \in Q(X_k)^* \times Q$ such that $\mu_k(\sigma)_{w,q} \neq \widetilde{0}$, let $\mu'_{k \cdot mx}(\sigma)_{w',q} = \mu_k(\sigma)_{w,q}$. Every other entry of $\mu'_{k \cdot mx}(\sigma)$ is $\widetilde{0}$. Note that $\mathcal{M}'$ need not be a bottom-up wtt because there may be $\mu'_l(\sigma)_{w,q} \neq \widetilde{0}$ with $\sigma \in \Gamma^{(l)}$ such that the order of the variables in w is not $x_1, \ldots, x_l$. However, by an appropriate reordering of the symbols $q(x_i)$ in w and the corresponding substitution variables z_j in

$\mu_l'(\sigma)_{w,q}$, we can turn $\mathcal{M}'$ into a bottom-up wtt $\mathcal{M}_2 = (Q', \Gamma, \Delta, S, \mu^2, F)$ such that $\tau_{\mathcal{M}'} = \tau_{\mathcal{M}_2}$. The proof of $\tau_{\mathcal{M}} = \tau_{\mathcal{M}_1}; \tau_{\mathcal{M}_2}$ can be done as follows. We define $h : T_\Sigma \to T_\Gamma$ for every Σ-tree $\xi = \sigma(\xi_1, \ldots, \xi_k)$ by

$$h(\xi) = \sigma\big(\underbrace{h(\xi_1), \ldots, h(\xi_1)}_{mx \text{ times}}, \ldots, \underbrace{h(\xi_k), \ldots, h(\xi_k)}_{mx \text{ times}}\big).$$

Clearly, $\tau_{\mathcal{M}_1}(\xi) = 1.h(\xi)$. Thus, it is sufficient to prove that $h_\mu(\xi)_q = h_{\mu^2}(h(\xi))_q$ for every $\xi \in T_\Sigma$ and $q \in Q$. Moreover, it is obvious that $\mathcal{M}_2$ is Boolean (linear, resp.), whenever $\mathcal{M}$ is so. This finishes the proof of (A). Finally, if $\mathcal{M}$ is a top-down wtt with regular look-ahead, then $\mathcal{M}'$ is a linear top-down wtt with regular look-ahead, and consequently, $\mathcal{M}_2$ is a linear bottom-up wtt, which proves (B). □

Putting Lemmata 5.11 and 5.12 together, we obtain the following characterizations of WTT(S) and TOP^R(S), which generalize the corresponding characterizations of gfst and top-down tree transducers with regular look-ahead that were obtained in [48], Theorems 5.10 and 5.15.

Theorem 5.13 ([102], Theorem 3). *For every combination* x *over* {l}*:*
(A) x-WTT(S) = xb-HOM(S); x-BOT(S).
(B) x-TOP^{R}(S) = xb-HOM(S); xl-BOT(S).

By definition, xl-WTT(S) = xl-TOP^{R}(S) for every combination x over {b, n}. Moreover, if the wtt $\mathcal{M}$ in the proof of Lemma 5.12(A) is linear, then the homomorphism top-down wtt $\mathcal{M}_1$ computes the identity. Hence, we obtain l-WTT(S) ⊆ l-BOT(S) and bl-WTT(S) ⊆ bl-BOT(S). These arguments verify the following characterization result, which generalizes [48], Theorem 5.13.

Theorem 5.14 ([102], Theorem 4). l-BOT(S) = l-TOP^{R}(S) *and* bl-BOT(S) = bl-TOP^{R}(S).

5.4.2 Results Concerning Bottom-up wtt

Here, we investigate composition results of the form C(S);D(S) ⊆ E(S), where C(S), D(S), and E(S) are subclasses of BOT(S). First, we prove that BOT(S) is closed under right-composition with db-BOT(S). This generalizes the corresponding result for bottom-up tree transformations, cf. [48], Theorem 4.6 and [4], Theorem 6, and the result BOT(S); b-HOM(S) ⊆ BOT(S) obtained in [53], Corollary 5.5.

Theorem 5.15 ([104], Theorem 24). *For every combination x over* {d, h, l, n} *we have* x-BOT(S); xdb-BOT(S) ⊆ x-BOT(S).

Proof. The proof immediately follows from Remark 5.8 and Lemma 5.10(a) because, by adding a dummy state, each bottom-up wtt can be turned into a total one computing the same tree series transformation. □

Next, we show a characterization of bottom-up wtt which generalizes Nivat's characterization of finite-state sequential machines [118], also cf. [6]. Note that Nivat's result was generalized for bottom-up tree transducers in [48]. To this end, we define finite-state relabeling bottom-up wtt in the way that we impose a further restriction on bottom-up wtt. A bottom-up wtt $\mathcal{M} = (Q, \Sigma, \Delta, S, \mu, F)$ is a *finite-state relabeling bottom-up wtt* if for every $k \geq 0$, $\sigma \in \Sigma^{(k)}$, $w = q_1(x_1) \ldots q_k(x_k) \in Q(X_k)^*$, and $q \in Q$, we have $\mathrm{supp}(\mu_k(\sigma)_{w,q}) \subseteq \{\delta(z_1, \ldots, z_k) \mid \delta \in \Delta^{(k)}\}$. We note that a wta bottom-up wtt is a particular finite-state relabeling bottom-up wtt. The class of all tree series transformations over S which are computable by finite-state relabeling bottom-up wtt, is denoted by $\mathrm{QREL}(S)$. Note that $\mathrm{QREL}(S) \subseteq \text{ln-BOT}(S) = \text{ln-TOP}(S)$, cf. Theorem 5.5.

Theorem 5.16 ([53], Theorem 5.7). *For every combination x over* $\{\mathrm{l}, \mathrm{n}\}$*:*

(A) $\text{x-BOT}(S) = \mathrm{QREL}(S); \text{xb-HOM}(S)$.
(B) $\text{x-BOT}(S) = \text{xl-BOT}(S); \text{xb-HOM}(S)$.

Moreover:

(C) $\mathrm{BOT}(S) = \text{l-TOP}(S); \text{b-HOM}(S)$.

Proof. We first prove (A). By Theorem 5.15, the right-hand side of the equation is a subset of its left-hand side. In order to prove the other inclusion, let $\mathcal{M} = (Q, \Sigma, \Delta, S, \mu, F)$ be a bottom-up wtt. We will construct a finite-state relabeling bottom-up wtt $\mathcal{M}_1$ and a Boolean homomorphism bottom-up wtt $\mathcal{M}_2$, such that, up to renaming of states, $\mathcal{M}$ is the syntactic composition of $\mathcal{M}_1$ and $\mathcal{M}_2$. For this, define the ranked alphabet $\Omega = \bigcup_{\sigma \in \Sigma} \Omega_\sigma$, where, for every $k \geq 0$, $\sigma \in \Sigma^{(k)}$, we let $\Omega_\sigma = \{[\sigma, w, q, \zeta]^{(k)} \mid w \in Q(X_k)^*, q \in Q, \zeta \in \mathrm{supp}(\mu_k(\sigma)_{w,q})\}$. Obviously, Ω is a finite set. Now let $\mathcal{M}_1 = (Q, \Sigma, \Omega, S, \mu^1, F)$ be such that for every $k \geq 0$, $\sigma \in \Sigma^{(k)}$, $w \in Q(X_k)^*$, $q \in Q$, and $\zeta' \in T_\Omega(Z_k)$,

$$(\mu_k^1(\sigma)_{w,q}, \zeta') = \begin{cases} (\mu_k(\sigma)_{w,q}, \zeta) & \text{if } \zeta' = [\sigma, w, q, \zeta](z_1, \ldots, z_k), \\ 0 & \text{otherwise.} \end{cases}$$

Let $\mathcal{M}_2 = (\{\star\}, \Omega, \Delta, S, \mu^2, \{\star\})$ be such that for every $k \geq 0$, $[\sigma, w, q, \zeta] \in \Omega^{(k)}$, and $\zeta' \in T_\Delta(Z_k)$, we have

$$(\mu_k^2([\sigma, w, q, \zeta])_{\star(x_1)\ldots\star(x_k),\star}, \zeta') = \begin{cases} 1 & \text{if } \zeta' = \zeta, \\ 0 & \text{otherwise.} \end{cases}$$

If $\mathcal{M}$ is nondeleting (resp., linear), then so is $\mathcal{M}_2$. Moreover, by identifying Q and $Q \times \{\star\}$, the bottom-up wtt $\mathcal{M}$ becomes the syntactic composition of $\mathcal{M}_1$ and $\mathcal{M}_2$. Thus, by Lemma 5.10(a), we have $\tau_\mathcal{M} = \tau_{\mathcal{M}_1}; \tau_{\mathcal{M}_2}$, which proves that the left-hand side of the equation is a subset of its right-hand side.

Now the inclusion $\subseteq$ for both (B) and (C) should be clear by (A) and the note we made above this theorem. Finally, by Theorem 5.15, the inclusion $\supseteq$ follows for (B), and by (B) and Theorem 5.14 the inclusion $\supseteq$ follows for (C).
□

Let us compare the equations of Theorem 5.13(B) and of Theorem 5.16(B) for x being the empty combination. We see that the classes $\mathrm{TOP}^{\mathrm{R}}(S)$ and $\mathrm{BOT}(S)$ can be characterized by the composition of the two classes l-BOT(S) and b-HOM(S) and that the difference between them lies in the order of their composition.

For further composition results on bottom-up wtt, we need another concept of syntactic composition. Let us explain why. For this, we consider two bottom-up wtt $\mathcal{M}_1 = (Q_1, \Sigma, \Delta, S, \mu^1, F_1)$ and $\mathcal{M}_2 = (Q_2, \Delta, \Gamma, S, \mu^2, F_2)$ and their composition $\mathcal{M}_1 \circ \mathcal{M}_2 = (Q_1 \times Q_2, \Sigma, \Gamma, S, \mu, F_1 \times F_2)$. We obtain the entry $\mu_k(\sigma)_{(p_1,q_1)(x_1)\dots(p_k,q_k)(x_k),(p,q)}$ by applying the homomorphism $h^{\overline{q}}_{\mu^2}$ to the entry $\mu^1_k(\sigma)_{p_1(x_1)\dots p_k(x_k),p}$, where $k \geq 0$, $\sigma \in \Sigma^{(k)}$, $p, p_1, \dots, p_k \in Q_1$, $q \in Q_2$, and $\overline{q} \in Q_2^{Z_k}$ with $q_i = \overline{q}_{z_i}$ for $1 \leq i \leq k$, and then selecting the q-component from the resulting Q-vector. As Lemma 5.10(a) states, the syntactic composition yields equality on the level of the semantics also, provided that $\mathcal{M}_2$ is total, deterministic, and Boolean. However, the following problem arises in the case when $\mathcal{M}_2$ does not have these properties. Let us suppose that $\mathcal{M}_1$ translates a tree $\xi \in T_\Sigma$ into an output tree $\zeta \in T_\Delta$ with weight $a \in S$ and that, during the translation, it deletes the translation $\zeta' \in T_\Delta$ with weight $a' \in S$ of a subtree ξ' of ξ. Still, due to the definition of IO-substitution of tree series, the weight a' of ζ' contributes to the weight a of ζ, whereas ζ' does not contribute to ζ. Furthermore, let us suppose that $\mathcal{M}_2$ transforms ζ into $\widetilde{\zeta} \in T_\Gamma$ with weight $b \in S$ and ζ' into $\widetilde{\zeta'} \in T_\Gamma$ with weight $b' \in S$. Since the input of $\mathcal{M}_2$ is ζ, it does not process the deleted tree ζ', and thus the weight b' does not contribute to b. However, when $\mathcal{M}_1 \circ \mathcal{M}_2$ processes the input tree ξ, it transforms its subtree ξ' into ζ' with weight a' using the family of rule mappings μ^1, and immediately also transforms ζ' into $\widetilde{\zeta'}$ with weight b' using the family of rule mappings μ^2. Then although $\mathcal{M}_1 \circ \mathcal{M}_2$ deletes the translation $\widetilde{\zeta'}$ of ζ', both a' and b' still contribute to the weight of the overall translation $\widetilde{\zeta}$, which contrasts the situation encountered when $\mathcal{M}_1$ and $\mathcal{M}_2$ run separately. In the case that $\mathcal{M}_2$ is Boolean, the weight b' can only be 0 or 1, so that one just has to avoid the case that $b' = 0$. This can be achieved by requiring that $\mathcal{M}_2$ is total and deterministic; see Lemma 5.10(a). However, we do not want to restrict $\mathcal{M}_2$ and, therefore, following [104], we propose another construction. Namely, we manipulate $\mathcal{M}_2$ such that it has a state $\diamond$, called a *blind state*, which is not a final state and which transforms each input tree into an output tree $\alpha \in \Delta^{(0)}$ with weight 1. Then we compose $\mathcal{M}_1$ and $\mathcal{M}_2$ by processing in state $\diamond$ the subtrees that $\mathcal{M}_1$ deletes. We note that the concept of blind state was introduced in [48], Theorem 2.8 (called e for *erasing* there) in order to construct a linear bottom-up tree transducer from a linear top-down tree transducer; it occurred already in the proof of Lemma 5.12.

A state $\diamond \in Q$ of a bottom-up wtt $\mathcal{M} = (Q, \Sigma, \Delta, S, \mu, F)$ is a *blind state* if $\diamond \notin F$, there is an $\alpha \in \Delta^{(0)}$ such that $\mu_k(\sigma)_{\diamond(x_1)\dots\diamond(x_k),\diamond} = 1.\alpha$, and $\mu_k(\sigma)_{q_1(x_1)\dots q_k(x_k),\diamond} \neq \widetilde{0}$ implies that $q_i = \diamond$ for every $1 \leq i \leq k$ (for every

$k \geq 0$, $\sigma \in \Sigma^{(k)}$, and $q_1, \ldots, q_k \in Q$). It is easy to prove that $h_\mu(\xi)_\diamond = 1.\alpha$ for every $\xi \in T_\Sigma$. Moreover, for every bottom-up wtt $\mathcal{M} = (Q, \Sigma, \Delta, S, \mu, F)$, we can construct a bottom-up wtt $\mathcal{M}' = (Q', \Sigma, \Delta, S, \mu', F)$ with blind state such that $\tau_\mathcal{M} = \tau_{\mathcal{M}'}$ in the following way. Let $Q' = Q \cup \{\diamond\}$, for every $k \geq 0$, $\sigma \in \Sigma^{(k)}$, and $q, q_1, \ldots, q_k \in Q$, let $\mu'_k(\sigma)_{\diamond(x_1)\ldots\diamond(x_k),\diamond} = 1.\alpha$, $\mu'_k(\sigma)_{q_1(x_1)\ldots q_k(x_k),q} = \mu_k(\sigma)_{q_1(x_1)\ldots q_k(x_k),q}$, and let all remaining entries be $\widetilde{0}$.

Now we formalize this new concept of syntactic composition, called bottom-up syntactic composition (cf. [104], Definition 17, where it was called just composition). Let $\mathcal{M}_1 = (Q_1, \Sigma, \Delta, S, \mu^1, F_1)$ and $\mathcal{M}_2 = (Q_2, \Delta, \Gamma, S, \mu^2, F_2)$ be bottom-up wtt such that $\diamond$ is a blind state of $\mathcal{M}_2$. The *bottom-up syntactic composition of* $\mathcal{M}_1$ *and* $\mathcal{M}_2$, denoted by $\mathcal{M}_1 \circ_{\mathrm{bu}} \mathcal{M}_2$ is the bottom-up wtt $(Q_1 \times Q_2, \Sigma, \Gamma, S, \mu, F)$, where $F = F_1 \times F_2$, and for every $k \geq 0$, $\sigma \in \Sigma^{(k)}$, $p, p_1, \ldots, p_k \in Q_1$, $q \in Q_2 \setminus \{\diamond\}$, and $\overline{q}, \overline{\diamond} \in Q_2^{Z_k}$ with $\overline{q}_{z_i} = q_i$ and $\overline{\diamond}_{z_i} = \diamond$ for $1 \leq i \leq k$, we have

$$\mu_k(\sigma)_{(p_1,q_1)(x_1)\ldots(p_k,q_k)(x_k),(p,q)} = h_{\mu^2}^{\overline{q}}\Bigg(\sum_{\substack{\zeta \in T_\Delta(Z_k),\\ (\forall 1 \leq i \leq k): z_i \notin \mathrm{var}(\zeta) \Leftrightarrow q_i = \diamond}} \big(\mu_k^1(\sigma)_{p_1(x_1)\ldots p_k(x_k),p}, \zeta\big).\zeta\Bigg)_q$$

and

$$\mu_k(\sigma)_{(p_1,\diamond)(x_1)\ldots(p_k,\diamond)(x_k),(p,\diamond)} = h_{\mu^2}^{\overline{\diamond}}\big(\mu_k^1(\sigma)_{p_1(x_1)\ldots p_k(x_k),p}\big)_\diamond .$$

All the remaining entries in μ are $\widetilde{0}$.

It should be clear that $\mathcal{M}_1 \circ_{\mathrm{bu}} \mathcal{M}_2$ does not always compute $\tau_{\mathcal{M}_1}; \tau_{\mathcal{M}_2}$ because the class of tree transformations computed by bottom-up tree transducers, i.e., by bottom-up wtt over $\mathbb{B}$ is not closed under composition; see [48], Theorem 2.5. However, the desired equality on the level of semantics holds for a linear $\mathcal{M}_1$, and we obtain the following generalization of [4], Theorem 6.

Lemma 5.17 ([104], Lemma 19). *Let* $\mathcal{M}_1$ *be a linear bottom-up wtt and* $\mathcal{M}_2$ *a bottom-up wtt with blind state. Then for every* $\xi \in T_\Sigma$, $p \in Q_1$, *and* $q \in Q_2$, *we have* $h_{\mu^2}(h_{\mu^1}(\xi)_p)_q = h_\mu(\xi)_{(p,q)}$ *and* $\tau_{\mathcal{M}_1 \circ_{\mathrm{bu}} \mathcal{M}_2} = \tau_{\mathcal{M}_1}; \tau_{\mathcal{M}_2}$.

Proof. Let $\diamond$ be the blind state of $\mathcal{M}_2$. The first equation can be proved by induction and case analysis, where the two cases are $q = \diamond$ and $q \neq \diamond$. In the first case, the proof uses the fact that $h_{\mu^2}(\zeta)_\diamond = 1.\alpha$ for every $\zeta \in \Delta$, and properties P5 and P7 of IO-substitution of tree series, while in the second case, it additionally uses Lemma 5.9 and property P2. Then the second equation follows straightforwardly from the first one. □

Now we can show that $\mathrm{BOT}(S)$ is closed under left-composition with l-$\mathrm{BOT}(S)$. This generalizes the corresponding result for bottom-up tree transformations, cf. [48], Theorem 4.5 and [4], Theorem 6.

Theorem 5.18 ([104], Theorem 20). *For every combination* x *over* $\{\mathrm{l}, \mathrm{n}\}$, *we have* $\mathrm{xl\text{-}BOT}(S); \mathrm{x\text{-}BOT}(S) \subseteq \mathrm{x\text{-}BOT}(S)$.

Proof. Let $\mathcal{M}_1$ and $\mathcal{M}_2$ be bottom-up wtt. We may assume without loss of generality that $\mathcal{M}_2$ has a blind state. Moreover, if $\mathcal{M}_1$ and $\mathcal{M}_2$ are linear (resp., nondeleting), then also $\mathcal{M}_1 \circ_{\mathrm{bu}} \mathcal{M}_2$ is linear (resp., nondeleting). Then the statement follows from Lemma 5.17. □

5.4.3 Results Concerning Top-down wtt

First, we show that TOP(S) is closed under right-composition with ln-TOP(S). This result generalizes [4], Theorem 1, in particular [4], Corollary 2(1).

Theorem 5.19 ([104], Theorem 26). *For every combination x over* $\{\mathrm{d}, \mathrm{l}, \mathrm{n}\}$, *we have* x-TOP($S$); xln-TOP($S$) $\subseteq$ x-TOP(S).

Proof. Let $\mathcal{M}_1$ be a top-down wtt and $\mathcal{M}_2$ a linear and nondeleting top-down wtt. By Theorem 5.5, there is a linear and nondeleting bottom-up wtt $\mathcal{M}_2'$ such that $\tau_{\mathcal{M}_2} = \tau_{\mathcal{M}_2'}$. By Lemma 5.10(b), we have $\tau_{\mathcal{M}_1}; \tau_{\mathcal{M}_2'} = \tau_{\mathcal{M}_1 \circ \mathcal{M}_2'}$. Since $\mathcal{M}_2'$ is linear and nondeleting, $\mathcal{M}_1 \circ \mathcal{M}_2'$ is a top-down wtt, cf. Remark 5.8. If $\mathcal{M}_1$ and $\mathcal{M}_2$ have property x, then $\mathcal{M}_1 \circ \mathcal{M}_2'$ has property x. □

One would expect from [4], Theorem 1, in particular [4], Corollary 3(2), that in the case x = d the linearity of $\mathcal{M}_2$ can be dropped. However, as shown in [53], Example 5.11, this is not the case.

The next result generalizes another case of [4], Theorem 1, viz. the case that $\mathcal{M}_1$ is total and deterministic, cf. [4], Corollary 2(4). However, linearity of $\mathcal{M}_2$ is still needed (for the same reason), and moreover, $\mathcal{M}_1$ must be Boolean as shown in [53], Example 5.12. It will be proved in Theorem 5.25 that $\mathcal{M}_2$ need not be linear if it is deterministic.

Theorem 5.20 ([104], Theorem 30). tdb-TOP(S); l-TOP(S) $\subseteq$ TOP(S).

Proof. Let $\mathcal{N}_1 = (Q_1, \Sigma, \Delta, S, \mu^1, \{q_1\})$ be a total, deterministic, and Boolean top-down wtt and $\mathcal{M} = (Q, \Delta, \Gamma, S, \mu, F)$ a linear top-down wtt. Let us apply the proof of Lemma 5.12(B) to $\mathcal{M}$. Since $\mathcal{M}$ is linear, the homomorphism top-down wtt $\mathcal{M}_1$ of that proof computes the identity, and so we obtain a linear bottom-up wtt $\mathcal{M}_2 = (Q_2, \Delta, \Gamma, S, \mu^2, F)$ such that $\tau_{\mathcal{M}} = \tau_{\mathcal{M}_2}$. Note that $Q_2 = Q \cup \{d\}$ and d is a blind state. Now let $\mathcal{M}_3 = (Q_1 \times Q_2, \Sigma, \Gamma, S, \mu, \{q_1\} \times F)$ be the syntactic composition of $\mathcal{N}_1$ and $\mathcal{M}_2$, i.e., $\mathcal{M}_3 = \mathcal{N}_1 \circ \mathcal{M}_2$. By Lemma 5.10(b), we have $\tau_{\mathcal{M}_3} = \tau_{\mathcal{N}_1}; \tau_{\mathcal{M}_2}$. Moreover, by Remark 5.8, $\mathcal{M}_3$ is a top-down wtt with regular look-ahead. Since $\mathcal{N}_1$ is total, deterministic, and Boolean, and since $\mathcal{M}_2$ is linear and d is a blind state of $\mathcal{M}_2$, the wtt $\mathcal{M}_3$ has the following properties:

(i) There is an $\alpha \in \Gamma^{(0)}$, such that $h_\mu(\xi)_{(p,d)} = 1.\alpha$ for every $\xi \in T_\Sigma$ and $p \in Q_1$.
(ii) For every $(p,q) \in Q_1 \times Q_2$, $k \geq 0$, $w = (p_1,q_1)(x_{i_1}) \dots (p_n,q_n)(x_{i_n}) \in (Q_1 \times Q_2)(X_k)^*$, $1 \leq j \leq n$, $\sigma \in \Sigma^{(k)}$, and $\zeta \in \mathrm{supp}(\mu_k(\sigma)_{w,(p,q)})$, we have $z_j \notin \mathrm{var}(\zeta)$ iff $q_j = d$.

This means that the look-ahead is trivial, and thus $\mathcal{M}_3$ can be transformed into an equivalent top-down wtt $\mathcal{M}'_3$ in the following way. For a tree $\zeta \in T_\Gamma(Z_n)$, which is linear in Z_n, let us denote by $\mathrm{norm}_n(\zeta)$ the linear and nondeleting tree in $T_\Gamma(Z_k)$ defined by $\mathrm{norm}_n(\zeta) = \zeta(\varphi(z_i) \mid 1 \leq i \leq n)$, where $\mathrm{var}(\zeta) = \{z_{i_1}, \ldots, z_{i_k}\}$ with $i_1 < \cdots < i_k$ and φ is any mapping $Z_n \to Z_k$ such that $\varphi(z_{i_j}) = z_j$ for all $1 \leq j \leq k$. Moreover, for a string $w \in (Q_1 \times Q_2)(X_k)^*$, let us denote by $\mathrm{del}(w)$ the string which is obtained from w by deleting all symbols of the form $(p, d)(x_i)$ from w. Now $\mathcal{M}'_3$ is obtained from $\mathcal{M}_3$ by changing μ to μ' in the following way: for every $k \geq 0$, $\sigma \in \Sigma^{(k)}$, $w' \in (Q_1 \times Q_2)(X_k)^*$, and $(p, q) \in Q_1 \times Q_2$,

$$\mu'_k(\sigma)_{w',(p,q)} = \sum_{\substack{w \in (Q_1 \times Q_2)(X_k)^* \\ \mathrm{del}(w) = w'}} \left(\sum_{\zeta \in T_\Gamma(Z_{|w|})} (\mu_k(\sigma)_{w,(p,q)}, \zeta)\mathrm{norm}_{|w|}(\zeta) \right).$$

Then we can prove that for every $\xi \in T_\Sigma$ and $(p, q) \in Q_1 \times Q_2$ such that $q \neq d$, we have $h_{\mu'}(\xi)_{(p,q)} = h_\mu(\xi)_{(p,q)}$. The proof is performed by induction on ξ, using properties (i) and (ii) of $\mathcal{M}_3$ and P2 of IO-substitution of tree series. Thus, $\tau_{\mathcal{M}_3} = \tau_{\mathcal{M}'_3}$ follows, which finishes the proof. □

Now we generalize [48], Theorem 3.6 and show that $\mathrm{TOP}(S)$ can be decomposed into bn-HOM(S) and l-TOP(S).

Lemma 5.21 ([53], Lemma 5.9). *For every combination* x *over* {t, d}, *we have* x-TOP(S) ⊆ bn-HOM(S); xl-TOP(S).

Proof. Let $\mathcal{M} = (Q, \Sigma, \Delta, S, \mu, F)$ be a top-down wtt. We construct a Boolean and nondeleting homomorphism top-down wtt $\mathcal{M}_1$ and a linear top-down wtt $\mathcal{M}_2$ such that $\tau_\mathcal{M} = \tau_{\mathcal{M}_1}; \tau_{\mathcal{M}_2}$. The proof is very similar to that of Lemma 5.12. In fact, $\mathcal{M}_1$ is constructed in the same way as in that proof and we construct $\mathcal{M}_2$ similarly to $\mathcal{M}'$. The main difference is that we do not need the extra state d because we do not need to force $\mathcal{M}_2$ to be a bottom-up wtt. Thus, the family of rule mappings underlying $\mathcal{M}_2$ slightly differs from that of $\mathcal{M}'$. Let $\mathcal{M}_2 = (Q, \Gamma, \Delta, S, \mu^2, F)$, and for every $(w, q) \in Q(X_k)^* \times Q$ such that $\mu_k(\sigma)_{w,q} \neq \widetilde{0}$, let $\mu^2_{k \cdot mx}(\sigma)_{\widetilde{w},q} = \mu_k(\sigma)_{w,q}$, where the string $\widetilde{w}$ is being defined as in the proof of Lemma 5.12. Every other entry of $\mu^2_{k \cdot mx}(\sigma)$ is $\widetilde{0}$. It should be clear that $\mathcal{M}_2$ is a linear top-down wtt. The proof of $\tau_\mathcal{M} = \tau_{\mathcal{M}_1}; \tau_{\mathcal{M}_2}$ can be done in the same way as in the proof of Lemma 5.12. Moreover, it is obvious that $\mathcal{M}_2$ inherits the properties x from $\mathcal{M}$. □

Next, we turn to further composition results for top-down wtt. For this, we define the top-down syntactic composition of two top-down wtt $\mathcal{M}_1$ and $\mathcal{M}_2$, denoted by $\mathcal{M}_1 \circ_{\mathrm{td}} \mathcal{M}_2$, thereby generalizing the corresponding concept defined for top-down tree transducers in [4], page 195. The wtt $\mathcal{M}_1 \circ_{\mathrm{td}} \mathcal{M}_2$ is obtained, as for wtt, by letting $\mathcal{M}_2$ work on the pieces of output produced by $\mathcal{M}_1$. In order to avoid too complex formulas, we consider only the case that $\mathcal{M}_1$ is deterministic, cf. [53], Definitions 5.13 and 5.14; [61], Definitions 5.2 and 5.3.

The pieces of output produced by $\mathcal{M}_1$ may contain variables from Z, which $\mathcal{M}_2$ cannot process. Therefore, just as in the case of the syntactic composition of bottom-up wtt, we first extend a top-down wtt $\mathcal{M} = (Q, \Sigma, \Delta, S, \mu, F)$ such that it can process input trees containing variables from Z_n for some $n \geq 0$. We define the $(\Sigma \cup Z_n)$-algebra $(S\langle T_\Delta(Q(Z_n))\rangle^Q, \mu_{\mathcal{M}}^{(n)})$ where every $z_j \in Z_n$ is a nullary symbol and $\mu_{\mathcal{M}}^{(n)}(z_j)()_q = 1.q(z_j)$ for every $q \in Q$. For every $k \geq 0$ and $\sigma \in \Sigma^{(k)}$, the k-ary operation $\mu_{\mathcal{M}}^{(n)}(\sigma) : S\langle T_\Delta(Q(Z_n))\rangle^Q \times \cdots \times S\langle T_\Delta(Q(Z_n))\rangle^Q \to S\langle T_\Delta(Q(Z_n))\rangle^Q$ is defined in the same way as $\mu_{\mathcal{M}}(\sigma)$. Let us denote the unique $(\Sigma \cup Z_n)$-algebra homomorphism from $T_{\Sigma \cup Z_n}$ to $S\langle T_\Delta(Q(Z_n))\rangle^Q$ by $h_\mu^{(n)}$.

For the definition of the family of rule mappings μ of $\mathcal{M}_1 \circ_{\text{td}} \mathcal{M}_2$, we need two more technical concepts: (a) substitution of particular strings and (b) linearization of a tree. For (a), let Q_1 and Q_2 be finite nonempty sets (e.g., state sets of top-down wtt), $k \geq 0$, $w \in Q_1(X_k)^*$ with $|w| = l$, and let $u \in Q_2(Z_l)^*$. Then $u\langle w\rangle$ is the string in $(Q_1 \times Q_2)(X_k)^*$ which is obtained from u by replacing, for every $q \in Q_2$ and $z_i \in Z_l$, the expression $q(z_i)$ by $(p, q)(x_j)$ where $p(x_j)$ is the ith symbol of w.

Let us now turn to (b). For a ranked alphabet Γ, the set of all trees in $T_\Gamma(Z_m)$, $m \geq 0$, which are both linear and nondeleting in Z_m and in which the order of the variables is $z_1, \ldots, z_m$, is denoted by $C_\Gamma^{(m)}$. Let $\xi \in T_\Gamma(H)$, where H is a set. The *linearization of* ξ *with respect to* H, denoted by $\text{lin}_H(\xi)$, is defined as the unique pair (ξ', u) where $\xi' \in C_\Gamma^{(m)}$ and $u = a_1 \ldots a_m \in H^*$ such that $\xi = \xi'(a_1, \ldots, a_m)$.

Now the *top-down syntactic composition* of a deterministic top-down wtt $\mathcal{M}_1 = (Q_1, \Sigma, \Delta, S, \mu^1, \{q_1\})$ and a top-down wtt $\mathcal{M}_2 = (Q_2, \Delta, \Gamma, S, \mu^2, F_2)$ is the top-down wtt $\mathcal{M}_1 \circ_{\text{td}} \mathcal{M}_2 = (Q_1 \times Q_2, \Sigma, \Gamma, S, \mu, \{q_1\} \times F_2)$, where the family of rule mappings μ is defined as shown in Fig. 2.

Using similar arguments as for the syntactic composition of a wtt with a bottom-up wtt, we will show that, under certain conditions, $\mathcal{M}_1 \circ_{\text{td}} \mathcal{M}_2$ computes $\tau_{\mathcal{M}_1}; \tau_{\mathcal{M}_2}$. For this, we first formulate a property of top-down wtt, namely, that h_μ distributes over substitutions $\xi(\xi_1, \ldots, \xi_l)$ for $\xi \in T_\Delta(Z_l)$ and $\xi_1, \ldots, \xi_l \in T_\Delta$. Note that this property corresponds to the one of (restricted) bottom-up wtt formulated in Lemma 5.9.

Lemma 5.22 ([61], Statement in the proof of Lemma 5.5). *Let* $\mathcal{M} = (Q, \Delta, \Gamma, S, \mu, F)$ *be a top-down wtt and* $l \geq 0$. *For every* $q \in Q$, $\xi \in T_\Delta(Z_l)$, *and* $\xi_1, \ldots, \xi_l \in T_\Delta$,

$$h_\mu(\xi(\xi_1, \ldots, \xi_l))_q = \sum_{\kappa=1}^{r} a_\kappa \zeta_\kappa \leftarrow \big(h_\mu(\xi_{i_{\kappa,1}})_{q_{\kappa,1}}, \ldots, h_\mu(\xi_{i_{\kappa,m_\kappa}})_{q_{\kappa,m_\kappa}}\big),$$

where $h_\mu^{(l)}(\xi)_q = a_1.\hat{\zeta}_1 + \cdots + a_r.\hat{\zeta}_r$ *for* $a_1, \ldots, a_r \in S \setminus \{0\}$ *and* $\hat{\zeta}_1, \ldots, \hat{\zeta}_r \in T_\Gamma(Q(Z_l))$, *and* $\text{lin}_{Q(Z_l)}(\hat{\zeta}_\kappa) = (\zeta_\kappa, q_{\kappa,1}(z_{i_{\kappa,1}}) \ldots q_{\kappa,m_\kappa}(z_{i_{\kappa,m_\kappa}}))$, $\zeta_\kappa \in C_\Gamma^{(m_\kappa)}$ *for every* $1 \leq \kappa \leq r$.

For every $\sigma \in \Sigma^{(k)}$ with $k \geq 0$,
for every $(w,p) \in Q_1(X_k)^* \times Q_1$ with $l = |w|$,
if $\mu_k^1(\sigma)_{w,p} = a.\zeta$ for $a \in S \setminus \{0\}$ and $\zeta \in T_\Delta(Z_l)$,
then { for every $q \in Q_2$,
if $h_{\mu^2}^{(l)}(\zeta)_q = a_1.\hat{\zeta}_1 + \cdots + a_r.\hat{\zeta}_r$ for $a_1, \ldots, a_r \in S \setminus \{0\}$
and $\hat{\zeta}_1, \ldots, \hat{\zeta}_r \in T_\Gamma(Q_2(Z_l))$
(by P4 also $h_{\mu^2}^{(l)}(\zeta)$ is polynomial),
then define, for every $1 \leq j \leq r$,
$\mu_k(\sigma)_{v_j,(p,q)} = \sum_{1 \leq i \leq r, v_i = v_j} (a \cdot a_i).\zeta_i$,
where, for every $1 \leq \kappa \leq r$,
$\text{lin}_{Q_2(Z_l)}(\hat{\zeta}_\kappa) = (\zeta_\kappa, u_\kappa)$, $\zeta_\kappa \in C_\Gamma^{(m_\kappa)}$,
$u_\kappa \in Q_2(Z_l)^*$, $|u_\kappa| = m_\kappa$ and $v_\kappa = u_\kappa\langle w\rangle$ }.

Moreover, for every $\sigma \in \Sigma^{(k)}$ with $k \geq 0$, $p \in Q_1$, $q \in Q_2$ and
$v \in (Q_1 \times Q_2)(X_k)^*$ not defined by the above conditions, let $\mu_k(\sigma)_{v,(p,q)} = \tilde{0}$.

Fig. 2. Definition of μ

Proof. The proof is performed by induction on ξ and it needs properties P5, P7, and a version of P8 which also assures associativity of tree series substitution, cf. [61], Corollary 2.6. □

The following sufficient conditions guarantee that $\mathcal{M}_1 \circ_{\text{td}} \mathcal{M}_2$ computes $\tau_{\mathcal{M}_1}; \tau_{\mathcal{M}_2}$, cf. Lemma 5.10.

Lemma 5.23 ([61], Lemma 5.5; [53], Lemma 5.17). *Let $\mathcal{M}_1$ be a total, deterministic, and Boolean top-down wtt and $\mathcal{M}_2$ a top-down wtt. If (a) $\mathcal{M}_1$ is a homomorphism top-down wtt or (b) $\mathcal{M}_2$ is deterministic, then for every $\xi \in T_\Sigma$, $p \in Q_1$, and $q \in Q_2$, we have $h_{\mu^2}(h_{\mu^1}(\xi)_p)_q = h_\mu(\xi)_{(p,q)}$ and $\tau_{\mathcal{M}_1 \circ_{\text{td}} \mathcal{M}_2} = \tau_{\mathcal{M}_1}; \tau_{\mathcal{M}_2}$.*

Proof. The first equation is proved by induction on ξ. The proof needs Lemma 5.22 and properties P5, P7, and P8 of IO-substitution of tree series. Then the second equation follows straightforwardly. □

The following theorem generalizes [48], Theorem 3.7.

Theorem 5.24 ([53], Theorem 5.18). *For every x over* $\{\text{t}, \text{d}\}$, *we have* x-TOP(S) = b-HOM(S); xl-TOP(S).

Proof. It follows from Lemma 5.21 and Lemma 5.23(a). □

By comparing the above equation and Theorem 5.16(C) for x being the empty combination, we observe that TOP(S) and BOT(S) can be characterized by the composition of the two classes b-HOM(S) and l-TOP(S). However, the orders of the subclasses in the two compositions are different.

Next, we generalize [133], Lemma 6.9 and [4], Corollary 3(3) (note that in that corollary the second *PDT* should be *DT*).

Theorem 5.25 ([53], Theorem 5.18). *For every x over* $\{\mathrm{t},\mathrm{l},\mathrm{n}\}$, *we have* $\text{xtdb-TOP}(S);\text{xd-TOP}(S) \subseteq \text{xd-TOP}(S)$.

Proof. It immediately follows from Lemma 5.23(b) and the fact that the top-down syntactic composition preserves the properties d, t, l, and n. □

5.5 The Inclusion Diagram of Some Fundamental wtt Classes

By the *inclusion diagram* of certain classes, we mean their Hasse diagram with respect to the partial order $\subseteq$, cf. [64], Sect. 2.2. In this subsection, we will be interested in the inclusion diagram of the classes $\text{WTT}(S)$, $\text{TOP}^{\text{R}}(S)$, $\text{TOP}(S)$, and $\text{BOT}(S)$ and their linear, and linear and nondeleting subclasses, where S is a proper semiring (altogether 12 classes of tree series transformations). We note that in [105] the same inclusion diagram was obtained for a positive semiring S. Moreover, by [107], the results of [105] can easily be generalized to the more general case that S is a proper semiring.

We already know that $\text{l-WTT}(S) = \text{l-TOP}^{\text{R}}(S)$ and $\text{ln-WTT}(S) = \text{ln-TOP}^{\text{R}}(S) = \text{ln-TOP}(S)$ by definition. Moreover, $\text{ln-TOP}(S) = \text{ln-BOT}(S)$ and $\text{l-BOT}(S) = \text{l-TOP}^{\text{R}}(S)$ by Theorems 5.5 and 5.14, respectively.

In Fig. 3, we visualize all the equalities and the inclusions among the involved 12 classes. In the rest of this subsection, we show that all inclusions are proper and that the unrelated classes are incomparable provided S is a proper semiring. For this, it is sufficient to verify the following four inequalities:

$$\text{TOP}(S) \setminus \text{BOT}(S) \neq \emptyset, \tag{9}$$

$$\text{BOT}(S) \setminus \text{TOP}^{\text{R}}(S) \neq \emptyset, \tag{10}$$

$$\text{l-BOT}(S) \setminus \text{TOP}(S) \neq \emptyset, \tag{11}$$

$$\text{l-TOP}(S) \setminus \text{ln-TOP}(S) \neq \emptyset. \tag{12}$$

First, we show that the above inequalities hold for $S = \mathbb{B}$ (and hence in this particular case the diagram in Fig. 3 is an inclusion diagram). Since tree transducers and wtt over $\mathbb{B}$ can be identified, cf. Sect. 5.2 and Theorem 5.6 for $S = \mathbb{B}$, we refer to the corresponding results in the theory of tree transducers. In fact, (9) follows from [48], Theorem 2.3, while (10) follows from [49], Corollary 2.4(1), and (11) from [48], Example 2.6. Finally, (12) is trivial: no nondeleting top-down tree transducer can translate, e.g., $\sigma(\alpha,\beta)$ to α.

Now we will lift these inequalities to every semiring S that is proper. For this, however, we need some preparation, cf. the end of Sect. 3.3. Let S' be another semiring and consider a mapping $f : S \to S'$. For every tree series transformation $\tau : T_\Sigma \to S\langle T_\Delta \rangle$, we define $f(\tau) : T_\Sigma \to S'\langle T_\Delta \rangle$ such that for every $\xi \in T_\Sigma$ we have $f(\tau)(\xi) = f \circ \tau(\xi)$. Moreover, for every wtt $\mathcal{M} = (Q, \Sigma, \Delta, S, \mu, F)$, we define the wtt $f(\mathcal{M}) = (Q, \Sigma, \Delta, S', \mu', F)$ over S' such that $\mu'_k(\sigma)_{w,q} = f(\mu_k(\sigma)_{w,q})$ for every $k \geq 0$, $\sigma \in \Sigma^{(k)}$, and $(w,q) \in Q(X_k)^* \times Q$. Then we can prove the following two statements, cf. Theorem 3.9.

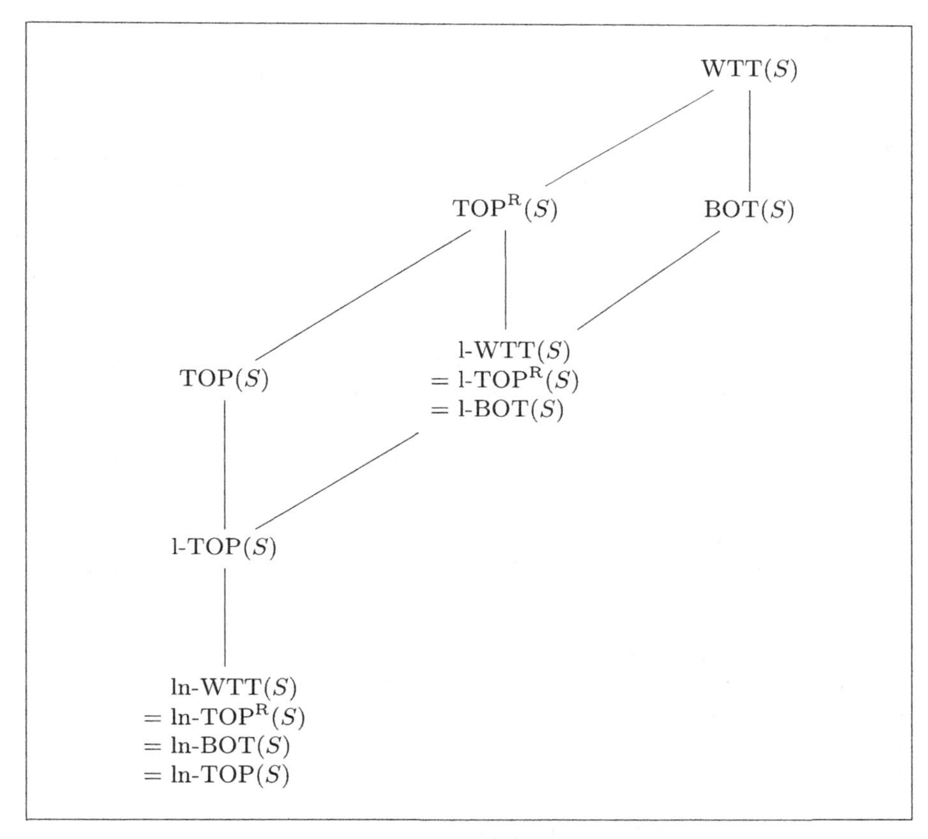

Fig. 3. The inclusion diagram of the classes $\mathrm{WTT}(S)$, $\mathrm{TOP}^{\mathrm{R}}(S)$, $\mathrm{TOP}(S)$, and $\mathrm{BOT}(S)$ and their linear, and linear and nondeleting subclasses, provided S is a proper semiring

Lemma 5.26 ([105], Lemma 2). *Let $\mathcal{M}$ be a wtt of any of the types in Fig. 3 and $f : S \to S'$ be a mapping such that $f(0) = 0$.*

(A) $f(\mathcal{M})$ is of the same type as $\mathcal{M}$.
(B) If f is a semiring homomorphism, then $\tau_{f(\mathcal{M})} = f(\tau_{\mathcal{M}})$.

Proof. Statement (A) can be proved by a direct inspection of each case. For statement (B), it can easily be shown by induction that for every $\xi \in T_\Sigma$, $\eta \in T_\Delta$, and $q \in Q$, we have $(h_{\mu'}(\xi)_q, \eta) = f((h_\mu(\xi)_q, \eta))$. Then $\tau_{f(\mathcal{M})} = f(\tau_{\mathcal{M}})$ follows easily. □

Now let $f : S \to S'$ be a semiring homomorphism and $\mathrm{C}(S)$ any of the tree series transformation classes in Fig. 3. By Lemma 5.26, $f(\mathrm{C}(S)) \subseteq \mathrm{C}(S')$, where, of course, $f(\mathrm{C}(S))$ denotes $\{f(\tau) \mid \tau \in \mathrm{C}(S)\}$. Now we can prove the following lemma.

Lemma 5.27 ([105], Lemma 2). *If $f : S \to S'$ is a surjective semiring homomorphism and* $\mathrm{C}(S)$ *is any of the classes in Fig. 3, then* $f(\mathrm{C}(S)) = \mathrm{C}(S')$.

Proof. Let $\mathcal{M} = (Q, \Sigma, \Delta, S', \mu, F)$ be a wtt of one of the types in Fig. 3. Define a mapping $g : S' \to S$ such that $g(0) = 0$ and for every $a \in S'$, we have $f(g(a)) = a$, this is possible because f is surjective. Then by Lemma 5.26(A), the wtt $g(\mathcal{M})$ is over S and of the same type as $\mathcal{M}$. Moreover, $f(g(\mathcal{M}))$ and $\mathcal{M}$ are syntactically the same, i.e., $f(g(\mathcal{M})) = \mathcal{M}$. Thus, by Lemma 5.26(B), we have $f(\tau_{g(\mathcal{M})}) = \tau_{f(g(\mathcal{M}))} = \tau_{\mathcal{M}}$. □

An easy computation verifies that for tree series transformations $\tau : T_\Sigma \to S\langle T_\Delta \rangle$ and $\tau' : T_\Delta \to S\langle T_\Gamma \rangle$ and semiring homomorphism $f : S \to S'$, we have $f(\tau; \tau') = f(\tau); f(\tau')$ (cf. [105], Lemma 3), and hence, $f(\mathrm{C}(S); \mathrm{D}(S)) = f(\mathrm{C}(S)); f(\mathrm{D}(S))$. Moreover, for $n \geq 1$, we denote the n-fold composition $\mathrm{C}(S); \ldots; \mathrm{C}(S)$ by $\mathrm{C}(S)^n$. Thus, $f(\mathrm{C}(S)^n) = f(\mathrm{C}(S))^n$ for every $n \geq 1$.

Lemma 5.28. *If S is a proper semiring, then for every two classes* $\mathrm{C}(S)$ *and* $\mathrm{D}(S)$ *in Fig. 3 and $m, n \geq 1$, the inequality* $\mathrm{C}(\mathbb{B})^m \setminus \mathrm{D}(\mathbb{B})^n \neq \emptyset$ *implies* $\mathrm{C}(S)^m \setminus \mathrm{D}(S)^n \neq \emptyset$.

Proof. We use a proof by contraposition. Assume $\mathrm{C}(S)^m \subseteq \mathrm{D}(S)^n$. By [137], Theorem 2.1, there is a (surjective) semiring homomorphism $f : S \to \mathbb{B}$. For this f, we have $f(\mathrm{C}(S)^m) \subseteq f(\mathrm{D}(S)^n)$. Furthermore, $f(\mathrm{C}(S)^m) = f(\mathrm{C}(S))^m$, which equals $\mathrm{C}(\mathbb{B})^m$ by Lemma 5.27. In the same way, we get $f(\mathrm{D}(S)^n) = \mathrm{D}(\mathbb{B})^n$. This implies $\mathrm{C}(\mathbb{B})^m \subseteq \mathrm{D}(\mathbb{B})^n$. □

Now we can state the main result of this subsection.

Theorem 5.29 ([105], Theorem 3). *If S is a proper semiring, then the diagram in Fig. 3 is the inclusion diagram of the depicted classes of tree series transformations.*

Proof. We saw that the inequalities (9)–(12) hold for $S = \mathbb{B}$. Then by Lemma 5.28, they also hold for every proper semiring S. □

5.6 Hierarchies

A *hierarchy* is a family $(K_n \mid n \geq 1)$, where K_n is a class such that $K_n \subseteq K_{n+1}$ for every $n \geq 1$. Recall that $\mathrm{C}(S)^n$ denotes the n-fold composition of a tree series transformation class $\mathrm{C}(S)$. Then for every class $\mathrm{C}(S)$ which we consider in this chapter, $(\mathrm{C}(S)^n \mid n \geq 1)$ is a hierarchy because $\mathrm{C}(S)$ contains the identity ι. In this subsection, we present the inclusion diagram consisting of the hierarchies $(\mathrm{TOP}(S)^n \mid n \geq 1)$ and $(\mathrm{BOT}(S)^n \mid n \geq 1)$, where S is a proper semiring. Hereby, we generalize the inclusion results concerning the n-fold compositions of the classes of top-down tree transformations and of bottom-up tree transformations, called hierarchy results, cf. [4], Theorem 13; [50], Theorem 3.14; and [68], Sect. 8. of Chap. IV. We note that such a generalization was made in [61] for top-down and bottom-up wtt over commutative, idempotent, and positive semirings, then it was shown in [105] that the idempotency is not necessary. By [107], these results hold even for proper semirings.

First, we show the generalization of [4], Theorem 13.

Lemma 5.30 ([61], Theorems 5.1 and 5.7). *For every $n \geq 1$, we have:*

(A) $\mathrm{TOP}(S)^n \subseteq \mathrm{BOT}(S)^{n+1}$.
(B) $\mathrm{BOT}(S)^n \subseteq \mathrm{TOP}(S)^{n+1}$.

Proof. Let us write T, B, and H for $\mathrm{TOP}(S)$, $\mathrm{BOT}(S)$, and $\mathrm{HOM}(S)$ for the sake of readability. Then using Theorems 5.24 and 5.16(C), we can compute as follows.

$$\begin{aligned} \mathrm{T}^n &\subseteq \text{l-T}; \mathrm{T}^n; \text{b-H} = \text{l-T}; (\text{b-H}; \text{l-T})^n; \text{b-H} \\ &= (\text{l-T}; \text{b-H})^{n+1} = \mathrm{B}^{n+1}, \\ \mathrm{B}^n &\subseteq \text{b-H}; \mathrm{B}^n; \text{l-T} = \text{b-H}; (\text{l-T}; \text{b-H})^n; \text{l-T} \\ &= (\text{b-H}; \text{l-T})^{n+1} = \mathrm{T}^{n+1}. \end{aligned}$$ □

In Fig. 4, we visualize the inclusions among the involved classes $\mathrm{TOP}(S)^n$ and $\mathrm{BOT}(S)^n$, $n \geq 1$. In the rest of this subsection, we show that all inclusions are proper and that the unrelated classes are incomparable provided S is a proper semiring. For this, it suffices to verify the following two inequalities for every $n \geq 1$:

$$\mathrm{TOP}(S)^n \setminus \mathrm{BOT}(S)^n \neq \emptyset, \tag{13}$$
$$\mathrm{BOT}(S)^n \setminus \mathrm{TOP}(S)^n \neq \emptyset. \tag{14}$$

Lemma 5.31 ([52]). *If* $\mathrm{TOP}(S)^n \subset \mathrm{TOP}(S)^{n+1}$ *for every* $n \geq 1$, *then both (13) and (14) hold.*

Proof. We again use the abbreviations T, B, and H introduced above. We first prove (13). For this, assume the opposite, i.e., that $\mathrm{T}^n \subseteq \mathrm{B}^n$. Then we obtain

$$\begin{aligned} \mathrm{T}^{n+2} &= \text{b-H}; \text{l-T}; \mathrm{T}^n; \text{b-H}; \text{l-T} \subseteq \text{b-H}; \text{l-B}; \mathrm{B}^n; \text{b-H}; \text{l-T} \\ &\subseteq \text{b-H}; \mathrm{B}^n; \text{l-T} \qquad\qquad\quad = \mathrm{T}^{n+1}, \end{aligned}$$

which contradicts the assumption of the lemma. In the first three steps of the computation, we used Theorem 5.24, Theorem 5.14, and Theorems 5.15 and 5.18, while the equality in the last step comes from the second computation in the proof of Lemma 5.30.

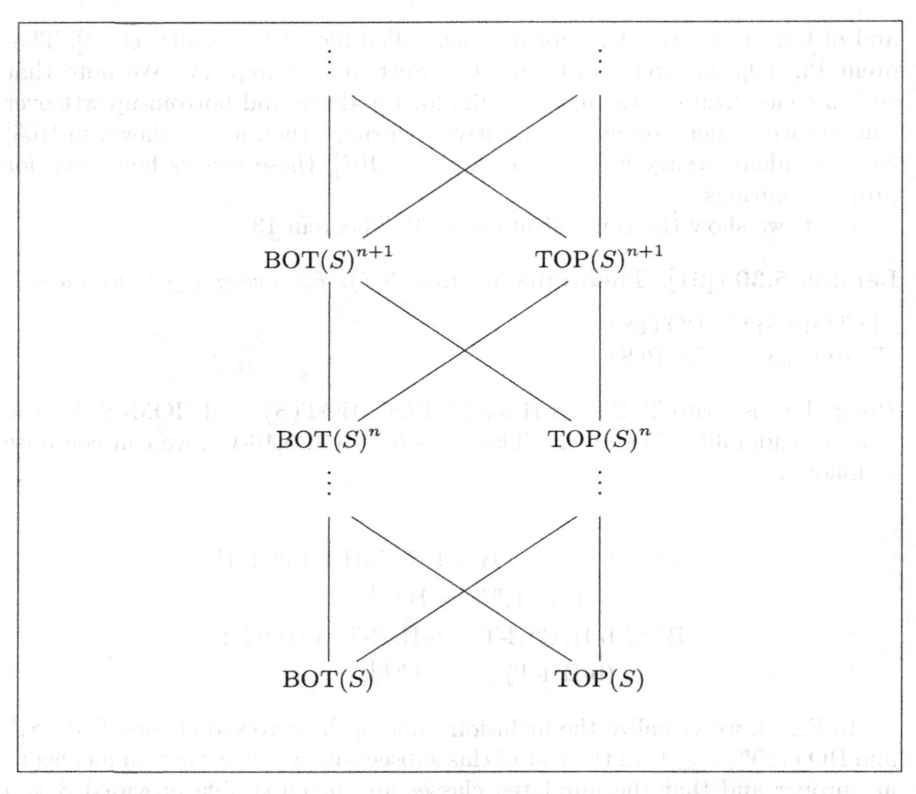

Fig. 4. The inclusion diagram of the classes $\mathrm{TOP}(S)^n$ and $\mathrm{BOT}(S)^n$, where $n \geq 1$ and S is a proper semiring

To prove (14), assume that $\mathrm{B}^n \subseteq \mathrm{T}^n$. Then we get

$$\begin{aligned}
\mathrm{T}^{n+1} &= \text{b-H}; \mathrm{B}^n; \text{l-T} && \subseteq \text{b-H}; \mathrm{T}^n; \text{l-T} \\
&= \text{b-H}; (\text{b-H}; \text{l-T})^n; \text{l-T} && \subseteq \text{b-H}; (\text{b-H}; \text{l-B})^n; \text{l-B} \\
&= (\text{b-H}; \text{l-B})^n && \subseteq \text{l-B}; (\text{b-H}; \text{l-B})^n; \text{b-H} \\
&= (\text{l-B}; \text{b-H})^{n+1} && = \mathrm{B}^{n+1},
\end{aligned}$$

which contradicts (13). In the first, third, and fourth steps, we used again the equality from Lemma 5.30, Theorem 5.24, and Theorem 5.14, respectively. In the fifth step, we used that both b-H and l-B are closed under composition; see Remark 5.8 and Lemma 5.10, and Theorem 5.18, respectively. Finally, the last step follows from Theorem 5.16(B). □

Now we can prove the main result of this subsection.

Theorem 5.32 ([105], Theorem 2). *If S is a proper semiring, then the diagram in Fig. 4 is the inclusion diagram of the depicted classes of tree series transformations.*

Proof. Since top-down tree transducers and top-down wtt over $\mathbb{B}$ can be identified, cf. Sect. 5.2, we obtain that $\mathrm{TOP}(\mathbb{B})^n \subset \mathrm{TOP}(\mathbb{B})^{n+1}$ for every $n \geq 1$, by [50], Theorem 3.14. Thus, by Lemmata 5.28 and 5.31, the inequalities (13) and (14) hold for every proper semiring S, which proves the theorem. □

5.7 Further Models of Weighted Tree Transducers

In this subsection, we will discuss other models of weighted tree transducers that occur in the literature. Actually, we would like to describe them as modifications of our wtt concept. However, an obstacle for this is that most of them have rule mappings of the type

$$\mu_k : \Sigma^{(k)} \to S\langle\langle T_\Delta(Z)\rangle\rangle^{Q(X_k)^* \times Q},$$

i.e., $\mu_k(\sigma)_{w,q}$ is not necessarily a polynomial. To remedy this problem, we first extend our wtt to so-called *infinite wtt* (for short: inf-wtt) that are defined exactly as wtt but the rule mapping μ_k has the above type. Moreover, we require that S is complete in order to have IO-substitution well defined. The tree transformation computed by an inf-wtt is defined in the same way as for wtt except that we use the Σ-algebra $(S\langle\langle T_\Delta\rangle\rangle^Q, \mu_\mathcal{M})$ and IO-substitution of arbitrary tree series. An inf-wtt is *polynomial* (for short: p) if the rule mapping μ_k maps into $S\langle T_\Delta(Z)\rangle^{Q(X_k)^* \times Q}$, i.e., a polynomial inf-wtt is the same as our wtt. The class of tree series transformations computed by certain restrictions of inf-wtt is denoted in the same way as the classes for the corresponding wtt except that we add 'inf' as index, like: $\text{l-TOP}(S)_{\text{inf}}$.

Bottom-up Inf-wtt with OI-Substitution

In [91], so-called *tree transducers* were defined. Such a tree transducer is a bottom-up inf-wtt $\mathcal{M} = (Q, \Sigma, \Delta, S, \mu, \nu)$ over a commutative and continuous semiring S with a root output $\nu : Q \to S\langle\langle T_\Delta(Z_1)\rangle\rangle$. As with our inf-wtt, a Σ-algebra is associated with $\mathcal{M}$, however, the operation $\mu_\mathcal{M}(\sigma)$ is defined in terms of the OI-substitution $\leftarrow_{\mathrm{OI}}$ for tree series (introduced in Sect. 3.3). Then the tree series transformation computed by $\mathcal{M}$ is defined as $\tau_\mathcal{M}(\xi) = \sum_{q \in Q} \nu(q) \leftarrow_{\mathrm{OI}} h_\mu(\xi)_q$ for every $\xi \in T_\Sigma$. In [91], it is shown that the polynomial versions of such tree transducers over $\mathbb{B}$ and the so-called *nondeterministically simple top-down tree transducers* are semantically equivalent. The latter are special top-down tree transducers with rules of the form $q(\sigma(x_1, \ldots, x_k)) \to \zeta(q_1(x_1), \ldots, q_k(x_k))$, where $k \geq 0$ and $\sigma \in \Sigma^{(k)}$, cf. [68], Exercise 4 in Chap. IV. A *recognizable tree transducer* is another restricted version of the tree transducer of [91]; in this model the tree series $\mu_k(\sigma)_{q_1(x_1)\ldots q_k(x_k),q}$ is in $\mathrm{Rec}(\Delta \cup Z_k, S)$ for every $k \geq 0$, $\sigma \in \Sigma^{(k)}$, and $q, q_1, \ldots, q_k \in Q$, and $\nu(q)$ has the form $a_q.z_1$ for every $q \in Q$. In [91], Corollary 17, it is shown that, for every linear, nondeleting, and recognizable tree transducer $\mathcal{M}$ and recognizable tree series $r \in S\langle\langle T_\Sigma\rangle\rangle$, the tree series $\tau_\mathcal{M}(r)$ is also recognizable. Finally, we mention that in [53], top-down inf-wtt and tree transducers of [91] are related.

Inf-wtt with OI-Substitution

In [102], the idea of [91] was generalized and inf-wtt (rather than bottom-up inf-wtt) with root output and OI-substitution over a complete semiring were investigated. We denote the class of tree series transformations computed by such inf-wtt by $\mathrm{WTT}_{\mathrm{OI}}(S)_{\mathrm{inf}}$; the restricted classes are denoted by applying the same system as in Sect. 5.3. In [102], Lemmas 5 and 6 it was proved that $\text{x-TOP}^{\mathrm{R}}_{\mathrm{OI}}(S)_{\mathrm{inf}} = \text{x-TOP}_{\mathrm{OI}}(S)_{\mathrm{inf}}$ for every combination $\mathrm{x} \in \{\mathrm{t}, \mathrm{d}, \mathrm{h}, \mathrm{p}, \mathrm{b}, \mathrm{l}, \mathrm{n}\}$ (where p stands for polynomial), and $\text{xp-TOP}^{\mathrm{R}}_{\mathrm{OI}}(S)_{\mathrm{inf}} = \text{xp-WTT}_{\mathrm{OI}}(S)_{\mathrm{inf}}$ for every combination $\mathrm{x} \in \{\mathrm{t}, \mathrm{d}, \mathrm{h}, \mathrm{b}, \mathrm{l}, \mathrm{n}\}$. Moreover, top-down wtt and polynomial top-down inf-wtt with OI-substitution have the same computation power, i.e., $\text{x-TOP}(S) = \text{xp-TOP}_{\mathrm{OI}}(S)_{\mathrm{inf}}$, for every combination $\mathrm{x} \in \{\mathrm{t}, \mathrm{d}, \mathrm{h}, \mathrm{b}, \mathrm{l}, \mathrm{n}\}$, cf. [102], Theorem 7.

Top-down Inf-wtt and Bottom-up Inf-wtt with IO°-Substitution

It was observed already in [53] that top-down and bottom-up inf-wtt do not generalize all fundamental properties of top-down and bottom-up tree transducers. For example, it was proved in [53], Proposition 3.14 that the computation power of homomorphism top-down wtt and of homomorphism bottom-up wtt over the semiring Nat are incomparable, while they are equal in the unweighted case, cf. [48], Lemma 3.2. Therefore, in [65], an alternative semantics, based on the so-called IO°-substitution of tree series, of top-down and bottom-up inf-wtt was suggested for consideration. Roughly speaking, the IO°-substitution differs from the IO-substitution defined on page 362 in that we take into account the number of occurrences of z_i in ξ for every $i \in I$ when computing the coefficient of a tree $\xi(\zeta_i \mid i \in I)$. Formally, the *IO°-substitution* of tree series $(s_i \mid i \in I)$ into r is defined by

$$r \leftarrow_{\mathrm{IO}^\circ} (s_i)_{i\in I} = \sum_{\substack{\xi\in T_\Delta(Z),\\ (\forall i\in I):\zeta_i\in \mathrm{supp}(s_i)}} \Big((r,\xi) \cdot \prod_{i\in I} (s_i,\zeta_i)^{|\xi|_{z_i}} \Big).\xi(\zeta_i \mid i \in I).$$

We note that [IO]-substitution as defined on page 347 takes into account whether z_i occurs in ξ or not but does not use the number of occurrences. Then top-down inf-wtt and bottom-up inf-wtt with IO°-substitution are defined in the same way as in this chapter except that we use IO°-substitution in the definition of $\mu_{\mathcal{M}}(\sigma)$ instead of IO-substitution. The classes of tree series transformations computed by top-down inf-wtt or bottom-up inf-wtt with IO°-substitution are denoted by indexing the corresponding notation with o, like: $\text{l-TOP}_{\mathrm{o}}(S)_{\mathrm{inf}}$. It turned out that the IO°-substitution does not provide anything new for top-down inf-wtt because, for every combination x over $\{\mathrm{t}, \mathrm{d}, \mathrm{h}, \mathrm{p}, \mathrm{b}, \mathrm{l}, \mathrm{n}\}$, we have $\text{x-TOP}(S)_{\mathrm{inf}} = \text{x-TOP}_{\mathrm{o}}(S)_{\mathrm{inf}}$, cf. [65], Theorem 5.2. However, for bottom-up inf-wtt it does because for every partially ordered semiring S with $1 \preceq 1 + 1$ and $\mathrm{x}, \mathrm{y} \in \{\mathrm{td}, \mathrm{d}, \mathrm{h}\}$, the classes x-BOT$(S)_{\mathrm{inf}}$ and y-BOT$_{\mathrm{o}}(S)_{\mathrm{inf}}$ are incomparable with respect to inclusion, cf. [110], Theorem 5.10. Also, for every combination x over $\{\mathrm{t},\mathrm{d},\mathrm{h},\mathrm{p},\mathrm{b},\mathrm{l},\mathrm{n}\}$, we have xln-BOT$(S)_{\mathrm{inf}}$ = xln-BOT$_{\mathrm{o}}(S)_{\mathrm{inf}}$ and xpb-BOT$(S)_{\mathrm{inf}}$ = xpb-BOT$_{\mathrm{o}}(S)_{\mathrm{inf}}$ provided S is idempotent, cf. [65], Theorems 5.5 and 5.8. Moreover, the following results of [65] generalize the corresponding ones concerning tree transducers. By Theorem 5.12, we have xh-TOP$(S)_{\mathrm{inf}}$ = xh-BOT$_{\mathrm{o}}(S)_{\mathrm{inf}}$ for every zero-divisor free semiring S. Moreover, xln-TOP$(S)_{\mathrm{inf}}$ = xln-BOT$_{\mathrm{o}}(S)_{\mathrm{inf}}$, cf. Theorem 5.5 and Proposition 5.30. Finally, for every combination x over $\{\mathrm{p},\mathrm{b}\}$, we have xl-TOP$(S)_{\mathrm{inf}} \subseteq$ xl-BOT$_{\mathrm{o}}(S)_{\mathrm{inf}}$, cf. Theorem 5.26.

Top-down and Bottom-up Inf-wtt with Term Rewrite Semantics

In this chapter, the semantics of top-down and bottom-up inf-wtt was defined in an algebraic framework, more precisely, as an initial algebra semantics. In [66] an alternative approach was suggested by introducing weighted tree transducers of which the semantics is defined in an operational style. A weighted tree transducer of [66] is a tree transducer in which each (term rewriting) rule is associated with a weight taken from S. Along a successful leftmost derivation, the weights of the involved rules are multiplied and, for every pair of input tree and output tree, the weights of its successful leftmost derivations are summed up. In [66], it is shown in a constructive way that the two approaches, i.e., weighted tree transducers with initial algebra semantics and weighted tree transducers with term rewrite semantics, are semantically equivalent for both, the top-down and the bottom-up case, cf. Theorems 6.9 and 5.10.

Deterministic Bottom-up Inf-wtt over Multiplicative Monoids

In [101], the concept of a *deterministic bottom-up weighted tree transducer* (for short: deterministic bu-w-tt) was defined in a similar way to our deterministic bottom-up inf-wtt, except for the following. Since in case of a deterministic bottom-up inf-wtt the "additive part" of S is needless, cf. [53], Proposition 3.12, the author defines deterministic bottom-up inf-wtt over a multiplicative monoid A with absorbing element 0. Every deterministic bottom-up inf-wtt over S is a deterministic bu-w-tt over the monoid $(S,\cdot,1)$. However, deterministic bu-w-tt are more general than our deterministic bottom-up inf-wtt because there exists a monoid $(A,\cdot,1)$ with absorbing element 0 for which there does not exist a semiring $(A,+,\cdot,0,1)$ (cf. [101], Observation 2.2). Deterministic bu-w-tt are defined with both IO-substitution and IO^{o}-substitution semantics and the restricted versions of both models are considered for every combination x over $\{\mathrm{t},\mathrm{h},\mathrm{l},\mathrm{n}\}$. In this way, there are 24 classes of tree series transformations computed by restricted deterministic bu-w-tt. Also, the underlying monoid is restricted, namely a nonperiodic monoid; a periodic, commutative and nonregular monoid; a periodic, commutative, and regular monoid; a commutative and idempotent monoid; and a periodic and commutative group is considered as the underlying monoid. For each kind of underlying monoid, the inclusion diagram of the 24 classes is presented, cf. [101], Theorems 4.8, 4.17, 4.20, 4.23, and 4.25, respectively.

Chapter 5 of [106] is a revised and extended version of [101]. The author considers deterministic bottom-up inf-wtt and deterministic top-down inf-wtt (over S). Thus, the bottom-up model is more restricted than the deterministic bu-w-tt of [101]. On the other hand, it is more general because deterministic bottom-up inf-wtt have final output tree series. Similarly, deterministic top-down inf-wtt have initial output tree series. For these models, results similar to those in [101] are obtained.

5.8 Further Results

It follows from Theorem 5.5 and Lemma 5.11(C) that the class ln-TOP(S) is closed under composition. In [89], Theorem 2.4 and [93], Theorem 3.7 this has been generalized to top-down inf-wtt in which the family of rule mappings has the property that $\mu_k(\sigma)_{w,q}$ is a recognizable tree series and algebraic tree series, respectively.

In [103] it was shown that bottom-up inf-wtt can be simulated by weighted tree automata over distributive multioperator monoids [90, 103, 62]. This model has already been discussed as concept (f) on pages 353 and 360.

Acknowledgments. We are indebted to Frank Drewes, Manfred Droste, Joost Engelfriet, and Andreas Maletti for their effort to help us in forming the final version of this chapter. We also thank Ferenc Gécseg, Johanna Högberg, Magnus Steinby, and Janis Voigtländer for their valuable comments.

References

1. A. Alexandrakis and S. Bozapalidis. Weighted grammars and Kleene's theorem. *Information Processing Letters*, 24(1):1–4, 1987.
2. D. Angluin. Learning regular sets from queries and counterexamples. *Information and Computation*, 75:87–106, 1987.
3. F. Baader and T. Nipkow. *Term Rewriting and All That.* Cambridge University Press, Cambridge, 1998.
4. B.S. Baker. Composition of top-down and bottom-up tree transductions. *Information and Control*, 41(2):186–213, 1979.
5. R. Bělohlávek. Determinism and fuzzy automata. *Information Sciences*, 143:205–209, 2002.
6. J. Berstel. *Transductions and Context-Free Languages.* Teubner, Stuttgart, 1979.
7. J. Berstel and C. Reutenauer. Recognizable formal power series on trees. *Theoretical Computer Science*, 18(2):115–148, 1982.

8. J. Berstel and C. Reutenauer. *Rational Series and Their Languages*, volume 2 of *Monographs in Theoretical Computer Science. An EATCS Series*, Springer, Berlin, 1988.
9. R.E. Block and G. Griffing. Recognizable formal series on trees and cofree coalgebraic systems. *Journal of Algebra*, 215:543–573, 1999.
10. S.L. Bloom and Z. Ésik. An extension theorem with an application to formal tree series. *Journal of Automata, Languages and Combinatorics*, 8:145–185, 2003.
11. B. Borchardt. The Myhill–Nerode theorem for recognizable tree series. In Z. Ésik and Z. Fülöp, editors, *Proceedings of the 7th International Conference on Developments in Language Theory (DLT), Szeged, Hungary*, volume 2710 of *Lecture Notes in Computer Science*, pages 146–158. Springer, Berlin, 2003.
12. B. Borchardt. A pumping lemma and decidability problems for recognizable tree series. *Acta Cybernetica*, 16(4):509–544, 2004.
13. B. Borchardt. Code selection by tree series transducers. In M. Domaratzki, A. Okhotin, K. Salomaa, and S. Yu, editors, *Proceedings of the 9th International Conference on Implementation and Application of Automata (CIAA), Kingston, Canada*, volume 3317 of *Lecture Notes in Computer Science*, pages 57–67. Springer, Berlin, 2005.
14. B. Borchardt. *The Theory of Recognizable Tree Series.* Verlag für Wissenschaft und Forschung, Berlin, 2005 (PhD thesis, TU Dresden, Germany, 2004).
15. B. Borchardt, Z. Fülöp, Z. Gazdag, and A. Maletti. Bounds for tree automata with polynomial costs. *Journal of Automata, Languages and Combinatorics*, 10:107–157, 2005.
16. B. Borchardt, A. Maletti, B. Šešelja, A. Tepavčevic, and H. Vogler. Cut sets as recognizable tree languages. *Fuzzy Sets and Systems*, 157:1560–1571, 2006.
17. B. Borchardt and H. Vogler. Determinization of finite state weighted tree automata. *Journal of Automata, Languages and Combinatorics*, 8(3):417–463, 2003.
18. S. Bozapalidis. Effective construction of the syntactic algebra of a recognizable series on trees. *Acta Informatica*, 28:351–363, 1991.
19. S. Bozapalidis. Representable tree series. *Fundamenta Informaticae*, 21:367–389, 1994.
20. S. Bozapalidis. Positive tree representations and applications to tree automata. *Information and Control*, 139(2):130–153, 1997.
21. S. Bozapalidis. Equational elements in additive algebras. *Theory of Computing Systems*, 32(1):1–33, 1999.
22. S. Bozapalidis. Context-free series on trees. *Information and Computation*, 169:186–229, 2001.
23. S. Bozapalidis and A. Alexandrakis. Représentations matricielles des séries d'arbre reconnaissables. *Theoretical Informatics and Applications, RAIRO*, 23(4):449–459, 1989.

24. S. Bozapalidis and O. Louscou-Bozapalidou. The rank of a formal tree power series. *Theoretical Computer Science*, 27:211–215, 1983.
25. S. Bozapalidis and G. Rahonis. On the closure of recognizable tree series under tree homomorphisms. *Journal of Automata, Languages and Combinatorics*, 10:185–202, 2005.
26. W.S. Brainerd. The minimalization of tree automata. *Information and Control*, 13:484–491, 1968.
27. P. Buchholz. Bisimulation relations for weighted automata. *Theoretical Computer Science*, 393(1–3):109–123, 2008.
28. J.R. Büchi. Weak second-order arithmetic and finite automata. *Zeitschrift für Mathematische Logik und Grundlagen der Mathematik*, 6:66–92, 1960.
29. H. Comon, M. Dauchet, R. Gilleron, F. Jacquemard, D. Lugiez, S. Tison, and M. Tommasi. Tree automata techniques and applications. Available on: http://www.grappa.univ-lille3.fr/tata, 1997.
30. B. Courcelle. Equivalences and transformations of regular systems—Applications to recursive program schemes and grammars. *Theoretical Computer Science*, 42:1–122, 1986.
31. B. Courcelle. Recursive applicative program schemes. In J. van Leeuwen, editor, *Handbook of Theoretical Computer Science, volume B*, pages 459–492. Elsevier, Amsterdam, 1990.
32. B. Courcelle. Basic notions of universal algebra for language theory and graph grammars. *Theoretical Computer Science*, 163:1–54, 1996.
33. J. Doner. Tree acceptors and some of their applications. *Journal of Computer and System Sciences*, 4:406–451, 1970.
34. F. Drewes. Computation by tree transductions. PhD thesis, University of Bremen, 1996.
35. F. Drewes. *Grammatical Picture Generation—A Tree-Based Approach, Texts in Theoretical Computer Science. An EATCS Series.* Springer, Berlin, 2006.
36. F. Drewes and H. Vogler. Learning deterministically recognizable tree series. *Journal of Automata, Languages and Combinatorics*, 12:333–354, 2007.
37. M. Droste and P. Gastin. Weighted automata and weighted logics. In L. Caires, G.F. Italiano, L. Monteiro, C. Palamidessi, and M. Yung, editors, *Proceedings of the 32nd International Colloquium on Automata, Languages and Programming (ICALP), Lisbon, Portugal*, volume 3580 of *Lecture Notes in Computer Science*, pages 513–525. Springer, Berlin, 2005.
38. M. Droste and P. Gastin. Weighted automata and weighted logics. *Theoretical Computer Science*, 380:69–86, 2007.
39. M. Droste and P. Gastin. Weighted automata and weighted logics. In this *Handbook*. Chapter 5. Springer, Berlin, 2009.
40. M. Droste and W. Kuich. Semirings and formal power series. In this *Handbook*. Chapter 1. Springer, Berlin, 2009.

41. M. Droste, Chr. Pech and H. Vogler. A Kleene theorem for weighted tree automata. *Theory of Computing Systems*, 38:1–38, 2005.
42. M. Droste and G. Rahonis. Weighted automata and weighted logics with discounting. In J. Holub and J. Zdárek, editors, *Proceedings of the 12th International Conference on Implementation and Application of Automata (CIAA), Prague*, volume 4783 of *Lecture Notes in Computer Science*, pages 73–84. Springer, Berlin, 2007.
43. M. Droste and H. Vogler. Weighted tree automata and weighted logics. *Theoretical Computer Science*, 366:228–247, 2006.
44. M. Droste and H. Vogler. Weighted logics for unranked tree automata. *Theory of Computing Systems*, 2009, in press.
45. S. Eilenberg. *Automata, Languages, and Machines—Volume A*, volume 59 of *Pure and Applied Mathematics*. Academic Press, San Diego, 1974.
46. C.C. Elgot. Decision problems of finite automata design and related arithmetics. *Transactions of the American Mathematical Society*, 98:21–52, 1961.
47. C.A. Ellis. Probabilistic tree automata. *Information and Control*, 19(5):401–416, 1971.
48. J. Engelfriet. Bottom-up and top-down tree transformations—A comparison. *Mathematical Systems Theory*, 9(3):198–231, 1975.
49. J. Engelfriet. Top–down tree transducers with regular look-ahead. *Mathematical Systems Theory*, 10:289–303, 1977.
50. J. Engelfriet. Three hierarchies of transducers. *Mathematical Systems Theory*, 15(2):95–125, 1982.
51. J. Engelfriet. Tree transducers and syntax-directed semantics. In *Proceedings of the 7th Colloquium on Trees in Algebra and Programming (CAAP), Lille, France, 1982.*
52. J. Engelfriet. On incomparability of tree series transformation classes. Personal communication, 2008.
53. J. Engelfriet, Z. Fülöp, and H. Vogler, Bottom-up and top-down tree series transformations. *Journal of Automata, Languages and Combinatorics*, 7:11–70, 2002.
54. J. Engelfriet and S. Maneth. A comparison of pebble tree transducers with macro tree transducers. *Acta Informatica*, 39(9):613–698, 2003.
55. J. Engelfriet and E.M. Schmidt. IO and OI.I. *Journal of Computer and System Sciences*, 15(3):328–353, 1977.
56. J. Engelfriet and E.M. Schmidt. IO and OI.II. *Journal of Computer and System Sciences*, 16(1):67–99, 1978.
57. Z. Ésik. Fixed point theory. In this *Handbook*. Chapter 2. Springer, Berlin, 2009.
58. Z. Ésik and W. Kuich. Formal tree series. *Journal of Automata, Languages and Combinatorics*, 8(2):219–285, 2003.
59. Z. Ésik and G. Liu. Fuzzy tree automata. *Fuzzy Sets and Systems*, 158:1450–1460, 2007.

60. I. Fichtner. Weighted picture automata and weighted logics. *Theory of Computing Systems*, 2009, in press.
61. Z. Fülöp, Z. Gazdag and H. Vogler, Hierarchies of tree series transformations. *Theoretical Computer Science*, 314:387–429, 2004.
62. Z. Fülöp, A. Maletti, and H. Vogler. A Kleene theorem for weighted tree automata over distributive multioperator monoids. *Theory of Computing Systems*, 44:455–499, 2009.
63. Z. Fülöp and S. Vágvölgyi. Congruential tree languages are the same as recognizable tree languages. *Bulletin of the European Association for Theoretical Computer Science*, 30:175–185, 1989.
64. Z. Fülöp and H. Vogler. *Syntax-Directed Semantics—Formal Models Based on Tree Transducers, Monographs in Theoretical Computer Science. An EATCS Series.* Springer, Berlin, 1998.
65. Z. Fülöp and H. Vogler. Tree series transformations that respect copying. *Theory of Computing Systems*, 36(3):247–293, 2003.
66. Z. Fülöp and H. Vogler. Weighted tree transducers. *Journal of Automata, Languages and Combinatorics*, 9:31–54, 2004.
67. Z. Fülöp and H. Vogler. Comparison of several classes of weighted tree automata. Technical report, TUD-FI06-08-Dez.2006, TU Dresden, 2006.
68. F. Gécseg and M. Steinby. *Tree Automata.* Akadémiai Kiadó, Budapest, 1984.
69. F. Gécseg and M. Steinby. Tree languages. In G. Rozenberg and A. Salomaa, editors, *Handbook of Formal Languages, volume 3*, pages 1–68. Springer, Berlin, 1997.
70. J.A. Goguen, J.W. Thatcher, E.G. Wagner, and J.B. Wright. Initial algebra semantics and continuous algebras. *Journal of the ACM*, 24:68–95, 1977.
71. J.S. Golan. *Semirings and Their Applications.* Kluwer Academic, Dordrecht, 1999.
72. J. Graehl and K. Knight. Training tree transducers. In *Human Language Technology Conference/North American Chapter of the Association for Computational Linguistics Annual Meeting (HLT-NAACL), Boston, Massachusetts, USA*, pages 105–112. Association for Computational Linguistics, Stroudsburg, 2004.
73. G. Grätzer. *Universal Algebra.* van Nostrand, Princeton, 1968.
74. I. Guessarian. Pushdown tree automata. *Theory of Computing Systems*, 16:237–263, 1983.
75. A. Habrard and J. Oncina. Learning multiplicity tree automata. In Y. Sakakibara, S. Kobayashi, K. Sato, T. Nishino, and E. Tomita, editors, *Proceedings of the 8th International Colloquium on Grammatical Inference: Algorithms and Applications (ICGI)*, volume 4201 of *Lecture Notes in Computer Science*, pages 268–280. Springer, Berlin, 2006.
76. U. Hebisch and H.J. Weinert. *Semirings—Algebraic Theory and Applications in Computer Science.* World Scientific, Singapore, 1998.

77. J. Högberg, A. Maletti, and J. May. Bisimulation minimisation of weighted tree automata. In T. Harju, J. Karhumäki, and A. Lepistö, editors, *Proceedings of the 11th International Conference on Developments in Language Theory (DLT)*, volume 4588 of *Lecture Notes in Computer Science*, pages 229–241. Springer, Berlin, 2007.
78. Y. Inagaki and T. Fukumura. On the description of fuzzy meaning of context-free languages. In L.A. Zadeh, editor, *Fuzzy Sets and Their Applications to Cognitive and Decision Processes*, pages 301–328. Academic Press, New York, 1975.
79. E.T. Irons. A syntax directed compiler for ALGOL 60. *Communications of the ACM*, 4(1):51–55, 1961.
80. D. Kirsten and I. Mäurer. On the determinization of weighted automata. *Journal of Automata, Languages and Combinatorics*, 10:287–312, 2005.
81. S.E. Kleene. Representation of events in nerve nets and finite automata. In C.E. Shannon and J. McCarthy, editors, *Automata Studies*, pages 3–42. Princeton University Press, Princeton, 1956.
82. J.W. Klop. Term rewrite systems. In S. Abramsky, D.M. Gabbay, T.S.E. Maibaum, editors, *Handbook of Logic in Computer Science, volume 2*, pages 1–116. Oxford Science Publications, Oxford, 1992.
83. K. Knight and J. Graehl. An overview of probabilistic tree transducers for natural language processing. In A. Gelbukh, editor, *Proceedings of the 6th International Conference on Computational Linguistics and Intelligent Text Processing (CICLing), Mexico City, Mexico*, volume 3406 of *Lecture Notes in Computer Science*, pages 1–24. Springer, Berlin, 2005.
84. K. Knight and J. May. Applications of weighted automata in natural language processing. In this *Handbook*. Chapter 14. Springer, Berlin, 2009.
85. H.-P. Kolb, J. Michaelis, U. Mönnich, and F. Morawietz. An operational and denotational approach to non-context-freeness. *Theoretical Computer Science*, 293:261–289, 2003.
86. D. Kozen. On the Myhill–Nerode theorem for trees. *Bulletin of the European Association for Theoretical Computer Science*, 47:170–173, 1992.
87. W. Kuich. Semirings and formal power series: Their relevance to formal languages and automata. In G. Rozenberg and A. Salomaa, editors, *Handbook of Formal Languages, volume 1*, Chapter 9, pages 609–677. Springer, Berlin, 1997.
88. W. Kuich. Formal power series over trees. In S. Bozapalidis, editor, *Proceedings of the 3rd International Conference on Developments in Language Theory (DLT), Thessaloniki, Greece*, pages 61–101. Aristotle University of Thessaloniki, Thessaloniki, 1998.
89. W. Kuich. Full abstract families of tree series I. In J. Karhumäki, H. Maurer, G. Paun, and G. Rozenberg, editors, *Jewels Are Forever*, pages 145–156. Springer, Berlin, 1999.
90. W. Kuich. Linear systems of equations and automata on distributive multioperator monoids. In D. Dorninger, G. Eigenthaler, and M. Goldstern, editors, *Contributions to General Algebra 12—Proceedings of the*

58th Workshop on General Algebra "58. Arbeitstagung Allgemeine Algebra", Vienna University of Technology, June 3–6, 1999, pages 1–10. Verlag Johannes Heyn, Klagenfurt, 1999.

91. W. Kuich. Tree transducers and formal tree series. *Acta Cybernetica*, 14:135–149, 1999.
92. W. Kuich. Formal series over algebras. In M. Nielsen and B. Rovan, editors, *Mathematical Foundations of Computer Science (MFCS), Bratislava, Slovakia*, volume 1893 of *Lecture Notes in Computer Science*, pages 488–496. Springer, Berlin, 2000.
93. W. Kuich. Full abstract families of tree series II. In R. Freund and A. Kelemenova, editors, *Proceedings of the International Workshop Grammar Systems 2000*, pages 347–358. Schlesische Universität Troppau 2000.
94. W. Kuich. Pushdown tree automata, algebraic tree systems, and algebraic tree series. *Information and Computation*, 165:69–99, 2001.
95. W. Kuich. Formal series over sorted algebras. *Discrete Mathematics*, 254:231–258, 2002.
96. W. Kuich and A. Salomaa. *Semirings, Automata, Languages*, volume 5 of *Monographs in Theoretical Computer Science. An EATCS Series* Springer, Berlin, 1986.
97. S. Lang. *Algebra*, 3rd edition. Addison–Wesley, Reading, 1993.
98. L. Libkin. Logics for unranked trees: An overview. In L. Caires, G.F. Italiano, L. Monteiro, C. Palamidessi, and M. Yung, editors, *Proceedings of the 32nd International Colloquium on Automata, Languages and Programming (ICALP), Lisbon, Portugal*, volume 3580 of *Lecture Notes in Computer Science*, pages 35–50. Springer, Berlin, 2005.
99. L. Libkin. Logics for unranked trees: An overview. *Logical Methods in Computer Science*, 2(3:2):1–31, 2006.
100. M. Magidor and G. Moran. Probabilistic tree automata and context free languages. *Israel Journal of Mathematics*, 8:340–348, 1970.
101. A. Maletti. Hasse diagrams for classes of deterministic bottom-up tree-to-tree-series transformations. *Theoretical Computer Science*, 339:200–240, 2005.
102. A. Maletti. The power of tree series transducers of type I and II. In C. De Felice and A. Restivo, editors, *Proceedings of the 9th International Conference on Developments in Language Theory (DLT), Palermo, Italy*, volume 3572 of *Lecture Notes in Computer Science*, pages 338–349. Springer, Berlin, 2005.
103. A. Maletti. Relating tree series transducers and weighted tree automata. *International Journal of Foundations of Computer Science*, 16(4):723–741, 2005.
104. A. Maletti. Compositions of tree series transformations. *Theoretical Computer Science*, 366:248–271, 2006.
105. A. Maletti. Hierarchies of tree series transformations revisited. In O.H. Ibarra and Z. Dang, editors, *Proceedings of the 10th International*

Conference on Developments in Language Theory (DLT), volume 4036 of *Lecture Notes in Computer Science*, pages 215–225. Springer, Berlin, 2006.

106. A. Maletti. *The Power of Tree Series Transducers.* Der Andere Verlag, Tönning, 2006 (PhD thesis, TU Dresden, Germany, 2006).
107. A. Maletti. Generalizing the inclusion and the hierarchy results from positive semirings to semirings which are not rings. Personal communication, 2007.
108. A. Maletti. Learning deterministically recognizable tree series—Revisited. In S. Bozapalidis and G. Rahonis, editors, *Proceedings of the 2nd International Conference on Algebraic Informatics*, volume 4728 of *Lecture Notes in Computer Science*, pages 218–235. Springer, Berlin, 2007.
109. A. Maletti. Myhill-Nerode theorem for recognizable tree series—Revisited. In E. Laber, C. Bornstein, L. Nogueira, and L. Faria, editors, *Proceedings of the 8th International Conference on Latin American Theoretical Informatics (LATIN)*, volume 4957 of *Lecture Notes in Computer Science*, pages 106–120. Springer, Berlin, 2008.
110. A. Maletti and H. Vogler. Incomparability results for classes of polynomial tree series transformations. *Journal of Automata, Languages and Combinatorics*, 10(4):535–568, 2005.
111. S. Maneth, A. Berlea, T. Perst, and H. Seidl. XML type checking with macro tree transducers. In *Proceedings of the Twenty-Fourth ACM SIGMOD–SIGACT–SIGART Symposium on Principles of Database Systems (PODS), Baltimore, Maryland*, pages 283–294. ACM, New York, 2005.
112. Chr. Mathissen. Definable transductions and weighted logics for texts. In T. Harju, J. Karhumäki, and A. Lepistö, editors, *Proceedings of the 11th International Conference on Developments in Language Theory (DLT), Turku*, volume 4588 of *Lecture Notes in Computer Science*, pages 324–336. Springer, Berlin, 2007.
113. Chr. Mathissen. Weighted logics for nested words and algebraic formal power series. In L. Aceto, I. Damgård, L.A. Goldberg, M.M. Halldórsson, A. Ingólfsdóttir, and I. Walukiewicz, editors, *Proceedings of the 35th International Colloquium on Automata, Languages and Programming (ICALP), Reykjavik*, volume 5126 of *Lecture Notes in Computer Science*, pages 221–232. Springer, Berlin, 2008.
114. J. May and K. Knight. Tiburon: A weighted tree automata toolkit. In O.H. Ibarra and H.-C. Yen, editors, *Proceedings of the 11th International Conference on Implementation and Application of Automata (CIAA)*, volume 4094 of *Lecture Notes in Computer Science*, pages 102–113. Springer, Berlin, 2006.
115. I. Meinecke. Weighted logics for traces. In D. Grigoriev, J. Harrison, and E.A. Hirsch, editors, *Proceedings of Computer Science—Theory and*

Applications, 1st CSR, St. Petersburg, volume 3967 of *Lecture Notes in Computer Science*, pages 235–246. Springer, Berlin, 2006.
116. J. Mezei and J.B. Wright. Algebraic automata and context-free sets. *Information and Control*, 11:3–29, 1967.
117. F. Neven. Automata, logic, and XML. In J. Bradfield, editor, *Computer Science Logic: 16th International Workshop (CSL), Edinburgh, Scotland, UK*, volume 2471 of *Lecture Notes in Computer Science*, pages 2–26. Springer, Berlin, 2002.
118. M. Nivat. Transduction des langages de Chomsky. *Université de Grenoble. Annales de l'Institut Fourier*, 18:339–456, 1968.
119. Chr. Pech. Kleene-type results for weighted tree automata. PhD thesis, TU Dresden, 2003.
120. Chr. Pech. Kleene's theorem for weighted tree-automata. In A. Lingas and B.J. Nilsson, editors, *Proceedings of the 14th International Symposium on Fundamentals of Computation Theory (FCT), Malmö, Sweden*, volume 2751 of *Lecture Notes in Computer Science*, pages 387–399. Springer, Berlin, 2003.
121. I. Petre and A. Salomaa. Algebraic systems and pushdown automata. In this *Handbook*. Chapter 7. Springer, Berlin, 2009.
122. G. Rahonis. Weighted Muller tree automata and weighted logics. *Journal of Automata, Languages, and Programming*, 12:455–483, 2007.
123. W.C. Rounds. Mappings and grammars on trees. *Theory of Computing Systems*, 4(3):257–287, 1970.
124. J. Sakarovitch. Rational and recognisable power series. In this *Handbook*. Chapter 4. Springer, Berlin, 2009.
125. A. Salomaa and M. Soittola. *Automata-Theoretic Aspects of Formal Power Series, Texts and Monographs in Computer Science*. Springer, Berlin, 1978.
126. T. Schwentick. Automata for XML—A survey. *Journal of Computer and System Sciences*, 73(3):289–315, 2007.
127. H. Seidl. Deciding equivalence of finite tree automata. *SIAM Journal on Computing*, 19(3):424–437, 1990.
128. H. Seidl. Finite tree automata with cost functions. *Theoretical Computer Science*, 126(1):113–142, 1994.
129. B. Šešelja and A. Tepavčević. Completion of ordered structures by cuts of fuzzy sets, an overview. *Fuzzy Sets and Systems*, 136(1):1–19, 2003.
130. M. Steinby. A theory of tree language varieties. In M. Nivat and A. Podelski, editors, *Tree Automata and Languages, Studies in Computer Science and Artificial Intelligence*, North-Holland, Amsterdam, 1992.
131. T. Stüber. Using the theorem of Cayley–Hamilton for proving tree series not to be recognizable. Personal communication, 2007
132. T. Stüber, H. Vogler, and Z. Fülöp. Decomposition of weighted multioperator tree automata. *International Journal Foundations of Computer Science*, 20(2):221–245, 2009.

133. J.W. Thatcher. Generalized^2 sequential machine maps. *Journal of Computer and System Sciences*, 4(4):339–367, 1970.
134. J.W. Thatcher. Tree automata: An informal survey. In A.V. Aho, editor, *Currents in the Theory of Computing*, pages 143–172. Prentice Hall, Englewood Cliffs, 1973.
135. J.W. Thatcher and J.B. Wright. Generalized finite automata theory with an application to a decision problem of second-order logic. *Theory of Computing Systems*, 2(1):57–81, 1968.
136. V. Vianu. A Web odyssey: From Codd to XML. In *Proceedings of the 20th ACM SIGACT–SIGMOD–SIGART Symposium on Principles of Database Systems (PODS), Santa Barbara, California, USA*, pages 1–15. ACM, New York, 2001.
137. H. Wang. On characters of semirings. *Houston Journal of Mathematics*, 23(3):391–405, 1997.
138. H. Wang. On rational series and rational languages. *Theoretical Computer Science*, 205(1–2):329–336, 1998.
139. W. Wechler. *Universal Algebra for Computer Scientists*, 1st edition, volume 25 of *Monographs in Theoretical Computer Science. An EATCS Series.* Springer, Berlin, 1992.

Chapter 10: Traces, Series-Parallel Posets, and Pictures: A Weighted Study

Ina Fichtner, Dietrich Kuske, and Ingmar Meinecke*

Institut für Informatik, Universität Leipzig, Leipzig, Germany
fichtner@informatik.uni-leipzig.de
kuske@informatik.uni-leipzig.de
meinecke@informatik.uni-leipzig.de

* The third author was supported by the Deutsche Forschungsgemeinschaft within the project "Gewichtete Automaten und gewichtete Logiken für diskrete Strukturen" DR 202/10-1.

M. Droste, W. Kuich, H. Vogler (eds.), *Handbook of Weighted Automata*, Monographs in Theoretical Computer Science. An EATCS Series, DOI 10.1007/978-3-642-01492-5_10, © Springer-Verlag Berlin Heidelberg 2009

1 Introduction

Words and finite automata over words are central notions in computer science. They model the sequential executions of systems. However, both the world in its own and computer science have to cope with another phenomenon as well: concurrency. Then not all events have to be ordered but some actions may appear independently. Not surprisingly, automata modeling concurrency are an important subject in theoretical computer science. On the other hand, words can be seen as one-dimensional objects. But certainly, we have to handle higher-dimensional objects like two-dimensional rectangles which we refer to as pictures. These generalizations of words can be seen as certain classes of partially ordered sets (short posets) or graphs. This chapter is concerned with weighted automata running on those special graphs. Hence, the behaviors of those automata will describe quantitative aspects of those posets or graphs accepted by the system.

In Sect. 2, we start with weighted distributed systems which are modeled as weighted asynchronous cellular automata (or wACA for short). Such an automaton is a collection of finitely many sequential automata where some of them depend on each other whereas others can proceed concurrently. Whenever a wACA executes a certain action a, the local process associated with a will change its state according to the states the dependent processes have adopted beforehand. The dependence relation induces a certain class of posets, called Mazurkiewicz traces or just traces. Traces stem originally from the behavior of certain Petri nets and have turned out to be one of the fundamental models in concurrency theory. A wealth of results on traces can be found in [12].

In this quantitative setting, a weight from a commutative semiring is affiliated with every transition of the automaton. Therefore, the concurrent behavior of a wACA is a function mapping traces to elements of the underlying semiring. We will characterize the behavior of a wACA by three other formalisms: presentations, logic formulas, and expressions. Here, presentations can be seen as weighted word automata satisfying additional features (known as diamond properties) due to the partial commutation of the actions. As logic formulas, we consider a fragment of weighted monadic second-order logic. For rational expressions, we have to restrict the application of iteration to monoalphabetic and connected subexpressions.

Branching automata provide another automaton model for concurrency; see [47] for relevant results. They realize concurrency by branching a process into several independent subprocesses which will be joined again in the future. In a branching automaton, two subruns are either composed sequentially or in parallel. Therefore, the partial order of the actions executed by the automaton can be built from singletons by the use of a sequential and a parallel product, giving rise to the class of series-parallel posets, or sp-posets for short. For weighted branching automata, which we present in Sect. 3, sequential and parallel composition are reflected in the weight structure by two different multiplications. Nondeterminism of the system is modeled by a third operation. Hence, the weight structure is not a semiring anymore but a bisemiring. A bisemiring can be understood as two semirings with the same additive structure (for nondeterminism) where the second product has to be commutative (due to the nature of commuting processes). The main result shown here will provide an equivalent formalism by way of expressions for weighted branching automata of bounded depth, i.e., those where only a bounded number of processes can be executed in parallel.

As suggested above, pictures (finite labeled grids) are another natural generalization of words; see [32] for an overview. Concerning pictures, in Sect. 4, we present weighted picture automata which examine for an action a at a certain position in the picture the states at the four poles of this position (east, south, west, and north). Again, such a transition or rule is equipped with a weight from a commutative semiring. We will characterize the behavior of these weighted picture automata as projections of the semantics of rational expressions and, moreover, as the semantics of a fragment of weighted monadic second-order logic.

Traces, sp-posets, and pictures are important and well-explored instances of posets and graphs. Other graph classes have been considered in the literature, likewise in a weighted setting. Related work as well as the history and bibliography concerning traces, sp-posets, and pictures are discussed at length at the end of the respective section.

2 Traces

Previous chapters of this handbook proved the equivalence of different formalisms for the description of functions $f : \Sigma^* \to S$ into some semiring S. The guiding idea was that a word $w \in \Sigma^*$ describes a sequence of actions to be performed by some system and $f(w)$ denotes the weight or cost associated with this sequence of actions. Because of the linear nature of words over Σ, this approach works fine for sequential systems. Here, we ask whether similar results hold for concurrent systems.

Concurrency means commutation of some actions within the system. Therefore, the respective weights have to commute, also. This is reflected by the commutativity of the semiring multiplication. *Hence, throughout this section, we assume a fixed commutative semiring S.*

2.1 Weighted Distributed Systems

Our model of a concurrent system will be that of an asynchronous cellular automaton. Such an automaton consists of finitely many finite automata each with its own alphabet. These finite automata reside in the nodes of a finite graph $(\mathbb{L}, D)$ and proceed asynchronously. Whenever the finite automaton present in a particular node executes an action, its subsequent state depends on the current one, the action performed, and on the states of those automata that reside in neighboring nodes.

Weighted Asynchronous Cellular Automata

Note that the architecture of the system is given by the graph $(\mathbb{L}, D)$. We therefore fix a nonempty and finite set $\mathbb{L}$ of *locations* and a symmetric and reflexive *dependence relation* $D \subseteq \mathbb{L} \times \mathbb{L}$. Its complement in $\mathbb{L}^2$ is called the *independence relation* I. A *distributed alphabet* is a tuple $\Sigma = (\Sigma_\ell)_{\ell \in \mathbb{L}}$ of nonempty and mutually disjoint alphabets. For $a \in \Sigma_\ell$, write $\mathrm{lc}(a)$ for the unique location of a, i.e., $\mathrm{lc}(a) = \ell$. Furthermore, we set $(a, b) \in D$ for $a, b \in \Sigma$ if and only if $(\mathrm{lc}(a), \mathrm{lc}(b)) \in D$ and $(a, b) \in I$ if and only if $(a, b) \notin D$.

Now we define our model of a concurrent system formally.

Definition 2.1. *A* weighted asynchronous cellular automaton *or* wACA *for short is a tuple* $\mathcal{A} = (\Sigma, (Q_m)_{m \in \mathbb{L}}, \mathrm{in}, (c_\ell)_{\ell \in \mathbb{L}}, \mathrm{out})$ *where:*

- Σ *is a distributed alphabet.*
- Q_m *is a finite set of local states for every* $m \in \mathbb{L}$.
- $c_\ell : (\prod_{m \in D(\ell)} Q_m) \times \Sigma_\ell \times Q_\ell \to S$ *is a local weight function for every* $\ell \in \mathbb{L}$ *where* $D(\ell) = \{m \in \mathbb{L} \mid (\ell, m) \in D\}$.
- $\mathrm{in}, \mathrm{out} : \prod_{m \in \mathbb{L}} Q_m \to S$ *are functions describing the cost for entering and leaving the system.*

Intuitively, the set of local states Q_m and the local weight function c_m describe the behavior of the automaton in location m. Next, we present three different behaviors of wACA.

Word Behavior of wACA

A *configuration* of $\mathcal{A}$ is an element of $\prod_{m \in \mathbb{L}} Q_m$, i.e., a tuple of local states, one for each location; Q denotes the set of configurations. The local weight functions c_ℓ define a *global weight function* $c : Q \times \Sigma \times Q \to S$. For this, let $a \in \Sigma_\ell$ be some action and let $p = (p_m)_{m \in \mathbb{L}}$ and $q = (q_m)_{m \in \mathbb{L}}$ be configurations. Then

$$c(p, a, q) = \begin{cases} c_\ell((p_m)_{m \in D(\ell)}, a, q_\ell) & \text{if } p_m = q_m \text{ for all } m \in \mathbb{L} \setminus \{\ell\}, \\ 0 & \text{otherwise.} \end{cases}$$

The mapping c can be understood as a matrix $A \in (S^\Sigma)^{Q \times Q}$ setting $A_{p,q}(a) = c(p, a, q)$. This way, the weighted asynchronous cellular automaton $\mathcal{A}$ defines a weighted finite automaton $\mathcal{B} = (Q, \mathrm{in}, A, \mathrm{out})$ over the semiring S and the alphabet $\bigcup_{\ell \in \mathbb{L}} \Sigma_\ell$ (cf. [21]). Then the *word behavior* $\|\mathcal{A}\|_W : \Sigma^+ \to S$ of the weighted asynchronous cellular automaton $\mathcal{A}$ is defined to be the restriction of the behavior $\|\mathcal{B}\| = \sum_{p,q \in Q} \mathrm{in}(p)(A^*)_{p,q}\mathrm{out}(q) : \Sigma^* \to S$ of $\mathcal{B}$ to the set of nonempty words.

Interleaving Behavior of wACA

Let $a \in \Sigma_k$ and $b \in \Sigma_\ell$ with $(k, \ell) \in I$ and let $p, q, r \in Q$ be configurations such that $c(p, a, q) \neq 0$ and $c(q, b, r) \neq 0$. Then one can check that $c(p, a, q) = c(q', a, r)$ and $c(q, b, r) = c(p, b, q')$ with $q'_\ell = r_\ell$ and $q'_m = p_m$ for $m \neq \ell$. Since the semiring S is commutative, this implies $\|\mathcal{A}\|_W(uabv) = \|\mathcal{A}\|_W(ubav)$, i.e., two words that only differ in the order of independent actions have the same weight. This observation leads to the following definition of Mazurkiewicz traces [55] (for more details, see [12]).

Let $\Sigma = (\Sigma_\ell)_{\ell\in\mathbb{L}}$ be a distributed alphabet. Then $\sim$ is the least congruence relation on the free semigroup Σ^+ with $ab \sim ba$ for all $a, b \in \Sigma$ with $(a, b) \in I$. The quotient $\mathbb{M}(\Sigma) = \Sigma^+/\sim$ is the Mazurkiewicz *trace semigroup generated by* Σ. Its elements are equivalence classes $[u]$ of words. A language $L \subseteq \Sigma^+$ is *trace closed* if $u \sim v$ and $v \in L$ imply $u \in L$, i.e., $L = \bigcup_{v\in L}[v]$. Similarly, a function $\mu : \Sigma^+ \to X$ to some set X is *trace closed* if $u \sim v$ implies $\mu(u) = \mu(v)$. Since we only deal with commutative semirings, the word behavior of every wACA $\mathcal{A}$ is trace closed. We can therefore define the *interleaving behavior* $\|\mathcal{A}\|_I$ of $\mathcal{A}$ as $\|\mathcal{A}\|_I : \mathbb{M}(\Sigma) \to S : [u] \mapsto \|\mathcal{A}\|_W(u)$.

Concurrent Behavior of wACA

Recall that elements of the trace semigroup are $\sim$-equivalence classes of words over Σ. These equivalence classes can be represented naturally as partial orders as follows. A *trace over* Σ is a finite and nonempty labeled poset $t = (V, \sqsubseteq, \lambda)$ with $\lambda : V \to \Sigma$ such that the following hold for all $x, y \in V$:

- If $(\lambda(x), \lambda(y)) \in D$, then $x \sqsubseteq y$ or $y \sqsubseteq x$.
- If $x \sqsubset y$ and there is no node in between, then $(\lambda(x), \lambda(y)) \in D$.

The set of (isomorphism classes of) traces over Σ is denoted $\mathbb{T}(\Sigma)$. For a trace $t = (V, \sqsubseteq, \lambda)$ and a node $x \in V$, let $\mathrm{lc}(x) = \mathrm{lc}(\lambda(x))$. Furthermore, $\mathrm{alph}(t) = \lambda(V)$ is the *alphabet of* t.

Let $t = (V, \sqsubseteq, \lambda)$ be a trace and $\ell \in \mathbb{L}$ a location. Then for every $U \subseteq V$ the set of nodes $x \in U$ with $\mathrm{lc}(x) = \ell$ is linearly ordered by $\sqsubseteq$, hence (if not empty) this set contains a largest element that we denote $\partial_\ell(U)$.

These notions lead to a truly concurrent definition of the behavior of the wACA $\mathcal{A} = (\Sigma, (Q_m)_{m\in\mathbb{L}}, \mathrm{in}, (c_\ell)_{\ell\in\mathbb{L}}, \mathrm{out})$: Let $t = (V, \sqsubseteq, \lambda) \in \mathbb{T}(\Sigma)$ be a trace. A function $r : V \to \bigcup_{\ell\in\mathbb{L}} Q_\ell$ is a *run* provided $r(x) \in Q_{\mathrm{lc}(x)}$ for all $x \in V$ – the idea is that $r(x)$ is the state of the finite automaton in location $\mathrm{lc}(x)$ *after* executing the event x. Let $\iota \in \prod_{\ell\in\mathbb{L}} Q_\ell$ be a configuration that the wACA starts from. Then $r^-_m(\iota, x)$ shall denote the state that the finite automaton in location m is in *before* executing the event x. To define this state formally, let $\Downarrow x = \{y \in V \mid y \sqsubset x\}$ and set $r^-_m(\iota, x) = r(\partial_m(\Downarrow x))$ if $\partial_m(\Downarrow x)$ is defined, and $r^-_m(\iota, x) = \iota_m$ otherwise for $x \in V$ and $m \in \mathbb{L}$. Then the wACA, when executing the event x in the run r started at ι, reads the states $r^-_m(\iota, x)$ for $m \in D(\mathrm{lc}(x))$ and moves into the local state $r(x)$. At the end of the run, the local automaton in location m is in state $\mathrm{final}_m(\iota, r, t)$ defined

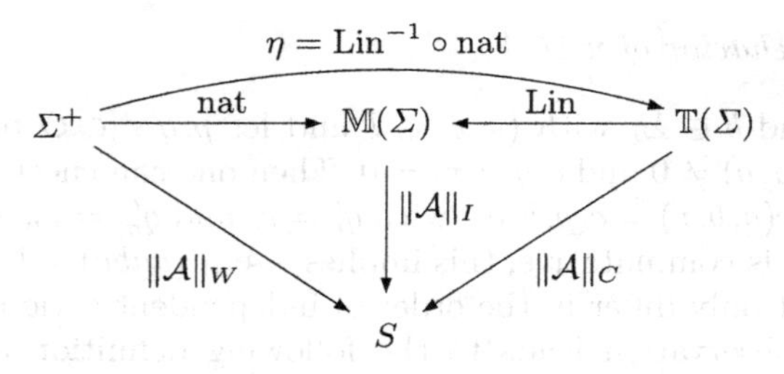

Fig. 1. Different behaviors of wACA

by $\mathrm{final}_m(\iota, r, t) = r(\partial_m(V))$ if $\partial_m(V)$ is defined, and $\mathrm{final}_m(\iota, r, t) = \iota_m$ otherwise. Finally, set $\mathrm{final}(\iota, r, t) = (\mathrm{final}_m(\iota, r, t))_{m \in \mathbb{L}}$. Then the *running weight of the run r starting in ι* is the product of the weights of all the transitions taken in this run, i.e.,

$$\mathrm{rwgt}(\iota, r, t) = \prod_{x \in V} c_{\mathrm{lc}(x)}\big((r_m^-(\iota, x))_{m \in D(\mathrm{lc}(x))}, \lambda(x), r(x)\big).$$

To obtain the total *weight of the run r starting in ι*, this quantity has to be multiplied with the weight for entering the system in configuration ι and leaving it in configuration $\mathrm{final}(\iota, r, t)$, i.e.,

$$\mathrm{wgt}(\iota, r, t) = \mathrm{in}(\iota) \cdot \mathrm{rwgt}(\iota, r, t) \cdot \mathrm{out}\big(\mathrm{final}(\iota, r, t)\big).$$

Then the *concurrent behavior* $\|\mathcal{A}\|_C : \mathbb{T}(\Sigma) \to S$ associates, with every trace $t = (V, \sqsubseteq, \lambda) \in \mathbb{T}(\Sigma)$, the sum of the weights of all possible runs:

$$\|\mathcal{A}\|_C(t) = \sum \Big(\mathrm{wgt}(\iota, r, t) \mid \iota \in \prod_{m \in \mathbb{L}} Q_m, r : V \to \bigcup_{m \in \mathbb{L}} Q_m \text{ run}\Big). \quad (1)$$

Relations Between These Behaviors

The three notions of behavior of a wACA are closely related (cf. Fig. 1).

Let $\mathrm{nat} : \Sigma^+ \to \mathbb{M}(\Sigma)$ denote the natural epimorphism defined by $\mathrm{nat}(u) = [u]$ for all $u \in \Sigma^+$. Then by the very definition of the word and the interleaving behavior of the wACA $\mathcal{A}$, we have $\|\mathcal{A}\|_I \circ \mathrm{nat} = \|\mathcal{A}\|_W$.

To relate the interleaving and the concurrent behavior of the wACA $\mathcal{A}$, we need a relation between the elements of $\mathbb{M}(\Sigma)$ and of $\mathbb{T}(\Sigma)$, i.e., between $\sim$-equivalence classes and traces. Let $t = (V, \sqsubseteq, \lambda)$ be a trace. A *linear extension* of t is a structure $(V, \preceq, \lambda)$ such that $\preceq$ is a linear order on the set V extending the partial order $\sqsubseteq$. Such a linear extension can naturally be considered as a word over Σ, hence we define $\mathrm{Lin}(t) \subseteq \Sigma^+$ as the set of all linear extensions of the trace t. Now a foundational result in trace theory asserts that Lin maps $\mathbb{T}(\Sigma)$ bijectively onto the trace semigroup $\mathbb{M}(\Sigma)$ generated by Σ (cf. [56, Theorem 1.4.8]). Then one obtains $\|\mathcal{A}\|_I \circ \mathrm{Lin} = \|\mathcal{A}\|_C$.

Thus, the diagram in Fig. 1 commutes. In particular, $\|\mathcal{A}\|_C \circ \eta = \|\mathcal{A}\|_W$ with $\eta = \mathrm{Lin}^{-1} \circ \mathrm{nat}$. The trace $\eta(w)$ can be described explicitly for $w = a_1 a_2 \dots a_n \in \Sigma^+$: Set $V = \{1, 2, \dots, n\}$ and $\lambda : V \to \Sigma : i \mapsto a_i$. Furthermore, $\sqsubseteq$ is the transitive closure of $E = \{(i,j) \in V^2 \mid i \leq j, (a_i, a_j) \in D\}$. Then $\eta(w) = (V, \sqsubseteq, \lambda)$.

The most interesting semantics of a wACA is the concurrent behavior $\|\mathcal{A}\|_C : \mathbb{M}(\Sigma) \to S$ since it considers the wACA as a truly concurrent device. Since $\mathrm{Lin} : \mathbb{T}(\Sigma) \to \mathbb{M}(\Sigma)$ is a bijection, we consider the set of traces $\mathbb{T}(\Sigma)$ as the underlying set of the trace semigroup $\mathbb{M}(\Sigma)$.

2.2 Other Formalisms: Presentations, Expressions, and Logics

This section introduces alternative formalisms for the description of functions from the trace semigroup $\mathbb{M}(\Sigma)$ into the semiring S.

Presentations

Recall that $\mathrm{nat} : \Sigma^+ \to \mathbb{M}(\Sigma)$ is the natural epimorphism. Hence, with a function $f : \mathbb{M}(\Sigma) \to S$, we can associate a function $g = f \circ \mathrm{nat} : \Sigma^+ \to S : u \mapsto f([u])$. Then g is recognizable as a formal power series [21], if it is the behavior of a weighted finite automaton without ε-moves $\mathcal{A} = (Q, R, A, P)$ where, in particular, Q is a finite set of states, $R \in S^{1 \times Q}$ is a row vector, $P \in S^{Q \times 1}$ a column vector, and $A \in (S^\Sigma)^{Q \times Q}$ a matrix. Setting $\mu(a)_{p,q} = A_{p,q}(a)$ for $a \in \Sigma$ and $p, q \in Q$, we obtain a mapping $\mu : \Sigma \to S^{Q \times Q}$ whose extension to a homomorphism $\Sigma^+ \to (S^{Q \times Q}, \cdot)$ we denote by μ, also. Then one has $g(u) = \sum_{p,q \in Q} R_p \cdot \mu(u)_{p,q} \cdot P_q$ for all $u \in \Sigma^+$, i.e., the vectors R and P together with the homomorphism μ represent the function $g = f \circ \mathrm{nat} : \Sigma^+ \to S$. This justifies to call a quadruple (Σ, R, μ, P) with $R \in S^{1 \times n}$, $\mu : \Sigma^+ \to S^{n \times n}$, and $P \in S^{n \times 1}$ satisfying $f([u]) = R \cdot \mu(u) \cdot P$ a *word series presentation* of f in which case we write $\|(\Sigma, R, \mu, P)\|$ for the function f.

The idea of presentations is to replace the free semigroup by the trace semigroup $\mathbb{M}(\Sigma)$, hereby following a more general concept of presentations of series over arbitrary monoids, cf. [63].

Definition 2.2. *An* n-dimensional presentation $(\Sigma, \lambda, \mu, \gamma)$ *comprises a distributed alphabet* Σ*, a row vector* $\lambda \in S^{1 \times n}$*, a homomorphism* $\mu : \mathbb{M}(\Sigma) \to (S^{n \times n}, \cdot)$ *from the trace semigroup into the multiplicative semigroup of* $n \times n$*-matrices over* S*, and a column vector* $\gamma \in S^{n \times 1}$*.*

The semantics *of the presentation is the function* $f = \|(\Sigma, \lambda, \mu, \gamma)\| : \mathbb{M}(\Sigma) \to S$ *defined by* $f(t) = \lambda \cdot \mu(t) \cdot \gamma = \sum_{1 \leq i,j \leq n} \lambda_i \cdot \mu(t)_{i,j} \cdot \gamma_j$ *for every* $t \in \mathbb{M}(\Sigma)$*.*

Expressions

An *expression* is a term using the constants sa with $s \in S$ and $a \in \Sigma$, the unary function symbols $^+$ and $(.)_A$ for $A \subseteq \Sigma$, and the binary function symbols $+$ and $\cdot$.

The *semantics* $[\![E]\!] : \mathbb{M}(\Sigma) \to S$ of an expression E is defined inductively as follows for any trace $t \in \mathbb{M}(\Sigma)$:

$$[\![sa]\!](t) = \begin{cases} s & \text{if } t = [a], \\ 0 & \text{otherwise,} \end{cases}$$

$$[\![E + F]\!](t) = [\![E]\!](t) + [\![F]\!](t),$$

$$[\![E \cdot F]\!](t) = \sum_{\substack{t_1, t_2 \in \mathbb{M}(\Sigma) \\ t = t_1 t_2}} [\![E]\!](t_1) \cdot [\![F]\!](t_2),$$

$$[\![E^+]\!](t) = \sum\Big(\prod_{1 \le i \le n} [\![E]\!](t_i) \mid n \in \mathbb{N},\ t_1, \ldots, t_n \in \mathbb{M}(\Sigma),\ t = t_1 t_2 \ldots t_n\Big),$$

$$[\![(E)_A]\!](t) = \begin{cases} [\![E]\!](t) & \text{if } \mathrm{alph}(t) = A, \\ 0 & \text{otherwise,} \end{cases}$$

where $A \subseteq \Sigma$ is arbitrary. Note that the sum in the definition of $[\![E^+]\!](t)$ is finite since there are only finitely many factorizations of the trace t. Furthermore, note that these definitions straightforwardly extend the rational operations on mappings $\Sigma^* \to S$ (see [18]).

Logic

Recall that the logical formalism from [16] describes the behavior of a weighted finite automaton. Here, we will use the same set of formulas but these formulas will be evaluated over a trace $(V, \sqsubseteq, \lambda)$ and not over a word. So, let var and VAR be disjoint infinite sets of *individual* and *set variables* and let Σ be some distributed alphabet. Then the syntax of our *weighted monadic second-order logic* wMSO is given by

$$\begin{aligned} \varphi ::= &\ s \mid P_a(x) \mid P_{\neg a}(x) \mid x \sqsubseteq y \mid x \not\sqsubseteq y \mid x \in X \mid x \notin X \mid \\ &\ \varphi \vee \varphi \mid \varphi \wedge \varphi \mid \exists x\, \varphi \mid \forall x\, \varphi \mid \exists X\, \varphi \mid \forall X\, \varphi \end{aligned}$$

where $s \in S$, $x, y \in \mathrm{var}$, $X \in \mathrm{VAR}$, and $a \in \Sigma$.

The *semantics* $[\![\varphi]\!]$ of a formula φ maps a trace $t = (V, \sqsubseteq, \lambda)$ together with an evaluation σ to an element of the semiring. Here, an *evaluation* maps individual variables from var to elements of V and set variables from VAR to subsets of V. Inductively, we define:

$$[\![s]\!](t, \sigma) = s$$

$$[\![P_a(x)]\!](t, \sigma) = \begin{cases} 1 & \text{if } \lambda(\sigma(x)) = a \\ 0 & \text{otherwise} \end{cases} \qquad [\![P_{\neg a}(x)]\!](t, \sigma) = \begin{cases} 0 & \text{if } \lambda(\sigma(x)) = a \\ 1 & \text{otherwise} \end{cases}$$

$$[\![x \sqsubseteq y]\!](t, \sigma) = \begin{cases} 1 & \text{if } \sigma(x) \sqsubseteq \sigma(y) \\ 0 & \text{otherwise} \end{cases} \qquad [\![x \not\sqsubseteq y]\!](t, \sigma) = \begin{cases} 0 & \text{if } \sigma(x) \sqsubseteq \sigma(y) \\ 1 & \text{otherwise} \end{cases}$$

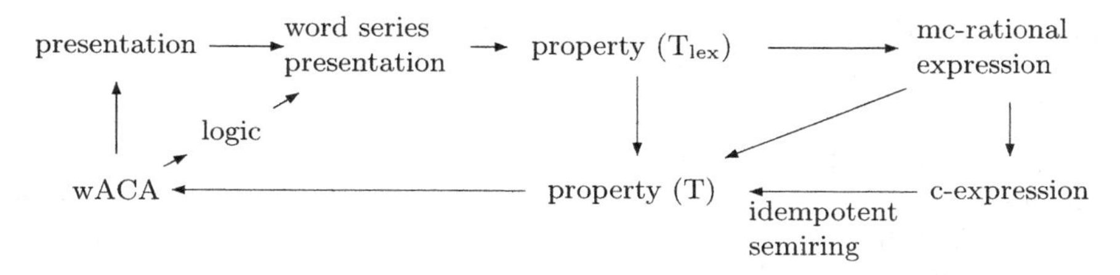

Fig. 2. Overview

$$[\![x \in X]\!](t,\sigma) = \begin{cases} 1 & \text{if } \sigma(x) \in \sigma(X) \\ 0 & \text{otherwise} \end{cases} \qquad [\![x \notin X]\!](t,\sigma) = \begin{cases} 0 & \text{if } \sigma(x) \in \sigma(X) \\ 1 & \text{otherwise} \end{cases}$$

$$[\![\varphi \vee \psi]\!](t,\sigma) = [\![\varphi]\!](t,\sigma) + [\![\psi]\!](t,\sigma) \qquad [\![\varphi \wedge \psi]\!](t,\sigma) = [\![\varphi]\!](t,\sigma) \cdot [\![\psi]\!](t,\sigma)$$

To define the semantics of quantifiers, it is convenient to write $\sigma \equiv_x \tau$ in case the valuations σ and τ differ at most in the value of the variable x. Then for a variable $x \in \mathrm{var} \cup \mathrm{VAR}$, we set

$$[\![\exists x\, \varphi]\!](t,\sigma) = \sum \left([\![\varphi]\!](t,\tau) \mid \tau \equiv_x \sigma\right),$$
$$[\![\forall x\, \varphi]\!](t,\sigma) = \prod \left([\![\varphi]\!](t,\tau) \mid \tau \equiv_x \sigma\right).$$

Since any trace has only finitely many nodes, these sums and products are finite. Since we consider only commutative semirings, the product in the semantics of the universal quantifier is independent of the order of the factors and, therefore, well defined. As usual, a formula where every variable is in the scope of a quantifier is called a *sentence*. If φ is a sentence, then $[\![\varphi]\!](t,\sigma) = [\![\varphi]\!](t,\tau)$ for all traces t and all evaluations σ and τ. Hence, we will understand $[\![\varphi]\!]$ for a sentence φ as a function that maps traces to elements of the semiring.

2.3 Relating the Formalisms

In this section, we show that all the formalisms introduced coincide in expressive power. As one can see from the plan of our proof in Fig. 2, the formalisms are connected by not-yet-defined properties (T) and (T_{lex}). Thus, before we can embark on the program of translating the formalisms into each other, we have to introduce these technical properties.

Property (T)

Let Σ and Γ be distributed alphabets. A function $\pi : \Gamma \to \Sigma$ is *location preserving* if $\mathrm{lc}(\pi(A)) = \mathrm{lc}(A)$ for all $A \in \Gamma$. We denote the natural extension of π to a semigroup homomorphism from Γ^+ to Σ^+ by π, also.

A function $f : \mathbb{M}(\Sigma) \to S$ *satisfies property (T)* if there exist a distributed alphabet Γ, a location preserving function $\pi : \Gamma \to \Sigma$, a homomorphism $c : \Gamma^+ \to (S, \cdot)$, and a language $L \subseteq \Gamma^+$ such that:

(T1) $[L] = \{V \in \Gamma^+ \mid \exists U \in L : U \sim V\}$ is regular.
(T2) $f([u]) = \sum(c(U) \mid U \in L \cap \pi^{-1}([u]))$ for all $u \in \Sigma^+$.
(T3) $\sum(c(U) \mid U \in [L] \cap \pi^{-1}(u)) = \sum(c(V) \mid V \in L \cap \pi^{-1}([u]))$ for all $u \in \Sigma^+$.

The strengthening ($\mathrm{T_{lex}}$) of property (T) needs linearly ordered distributed alphabets. So, we will assume for every distributed alphabet Σ a linear order $\preceq_\Sigma$ on Σ. A function $\pi : \Gamma \to \Sigma$ is *order preserving* if $A \preceq_\Gamma B$ implies $\pi(A) \preceq_\Sigma \pi(B)$ for all $A, B \in \Gamma$. For $u \in \Sigma^+$, the *lexicographic normal form* $\mathrm{lnf}(u)$ is the lexicographically minimal word from the equivalence class $[u]$. We let $\mathrm{LNF}(\Sigma)$ denote the set of all words $\mathrm{lnf}(u)$ with $u \in \Sigma^+$.

Now a function $f : \mathbb{M}(\Sigma) \to S$ *satisfies property* (T_{lex}) if it satisfies property (T) where, in addition, we assume π to be order-preserving and the language L to consist of words in lexicographic normal form, only (i.e., $L \subseteq \mathrm{LNF}(\Gamma)$).

2.3.1 wACA, (Word Series) Presentations, and Property (T)

From wACA to Presentations

Let $\mathcal{A}$ be a wACA and let Q be the set of configurations of $\mathcal{A}$. Recall the global cost function $c : Q \times \Sigma \times Q \to S$ that we used to define the word behavior $\|\mathcal{A}\|_W$. For $a \in \Sigma$ and $p, q \in Q$, set $\mu([a])_{p,q} = c(p, a, q)$. Then $\mu([a]) \cdot \mu([b]) = \mu([b]) \cdot \mu([a])$ for $(a, b) \in I$. Since $\mathbb{M}(\Sigma) = \Sigma^+/\!\sim$ and $\sim$ is the least congruence identifying ab and ba for $(a, b) \in I$, the mapping μ can uniquely be extended to a homomorphism $\mu : \mathbb{M}(\Sigma) \to (S^{Q\times Q}, \cdot)$. Understanding the mappings in, out $: Q \to S$ as row and column vector, respectively, $(\Sigma, \mathrm{in}, \mu, \mathrm{out})$ is a $|Q|$-dimensional presentation. The semantics $\|(\Sigma, \mathrm{in}, \mu, \mathrm{out})\|$ of this presentation coincides with the concurrent behavior $\|\mathcal{A}\|_C$ of the wACA $\mathcal{A}$ we started with. Hence, we get the following theorem.

Theorem 2.3 ([43, Proposition 4.5]). *For every weighted asynchronous cellular automaton $\mathcal{A}$, there exists a presentation $(\Sigma, \lambda, \mu, \gamma)$ such that $\|\mathcal{A}\|_C = \|(\Sigma, \lambda, \mu, \gamma)\|$.*

From Presentations to Word Series Presentations

Now let $(\Sigma, \lambda, \mu, \gamma)$ be an n-dimensional presentation of $f : \mathbb{M}(\Sigma) \to S$. Then $\mu \circ \mathrm{nat} : \Sigma^+ \to S$ is a homomorphism and $(\Sigma, \lambda, \mu \circ \mathrm{nat}, \gamma)$ is a word series presentation of f since $f([u]) = \lambda\mu([u])\gamma = \lambda(\mu \circ \mathrm{nat})(u)\gamma$. Hence, we have the following theorem.

Theorem 2.4. *Let $(\Sigma, \lambda, \mu, \gamma)$ be a presentation. Then $(\Sigma, \lambda, \mu \circ \mathrm{nat}, \gamma)$ is a word series presentation of $\|(\Sigma, \lambda, \mu, \gamma)\|$.*

From Word Series Presentations to Property (T_{lex})

Now let $(\Sigma, \lambda, \mu, \gamma)$ be an n-dimensional word series presentation and set $Q = \{1, \dots, n\}$. Using a normalization procedure (cf. [21]), we can assume $\lambda_p, \gamma_p \in \{0, 1\}$ for all $p \in Q$. Then we define a distributed alphabet Γ setting $\Gamma_\ell = Q \times \Sigma_\ell \times Q$ for $\ell \in \mathbb{L}$. Now the mapping $\pi : \Gamma \to \Sigma : (p, a, q) \mapsto a$ is location preserving. Define linear orders $\preceq_\Sigma$ and $\preceq_\Gamma$ on the distributed alphabets Σ and Γ such that π is also order preserving. Let $c : \Gamma^+ \to (S, \cdot)$ be the homomorphism defined by $c(p, a, q) = \mu(a)_{p,q}$.

A *path from* p_1 *to* p_{n+1} is a word $U \in \Gamma^+$ of the form $U = (p_i, a_i, p_{i+1})_{1 \le i \le n}$ with $\mu(a_i)_{p_i, p_{i+1}} \neq 0$ for all $1 \le i \le n$. Let the language $L \subseteq \Gamma^+$ consist of all paths $U \in \mathrm{LNF}(\Gamma)$ from some $p \in Q$ with $\lambda_p = 1$ to some $q \in Q$ with $\gamma_q = 1$.

The language L is regular since it is the intersection of the regular language of all paths from some initial to some final state with the regular language $\mathrm{LNF}(\Gamma)$. By [60], $[L]$ is regular since it is the $\sim$-closure of a regular language of lexicographic normal forms; hence, (T1) holds. Since π is location and order preserving, $U \in \mathrm{LNF}(\Gamma)$ if and only if $\pi(U) \in \mathrm{LNF}(\Sigma)$ for all $U \in \Gamma^+$. Hence, for $u \in \Sigma^+$, we have

$$\begin{aligned}
\|(\Sigma, \lambda, \mu, \gamma)\|([u]) &= \|(\Sigma, \lambda, \mu, \gamma)\|\big([\mathrm{lnf}(u)]\big) \\
&= \sum_{\lambda_p = 1} \sum_{\gamma_q = 1} \big(\mu(\mathrm{lnf}(u))\big)_{p,q} \\
&= \sum_{\lambda_p = 1} \sum_{\gamma_q = 1} \sum \big(c(U) \mid U \text{ path from } p \text{ to } q,\ \pi(U) = \mathrm{lnf}(u)\big) \\
&= \sum \big(c(U) \mid U \in L,\ \pi(U) = \mathrm{lnf}(u)\big) \\
&= \sum \big(c(U) \mid U \in L \cap \pi^{-1}([u])\big)
\end{aligned}$$

proving (T2). Finally, to prove (T3), let $u \in \Sigma^+$ be arbitrary. Then the mapping $h = \mathrm{lnf}\restriction_{[L] \cap \pi^{-1}(u)}$ maps $[L] \cap \pi^{-1}(u)$ to $L \cap \pi^{-1}([u]) \subseteq \mathrm{LNF}(\Gamma)$ such that $c(U) = c(h(U))$ since $h(U) = \mathrm{lnf}(U)$ is some reordering of the word U.

To verify injectivity of h, let $U, V \in [L] \cap \pi^{-1}(u)$ with $h(U) = h(V)$. Then $U \sim h(U) = h(V) \sim V$ and $\pi(U) = \pi(V)$. Since the first letters of $\pi(U)$ and $\pi(V)$ are equal, the first letters of U and V belong to the same location; hence, they are dependent. Now their equality follows from $U \sim V$. The proof then proceeds by induction.

To verify surjectivity of h, let $V \in L \cap \pi^{-1}([u])$. Then $v = \pi(V) \sim u$ is obtained from u by successively permuting adjacent and independent letters. Since π is location preserving, these swaps can be simulated in V resulting in a word $U \in [V] \cap \pi^{-1}(u) \subseteq [L] \cap \pi^{-1}(u)$. Hence, h is a bijection which proves (T3) and, therefore, we have the following theorem.

Theorem 2.5. *Let $(\Sigma, \lambda, \mu, \gamma)$ be a word series presentation. Then the function $\|(\Sigma, \lambda, \mu, \gamma)\|$ satisfies property T_{lex}.*

From Property (T) to wACA

Suppose the function $f : \mathbb{M}(\Sigma) \to S$ satisfies property (T), i.e., there are Γ, π, c, and L as required that satisfy (T1)–(T3). By (T1) and [67], the language $[L]$ can be accepted by some deterministic asynchronous cellular automaton. Translated into our context, there is a wACA $\mathcal{A} = (\Gamma, (Q_m)_{m\in\mathbb{L}}, \text{in}, (c_\ell)_{\ell\in\mathbb{L}}, \text{out})$ whose word behavior $\|\mathcal{A}\|_W$ is the characteristic function $\chi_{[L]} : \Gamma^+ \to S$ of $[L]$. We now define another wACA $\mathcal{A}' = (\Sigma, (Q_m)_{m\in\mathbb{L}}, \text{in}, (c'_\ell)_{\ell\in\mathbb{L}}, \text{out})$ setting

$$c'_\ell\big((p_m)_{m\in D(\ell)}, a, q_\ell\big) = \sum \big(c(A) \cdot c_\ell\big((p_m)_{m\in D(\ell)}, A, q_\ell\big) \bigm| a \in \pi^{-1}(A)\big)$$

for $a \in \Sigma_\ell$, $p_m \in Q_m$, and $q_\ell \in Q_\ell$. For $u \in \Sigma^+$ we then have

$$\begin{aligned}\|\mathcal{A}'\|_C([u]) = \|\mathcal{A}'\|_W(u) &= \sum \big(c(U) \cdot \|\mathcal{A}\|_W(U) \bigm| U \in \pi^{-1}(u)\big)\\ &= \sum \big(c(U) \bigm| U \in [L] \cap \pi^{-1}(u)\big)\\ &= \sum \big(c(V) \bigm| V \in L \cap \pi^{-1}([u])\big) \quad \text{by (T3)}\\ &= f([u]) \quad \text{by (T2)}.\end{aligned}$$

This proves

Theorem 2.6. *If $f : \mathbb{M}(\Sigma) \to S$ satisfies property (T), then it is the concurrent behavior of some wACA.*

2.3.2 wACA, Logic, and Word Series Presentations

A Fragment of Weighted Monadic Second-Order Logic

Already in the setting of words, the full weighted monadic second-order logic exceeds the class of behaviors of weighted automata. In [14, 16], Droste and Gastin define two different fragments of wMSO that are both expressively equivalent to weighted finite automata on words. Here, we adapt the approach from [14] to our setting. To this end, a formula φ is a *definable step formula* if there exist $n \in \mathbb{N}$, $s_i \in S$, and wMSO-formulas β_i over the Boolean semiring $\mathbb{B}$ (for $1 \le i \le n$) such that, for every trace $t \in \mathbb{M}(\Sigma)$ and every evaluation σ, we have $[\![\varphi]\!](t,\sigma) = \sum_{1\le i\le n} s_i [\![\beta_i]\!](t,\sigma)$ (where we identify the elements 0 and 1 of the Boolean semiring $\mathbb{B}$ with the corresponding elements of the semiring S).[2]

Now $\varphi \in$ wMSO is *restricted* if φ does not contain any second-order universal quantification, and for every subformula $\forall x\, \psi$ of φ, the formula ψ is a definable step formula. The class of all restricted formulas is denoted by wRMSO.

[2] The formally correct statement would be $[\![\varphi]\!](t,\sigma) = \sum_{1\le i\le n} s_i f_i(t,\sigma)$ where f_i is the characteristic function of the support of $[\![\beta_i]\!]$ (whose values belong to the Boolean semiring). But f_i takes values in the semiring S.

The class wREMSO comprises the existential formulas of wRMSO, i.e., those of the form $\exists X_1 \ldots \exists X_m \psi$ where $\psi \in$ wRMSO does not contain any second-order quantification.

From wACA to Logic

It is generally known that the behavior of finite automata can be described by sentences of existential monadic second-order logic. In the weighted setting, we have to be careful in adapting this approach. Whereas on the one hand the weights of the transitions have to be included, on the other hand formulas describing only qualitative properties of the automaton have to preserve a "Boolean" 0–1-semantics. Therefore, we introduce *unambiguous formulas.* All atomic formulas apart from s for $s \in S$ are unambiguous. Let φ and ψ be unambiguous formulas. Then so are $\varphi \wedge \psi$, $\forall x\, \varphi$, and $\forall X\, \varphi$. The formula $\varphi \vee \psi$ is unambiguous if, for all traces t and all valuations σ, at most one of $[\![\varphi]\!](t, \sigma)$ and $[\![\psi]\!](t, \sigma)$ is nonzero. Finally, the formula $\exists z\, \varphi$ with $z \in \text{var} \cup \text{VAR}$ is unambiguous if, for every trace t and every valuation σ, there is at most one valuation τ with $\tau \equiv_z \sigma$ and $[\![\varphi]\!](t, \tau) \neq 0$. Then $[\![\varphi]\!](t, \tau) \in \{0, 1\}$ for any unambiguous formula φ, any trace t, and any valuation τ.

In the setting of traces, we can find for every first-order formula φ over the Boolean semiring an unambiguous one φ^+ over the semiring S such that $[\![\varphi]\!]$ and $[\![\varphi^+]\!]$ have the same support, i.e., $[\![\varphi]\!](t, \sigma) = 1$ in the Boolean semiring if and only if $[\![\varphi^+]\!](t, \sigma) = 1$ in S for all $t \in \mathbb{M}(\Sigma)$ and all evaluations σ. This is an easy exercise for quantifier-free formulas. With regard to $\exists x\, \varphi$, we can single out a unique node (if there is at least one) for x such that φ is satisfied. This can be done as follows: fix a linear order $\preccurlyeq$ on Σ and define a strict linear order on the nodes of any trace t by the unambiguous formula

$$\omega(x, y) = \bigvee_{(a,b) \in \prec} (P_a(x) \wedge P_b(y)) \vee \bigvee_{a \in \Sigma} (P_a(x) \wedge P_a(y) \wedge x \sqsubset y)$$

and amend φ appropriately; cf. [59] for details.

The transitions of a wACA $\mathcal{A} = (\Sigma, (Q_m)_{m \in \mathbb{L}}, \text{in}, (c_\ell)_{\ell \in \mathbb{L}}, \text{out})$ do not use the partial order $\sqsubseteq$ but the node $\partial_m(\Downarrow x)$ for a node x and $m \in \mathbb{L}$. Take $P_m(x)$ as a shorthand for $\bigvee_{a \in \Sigma_m} P_a(x)$. Then $y = \partial_m(\Downarrow x)$ if and only if x and y satisfy the formula

$$\beta(x, y) = P_m(y) \wedge (y \sqsubset x) \wedge \forall z \big[(y \sqsubseteq z \sqsubset x \wedge P_m(z)) \implies y = z \big].$$

Note that β is in wRMSO. Now we follow the usual way (cf., e.g., [65]) of transforming an automaton into a formula where a valuation of second-order variables reflects an assignment of nodes to states (cf. [9] for more details in a more general setting). Therefore, we introduce variables X_q for every $q \in Q$. Then we build formulas with free variables X_q for the following statements:

- The sets $\{X_q \mid q \in Q\}$ form a partition of the nodes of the trace.

- The node x and its local neighborhood is matched by the transition $((p_m)_{m\in D(\ell)}, a, q_\ell)$ (here the formula β from above is used).
- The node x is the last node of a location m.

All these formulas describe qualitative statements and can be made unambiguous according to what was shown above. Finally, we add the appropriate weights by conjunction to the particular formulas. Altogether, we obtain the following theorem.

Theorem 2.7. *For every wACA $\mathcal{A}$, there is a sentence $\varphi \in$ wREMSO such that $[\![\varphi]\!] = \|\mathcal{A}\|_C$.*

From Logic to Word Series Presentations

Consider a sentence $\varphi \in$ wRMSO evaluated over traces from $\mathbb{M}(\Sigma)$. Our strategy is to transform φ into $\widetilde{\varphi} \in$ wRMSO evaluated over words. The logic wRMSO over words is defined in the same way as for traces, but this time with a linear order $\leq$ instead of the partial order $\sqsubseteq$ as used for traces (cf. [16]).

We wish to construct $\widetilde{\varphi}$ with $[\![\widetilde{\varphi}]\!] = [\![\varphi]\!] \circ \text{nat}$ where $\text{nat} : \Sigma^+ \to \mathbb{M}(\Sigma)$ is the natural epimorphism. For this, we replace every atomic formula $x \sqsubseteq y$ in φ by an unambiguous version of the formula

$$\exists x_0, \ldots, x_{|\Sigma|} : x = x_0 \wedge \bigwedge_{i=0,\ldots,|\Sigma|-1} (x_i \leq x_{i+1} \wedge (x_i, x_{i+1}) \in D) \wedge x_{|\Sigma|} = y$$

where $(x_i, x_{i+1}) \in D$ is a shorthand for $\bigvee_{(a,b)\in D}(P_a(x) \wedge P_b(y))$. Then $[\![\widetilde{\varphi}]\!](w) = [\![\varphi]\!](\text{nat}(w)) = [\![\varphi]\!] \circ \text{nat}(w)$ for all $w \in \Sigma^+$. It turns out that $\widetilde{\varphi}$ is in wRMSO. In fact, we can also show the other direction (see [59] for details) and obtain the following proposition.

Proposition 2.8. *For $f : \mathbb{M}(\Sigma) \to S$ the following are equivalent:*

1. *f is definable in* wRMSO *over traces.*
2. *$f \circ \text{nat} : \Sigma^+ \to S$ is definable in* wRMSO *over words.*

Since $[\![\varphi]\!] \circ \text{nat}$ is definable in wRMSO, it is the behavior of a weighted finite automaton $\mathcal{A}$ with state set Q (see [16]). Then by what we saw on page 411, $\mathcal{A}$ can be transformed into a word series presentation of $[\![\varphi]\!]$. This proves the following result.

Theorem 2.9. *Let $\varphi \in$ wRMSO be a sentence. Then $[\![\varphi]\!]$ admits a word series presentation.*

2.3.3 Expressions and Property (T)

From Property (T_{lex}) to Expressions

Suppose the function $f : \mathbb{M}(\Sigma) \to S$ satisfies property ($\mathrm{T_{lex}}$), i.e., there are Γ, π, c, and $L \subseteq \mathrm{LNF}(\Gamma)$ as required that satisfy (T1)–(T3). Since the language $[L]$ is regular, so is $L = \{\mathrm{lnf}(U) \mid U \in [L]\} = [L] \cap \mathrm{LNF}(\Gamma)$. Hence, there is a rational expression E whose language $\mathcal{L}(E)$ equals L. As for every rational expression, we can assume (cf. [19]) that $\mathcal{L}(F)$ is *monoalphabetic* for every subexpression F^+ of E (i.e., $\mathrm{alph}(U) = \mathrm{alph}(V)$ for all $U, V \in \mathcal{L}(F)$). Since $L \subseteq \mathrm{LNF}(\Gamma)$, the language $\mathcal{L}(F)$ is *connected* for every subexpression F^+ of E (i.e., $\mathrm{alph}(U)$ is a connected subgraph of (Γ, D) for every $U \in \mathcal{L}(F)$) [60].

Now replace every appearance of a letter $A \in \Gamma$ in E with $c(A)\pi(A)$. This results in an expression G. It can be checked that the function f is the semantics $[\![G]\!]$ of the expression G (see [42, proof of Lemma 4.1]). Since E is a rational expression, the expression G does not use the construct $(.)_A$. Furthermore, by what we saw above, if H^+ is a subexpression of G, and $[\![H]\!](t) \neq 0 \neq [\![H]\!](t')$, then $\mathrm{alph}(t) = \mathrm{alph}(t')$ is connected, i.e., the expression G is *mc-rational* as defined in [13].

Theorem 2.10 ([42]). *If $f : \mathbb{M}(\Sigma) \to S$ satisfies property T_{lex}, then it is the semantics of some mc-rational expression G.*

From Expressions to Property (T)

We now aim at the converse, i.e., we want to show for each suitable expression E that $[\![E]\!]$ satisfies property (T). The starting idea is extremely simple: To construct the language L, understand the constant sa in an expression E as the letter (s, a) from some distributed alphabet Γ with $\Gamma_\ell \subseteq S \times \Sigma_\ell$. In this construction of L, we let $(F)_A$ denote the intersection of the language denoted by F with the set of words U with $\pi(\mathrm{alph}(U)) = A$ (where $\pi(s, a) = a$). The homomorphism $c : \mathbb{M}(\Gamma) \to (S, \cdot)$ is then given by $c(s, a) = s$.

To verify (T3), let $u \in \Sigma^+$ and $V \in L \cap \pi^{-1}([u])$. Then $\pi(V) \sim u$ implies the existence of some word U with $V \sim U$ and $\pi(U) = u$ and, therefore, $U \in [L] \cap \pi^{-1}(u)$. Hence, there is a function $f_u : L \cap \pi^{-1}([u]) \to [L] \cap \pi^{-1}(u)$ with $f_u(V) \sim V$ and, therefore, $c(f_u(V)) = c(V)$. This function is even surjective: if $U \in [L] \cap \pi^{-1}(u)$, then there exists at least one word $V \in L$ with $U \sim V$. Since π is location-preserving, this implies $\pi(V) \sim \pi(U) = u$. Hence, we have $f_u(V) \sim V \sim U$ and $\pi(f_u(V)) = u = \pi(U)$ which implies $U = f_u(V)$. Since f_u is in general not injective, (T3) holds provided the semiring S is idempotent. In this case, we can also verify (T2) inductively along the construction of the expression E. Finally, consider (T1), i.e., the regularity of the closure of the language L. From [60], we know that this is the case at least if, for every subexpression F^+ of E, the language denoted by F is connected (without this assumption, the regularity of $[L]$ is not guaranteed and even undecidable [62]). We enforce this condition by the following syntactic restriction on E: whenever

F^+ is a subexpression of E, then $F = \sum(G)_A$ for some expression G where the sum extends over all sets $A \subseteq \Sigma$ such that (A, D) is connected. Calling these expressions *c-rational* (since iteration is applied to connected functions, only), we therefore obtain the following theorem.

Theorem 2.11 ([42]). *If E is a c-rational expression and the semiring S is idempotent, then the semantics $[\![E]\!]$ of E satisfies property (T).*

The expression $(1a + 1b)^+$ is equivalent to a c-rational expression. Now assume $(a, b) \in I$ and consider the natural semiring $(\mathbb{N}, +, \cdot, 0, 1)$. Then the function $[\![(1a+1b)^+]\!]$ does not have a presentation (by [13, Example 39]) and, therefore, violates property (T) by Theorems 2.6 and 2.3. Hence, idempotency is necessary in the above theorem.

Recall that S was assumed idempotent since the function f_u is surjective but not necessarily injective. Here is an example: consider the expression $E = (1a + 1b) \cdot (1a + 1b)$ and assume $(a, b) \in I$. Then the above proof yields $L = \{A^2, AB, BA, B^2\} = [L]$ where $A = (1, a)$ and $B = (1, b)$. With $u = ab$, we then have $[L] \cap \pi^{-1}(u) = \{AB\}$ and $L \cap \pi^{-1}([u]) = \{AB, BA\}$, i.e., there cannot be a bijection between these two sets. The problem arises since the two occurrences of a in E cannot be distinguished. If, in the construction of the language L, we replace the two occurrences of a by distinct letters A_1 and A_2, then this problem vanishes. But there is a related one: let $E = (1a+1b)^+$. Then as above, there cannot be a bijection between $L \cap \pi^{-1}([ab])$ and $[L] \cap \pi^{-1}(ab)$ although every letter appears only once in E. Here, we replace the expression E with the equivalent one

$$(1a + 1b) + \big((1a + 1b)(1a + 1b)\big)^+ + \big((1a + 1b)(1a + 1b)\big)^+(1a + 1b)$$

and then replace the occurrences of $1a$ and $1b$ by mutually distinct letters A_i and B_i for $1 \le i \le 6$. But then (T1) will not hold since the middle term gives rise to the language $L = (\{A_2, B_2\} \cdot \{A_3, B_3\})^+$ whose closure $[L]$ is not regular (since the word A_2B_3 is not connected). This latter problem cannot arise if E is mc-rational. Restricting to these mc-rational expressions E, we therefore obtain property (T) as follows: first, replace every subexpression F^+ with $F + (F \cdot F)^+ + (F \cdot F)^+ \cdot F$. In a second step, replace the constants sa in the resulting expression with mutually distinct letters (s, a, i) for some distributed alphabet Γ with $\Gamma_\ell \subseteq S \times \Sigma_\ell \times \mathbb{N}$. Setting $\pi(s, a, i) = a$ and $c(s, a, i) = s$ finishes the construction.

Theorem 2.12 ([42]). *If E is an mc-rational expression, then the semantics $[\![E]\!]$ of E satisfies property (T).*

2.3.4 The Characterization Theorem

The last result completes the picture of Fig. 2 showing the equivalence of wACA, presentations, logics, and expressions.

Theorem 2.13 ([13, 59, 43]). *Let Σ be a distributed alphabet, S a commutative semiring, and $f : \mathbb{M}(\Sigma) \to S$ a function. Then the following are equivalent:*

1. *f is the concurrent behavior of some wACA.*
2. *f admits a presentation.*
3. *f is the semantics of some sentence from wRMSO.*
4. *f is the semantics of some mc-expression.*

If the semiring is idempotent, any of the above statements is equivalent to

5. *f is the semantics of some c-expression.*

It can be checked that all our proofs are effective. This makes it possible to intertranslate the different formalisms for describing functions from $\mathbb{M}(\Sigma)$ to S.

2.4 History and Overview

Every arrow in Fig. 2 from formalism A to formalism B denotes that we proved that formalism A can equivalently be transformed into formalism B; thus, all these formalisms are equivalent. We call mappings $\mathbb{M}(\Sigma) \to S$ that can be presented in any of these formalisms *recognizable trace series*. An early study of these series can be found in the work of Fliess [25, 26] who defined recognizability by means of presentations. Most of his results turned out to hold for arbitrary monoids (cf. [64, 63]) and are therefore beyond the scope of this article.

Whereas Fliess' work was motivated from combinatorics, traces attracted growing interest as a model for the behavior of concurrent systems, especially Petri nets. A comprehensive theory was developed; see [12]. In this light, the study of recognizable series for partially commuting variables was revived by Droste and Gastin [13] who showed that a trace series admits a presentation if and only if it is mc-rational (if and only if it is c-rational in case the semiring is idempotent), i.e., the equivalence of 2 and 4 (resp. 5) in Theorem 2.13.

Consider again the equivalent formalisms stated by the arrows in Fig. 2. Using mc-rational expressions, it follows that the class of recognizable series is closed under the rational operations (iteration is restricted to monoalphabetic and connected functions)—this is directly shown in [13] in order to prove that every mc-rational mapping has a presentation (for very special semirings, closure under the Cauchy-product was known from [26]). Yet another proof of this fact was given in [7] by Berstel and Reutenauer who used algebraic means. A construction of equivalent expressions from a presentation was firstly given in [13], the proof via property ($\mathrm{T}_{\mathrm{lex}}$) can be seen as a formalization of this construction. The proof from [13] was extended by Mathissen [49] allowing deflation parameters in the style of [20]. Every recognizable series $f : \mathbb{M}(\Sigma) \to S$ can be defined in wRMSO and wREMSO; Meinecke [59] derived this result as a corollary to the corresponding result on formal power series

$f : \Sigma^+ \to S$ from [14]. The proof we present here is due to [9] where it can be found for message sequence charts and certain dags. Our proof that every function of the form $[\![\varphi]\!]$ satisfies property (T_{lex}) formalizes the construction of a presentation from [59]. Also, the construction of a wACA from a presentation via property (T) formalizes the original proof by Kuske [43]; a consequence of that proof is that we can require presentations to have certain natural properties ("I-consistency") without restricting the expressive power.

This article did not consider *aperiodic* trace series, i.e., functions $\mathbb{M}(\Sigma) \to S$ that admit a presentation $(\Sigma, \lambda, \mu, \gamma)$ and an index $n \in \mathbb{N}$ such that $\mu(t^n) = \mu(t^{n+1})$ for all traces $t \in \mathbb{M}(\Sigma)$. For idempotent semirings with Burnside matrix monoids, Droste and Gastin [15] show the coincidence of aperiodic trace series and trace series defined by *starfree* expressions. For commutative, weakly bi-aperiodic semirings, Meinecke [59] lifts a result from [14] and shows that first-order formulas define exactly the class of aperiodic trace series. A characterization in terms of wACA is not known nor does it seem likely that there is a property-(T)-like characterization since these trace series are not closed under projections.

Concerning logics, some decidability issues arise [59, 14]. If the semiring S is a *computable field*, then it is decidable whether a wMSO-formula φ is in wRMSO. Moreover, satisfiability and equivalence can be decided. For a *locally finite* semiring S, the semantics of every wMSO formula φ is recognizable. Then universality and equivalence turn out to be decidable.

Traces appear naturally as *heaps of pieces*, i.e., monoids generated by solid blocks in rectangular form as known from the Tetris game. The concatenation is determined by piling up those pieces. This yields a nice graphical interpretation for traces. In [29], Gaubert and Mairesse described the timed behavior of safe timed Petri nets by functions from the associated heap of pieces into the $(\mathsf{max}, +)$-semiring. It can be computed by a $(\mathsf{max}, +)$-automaton. This correspondence is exploited for performance evaluation [27, 29] and asymptotic analysis [28] using spectral theory techniques.

The model checking of probabilistic systems also benefits from distributed alphabets: [4, Definition 5.6] turns a Markov decision process into a probabilistic presentation that allows to apply partial-order reductions to speed up the model checking of LTL-properties.

Probabilistic asynchronous automata running on traces were introduced by Jesi, Pighizzini, and Sabadini [37]. They showed that probabilistic asynchronous automata and probabilistic presentations (which they call probabilistic M-automata) define the same class of behaviors. However, they have to restrict themselves to distributed alphabets with an acyclic dependence relation. It is open whether the techniques presented here can be used to derive the general result.

Quantitative aspects are important for communication scenarios and distributed systems. However, the trace model has natural limitations. To describe the behavior of message passing systems the model of *message sequence charts*, MSCs for short, is more appropriate. Bollig and Meinecke [9] proved an equivalence result between weighted asynchronous cellular automata and some weighted MSO logic in this setting. A robust theory of weighted MSCs still has to be developed. We would like to exploit techniques using trace theory to obtain new results also for weighted systems over MSCs. Such techniques were initiated by Arnold [3] and elaborated by several authors in different settings [17, 45, 40, 30].

3 Series-Parallel Posets

In the previous section, the behavior of certain concurrent systems was described by means of traces. However, the weights of transitions appearing concurrently were handled in the same way as those of transitions executed sequentially. In either case, the weights were multiplied within the given semiring. This does not seem reasonable when considering the execution time for example. Duration times should be added for events executed sequentially whereas for parallel execution one would suggest taking the maximum of the respective execution times. Here, we give an automaton model for such concurrent systems and provide a characterization of its behavior.

3.1 Series-Parallel Posets and Bisemirings

Series-Parallel Posets

We first define the class of *series-parallel posets* or *sp-posets* for short, which will serve for the description of executions of the concurrent systems considered [35, 61]. Let Σ be a finite alphabet. Let $t_1 = (V_1, \leq_1, l_1)$ and $t_2 = (V_2, \leq_2, l_2)$ be two Σ-labeled posets, i.e., $l_i : V_i \to \Sigma$ for $i = 1, 2$. Assume $V_1 \cap V_2 = \emptyset$. Then define the sequential product $t_1 \cdot t_2 = (V, \leq, l)$ by $V = V_1 \cup V_2$, $\leq \, = \, \leq_1 \cup (V_1 \times V_2) \cup \leq_2$, and $l = l_1 \cup l_2$. Similarly, the parallel product $t_1 \parallel t_2 = (V, \leq, l)$ is defined by $V = V_1 \cup V_2$, $\leq \, = \, \leq_1 \cup \leq_2$, and $l = l_1 \cup l_2$.

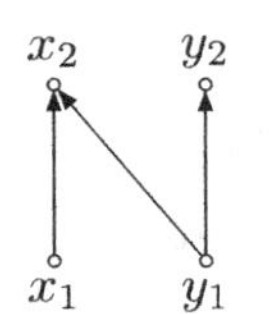

The class of (finite and nonempty) *sp-posets* $\mathrm{SP}(\Sigma)$ is the smallest class of Σ-labeled posets containing the singleton posets and closed under sequential and parallel product. The sp-posets can be characterized as the so-called N-free posets, i.e., those finite posets in which the poset $(N, \leq_N)$ (on the left) cannot be embedded; cf. [35]. The absence of a subposet isomorphic to $(N, \leq_N)$ prohibits a modeling of any kind of message passing by sp-posets. The set of all sp-posets $(\mathrm{SP}(\Sigma), \cdot, \parallel)$ together with sequential and parallel product is the free sp-algebra. Note that both operations are associative and, moreover, the parallel product is commutative. Any $t \in \mathrm{SP}(\Sigma)$ allows for a unique maximal sequential decomposition $t = t_1 \cdot \dots \cdot t_m$ where every $t_i \neq t't''$ for any $t', t'' \in \mathrm{SP}(\Sigma)$. Similarly, t has a unique (up to commutativity) maximal parallel decomposition $t = t_1 \parallel \dots \parallel t_n$.

Bisemirings

Usually, weights for finite automata stem from semirings. In our setting, we are in need of an additional second multiplication to handle parallel composition.

Definition 3.1. *A* bisemiring $(S, +, \cdot, \diamond, 0, 1)$ *is a set* S *together with three binary operations called addition* $+$, *sequential multiplication* $\cdot$, *and parallel multiplication* $\diamond$, *and two constants* 0 *and* 1 *such that:*

- $(S, +, 0)$ *is a commutative monoid,* $(S, \cdot, 1)$ *is a monoid, and* $(S, \diamond)$ *is a commutative semigroup.*
- *Both* $\cdot$ *and* $\diamond$ *distribute over* $+$.
- 0 *is absorbing for* $\cdot$ *and* $\diamond$, *i.e.,* $s \cdot 0 = 0 \cdot s = s \diamond 0 = 0$ *for all* $s \in S$.

Broadly speaking, a bisemiring S consists of two semirings with the same basic set and the same additive structure where parallel multiplication has to be commutative but may miss a unit. Next, we collect some examples of bisemirings.

Example 3.2. The structure $S = (\mathbb{N} \cup \{+\infty\}, \mathsf{min}, +, \mathsf{max}, +\infty, 0)$ is a bisemiring. The number 0 is the unit for the sequential multiplication $+$ and $+\infty$ is the absorbing zero. Let $a \in \Sigma$ be some action with execution time $\text{time}(a)$. If a cannot be performed, put $\text{time}(a) = +\infty$. Now we define $\text{time} : \text{SP}(\Sigma) \to S$ inductively by $\text{time}(t_1 \cdot t_2) = \text{time}(t_1) + \text{time}(t_2)$ and $\text{time}(t_1 \parallel t_2) = \mathsf{max}(\text{time}(t_1), \text{time}(t_2))$. Then time is an sp-algebra homomorphism and gives the execution time of the sp-poset t.

Example 3.3. $S = (\mathbb{N} \cup \{-\infty, +\infty\}, \mathsf{max}, \mathsf{min}, +, -\infty, +\infty)$ where $(-\infty) + (+\infty) = -\infty$ is a bisemiring. By help of this bisemiring, we can model a series-parallel channel system. The weights give the channel capacity. Along a sequence of channels, the minimal capacity determines the overall capacity, but for parallel channels we can add the capacities.

Example 3.4. Let M be a set and R_M the set of binary relations on M. Let $\circ$ denote the usual relational product, i.e., for $A, B \in R_M$

$$A \circ B = \big\{(a, c) \in M^2 \bigm| \exists\, b \in M : (a, b) \in A \text{ and } (b, c) \in B\big\}.$$

By Δ, we denote the diagonal relation: $\Delta = \{(a, a) \mid a \in M\}$. Then $(R_M, \cup, \circ, \cap, \emptyset, \Delta)$ is a bisemiring which can be interpreted as follows: Let Σ be a finite set of atomic conditions and M an arbitrary set of ports. A condition $a \in \Sigma$ states which ports are connected via the relation $\text{link}(a) \subseteq M \times M$. Now some conditions may emerge sequentially and others at the same time. For $t = t_1 \cdot t_2$, we put $\text{link}(t) = \text{link}(t_1) \circ \text{link}(t_2)$. Hence, $\text{link}(t)$ indicates the connected ports after a sequence of conditions occurred. On the other hand, if $t = t_1 \parallel t_2$, then $\text{link}(t) = \text{link}(t_1) \cap \text{link}(t_2)$, i.e., only those port connections are still enabled that satisfy both conditions t_1 and t_2 simultaneously.

Any semiring S' can be understood as a bisemiring by defining a trivial parallel multiplication $s \diamond s' = 0$ for $s, s' \in S'$. If $S' \neq \{0\}$, this is an example of a bisemiring without unit for $\diamond$. If S' is commutative, i.e., multiplication commutes, then also $(S', +, \cdot, \cdot, 0, 1)$ is a bisemiring. We refer to these latter bisemirings as *doubled semirings.* One of this kind is the *Boolean bisemiring* $(\mathbb{B}, \vee, \wedge, \wedge, 0, 1)$.

3.2 Weighted Branching Automata

We fix a finite alphabet Σ and a bisemiring S. If Q is a set and $m \in \mathbb{N}$, we denote by $\binom{Q}{m}$ the collection of all subsets of Q with cardinality m.

Definition 3.5. *A* weighted branching automaton *over the alphabet Σ and the bisemiring S, or a* wBA *for short, is a 6-tuple $\mathcal{A} = (Q, \mu_{\text{seq}}, \mu_{\text{fork}}, \mu_{\text{join}}, \lambda, \gamma)$ where:*

- *Q is a finite set of* states.
- *$\mu_{\text{seq}} : Q \times \Sigma \times Q \to S$ is the* sequential transition function.
- *$\mu_{\text{fork}} = (\mu_{\text{fork}}^m : Q \times \binom{Q}{m} \to S \mid m = 2, \dots, |Q|)$ is the family of* fork transition functions.
- *$\mu_{\text{join}} = (\mu_{\text{join}}^m : \binom{Q}{m} \times Q \to S \mid m = 2, \dots, |Q|)$ is the family of* join transition functions.
- *$\lambda, \gamma : Q \to S$ are the* initial *and the* final weight function, *respectively.*

We write $p \xrightarrow{a}_s q$ if $\mu_{\text{seq}}(p, a, q) = s \neq 0$ and call it a *sequential transition* from p to q with action a and weight s. If we do not care about the weight $s \neq 0$, then we just state $p \xrightarrow{a} q$. We also write $p \to_s \{p_1, \dots, p_m\}$ and $p \to \{p_1, \dots, p_m\}$ if $\mu_{\text{fork}}^m(p, \{p_1, \dots, p_m\}) = s \neq 0$. In the same way, $\{q_1, \dots, q_m\} \to_s q$ and $\{q_1, \dots, q_m\} \to q$ are to be understood as $\mu_{\text{join}}^m(\{q_1, \dots, q_m\}, q) = s \neq 0$. In these cases, we speak of a *fork transition* from p to $\{p_1, \dots, p_m\}$ with weight s and of a *join transition* from $\{q_1, \dots, q_m\}$ to q with weight s, respectively. The integer m is called the *arity* of the fork and the join transition, respectively.[3] A state $q \in Q$ is *initial* if $\lambda(q) \neq 0$. Dually, q is a *final* state if $\gamma(q) \neq 0$.

A weighted branching automaton $\mathcal{A}$ can be graphically represented similarly to classical finite automata (cf. Fig. 3): sequential transitions are labeled with actions and weights, fork and join transitions are depicted by directed edges connected with a semicircle labeled with the weight of the transition only. In Fig. 3, the states p and p_3 are initial with entry weights of 1 and 2, respectively, and r and q_3 are final with weights 2 and 1. There are four sequential transitions, e.g., from p_1 to q_1 with label a and weight 1, one fork transition $p \to_1 \{p_1, p_2, p_3\}$, and one join transition $\{q_1, q_2, q_3\} \to_2 q$.

To define the behavior of some wBA $\mathcal{A}$, we have to clarify the notion of a *run.* In classical finite automata, a run can be seen as a very special

[3] Note that we fork into or join always different states, i.e., we fork into a set not a multiset. This is just a technical trifle and actually no restriction of the model.

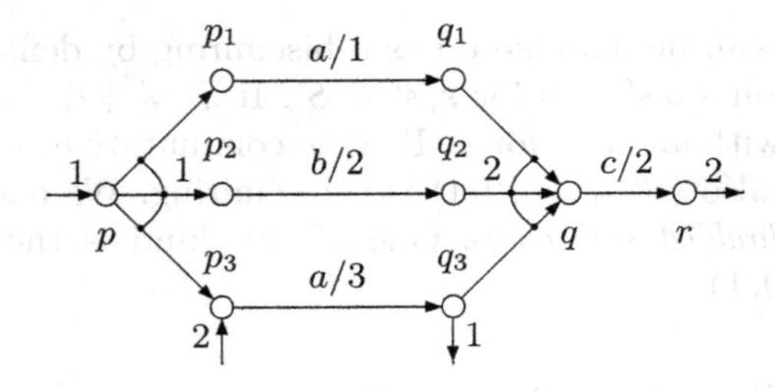

Fig. 3. Graphical representation of wBA

graph obtained by concatenating matching basic graphs, i.e., the transitions. Here, we consider a broader class of graphs. Let $\mathcal{A} = (Q, \mu_{\text{seq}}, \mu_{\text{fork}}, \mu_{\text{join}}, \lambda, \gamma)$ be a wBA. Then we consider finite directed graphs G which have a unique source $\text{src}(G)$ and a unique sink $\text{sk}(G)$, whose nodes are Q-labeled, and whose edges are partially Σ-labeled. We introduce two products for such graphs. The sequential product $G = G_1G_2$ of G_1 and G_2 is defined if $\text{sk}(G_1)$ and $\text{src}(G_2)$ are labeled with the same state, and then we take G as the disjoint union of G_1 and G_2, but fuse $\text{sk}(G_1)$ and $\text{src}(G_2)$ to one node with the same label as before. Now take graphs $G_1, \ldots, G_m$ of our class. Let p_i be the label of $\text{src}(G_i)$ and q_i the label of $\text{sk}(G_i)$. For $p, q \in Q$ the p–q-parallel product of $G_1, \ldots, G_m$ is defined if $p \to \{p_1, \ldots, p_m\}$ is a fork and $\{q_1, \ldots, q_m\} \to q$ is a join. Then $G = \|_{p,q}(G_1, \ldots, G_m)$ is the disjoint union of $G_1, \ldots, G_m$ supplemented by two new nodes labeled with p and q, respectively. Moreover, from the p-node, there is an unlabeled edge to $\text{src}(G_i)$ for every i, and dually, an unlabeled edge from every $\text{sk}(G_i)$ to the q-node (cf. Fig. 4).

Now we are ready to define the runs of $\mathcal{A}$. If $p \xrightarrow{a} q$ is a sequential transition of $\mathcal{A}$, then $p \xrightarrow{a} q$ is an atomic run. The set of all runs $\mathcal{R}(\mathcal{A})$ is the smallest set of graphs that contains all atomic runs and that is closed under sequential product and all p–q-parallel products for $p, q \in Q$ as defined above.

Next, we define the label $\text{lab}(G)$ for any run G as an sp-poset over Σ and the weight $\text{wgt}(G)$ as an element of S inductively as follows: For an atomic run $G : p \xrightarrow{a} q$, we put $\text{lab}(G) = a$ and $\text{wgt}(G) = \mu_{\text{seq}}(p, a, q)$. If $G = G_1G_2$,

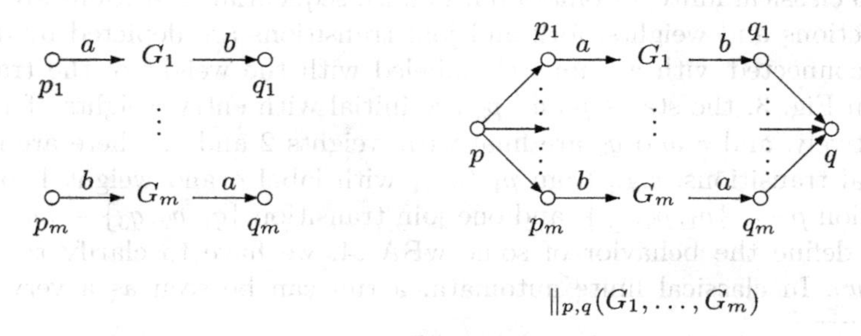

Fig. 4. The p–q-parallel product of $G_1, \ldots, G_m$

then $\text{lab}(G) = \text{lab}(G_1) \cdot \text{lab}(G_2)$ and $\text{wgt}(G) = \text{wgt}(G_1) \cdot \text{wgt}(G_2)$. Now let $G = \|_{p,q}(G_1, \ldots, G_n)$ $(n \geq 2)$ be the parallel decomposition of G where p_i is the state label of $\text{src}(G_i)$ and q_i is the state label of $\text{sk}(G_i)$ for $i = 1, \ldots, n$. Then $\text{lab}(G) = \text{lab}(G_1) \parallel \cdots \parallel \text{lab}(G_n)$ and $\text{wgt}(G)$ equals

$$\mu^n_{\text{fork}}(p, \{p_1, \ldots, p_n\}) \cdot [\text{wgt}(G_1) \diamond \cdots \diamond \text{wgt}(G_n)] \cdot \mu^n_{\text{join}}(\{q_1, \ldots, q_n\}, q) .$$

The weight of a parallel run can be interpreted as follows: Firstly, a weight for branching the process emerges, then the weights for the n subprocesses, and finally, the weight for joining the subprocesses. These weights for branching and joining occur consecutively, and are multiplied sequentially. On the other hand, the weights of the n subprocesses are multiplied in parallel.

If a run G has label $t \in \text{SP}(\Sigma)$, then we say that G is a run on t. If $\text{src}(G)$ is labeled with p and $\text{sk}(G)$ with q, then we write $G : p \xrightarrow{t} q$. For any $t \in \text{SP}(\Sigma)$, there are only finitely many runs G of $\mathcal{A}$ on t. Now we put $\text{wgt}(p, t, q) = \sum(\text{wgt}(G) \mid G \in \mathcal{R}(\mathcal{A}) \;\&\; G : p \xrightarrow{t} q)$ and

$$\|\mathcal{A}\|(t) = \sum_{p,q \in Q} \lambda(p) \cdot \text{wgt}(p, t, q) \cdot \gamma(q).$$

The function $\|\mathcal{A}\| : \text{SP}(\Sigma) \to S$ is the *behavior* of the wBA $\mathcal{A}$. If we assume $S = (\mathbb{N} \cup \{+\infty\}, \mathsf{min}, +, \mathsf{max}, +\infty, 0)$, then the wBA $\mathcal{A}$ from Fig. 3 on page 426 has the behavior $\|\mathcal{A}\|(a) = 6$, $\|\mathcal{A}\|((a \parallel b \parallel a)\, c) = 11$, and $\|\mathcal{A}\|(t) = +\infty$ for $t \notin \{a, (a \parallel b \parallel a)\, c\}$.

Remark 3.6. Note that parallel runs may differ in their branching structure. In Fig. 3 on page 426, there is a run on $a \parallel b \parallel a$ which forks into three subprocesses immediately. On the other hand, two runs on $a \parallel b \parallel c$ which realize $a \parallel b \parallel c$ by a branching in cascades are depicted in Fig. 6 on page 434. One run forks into a and $b \parallel c$ firstly before the subprocess executing $b \parallel c$ forks into b and c. The other run forks into $a \parallel b$ and c firstly and only then into a and b.

A bisemiring S is *positive* if it is zero-sum-free and zero-divisor-free for both products, i.e., $s * s' = 0$ implies $s = 0$ or $s' = 0$ for $* \in \{+, \cdot, \diamond\}$. Recall that a run is composed of transitions with nonzero weight. Hence, for a wBA $\mathcal{A}$ over a positive bisemiring, we have $\|\mathcal{A}\| = 0$ if and only if there is no run from an initial to a final state. Exploiting this property we get the following theorem.

Theorem 3.7 ([57, Lemma 5.14]). *Let S be a positive bisemiring. Then for a wBA $\mathcal{A}$, it is decidable whether there is a $t \in \text{SP}(\Sigma)$ with $\|\mathcal{A}\|(t) \neq 0$.*

3.3 Rationality

In this section, we will characterize the behavior of wBA by means of rational expressions.

Rational Expressions

A *rational expression* over an alphabet Σ and a bisemiring S is a term using the constants a with $a \in \Sigma$, the unary function symbols $^+$, $^\boxplus$, $s\cdot$, and $\cdot s$ for $s \in S$, and the binary function symbols $+$, $\cdot$ and $\|$. A *series-rational expression* is a rational expression not using the function symbol $^\boxplus$.

The *semantics* $[\![E]\!] : \mathrm{SP}(\Sigma) \to S$ of a rational expression E is defined inductively. We put for every $t \in \mathrm{SP}(\Sigma)$:

$$
\begin{aligned}
[\![a]\!](t) &= \begin{cases} 1 & \text{if } t = a, \\ 0 & \text{otherwise,} \end{cases} \\
[\![sE]\!](t) &= s \cdot [\![E]\!](t), \\
[\![Es]\!](t) &= [\![E]\!](t) \cdot s, \\
[\![E_1 + E_2]\!](t) &= [\![E_1]\!](t) + [\![E_2]\!](t), \\
[\![E_1 \cdot E_2]\!](t) &= \sum_{\substack{t = t_1 \cdot t_2 \\ t_1, t_2 \in \mathrm{SP}(\Sigma)}} [\![E_1]\!](t_1) \cdot [\![E_2]\!](t_2), \\
[\![E_1 \parallel E_2]\!](t) &= \sum_{\substack{t = t_1 \| t_2 \\ (t_1, t_2) \in \mathrm{SP}(\Sigma)^2}} [\![E_1]\!](t_1) \diamond [\![E_2]\!](t_2), \\
[\![E^+]\!](t) &= \sum_{\substack{t = t_1 \cdot \cdots \cdot t_m \\ t_1, \ldots, t_m \in \mathrm{SP}(\Sigma)}} [\![E]\!](t_1) \cdot \cdots \cdot [\![E]\!](t_m), \\
[\![E^\boxplus]\!](t) &= \sum_{\substack{t = t_1 \| \cdots \| t_m \\ (t_1, \ldots, t_m) \in \mathrm{SP}(\Sigma)^m}} [\![E]\!](t_1) \diamond \cdots \diamond [\![E]\!](t_m).
\end{aligned}
$$

Note that the sums in the definition of $[\![E^+]\!]$ and $[\![E^\boxplus]\!]$ are finite since there are only finitely many sequential and parallel decompositions of any sp-poset t. In the definition of $[\![E_1 \parallel E_2]\!]$ the sum is taken over all pairs $(t_1, t_2) \in \mathrm{SP}(\Sigma)^2$, i.e., for $t_1 \neq t_2$ both $[\![E_1]\!](t_1) \diamond [\![E_2]\!](t_2)$ and $[\![E_1]\!](t_2) \diamond [\![E_2]\!](t_1)$ contribute to the sum (similarly for $[\![E^\boxplus]\!]$). In the sequel, we will use the function symbols of expressions also as operations of functions $f : \mathrm{SP}(\Sigma) \to S$ with the semantics as defined above.

From Rational Expressions to Weighted Branching Automata

By standard constructions as in the case of words, we get the following proposition.

Proposition 3.8. *Let $\mathcal{A}$ and $\mathcal{A}'$ be two wBA and $s \in S$. Then there are wBA $\mathcal{A}_{s\cdot}$, $\mathcal{A}_{\cdot s}$, and $\mathcal{A}_+$ with $\|\mathcal{A}_{s\cdot}\| = s \cdot \|\mathcal{A}\|$, $\|\mathcal{A}_{\cdot s}\| = \|\mathcal{A}\| \cdot s$, and $\|\mathcal{A}_+\| = \|\mathcal{A}\| + \|\mathcal{A}'\|$, respectively.*

The next goal is to construct wBA $\mathcal{A}_\|$ and $\mathcal{A}_\boxplus$ with $\|\mathcal{A}_\|\| = \|\mathcal{A}\| \parallel \|\mathcal{A}'\|$ and $\|\mathcal{A}_\boxplus\| = \|\mathcal{A}\|^\boxplus$, respectively. For this purpose, we have to normalize $\mathcal{A}$ and

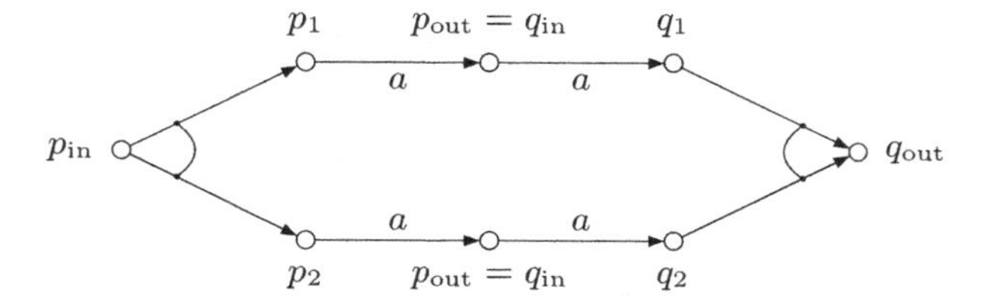

Fig. 5. A problematic run in the classical product construction

$\mathcal{A}'$ at first. A wBA $\mathcal{A} = (Q, \mu_{\text{seq}}, \mu_{\text{fork}}, \mu_{\text{join}}, \lambda, \gamma)$ is *normalized* if there are unique initial and final states p_{in} and p_{out} such that: $\lambda(p_{\text{in}}) = 1$ and $\lambda(p) = 0$ for all $p \neq p_{\text{in}}$, $\gamma(p_{\text{out}}) = 1$ and $\gamma(p) = 0$ for all $p \neq p_{\text{out}}$, and there are no transitions entering p_{in}, and no transitions leaving p_{out} (neither sequential, fork, nor join). Every wBA $\mathcal{A}$ can be normalized by adding the new initial state p_{in} and the new final state p_{out} and by summarizing transitions from old initial states to a new one starting in p_{in} with appropriate weight, e.g., $p_{\text{in}} \xrightarrow{a}_s q$ where $s = \sum_{p \in Q} \lambda(p) \cdot \mu_{\text{seq}}(p, a, q)$. Similarly, we proceed for the new final state; see [44] for details.

Proposition 3.9. *Let $\mathcal{A}$ and $\mathcal{A}'$ be two wBA. There is a wBA $\mathcal{A}_{\|}$ with $\|\mathcal{A}_{\|}\| = \|\mathcal{A}\| \parallel \|\mathcal{A}'\|$.*

Proof. Assume $\mathcal{A}$ and $\mathcal{A}'$ to be normalized with p_{in}, p_{out} and p'_{in}, p'_{out} the respective unique initial and final states. Now $\mathcal{A}_{\|}$ is built by taking the disjoint union of $\mathcal{A}$ and $\mathcal{A}'$, adding two new states r_{in} and r_{out} and, moreover, a fork $r_{\text{in}} \rightarrow_1 \{p_{\text{in}}, p'_{\text{in}}\}$ and a join $\{p_{\text{out}}, p'_{\text{out}}\} \rightarrow_1 r_{\text{out}}$. We put $\lambda(r_{\text{in}}) = 1$ and $\gamma(r_{\text{out}}) = 1$. All other initial and final weights are equal to 0. By distributivity of $\diamond$ over $+$, one can show that $\|\mathcal{A}_{\|}\| = \|\mathcal{A}\| \parallel \|\mathcal{A}'\|$. □

Using a similar construction, [57] proves the following result.

Proposition 3.10. *Let $\mathcal{A}$ be a wBA. There is a wBA $\mathcal{A}_{\boxplus}$ with $\|\mathcal{A}_{\boxplus}\| = \|\mathcal{A}\|^{\boxplus}$.*

Most difficulties arise from sequential product and iteration. It is tempting to use the construction as known for finite automata: normalize $\mathcal{A}$ and $\mathcal{A}'$ and "fuse" the unique final state of $\mathcal{A}$ and the unique initial one of $\mathcal{A}'$. But already for the Boolean bisemiring $\mathbb{B}$, serious difficulties arise as noted by Lodaya and Weil [47] firstly. Consider $\mathcal{A}$ which has just the following transitions: a fork $p_{\text{in}} \rightarrow_1 \{p_1, p_2\}$ and two sequential transitions $p_1 \xrightarrow{a}_1 p_{\text{out}}$ and $p_2 \xrightarrow{a}_1 p_{\text{out}}$. Here, p_{in} is initial and p_{out} is final. On the other hand, $\mathcal{A}'$ consists of two sequential transitions $q_{\text{in}} \xrightarrow{a}_1 q_1$ and $q_{\text{in}} \xrightarrow{a}_1 q_2$ and a join $\{q_1, q_2\} \rightarrow_1 q_{\text{out}}$ with q_{in} initial and q_{out} final. Then for every $t \in \text{SP}(\Sigma)$, we have $\|\mathcal{A}\|(t) = \|\mathcal{A}'\|(t) = 0$ and, therefore, $\|\mathcal{A}\| \cdot \|\mathcal{A}'\|(t) = 0$. The classical construction would suggest a wBA which would have the successful run depicted in Fig. 5, a contradiction. The problem is that in the usual construction the new automaton can switch in parallel subruns from $\mathcal{A}$ to $\mathcal{A}'$. Therefore, we have to ensure that the switch from $\mathcal{A}$ to $\mathcal{A}'$ can be done only on the most upper level of the computation, i.e., when all fork transitions of $\mathcal{A}$ are closed again by matching joins from $\mathcal{A}$.

Proposition 3.11. *Let $\mathcal{A}$ and $\mathcal{A}'$ be two wBA. There is a wBA $\mathcal{A}_\bullet$ such that $\|\mathcal{A}_\bullet\| = \|\mathcal{A}\| \cdot \|\mathcal{A}'\|$.*

Proof. Given are the two wBA $\mathcal{A} = (Q, \mu_{\text{seq}}, \mu_{\text{fork}}, \mu_{\text{join}}, \lambda, \gamma)$ and $\mathcal{A}' = (Q', \mu'_{\text{seq}}, \mu'_{\text{fork}}, \mu'_{\text{join}}, \lambda', \gamma')$. The idea for the construction of $\mathcal{A}_\bullet$ is to send a signal along the runs of $\mathcal{A}$ which is propagated to one branch only when the run takes a fork. We are allowed to switch to $\mathcal{A}'$ only if the signal is present. Hence, we cannot switch to $\mathcal{A}'$ in all parallel subruns simultaneously and, therefore, parallel subruns have to terminate by a join in $\mathcal{A}$ firstly before continuing the run in $\mathcal{A}'$. Hence, the first step in our construction is to build a wBA $\widetilde{\mathcal{A}}$ with $\|\widetilde{\mathcal{A}}\| = \|\mathcal{A}\|$ which is done in the following way (see [44] for details): We put $\widetilde{Q} = Q \times \{0, 1\}$. We refer to the second component of a state from $\widetilde{Q}$ as a signal. Along sequential transitions the signal is propagated, i.e., $\widetilde{\mu}_{\text{seq}}((p, x), a, (q, y)) = \mu_{\text{seq}}(p, a, q)$ if $x = y$ and otherwise it equals 0. To handle branching, we fix a linear order on the state set Q and transmit the signal (if there is one) by a fork to the smallest state of the target states of the fork transition. On the other hand, a join transition propagates a signal if there is at least one signal in the different branches. Certainly, in initial and final states, the signal has to be present. The appropriate weights of the transitions carry over from $\mathcal{A}$. Note that the construction of $\widetilde{\mathcal{A}}$ preserves normalization.

Now we can apply to $\widetilde{\mathcal{A}}$ and $\mathcal{A}'$ the classical construction to obtain $\mathcal{A}_\bullet$, i.e., we take the disjoint union of $\widetilde{\mathcal{A}}$ and $\mathcal{A}'$ and fuse the unique final state of $\widetilde{\mathcal{A}}$ and the unique initial one of $\mathcal{A}'$ to one state where the automaton switches from $\widetilde{\mathcal{A}}$ to $\mathcal{A}'$. By the construction above, we can ensure that a run in $\mathcal{A}_\bullet$ is the sequential product of a run in $\widetilde{\mathcal{A}}$ and a run in $\mathcal{A}'$. Hence, $\|\mathcal{A}_\bullet\| = \|\widetilde{\mathcal{A}}\| \cdot \|\mathcal{A}'\| = \|\mathcal{A}\| \cdot \|\mathcal{A}'\|$. □

For the sequential iteration, one has to be even more careful. Now it is not sufficient to prevent the switching to an initial state of $\mathcal{A}$ in at least one of the parallel subruns. This time we have to ensure this property for every parallel subrun. This can be achieved by sending two signals which separate in a fork transition and travel along different ways. Then we allow a jump from a final to an initial state only in case both signals are present. This way we get the following proposition.

Proposition 3.12 ([44]). *For every wBA $\mathcal{A}$, there is a wBA $\mathcal{A}_+$ satisfying $\|\mathcal{A}_+\| = \|\mathcal{A}\|^+$.*

Certainly, there is a wBA $\mathcal{A}$ with $\|\mathcal{A}\| = [\![a]\!]$ for every $a \in \Sigma$. Then Propositions 3.8, 3.9, 3.10, 3.11, and 3.12 imply the following theorem.

Theorem 3.13 ([44]). *For every rational expression E, there is a wBA $\mathcal{A}$ such that $[\![E]\!] = \|\mathcal{A}\|$.*

From Weighted Branching Automata to Expressions

In contrast to weighted automata over words, rational expressions are too weak to describe the behavior of all wBA as already noted by Lodaya and Weil [47] for the Boolean bisemiring.

Nevertheless, we will be able to characterize a subclass of wBA by series-rational expressions. We specify this class in two different ways: firstly, by purely syntactic restrictions of the wBA, and secondly by the bounded-width property of the behavior.

In order to construct a series-rational expression from a wBA $\mathcal{A}$, we have to arrange the fork and join transitions in a certain hierarchy. Therefore, we define a *depth function* $\mathrm{dp} : \mathcal{R}(\mathcal{A}) \to \mathbb{N}$ for every run of $\mathcal{A}$ inductively:

- Every atomic run $G : p \xrightarrow{a} q$ is of depth 0.
- If $G = G_1 \cdot \dots \cdot G_m$, then $\mathrm{dp}(G) = \max\{\mathrm{dp}(G_i) \mid i = 1, \dots, m\}$.
- If $G = \|_{p,q}(G_1, \dots, G_m)$, then $\mathrm{dp}(G) = 1 + \max\{\mathrm{dp}(G_i) \mid i = 1, \dots, m\}$.

The depth of a run measures the nesting of branchings within this run. A wBA $\mathcal{A}$ is of *bounded depth* if there is a $d \in \mathbb{N}$ such that $\mathrm{dp}(G) \leq d$ for every run $G \in \mathcal{R}(\mathcal{A})$. "Bounded depth" is a property of the runs of the automaton. Here, the weights need not be considered. By analyzing the "unweighted" transition relations of $\mathcal{A}$, one can show the following theorem.

Theorem 3.14 ([57]). *Let S be an arbitrary bisemiring. It is decidable whether a given wBA $\mathcal{A}$ is of bounded depth.*

With this restriction, we can describe $\mathcal{A}$ by an expression.

Theorem 3.15 ([44]). *Let $\mathcal{A}$ be a wBA of bounded depth. Then there is a series-rational expression E with $[\![E]\!] = \|\mathcal{A}\|$.*

Proof. Let f be a fork and j a join transition of $\mathcal{A}$. We say that (f, j) is a *matching pair* if there is a run G of $\mathcal{A}$ that starts with fork f and ends with join j. Then we say G is limited by (f, j). Let M be the set of all matching pairs of $\mathcal{A}$. We define a binary relation $\prec$ on M by $(f, j) \prec (f', j')$ if there is a run $G' \in \mathcal{R}(\mathcal{A})$ limited by (f', j') that contains a proper subrun (defined in the obvious manner) G limited by (f, j). Due to bounded depth of $\mathcal{A}$, the relation $\prec$ is irreflexive. Obviously, $\prec$ is transitive. This implies that the reflexive closure $\preceq$ is a partial order which can be extended to a linear one $\sqsubseteq$.

For $J \subseteq M$ and $p, q \in Q$, let $\|\mathcal{A}\|^J_{p,q} : \mathrm{SP}(\Sigma) \to S$ denote the function $\|\mathcal{A}\|^J_{p,q}(t) = \sum \mathrm{wgt}(G)$ where $t \in \mathrm{SP}(\Sigma)$ and where the sum extends over all runs G from p to q with label t such that only matching pairs from J are used. We show that $\|\mathcal{A}\|^J_{p,q}$ can be defined by a series-rational expression $E^J_{p,q}$ for every initial segment $J = \{(f', j') \in M \mid (f', j') \sqsubseteq (f, j)\}$ with $(f, j) \in M$. We proceed by induction on $|J|$.

If $|J| = 0$, then no forks or joins are used and $\|\mathcal{A}\|^{\emptyset}_{p,q}$ is defined by a series-rational expression $E^{\emptyset}_{p,q}$ using a classical result of Schützenberger; cf. [21].

In fact, here we deal with words and the parallel product is not used at all. Now let $J = \{(f',j') \mid (f',j') \sqsubseteq (f,j)\}$ with $f : o \to_s \{o_1,\dots,o_m\}$ and $j : \{r_1,\dots,r_m\} \to_{\bar{s}} r$. Let $\|\mathcal{A}\|_{(f,j)}$ be such that for every $t \in \mathrm{SP}(\Sigma)$

$$\|\mathcal{A}\|_{(f,j)}(t) = \sum_{\substack{G:o\xrightarrow{t}r \\ G \text{ limited by } (f,j)}} \mathrm{wgt}(G).$$

Let $G : o \xrightarrow{t} r$ be limited by (f,j). Then $G = \|_{o,r}(H_1,\dots,H_m)$ with H_1, $\dots, H_m$ containing only matching pairs from $J' = J \setminus \{(f,j)\}$. Hence, the induction hypothesis can be applied to $H_1,\dots,H_m$. Let S_m denote the symmetric group on $\{1,\dots,m\}$. Then the series-rational expression

$$E_{(f,j)} = s \cdot \left(\sum_{\pi \in S_m} \left[E^{J'}_{o_1,r_{\pi(1)}} \parallel \cdots \parallel E^{J'}_{o_m,r_{\pi(m)}} \right] \right) \cdot \bar{s}$$

has the semantics $[\![E_{(f,j)}]\!] = \|\mathcal{A}\|_{(f,j)}$. Now, also $\|\mathcal{A}\|^J_{p,q}$ can be described by a series-rational expression. Unfortunately, one has to distinguish several cases. Here we consider $p \neq o$, $r \neq q$, and $o \neq r$ only. In this case we put

$$E^J_{p,q} = E^{J'}_{p,q} + E^{J'}_{p,o} \cdot E_{(f,j)} \cdot E^{J'}_{r,q} + E^{J'}_{p,o} \cdot E_{(f,j)} \cdot \left(E^{J'}_{r,o} \cdot E_{(f,j)}\right)^+ \cdot E^{J'}_{r,q}$$

and get $[\![E^J_{p,q}]\!] = \|\mathcal{A}^J_{p,q}\|$. The first expression $E^{J'}_{p,q}$ of this sum covers all runs that do not use (f,j) at the upper level. The second one covers all runs $G_1 \cdot G_2 \cdot G_3$ such that G_2 is limited by (f,j), but neither in G_1 nor in G_3 (f,j) appears as a matching pair. The third subexpression covers all runs where we have more than one such subrun limited by (f,j).

Finally, we put $E = \sum_{p,q \in Q} \lambda(p) \cdot E^M_{p,q} \cdot \gamma(q)$ which is a series-rational expression and satisfies $[\![E]\!] = \|\mathcal{A}\|$. □

Bounded Depth and Bounded Width

The notion of bounded depth is a property of the automaton not referring to its behavior. Next, we define a semantic property of its behavior. For every $t \in \mathrm{SP}(\Sigma)$, we define the *width* of t inductively as follows: for $a \in \Sigma$ we have $\mathrm{wd}(a) = 1$, $\mathrm{wd}(t_1 \cdot t_2) = \max(\mathrm{wd}(t_1), \mathrm{wd}(t_m))$, and finally $\mathrm{wd}(t_1 \parallel t_2) = \mathrm{wd}(t_1) + \mathrm{wd}(t_2)$. In terms of computer science, $\mathrm{wd}(t)$ gives the minimal number of processors it takes for t to be executed. Now let $f : \mathrm{SP}(\Sigma) \to S$ be a function. We put $\mathrm{wd}(f) = \sup\{\mathrm{wd}(t) \mid f(t) \neq 0\}$. Note that $\mathrm{wd}(f)$ may be infinite. We say f is of *bounded width* if $\mathrm{wd}(f)$ is finite.

Let $\mathcal{A}$ be a wBA of bounded depth with $\mathrm{dp}(G) \leq d$ for all runs G, and let B be the highest arity of all fork and join transitions. Then it is easy to show that $\mathrm{wd}(\mathrm{lab}(G)) \leq B^d$ for all $G \in \mathcal{R}(\mathcal{A})$. Hence, $\|\mathcal{A}\|$ is of bounded width.

Conversely, let $\mathcal{A}$ be a wBA such that $\|\mathcal{A}\|$ is of bounded width. The aim is to construct a wBA $\mathcal{A}'$ of bounded depth with $\|\mathcal{A}'\| = \|\mathcal{A}\|$. At first glance, one would try to implement a depth counter and to restrict the reachable depth by $\mathrm{wd}(\|\mathcal{A}\|)$. But this approach works for certain bisemirings only. In general, $t \in \mathrm{SP}(\Sigma)$ may be executed in $\mathcal{A}$ by runs of different depth. If S is not zero-sum-free, i.e., there are $s, s' \neq 0$ with $s + s' = 0$, then the weights of runs of different depth may sum up to zero. Hence, if we stop the computation at a certain depth, we may make a mistake. An elaborated example of such a wBA is given in [57]. Therefore, we have to refine the construction to overcome this problem. The new wBA constructed will keep track of the width of the poset computed so far. This is realized by a stack where the widths encountered up to the last fork transition are stored. Hereby, we can limit the size of the stack and the numbers stored therein to $\mathrm{wd}(\|\mathcal{A}\|)$ whereby we stay within the realm of finite-state systems. The involved construction can be found in [44, 57] and yields the following result.

Theorem 3.16. *Let $\mathcal{A}$ be a wBA such that $\|\mathcal{A}\|$ is of bounded width. There is a wBA $\mathcal{A}'$ of bounded depth with $\|\mathcal{A}'\| = \|\mathcal{A}\|$.*

If $\mathcal{A}$ is of bounded depth, then $\|\mathcal{A}\|$ is of bounded width. If $\mathcal{A}$ is not depth-bounded, we cannot be sure whether $\|\mathcal{A}\|$ is of unbounded width. But for positive bisemirings, this is the case. Thus, we can prove by the help of Theorem 3.14 the following theorem.

Theorem 3.17. *Let S be a positive bisemiring. Then it is decidable whether the behavior $\|\mathcal{A}\|$ of a wBA $\mathcal{A}$ is of bounded width.*

The Characterization Theorem

Now all ingredients are allocated for the main theorem.

Theorem 3.18 ([44]). *Let S be an arbitrary bisemiring and $f : \mathrm{SP}(\Sigma) \to S$. Then the following are equivalent:*

1. *f is of bounded width and the behavior of some wBA.*
2. *f is the behavior of some wBA of bounded depth.*
3. *f is the semantics of some series-rational expression.*

Proof. If $f = \|\mathcal{A}\|$ for some wBA $\mathcal{A}$ and f is of bounded width, then there is a wBA $\mathcal{A}'$ of bounded depth with $f = \|\mathcal{A}'\|$ by Theorem 3.16. Due to Theorem 3.15, we can construct from $\mathcal{A}'$ a series-rational expression E with $[\![E]\!] = \|\mathcal{A}'\| = f$. For every series-rational expression E, there is a wBA $\mathcal{A}$ with $[\![E]\!] = \|\mathcal{A}\|$ as stated by Theorem 3.13. Moreover, a simple inductive argument shows that $\|\mathcal{A}\| = [\![E]\!]$ is of bounded width. □

3.4 The Hadamard Product

The *Hadamard product* $f \odot g$ of two functions $f, g : \mathrm{SP}(\Sigma) \to S$ is the pointwise sequential product, i.e., $(f \odot g)(t) = f(t) \cdot g(t)$ for every $t \in \mathrm{SP}(\Sigma)$. If S is the Boolean bisemiring $\mathbb{B}$, it models the intersection of the supports of f and g. Here, we raise the question whether the Hadamard product of the

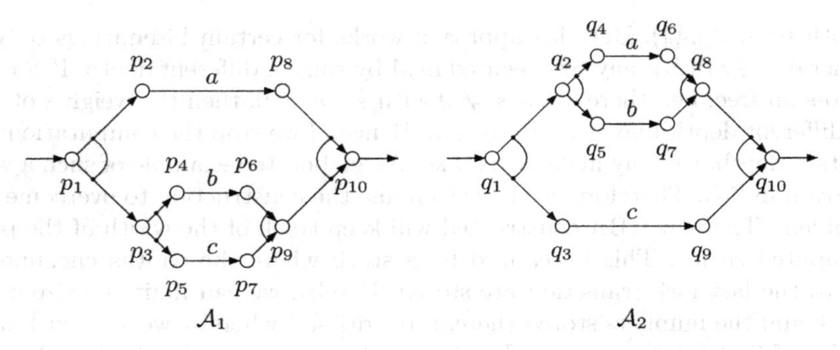

Fig. 6. Two wBA with same behavior but different branching structure

behaviors of two wBA $\mathcal{A}_1$ and $\mathcal{A}_2$ is again the behavior of a wBA $\mathcal{A}$. For weighted automata over words, this question can be affirmed by the usual product automaton construction (see [63, Theorem 4.13]). But in our setting, the situation turns out to be more challenging.

Consider the two wBA over the Boolean bisemiring $\mathbb{B}$ depicted in Fig. 6. They have the same behavior but a different branching structure. In a product automaton $\mathcal{A}$ of $\mathcal{A}_1$ and $\mathcal{A}_2$, there would be the following sequential transitions: $(p_2, q_4) \xrightarrow{a} (p_8, q_6)$, $(p_4, q_5) \xrightarrow{b} (p_6, q_7)$, and $(p_5, q_3) \xrightarrow{c} (p_7, q_9)$. How should we define the new fork transitions starting in (p_1, q_1)? From a permissive point of view, we could allow $(p_1, q_1) \rightarrow \{(p_2, q_2), (p_3, q_3)\}$ and $(p_1, q_1) \rightarrow \{(p_2, q_3), (p_3, q_2)\}$ in $\mathcal{A}$. Similarly, there may exist forks $(p_3, q_2) \rightarrow \{(p_4, q_4), (p_5, q_5)\}$ and $(p_3, q_2) \rightarrow \{(p_4, q_5), (p_5, q_4)\}$. But even with these four fork transitions, we cannot branch into the states (p_2, q_4) and (p_5, q_3) where two of the three sequential transitions start. Therefore, $\|\mathcal{A}\|$ would map also $a \parallel b \parallel c$ to 0. Hence, $\|\mathcal{A}\| \neq \|\mathcal{A}_1\| \odot \|\mathcal{A}_2\|$. To handle this problem, we will disallow such a branching in cascades as it is done by $\mathcal{A}_1$ and $\mathcal{A}_2$. Therefore, we turn these wBA into wBA which branch at once, i.e., there would be a fork of arity 3 in our example.

Moreover, we have to impose heavy restrictions on the underlying bisemiring S. We call S *idempotent* if the addition is idempotent, i.e., $s + s = s$ for all $s \in S$. Moreover, remember that for a doubled semiring the two multiplications are equal and, therefore, commutative.

Theorem 3.19 ([58]). *Let $\mathcal{A}_1$ and $\mathcal{A}_2$ be two wBA of bounded depth over an idempotent doubled semiring S. Then there is a wBA $\mathcal{A}$ of bounded depth over S with $\|\mathcal{A}\| = \|\mathcal{A}_1\| \odot \|\mathcal{A}_2\|$.*

Proof. Firstly, we can transform $\mathcal{A}_1$ and $\mathcal{A}_2$ into wBA where the branching within a run happens always at once, i.e., a fork transition is not immediately followed by another fork. For this, we use Theorem 3.18, i.e., the wBA are transferred to equivalent expressions. From the expressions, we can build wBA with the required property. In doing so, we have to give another construction for the parallel product. We succeed by fusing fork and join transitions to such of higher arity and multiply the affected weights (which is possible since S is a doubled semiring); cf. [57, Proposition 7.5] and [58, Proposition 4.6]. For those new wBA, one can apply a product automaton construction. However, it is possible that some runs are duplicated. Here, the idempotency of S is applied [58, Theorem 5.4]. Thus, we succeed in constructing a wBA $\mathcal{A}$ such that $\|\mathcal{A}\| = \|\mathcal{A}_1\| \odot \|\mathcal{A}_2\|$. □

Instances of idempotent doubled semirings are $(\mathbb{N} \cup \{+\infty\}, \mathsf{min}, +, +, +\infty, 0)$ or the Boolean bisemiring. For bisemirings S with two different multiplications, one can find counter-examples to the statement of the last theorem even when we endow S with rich properties [58, Example 5.3].

3.5 History and Overview

Series-parallel posets were proposed as a model for the behavior of concurrent processes by Grabowski [35], Pratt [61], and Gischer [34]. Lodaya and Weil [47] introduced branching automata as a device for accepting languages of sp-posets and characterized the behavior of these automata by expressions and algebraically in the case of bounded width. This work was extended by Kuske [41] who considered infinite sp-posets and gave a logical characterization.

Bisemirings and weighted branching automata were proposed by Kuske and Meinecke in [44] where the Kleene-type theorem was shown. Different regularity concepts concerning the running modi of wBA are investigated in [58, 57]. These results were used for the closure of behaviors of wBA under Hadamard product. An open question is whether some results generalize to sp-posets of unbounded width. For the Boolean bisemiring $\mathbb{B}$, Lodaya and Weil [48] gave a characterization of the behaviors of general wBA by exploiting another concept of rationality using ζ-substitution and ζ-exponentiation. This concept may carry over for doubled semirings, but it is by far not understood how to handle general bisemirings.

Moreover, we miss a logical and an algebraic characterization of the behavior of wBA. Mathissen showed the equivalence of weighted automata and logic in the setting of texts and sp-biposets [50] as well as of nested words [51]. But it is not clear whether this result can be transferred to sp-posets. For an algebraic characterization along the lines of semimodules as stated in [6] (see [21, 63]), we even miss an algebraic analogon to semimodules that can handle two different products.

Infinite sp-posets have been explored so far only in the Boolean setting [41, 5]. It would be of interest to define an appropriate weighted setting for infinite sp-posets and to show the equivalence of different formalisms.

4 Pictures

In this section, we extend one-dimensional words to two-dimensional structures called pictures. We will investigate picture series characterized in terms of automata, rational expressions, and weighted logics. It provides an extension of both the case of formal power series on words and the theory of two-dimensional languages.

Throughout this section, we assume a fixed commutative semiring S.

4.1 Pictures and Weighted Picture Automata

We set $\mathbb{N}_1 := \mathbb{N} \setminus \{0\}$ and $[n] = \{1, \ldots, n\}$ for $n \in \mathbb{N}_1$. Let $m, n \in \mathbb{N}_1$ and let Σ be an alphabet. A *picture* of *size* (m, n) over Σ is an $m \times n$-matrix over Σ, i.e., a mapping $p : [m] \times [n] \to \Sigma$. We write $p(i, j)$ or $p_{i,j}$ for the component of p at position (i, j). Furthermore, the *height* (or *vertical length*) and *width* (or *horizontal length*) of p are $l_v(p) = m$ and $l_h(p) = n$, respectively. We denote by Σ^{++} the set of all pictures over Σ. The set $\Sigma^{m\times n} \subseteq \Sigma^{++}$ comprises all pictures of size (m, n). A *picture language* is a set of pictures.

It is the aim of this section to consider quantitative properties of pictures, i.e., functions from Σ^{++} into some commutative semiring S that we call *picture series*. For motivation, we give two examples of such functions $r : \Sigma^{++} \to \mathbb{R} \cup \{\infty\}$ and $t : \Sigma^{++} \to \mathbb{N}$.

Example 4.1. Let $D \subset [0, 1]$ be a finite set of rational values and let $L \subseteq D^{++}$ be any picture language over D. Let furthermore $r : D^{++} \to \mathbb{R} \cup \{\infty\}$ be defined by $r(p) = \sum_{i,j} p_{i,j}$ for $p \in L$ and ∞ for $p \in \Sigma^{++} \setminus L$. We could interpret values in D as different levels of gray, then for each picture p in L, the function r provides the total value $r(p)$ of light of p.

Example 4.2. Let C be a finite set of colors and consider the function $t : C^{++} \to \mathbb{N}$ where $t(p)$ is the largest area of a monochromatic rectangle in the picture p.

One way of defining such functions is via weighted picture automata that we present next. These devices were first introduced by Bozapalidis and Grammatikopoulou in [10] and named weighted quadrapolic picture automata (extending Wang systems by de Prophetis and Varricchio from [11]).

Definition 4.3 ([10]). *A* weighted picture automaton *or* wPA *is a 6-tuple* $\mathcal{A} = (Q, R, F_w, F_n, F_e, F_s)$ *consisting of a finite set* Q *of states, a finite set of rules* $R \subseteq \Sigma \times S \times Q^4$ *as well as four* poles of acceptance $F_w, F_n, F_e, F_s \subseteq Q$.

Given a rule $t = (a, \bar{s}, q_w, q_n, q_e, q_s) \in R$, let $\text{label}(t) = a$ be its (input) label, $\text{wgt}(t) = \bar{s}$ its weight, $\sigma_x(t) = q_x$ for $x \in \{w, n, e, s\}$ its four poles. We extend the functions label and wgt to pictures by setting for a picture $c \in R^{l\times r}$ over the set of rules: $\text{label}(c) : [l] \times [r] \to \Sigma : (i, j) \mapsto \text{label}(c_{i,j})$ and

$\mathrm{wgt}(c) = \prod_{(i,j)\in[l]\times[r]} \mathrm{wgt}(c_{i,j}) \in S$. We call $\mathrm{label}(c) \in \Sigma^{l\times r}$ the *label of* c and $\mathrm{wgt}(c) \in S$ the *weight of* c.

A picture $c \in R^{l\times r}$ is a *run of* $\mathcal{A}$ if $\sigma_s(c_{i,j}) = \sigma_n(c_{i+1,j})$ for all $(i,j) \in [l-1, r]$ and $\sigma_e(c_{i,j}) = \sigma_w(c_{i,j+1})$ for all $(i,j) \in [l] \times [r-1]$. It is *successful* if it has its (outer) pole states in the respective poles of acceptance, i.e., $\sigma_w(c_{i,1}) \in F_w$, $\sigma_n(c_{1,j}) \in F_n$, $\sigma_e(c_{i,r}) \in F_e$, and $\sigma_s(c_{l,j}) \in F_s$ for all $i \in [l]$ and $j \in [r]$. For every $p \in \Sigma^{++}$, we denote the set of successful runs c with $\mathrm{label}(c) = p$ by $\mathrm{Succ}(p)$.

Then the *behavior* of the weighted picture automaton $\mathcal{A}$ is the picture series $[\![\mathcal{A}]\!] : \Sigma^{++} \to S$ with

$$[\![\mathcal{A}]\!](p) = \sum \big(\mathrm{wgt}(c) \mid c \in \mathrm{Succ}(p)\big)$$

for all $p \in \Sigma^{++}$. A picture series $r : \Sigma^{++} \to S$ is *recognizable* if it is the behavior of some weighted picture automaton.

Let us consider again Examples 4.1 and 4.2. Assume the picture language $L \subseteq D^{++}$ in Example 4.1 to be recognizable (i.e., in our terminology, it is the support of the behavior of some weighted picture automaton $\mathcal{A}$ over the Boolean semiring $\mathbb{B}$). If $(a, 1, q_w, q_n, q_e, q_s)$ is a rule of the automaton $\mathcal{A}$, then let $(a, a, q_w, q_n, q_e, q_s)$ be a rule of the weighted picture automaton $\mathcal{B}$ over the tropical semiring $\mathbb{T} = (\mathbb{R}_+ \cup \{\infty\}, \mathsf{min}, +, \infty, 0)$. Then the behavior of $\mathcal{B}$ equals the picture series r. The picture series t from Example 4.2 is the behavior of a weighted picture automaton over the max-plus semiring $(\mathbb{N} \cup \{-\infty\}, \mathsf{max}, +, -\infty, 0)$. This automaton has one successful run for every monochromatic rectangle q in p. We weight the rules that overlay q with 1. Then the weight of this particular run is $l_v(q) \cdot l_h(q)$. Since we get the behavior by adding (which, in our semiring, means taking the maximum) the weights for successful runs reading p, the maximal size is extracted.

Recall that for a regular word language, the characteristic series over an arbitrary semiring is recognizable. In contrast, the following example shows that this is not the case for picture series. For this, consider the characteristic series 1_L of the set L of pictures containing two equal rows. Then:

1. There is a wPA over the Boolean semiring whose behavior equals 1_L.
2. There is no wPA over the semiring $(\mathbb{N}, +, \cdot, 0, 1)$ whose behavior is 1_L [2].

Weighted 2-Dimensional On-line Tessellation Automata

These are the weighted counterpart of on-line tessellation automata that were introduced by Inoue and Nakamura in [36]. In our context, we can understand them as a variant of wPA where:

1. For each rule $(a, \bar{s}, q_w, q_n, q_e, q_s)$, one has $q_e = q_s$.
2. The success of a run is determined as for wPA in the west and the north, but concerning the south and the east, we only require the states at the south-east corner to be accepting.

In [23], it is shown that this automaton model creates the same behaviors as wPA.

Weighted Tiling Systems

A *tile* is a square picture of dimension 2×2 and a *weighted tiling system* is a function that assigns weights to tiles. Then the weight of an arbitrary picture is obtained by multiplying all the weights of 2×2-subpictures according to the weighted tiling system (more precisely: one first adds a border marked # and then considers the subpictures). Since weighted tiling systems lack states that are present in wPA, their expressive power is strictly less than that of wPA. But the projections of their behaviors coincide with the behaviors of wPA [54].

4.2 Other Formalisms: Expressions and Logics

Rational Picture Series

Another way of specifying picture series is via rational expressions that we introduce next: a *rational expression* is a term using the constants a for $a \in \Sigma$, the unary function symbols $s\cdot$ for $s \in S$, $^{\oplus^+}$, and $^{\ominus^+}$, and the binary function symbols $+$, $\oplus$, $\ominus$, and $\odot$.

Before we can define the semantics $[\![E]\!] : \Sigma^{++} \to S$ of a rational expression E, we need the column and row concatenation of pictures. The *column concatenation* $p \oplus q$ juxtaposes two pictures next to each other provided they have the same height; the *row concatenation* $p \ominus q$ places p on top of q provided they have the same width:

$$p \oplus q = \boxed{\;p\;|\;q\;} \quad \text{and} \quad p \ominus q = \begin{array}{|c|}\hline p \\ \hline q \\ \hline\end{array}\,.$$

Now we set inductively

$$[\![a]\!](p) = \begin{cases} 1 & \text{if } p \in \Sigma^{1\times 1} \text{ and } p_{1,1} = a, \\ 0 & \text{otherwise,} \end{cases}$$

$$[\![s \cdot E]\!](p) = s \cdot [\![E]\!](p),$$

$$[\![E + F]\!](p) = [\![E]\!](p) + [\![F]\!](p),$$

$$[\![E \odot F]\!](p) = [\![E]\!](p) \cdot [\![F]\!](p),$$

$$[\![E \oplus F]\!](p) = \sum_{\substack{p_1,p_2 \in \Sigma^{++} \\ p = p_1 \oplus p_2}} [\![E]\!](p_1) \cdot [\![F]\!](p_2),$$

$$[\![E^{\oplus^+}]\!](p) = \sum \left(\prod_{1 \le i \le n} [\![E]\!](p_i) \;\middle|\; \begin{array}{l} n \in \mathbb{N},\ p_1, \ldots, p_n \in \Sigma^{++}, \\ p = p_1 \oplus \cdots \oplus p_n \end{array} \right),$$

and the semantics of $E \ominus F$ and $E^{\ominus^+}$ are defined analogously. Then a picture series $f : \Sigma^{++} \to S$ is called *rational* if there exists a rational expression E with $[\![E]\!] = f$.

Logic

We proceed similarly to the case of words (see [16]) and traces (page 412). The only difference is the set of atomic formulas since we want to speak about pictures and not about labeled (partial) orders. So, let var and VAR be disjoint infinite sets of *individual* and *set variables* and let Σ be some alphabet. Then the syntax of weighted monadic second-order logic (for short: wMSO) is given by

$$\varphi ::= s \mid P_a(x) \mid P_{\neg a}(x) \mid xS_vy \mid x\not{S}_vy \mid xS_hy \mid x\ \not{S}_hy \mid x = y \mid x \neq y \mid \\ x \in X \mid x \notin X \mid \varphi \vee \varphi \mid \varphi \wedge \varphi \mid \exists x\, \varphi \mid \forall x\, \varphi \mid \exists X\, \varphi \mid \forall X\, \varphi,$$

where $s \in S$, $x, y \in$ var, $X \in$ VAR, and $a \in \Sigma$.

The *semantics* $[\![\varphi]\!]$ of a formula φ maps a picture $p \in \Sigma^{m \times n}$ together with an evaluation σ to an element of the semiring. Here, an *evaluation* maps individual variables from var to elements of $[m] \times [n]$ and set variables from VAR to subsets of $[m] \times [n]$. Then the semantics is defined as in the case of traces based on the following evaluation of atomic formulas (here $x, y \in$ var, $X \in$ VAR, and $a \in \Sigma$) where in the semantics of xS_vy and xS_hy we use S_v and S_h also to denote the vertical and horizontal successor relations, respectively: S_v is the set of pairs $((i,j),(i+1,j))$ and S_h that of pairs $((i,j),(i,j+1))$:

$$[\![P_a(x)]\!](p,\sigma) = \begin{cases} 1 & \text{if } p(\sigma(x)) = a \\ 0 & \text{otherwise} \end{cases} \qquad [\![xS_vy]\!](p,\sigma) = \begin{cases} 1 & \text{if } \sigma(x)S_v\sigma(y) \\ 0 & \text{otherwise} \end{cases}$$

$$[\![xS_hy]\!](p,\sigma) = \begin{cases} 1 & \text{if } \sigma(x)S_h\sigma(y) \\ 0 & \text{otherwise} \end{cases} \qquad [\![x = y]\!](p,\sigma) = \begin{cases} 1 & \text{if } \sigma(x) = \sigma(y) \\ 0 & \text{otherwise} \end{cases}$$

In addition, if φ is the negated version of the atomic formula ψ, then the rôles of 0 and 1 in the above definitions are interchanged:

$$[\![P_{\neg a}(x)]\!](p,\sigma) = \begin{cases} 0 & \text{if } p(\sigma(x)) = a \\ 1 & \text{otherwise} \end{cases} \qquad [\![x\not{S}_vy]\!](p,\sigma) = \begin{cases} 0 & \text{if } \sigma(x)S_v\sigma(y) \\ 1 & \text{otherwise} \end{cases}$$

$$[\![x\not{S}_hy]\!](p,\sigma) = \begin{cases} 0 & \text{if } \sigma(x)S_h\sigma(y) \\ 1 & \text{otherwise} \end{cases} \qquad [\![x \neq y]\!](p,\sigma) = \begin{cases} 0 & \text{if } \sigma(x) = \sigma(y) \\ 1 & \text{otherwise} \end{cases}$$

It is easily shown that $[\![\varphi]\!](p,\sigma) = [\![\varphi]\!](p,\tau)$ provided σ and τ coincide on those variables that occur freely in the formula $\varphi \in$ wMSO. If, in the extreme, $\varphi \in$ wMSO is a *sentence* (i.e., does not have any free variables), then it therefore defines a picture series $f : \Sigma^{++} \to S$ by $f(p) = [\![\varphi]\!](p,\sigma)$ for an arbitrary evaluation σ, and we denote this picture series by $[\![\varphi]\!]$, also. Any such picture series is called wMSO-*definable*; if $\mathcal{L} \subseteq$ wMSO and $\varphi \in \mathcal{L}$ is a sentence, then the picture series $[\![\varphi]\!]$ is called $\mathcal{L}$-*definable*.

4.3 Relating the Formalisms

Here, we will prove that for any commutative semiring the family of behaviors of wPA coincides with the class of projections of rational picture series (Theorems 4.4 and 4.5) and equals the family of picture series defined in terms of wRMSO and wREMSO logic (Theorems 4.6 and 4.10), respectively. This indicates that the class of recognizable picture series is robust.

4.3.1 Recognizable and Rational Picture Series

For words (see [21, 63]), traces, and sp-posets (see earlier sections), the behavior of every automaton can be constructed using a fixed set of rational operations. Marking another difference to picture series, this does not hold in the current setting: the characteristic series of all square pictures (over the Boolean semiring $\mathbb{B}$) is recognizable, but not rational [54] (based on results from [32, 10]). But, as we will show next, recognizable picture series are precisely the "projections" of rational picture series. Any such *projection* is given by a function $\pi : \Gamma \to \Sigma$ for two alphabets Γ and Σ. Then π can naturally be extended to a function $\pi' : \Gamma^{++} \to \Sigma^{++}$ setting $\pi'(p_\Gamma) = \pi \circ p_\Gamma$ for every $p_\Gamma \in \Gamma^{++}$. This function can then be lifted to a function $\pi'' : S^{\Gamma^{++}} \to S^{\Sigma^{++}}$ by $\pi''(f)(p_\Sigma) = \sum(f(p_\Gamma) \mid p_\Gamma \in \Gamma^{++}, \pi'(p_\Gamma) = p_\Sigma)$ for any $f : \Gamma^{++} \to S$ and $p_\Sigma \in \Sigma^{++}$. To simplify the notation, we will denote the derived functions π' and π'' also by π.

Weighted picture automata with behavior $[\![a]\!]$ are built easily. Generalizing the constructions from [21] from words to pictures, Bozapalidis and Grammatikopoulou [10] showed that the class of recognizable picture series is closed under the rational operations as well as under projections. This yields the following theorem.

Theorem 4.4 ([10]). *If f is the projection of a rational picture series, then there is a wPA $\mathcal{A}$ with $f = [\![\mathcal{A}]\!]$.*

To demonstrate the converse implication, let $\mathcal{A} = (Q, R, F_w, F_n, F_e, F_s)$ be a weighted picture automaton. Then let $L_h \subseteq R^+$ denote the set of words $r_1 r_2 \ldots r_n$ over the alphabet of rules such that:

(i) $\sigma_w(r_1) \in F_w$.
(ii) $\sigma_e(r_i) = \sigma_w(r_{i+1})$ for $1 \le i < n$.
(iii) $\sigma_e(r_n) \in F_e$.

In other words, L_h is the set of rows that can appear in successful runs of $\mathcal{A}$. Then define a function $f : R^+ \to S$ by

$$f(r_1 r_2 \ldots r_n) = \begin{cases} \prod(\mathrm{wgt}(r_i) \mid 1 \le i \le n) & \text{if } r_1 r_2 \ldots r_n \in L_h, \\ 0 & \text{otherwise.} \end{cases}$$

Understanding words over R as pictures from $R^{1\times n}$, the function f extends naturally to a picture series $r_h : R^{++} \to S$ that maps all pictures with at least two rows to 0. Similarly, we can define the set $L_v \in R^+$ that comprises all columns that can appear in a successful run of $\mathcal{A}$ (just replace w by n and e by s in (i), (ii), (iii) above). We then understand the words from L_v as columns and let r_v denote the characteristic series of L_v. Since the languages L_h and L_v are regular, they are rational by Kleene's theorem [39]. Hence, $f : R^+ \to S$ and the characteristic function g of L_v are rational formal power series. This implies the rationality of r_h and r_v, i.e., there are rational expressions E_h and E_v with $[\![E_h]\!] = r_h$ and $[\![E_v]\!] = r_v$. Finally, note that the behavior of $\mathcal{A}$ equals the projection under label : $R \to \Sigma$ of the rational picture series $[\![E_h^{\ominus^+} \odot E_v^{\oplus^+}]\!]$. This proves the converse implication of Theorem 4.4.

Theorem 4.5 ([54]). *If $\mathcal{A}$ is a wPA, then $[\![\mathcal{A}]\!]$ is the projection of some rational picture series.*

4.3.2 Recognizable Picture Series and Logics

For traces, we used the notion of a definable step formula. In the setting of pictures, we need a further restriction to FO-definable step formulas: a formula φ is a *first-order definable step formula* if there exist $n \in \mathbb{N}$, $s_1, \ldots, s_n \in S$, and wMSO-formulas $\beta_1, \ldots, \beta_n$ over the Boolean semiring $\mathbb{B}$ *without set quantification* such that, for every picture $p \in \Sigma^{++}$ and every evaluation σ, we have $[\![\varphi]\!](p,\sigma) = \sum_{1\le i\le n} s_i f_i(p,\sigma)$ where f_i is the characteristic function (in the semiring S) of the support of $[\![\beta_i]\!]$. *Unambiguous formulas*, the set wRMSO of *restricted formulas*, and the set wREMSO of *restricted existential formulas* are defined in the same way as for traces (cf. page 416 et seq.; here, $\forall x\, \alpha$ is restricted if α is a FO-definable step formula). For every positive Boolean combination φ of (negated) atomic formulas, one can easily find an equivalent unambiguous formula φ^+ (this does not extend to first-order formulas over the Boolean semiring since one cannot handle the existential quantifier due to the local nature of the relations S_v and S_h).

Now we follow the usual way (cf. [65]) of transforming an automaton into a formula where an evaluation of second-order variables reflects an assignment of positions to rules. Therefore, we introduce variables X_r for every $r \in R$ and build formulas for the following statements:

- The sets $\{X_r \mid r \in R\}$ form a partition of the positions of the picture.
- If a position belongs to X_r, then its label equals that of the rule r, i.e., $\forall x \bigwedge_{r\in R}(x \notin X_r \vee P_{\text{label}(r)}(x))$.
- Every position x and its local neighborhood satisfies the condition on a run, e.g., $\forall x \forall y (x \not S_h y \vee \bigvee_{r,r'\in R, \sigma_w(r)=\sigma_e(r')}(x \in X_r \wedge y \in X_{r'}))$ in conjunction with the analogous formula for the vertical direction.
- All positions on the left border have a state from F_w at their left pole, e.g., $\forall x(\exists y(y S_h x) \vee \bigvee_{r\in R, \sigma_w(r)\in F_w} x \in X_r)$ and similarly for the other borders.

All these formulas describe qualitative statements and we can write down equivalent unambiguous formulas. Finally, we take the conjunction of all these unambiguous formulas and the formula $\forall x(\bigvee_{r\in R} \mathrm{wgt}(r) \wedge x \in X_r)$. This construction proves the following theorem.

Theorem 4.6. *For every wPA $\mathcal{A}$, there is a sentence $\varphi \in$ wREMSO such that $[\![\varphi]\!] = [\![\mathcal{A}]\!]$.*

We now want to prove the converse implication. As in the case of words, this is done by induction over the construction of a formula $\varphi \in$ wRMSO. Hence, we have to handle free variables, i.e., pictures p together with an evaluation σ of the variables. To this aim, let $\mathcal{V} \subseteq \mathrm{var} \cup \mathrm{VAR}$ be a finite set of variables, $p \in \Sigma^{m\times n}$, and σ a evaluation. Then $\Sigma_{\mathcal{V}} = \Sigma \times \{0,1\}^{\mathcal{V}}$ is an enriched alphabet where each letter is extended by one bit per variable from $\mathcal{V}$. We further define $r = (p,\sigma)_{\mathcal{V}} \in \Sigma_{\mathcal{V}}^{m\times n}$ by $r_{i,j} = (p_{i,j}, (b_v)_{v\in\mathcal{V}})$ with $b_x = 1$ if and only if $\sigma(x) = (i,j)$ for $x \in \mathcal{V} \cap \mathrm{var}$ and $b_X = 1$ if and only if $(i,j) \in \sigma(X)$ for $X \in \mathcal{V} \cap \mathrm{VAR}$. Note that encodings r which we obtain this way have the following property: for every $x \in \mathcal{V} \cap \mathrm{var}$, there is a unique position in $[m] \times [n]$ such that the bit b_x at this position equals 1. Arbitrary pictures over the enriched alphabet with this property are called *valid*.

Conversely, let r be a valid picture and p be its projection to Σ. Then there exists an evaluation σ such that r is obtained from p and σ as above. If now the free variables of the formula $\varphi \in$ wMSO are contained in the set $\mathcal{V}$, we define a new picture series $[\![\varphi]\!]_{\mathcal{V}} : \Sigma_{\mathcal{V}}^{++} \to S$ as follows: for $r = (p,\sigma)_{\mathcal{V}}$, set $[\![\varphi]\!]_{\mathcal{V}}(r) = [\![\varphi]\!](p,\sigma)$.[4] Extend this to a total function by setting $[\![\varphi]\!]_{\mathcal{V}}(r) = 0$ if r is not valid.

As mentioned above, we construct a wPA for $[\![\varphi]\!]$ by structural induction over φ. The base case as well as disjunction, conjunction, and existential quantification are first dealt with in the following lemma.

Lemma 4.7. *Let $\varphi, \psi \in$ wMSO and $\mathcal{V}$ be a finite set of variables containing all the free variables of φ and ψ. Then the following hold:*

1. *If φ is an atomic formula (positive or negative), then $[\![\varphi]\!]_{\mathcal{V}}$ is recognizable.*
2. *If $[\![\varphi]\!]_{\mathcal{V}}$ and $[\![\psi]\!]_{\mathcal{V}}$ are recognizable, then so are $[\![\varphi \vee \psi]\!]_{\mathcal{V}}$ and $[\![\varphi \wedge \psi]\!]_{\mathcal{V}}$.*
3. *If $z \in \mathrm{var} \cup \mathrm{VAR}$ and $[\![\varphi]\!]_{\mathcal{V}}$ is recognizable, then so is $[\![\exists z\varphi]\!]_{\mathcal{V}\setminus\{z\}}$.*

Proof. (1) Weighted picture automata with the necessary behaviors can easily be constructed explicitly.

(2) Recall that $[\![\varphi \vee \psi]\!]_{\mathcal{V}} = [\![\varphi]\!]_{\mathcal{V}} + [\![\psi]\!]_{\mathcal{V}}$ and $[\![\varphi \wedge \psi]\!]_{\mathcal{V}} = [\![\varphi]\!]_{\mathcal{V}} \odot [\![\psi]\!]_{\mathcal{V}}$. From wPA for $[\![\varphi]\!]_{\mathcal{V}}$ and $[\![\psi]\!]_{\mathcal{V}}$, wPA for these series have been constructed in [10].

(3) Since $[\![\exists z\,\varphi]\!]_{\mathcal{V}\setminus\{z\}}$ is the projection of $[\![\varphi]\!]_{\mathcal{V}}$, a wPA for $[\![\exists z\,\varphi]\!]_{\mathcal{V}\setminus\{z\}}$ can be constructed from one for $[\![\varphi]\!]_{\mathcal{V}}$ by [10]. □

[4] The valid picture r does not completely define the evaluation σ. Since the free variables of φ all belong to $\mathcal{V}$, any two evaluations σ and τ with $r = (p,\sigma) = (p,\tau)$ satisfy $[\![\varphi]\!](p,\sigma) = [\![\varphi]\!](p,\tau)$ which proves that $[\![\varphi]\!]_{\mathcal{V}}$ is nevertheless well defined.

As in the case of words, the most difficult step is universal quantification. For this step, we need the following notion and result: a wPA $\mathcal{A}$ is *unambiguous* if, for every picture p, there is at most one successful run of $\mathcal{A}$ on p.

Proposition 4.8 ([53]). *Let $\varphi \in$ wMSO be a first-order formula over the Boolean semiring $\mathbb{B}$ with free variables in $\mathcal{V}$. Then $[\![\varphi]\!]_{\mathcal{V}}$ is the behavior of some unambiguous wPA.*

Proof. By [33, Proposition 5], the support of $[\![\varphi]\!]_{\mathcal{V}}$ is a locally threshold testable (LTT) picture language. Any such set of pictures is a finite disjoint union of strictly LTT languages. These languages can be accepted by unambiguous picture automata in the sense of [33], i.e., in our terminology, their characteristic function over $\mathbb{B}$ is the behavior of some unambiguous wPA. Since the behaviors of unambiguous wPA are closed under addition (provided the supports are disjoint), the result follows. □

Lemma 4.9. *Let $\varphi \in$ wMSO be a FO-definable step formula with free variables in $\mathcal{V}$ and let $x \in$ var. Then $[\![\forall x\, \varphi]\!]_{\mathcal{V}\setminus\{x\}}$ is the behavior of some wPA.*

Proof. Since φ is a FO-definable step formula, there are $n \in \mathbb{N}$, $s_\ell \in S$, and wMSO-formulas β_ℓ without set variables over the Boolean semiring such that $[\![\varphi]\!](p,\sigma) = \sum_{1\leq \ell \leq n} s_\ell f_\ell(p,\sigma)$ where f_ℓ is the characteristic function (in the semiring S) of the support L_ℓ of $[\![\beta_\ell]\!]$. One can assume that the languages L_ℓ form a partition.

Next, let $\widetilde{L}$ denote the set of triples (p,ν,σ) with $p \in \Sigma^{++}$ a picture, σ an evaluation, and $\nu : \mathrm{Dom}(p) \to [n]$ such that

$$\nu(i,j) = \ell \quad \Longleftrightarrow \quad \big(p, \sigma[x \to (i,j)]\big) \in L_\ell$$

where $\sigma[x \to (i,j)]$ denotes the evaluation τ with $\tau(x) = (i,j)$ that coincides with σ on all the variables distinct from x. The pairs (p,ν) can naturally be understood as pictures over the extended alphabet $\widetilde{\Sigma} = \Sigma \times [n]$.

From the formulas β_ℓ, one then builds a formula β without set variables over the Boolean semiring such that $[\![\beta]\!]_{\mathcal{V}}$ is the characteristic function of $\{((p,\nu),\sigma)_{\mathcal{V}} \mid (p,\nu,\sigma) \in \widetilde{L}\}$. Then by Proposition 4.8, there is an unambiguous wPA over $\mathbb{B}$ whose behavior equals $[\![\beta]\!]$. Replacing all rules $((a,\ell,(b_v)_{v\in\mathcal{V}}),1,q_w,q_n,q_e,q_s)$ by $((a,(b_v)_{v\in\mathcal{V}\setminus\{x\}}),s_\ell,q_w,q_n,q_e,q_s)$ yields a wPA, and one can show that its behavior is $[\![\forall x\,\varphi]\!]_{\mathcal{V}\setminus\{x\}}$. □

From Lemmas 4.7 and 4.9, we finally obtain the converse implication of Theorem 4.6.

Theorem 4.10 ([53]). *For every $\varphi \in$ wRMSO, there exists a wPA $\mathcal{A}$ with $[\![\varphi]\!] = [\![\mathcal{A}]\!]$.*

As in the case of words, the existential fragment of wMSO can express behaviors that are beyond the abilities of wPA. An interesting problem is to find properties of the semiring that prohibit this. In the following, we sketch one such example. A monoid M is called *weakly aperiodic* if for every $x \in M$ there exists $m_x \in \mathbb{N}$ with $x^{m_x} = x^{m_x+1}$. A semiring S is *weakly bi-aperiodic* if both $(S, +, 0)$ and $(S, \cdot, 1)$ are weakly aperiodic monoids. This condition allows us to prove that every formula from wMSO without set quantification is a FO-definable step formula. Hence, from Theorems 4.6 and 4.10, we obtain the following theorem.

Theorem 4.11 ([22, Proposition 7.28]). *Let S be weakly bi-aperiodic and $f : \Sigma^{++} \to S$. Then f is the behavior of some existential formula if and only if it is the behavior of some wPA.*

4.3.3 The Characterization Theorem

In summary, we obtain the equivalence of all the formalisms introduced. This indicates that the class of recognizable picture series is robust.

Theorem 4.12 ([10, 23, 54]). *Let S be a commutative semiring and $f : \Sigma^{++} \to S$ a picture series. Then the following are equivalent:*

1. *f is the behavior of some wPA.*
2. *f is the projection of a rational picture series.*
3. *f is the semantics of some sentence from wRMSO.*

Proof. Equivalence of the first two statements follows from Theorems 4.4 and 4.5, that of the first and last statement from Theorems 4.6 and 4.10. □

4.4 Decidability Issues

Recall the following result from the one-dimensional setting of words [14]: Let S be a computable field. Then it is decidable whether a weighted formula φ is restricted. In case φ is restricted, one can effectively compute a weighted finite automaton for $[\![\varphi]\!]$. Due to the two dimensions in pictures, this result breaks down completely.

Proposition 4.13 ([22, Proposition 8.1]). *It is undecidable whether a formula φ from* wMSO *is restricted or not.*

Proof. Let $w = a_1 a_2 \dots a_n \in \Gamma^+$ be a word and $h : \Gamma^+ \to \Delta^+$ a semigroup homomorphism. The picture $p_h(w)$ over $\Gamma \cup \Delta \cup \{\square\}$ contains, in the top row, the word $h(w)$ and in the bottom row a word from $w\square^*$. The intermediate rows are used to construct inductively the words $h(a_1 \dots a_i) a_{i+1} \dots a_n$.

Now let (h, g) be an instance of Post's correspondence problem PCP, i.e., two homomorphisms $h, g : \Gamma^+ \to \Delta^+$. Then we consider the set L of pictures that are overlays of two pictures $p_h(w)$ and $p_g(w)$ that coincide in the top and bottom rows. Hence, L encodes the solutions of the PCP-instance (h, g). Now consider the projection K of L to a unary alphabet Σ. Only in the trivial case of $K = \emptyset$ we can express membership in K by a first-order formula. But this is always possible in monadic second-order logic. Therefore, one can construct a wMSO-formula φ such that $[\![\varphi]\!]$ is the characteristic function for K. It follows that $\forall x\, \varphi$ is restricted if and only if φ is a FO-definable step formula if and only if $K = \emptyset$ if and only if (h, g) has no solution. Since this is undecidable, the result follows. □

Note that the previous proposition holds for any commutative semiring. With the same generality, using similar ideas, we get the following proposition.

Proposition 4.14 ([22, Corollary 8.3]). *It is undecidable whether two given* wMSO*-formulas φ and ψ satisfy $[\![\varphi]\!] = [\![\psi]\!]$. It is also undecidable whether the behavior of a given wPA is the semantics of some FO-definable step formula.*

Note that in a locally finite semiring, the image of the behavior of every wPA is finite. Now let S not be locally finite. Using distributivity, one obtains an element s of S that has infinite order in $(S, \cdot, 1)$ or in $(S, +, 0)$ (we consider the first case, the second is dealt with similarly). With the notions from the proof of Proposition 4.13, we have that (f, g) has a solution if and only if (f, g) has infinitely many solutions if and only if K is infinite. In addition, the picture series f with $f(p) = s^{mn}$ for $p \in K$ of dimension $m \times n$ and 0 for $p \notin K$ is recognizable. Since s has infinite order in $(S, \cdot, 1)$, this picture series f has finite image if and only if K is empty which is undecidable. Thus, we have proved the following proposition.

Proposition 4.15 ([22, Proposition 8.4]). *Let S not be locally finite. Then it is undecidable whether the behavior of a given wPA has finite image.*

4.5 History and Overview

In the literature, one finds a wealth of proposals of formal models for recognizing or generating two-dimensional arrays of symbols, e.g., three-way and four-way machines as well as two-dimensional Turing machines or marker automata [8, 36, 46]. Finally, in [31, 32], Restivo and Giammarresi introduced a new definition of finite-state recognizability for two-dimensional languages: recognizable picture languages. This definition takes as starting point a well-known characterization of recognizable string languages in terms of local languages and projections. Besides its conceptual simplicity, this device turned out to be very robust since it admits also elegant connections with other two-dimensional machine-based models and logical formalisms; more precisely, recognizable picture languages correspond to EMSO logic, which for pictures is strictly weaker than the full MSO logic (even more, the monadic second-order quantifier alternation hierarchy over the class of pictures is strict [52]).

Bozapalidis and Grammatikopoulou [10] based their extension to a quantitative setting on this notion of recognizability for picture languages introducing the model of a weighted picture automaton. Their main result asserts that picture series computed by weighted picture automata are closed under rational operations and projections.

Picture languages and picture series can be related via the notion of the support of a series. If the semiring S is not a ring, then every recognizable picture language is the support of some recognizable picture series over S. Conversely, one considers so called cut languages: Let $f : \Sigma^{++} \to \mathbb{N}$ and let $n \in \mathbb{N}$. Then the *cut language* L_n is the set of pictures whose value under f exceeds n. Then every cut language of a recognizable picture series is effectively recognizable—a result that holds for many more semirings [22, Proposition 9.5].

Let us discuss some extensions. Surely, two dimensions are not the end of the story: we could consider objects of higher dimension or other regular structures. Probably some of the results presented in this section carry over to higher dimensions.

An extension of picture series to series over infinite pictures is another interesting problem to be considered. There are results and different automata models to recognize languages of infinite pictures; see, e.g., [1, 24]. In particular, there is a hierarchy of acceptance conditions as known from the theory of ω-languages together with many new combinatorial phenomena. Can we extend these results and new proof techniques to series on infinite pictures?

Moreover, there is a well-established theory of probabilistic cellular automata working on (dynamic) pictures, e.g., [38]. Concentrating on their stable configurations, we can comprehend their behavior on infinite (static) pictures within some quantitative setting. Are there similarities between probabilistic cellular automata seen on infinite pictures in this way and series on infinite pictures?

All objects of this chapter are in some sense finite directed labeled graphs having particular structure. What we would like to have is a general concept of weighted graph devices covering the presented quantitative settings for these structures. Thomas [66] introduced graph acceptors working on graphs of bounded degree and revealed their relation to logical definability. This concept at least covers the unweighted versions of the automata models considered in this chapter and could therefore serve as a starting point for a general theory (for a first attempt, see [22]).

References

1. J. Altenbernd, W. Thomas, and S. Wöhrle. Tiling systems over infinite pictures and their acceptance conditions. In M. Ito and M. Toyama, editors, *DLT 2002*, volume 2450 of *Lecture Notes in Computer Science*, pages 297–306. Springer, Berlin, 2002.
2. M. Anselmo, D. Giammarresi, M. Madonia, and A. Restivo. Unambiguous recognizable two-dimensional languages. *Informatique Théorique et Applications, RAIRO*, 40:277–293, 2006.

3. A. Arnold. An extension of the notions of traces and of asynchronous automata. *Informatique Théorique et Applications, RAIRO*, 25:355–393, 1991.
4. C. Baier, M. Größer, and F. Ciesinski. Model checking linear-time properties of probabilistic systems. In this *Handbook*. Chapter 13. Springer, Berlin, 2009.
5. N. Bedon and C. Rispal. Series-parallel languages on scattered and countable posets. In L. Kucera and A. Kucera, editors, *MFCS 2007*, volume 4708 of *Lecture Notes in Computer Science*, pages 477–488. Springer, Berlin, 2007.
6. J. Berstel and C. Reutenauer. *Rational Series and Their Languages*, volume 12 of *Monographs in Theoretical Computer Science. An EATCS Series*. Springer, Berlin, 1988.
7. J. Berstel and C. Reutenauer. Extension of Brzozowski's derivation calculus of rational expressions to series over the free partially commutative monoids. *Theoretical Computer Science*, 400:144–158, 2008.
8. M. Blum and C. Hewitt. Automata on a 2-dimensional tape. In *IEEE Symposium on Switching and Automata Theory*, pages 155–160, 1967.
9. B. Bollig and I. Meinecke. Weighted distributed systems and their logics. In S.N. Artemov and A. Nerode, editors, *LFCS 2007*, volume 4514 of *Lecture Notes in Computer Science*, pages 54–68. Springer, Berlin, 2007.
10. S. Bozapalidis and A. Grammatikopoulou. Recognizable picture series. *Journal of Automata, Languages and Combinatorics*, 10(2/3):159–183, 2005.
11. L. de Prophetis and S. Varricchio. Recognizability of rectangular pictures by Wang systems. *Journal of Automata, Languages and Combinatorics*, 2(4):269–288, 1997.
12. V. Diekert and G. Rozenberg, editors. *The Book of Traces*. World Scientific, Singapore, 1995.
13. M. Droste and P. Gastin. The Kleene–Schützenberger theorem for formal power series in partially commuting variables. *Information and Computation*, 153:47–80, 1999.
14. M. Droste and P. Gastin. Weighted automata and weighted logics. *Theoretical Computer Science*, 380:69–86, 2007.
15. M. Droste and P. Gastin. On aperiodic and star-free formal power series in partially commuting variables. *Theory of Computing Systems*, 42:608–631, 2008.
16. M. Droste and P. Gastin. Weighted automata and weighted logics. In this *Handbook*. Chapter 5. Springer, Berlin, 2009.
17. M. Droste, P. Gastin, and D. Kuske. Asynchronous cellular automata for pomsets. *Theoretical Computer Science*, 247:1–38, 2000.
18. M. Droste and W. Kuich. Semirings and formal power series. In this *Handbook*. Chapter 1. Springer, Berlin, 2009.
19. M. Droste and D. Kuske. Recognizable languages in divisibility monoids. *Mathematical Structures in Computer Science*, 11:743–770, 2001.

20. M. Droste and D. Kuske. Skew and infinitary formal power series. *Theoretical Computer Science*, 366:199–227, 2006.
21. Z. Esik and W. Kuich. Finite automata. In this *Handbook*. Chapter 3. Springer, Berlin, 2009.
22. I. Fichtner. Characterizations of recognizable picture series. Dissertation, Universität Leipzig, 2007.
23. I. Fichtner. Weighted picture automata and weighted logics. *Theory of Computing Systems*, 2009, accepted. Extended abstract in B. Durand and W. Thomas, editors, *STACS 2006*, volume 3884 of *Lecture Notes in Computer Science*, pages 313–324. Springer, Berlin, 2006.
24. O. Finkel. On recognizable languages of infinite pictures. *International Journal of Foundations of Computer Science*, 15(6):823–840, 2004.
25. M. Fliess. Séries reconnaissables, rationelles et algébriques. *Bulletin des Sciences Mathématiques*, 94:231–239, 1970.
26. M. Fliess. Matrices de Hankel. *Journal de Mathématiques Pures et Appliquées*, 53:197–224, 1974.
27. S. Gaubert and J. Mairesse. Task resource models and (max,+) automata. In J. Gunawardena, editor, *Idempotency*, pages 133–144. Cambridge University Press, Cambridge, 1997.
28. S. Gaubert and J. Mairesse. Asymptotic analysis of heaps of pieces and application to timed Petri nets. In P. Buchholz and M. Silva, editors, *Petri Nets and Performance Models (PNPM'99)*, pages 158–169. IEEE Computer Society, Los Alamitos, 1999.
29. S. Gaubert and J. Mairesse. Modeling and analysis of timed Petri nets using heaps of pieces. *IEEE Transactions on Automatic Control*, 44(4):683–698, 1999.
30. B. Genest, D. Kuske, and A. Muscholl. A Kleene theorem and model checking algorithms for existentially bounded communicating automata. *Information and Computation*, 204:920–956, 2006.
31. D. Giammarresi and A. Restivo. Recognizable picture languages. *International Journal of Pattern Recognition and Artificial Intelligence*, 6 (2–3):241–256, 1992.
32. D. Giammarresi and A. Restivo. Two-dimensional languages. In G. Rozenberg and A. Salomaa, editors, *Handbook of Formal Languages, volume 3*, pages 215–267. Springer, Berlin, 1997.
33. D. Giammarresi, A. Restivo, S. Seibert, and W. Thomas. Monadic second-order logic over rectangular pictures and recognizability by tiling systems. *Information and Computation*, 125:32–45, 1996.
34. J. Gischer. The equational theory of pomsets. *Theoretical Computer Science*, 61:199–224, 1988.
35. J. Grabowski. On partial languages. *Fundamenta Informaticae*, 4(2):427–498, 1981.
36. K. Inoue and A. Nakamura. Some properties of two-dimensional on-line tessellation acceptors. *Information and Computation*, 13:95–121, 1977.

37. S. Jesi, G. Pighizzini, and N. Sabadini. Probabilistic asynchronous automata. *Theory of Computing Systems*, 29(1):5–31, 1996.
38. J. Kari. Theory of cellular automata: A survey. *Theoretical Computer Science*, 334(1–3):3–33, 2005.
39. S. Kleene. Representation of events in nerve nets and finite automata. In C. Shannon and J. McCarthy, editors, *Automata Studies*, volume 34 of *Annals of Mathematics Studies*, pages 3–40. Princeton University Press, Princeton, 1956.
40. D. Kuske. Regular sets of infinite message sequence charts. *Information and Computation*, 187:80–109, 2003.
41. D. Kuske. Towards a language theory for infinite N-free pomsets. *Theoretical Computer Science*, 299:347–386, 2003.
42. D. Kuske. Weighted and unweighted trace automata. *Acta Cybernetica*, 2009, in press.
43. D. Kuske. Weighted asynchronous cellular automata. *Theoretical Computer Science*, 374:127–148, 2007.
44. D. Kuske and I. Meinecke. Branching automata with costs—A way of reflecting parallelism in costs. *Theoretical Computer Science*, 328:53–75, 2004.
45. D. Kuske and R. Morin. Pomsets for local trace languages: Recognizability, logic and Petri nets. *Journal of Automata, Languages and Combinatorics*, 7:187–224, 2002.
46. M. Latteux and D. Simplot. Recognizable picture languages and domino tiling. *Theoretical Computer Science*, 178:275–283, 1997.
47. K. Lodaya and P. Weil. Series-parallel languages and the bounded-width property. *Theoretical Computer Science*, 237:347–380, 2000.
48. K. Lodaya and P. Weil. Rationality in algebras with a series operation. *Information and Computation*, 171:269–293, 2001.
49. C. Mathissen. Verhalten gewichteter Automaten mit Deflationsparameter über Spurmonoiden. Diplomarbeit, Technische Universität Dresden, 2005.
50. C. Mathissen. Definable transductions and weighted logics for texts. In T. Harju, J. Karhumäki, and A. Lepistö, editors, *DLT 2007*, volume 4588 of *Lecture Notes in Computer Science*, pages 324–336. Springer, Berlin, 2007.
51. C. Mathissen. Weighted logics for nested words and algebraic formal power series. In L. Aceto, I. Damård, L.A. Goldberg, M.M. Halldórsson, A. Ingólfsdóttir, and I. Walukiewicz, editors, *ICALP 2008*, volume 5126 of *Lecture Notes in Computer Science*, pages 221–232. Springer, Berlin, 2008.
52. O. Matz, N. Schweikardt, and W. Thomas. The monadic quantifier alternation hierarchy over grids and graphs. *Information and Computation*, 179(2):356–383, 2002.

53. I. Mäurer. Weighted picture automata and weighted logics. In B. Durand and W. Thomas, editors, *STACS 2006*, volume 3884 of *Lecture Notes in Computer Science*, pages 313–324. Springer, Berlin, 2006.
54. I. Mäurer. Characterizations of recognizable picture series. *Theoretical Computer Science*, 374:214–228, 2007.
55. A. Mazurkiewicz. Concurrent program schemes and their interpretation. Technical report, DAIMI Report PB-78, Aarhus University, 1977.
56. A. Mazurkiewicz. Introduction to trace theory. In V. Diekert and G. Rozenberg, editors, *The Book of Traces.* World Scientific, Singapore, 1995.
57. I. Meinecke. Weighted branching automata—Combining concurrency and weights. Dissertation, Technische Universität Dresden, 2004.
58. I. Meinecke. The Hadamard product of sequential-parallel series. *Journal of Automata, Languages and Combinatorics*, 10(2/3):313–346, 2005.
59. I. Meinecke. Weighted logics for traces. In D. Grigoriev, J. Harrison, and E.A. Hirsch, editors, *CSR 2006*, volume 3976 of *Lecture Notes in Computer Science*, pages 235–246. Springer, Berlin, 2006.
60. E. Ochmański. Regular behaviour of concurrent systems. *Bulletin of the European Association for Theoretical Computer Science*, 27:56–67, 1985.
61. V. Pratt. Modelling concurrency with partial orders. *International Journal of Parallel Programming*, 15:33–71, 1986.
62. J. Sakarovitch. The "last" decision problem for rational trace languages. In I. Simon, editor, *LATIN 1992*, volume 583 of *Lecture Notes in Computer Science*, pages 460–473. Springer, Berlin, 1992.
63. J. Sakarovitch. Rational and recognisable power series. In this *Handbook.* Chapter 4. Springer, Berlin, 2009.
64. A. Salomaa and M. Soittola. *Automata-Theoretic Aspects of Formal Power Series.* Springer, Berlin, 1978.
65. W. Thomas. Automata on infinite objects. In J. van Leeuwen, editor, *Handbook of Theoretical Computer Science*, pages 133–191. Elsevier, Amsterdam, 1990.
66. W. Thomas. On logics, tilings, and automata. In J. Leach, editor, *ICALP 1991*, volume 510 of *Lecture Notes in Computer Science*, pages 441–453. Springer, Berlin, 1991.
67. W. Zielonka. Notes on finite asynchronous automata. *Informatique Théorique et Applications, RAIRO*, 21:99–135, 1987.

Part IV

Applications

Chapter 11: Digital Image Compression

Jürgen Albert[1] and Jarkko Kari[2]

[1] Informatik II, Universität Würzburg, 97074 Würzburg, Germany
albert@informatik.uni-wuerzburg.de
[2] Department of Mathematics, University of Turku, 20014 Turku, Finland
jkari@utu.fi

1 Introduction

Regular languages have a self-similar structure, given explicitly in their finite automata descriptions. This fractal nature can be observed visually by using a suitable interpretation of words of the language as addresses of black image pixels.

Under a similar addressing scheme, weighted languages and automata define grayscale images: the weight of a word gives the intensity of the corresponding image pixel. It turns out that weighted finite automata (WFA) are already powerful enough to describe both fractal-like and smooth images. Simple inference algorithms exist for constructing WFA representations of given images, and standard automata minimization techniques can be used to minimize the number of states. This naturally leads to the idea of using

M. Droste, W. Kuich, H. Vogler (eds.), *Handbook of Weighted Automata*, Monographs in Theoretical Computer Science. An EATCS Series,
DOI 10.1007/978-3-642-01492-5_11,

such automata as compressed representations of images. This concept was introduced in the early 1990s [7, 9–11], and it soon led to further improved and generalized algorithms; see, e.g., [3, 4, 16, 21, 22, 25, 26]. See also [24] for a tutorial and further examples. In this article, the basic concepts and algorithms are presented.

We start in Sect. 2 by discussing our addressing scheme of pixels using words over a four-letter alphabet. Infinite languages define infinitely sharp images, so we next briefly discuss the concept of multi-resolution images. A multi-resolution image is simply a formal power series, that is, a function that assigns colors to words. In case the color set is a semiring, weighted finite automata can be used to describe power series. This is the topic of Sect. 3. Most of the time, the semiring used is $\mathbb{R}$; the set of real numbers under normal addition and multiplication. In Sect. 4, we discuss how one can efficiently draw the image specified by a given WFA. Next, we turn our attention to the converse problem: Sect. 5 concentrates on the problem of inferring a WFA that represents a given input image. We provide an efficient algorithm for this task. The algorithm is guaranteed to produce the minimum state WFA. In Sect. 6, we outline the ideas used to transform the theoretical inference algorithm into a practical image compression technique. The details of these ideas are skipped and the interested reader is referred to the more technical descriptions [10, 23]. We compare our algorithm with the image compression standard JPEG [31] that is based on the discrete cosine transform. Using several test images, we show the difference in performance, and we demonstrate that the WFA technique performs especially well on images with sharp edges. Then in Sect. 7, we briefly discuss one nice aspect of the WFA representation of images: several natural image transformations can be implemented directly in the compressed form, without the need to decode the image first. We show several examples of image operations that can be defined using weighted finite transducers (WFT). Finally, we stress the fractal nature of WFA-concepts in Sect. 8 by generalizing from WFA to higher-dimensional parametric weighted finite automata, which can, e.g., simulate any iterated function system easily.

2 Image Types

A *finite resolution* image is a $w \times h$ rectangular array of pixels, each with its own color. In the context of this work, squares of dimensions that are powers of two come up naturally, so that

$$w = h = 2^k \quad \text{for some } k \in \mathbb{Z}_+.$$

Definition 2.1. *The image is then just a function*

$$\{1, 2, \ldots, w\} \times \{1, 2, \ldots, h\} \to S$$

that assigns to each pixel a color from a set S of possible colors. The color set depends on the type of the image. It can be:

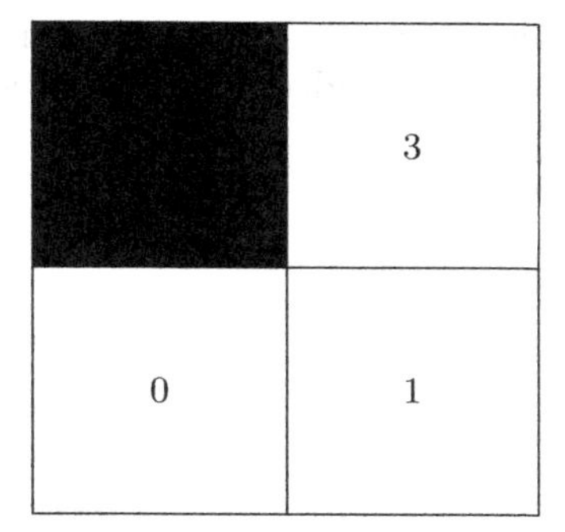

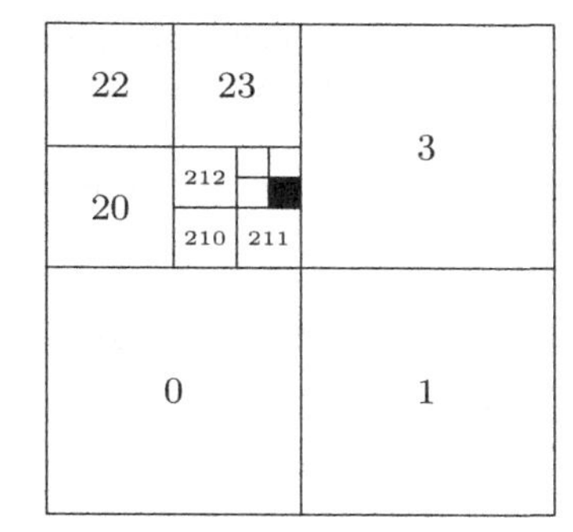

Fig. 1. Pixels with addresses 2 and 2131 respectively

- $S = \mathbb{B} = \{0, 1\}$ *on bi-level or binary images, where* 0 *and* 1 *represent white and black, respectively.*
- $S = \mathbb{R}$, *the set of reals, on grayscale images. The color value represents the intensity of the pixel. Sometimes only non-negative real numbers* $S = \mathbb{R}_+$ *are allowed, sometimes the values are restricted to an interval, e.g.,* $S = [0, 1]$. *In digital images the intensity values are quantized in a finite number of intensity levels. However, in this article, we always use* $S = \mathbb{R}$.
- $S = \mathbb{R}^3$, *vectors of three real numbers, on color images. Three values represent the intensities of three color components. Also,* $\mathbb{R}^3_+$ *or* $[0, 1]^3$ *may be used.*

Rather than addressing the pixels by their x- and y-coordinates, let us address the pixels using words over the four-letter alphabet $\Sigma = \{0, 1, 2, 3\}$. Pixels of a $2^k \times 2^k$ image are addressed by words of length k as follows.

Definition 2.2. *The image is first divided into four quadrants. The first letter of the address of a pixel is determined by the quadrant that contains the pixel.*

The rest of the address is defined recursively as the word of length $k - 1$, *that is, the address of the pixel in the quadrant when the quadrant is viewed as a* $2^{k-1} \times 2^{k-1}$ *image.*

The only pixel of a 1×1 resolution "image" has the empty word ε as address.

Example 2.3. Figure 1 shows the locations of the pixels addressed by 2 at resolution 2×2 and 2131 at resolution 16×16, respectively.

Our addressing scheme thus defines a *quadtree* structure on the pixels of the images. Quadtrees (also octtrees, etc.) can be embedded into *bintrees* in a canonical way when encoding the alphabet symbols $0, 1, 2, 3$ by their binary number-representations $00, 01, 10, 11$. For addresses of odd length, this opens a possibility to address also images and sub-images of sizes $2^{k+1} \times 2^k$. Bintrees have been used successfully for fine-tuning the encoding algorithm presented in the section on practical image compression (see, e.g., [15]). For simplicity of the arguments, we will stick to the common quadtree representations for the following.

A $2^k \times 2^k$ resolution image is now a function

$$r_k : \Sigma^k \to S$$

that assigns a color to each pixel.

Definition 2.4. *A multi-resolution image is a function that assigns a color to each pixel in resolutions $2^k \times 2^k$ for all $k \geq 0$. In other words, it is a formal power series $r \in S\langle\langle \Sigma^* \rangle\rangle$ where $\Sigma = \{0, 1, 2, 3\}$ and S is the color set. Coefficient (r, w) is the color of pixel w.*

A multi-resolution image can be viewed as a sequence $r_0, r_1, r_2, \ldots$ of images where r_k, the restriction of r to words of length k, is a $2^k \times 2^k$ image.

Note that without any additional constraint on a multi-resolutions image r, the images r_k may be completely unrelated. However, one would like them to represent approximations of the same image at different resolutions. This constraint is captured into the concept of average preservation. Let us consider grayscale images, i.e., the case $S = \mathbb{R}$. A simple way to interpolate a $2^k \times 2^k$ resolution grayscale image into a $2^{k-1} \times 2^{k-1}$ resolution image is to average the intensity values on 2×2 blocks of pixels.

Definition 2.5. *Let r be a multi-resolution grayscale image. If*

$$(r, w) = \frac{(r, w0) + (r, w1) + (r, w2) + (r, w3)}{4}$$

for every $w \in \Sigma^*$, then r_{k-1} is the interpolation of the next sharper image r_k, for every $k = 1, 2, \ldots$. In this case, we say that r is average preserving, or ap for short.

In the quadtree form, this property simply states that each node is the average of its four children. If r is average preserving, the images in the sequence $r_0, r_1, r_2, \ldots$ form sharper and sharper approximations of some infinitely sharp grayscale image.

In image processing algorithms, we typically rely on efficient algorithms of linear algebra. These are available because $\mathbb{R}\langle\langle \Sigma^* \rangle\rangle$ is a linear space over $\mathbb{R}$ under the usual pointwise sum and scalar product. Note that the set of average preserving multi-resolution images is a linear subspace.

In an ideal situation, multi-resolution images are used. In practical applications in digital image processing, one only has finite resolution $2^k \times 2^k$ images available. They also form a linear space that is isomorphic to $\mathbb{R}^{4^k}$. Here, we can find an expression of a given finite resolution image r_k as a linear combination of finite resolution images $\psi_1, \ldots, \psi_n$, if such an expression

$$r_k = c_1 \cdot \psi_1 + c_2 \cdot \psi_2 + \cdots + c_n \cdot \psi_n$$

with $c_1, c_2, \ldots, c_n \in \mathbb{R}$ exists.

3 Weighted Finite Automata and Multi-resolution Images

Suppose the color set S is a semiring. This is the case in all our examples where $S = \mathbb{B}$ (bi-level image), $S = \mathbb{R}$ (grayscale image) or $S = \mathbb{R}^3$ (color image). Then multi-resolution images can be described using weighted automata.

Weighted finite automata (WFA) are finite automata whose transitions are weighted by elements of the coloring semiring S, and whose states have two associated weights called the initial and the final distribution values. In our applications, the input alphabet is $\Sigma = \{0, 1, 2, 3\}$, and the semiring S is $\mathbb{B}$, $\mathbb{R}$, or $\mathbb{R}^3$.

Definition 3.1. *A WFA is specified by:*

- *The finite state set* Q
- *Four transition matrices* $A_0, A_1, A_2, A_3 \in S^{Q\times Q}$
- *A final distribution vector* $F \in S^{Q\times 1}$
- *An initial distribution vector* $I \in S^{1\times Q}$

The WFA defines a multi-resolution image r *as follows:*

$$(r, a_1 a_2 \ldots a_k) = I A_{a_1} A_{a_2} \cdots A_{a_k} F$$

for all $a_1 a_2 \ldots a_k \in \Sigma^*$. *Let us use the following shorthand notation: For every* $w = a_1 a_2 \ldots a_k \in \Sigma^*$, *let*

$$A_w = A_{a_1} A_{a_2} \cdots A_{a_k}$$

be the product of the matrices corresponding to the letters of the word w. *Then* $(r, w) = I A_w F$.

Usually a WFA is drawn as a labeled, weighted directed graph. The vertex set is the state set Q, and there is an edge from vertex $i \in Q$ to vertex $j \in Q$ with label $a \in \Sigma$ and weight $s \in S$ iff $(A_a)_{ij} = s$. Edges with weight 0 are usually not drawn. The initial and final distribution values are marked inside the vertices.

Example 3.2. Let $S = \mathbb{B}$, $|Q| = 2$ and

$$I = (1\ 0) \qquad A_0 = \begin{pmatrix} 0 & 1 \\ 0 & 1 \end{pmatrix} \qquad A_1 = \begin{pmatrix} 1 & 0 \\ 0 & 1 \end{pmatrix}$$

$$F = \begin{pmatrix} 1 \\ 1 \end{pmatrix} \qquad A_2 = \begin{pmatrix} 1 & 0 \\ 0 & 1 \end{pmatrix} \qquad A_3 = \begin{pmatrix} 0 & 0 \\ 0 & 1 \end{pmatrix}.$$

Weighted finite automata over $\mathbb{B}$ are classical finite automata. The directed graph representation of the sample automaton is

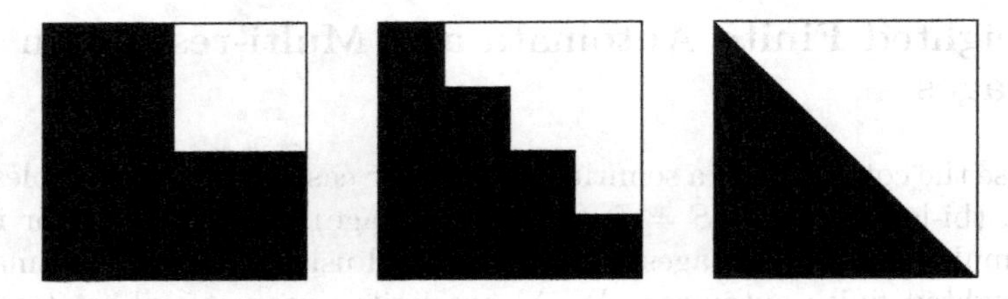

Fig. 2. Finite resolution images defined by the finite automaton of Example 3.2

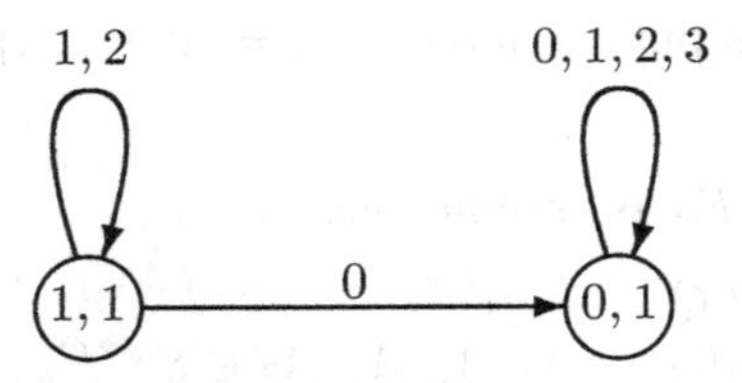

Here, all edges are with weight 1. A regular expression for the accepted language is $(1 + 2)^* + (1 + 2)^*0\Sigma^*$. Figure 2 shows the corresponding finite resolution images at resolutions 2×2, 4×4, and 256×256.

Example 3.3. Let $S = \mathbb{R}$, $|Q| = 2$ and

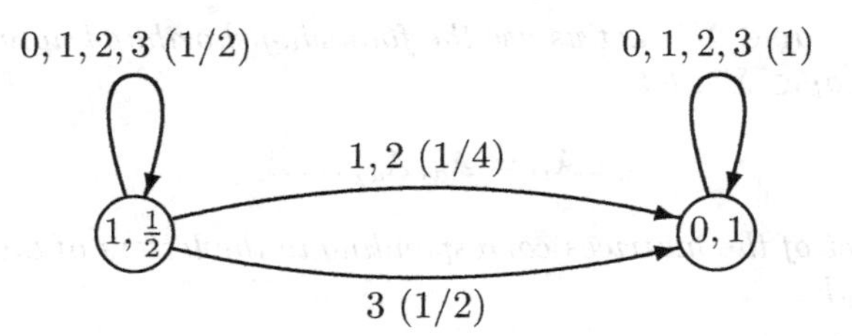

The weight of each edge is given in parentheses after the label.

The finite resolution images defined by this WFA at resolutions 2×2, 4×4, and 256×256 are shown in Fig. 3. Here (and in all our grayscale examples), value 0 is drawn black, value 1 in white, and intermediate values give different shades of gray.

Fig. 3. Finite resolution images defined by the weighted finite automaton of Example 3.3

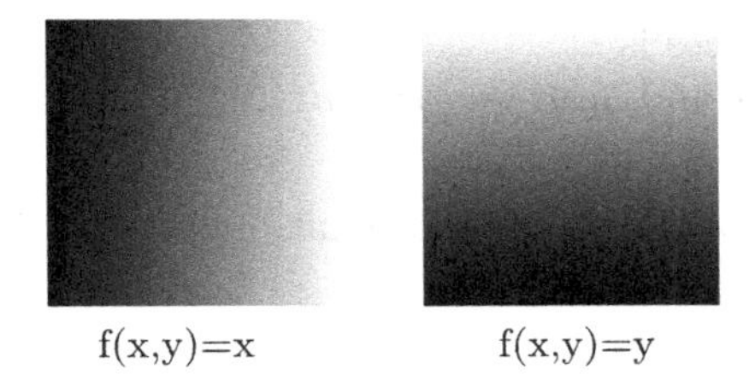

Fig. 4. High resolution images generated by the two state WFAs of Example 3.4

Let $S = \mathbb{R}$. If

$$(A_0 + A_1 + A_2 + A_3) \cdot F = 4F \tag{1}$$

holds then

$$(r, w0) + (r, w1) + (r, w2) + (r, w3) = 4(r, w)$$

for all $w \in \Sigma^\star$. In other words, the multi-resolution image specified by the automaton is then average preserving. Therefore, a WFA is called *average preserving* if (1) is satisfied. Note that condition (1) states that number 4 is an eigenvalue of the matrix $A_0 + A_1 + A_2 + A_3$, and F is a corresponding eigenvector.

Given an n-state WFA A and an m-state WFA B that compute multi-resolution functions r_1 and r_2, respectively, it is easy to construct:

- An $(n + m)$-state WFA that computes the sum $r_1 + r_2$
- An n-state WFA that computes the point-wise scalar multiple sr_1 of r_1 by $s \in \mathbb{R}$
- An nm-state WFA that computes the Hadamard product $r_1 \odot r_2$

Example 3.4. The 2-state WFAs

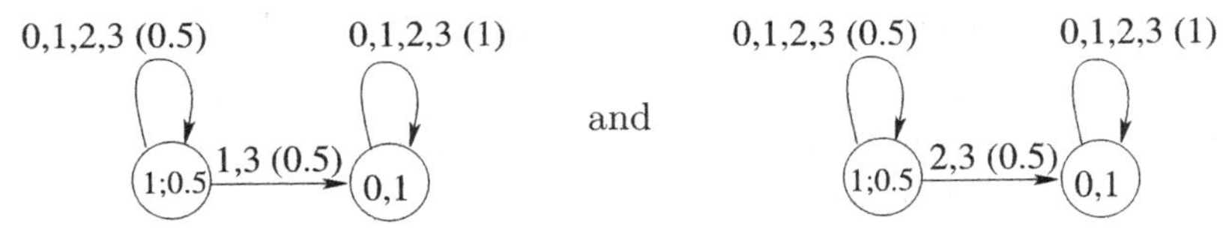

generate the linear functions of Fig. 4. From these and the constant function 1, one can build a WFA for any polynomial.

4 Drawing WFA Images

In this section, we consider the following *decoding problem*: Given a WFA, draw the corresponding image at some specified finite resolution $2^k \times 2^k$. Decoding at resolution $2^k \times 2^k$ involves forming the matrix products IA_wF for all $w \in \Sigma^k$. Note that the number of multiplications (and additions) required by the trivial algorithm to compute the product of:

- Two $n \times n$ matrices is n^3
- An $n \times n$ matrix and an n-vector is n^2
- Two n-vectors is n

Observe that it is naturally better to multiply vectors than matrices.

Decoding Algorithm #1

1. Form the products IA_w for all non-empty words $w \in \Sigma^{\leq k}$ in the order of increasing length of w. Because $IA_{ua} = IA_u A_a$, we need one product of a vector and a matrix for each word $w = ua$.
2. In the end, for every $w \in \Sigma^k$, multiply vectors IA_w and F.

Let us analyze the complexity of the algorithm. Let $N = 4^k$ be the number of pixels, and let n be the number of states. The number of non-empty words of length $\leq k$ is

$$4 + 16 + \cdots + 4^k \leq N\left(1 + \frac{1}{4} + \frac{1}{16} + \cdots\right) = \frac{4}{3}N.$$

Step 1 of the algorithm requires then at most $\frac{4}{3}Nn^2$ multiplications. Step 2 requires Nn multiplications so the total number of multiplications is at most $Nn(1 + \frac{4n}{3})$. Let us consider then a more efficient alternative.

Decoding Algorithm #2

1. Use the first step of the decoding algorithm #1 to compute the products IA_u for words u of length $k/2$ and products $A_v F$ for words v of length $k/2$. If k is odd, then we round the lengths so that u has length $\lfloor k/2 \rfloor$ and v has length $\lceil k/2 \rceil$.
2. Form all possible products $(IA_u)(A_v F)$ for all $u, v \in \Sigma^{k/2}$.

Step 1 requires at most $2 \times \frac{4}{3}4^{k/2}n^2 = \frac{8}{3}\sqrt{N}n^2$ multiplications. Step 2 requires Nn multiplications. Now the total is $Nn(1 + \frac{8n}{3\sqrt{N}})$ multiplications. This is considerably better than algorithm #1, especially when N, the number of pixels, is very large.

5 An Encoding Algorithm

In this chapter, we assume that $S = \mathbb{R}$. For $r \in S\langle\langle \Sigma^* \rangle\rangle$ and $w \in \Sigma^*$ we denote by $w^{-1}r$ the left quotient of r by w. It is the power series defined as follows:

$$(w^{-1}r, u) = (r, wu)$$

for all $u \in \Sigma^*$. Intuitively speaking, $w^{-1}r$ is the image obtained from image r by zooming into the sub-square whose address is w.

In a WFA A, the transitions A_0, A_1, A_2, and A_3 and the final distribution F define a multi-resolution image ψ_i for every state i: Image ψ_i is the multi-resolution image computed by the WFA that is obtained from A by changing the initial distribution so that the initial distribution value of state i is 1, and all other states have initial distribution 0. In other words, for every $w \in \Sigma^*$, we have

$$(\psi_i, w) = (A_w F)_i$$

where we use the notation that $(A_w F)_i$ is the ith component of the vector $A_w F$. Multi-resolution ψ_i is called the image of state i. It is average preserving if the WFA is.

The WFA gives a mutually recursive definition of the ψ_i multi-resolutions:

- $(\psi_i, \varepsilon) = F_i$, that is, the final distribution values are the average intensities of the state images.
- For every $a \in \Sigma, w \in \Sigma^\star$

  $$\begin{aligned}(a^{-1}\psi_i, w) &= (\psi_i, aw) = (A_{aw}F)_i = [A_a(A_w F)]_i = \sum_{j=1}^{n}(A_a)_{ij}(A_w F)_j \\ &= s_1(\psi_1, w) + s_2(\psi_2, w) + \cdots + s_n(\psi_n, w)\end{aligned}$$

  where $s_j = (A_a)_{ij}$ is the weight of the transition from state i into state j with label a. Since the coefficients $s_j = (A_a)_{ij}$ are independent of w, we have

  $$a^{-1}\psi_i = s_1\psi_1 + s_2\psi_2 + \cdots + s_n\psi_n$$

  in the vector space of multi-resolution images. In other words, the ith row of the transition matrix A_a tells how the quadrant a of the state image ψ_i is expressed as a linear combination of state images $\psi_1, \ldots, \psi_n$. See Fig. 5 for an illustration.
- The initial distribution I tells how the multi-resolution image r computed by the WFA is composed of state images $\psi_1, \psi_2, \ldots, \psi_n$:

  $$r = I_1 \cdot \psi_1 + I_2 \cdot \psi_2 + \cdots + I_n \cdot \psi_n.$$

The final distribution values and the transition matrices define state images $\psi_1, \ldots, \psi_n$ uniquely.

The following algorithm infers a minimum state WFA for a given multi-resolution function r:

Encoding Algorithm for Grayscale Images

Input: multi-resolution image r

Variables: n : number of states so far
i : first non-processed state
ψ_j : Image of state j

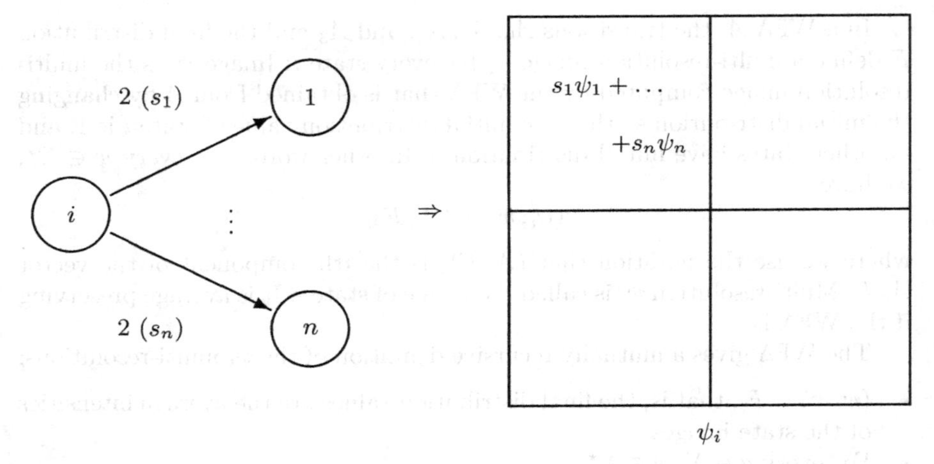

Fig. 5. The transitions from state i with label a specify how the quadrant a of the state image ψ_i is expressed as a linear combination of the state images $\psi_1, \ldots, \psi_n$

1. $n \leftarrow 1, i \leftarrow 1, \psi_1 \leftarrow r$
2. For quadrants $a = 0, 1, 2, 3$ do
   (a) If $\exists s_1, \ldots, s_n \in \mathbb{R}$ such that $a^{-1}\psi_i = s_1\psi_1 + \cdots + s_n\psi_n$, then add the transitions

   $$(i) \xrightarrow{a\ (s_j)} (j)$$

   for all $j = 1, 2, \ldots, n$
   (b) Else create a new state: Set $n \leftarrow n + 1$, $\psi_n \leftarrow a^{-1}\psi_i$, and add the transition

   $$(i) \xrightarrow{a\ (1)} (n)$$

3. $i \leftarrow i + 1$. If $i \leq n$, then goto 2
4. Initial distributi.on: $I_1 = 1, I_i = 0$ for $i = 2, 3, \ldots, n$
   Final distribution: $F_i = (\psi_i, \epsilon)$ for $i = 1, 2, \ldots, n$

Theorem 5.1 ([11]). *Let $r \in \mathbb{R}\langle\langle \Sigma^* \rangle\rangle$.*

- *Multi-resolution image r can be generated by a WFA if and only if the multi-resolution images*

$$u^{-1}r, \quad \textit{for all } u \in \Sigma^\star,$$

  generate a finite dimensional vector space. The dimension of the vector space is the same as the smallest possible number of states in any WFA that generates r.
- *If r can be generated by a WFA, then the algorithm above produces a WFA with the minimum number of states.*

- *If r is average preserving then the algorithm produces an average preserving WFA.*

The setup of the encoding algorithm above is theoretical: the input r is an infinite object but the input of a concrete algorithm should have a finite presentation. In practice, the input to the algorithm is a finite resolution image. When encoding a $2^k \times 2^k$ size image, the subtrees $w^{-1}r$ are only known up to depth $k - |w|$. When forming linear combinations, the values deeper in the tree are "do not care" nodes, that is, it is enough to express the known part of each subtree as a linear combination of other trees and the do not care nodes may get arbitrary values. Fortunately, as the process is done in the breadth-first order, the sub-images are processed in the decreasing order of size. This means that all previously created states have trees assigned to them that are known at least to the depth of the current subtree. Hence, the don't care values of prior images only can affect the don't care values of the present image, and the linear expressions are precise at the known part of the tree.

6 Practical Image Compression Using WFA

In the previous section we saw how we can find a minimum state WFA for a given grayscale image. If the WFA is small in size, then the WFA can be used as a compressed representation of the image. However, the encoding algorithm of the previous section as such is ill suited for image compression purposes. Even though the resulting WFA is minimal with respect to the number of states, it may have a very large number of edges. The algorithm also represents the image exactly, and the last details of the image often require a very large increase in the size of the WFA. This is called *lossless* image compression. More often one is interested in *lossy* compression where small errors are allowed in the regenerated image, if this admits sufficient decrease in the size of the compressed image file. In this section, we outline how the encoding algorithm can be modified into a practical image compression method. Note that the techniques are heuristic in nature and no guarantees of optimality exist. We can only use compression experiments and comparisons with other algorithms to show their usefulness.

Recall step 2 of the encoding algorithm from the previous section:

(a) If $\exists s_1, \ldots, s_n \in \mathbb{R}$ such that $a^{-1}\psi_i = s_1\psi_1 + \cdots + s_n\psi_n$, then for all $j = 1, 2, \ldots, n$ add the transitions

$$(i) \xrightarrow{a\ (s_j)} (j)$$

(b) Else create a new state: Set $n \leftarrow n+1$, $\psi_n \leftarrow a^{-1}\psi_i$, and add the transition

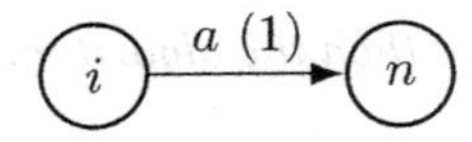

It turns out that better results are obtained in practice if we try both alternatives (a) and (b), and choose the one that gives the smaller automaton. It may namely happen that the linear combination in (a) contains so many coefficients that it is better to create a new state in (b) and process its quadrants.

In order to make a fair comparison between (a) and (b), we need to process the new state created in (b) completely before we can determine which choice to make. Therefore, we change the order of processing the states from the breadth-first order into the depth-first order: instead of processing all four quadrants of ψ_i before advancing to the next state image, all new states created by a quadrant are processed before advancing to the next quadrant.

Let us first measure the size of the automaton by the quantity

$$|E| + P \cdot |V|$$

where $P \in \mathbb{R}_+$ is a given constant, $|E|$ is the number of transitions, and $|V|$ is the number of states of the automaton. Constant P is a Lagrange multiplier that formulates the relative cost of states vs. edges in the automaton. If we want to minimize the number of edges, we set $P = 0$. The goal of the next inference algorithm is to find for a given multi-resolution image ψ a small WFA in the sense that the value $|E| + P \cdot |V|$ is small.

Because we now process the quadtree in the depth-first order, it is natural to make the algorithm recursive. The new encoding algorithm consists of a recursive routine *make_wfa(ψ_i,max)* that adds new states and edges to the WFA constructed so far, with the goal of representing the state image ψ_i in such a way that the value $\Delta_E + P \cdot \Delta_V$ is small, where Δ_E and Δ_V are the numbers of new transitions and states added by the recursive call. If $\Delta_E + P \cdot \Delta_S \leq max$ then the routine returns the value $\Delta_E + P \cdot \Delta_V$, otherwise it returns ∞ to indicate that no improvement over the target value *max* was obtained.

Encoding Algorithm #2 [10]

Input: Multi-resolution image r and a positive real number P
Global variables used: n : number of states
ψ_j : image of state j

1. $n \leftarrow 1, \psi_1 \leftarrow r$
2. *make_wfa*(ψ_1, ∞)

make_wfa(ψ_i,max) :

1. If $max < 0$ then return(∞)
2. Set $cost \leftarrow 0$

3. For quadrants $a = 0, 1, 2, 3$ do
   (a) If $\exists s_1, \dots, s_n \in \mathbb{R}$ such that $a^{-1}\psi_i = s_1\psi_1 + \cdots + s_n\psi_n$ then

$$\textit{cost1} \leftarrow \text{number of non-zero coefficients } s_j$$

   else

$$\textit{cost1} \leftarrow \infty$$

   (b) Set $n_0 \leftarrow n$, $n \leftarrow n + 1$, $\psi_n \leftarrow a^{-1}\psi_i$ and add the transition

$$(i) \xrightarrow{a\ (1)} (n)$$

   (c) Set $\textit{cost2} \leftarrow P + \textit{make\_wfa}(\psi_n, \textit{min}\{\textit{max} - \textit{cost} - P, \textit{cost1} - P\})$
   (d) If $\textit{cost2} \leq \textit{cost1}$ then
   - $\textit{cost} \leftarrow \textit{cost} + \textit{cost2}$

   else
   - $\textit{cost} \leftarrow \textit{cost} + \textit{cost1}$
   - remove all transitions from states $n_0 + 1, \dots, n$, and set $n \leftarrow n_0$
   - remove the transition

$$(i) \xrightarrow{a\ (1)} (n)$$

   - add transitions

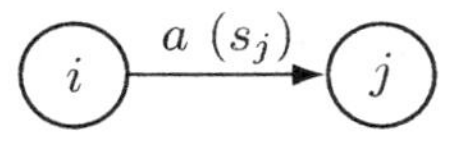

   for $s_j \neq 0$
4. If $\textit{cost} \leq \textit{max}$ then return($\textit{cost}$) else return(∞)

A few words to explain the algorithm: The main step is line 3 where we try to find a WFA representation for each of the four quadrants of ψ_i. For each quadrant, we try two alternatives: (a) to express the quadrant as a linear combination of existing states, and (b) to create a new state whose state image is the quadrant, and to recursively process the new state. In variable *cost1,* we store the cost of alternative (a), i.e., the number of transitions created in the automaton, and in *cost2* we store the cost of alternative (b), i.e., the sum of P (for the new state created) and the cost returned from the recursive call to process the new state. In step 3(d), the algorithm chooses the better of the two alternatives.

The algorithm above is still lossless. The WFA represents the input image precisely, without any loss of detail. Much better compression is obtained if we allow small errors in the regenerated image whenever that helps to compress the image more. There is a trade-off between the amount of image degradation and the size of the compressed image. Let us measure the amount of degradation by the square difference metric $d(\cdot, \cdot)$. For two images ψ and ϕ at resolution $2^k \times 2^k$, this metric defines the image difference value as

$$d(\psi, \phi) = \sum_{w \in \Sigma^k} \left((\psi, w) - (\phi, w)\right)^2.$$

This is a reasonable measure for the reconstruction error. It is also convenient for our linear combinations approach since it is the square of the normal Euclidean distance.

We also introduce a new Lagrange multiplier G that controls the trade-off between the image size and the reconstruction error. Parameter G is given as an input by the user, and the algorithm will produce a WFA that generates an image ψ' such that the value of

$$d(\psi, \psi') + G \cdot S$$

is small, where ψ is the image to be compressed and S is the size of the WFA constructed by the algorithm. We may continue using

$$S = |E| + P \cdot |V|,$$

but in practical image compression it is better to define S as the actual number of bits required to store the WFA in a file. The WFA is stored using suitable entropy coder, e.g., an arithmetic coder. See [23] for details on how the different items such that states, transitions, and weights are stored using arithmetic coding.

The Lagrange multiplier G is the parameter that the user can change to adjust the file size and the image quality:

Small $G \Rightarrow$ big automaton, small error
Large $G \Rightarrow$ small automaton, big error

The following modifications to the algorithm were also made:

- Edges back to states that have not yet been completely processed are problematic in lossy coding, as the actual image of those states is not yet precisely known. Therefore, we opted to only allow edges to states already completely processed. Note that this prevents the creation of any loops in the WFA.
- In order to have at least one processed state to begin with, we introduce an initial base: Before calling *make_wfa* the first time, we set $n \leftarrow N$ with some fixed images $\psi_1, \ldots, \psi_N$. In our tests below, we used $N = 6$ base images that were linearly independent quadratic polynomials, i.e., functions 1, x, y, x^2, y^2, and xy.

With these modifications, we have a practical encoding algorithm that compares favorably with other compression techniques. Let us compare WFA compression with the JPEG image compression standard using the test image of Fig. 6. This color image contains large smooth areas, and is therefore well suited for JPEG compression. A color image consists of three color layers, each

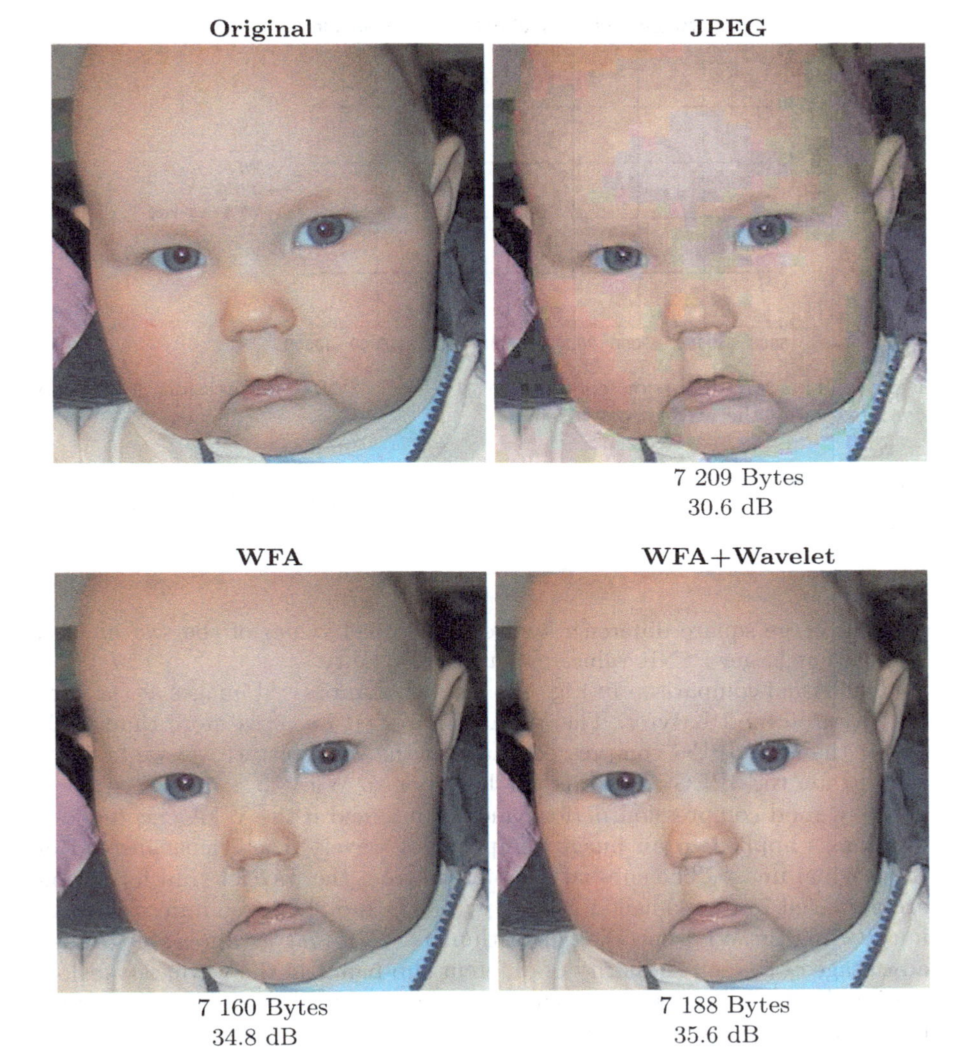

Fig. 6. A comparison of JPEG and the WFA compression at a low bitrate

of which is compressed as a grayscale image. However, only one automaton is built, that is, different color components can refer to the sub-images of other color layers.

In the compression experiment, the reconstruction errors are reported as the peak signal-to-noise ratio (PSNR). The units of this measure are decibels (dB). The PSNR value is directly obtained from the square difference $d(\cdot,\cdot)$ as follows:

$$\mathrm{PSNR}(\cdot,\cdot) = 10\log_{10}\left(\frac{A^2}{\sigma}\right)$$

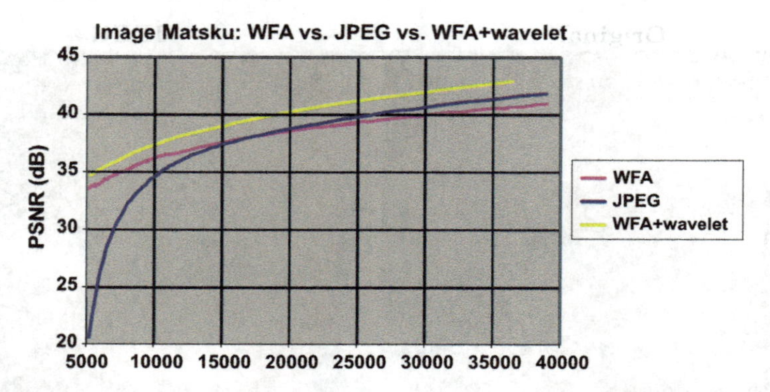

Fig. 7. The rate-distortion comparison of JPEG, WFA compression, and WFA compression of the wavelet coefficients

where A is the maximum intensity value ($A = 255$ in our 8 bits-per-pixel image) and

$$\sigma = \frac{d(\cdot, \cdot)}{\#\text{ of pixels}}$$

is the average square difference between the pixel values of the two images. Notice that larger PSNR values mean better quality.

Our visual comparison in Fig. 6 is between compressed images at the very low bitrate of 7.2 KBytes. The WFA compressed image is more than 4 dB better than the JPEG compressed image at the same bitrate. As we increase the bitrate, the JPEG algorithm catches up with WFA.

Very good compression performance is obtained if the WFA encoding algorithm is applied to an image composed of wavelet coefficients instead of the original image. The sub-bands obtained from the wavelet transformation are arranged into a so-called Mallat pyramid, and this is compressed as an image using the WFA encoding algorithm. The WFA algorithm is able to take advantage of the self-similarity of different sub-bands in the wavelet transformation. The subdivision into quadrants used by our algorithm matches with the organization of the sub-bands in the Mallat form. A result is provided in Fig. 6 for visual comparison. In our tests, the Daubechies W6 wavelets are used.

Figure 7 summarizes numerically the rate-distortion performances of JPEG, WFA, and WFA with wavelets for the test image of Fig. 6. All three algorithms were used at various bitrates, and the bitrate versus image quality values were plotted. Note how JPEG surpasses WFA at 16 KByte compression for this type of image characteristics.

WFA (without wavelets) compress very well images with sharp edges. As a second experiment, let us compare WFA compression and JPEG on the test image shown in Fig. 8. The difference between JPEG and WFA is clear even at high quality setting, as seen in Fig. 8.

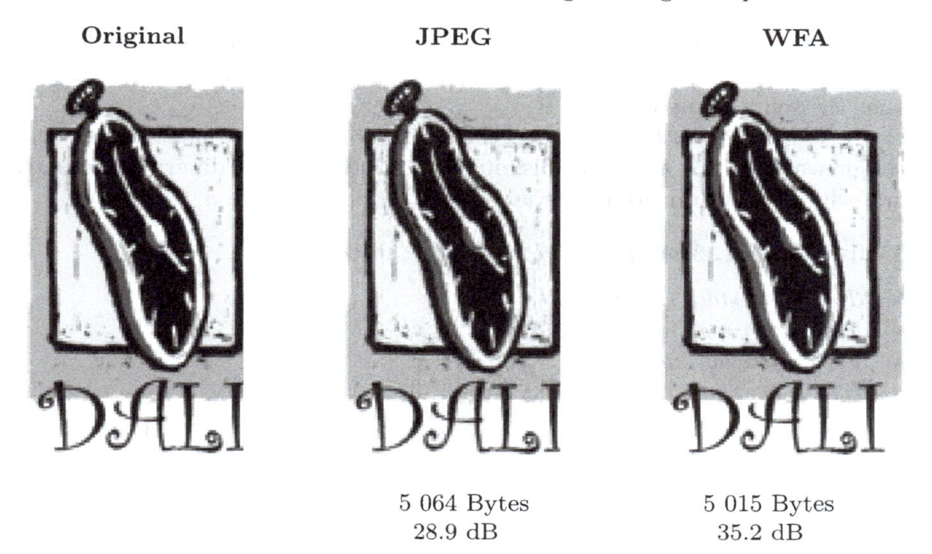

Fig. 8. The second test image that contains sharp edges, and the compression results using JPEG and WFA compression at 5 KBytes

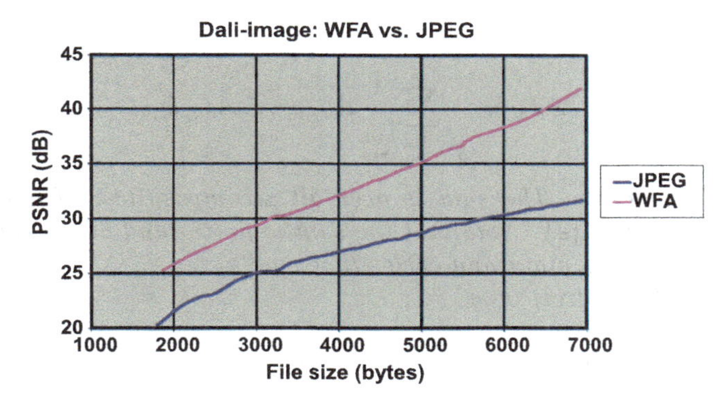

Fig. 9. The rate-distortion comparison of JPEG and WFA compression on the second test image

Numerical comparisons in Fig. 9 indicate that WFA compression remains superior through all bitrates.

7 Weighted Finite Transducers (WFT)

A nice feature of WFA image representations is the property that one can perform interesting and useful image operations directly in the WFA form. Bi-level images and regular languages can be transformed using finite state transducers. Analogously, grayscale images and WFA are transformed using finite state transducers with real weights. More details of the examples and results in this section can be found in [8]. We assume throughout this chapter that the semiring used is $S = \mathbb{R}$.

A *Weighted Finite Transducer (WFT)* is obtained by introducing edge weights and initial and final distribution values to an ordinary finite state transducer. The transitions are labeled by pairs a/b where $a, b \in \Sigma \cup \{\varepsilon\}$.

Definition 7.1. *More precisely, a WFT is specified by:*

- *The finite state set Q*
- *Transition matrices $A_{a,b} \in \mathbb{R}^{Q \times Q}$ for all $a, b \in \Sigma \cup \{\varepsilon\}$*
- *Final distribution vector $F \in \mathbb{R}^{Q \times 1}$*
- *Initial distribution vector $I \in \mathbb{R}^{1 \times Q}$*

The WFT is called ε-free if the weight matrices $A_{a,\varepsilon}$, $A_{\varepsilon,b}$, and $A_{\varepsilon,\varepsilon}$ are zero matrices for all $a, b \in \Sigma$. The WFT defines a function

$$\rho : \Sigma^* \times \Sigma^* \to \mathbb{R}$$

called a weighted relation as follows: For every $u, v \in \Sigma^*$, we have

$$\rho(u, v) = I A_{u,v} F$$

where

$$A_{u,v} = \sum_{\substack{a_1 \ldots a_m = u \\ b_1 \ldots b_m = v}} A_{a_1,b_1} \ldots A_{a_m,b_m}$$

if the sum converges. The sum is over all decompositions of u and v into symbols $a_i, b_i \in \Sigma \cup \{\varepsilon\}$. Note that the sum is finite (and hence converges) if the WFT does not contain any cycles that read ε/ε.

If the WFT is ε-free, then

$$\rho(a_1 \ldots a_k, b_1 \ldots b_k) = I A_{a_1,b_1} \ldots A_{a_k,b_k} F$$

where all $a_i, b_i \in \Sigma$, and $\rho(u, v) = 0$ when $|u| \neq |v|$.

Next, we define the action of a weighted relation ρ on a multi-resolution image f. The result is a new multi-resolution function $g = \rho(f)$, defined by

$$g(w) = \sum_{u \in \Sigma^*} f(u) \rho(u, w), \quad \text{for all } w \in \Sigma^*,$$

provided the sum converges. The sum is finite if the WFT is ε-free, or more generally, if the weight matrices $A_{a,\varepsilon}$ are zero, for all $a \in \Sigma \cup \{\varepsilon\}$. In this case, the sum is over all words u whose length is not greater than the length of w.

It is easy to see that the operator

$$\rho : \mathbb{R}\langle\langle \Sigma^* \rangle\rangle \to \mathbb{R}\langle\langle \Sigma^* \rangle\rangle$$

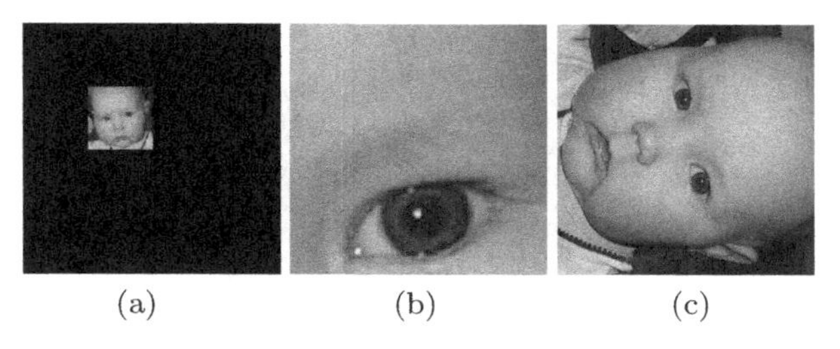

(a) (b) (c)

Fig. 10. The output of transforming the test image using the WFT of Examples 7.2, 7.3, and 7.4

is linear, that is, for arbitrary multi-resolution functions f and g, and arbitrary real numbers x and y we have

$$\rho(xf + yg) = x\rho(f) + y\rho(g).$$

Many interesting and natural linear image transformations can be implemented as a WFT. In the following, we see several examples.

Example 7.2. Let $w \in \Sigma^*$ be a fixed word. The WFT

1,0 —ε /w (1)→ 0,1 ⟲ 0/0, 1/1, 2/2, 3/3 (1)

computes the weighted relation $\rho(u, wu) = 1$ for every $u \in \Sigma^*$, and $\rho(u, v) = 0$ if $v \neq wu$. The effect is to shrink the input image and place it at the sub-square addressed by w. For example, Fig. 10(a) shows the result of our test image for $w = 21$.

Example 7.3. Consider then the WFT

1,0 —w/ε (1)→ 0,1 ⟲ 0/0, 1/1, 2/2, 3/3 (1)

It computes the weighted relation $\rho(wu, u) = 1$ for every $u \in \Sigma^*$, and $\rho(v, u) = 0$ if $v \neq wu$. Now the effect is to zoom the sub-square of the input image whose address is w. For example, with $w = 30$, our test image is mapped into the image shown in Fig. 10(b).

Example 7.4. The WFT

1,1 ⟲ 0/2, 2/3, 3/1, 1/0 (1)

rotates the image 90°, as shown in Fig. 10(c).

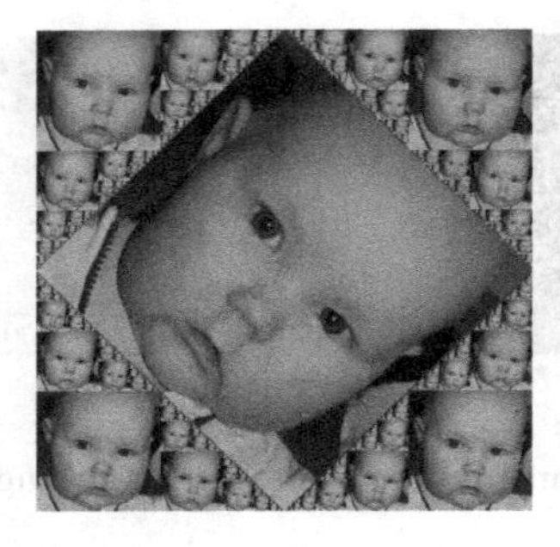

Fig. 11. The output of transforming the test image using an 11 state WFT

Also, fractal-like transformations can be defined. For example, 11 states are enough to implement a transformation that produces Fig. 11.

Next, we define the action of WFTs on WFAs.

Definition 7.5. *An application of an ε-free n-state WFT M to an m-state WFA A is the mn state WFA $M(A)$ whose states are the pairs (p,q) of states of A and M, the initial and final distribution values are obtained by multiplying the corresponding distributions of A and M, and the weight of the transition*

$$(p,q) \xrightarrow{a} (s,t)$$

is

$$\sum_{x \in \Sigma} (A_x)_{p,s} (M_{x,a})_{q,t}$$

where A_x and $M_{x,a}$ are the weight matrices of A and M, respectively.

This is a straightforward generalization of the usual applications of a finite letter-to-letter transducer on a finite automaton. It is easy to see that $M(A)$ generates the multi-resolution function $\rho(f)$ where ρ is the weighted relation of M and f is the multi-resolution determined by A.

Applying WFT M directly to a WFA A has the advantages that:

- It is fast if A and M are small. There is no need to decode the image into the usual pixel form before applying the operation.
- The result is correct at every resolution.

A concept of average preservation can also defined for WFT; see [8] for details and for more examples.

8 Parametric Weighted Finite Automata (PWFA)

We will generalize now the way input words can define pixel positions in WFA. Instead of using a fixed binary (or k-ary) representation of addresses, we allow that the pixel positions are also computed by some WFA [1]. We call these automata *parametric weighted finite automata*, because the input string acts as a parameter binding the functions for different dimensions together. Instead of computing single values for the input strings, in parametric weighted finite automata, we get points of higher dimensional spaces. This requires only one change in the definition for WFA, namely our initial distribution vector I becomes an initial distribution matrix of size $d \times Q$ for some $d > 0$. For these parametric WFA, we will only consider weights and vectors over $S = \mathbb{R}$.

Definition 8.1. *A Parametric Weighted Finite Automaton (PWFA) is thus specified by:*

- *The finite state set* Q
- *The input alphabet* $\Sigma = \{0, 1, \ldots, m-1\}$
- *The weight matrices for transitions* $A = (A_0, A_1, \ldots, A_{m-1}),\ A_i \in \mathbb{R}^{Q \times Q}$
- *The final distribution* $F \in \mathbb{R}^{Q \times 1}$
- *The initial distribution matrix* $I \in \mathbb{R}^{d \times Q}$

In the transition diagrams for PWFAs now inside every node d, initial distribution values and one final distribution value are inserted. And computing (r, w) with an initial distribution matrix I of size $d \times Q$ looks exactly as in the case of WFA: $(r, w) = IA_wF$.

Definition 8.2. *For the given PWFA P over the alphabet Σ, let $R_n(P)$ denote the set of points computed by P on inputs of length n, and $R_{\geq n}(P)$ the set of points computed on inputs of length at least n:*

$$R_n(P) = \{(r, w) \mid w \in \Sigma^n\}, \qquad R_{\geq n}(P) = \bigcup_{i=n}^{\infty} R_i(P).$$

Now a topologically closed set $R(P)$ can be associated with a PWFA P:

$$R(P) = \bigcap_{n=0}^{\infty} \overline{R_{\geq n}(P)}$$

where $\overline{R_{\geq n}(P)}$ is the topological closure of $R_{\geq n}(P)$.

In other words, a point $\overline{x} \in \mathbb{R}^d$ is in $R(P)$ if either (i) there exist infinitely many words w such that $(r, w) = \overline{x}$, or (ii) there exist points $(r, w) \neq \overline{x}$ arbitrarily close to $\overline{x}$.

Multi-dimensional sets $R(P)$ given by a PWFA P can be interpreted as relations or images in many different ways. If $d = 2$, it is natural to interpret pairs (x, y) as points of the Euclidean plane, so $R(P)$ becomes a bi-level image. In case $d = 3$, we might have a set of points (x, y, z) of a 3D object or a description of pixel locations x, y and intensities z of a 2D image, or a description of a moving 2D bi-level object where the third dimension is interpreted as the time coordinate. Case $d = 4$ could be a description of a 3D

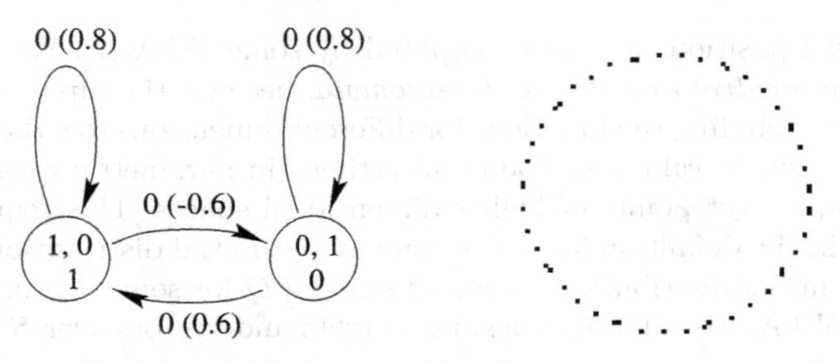

Fig. 12. **(left)** Rotation by $\cos^{-1}(0.8)$. **(right)** First 50 points of the circle

grayscale object, or a 2D grayscale video, etc. In all cases, PWFA-decoding computes d-dimensional points, followed by their interpretation. This new degree of freedom by separating the generation of real values from their interpretation yields a high descriptional power. For example, PWFA are used as representations of multi-dimensional wavelets [27], shapes of figures [30], spline-curves, and textured 3D patches [2].

PWFA over Unary Alphabet

WFA over a unary alphabet do not define useful functions, but from PWFA over a single input symbol 0 one can derive already interesting structures.

Example 8.3. Consider the following one-symbol, two-states PWFA C in Fig. 12.

The corresponding weight matrix A_0 defines a rotation of the plane $\mathbb{R}^2$ by the angle $\alpha = \cos^{-1}(0.8)$. The ratio of α to π is irrational. To see this, we can use the following nice, short proof from [20]. We have $\cos\alpha = \frac{4}{5}$. Using repeatedly the formula $\cos 2\alpha = 2\cos^2\alpha - 1$, we easily obtain

$$\cos 2^k\alpha = \frac{a_k}{5^{2^k}}$$

for all k, where a_k is an integer not divisible by 5. Hence, all $\cos 2^k\alpha$ are distinct, so the set $\{\cos i\alpha \mid i \in \mathbb{Z}\}$ is infinite. This means that α/π is irrational.

The orbit of a point under the iterated rotation by an irrational angle defines a dense subset of a circle, so

$$R(C) = \{(x, y) \mid x^2 + y^2 = 1\} = \{(\cos(t), \sin(t)) \mid t \in \mathbb{R}\}.$$

The unary alphabet PWFAs have been characterized by decidability results and closure properties [28, 19]. Furthermore, it is not hard to prove that two symbols in Σ actually suffice. Please note, that on the other hand the number of states gives rise to an infinite hierarchy, as can be concluded from the facts about polynomials of degree m.

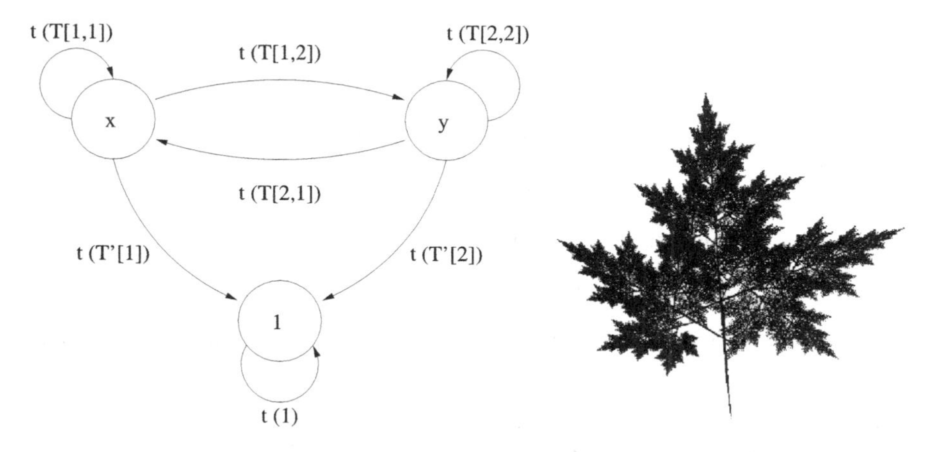

Fig. 13. Simulating the IFS for the maple leaf

PWFA and Iterated Function Systems

Consider any Iterated Function System (IFS) with k contractive affine maps of $\mathbb{R}^2$ [5]. A PWFA simulating the IFS just needs k symbols. Each symbol t corresponds to one affine transform represented by some 2×2 matrix T for scaling and/or rotating the plane and a 2×1 translation vector T'.

Two states are needed to represent the x- and y-coordinates and one state for the constant 1. The weights for transitions between states are assigned for each symbol t in a straightforward manner from the matrix- and vector-entries of T and T'.

Example 8.4. The fractal maple leaf is generated this way by a 3-state, 4-symbols PWFA (see Fig. 13):

PWFA and Polynomial Curves

If each of the d functions computed by a PWFA is a polynomial, we can produce a very compact automaton for the corresponding polynomial curve in $\mathbb{R}^d$.

Example 8.5. Consider the 4-state, 2-symbols, 2-dimensional PWFA P as given in Fig. 14. It has initial distributions $(1, -1, 0, 0)$ and $(0, 1, -1, 0)$, if the states are numbered from left to right.

In the bintree representation of a WFA, the four states as such—from left to right—would compute the functions t^3, t^2, t, and 1, respectively, over the interval $[0, 1)$. Let us interpret now the two dimensions of the PWFA P as the x- and y-coordinates of points. Then the given PWFA computes the points of the curve segment shown in the middle of Fig. 14. The second image is also

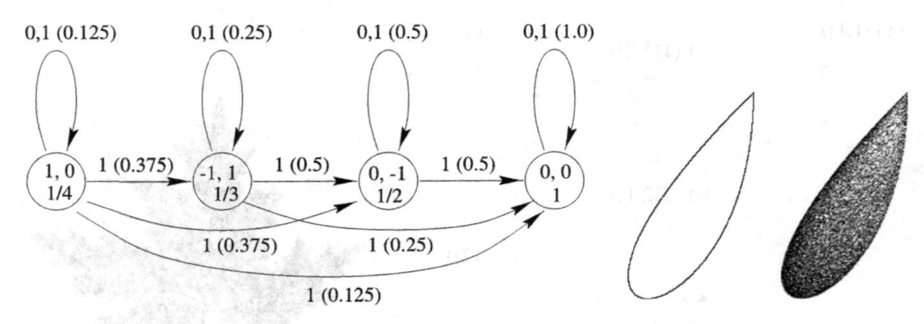

Fig. 14. PWFA for $\{(t^3 - t^2, t^2 - t) \mid 0 \leq t \leq 1\}$, curve and filling the interior

generated by some PWFA, which essentially holds two copies of the PWFA P and three extra symbols plus two helper states. Then for any two points on the curve of P, the line between those points is gradually filled with black pixels in some random fashion. For the displayed picture, the computation was stopped intentionally to leave some pixels of the interior unchanged.

We had mentioned already that any polynomial $p(x)$ of degree m can be computed by a standard one-dimensional WFA with $m + 1$ states, as shown already in [6]. Therefore, any d-dimensional curve

$$\{(p_1(t), p_2(t), \ldots, p_d(t)) \mid 0 \leq t \leq 1\}$$

with parametric representation using the polynomials $p_1(t), p_2(t), \ldots, p_d(t)$ is computable by a PWFA. Furthermore, if the highest degree of the polynomials $p_1(t), p_2(t), \ldots, p_d(t)$ is m, the PWFA will only need $m + 1$ states again. It is worthwhile noting that these polynomial curves include, e.g., the square root (t^2, t), which can only be approximated by WFA, but is not computable with arbitrary precision [14, 13]. For many practical purposes, these polynomial curves are used to approximate figures in the plane like shapes of, e.g., font letters. This has been studied for WFA and chain-code languages in [12] and in [29, 30] for several types of splines and also for textured 3D Bezier-spline surfaces [2].

9 Conclusions and Open Problems

For weighted finite automata and their extensions to, e.g., weighted finite transducers or parametric WFA several attractive possibilities of image generation and image compression have been demonstrated. The WFA inference algorithm works very well, especially for the combination with wavelet-transforms as preprocessing. This seems to indicate that there is still quite a bit of potential to improve the WFA compression rates in hybrid WFA image compression heuristics.

WFT and PWFA have been explored up to now mostly with respect to inclusion properties and decidability questions and a small number of interesting "hand-made" examples have been provided. For practical applications, an important question is whether the WFA inference algorithm can be extended to WFT or PWFA, e.g., to PWFA-representation of 3D spline-patches.

For some of the published examples of PWFA, it seemed essential that irrational weights can be employed. In a strict sense, it is arguable here, whether the attribute "finite" is indeed justified for those PWFA, since we do not generate the irrational number by some kind of finite state device. There are results on language families and decidability questions for integer weighted finite automata by Halava and Harju [17, 18], but PWFA with rational weights are still to be studied in detail.

References

1. J. Albert and J. Kari. Parametric weighted finite automata and iterated function systems. In M. Dekking et al., editors, *Proc. Fractals: Theory and Applications in Engineering, Delft*, pages 248–255, 1999.
2. J. Albert and G. Tischler. On generalizations of weighted finite automata and graphics applications. In S. Bozapalidis and G. Rahonis, editors, *Proc. CAI 2007*, volume 4728 of *Lecture Notes in Computer Science*, pages 1–22. Springer, Berlin, 2007.
3. S. Bader, S. Hölldobler, and A. Scalzitti. Semiring artificial neural networks and weighted automata—And an application to digital image encoding. In *Advances in Artificial Intelligence*, volume 3238 of *Lecture Notes in Computer Science*, pages 281–294. Springer, Berlin, 2004.
4. P. Bao and X.L. Wu. L-infinity-constrained near-lossless image compression using weighted finite automata encoding. *Computers & Graphics*, 22:217–223, 1998.
5. M. Barnsley. *Fractals Everywhere*, 2nd edition. Academic Press, San Diego, 1993.
6. K. Culik II and J. Karhumäki. Finite automata computing real functions. *SIAM Journal on Computing*, 23(4):789–814, 1994.
7. K. Culik II and J. Kari. Digital images and formal languages. In A. Salomaa and G. Rozenberg, editors, *Handbook of Formal Languages, volume 3*, pages 599–616. Springer, Berlin, 1997.
8. K. Culik II and J. Kari. Finite state transformations of images. *Computers & Graphics*, 20:125–135, 1996.
9. K. Culik II and J. Kari. Finite state methods for compression and manipulation of images. In J.A. Storer and M. Cohen, editors, *Proceedings of the Data Compression Conference*, pages 142–151. IEEE Computer Society Press, Los Alamitos, 1995.
10. K. Culik II and J. Kari. Image-data compression using edge-optimizing algorithm for WFA inference. *Information Processing & Management*, 30:829–838, 1994.

11. K. Culik II and J. Kari. Image compression using weighted finite automata. *Computers & Graphics*, 17:305–313, 1993.
12. K. Culik II, V. Valenta, and J. Kari. Compression of silhouette-like images based on WFA. *Journal of Universal Computer Science*, 3(10):1100–1113, 1997.
13. D. Derencourt, J. Karhumäki, M. Latteux, and A. Terlutte. On computational power of weighted finite automata. In *Proceedings of MFCS'92*, volume 629 of *Lecture Notes in Computer Science*, pages 236–245. Springer, Berlin, 1992.
14. M. Droste, J. Kari, and P. Steinby. Observations on the smoothness properties of real functions computed by weighted finite automata. *Fundamenta Informaticae*, 73(1,2):99–106, 2006.
15. U. Hafner. Refining image compression with weighted finite automata. In J.A. Storer and M. Cohn, editors, *Proc. Data Compression Conference*, pages 359–368, 1996.
16. U. Hafner, J. Albert, S. Frank, and M. Unger. Weighted finite automata for video compression. *IEEE Journal on Selected Areas in Communication*, 16:108–119, 1998.
17. V. Halava and T. Harju. Undecidability in integer weighted finite automata. *Fundamenta Informaticae*, 38(1,2):189–200, 1999.
18. V. Halava and T. Harju. Languages accepted by integer weighted finite automata. In J. Karhumäki, H. Maurer, G. Paun, and G. Rozenberg, editors, *Jewels Are Forever*, pages 123–134. Springer, Berlin, 1999.
19. V. Halava, T. Harju, M. Hirvensalo, and J. Karhumäki. Skolems problem—On the border between decidability and undecidability. Technical Report 683, Turku Centre for Computer Science, 2005
20. J. Jahnel. When is the (co)sine of a rational angle equal to a rational number? Unpublished, available online at www.uni-math.gwdg.de/jahnel/Preprints/cos.pdf, 2005.
21. Z.H. Jiang, O. de Vel, and B. Litow. Unification and extension of weighted finite automata applicable to image compression. *Theoretical Computer Science*, 302:275–294, 2003.
22. Z.H. Jiang, B. Litow, and O. de Vel. Similarity enrichment in image compression through weighted finite automata. In *Computing and Combinatorics*, volume 1858 of *Lecture Notes in Computer Science*, pages 447–456. Springer, Berlin, 2000.
23. J. Kari and P. Fränti. Arithmetic coding of weighted finite automata. *Informatique Théorique et Applications, RAIRO*, 28:343–360, 1994.
24. J. Kari. Image processing using finite automata. In Z. Esik, C. Martin-Vide, and V. Mitrana, editors, *Recent Advances in Formal Languages and Applications*, volume 25 of *Studies in Computational Intelligence*, pages 171–208. Springer, Berlin, 2006.
25. F. Katritzke, W. Merzenich, and M. Thomas. Enhancements of partitioning techniques for image compression using weighted finite automata. *Theoretical Computer Science*, 313:133–144, 2004.

26. S.V. Ramasubramanian and K. Krithivasan. Finite automata digital images. *International Journal of Pattern Recognition and Artificial Intelligence*, 14:501–524, 2000.
27. G. Tischler, J. Albert, and J. Kari. Parametric weighted finite automata and multidimensional dyadic wavelets. In *Proc. Fractals in Engineering, Tours, France*. Published on CD-ROM, 2005.
28. G. Tischler. Theory and applications of parametric weighted finite automata. Dissertation, Würzburg, 2008.
29. G. Tischler. Properties and applications of parametric weighted finite automata. *Journal of Automata, Languages and Combinatorics*, 10(2/3):347–365, 2005.
30. G. Tischler. Parametric weighted finite automata for figure drawing. In *Implementation and Application of Automata*, volume 3317 of *Lecture Notes in Computer Science*, pages 259–268. Springer, Berlin, 2004.
31. G.K. Wallace. The JPEG still picture compression standard. *Communications of the ACM*, 34(4):30–44, 1991.

Chapter 12: Fuzzy Languages

George Rahonis

Department of Mathematics, Aristotle University of Thessaloniki,
54124 Thessaloniki, Greece
grahonis@math.auth.gr

1 Introduction

Classical logic, as founded by the Greek philosopher Aristotle, is based on the *principle of bivalence* which states that every proposition can be assigned exactly one of the logical values *true* or *false*. However, Aristotle himself observed that this principle cannot describe the status of *all* propositions especially the ones which refer to *future contingents*. In his treatise *On Interpretation 9*, the philosopher formulated the famous sentence "There will be a sea-battle tomorrow", which is actually neither true nor false. Clearly, (at least) a *third* logical value is required in order to describe such situations. Actually this third value spoils the principle of bivalence. Nevertheless, despite the efforts

M. Droste, W. Kuich, H. Vogler (eds.), *Handbook of Weighted Automata*,
Monographs in Theoretical Computer Science. An EATCS Series,
DOI 10.1007/978-3-642-01492-5_12,

of philosophers and mathematicians in the Middle Ages, a *three-valued propositional logic* was successfully established by Łukasiewicz and Post only in 1920 (see [48, 59]). However, it came up that even that three-valued logic was not sufficient enough to describe the logical status of real world statements. Therefore, the three-valued logic has been extended to *multi-valued* (or *many-valued*) *logic* by considering (finitely or infinitely) many logical values. For textbooks on multi-valued logic, we refer the reader to [50, 64] (see also [66] for historical details for the multi-valued logic's progress and the contribution of Gr. Moisil and A. Salomaa to this field).

On the other hand, Zadeh [78] introduced in 1965 the concept of *fuzzy sets*. He was motivated by the real world where sentences like "the class of real numbers that are much greater than 1" or "the class of beautiful women" are naturally imprecise, and they do not determine sets in the usual mathematical sense. In 1973, Zadeh founded his *fuzzy logic* as a multi-valued logic over the interval $[0,1] \subseteq \mathbb{R}$, enriched with further fuzzy quantifiers like *most*, *few*, *many*, and *several*. In the meantime, Wee [75] introduced the *fuzzy automaton* as a model of learning systems. The fuzzy automaton model is the natural *fuzzification* of the classical finite automaton and it is actually a weighted automaton model (over the fuzzy semiring $\langle [0,1], \mathsf{max}, \mathsf{min}, 0, 1 \rangle$) in the sense of [24]. Since then, fuzzy automata theory has been extended to more general structures like lattices, residuated lattices, and ℓ-monoids. However, in all these cases, the corresponding fuzzy automata act on semirings induced by the original structures. Therefore, all the well-known results for recognizable formal power series over semirings hold in particular for *fuzzy recognizable languages* accepted by fuzzy automata. More specific results can be obtained for fuzzy automata and their behaviors due to the special properties of their underlying semirings inherited by the original structures. For instance, the determinization problem is effectively solved for fuzzy automata and the equality is decidable for fuzzy recognizable languages over bounded distributive lattices.

Fuzzy structures and fuzzy logic contribute to a wide range of real world applications because they can effectively incorporate the impreciseness of practical problems. It is the purpose of this chapter, to present the theory of fuzzy recognizable languages as a paradigm of recognizable formal power series. Our fuzzy languages are defined over bounded distributive lattices. This is a more general case than the very first definition of fuzzy languages over the interval $[0,1]$, but still almost all the recognizability properties remain valid. In our development, we refer only briefly to those results which are inherited from the general theory of weighted automata and power series. Instead, we focus on results which do not hold for power series over arbitrary semirings. More precisely, our fuzzy recognizable languages are obtained as behaviors of multi-valued automata. We show that for every such multi-valued automaton we can effectively construct an equivalent trim deterministic one which moreover has a minimum counterpart. Furthermore, the equivalence problem for multi-valued automata is decidable and a pumping lemma holds for fuzzy recognizable languages. The equivalence problem turns out also to be decidable for multi-valued automata over infinite words. Our treatment of fuzzy recognizable languages is based on automata-theoretic techniques. It is worth noting that fuzzy recognizability over finite words, especially over the fuzzy semiring, has been also defined by means of finite monoid representations, syntactic congruences, syntactic monoids, and (left and right) derivatives (see [8, 9, 36, 47, 54, 57]). On the other hand, several authors have fuzzified notions like monoids [36], trees [22, 28], and algebras [46, 70–72, 74].

In the sequel, we briefly describe the contents of the chapter. First, we introduce basic notions like (bounded distributive) lattices and the more particular class of De Morgan algebras. We show that the collection of De Morgan algebras coincides with the family of semirings with complement function. We define the concept of fuzzy languages as formal power series over bounded distributive lattices. Then we deal with fuzzy recognizable languages over finite (resp. infinite) words obtained as behaviors of multi-valued (resp. multi-valued Büchi and Muller) automata. An MSO logic characterization of fuzzy recognizable languages is also provided. Next, we briefly investigate fuzzy languages over bounded ℓ-monoids and residuated lattices. These are the most general classes of fuzzy languages, but still they are special cases of power series. Finally, we refer to practical applications of fuzzy languages. The material concerning multi-valued automata over infinite words, De Morgan algebras, and the MSO logic is contained in [22].

For monographs presenting fuzzy logic, fuzzy languages, and fuzzy automata, we refer the reader to [32, 35, 54, 74]. Our list of references includes only these ones which are connected with the context of the chapter. In [1, 35, 54], there are extended lists of references until 2002. Also, the journal *Fuzzy Sets and Systems* publishes periodically an article entitled *Recent Literature*, and it presents the latest developments in fuzzy theory (see for instance volume 159 (2008), pages 857–865).

2 Lattices and Fuzzy Languages

A partially ordered set $(L, \leq)$ is called a *lattice* if the supremum (called also least upper bound or join) $a \vee b$ and the infimum (called also greatest lower bound or meet) $a \wedge b$ exist in L for every $a, b \in L$ (see [21]). A lattice $(L, \leq)$ (which is simply denoted by L if the order relation is understood) is *distributive* if it satisfies the equation $a \wedge (b \vee c) = (a \wedge b) \vee (a \wedge c)$ (which in turn implies $a \vee (b \wedge c) = (a \vee b) \wedge (a \vee c)$) for every $a, b, c \in L$. The supremum (resp. the infimum) of every $A \subseteq L$ is denoted (if it exists in L) by $\vee A$ (resp. $\wedge A$). If $A = (a_i \mid i \in I)$, then we also use the notation $\bigvee_{i \in I} a_i$ (resp. $\bigwedge_{i \in I} a_i$). A lattice L is *bounded* if it contains two distinguished elements $0, 1 \in L$ such that $0 \leq a \leq 1$ for every $a \in L$. Furthermore, a lattice L is called *complete* if $\vee A$ and $\wedge A$ exist for every $A \subseteq L$. Observe that a complete lattice is also bounded with $0 = \vee\emptyset$ and $1 = \wedge\emptyset$. It is well known that if L is any distributive lattice and $A \subseteq L$ a finite subset, then the sub-lattice L_A of L generated by A is finite. In fact, if $A' = \{\wedge B \mid B \subseteq A\}$, then we have $L_A = \{\vee C \mid C \subseteq A'\}$ due to the distributivity law. Obviously, every finite lattice L is complete. A bounded distributive lattice L forms a semiring $\langle L, \vee, \wedge, 0, 1\rangle$ whose operations are both idempotent. An element $a \neq 0$ of a lattice L is called *join-irreducible* if $a = b \vee c$ implies $a = b$ or $a = c$ for every $b, c \in L$. We denote by $J(L)$ the set of all join-irreducible elements of L. If the lattice L is finite, then

$$a = \vee\{b \in J(L) \mid b \leq a\}$$

for every $a \in L$. Moreover, if L is distributive, then every join-irreducible element $a \in L$ is *prime*, i.e., whenever $a \leq b \vee c$ with $b, c \in L$, then $a \leq b$ or $a \leq c$ (cf. [6, 16]).

Let $(L, \leq)$ be a bounded distributive lattice and $^- : L \to L$ be any function with $\overline{0} = 1$ and $\overline{1} = 0$. Then we call $^-$ a (*general*) *negation function* and $(L, \leq, ^-)$ a *bounded distributive lattice with negation function.* Note that every bounded distributive lattice L can be equipped with a negation function $^-$ by letting for instance $\overline{0} = 1$ and $\overline{x} = 0$ for every $x \in L \setminus \{0\}$. De Morgan algebras, Heyting algebras, and variants of pseudo-complemented lattices are well-investigated classes of distributive lattices with negation function (see [3, 16]). Recently, De Morgan algebras have been investigated intensively for multi-valued model checking (see [13, 31, 39]). More precisely, a *De Morgan* (or *quasi-Boolean*) *algebra* is a distributive lattice $(L, \leq, ^-)$ with a *complement mapping* $^-$ satisfying the involution $\overline{\overline{a}} = a$ and De Morgan laws, i.e., $\overline{a \vee b} = \overline{a} \wedge \overline{b}$ and $\overline{a \wedge b} = \overline{a} \vee \overline{b}$ for every $a, b \in L$. Then $a \leq b$ implies $\overline{b} \leq \overline{a}$ for every $a, b \in L$. Furthermore, if L is bounded, then $\overline{0} = 1$ and $\overline{1} = 0$, i.e., the function $^-$ is a negation function. Moreover, the mapping $^- : (L, \leq) \to (L, \geq)$ is an order-isomorphism. Hence, if $(a_i \mid i \in I) \subseteq L$ is a family of elements of L for which $\bigvee_{i \in I} a_i$ exists, then $\overline{\bigvee_{i \in I} a_i} = \bigwedge_{i \in I} \overline{a_i}$. For instance, the lattice $([0,1], \leq, ^-)$ with $\leq$ the usual order of real numbers, and $\overline{a} = 1 - a$ for every $a \in [0,1]$ is a De Morgan algebra. The induced semiring $\langle [0,1], \mathsf{max}, \mathsf{min}, 0, 1\rangle$ is referred to as the *fuzzy semiring.* In the sequel, without any further notation, for every De Morgan algebra $(L, \leq, ^-)$, we require the lattice L to be bounded. On the other hand, every bounded distributive lattice can be endowed with a negation function, therefore, lattices with negation function constitute a much larger class than De Morgan algebras. In particular, any bounded distributive lattice which is not anti-isomorphic to itself, does not have a complement operation, and thus cannot be structured to a De Morgan algebra.

Next, we investigate the relationship between De Morgan algebras and semirings. Given a semiring $\langle S, +, \cdot, 0, 1\rangle$, a mapping $f : S \to S$ is called a *complement function*, if it satisfies the following statements:

(i) f is an *involution*, i.e., $f(f(a)) = a$ for every $a \in S$.
(ii) f is a monoid morphism from $\langle S, +, 0\rangle$ to $\langle S, \cdot, 1\rangle$, i.e., $f(0) = 1$ and $f(a + b) = f(a) \cdot f(b)$ for every $a, b \in S$.

It is easily seen that $f(1) = 0$ and $f(a \cdot b) = f(a) + f(b)$ for every $a, b \in S$, hence f is a monoid isomorphism from $\langle S, +, 0\rangle$ to $\langle S, \cdot, 1\rangle$ and from $\langle S, \cdot, 1\rangle$ to $\langle S, +, 0\rangle$. Every De Morgan algebra $(L, \leq, ^-)$ induces a semiring $\langle L, \vee, \wedge, 0, 1\rangle$ with complement mapping $^-$, therefore, the following result concludes that De Morgan algebras and semirings with complement function coincide. This indicates the relation between the MSO logic over De Morgan algebras (see Sect. 3.3) and semirings (see [18, 23, 20]).

Proposition 2.1 ([22]). *Let $\langle S, +, \cdot, 0, 1\rangle$ be a semiring with complement function f. For every $a, b \in S$, we put $a \leq b$ iff $a + b = b$. Then $(S, \leq, f)$ is a De Morgan algebra.*

Proof. We have $f(0) = 1$, f is an involution, $f(a + b) = f(a) \cdot f(b)$ and $f(a \cdot b) = f(a) + f(b)$. Hence, $0 \cdot 0 = 0$ implies $1 + 1 = 1$, so $\langle S, +, 0\rangle$, and hence also $\langle S, \cdot, 1\rangle$ are idempotent. Thus, $\leq$ is a partial order on S (see Proposition 20.19 in [30]) and $a + b$ is the supremum of a and b in this partial order. Moreover, $0 \leq a$ for every $a \in S$, and $a \cdot 0 = 0$ implies $f(a) + 1 = 1$, so $f(a) \leq 1$, showing also $a \leq 1$ for every $a \in S$.

Next, observe that if $a \leq b$, then by distributivity we have $a \cdot c \leq b \cdot c$ for every $a, b, c \in S$. We show that $a \cdot b$ is the infimum of a and b in $(S, \leq)$ for every $a, b \in S$. Since $a \leq 1$, the previous remark implies $a \cdot b \leq b$ and similarly $a \cdot b \leq a$. Now if $c \in S$ with $c \leq a$ and $c \leq b$, then $c = c \cdot c \leq a \cdot c \leq a \cdot b$, proving that $a \cdot b = a \wedge b$. Hence, $(S, \leq)$ is a distributive lattice with $+$ being the operation supremum and $\cdot$ being the infimum. Moreover, $(S, \leq)$ is bounded, and f is a complement mapping satisfying De Morgan laws. Thus, the proof is completed. □

The interested reader should find further characterizations of bounded distributive lattices by means of semirings in Example 1.5 of [30].

Given two lattices $(L, \leq)$ and $(L', \leq)$, a mapping $f : L \to L'$ is a *lattice morphism* if it preserves suprema and infima, i.e., for every $a, b \in L$

$$f(a \vee b) = f(a) \vee f(b) \quad \text{and} \quad f(a \wedge b) = f(a) \wedge f(b).$$

Then $a \leq b$ implies $f(a) \leq f(b)$ for every $a, b \in L$. Furthermore, if $(L, \leq)$ and $(L', \leq)$ are bounded distributive lattices, then a lattice morphism $f : L \to L'$ satisfying $f(0) = 0$ and $f(1) = 1$ is a semiring morphism from $\langle L, \vee, \wedge, 0, 1\rangle$ to $\langle L', \vee, \wedge, 0, 1\rangle$.

Now we turn to fuzzy sets originally introduced by Zadeh in [78]. Given a non-empty set X, a *fuzzy set A in X* (or a *fuzzy subset A of X*) is defined by a membership function

$$f_A : X \to [0, 1].$$

A fuzzy subset of a free monoid is called a *fuzzy language* [40]. Thus, a fuzzy language is nothing else but a formal power series over the fuzzy semiring $\langle [0, 1], \mathsf{max}, \mathsf{min}, 0, 1\rangle$. So far, the term fuzzy language has been also used for power series over lattices, residuated lattices, and ℓ-monoids (see Sect. 4).

Here, we deal with fuzzy languages over bounded distributive lattices. More precisely, let S be a set and L be a bounded distributive lattice. A *formal power series* (*over S and L*) is a mapping $r : S \to L$. Such a series is called (*finitary*) *fuzzy language* (resp. *infinitary fuzzy language*) over some finite alphabet Σ if $S = \Sigma^*$ (resp. $S = \Sigma^\omega$, i.e., the set of all infinite words over Σ). Subsequently, we will only need the cases where $S = \Sigma^*$ or $S = \Sigma^\omega$. The support $\mathrm{supp}(r)$ of a series r over S and L is defined as usually by $\mathrm{supp}(r) = \{s \in S \mid (r, s) \neq 0\}$, and the *image* of r is the set $\{l \in L \mid \exists s \in S : (r, s) = l\}$. The collection $L\langle\langle S\rangle\rangle$ of all power series over S and L is itself a bounded distributive lattice $(L\langle\langle S\rangle\rangle, \leq)$; for $r, r' \in L\langle\langle S\rangle\rangle$ the partial order $\leq$ is determined by $r \leq r'$ iff $(r, s) \leq (r', s)$ for every $s \in S$. Then the supremum $r \vee r'$ and the infimum $r \wedge r'$ are defined elementwise by $(r \vee r', s) = (r, s) \vee (r', s)$ and $(r \wedge r', s) = (r, s) \wedge (r', s)$ for every $s \in S$. Furthermore, for every $k \in L$, the scalar infimum $k \wedge r$ is determined by $(k \wedge r, s) = k \wedge (r, s)$ for every $s \in S$. If $(L, \leq, ^-)$ is a bounded distributive lattice with negation function (resp. a De Morgan algebra), then $(L\langle\langle S\rangle\rangle, \leq, ^-)$ constitutes also a bounded distributive lattice with negation function (resp. a De Morgan algebra); for every $r \in L\langle\langle S\rangle\rangle$ its negation $\overline{r} \in L\langle\langle S\rangle\rangle$ is defined by $(\overline{r}, s) = \overline{(r, s)}$ for every $s \in S$.

Assume that $(L, \leq)$ and $(L', \leq)$ are two distributive lattices, and let $f : L \to L'$ be any mapping. Then f is extended to a mapping $f : L\langle\langle S\rangle\rangle \to L'\langle\langle S\rangle\rangle$ in the following way. For every $r \in L\langle\langle S\rangle\rangle$, the series $f(r) \in L'\langle\langle S\rangle\rangle$ is determined by $(f(r), s) = f((r, s))$ for every $s \in S$.

3 Fuzzy Recognizability over Bounded Distributive Lattices

We consider the concept of fuzzy recognizable languages obtained as behaviors of weighted automata over bounded distributive lattices. Such automata are called *multi-valued,* and they have recently contributed to multi-valued logics [22] and multi-valued model checking employing distributive lattices [10, 39]. Several other names occur in the literature for automata over lattices, like fuzzy automaton, max-min automaton, L-fuzzy automaton, and lattice automaton depending on the properties of the underlying lattice (see, for instance, [43, 54]).

First, we deal with fuzzy recognizable languages over finite words. For these languages a Kleene–Schützenberger theorem is obtained as a special case of the corresponding theorem for recognizable series over commutative semirings. Then we show that fuzzy recognizable languages have an elegant characterization, namely they are written as fuzzy recognizable step languages. This enables us to give short proofs for well-known results concerning multi-valued automata. More precisely, we show that:

(i) For every multi-valued automaton, we can effectively construct an equivalent minimum trim deterministic one.

(ii) The equivalence problem for multi-valued automata is decidable.
(iii) A pumping lemma holds for fuzzy recognizable languages.
(iv) A fuzzy language is recognizable iff it has finite image and each of its cut languages is recognizable.

It is worth noting that these results do not hold in general for weighted automata over arbitrary semirings. Next, we introduce Büchi and Muller multi-valued automata working on infinite words. As in the finitary case, we show that fuzzy Büchi recognizable languages can be written as fuzzy Büchi recognizable step languages. Using this simple characterization, we give elegant proofs for two important results. Namely, the classes of fuzzy Büchi and fuzzy Muller recognizable languages coincide, and a Kleene theorem holds for them. Moreover, we introduce a multi-valued MSO logic and we show the fundamental theorem of Büchi, i.e., fuzzy definable languages over infinite words coincide with fuzzy Büchi recognizable languages. For the rest of this section, Σ will denote an arbitrary finite alphabet and L an arbitrary bounded distributive lattice.

3.1 Fuzzy Recognizability over Finite Words

We start with the concept of multi-valued automata.

Definition 3.1. *A* multi-valued automaton *(MVA for short)* over Σ and L *is a quadruple* $\mathcal{A} = (Q, \mathrm{in}, \mathrm{wt}, \mathrm{out})$, *where* Q *is the* finite state set, $\mathrm{in} : Q \to L$ *is the* initial distribution, $\mathrm{wt} : Q \times \Sigma \times Q \to L$ *is the mapping assigning* weights *to the transitions of the automaton, and* $\mathrm{out} : Q \to L$ *is the* final distribution.

Let $w = a_0 \dots a_{n-1} \in \Sigma^*$ where $a_0, \dots, a_{n-1} \in \Sigma$. A *path of* $\mathcal{A}$ *over* w is a sequence $P_w = (t_i)_{0 \le i \le n-1}$ of transitions, such that $t_i = (q_i, a_i, q_{i+1}) \in Q \times \Sigma \times Q$ for every $0 \le i \le n-1$. The *weight of* P_w is defined by

$$\mathrm{weight}(P_w) = \mathrm{in}(q_0) \wedge \bigwedge_{0 \le i \le n-1} \mathrm{wt}(t_i) \wedge \mathrm{out}(q_n).$$

We shall denote by $L_{\mathcal{A}}$ the finite sub-lattice of L generated by $\{0,1\} \cup \{\mathrm{in}(q) \mid q \in Q\} \cup \{\mathrm{out}(q) \mid q \in Q\} \cup \{\mathrm{wt}(t) \mid t \in Q \times \Sigma \times Q\}$. Clearly $\mathrm{weight}(P_w) \in L_{\mathcal{A}}$. The *behavior* of $\mathcal{A}$ is the fuzzy language

$$\|\mathcal{A}\| : \Sigma^* \to L$$

which is defined by

$$(\|\mathcal{A}\|, w) = \bigvee_{P_w} \mathrm{weight}(P_w)$$

for $w \in \Sigma^*$, where the supremum is taken over all paths P_w of $\mathcal{A}$ over w. It should be clear that $(\|\mathcal{A}\|, \varepsilon) = \bigvee_{q \in Q} \mathrm{in}(q) \wedge \mathrm{out}(q)$. Again, $(\|\mathcal{A}\|, w) \in L_{\mathcal{A}}$ for every $w \in \Sigma^*$.

Two multi-valued automata $\mathcal{A}$ and $\mathcal{A}'$ over Σ and L are called *equivalent* if they have the same behavior, i.e., $\|\mathcal{A}\| = \|\mathcal{A}'\|$.

A fuzzy language $r \in L\langle\langle \Sigma^* \rangle\rangle$ is said to be *fuzzy recognizable over Σ and L* if there is an MVA $\mathcal{A}$ such that $r = \|\mathcal{A}\|$. We denote the family of all fuzzy recognizable languages over Σ and L by $L^{\text{rec}}\langle\langle \Sigma^* \rangle\rangle$. The reader should observe that an MVA over Σ and L is just a weighted automaton over Σ and the semiring $\langle L, \vee, \wedge, 0, 1 \rangle$ in the sense of [24, 38, 67] (see also Theorems 2.2 and 3.6 in [27]). Thus, the class $L^{\text{rec}}\langle\langle \Sigma^* \rangle\rangle$ coincides with the collection of all recognizable series over Σ and the semiring L. Therefore, as a consequence of the general Kleene–Schützenberger theorem for series over arbitrary semirings (see, for instance, [65]), we immediately obtain its reformulation for fuzzy languages as follows. Let us first reconsider the rational operations of formal power series in the setting of fuzzy languages. Let $r, r' \in L\langle\langle \Sigma^* \rangle\rangle$. The *Cauchy product rr' of r and r'* is a fuzzy language in $L\langle\langle \Sigma^* \rangle\rangle$ which is determined by $(rr', w) = \bigvee_{uu'=w} (r, u) \wedge (r', u')$ for every $w \in \Sigma^*$. If r is *proper*, i.e., $(r, \varepsilon) = 0$, then we define the *star $r^* \in L\langle\langle \Sigma^* \rangle\rangle$ of r* by $(r^*, w) = \vee\{(r, u_1) \wedge \cdots \wedge (r, u_n) \mid u_1 \ldots u_n = w, u_1, \ldots, u_n \in \Sigma^*\}$ for every $w \in \Sigma^*$. The *rational operations of* fuzzy languages in $L\langle\langle \Sigma^* \rangle\rangle$ are the supremum, the Cauchy product, and the star. We denote by $L^{\text{rat}}\langle\langle \Sigma^* \rangle\rangle$ the least class of fuzzy languages from $L\langle\langle \Sigma^* \rangle\rangle$ which contains the polynomials, i.e., the fuzzy languages with finite support, and is closed under the rational operations.

Theorem 3.2 (Kleene–Schützenberger). *Let Σ be an alphabet and L be a bounded distributive lattice. Then $L^{\text{rec}}\langle\langle \Sigma^* \rangle\rangle = L^{\text{rat}}\langle\langle \Sigma^* \rangle\rangle$.*

Let Σ, Δ be alphabets and $h : \Sigma^* \to \Delta^*$ be any morphism. Then we can define the mapping $h^{-1} : L\langle\langle \Delta^* \rangle\rangle \to L\langle\langle \Sigma^* \rangle\rangle$ (see [21]); if L is a complete lattice or h is non-deleting, then the mapping $h : L\langle\langle \Sigma^* \rangle\rangle \to L\langle\langle \Delta^* \rangle\rangle$ is also well defined.

Proposition 3.3 ([24]). *Let Σ, Δ be two alphabets and $h : \Sigma^* \to \Delta^*$ be any morphism. Then:*

(i) $h^{-1} : L\langle\langle \Delta^* \rangle\rangle \to L\langle\langle \Sigma^* \rangle\rangle$ preserves fuzzy recognizability.
(ii) If h is non-deleting, then $h : L\langle\langle \Sigma^* \rangle\rangle \to L\langle\langle \Delta^* \rangle\rangle$ preserves fuzzy recognizability.

Recall that for every language $R \subseteq \Sigma^*$, its characteristic series $1_R \in L\langle\langle \Sigma^* \rangle\rangle$ is defined by $(1_R, w) = 1$ if $w \in R$, and 0 otherwise, for every $w \in \Sigma^*$. Here, we call 1_R the *characteristic fuzzy language of R*. Every unweighted finite automaton with input alphabet Σ can be considered in the obvious way, as an MVA over Σ and L with weights only 0 and 1. Therefore, for every recognizable language R, its characteristic language 1_R is fuzzy recognizable. Assume now that $R_1, \ldots, R_n \subseteq \Sigma^*$ are recognizable languages and $k_1, \ldots, k_n \in L$. Clearly, the fuzzy language $k_i \wedge 1_{R_i}$ is recognizable for every $1 \leq i \leq n$. Then by Theorem 3.2, the fuzzy language

$$r = \bigvee_{1 \le i \le n} k_i \wedge 1_{R_i}$$

is also recognizable. Such a language r is called a *fuzzy recognizable step language.* The class of recognizable languages is closed under the Boolean operations; hence, for every fuzzy recognizable step language $r = \bigvee_{1 \le i \le n} k_i \wedge 1_{R_i}$, we may assume that the family $(R_i \mid 1 \le i \le n)$ forms a partition of Σ^*. Next, we show that fuzzy recognizable languages and fuzzy recognizable step languages coincide. This important result has been firstly proved in [19] for power series over locally finite semirings. Therefore, it can be applied to the class of fuzzy languages over bounded distributive lattices (recall that for every bounded distributive lattice L, the semiring $\langle L, \vee, \wedge, 0, 1 \rangle$ is locally finite). However, here we give an alternative proof based on lattices. The same proof has been also used in [22] for the corresponding result for infinitary fuzzy languages (see Sect. 3.2). We shall need the following lemma which is easily proved by a standard automata construction.

Lemma 3.4. *Let $(L, \le)$ and $(L', \le)$ be two bounded distributive lattices and $f : L \to L'$ be a lattice morphism. Then for every fuzzy recognizable language r in $L\langle\langle \Sigma^* \rangle\rangle$, the fuzzy language $f(r) \in L'\langle\langle \Sigma^* \rangle\rangle$ is again recognizable.*

Theorem 3.5. *Let Σ be an alphabet and L be a bounded distributive lattice. Then a fuzzy language $r \in L\langle\langle \Sigma^* \rangle\rangle$ is recognizable iff it is a fuzzy recognizable step language.*

Proof. Let r be fuzzy recognizable and $\mathcal{A} = (Q, \mathrm{in}, \mathrm{wt}, \mathrm{out})$ be an MVA over Σ and L such that $r = \|\mathcal{A}\|$ and $L_{\mathcal{A}} = \{k_1, \ldots, k_n\}$. We set $R_i = \{w \in \Sigma^* \mid (r, w) = k_i\}$ for every $1 \le i \le n$. Then

$$r = \bigvee_{1 \le i \le n} k_i \wedge 1_{R_i}.$$

We shall show that the languages R_i $(1 \le i \le n)$ are recognizable. Let $\mathbb{B} = (\{0, 1\}, \le)$ be the two-valued Boolean lattice. For every join-irreducible element p of $L_{\mathcal{A}}$, we define a mapping $f_p : L_{\mathcal{A}} \to \{0, 1\}$ by putting

$$f_p(a) = \begin{cases} 1 & \text{if } p \le a, \\ 0 & \text{otherwise} \end{cases}$$

for every $a \in L$.

We claim that f_p is a lattice morphism. Indeed, $p \neq 0$; hence, $f_p(0) = 0$ and $f_p(1) = 1$. Next, note that if $a, a' \in L_{\mathcal{A}}$ and $f_p(a \vee a') = 1$, then $p \le a \vee a'$ which implies $p \le a$ or $p \le a'$ since p is prime, proving $f_p(a \vee a') = f_p(a) \vee f_p(a')$. Clearly, $f_p(a \wedge a') = f_p(a) \wedge f_p(a')$. By Lemma 3.4, the fuzzy language $f_p(r)$ of $\mathbb{B}\langle\langle \Sigma^* \rangle\rangle$ is recognizable and, therefore, the language $\mathrm{supp}(f_p(r)) = \{w \in \Sigma^* \mid p \le (r, w)\}$ is recognizable. Now let $1 \le i \le n$. Since the element k_i of $L_{\mathcal{A}}$ is the supremum of the join-irreducible elements of $L_{\mathcal{A}}$ below k_i, the language R_i is obtained as the intersection of the languages $\mathrm{supp}(f_p(r))$ ($p \leq k_i$ and join-irreducible) and of the complements of the languages $\mathrm{supp}(f_p(r))$ ($p \not\leq k_i$ and join-irreducible). The class of recognizable languages is closed under the Boolean operations. Therefore, we conclude that R_i is a recognizable language, as required.

The converse is also true as already noted. □

Observe that the proof of the above theorem is effective. Indeed, starting from the weights of the multi-valued automaton $\mathcal{A}$, we compute the sub-lattice $L_{\mathcal{A}}$ in finitely many steps. Then following our proof, we obtain finite automata for the languages R_i ($1 \leq i \leq n$).

Due to Theorem 3.5, in the sequel, we write every fuzzy recognizable language as a fuzzy recognizable step language. This has very interesting consequences. Firstly, generalizing Lemma 3.4, we show that fuzzy recognizability is preserved even by arbitrary mappings between lattices.

Proposition 3.6. *Let* $(L, \leq)$ *and* $(L', \leq)$ *be two bounded distributive lattices and* $f : L \to L'$ *be any mapping. Then for every fuzzy recognizable language* $r \in L\langle\langle \Sigma^* \rangle\rangle$ *the language* $f(r) \in L'\langle\langle \Sigma^* \rangle\rangle$ *is again fuzzy recognizable.*

Proof. Let $r = \bigvee_{1 \leq i \leq n} k_i \wedge 1_{R_i}$. Then $f(r) = \bigvee_{1 \leq i \leq n} f(k_i) \wedge 1_{R_i}$ and so $f(r)$ is fuzzy recognizable. □

Next, we get a classical result from fuzzy language theory. More precisely, given a fuzzy language $r \in L\langle\langle \Sigma^* \rangle\rangle$ and $l \in L$, the *l-cut of r* is the language $r_{\geq l} = \{w \in \Sigma^* \mid (r, w) \geq l\}$. Furthermore, for every $l \in L$, we let $r_{=l} = r^{-1}(l) = \{w \in \Sigma^* \mid (r, w) = l\}$.

Proposition 3.7 ([43]). *For every fuzzy language* $r \in L\langle\langle \Sigma^* \rangle\rangle$*, the following statements are equivalent:*

(i) r *is fuzzy recognizable.*
(ii) r *has finite image, and for every* $l \in L$, $r_{=l}$ *is a recognizable language.*
(iii) r *has finite image, and for every* $l \in L$, $r_{\geq l}$ *is a recognizable language.*

Proof. The equivalence of (i) and (ii) is immediate by Theorem 3.5. We show the implication (i) ⇒ (iii). Let $r = \bigvee_{1 \leq i \leq n} k_i \wedge 1_{R_i}$ with pairwise disjoint recognizable languages R_i. Consider an $l \in L$. If there is no $i \in \{1, \ldots, n\}$ such that $k_i \geq l$, then $r_{\geq l} = \emptyset$. Otherwise, let $k_{i_1}, \ldots, k_{i_m}$ ($1 \leq i_1 < \cdots < i_m \leq n$) be all the values of r with $k_{i_1}, \ldots, k_{i_m} \geq l$. Then $r_{\geq l} = R_{i_1} \cup \cdots \cup R_{i_m}$, hence $r_{\geq l}$ is recognizable. Finally, assume that statement (iii) is true. For every $l \in L$, we have $r_{=l} = r_{\geq l} \setminus \bigcup_{l' \in L, l < l'} r_{\geq l'}$, and thus $r_{=l}$ is recognizable which concludes our proof. □

In the sequel, we deal with the determinization and minimization problems of multi-valued automata. These problems do not always have a solution for weighted automata over arbitrary semirings (see [12, 34, 52]) or even over residuated lattices and ℓ-monoids (see [33, 42]). However, due to the local finiteness property of distributive lattices, we show that for every multi-valued automaton we can effectively construct an equivalent trim deterministic one. The determinization problem for fuzzy automata over $([0,1], \leq)$ was first solved in [51]. Borchardt in [7] showed that weighted tree automata over locally finite semirings can be effectively determinized (see [34] for the word case). A reformulation of the same method is used in [42] for fuzzy automata over bounded ℓ-monoids, and in [5, 43] (resp. in [33]) for the special case of fuzzy automata over bounded distributive lattices (resp. over residuated lattices). In all the aforementioned papers, the authors followed the well-known subset construction (or even the accessible subset construction in [33]) in the weighted setting. Here, we use the result of Theorem 3.5 and we reduce the determinization of multi-valued automata to the determinization of classical finite automata (as indicated in [19]). Then we minimize the trim deterministic multi-valued automaton. For this minimization procedure, we use the classical *reduction algorithm* (see [24]). In [4, 68] (resp. in [58]), the size (number of states) of a non-deterministic fuzzy automaton (over the fuzzy semiring) is reduced by means of equivalences (resp. congruences) on the set of states.

A *deterministic multi-valued automaton* (DMVA for short) *over Σ and L* is an MVA $\mathcal{A} = (Q, \text{in}, \text{wt}, \text{out})$ such that the following two conditions hold:

(i) There is exactly one $q_0 \in Q$ such that $\text{in}(q_0) = 1$ and for every $p \in Q$ with $p \neq q_0$ we have $\text{in}(p) = 0$.
(ii) For every $q \in Q$ and $\sigma \in \Sigma$, there is at most one state $q' \in Q$ such that $\text{wt}(q, \sigma, q') = 1$ and for every $p \in Q$ with $p \neq q'$ we have $\text{wt}(q, \sigma, p) = 0$.

Clearly, for a DMVA $\mathcal{A}$, the function wt can be equivalently expressed by a (partial) function $\delta : Q \times \Sigma \to Q$ in the obvious way. Therefore, we will denote in the sequel a DMVA by $(Q, q_0, \delta, \text{out})$ with $q_0 \in Q$ and $\delta : Q \times \Sigma \to Q$ as a partial function. Thus, a DMVA $\mathcal{A}$ can be considered as a classical deterministic automaton with weights attached only to the final states. The DMVA $\mathcal{A} = (Q, q_0, \delta, \text{out})$ is called *accessible* if for every state $q \in Q$ there exists a word $w \in \Sigma^*$ such that $\delta(q_0, w) = q$. Furthermore, $\mathcal{A}$ is *co-accessible* if for every $q \in Q$ there exists $w \in \Sigma^*$ such that $\text{out}(\delta(q, w)) > 0$. A DMVA is called *trim* if it is accessible and co-accessible. Observe that in a DMVA $\mathcal{A}$, for every word $w = a_0 \ldots a_{n-1} \in \Sigma^*$ and for every path $P_w = (p_i, a_i, p_{i+1})_{0 \leq i \leq n-1}$ of $\mathcal{A}$ over w such that $\delta(p_i, a) = p_{i+1}$, we have

$$\text{weight}(P_w) = \begin{cases} \text{out}(p_n) & \text{if } p_0 = q_0, \\ 0 & \text{otherwise.} \end{cases}$$

Theorem 3.8. *Let Σ be an alphabet and L be a bounded distributive lattice. For every MVA $\mathcal{A} = (Q, \text{in}, \text{wt}, \text{out})$ over Σ and L, we can effectively construct a trim DMVA $\mathcal{A}'$ over Σ and L such that $\|\mathcal{A}'\| = \|\mathcal{A}\|$.*

Proof. Let $\|\mathcal{A}\| = \bigvee_{1\le i\le n} k_i \wedge 1_{R_i}$ with pairwise disjoint recognizable languages R_i. Clearly, we may assume that $k_i \neq 0$ for every $1 \le i \le n$. Let $\mathcal{A}_i = (Q_i, \Sigma, q_{0i}, \delta_i, F_i)$ $(1 \le i \le n)$ be a complete deterministic (i.e., δ_i is a total mapping) finite automaton accepting R_i. Now we perform a classical construction of an automaton accepting a union of languages. Consider the finite automaton $\widetilde{\mathcal{A}} = (\widetilde{Q}, \Sigma, q_0, \widetilde{\delta}, \widetilde{F})$ with $\widetilde{Q} = Q_1 \times \cdots \times Q_n$, $q_0 = (q_{01}, \ldots, q_{0n})$, and $\widetilde{F} = \bigcup_{1\le i\le n} Q_1 \times \cdots \times Q_{i-1} \times F_i \times Q_{i+1} \times \cdots \times Q_n$. The (total) mapping $\widetilde{\delta} : \widetilde{Q} \times \Sigma \to \widetilde{Q}$ is determined by $\widetilde{\delta}((q_1, \ldots, q_n), a) = (\delta_1(q_1, a), \ldots, \delta_n(q_n, a))$ for every $(q_1, \ldots, q_n) \in \widetilde{Q}, a \in \Sigma$. Obviously, $\widetilde{\mathcal{A}}$ is deterministic with behavior $R_1 \cup \cdots \cup R_n$. Now let $\overline{\mathcal{A}} = (Q, \Sigma, q_0, \delta, F)$ be the trim part of $\widetilde{\mathcal{A}}$ (see [24]). We consider the DMVA $\mathcal{A}' = (Q, q_0, \delta, \mathrm{out})$ over Σ and L with $\mathrm{out}((q_1, \ldots, q_n)) = \bigvee_{1\le i\le n} \mathrm{out}_i(q_i)$ where

$$\mathrm{out}_i(q_i) = \begin{cases} k_i & \text{if } q_i \in F_i, \\ 0 & \text{otherwise.} \end{cases}$$

The finite automaton $\overline{\mathcal{A}}$ is accessible, and thus the DMVA $\mathcal{A}'$ is also accessible. Moreover, $\mathcal{A}'$ is trim. Indeed, let $(q_1, \ldots, q_n) \in Q$. Since $\overline{\mathcal{A}}$ is co-accessible there is a $w \in \Sigma^*$ such that $\delta((q_1, \ldots, q_n), w) \in F$, i.e., there is an index $1 \le i \le n$ such that $\delta_i(q_i, w) \in F_i$ which in turn implies that $\mathrm{out}_i(\delta_i(q_i, w)) = k_i$. In fact, since the languages R_i are pairwise disjoint, there is exactly one index i with this property, and for every other $1 \le j \le n$ with $j \neq i$, we have $\delta_j(q_j, w) \in Q_j \setminus F_j$. Hence, $\mathrm{out}(\delta((q_1, \ldots, q_n), w)) = k_i > 0$. Now for every $w \in \Sigma^*$,

$$\begin{aligned}(\|\mathcal{A}'\|, w) &= \mathrm{out}\big(\delta(q_0, w)\big) = \mathrm{out}\big((\delta_1(q_{01}, w), \ldots, \delta_n(q_{0n}, w))\big) \\ &= \bigvee_{1\le i\le n} \mathrm{out}_i(\delta_i(q_{0i}, w)) = \bigvee_{1\le i\le n} (k_i \wedge 1_{R_i}, w)\end{aligned}$$

i.e., $\|\mathcal{A}'\| = \|\mathcal{A}\|$ as required. □

Let $\mathcal{A} = (Q, q_0, \delta, \mathrm{out})$ and $\mathcal{A}' = (Q', q_0', \delta', \mathrm{out}')$ be two DMVA over Σ and L, and let $\varphi : Q \to Q'$ be a mapping such that:

(i) $\varphi(q_0) = q_0'$.
(ii) If $\delta(q, a)$ exists, then $\delta'(\varphi(q), a)$ exists and $\varphi(\delta(q, a)) = \delta'(\varphi(q), a)$ for every $q \in Q, a \in \Sigma$.

Then φ is called a *homomorphism from* $\mathcal{A}$ *to* $\mathcal{A}'$ and is denoted by $\varphi : \mathcal{A} \to \mathcal{A}'$. If $\mathrm{out}'(\varphi(q)) = \mathrm{out}(q)$ for every $q \in Q$, then φ is termed a *strong homomorphism.* A bijective strong homomorphism φ is an *isomorphism.*

Lemma 3.9. *Let* $\mathcal{A} = (Q, q_0, \delta, \mathrm{out})$ *and* $\mathcal{A}' = (Q', q_0', \delta', \mathrm{out}')$ *be two equivalent trim DMVA. Then there is at most one homomorphism* $\varphi : \mathcal{A} \to \mathcal{A}'$*. Every such homomorphism is surjective and strong.*

Proof. Assume that there are two homomorphisms $\varphi : \mathcal{A} \to \mathcal{A}'$ and $\psi : \mathcal{A} \to \mathcal{A}'$. For every $q \in Q$, there exists a word $w \in \Sigma^*$ such that $\delta(q_0, w) = q$. Then $\varphi(q) = \varphi(\delta(q_0, w)) = \delta'(\varphi(q_0), w) = \delta'(q_0', w)$. Similarly, we show that $\psi(q) = \delta'(q_0', w)$, and thus $\varphi(q) = \psi(q)$, i.e., $\varphi = \psi$. Next, we show that φ, whenever it exists, is surjective and strong. Consider $q' \in Q'$. Since $\mathcal{A}'$ is accessible, there is $w \in \Sigma^*$ with $q' = \delta'(q_0', w)$. Moreover, there exists $w' \in \Sigma^*$ such that $0 < \text{out}'(\delta'(q', w')) = (\|\mathcal{A}'\|, ww') = (\|\mathcal{A}\|, ww')$. So, there exists $q \in Q$ with $q = \delta(q_0, w)$. Therefore, $\varphi(q) = \varphi(\delta(q_0, w)) = \delta'(\varphi(q_0), w) = \delta'(q_0', w) = q'$, showing that φ is surjective. Keeping the same notations, we have $\text{out}'(q') = (\|\mathcal{A}'\|, w) = (\|\mathcal{A}\|, w) = \text{out}(q)$ yielding that φ is a strong homomorphism. □

For every $r \in L^{\text{rec}}\langle\langle \Sigma^* \rangle\rangle$, let $\text{TR}(r)$ be the collection of all trim DMVA accepting r. We define a pre-order $\leq$ in $\text{TR}(r)$; for every $\mathcal{A}, \mathcal{A}' \in \text{TR}(r)$, we set $\mathcal{A}' \leq \mathcal{A}$ iff there exists an homomorphism $\varphi : \mathcal{A} \to \mathcal{A}'$. We show that if $\mathcal{A} \leq \mathcal{A}'$ and $\mathcal{A}' \leq \mathcal{A}$, then $\mathcal{A}$ and $\mathcal{A}'$ are isomorphic. Indeed, $\mathcal{A} \leq \mathcal{A}'$ and $\mathcal{A}' \leq \mathcal{A}$ imply that there exist homomorphisms $\varphi' : \mathcal{A}' \to \mathcal{A}$, $\varphi : \mathcal{A} \to \mathcal{A}'$. Then $\varphi' \circ \varphi : \mathcal{A} \to \mathcal{A}$, $\varphi \circ \varphi' : \mathcal{A}' \to \mathcal{A}'$ are also homomorphisms and by Lemma 3.9, $\varphi' \circ \varphi = 1_{\mathcal{A}}$ and $\varphi \circ \varphi' = 1_{\mathcal{A}'}$ where $1_{\mathcal{A}}$ and $1_{\mathcal{A}'}$ are the identity isomorphisms of $\mathcal{A}$ and $\mathcal{A}'$, respectively, and φ is strong. So, φ is an isomorphism. We conclude that the collection of the isomorphism classes of all trim DMVA accepting r forms a partial order. Clearly, the question of the existence (up to an isomorphism) of a minimum trim DMVA accepting r arises. Here, minimum refers to a trim DMVA in $\text{TR}(r)$ which has as few states as any other automaton in $\text{TR}(r)$. In the following, we show that such a minimum trim DMVA accepting r, always can be constructed and is unique up to isomorphism.

Given a fuzzy language $r \in L\langle\langle \Sigma^* \rangle\rangle$, we define an equivalence relation $\equiv_r$ on Σ^* as follows. For every $w_1, w_2 \in \Sigma^*$, $w_1 \equiv_r w_2$ iff $(r, w_1 w) = (r, w_2 w)$ for every $w \in \Sigma^*$. It is clear that $\equiv_r$ is a right congruence.

Proposition 3.10. *The fuzzy language $r \in L\langle\langle \Sigma^* \rangle\rangle$ is recognizable iff the right congruence $\equiv_r$ has finite index.*

Proof. Assume first that r is accepted by a trim DMVA $\mathcal{A} = (Q, q_0, \delta, \text{out})$. We define an equivalence relation $\equiv_{\mathcal{A}}$ on Σ^* as follows. For every $w_1, w_2 \in \Sigma^*$, $w_1 \equiv_{\mathcal{A}} w_2$ iff $\delta(q_0, w_1) = \delta(q_0, w_2)$. Obviously, $\equiv_{\mathcal{A}}$ is a right congruence, i.e., $w_1 \equiv_{\mathcal{A}} w_2$ implies $w_1 w \equiv_{\mathcal{A}} w_2 w$ for every $w \in \Sigma^*$, and thus $(r, w_1 w) = (r, w_2 w)$; therefore, $\equiv_{\mathcal{A}} \subseteq \equiv_r$. Since Q is finite, $\equiv_{\mathcal{A}}$ has finite index, hence $\equiv_r$ has also finite index.

Conversely, assume that $\equiv_r$ has finite index and let $[w]$ denote the equivalence class of $w \in \Sigma^*$. We construct the accessible DMVA $\mathcal{A}' = (Q', [\varepsilon], \delta_r, \text{out}_r)$ with $Q' = \{[w] \mid w \in \Sigma^*\}$. The function δ_r is determined by $\delta_r([w], a) = [wa]$ for every $[w] \in Q', a \in \Sigma$, and $\text{out}_r([w]) = (r, w)$ for every $[w] \in Q'$. Then $\|\mathcal{A}'\| = r$ and thus r is fuzzy recognizable. By letting $Q_r = \{[w] \in Q' \mid \exists u \in \Sigma^* : (r, wu) > 0\}$, we get an equivalent trim DMVA $\mathcal{A}_r = (Q_r, [\varepsilon], \delta_r, \text{out}_r)$. □

Keeping the notations of the previous proof, assume now that $r \in L^{\mathrm{rec}}\langle\langle \Sigma^* \rangle\rangle$ and let $\mathcal{A} = (Q, q_0, \delta, \mathrm{out})$ be a trim DMVA accepting r. Then for every $q \in Q$ there exists $w_q \in \Sigma^*$ such that $q = \delta(q_0, w_q)$. If w'_q is another word such that $q = \delta(q_0, w'_q)$, then $[w_q] = [w'_q]$. We define a mapping $\varphi_r : Q \to Q_r$ by $\varphi(q) = [w_q]$. Then $\varphi_r : \mathcal{A} \to \mathcal{A}_r$ is a homomorphism. Indeed, $\delta(q_0, \varepsilon) = q_0$, and thus $\varphi_r(q_0) = [\varepsilon]$. Furthermore, let $q \in Q$ and $w \in \Sigma^*$ such that $\delta(q, w)$ exists. Then $\varphi_r(\delta(q, w)) = \varphi_r(\delta(\delta(q_0, w_q), w)) = \varphi_r(\delta(q_0, w_q w)) = [w_q w] = \delta'([\varepsilon], w_q w) = \delta'([w_q], w) = \delta'(\varphi_r(q), w)$ proving our claim. Hence, we have obtained the following result.

Theorem 3.11. *Let Σ be an alphabet and L be a bounded distributive lattice. For every fuzzy recognizable language $r \in L^{\mathrm{rec}}\langle\langle \Sigma^* \rangle\rangle$, there exists a minimum trim DMVA $\mathcal{A}_r$ with $\|\mathcal{A}_r\| = r$.*

Any trim DMVA $\mathcal{A}$ which is isomorphic to $\mathcal{A}_r$ will be also called a *minimum* automaton for r. Next, we show that for every fuzzy recognizable language $r \in L^{\mathrm{rec}}\langle\langle \Sigma^* \rangle\rangle$, we can effectively construct a minimum automaton accepting r. Let us assume that $\mathcal{A} = (Q, q_0, \delta, \mathrm{out})$ is a trim DMVA with behavior $\|\mathcal{A}\| = r$. We define an equivalence relation $\equiv$ on Q as follows: for every $q, q' \in Q$, $q \equiv q'$ iff $\mathrm{out}(\delta(q, w)) = \mathrm{out}(\delta(q', w))$ for every $w \in \Sigma^*$. Then $\mathcal{A}$ is called *reduced* if $q \equiv q'$ implies $q = q'$ for every $q, q' \in Q$. It is easy to see that $\mathcal{A}$ is reduced iff the strong homomorphism $\varphi_r : \mathcal{A} \to \mathcal{A}_r$ is injective. Since φ_r is also surjective, we conclude the following proposition.

Proposition 3.12. *A DMVA $\mathcal{A}$ accepting $r \in L^{\mathrm{rec}}\langle\langle \Sigma^* \rangle\rangle$ is minimum iff it is trim and reduced.*

The previous proposition actually points out a way to construct a minimum DMVA accepting r: we start from a trim DMVA $\mathcal{A} = (Q, q_0, \delta, \mathrm{out})$ with $\|\mathcal{A}\| = r$ and we merge its equivalent states. Therefore, we prove that the equivalence $q \equiv q'$ is decidable for every pair of states $q, q' \in Q$, and we give an algorithm which uses at most $\mathrm{card}(Q)$ iterations. To this end, we introduce the equivalence relations $\equiv_n$ $(n \geq 0)$ on Q, given by $q \equiv_n q'$ iff $\mathrm{out}(\delta(q, w)) = \mathrm{out}(\delta(q', w))$ for every $w \in \bigcup_{0 \leq k \leq n} \Sigma^k$. Obviously, $\equiv_0 \supseteq \equiv_1 \supseteq \cdots \supseteq \equiv_n \supseteq \cdots$ hence $\equiv = \bigcap_{n \geq 0} \equiv_n$. We show that if there exists an $n \geq 0$ such that $\equiv_n = \equiv_{n+1}$, then $\equiv_{n+1} = \equiv_{n+l}$ for every $l \geq 2$. Indeed, assume that $\equiv_n = \equiv_{n+1}$. Then for every $q, q' \in Q$,

$$
\begin{aligned}
& q \equiv_{n+1} q' \\
\iff & \mathrm{out}\big(\delta(q, w)\big) = \mathrm{out}\big(\delta(q', w)\big) \quad \text{for every } w \in \bigcup_{0 \leq k \leq n+1} \Sigma^k \\
\iff & \mathrm{out}\big(\delta(q, au)\big) = \mathrm{out}\big(\delta(q', au)\big) \quad \text{for every } a \in \Sigma,\ u \in \bigcup_{0 \leq k \leq n} \Sigma^k
\end{aligned}
$$

$$\begin{aligned}
&\Longleftrightarrow \quad \mathrm{out}\big(\delta(\delta(q,a),u)\big) = \mathrm{out}\big(\delta\big(\delta(q',a),u\big)\big)\\
&\qquad \text{for every } a \in \Sigma,\ u \in \bigcup_{0\leq k\leq n} \Sigma^k\\
&\Longleftrightarrow \quad \delta(q,a) \equiv_n \delta(q',a) \quad \text{for every } a \in \Sigma\\
&\Longleftrightarrow \quad \delta(q,a) \equiv_{n+1} \delta(q',a) \quad \text{for every } a \in \Sigma \quad \text{(by hypothesis)}\\
&\Longleftrightarrow \quad \mathrm{out}\big(\delta\big(\delta(q,a),u\big)\big) = \mathrm{out}\big(\delta\big(\delta(q',a),u\big)\big)\\
&\qquad \text{for every } a \in \Sigma, u \in \bigcup_{0\leq k\leq n+1} \Sigma^k\\
&\Longleftrightarrow \quad q \equiv_{n+2} q'.
\end{aligned}$$

Therefore, by induction, we have $\equiv_{n+1} \; = \; \equiv_{n+l}$ for every $l \geq 2$. Now we let $e_0, e_1, \ldots$ denote the numbers of the equivalence classes of $\equiv_0, \equiv_1, \ldots$, respectively. Then $e_0 \leq e_1 \leq \cdots \leq \mathrm{card}(Q)$. Thus, there exists an $n \leq \mathrm{card}(Q)$ such that $e_n = e_{n+1}$, hence $\equiv_n \; = \; \equiv_{n+1}$ and so $\equiv \; = \; \equiv_n$. We conclude that the equivalence $q \equiv q'$ is decidable in at most $\mathrm{card}(Q)$ iterations.

We complete this subsection with two further important consequences of Theorem 3.5. First, a *pumping lemma* is valid within the class $L^{\mathrm{rec}}\langle\langle \Sigma^* \rangle\rangle$.

Proposition 3.13. *Let $r \in L^{\mathrm{rec}}\langle\langle \Sigma^* \rangle\rangle$. There exists an integer $m > 0$ such that for every $w \in \Sigma^*$ with $|w| > m$, the word w can be written as $w = w_1 u w_2$ with $|u| > 0$ and $|w_1 w_2| < m$, and $(r, w_1 u^k w_2) = (r, w)$ for every $k \geq 0$.*

Proof. Let $r = \bigvee_{1\leq i\leq n} k_i \wedge 1_{R_i}$. Then the pumping lemma holds for every recognizable language R_i $(1 \leq i \leq n)$, and let m_i be the corresponding integer for R_i. We conclude our proof by letting $m = \max\{m_1, \ldots, m_n\}$. □

A pumping lemma for fuzzy recognizable languages over the interval $[0, 1]$, has been proved in [8] by means of fuzzy monoid recognizability.

Now we show that the *equivalence problem* is decidable for multi-valued automata over Σ and L. In fact, we prove the following stronger result.

Theorem 3.14. *Let Σ be an alphabet and L be a bounded distributive lattice. For every two fuzzy recognizable languages $r, r' \in L^{\mathrm{rec}}\langle\langle \Sigma^* \rangle\rangle$, the relations $r \leq r'$ and $r = r'$ are decidable.*

Proof. Let $r = \bigvee_{1\leq i\leq n} k_i \wedge 1_{R_i}$ with pairwise disjoint recognizable languages R_i and $r' = \bigvee_{1\leq j\leq m} k'_j \wedge 1_{R'_j}$ with pairwise disjoint recognizable languages R'_j. Clearly, our decidability problems reduce to well-known decidability problems for recognizable languages. For instance in case of equality, we check that whenever $R_i \cap R'_j \neq \emptyset$ then $k_i = k'_j$. □

3.2 Fuzzy Recognizability over Infinite Words

In this subsection, we introduce Büchi and Muller multi-valued automata consuming infinite words. We show that both models accept the same class of infinitary fuzzy languages, and a Kleene-type theorem holds for this class. The material of the present and the next subsection is based on [22].

Definition 3.15.

(a) A multi-valued Muller automaton *(MVMA for short)* over Σ and L *is a quadruple* $\mathcal{A} = (Q, \mathrm{in}, \mathrm{wt}, \mathcal{F})$, *where* Q *is the* finite state set, $\mathrm{in} : Q \to L$ *is the* initial distribution, $\mathrm{wt} : Q \times \Sigma \times Q \to L$ *is the mapping assigning* weights *to the transitions of the automaton, and* $\mathcal{F} \subseteq \mathcal{P}(Q)$ *is the* family of final state sets.

(b) An MVMA $\mathcal{A}$ *is a* multi-valued Büchi automaton *(MVBA for short) if there is a set* $F \subseteq Q$ *such that* $\mathcal{F} = \{S \subseteq Q \mid S \cap F \neq \emptyset\}$.

Let $w = a_0 a_1 \ldots \in \Sigma^\omega$. A *path of* $\mathcal{A}$ *over* w is an infinite sequence of transitions $P_w = (t_i)_{i \geq 0}$, so that $t_i = (q_i, a_i, q_{i+1}) \in Q \times \Sigma \times Q$ for every $i \geq 0$. The *weight of* P_w is defined by

$$\mathrm{weight}(P_w) = \mathrm{in}(q_0) \wedge \bigwedge_{i \geq 0} \mathrm{wt}(t_i).$$

Observe that $\mathrm{weight}(P_w)$ is well-defined since $\mathrm{weight}(P_w) \in L_{\mathcal{A}}$, where $L_{\mathcal{A}}$ is the finite sub-lattice of L generated by $\{0,1\} \cup \{\mathrm{in}(q) \mid q \in Q\} \cup \{\mathrm{wt}(t) \mid t \in Q \times \Sigma \times Q\}$. The path P_w is called *successful* if the set of states that appear infinitely often along P_w constitutes a final state set. The *behavior* of $\mathcal{A}$ is the infinitary fuzzy language

$$\|\mathcal{A}\| : \Sigma^\omega \to L$$

which is defined by

$$(\|\mathcal{A}\|, w) = \bigvee_{P_w} \mathrm{weight}(P_w)$$

for $w \in \Sigma^\omega$, where the supremum is taken over all successful paths P_w of $\mathcal{A}$ over w. Since $L_{\mathcal{A}}$ is finite, $(\|\mathcal{A}\|, w)$ exists and $(\|\mathcal{A}\|, w) \in L_{\mathcal{A}}$ for every $w \in \Sigma^\omega$.

An infinitary fuzzy language $r \in L\langle\langle \Sigma^\omega \rangle\rangle$ is said to be *fuzzy Muller recognizable* (resp. *fuzzy Büchi recognizable* or *fuzzy* ω*-recognizable*) if there is an MVMA (resp. an MVBA) $\mathcal{A}$ so that $r = \|\mathcal{A}\|$. We denote the family of all fuzzy Muller recognizable (resp. fuzzy ω-recognizable) languages over Σ and L by $L^{\text{M-rec}}\langle\langle \Sigma^\omega \rangle\rangle$ (resp. $L^{\omega\text{-rec}}\langle\langle \Sigma^\omega \rangle\rangle$). It should be clear that the class $L^{\text{M-rec}}\langle\langle \Sigma^\omega \rangle\rangle$ (resp. $L^{\omega\text{-rec}}\langle\langle \Sigma^\omega \rangle\rangle$) coincides with the class of Muller recognizable (resp. ω-recognizable) series over Σ and the semiring $\langle L, \vee, \wedge, 0, 1\rangle$ (see [22, 23]). Clearly $L^{\omega\text{-rec}}\langle\langle \Sigma^\omega \rangle\rangle \subseteq L^{\text{M-rec}}\langle\langle \Sigma^\omega \rangle\rangle$. Later on, we shall prove that in fact the two classes coincide.

Two multi-valued Muller (resp. Büchi) automata $\mathcal{A}$ and $\mathcal{A}'$ over Σ and L are called *equivalent* if $\|\mathcal{A}\| = \|\mathcal{A}'\|$.

Given a language $R \subseteq \Sigma^\omega$, its characteristic infinitary fuzzy language $1_R \in L\langle\langle \Sigma^\omega \rangle\rangle$ is defined in a similar way as for finitary languages. Obviously, every unweighted Büchi automaton with input alphabet Σ can be considered as an

MVBA over Σ and L with weights only 0 and 1. Therefore, we immediately obtain the next proposition.

Proposition 3.16 ([23]). *Let $R \subseteq \Sigma^\omega$ be an ω-recognizable language. Then the characteristic infinitary fuzzy language $1_R \in L\langle\langle \Sigma^\omega \rangle\rangle$ is ω-recognizable.*

Assume now that $R_1, \ldots, R_n \subseteq \Sigma^\omega$ are ω-recognizable languages, $k_1, \ldots, k_n \in L$, and let

$$r = \bigvee_{1 \leq i \leq n} k_i \wedge 1_{R_i}.$$

Such a language r is called *fuzzy ω-recognizable step language* [23]. Actually a fuzzy ω-recognizable step language is fuzzy ω-recognizable. Indeed, let us assume that for every $1 \leq i \leq n$ we are given a Büchi automaton $\mathcal{A}_i = (Q_i, I_i, \Delta_i, \mathcal{F}_i)$ accepting R_i (see [56]). We fix an $1 \leq i \leq n$. Then as already noted above, $\mathcal{A}_i$ can be considered as an MVBA $(Q_i, \mathrm{in}_i, \mathrm{wt}_i, \mathcal{F}_i)$ over Σ and L with behavior 1_{R_i}. We consider the MVBA $\overline{\mathcal{A}_i} = (Q_i, k_i \wedge \mathrm{in}_i, \mathrm{wt}_i, \mathcal{F}_i)$. Obviously $\|\overline{\mathcal{A}_i}\| = k_i \wedge 1_{R_i}$. Now let $\mathcal{A}$ be the MVBA obtained as the disjoint union of all $\overline{\mathcal{A}_i}$ $(1 \leq i \leq n)$. Clearly, $\|\mathcal{A}\| = r$ proving our claim.

Theorem 3.17 ([22]). *Let Σ be an alphabet and L be a bounded distributive lattice. Then the following statements are equivalent for every infinitary fuzzy language $r \in L\langle\langle \Sigma^\omega \rangle\rangle$:*

(i) r is fuzzy Muller recognizable.
(ii) r is fuzzy ω-recognizable.
(iii) r is a fuzzy ω-recognizable step language.

Proof. We show that (i) implies (iii). Let $r \in L^{\text{M-rec}}\langle\langle \Sigma^\omega \rangle\rangle$ and $\mathcal{A}$ be an MVMA accepting r. Then $r = \bigvee_{1 \leq i \leq n} k_i \wedge 1_{R_i}$ where $L_\mathcal{A} = \{k_1, \ldots, k_n\}$ and $R_i = \{w \in \Sigma^\omega \mid (r, w) = k_i\}$ for every $1 \leq i \leq n$. Following the proof of Theorem 3.5, we can show that the languages R_i $(1 \leq i \leq n)$ are Muller recognizable, and thus ω-recognizable, which in turn implies that r is a fuzzy ω-recognizable step language.

The implications (iii) $\Rightarrow$ (ii) and (ii) $\Rightarrow$ (i) are also true as already shown. □

Observe that our proof above is effective (recall the discussion after Theorem 3.5). In the sequel without any further notation, we write every fuzzy ω-recognizable language r over Σ and L as $r = \bigvee_{1 \leq i \leq n} k_i \wedge 1_{R_i}$.

Theorem 3.17 has very interesting consequences. Firstly, we can easily obtain closure properties of fuzzy ω-recognizable languages.

Proposition 3.18. *The class $L^{\omega\text{-rec}}\langle\langle \Sigma^\omega \rangle\rangle$ of fuzzy ω-recognizable languages is closed under supremum, infimum, and scalar infimum.*

Proof. Closure under supremum is immediate and closure under scalar infimum is obtained by distributivity of L. Furthermore, for the closure under infimum one has to recall that the class of ω-recognizable languages is closed under intersection. □

Consider now two alphabets Σ, Δ and a non-deleting homomorphism $h : \Sigma^* \to \Delta^*$. Then h can be extended to a mapping $h : \Sigma^\omega \to \Delta^\omega$ by setting $h(a_0a_1\ldots) = h(a_0)h(a_1)\ldots$ for every infinite word $a_0a_1\ldots \in \Sigma^\omega$. Let $r \in L\langle\langle\Sigma^\omega\rangle\rangle$ be an infinitary fuzzy language having finite image, and $R \subseteq \Sigma^\omega$. We define the infinitary fuzzy language $h_R(r) \in L\langle\langle\Delta^\omega\rangle\rangle$ by

$$(h_R(r), u) = \bigvee_{w \in h^{-1}(u) \cap R} (r, w)$$

for every $u \in \Delta^\omega$. We denote the mapping h_{Σ^ω} simply by h. Furthermore, if $s \in L\langle\langle\Delta^\omega\rangle\rangle$, then the fuzzy language $h^{-1}(s) \in L\langle\langle\Sigma^\omega\rangle\rangle$ is specified by

$$(h^{-1}(s), w) = (s, h(w))$$

for every $w \in \Sigma^\omega$.

Proposition 3.19.

(i) Let $(L, \leq)$ and $(L', \leq)$ be two bounded distributive lattices and $f : L \to L'$ be any mapping. Then for every fuzzy ω-recognizable language r in $L\langle\langle\Sigma^\omega\rangle\rangle$ the fuzzy language $f(r) \in L'\langle\langle\Sigma^\omega\rangle\rangle$ is again ω-recognizable.

(ii) Let $h : \Sigma^\omega \to \Delta^\omega$ be a non-deleting homomorphism and $R \subseteq \Sigma^\omega$ be an ω-recognizable language. Then $h_R : L\langle\langle\Sigma^\omega\rangle\rangle \to L\langle\langle\Delta^\omega\rangle\rangle$ and $h^{-1} : L\langle\langle\Delta^\omega\rangle\rangle \to L\langle\langle\Sigma^\omega\rangle\rangle$ preserve the ω-recognizability property of fuzzy languages.

Proof. Statement (i) can be shown as Proposition 3.6, using Theorem 3.17. Now let $r \in L^{\omega\text{-rec}}\langle\langle\Sigma^\omega\rangle\rangle$ with $r = \bigvee_{1\leq i\leq n} k_i \wedge 1_{R_i}$. For every $u \in \Delta^\omega$, we have

$$(h_R(r), u) = \bigvee_{w \in h^{-1}(u) \cap R} (r, w) = \bigvee_{1\leq i\leq n} \Big(k_i \wedge \bigvee_{w \in h^{-1}(u) \cap R} (1_{R_i}, w) \Big)$$

which is equal to $\bigvee_{1\leq i\leq n} k_i \wedge (1_{h(R_i \cap R)}, u)$. Hence,

$$h_R(r) = \bigvee_{1\leq i\leq n} k_i \wedge 1_{h(R_i \cap R)}.$$

Since the class of ω-recognizable languages is closed under non-deleting homomorphisms [56], we obtain that the fuzzy language $h_R(r)$ is ω-recognizable.

Finally, assume that $s = \bigvee_{1\leq j\leq m} k'_j \wedge 1_{R'_j}$. Then

$$h^{-1}(s) = \bigvee_{1\leq j\leq m} k'_j \wedge 1_{h^{-1}(R'_j)}.$$

The class of ω-recognizable languages is closed under inverse non-deleting homomorphisms [56], therefore, $h^{-1}(s)$ is fuzzy ω-recognizable and our proof is completed. □

As an immediate consequence of Proposition 3.19(i), we obtain the closure of fuzzy ω-recognizable languages under negation functions.

Corollary 3.20. *Let $(L, \leq, ^{-})$ be a bounded distributive lattice with negation function, and $r \in L^{\omega\text{-rec}}\langle\langle \Sigma^\omega \rangle\rangle$. Then also $\overline{r} \in L^{\omega\text{-rec}}\langle\langle \Sigma^\omega \rangle\rangle$.*

By Theorem 3.17, we get the statements of Proposition 3.7 in the setting of infinitary fuzzy languages. More precisely, for every $r \in L\langle\langle \Sigma^\omega \rangle\rangle$ and $l \in L$, we consider the infinitary languages $r_{\geq l} = \{w \in \Sigma^\omega \mid (r, w) \geq l\}$ and $r_{=l} = r^{-1}(l) = \{w \in \Sigma^\omega \mid (r, w) = l\}$.

Proposition 3.21. *For every fuzzy language $r \in L\langle\langle \Sigma^\omega \rangle\rangle$, the following statements are equivalent:*

(i) r is fuzzy ω-recognizable.
(ii) r has finite image, and for every $l \in L$, $r_{=l}$ is an ω-recognizable language.
(iii) r has finite image, and for every $l \in L$, $r_{\geq l}$ is an ω-recognizable language.

As a further consequence of Theorem 3.17, we prove that the *equivalence problem* is decidable for multi-valued Muller (resp. Büchi) automata over Σ and L. In fact, we get the subsequent stronger decidability result.

Theorem 3.22. *Let Σ be an alphabet and L be a bounded distributive lattice. For every two fuzzy ω-recognizable languages $r, r' \in L^{\omega\text{-rec}}\langle\langle \Sigma^\omega \rangle\rangle$, the relations $r \leq r'$ and $r = r'$ are decidable.*

Proof. See the proof of Theorem 3.14. □

Finally, we show that a Kleene theorem holds for fuzzy ω-recognizable languages. We firstly recall the ω-rational operations of fuzzy languages (see [37, 63, 27]). Let $r \in L\langle\langle \Sigma^* \rangle\rangle$ and $r' \in L\langle\langle \Sigma^\omega \rangle\rangle$. Then the *Cauchy product $rr' \in L\langle\langle \Sigma^\omega \rangle\rangle$ of r and r'* is defined by $(rr', w) = \vee\{(r, u) \wedge (r', u') \mid w = uu', u \in \Sigma^*, u' \in \Sigma^\omega\}$ for every $w \in \Sigma^\omega$. Furthermore, whenever r is proper, i.e., $(r, \varepsilon) = 0$, we define the *ω-star $r^\omega \in L\langle\langle \Sigma^\omega \rangle\rangle$ of r* as follows: $(r^\omega, w) = \vee\{\wedge\{(r, w_1), (r, w_2), \ldots\} \mid w = w_1 w_2 \ldots$ with $w_1, w_2, \ldots \in \Sigma^*\}$ for every $w \in \Sigma^\omega$. Now the class of *fuzzy ω-rational languages over Σ and L*, denoted by $L^{\omega\text{-rat}}\langle\langle \Sigma^\omega \rangle\rangle$, is the least class of infinitary fuzzy languages generated by the finitary fuzzy languages (over Σ and L) with finite support, applying finitely many times the operations of supremum, Cauchy product, star, and ω-star. Every fuzzy ω-recognizable language $r = \bigvee_{1 \leq i \leq n} k_i \wedge 1_{R_i}$ over Σ and L is ω-rational. Indeed, for every $1 \leq i \leq n$ the language R_i is ω-rational and thus 1_{R_i} is a fuzzy ω-rational language with values 0 and 1. Then $k_i \wedge 1_{R_i}$ is just the Cauchy product of the series $k_i\varepsilon$ and 1_{R_i}, where the series $k_i\varepsilon$ is defined by $(k_i\varepsilon, w) = 1$ if $w = \varepsilon$ and $(k_i\varepsilon, w) = 0$ otherwise, for every $w \in \Sigma^*$. Conversely, we claim that $L^{\omega\text{-rat}}\langle\langle \Sigma^\omega \rangle\rangle \subseteq L^{\omega\text{-rec}}\langle\langle \Sigma^\omega \rangle\rangle$. For this, it suffices to show that for every $r \in L^{\text{rec}}\langle\langle \Sigma^* \rangle\rangle, r' \in L^{\omega\text{-rec}}\langle\langle \Sigma^\omega \rangle\rangle$ the fuzzy language $rr' \in L^{\omega\text{-rec}}\langle\langle \Sigma^\omega \rangle\rangle$, and for every proper fuzzy language $r \in L^{\text{rec}}\langle\langle \Sigma^* \rangle\rangle$, the ω-star $r^\omega \in L^{\omega\text{-rec}}\langle\langle\Sigma^\omega\rangle\rangle$. Once again by using Theorems 3.5 and 3.17, this is reduced to the well-known closure properties of recognizable languages under the ω-rational operations (see, for instance, [56]). Therefore, we get the subsequent Kleene theorem for infinitary fuzzy languages.

Theorem 3.23. *Let Σ be an alphabet and L be a bounded distributive lattice. Then $L^{\omega\text{-rec}}\langle\langle\Sigma^\omega\rangle\rangle = L^{\omega\text{-rat}}\langle\langle\Sigma^\omega\rangle\rangle$.*

For a Kleene theorem for infinitary formal power series over a larger class of semirings than bounded distributive lattices, we refer the reader to [25–27].

3.3 Multi-valued MSO Logic

Following [22], we introduce a multi-valued monadic second-order logic (multi-valued MSO logic, for short) over infinite words, and we state a multi-valued version of Büchi's theorem [11] for fuzzy languages over bounded distributive lattices with negation function. A corresponding theory for finite words has been obtained as an application of weighted logics over locally finite semirings (see [18, 20]). Throughout this subsection, we assume that $(L, \leq, ^-)$ is a bounded distributive lattice with negation function.

Every word $w = a_0 a_1 \ldots \in \Sigma^\omega$, with $a_0, a_1, \ldots \in \Sigma$, is also written as $w = w(0)w(1)\ldots$ with $w(i) = a_i$ for $i \geq 0$. Then every $w \in \Sigma^\omega$ is represented by the structure $(\omega, \leq, (R_a)_{a\in\Sigma})$ where $R_a = \{i \mid w(i) = a\}$ for $a \in \Sigma$. Given a finite set $\mathcal{V}$ of first- and second-order variables, a $(w, \mathcal{V})$-*assignment* σ is a mapping assigning elements of ω to first-order variables from $\mathcal{V}$, and subsets of ω to second-order variables from $\mathcal{V}$. If x is a first-order variable and $i \in \omega$, then $\sigma[x \to i]$ denotes the $(w, \mathcal{V} \cup \{x\})$-assignment which assigns i to x and acts as σ on $\mathcal{V} \setminus \{x\}$. For a second-order variable X and $I \subseteq \omega$, the notation $\sigma[X \to I]$ has a similar meaning.

By using the extended alphabet $\Sigma_\mathcal{V} = \Sigma \times \{0,1\}^\mathcal{V}$, we encode pairs (w, σ) for every $w \in \Sigma^\omega$ and every $(w, \mathcal{V})$-assignment σ. Every word in $\Sigma_\mathcal{V}^\omega$ is considered as a pair (w, σ) where w is the projection over Σ, and σ is the projection over $\{0,1\}^\mathcal{V}$. Then σ is a *valid* $(w, \mathcal{V})$-assignment if for every first-order variable $x \in \mathcal{V}$ the x-row contains exactly one 1. In this case, we identify σ with the $(w, \mathcal{V})$-assignment so that for every first-order variable $x \in \mathcal{V}$, $\sigma(x)$ is the position of the 1 on the x-row, and for every second-order variable $X \in \mathcal{V}$, $\sigma(X)$ is the set of positions labeled with 1 along the X-row. By standard automata constructions, it can be shown that the language

$$N_\mathcal{V} = \{(w, \sigma) \in \Sigma_\mathcal{V}^\omega \mid \sigma \text{ is a valid } (w, \mathcal{V})\text{-assignment}\}$$

is ω-recognizable.

Definition 3.24. *The set of all* MSO(L, Σ)*-formulas of the multi-valued MSO logic over Σ and L is defined to be the smallest set F such that:*

- *F contains all* atomic formulas $k, P_a(x), x \leq y, x \in X$.
- *If* $\varphi, \psi \in F$, *then also* $\neg\varphi, \varphi \vee \psi, \varphi \wedge \psi, \exists x \centerdot \varphi, \exists X \centerdot \varphi, \forall x \centerdot \varphi, \forall X \centerdot \varphi \in F$

where $k \in L$, $a \in \Sigma$, x, y *are first-order variables and* X *is a second-order variable.*

We represent the semantics of the formulas in $\mathrm{MSO}(L, \Sigma)$ as infinitary fuzzy languages over the extended alphabet $\Sigma_{\mathcal{V}}$ and the lattice L. Here, our definition of semantics is more general that the one used in [18, 23, 20]. There, the authors assigned to every atomic formula $P_a(x), x \leq y$, or $x \in X$, respectively, the characteristic series of its associated MSO-language. These series take on only $0, 1$. Here, we assume that there is a function f assigning to every atomic formula φ of the form $P_a(x), x \leq y$, or $x \in X$, respectively, an infinitary fuzzy language $f(\varphi)$ in $L\langle\langle \Sigma_\varphi^\omega \rangle\rangle$ (where Σ_φ stands for $\Sigma_{\mathrm{Free}(\varphi)}$). This generalization has been already used in other logics. For instance, in many-valued predicate logic, every object variable is being assigned a value from an L-structure M, where L is a BL-algebra (see Sect. 5 in [32]). In [39], the atomic propositions of the multi-valued LTL take values from a subset of the underlying De Morgan algebra. Our assignment f here is called *ω-recognizable* if the fuzzy language $f(\varphi)$ is ω-recognizable for every atomic formula φ. Later on, we always require that f is an ω-recognizable assignment. Thus, the language $f(\varphi)$ will be taking on only finitely many values, for every atomic formula φ. Therefore, we will call f a *multi-valued atomic assignment over* Σ, if $f(\varphi)$ takes on only finitely many values, for every atomic formula φ.

Definition 3.25. *Let* $\varphi \in \mathrm{MSO}(L, \Sigma)$, $\mathcal{V}$ *be a finite set of variables containing* $\mathrm{Free}(\varphi)$, *and* f *be a multi-valued atomic assignment over* Σ. *We define the* f-semantics *of* φ *to be an infinitary fuzzy language* $\|\varphi\|_{\mathcal{V}}^f \in L\langle\langle \Sigma_{\mathcal{V}}^\omega \rangle\rangle$ *in the following way. Let* $(w, \sigma) \in \Sigma_{\mathcal{V}}^\omega$. *If* σ *is not a valid* $(w, \mathcal{V})$*-assignment, then we put* $(\|\varphi\|_{\mathcal{V}}^f, (w, \sigma)) = 0$. *Otherwise, we inductively define* $(\|\varphi\|_{\mathcal{V}}^f, (w, \sigma)) \in L$ *as follows:*

- $(\|k\|_{\mathcal{V}}^f, (w, \sigma)) = k$
- $(\|\varphi\|_{\mathcal{V}}^f, (w, \sigma)) = (f(\varphi), (w, \sigma|_{\mathrm{Free}(\varphi)}))$ *if* φ *is an atomic formula of the form* $P_a(x), x \leq y$, *or* $x \in X$
- $(\|\neg\varphi\|_{\mathcal{V}}^f, (w, \sigma)) = \overline{(\|\varphi\|_{\mathcal{V}}^f, (w, \sigma))}$
- $(\|\varphi \vee \psi\|_{\mathcal{V}}^f, (w, \sigma)) = (\|\varphi\|_{\mathcal{V}}^f, (w, \sigma)) \vee (\|\psi\|_{\mathcal{V}}^f, (w, \sigma))$
- $(\|\varphi \wedge \psi\|_{\mathcal{V}}^f, (w, \sigma)) = (\|\varphi\|_{\mathcal{V}}^f, (w, \sigma)) \wedge (\|\psi\|_{\mathcal{V}}^f, (w, \sigma))$
- $(\|\exists x \centerdot \varphi\|_{\mathcal{V}}^f, (w, \sigma)) = \bigvee_{i \in \omega} (\|\varphi\|_{\mathcal{V} \cup \{x\}}^f, (w, \sigma[x \to i]))$
- $(\|\exists X \centerdot \varphi\|_{\mathcal{V}}^f, (w, \sigma)) = \bigvee_{I \subseteq \omega} (\|\varphi\|_{\mathcal{V} \cup \{X\}}^f, (w, \sigma[X \to I]))$
- $(\|\forall x \centerdot \varphi\|_{\mathcal{V}}^f, (w, \sigma)) = \bigwedge_{i \in \omega} (\|\varphi\|_{\mathcal{V} \cup \{x\}}^f, (w, \sigma[x \to i]))$
- $(\|\forall X \centerdot \varphi\|_{\mathcal{V}}^f, (w, \sigma)) = \bigwedge_{I \subseteq \omega} (\|\varphi\|_{\mathcal{V} \cup \{X\}}^f, (w, \sigma[X \to I]))$.

It should be clear that in Definition 3.25 all the occurring infinite suprema and infima exist in L (without any further completeness assumption). More precisely, one can show by induction on the structure of formulas φ that $\|\varphi\|^f_{\mathcal{V}}$ takes on only finitely many values. Indeed, for atomic formulas, this is clear by assumption, and the property is preserved by negation, disjunction, and conjunction. Since L is a lattice, the property is also preserved by infinite suprema and infima, proving our claim.

If the multi-valued atomic assignment is well-known, then we omit the superscript f from $\|\varphi\|^f_{\mathcal{V}}$. Furthermore, we simply write $\|\varphi\|$ for $\|\varphi\|_{\mathrm{Free}(\varphi)}$. If φ has no free variables, i.e., if it is a sentence, then $\|\varphi\| \in L\langle\langle\Sigma^\omega\rangle\rangle$.

An infinitary fuzzy language $r \in L\langle\langle\Sigma^\omega\rangle\rangle$ is called *MSO-f-definable* if there is a sentence $\varphi \in \mathrm{MSO}(L, \Sigma)$ such that $r = \|\varphi\|^f$. We let $L^{f\text{-mso}}\langle\langle\Sigma^\omega\rangle\rangle$ comprise all fuzzy languages from $L\langle\langle\Sigma^\omega\rangle\rangle$ which are f-definable by some sentence in $\mathrm{MSO}(L, \Sigma)$. In the sequel, we show that the classes $L^{f\text{-mso}}\langle\langle\Sigma^\omega\rangle\rangle$ and $L^{\omega\text{-rec}}\langle\langle\Sigma^\omega\rangle\rangle$ coincide.

Let us first give an example of possible interpretations of multi-valued MSO-formulas. The reader can find more examples in [18, 22, 23, 20].

Example 3.26 ([22]). We consider the bounded distributive lattice $(\mathbb{N} \cup \{\infty\}, \leq, ^-)$ (where $\mathbb{N}$ is the set of natural numbers and $^-$ is an arbitrary negation function). Let $\Sigma = \{a, b, c\}$ and f be the multi-valued atomic assignment over Σ, determined in the following way. For every $w \in \Sigma^\omega$ and every valid $(w, \{x\})$-assignment σ, we set:

- $(f(P_a(x)), (w, \sigma)) = 0$
- $(f(P_b(x)), (w, \sigma)) = \begin{cases} 1 & \text{if } w(\sigma(x)) = b, \\ 0 & \text{otherwise} \end{cases}$
- $(f(P_c(x)), (w, \sigma)) = \begin{cases} 2 & \text{if } w(\sigma(x)) = c, \\ 0 & \text{otherwise.} \end{cases}$

For every other atomic formula φ, $f(\varphi)$ is the fuzzy language with image $\{0\}$. Let $\varphi = \forall x.(P_a(x) \vee P_b(x) \vee P_c(x))$. In fact, φ is a sentence, and for every word $w \in \Sigma^\omega$ the semantics $\|\varphi\|^f$ returns the value 0 if the letter a occurs at least once in w, the value 1 if no a appears in w but b occurs at least once, and it returns the value 2 if $w = c^\omega$.

The reader should observe that the above definition of semantics is valid for every formula $\varphi \in \mathrm{MSO}(L, \Sigma)$ and every finite set $\mathcal{V}$ of variables containing $\mathrm{Free}(\varphi)$. The following proposition states that the f-semantics $\|\varphi\|^f_{\mathcal{V}}$ is in fact independent of the set $\mathcal{V}$; it depends only on $\mathrm{Free}(\varphi)$. For a proof, we refer the reader to [18, 20].

Proposition 3.27. *For every $\varphi \in \mathrm{MSO}(L, \Sigma)$, every finite set $\mathcal{V}$ of variables with $\mathrm{Free}(\varphi) \subseteq \mathcal{V}$, and every multi-valued atomic assignment f over Σ, it holds that*

$$\left(\|\varphi\|_{\mathcal{V}}^{f},(w,\sigma)\right)=\left(\|\varphi\|^{f},(w,\sigma|_{\mathrm{Free}(\varphi)})\right)$$

for every $(w,\sigma)\in\Sigma_{\mathcal{V}}^{\omega}$, *where* σ *is a valid* $(w,\mathcal{V})$*-assignment. Furthermore, the fuzzy language* $\|\varphi\|^{f}$ *is* ω*-recognizable iff the fuzzy language* $\|\varphi\|_{\mathcal{V}}^{f}$ *is* ω*-recognizable.*

The next lemma states a further closure property of the class of fuzzy ω-recognizable languages.

Lemma 3.28. *Let* $h:\Sigma^{\omega}\to\Delta^{\omega}$ *be a non-deleting homomorphism,* $R\subseteq\Sigma^{\omega}$ *be an* ω*-recognizable language, and* $r\in L^{\omega\text{-rec}}\langle\langle\Sigma^{\omega}\rangle\rangle$ *be a fuzzy* ω*-recognizable language. Then the language* $\bigwedge_{h,R}(r)\in L\langle\langle\Delta^{\omega}\rangle\rangle$ *defined by* $(\bigwedge_{h,R}(r),u)=\bigwedge_{w\in h^{-1}(u)\cap R}(r,w)$ *is fuzzy* ω*-recognizable.*

Proof. Let $(L^{d},\leq^{d})=(L,\geq)$ be the dual lattice of L, which is obtained by interchanging suprema and infima. Since r takes on only finitely many values and each value on an ω-recognizable language, r is also fuzzy ω-recognizable over L^{d}. Consider the transformation $h_{R}^{d}:L^{d}\langle\langle\Sigma^{\omega}\rangle\rangle\to L^{d}\langle\langle\Delta^{\omega}\rangle\rangle$. By Proposition 3.19(ii), we obtain $h_{R}^{d}(r)\in(L^{d})^{\omega\text{-rec}}\langle\langle\Delta^{\omega}\rangle\rangle$ which in turn means that $h_{R}^{d}(r)\in L^{\omega\text{-rec}}\langle\langle\Delta^{\omega}\rangle\rangle$. Since suprema in L^{d} equal infima in L, we have $h_{R}^{d}(r)=\bigwedge_{h,R}(r)$ and our proof is completed. □

Proposition 3.29. *Let* $\varphi,\psi\in\mathrm{MSO}(L,\Sigma)$ *such that* $\|\varphi\|_{\mathcal{V}}^{f},\|\psi\|_{\mathcal{V}}^{f}$ *are fuzzy* ω*-recognizable languages where* f *is a multi-valued atomic assignment, and* $\mathcal{V}$ *is a finite set of variables with* $\mathrm{Free}(\varphi)\cup\mathrm{Free}(\psi)\subseteq\mathcal{V}$*. Then the languages* $\|\neg\varphi\|_{\mathcal{V}}^{f}$, $\|\varphi\vee\psi\|_{\mathcal{V}}^{f}$, $\|\varphi\wedge\psi\|_{\mathcal{V}}^{f}$, $\|\exists x\,.\,\varphi\|_{\mathcal{V}}^{f}$, $\|\exists X\,.\,\varphi\|_{\mathcal{V}}^{f}$, $\|\forall x\,.\,\varphi\|_{\mathcal{V}}^{f}$, *and* $\|\forall X\,.\,\varphi\|_{\mathcal{V}}^{f}$ *are fuzzy* ω*-recognizable.*

Proof. The f-semantics of the negation of φ is fuzzy ω-recognizable by Corollary 3.20. The f-semantics of disjunction and conjunction of φ and ψ are fuzzy ω-recognizable by Proposition 3.18. Next, we deal with existential and universal quantifiers. By assumption, $\|\varphi\|_{\mathcal{V}}^{f}$ is fuzzy ω-recognizable. Let $\|\varphi\|_{\mathcal{V}}^{f}=\bigvee_{1\leq i\leq n}k_{i}\wedge 1_{R_{i}}$, and

$$h:\Sigma_{\mathcal{V}\cup\{x\}}^{\omega}\to\Sigma_{\mathcal{V}}^{\omega}\quad\text{and}\quad h':\Sigma_{\mathcal{V}\cup\{X\}}^{\omega}\to\Sigma_{\mathcal{V}}^{\omega}$$

be the non-deleting homomorphisms erasing the x-row and the X-row, respectively. Clearly,

$$\begin{aligned}\|\exists x\,.\,\varphi\|_{\mathcal{V}}^{f}&=h\left(\|\varphi\|_{\mathcal{V}\cup\{x\}}^{f}\right) & \|\exists X\,.\,\varphi\|_{\mathcal{V}}^{f}&=h'\left(\|\varphi\|_{\mathcal{V}\cup\{X\}}^{f}\right)\\ \|\forall x\,.\,\varphi\|_{\mathcal{V}}^{f}&=\bigwedge\nolimits_{h,N_{\mathcal{V}\cup\{x\}}}\left(\|\varphi\|_{\mathcal{V}\cup\{x\}}^{f}\right) & \|\forall X\,.\,\varphi\|_{\mathcal{V}}^{f}&=\bigwedge\nolimits_{h',\Sigma_{\mathcal{V}\cup\{X\}}^{\omega}}\left(\|\varphi\|_{\mathcal{V}\cup\{X\}}^{f}\right).\end{aligned}$$

We conclude our proof by applying Proposition 3.19(ii) and Lemma 3.28. □

Proposition 3.30. *Let* f *be any* ω*-recognizable multi-valued atomic assignment. Then* $L^{f\text{-mso}}\langle\langle\Sigma^{\omega}\rangle\rangle\subseteq L^{\omega\text{-rec}}\langle\langle\Sigma^{\omega}\rangle\rangle$.

Proof. We apply induction on the structure of $\mathrm{MSO}(L, \Sigma)$-formulas using Proposition 3.29. □

Next, we define the *crisp atomic assignment* cf for atomic formulas. More precisely, let φ be an atomic formula of the form $P_a(x), x \leq y$, or $x \in X$. Then for every $(w, \sigma) \in \Sigma_\varphi^\omega$ with σ a valid assignment, we set:

- $(cf(P_a(x)), (w, \sigma)) = \begin{cases} 1 & \text{if } w(\sigma(x)) = a, \\ 0 & \text{otherwise} \end{cases}$
- $(cf(x \leq y), (w, \sigma)) = \begin{cases} 1 & \text{if } \sigma(x) \leq \sigma(y), \\ 0 & \text{otherwise} \end{cases}$
- $(cf(x \in X), (w, \sigma)) = \begin{cases} 1 & \text{if } \sigma(x) \in \sigma(X), \\ 0 & \text{otherwise.} \end{cases}$

Note that if φ is an atomic formula of this form, then $(\|\neg\varphi\|^{cf}, (w, \sigma)) = \overline{(cf(\varphi), (w, \sigma))}$ for every $(w, \sigma) \in N_\varphi$, and by the property of $^-$ that $\overline{1} = 0$ and $\overline{0} = 1$, the semantics of $\neg\varphi$ coincides with the one given in [23, 20]. Furthermore, the crisp atomic assignment is ω-recognizable [23]. We denote the class $L^{cf\text{-mso}}\langle\langle \Sigma^\omega \rangle\rangle$ simply by $L^{\mathrm{mso}}\langle\langle \Sigma^\omega \rangle\rangle$.

Now we can state our Büchi-type characterization of the class $L^{\omega\text{-rec}}\langle\langle \Sigma^\omega \rangle\rangle$.

Theorem 3.31. *Let Σ be an alphabet and L be a bounded distributive lattice with any negation function. Then*

$$L^{\omega\text{-rec}}\langle\langle \Sigma^\omega \rangle\rangle = \bigcup_f L^{f\text{-mso}}\langle\langle \Sigma^\omega \rangle\rangle = L^{\mathrm{mso}}\langle\langle \Sigma^\omega \rangle\rangle$$

where the union is taken over all ω-recognizable multi-valued atomic assignments.

Proof. Let $r \in L^{\omega\text{-rec}}\langle\langle \Sigma^\omega \rangle\rangle$ with $r = \bigvee_{1 \leq i \leq n} k_i \wedge 1_{R_i}$. We fix an $1 \leq i \leq n$. By Büchi's theorem [11], R_i is definable by a classical MSO-sentence φ_i. Clearly, φ_i can be considered as a multi-valued sentence over Σ and L. Then $\|\varphi_i\| = 1_{R_i}$ which in turn implies that $\| \bigvee_{1 \leq i \leq n} k_i \wedge \varphi_i \| = r$. Thus, $L^{\omega\text{-rec}}\langle\langle \Sigma^\omega \rangle\rangle \subseteq L^{\mathrm{mso}}\langle\langle \Sigma^\omega \rangle\rangle$. Now Proposition 3.30 completes our proof. □

This result shows that for every formula $\varphi \in \mathrm{MSO}(L, \Sigma)$, whose semantics is defined with any ω-recognizable multi-valued atomic assignment, we can construct an equivalent $\mathrm{MSO}(L, \Sigma)$-formula with the crisp atomic assignment. In case of De Morgan algebras, an alternative simpler syntax of formulas of multi-valued MSO logic can be given by the grammar

$$\varphi ::= k \mid P_a(x) \mid x \leq y \mid x \in X \mid \neg\varphi \mid \varphi \vee \varphi \mid \exists x \centerdot \varphi \mid \exists X \centerdot \varphi.$$

We define the semantics $\|\varphi\|$ of formulas φ of this syntax exactly as in Definition 3.25. Given a multi-valued atomic assignment f, let $L^{\mathrm{dm}\text{-}f\text{-mso}}\langle\langle \Sigma^\omega \rangle\rangle$ be the collection of all infinitary fuzzy languages definable in this logic. Then conjunction and universal quantifiers are determined by:

- $\varphi \wedge \psi = \neg(\neg\varphi \vee \neg\psi)$
- $\forall x \,.\, \varphi = \neg(\exists x \,.\, \neg\varphi)$
- $\forall X \,.\, \varphi = \neg(\exists X \,.\, \neg\varphi)$

for every $\varphi, \psi \in \mathrm{MSO}(L, \Sigma)$. By using the De Morgan laws, we have the following equalities for every $(w, \sigma) \in \Sigma_{\mathcal{V}}^{\omega}$ where σ is a valid assignment:

- $(\|\varphi \wedge \psi\|_{\mathcal{V}}^{f}, (w, \sigma)) = (\|\varphi\|_{\mathcal{V}}^{f}, (w, \sigma)) \wedge (\|\psi\|_{\mathcal{V}}^{f}, (w, \sigma))$
- $(\|\forall x \,.\, \varphi\|_{\mathcal{V}}^{f}, (w, \sigma)) = \bigwedge_{i \in \omega}(\|\varphi\|_{\mathcal{V}\cup\{x\}}^{f}, (w, \sigma[x \to i]))$
- $(\|\forall X \,.\, \varphi\|_{\mathcal{V}}^{f}, (w, \sigma)) = \bigwedge_{I \subseteq \omega}(\|\varphi\|_{\mathcal{V}\cup\{X\}}^{f}, (w, \sigma[X \to I]))$.

The crisp atomic assignment cf is also defined as before, and we denote again the class $L^{\text{dm-}cf\text{-mso}}\langle\langle \Sigma^{\omega} \rangle\rangle$ simply by $L^{\text{dm-mso}}\langle\langle \Sigma^{\omega} \rangle\rangle$. Then the next result is an immediate consequence of Theorem 3.31 and the above equalities.

Corollary 3.32. *Let Σ be an alphabet and $(L, \leq, ^{-})$ be a De Morgan algebra. Then*

$$L^{\omega\text{-rec}}\langle\langle \Sigma^{\omega} \rangle\rangle = \bigcup_{f} L^{\text{dm-}f\text{-mso}}\langle\langle \Sigma^{\omega} \rangle\rangle = L^{\text{dm-mso}}\langle\langle \Sigma^{\omega} \rangle\rangle$$

where the union is taken over all ω-recognizable multi-valued atomic assignments.

4 Fuzzy Languages: An Overview

In the previous section, we have focused on fuzzy languages over bounded distributive lattices. Several other concepts of fuzzy languages occur in the literature, and they mainly differ in their underlying structure. The most general cases are covered by fuzzy languages over bounded ℓ-monoids and residuated lattices. Since every residuated lattice is a bounded ℓ-monoid, fuzzy languages over residuated lattices constitute a subclass of fuzzy languages over bounded ℓ-monoids. Fuzzy automata over these two concepts have been investigated recently. Actually, they are weighted automata over the corresponding induced semirings. Therefore, the properties of their behaviors mostly follow from the general theory of recognizable formal power series. Further properties of fuzzy recognizable languages over residuated lattices and bounded ℓ-monoids require specific restrictions for their underlying structures. In this section, we only highlight the most interesting results (without proofs) for fuzzy recognizable languages over bounded ℓ-monoids and residuated lattices. We also succinctly refer to fuzzy automata with outputs, to families of fuzzy languages, and to fuzzy tree languages. For the rest of this section, Σ will denote an arbitrary finite alphabet.

4.1 Fuzzy Languages over ℓ-Monoids

A *lattice-ordered monoid* (or *ℓ-monoid* for short) is a lattice $(L, \leq)$ equipped with an operation $\cdot$ and a distinguished element $e \in L$ such that the following conditions hold:

(i) $\langle L, \cdot, e\rangle$ is a monoid
(ii) $a \cdot (b \vee c) = a \cdot b \vee a \cdot c$ and $(a \vee b) \cdot c = a \cdot c \vee b \cdot c$

for every $a, b, c \in L$ (see [6]). Note that then $a \leq b$ implies $a \cdot c \leq b \cdot c$ and $c \cdot a \leq c \cdot b$, for every $a, b, c \in L$.

The ℓ-monoid defined above is denoted by $(L, \vee, \cdot)$, and is called *bounded* if the lattice $(L, \leq)$ is bounded and

(iii) $a \cdot 0 = 0 \cdot a = 0$

for every $a \in L$. Furthermore, if $(L, \leq)$ is a complete lattice satisfying

(iv) $a \cdot (\bigvee_{i\in I} b_i) = \bigvee_{i\in I}(a \cdot b_i)$ and $(\bigvee_{i\in I} b_i) \cdot a = \bigvee_{i\in I}(b_i \cdot a)$

for every $a \in L$ and every countable family $(b_i \mid i \in I) \subseteq L$ of elements of L, then $(L, \vee, \cdot)$ is called *countably distributive*. Every bounded distributive lattice is a bounded ℓ-monoid with $\cdot = \wedge$ and $e = 1$. A further example of a bounded ℓ-monoid with $e = 1$ is given by any residuated lattice $\mathcal{L}$ (see Sect. 4.2 below). Given a bounded ℓ-monoid $(L, \vee, \cdot)$, the structure $\langle L, \vee, \cdot, 0, e\rangle$ is a semiring. An *L-valued language r over Σ and* $(L, \vee, \cdot)$ is a formal power series $r \in L\langle\langle \Sigma^* \rangle\rangle$. Automata over bounded ℓ-monoids, called L-fuzzy automata, were introduced in [42]. More precisely, an *L-fuzzy automaton over Σ and* $(L, \vee, \cdot)$ is just a weighted automaton over Σ and the semiring $\langle L, \vee, \cdot, 0, e\rangle$. A Kleene theorem for L-valued recognizable languages is stated in [42] under the assumption that the ℓ-monoid $(L, \vee, \cdot)$ is countably distributive or $e = 1$. In fact, this is actually an application of the Kleene–Schützenberger theorem for recognizable series (see for instance [27, 65]). A *deterministic L-fuzzy automaton over Σ and* $(L, \vee, \cdot)$ is defined in the same way as the deterministic multi-valued automaton (see Sect. 3.1). The subsequent theorem indicates the requirements for the determinization of L-fuzzy automata over bounded l-monoids.

Theorem 4.1 ([42]). *Let $(L, \vee, \cdot)$ be a bounded ℓ-monoid. Then for every L-fuzzy automaton over $(L, \vee, \cdot)$, there exists an equivalent deterministic L-fuzzy automaton over $(L, \vee, \cdot)$ iff the semiring $\langle L, \vee, \cdot, 0, e\rangle$ is locally finite.*

For the "if" part, we refer the reader also to [7, 19] where corresponding statements for arbitrary locally finite semirings are shown. For the "only if" part, we consider an arbitrary finite subset A of L, and we construct an L-fuzzy automaton $\mathcal{A}$ having A as its set of weights and the submonoid L_A of $\langle L, \cdot, e\rangle$ generated by A as the image of its behavior (see [42]). Since every deterministic L-fuzzy automaton takes on only finitely many values and $\mathcal{A}$

can be determinized, L_A is finite. Since the monoid $\langle L, \vee, 0\rangle$ is clearly locally finite, it follows that the semiring $\langle L, \vee, \cdot, 0, e\rangle$ is locally finite.

If the semiring $\langle L, \vee, \cdot, 0, e\rangle$ is locally finite, then by [19] we get that an L-valued language $r \in L\langle\langle \Sigma^* \rangle\rangle$ is recognizable iff it is an L-valued recognizable step language (in the sense of Sect. 3 and [19]). Then following our constructions in Sect. 3.1, we get similar results as for fuzzy recognizable languages over bounded distributive lattices. We collect these results in the subsequent theorem.

Theorem 4.2. *Let Σ be an alphabet and $(L, \vee, \cdot)$ be a bounded ℓ-monoid such that the semiring $\langle L, \vee, \cdot, 0, e\rangle$ is locally finite. Then:*

- *For every L-valued recognizable language $r \in L\langle\langle \Sigma^* \rangle\rangle$ we can effectively construct a minimum trim deterministic L-fuzzy automaton with behavior r.*
- *An L-valued language $r \in L\langle\langle \Sigma^* \rangle\rangle$ is recognizable iff r has finite image and each of its cut languages is recognizable.*[1]
- *A pumping lemma holds for L-valued recognizable languages in $L\langle\langle \Sigma^* \rangle\rangle$.*
- *The equivalence problem is decidable for L-fuzzy automata over Σ and $(L, \vee, \cdot)$.*

4.2 Fuzzy Languages over Residuated Lattices

Now we turn to residuated lattices. A *residuated lattice* is an algebra $\mathcal{L} = \langle L, \vee, \wedge, \otimes, \rightarrow, 0, 1\rangle$ where $(L, \leq)$ is a bounded lattice equipped with two operations $\otimes, \rightarrow$ such that the following conditions hold:

(i) $\langle L, \otimes, 1\rangle$ is a commutative monoid
(ii) $\otimes$ and $\rightarrow$ form an *adjoint pair*, i.e., $a \otimes b \leq c \Leftrightarrow a \leq b \rightarrow c$

for every $a, b, c \in L$ (see [73]).

If the lattice L is complete, then $\mathcal{L}$ is called a *complete residuated lattice*. Examples of (complete) residuated lattices are provided by the fuzzy semiring $\langle [0,1], \mathsf{max}, \mathsf{min}, 0, 1\rangle$ equipped with operations $\otimes$ and $\rightarrow$ defined, respectively, as follows. For every $a, b \in [0,1]$, let:

- $a \otimes b = \mathsf{max}(a + b - 1, 0)$, and $a \rightarrow b = \mathsf{min}(1 - a + b, 1)$ (the *Łukasiewicz structure*)
- $a \otimes b = a \cdot b$, and $a \rightarrow b = 1$ if $a \leq b$ and $a \rightarrow b = b/a$ (the usual quotient of real numbers) otherwise (the *product structure*)
- $a \otimes b = \mathsf{min}(a, b)$, and $a \rightarrow b = 1$ if $a \leq b$ and $a \rightarrow b = b$ otherwise (the *Gödel structure*).

Next, we claim that if $(a_i \mid i \in I) \subseteq L$ is a family of elements of L such that $\bigvee_{i \in I} a_i$ exists, then for every $a \in L$ we have

[1] This statement has been also derived in [42].

$$\left(\bigvee_{i\in I} a_i\right) \otimes a = \bigvee_{i\in I} (a_i \otimes a).$$

Indeed, for every $j \in I$, we have

$$a_j \leq \bigvee_{i\in I} a_i \leq \left(a \to \left(\left(\bigvee_{i\in I} a_i\right) \otimes a\right)\right),$$

so $a_j \otimes a \leq (\bigvee_{i\in I} a_i) \otimes a$. Now let $c \in L$ such that $a_i \otimes a \leq c$ for every $i \in I$, hence $a_i \leq a \to c$. Then we get $\bigvee_{i\in I} a_i \leq a \to c$, and thus $(\bigvee_{i\in I} a_i) \otimes a \leq c$ proving our claim.

By (ii), we obtain also $0 \otimes a = 0$ for every $a \in L$.

Clearly, for every residuated lattice $\mathcal{L} = \langle L, \vee, \wedge, \otimes, \to, 0, 1\rangle$, the triple $(L, \vee, \otimes)$ is a bounded ℓ-monoid (with $e = 1$). Moreover, $\langle L, \vee, \otimes, 0, 1\rangle$ is a commutative semiring which is called the *semiring reduct of* $\mathcal{L}$ and is denoted by $\mathcal{L}^*$. Obviously, the semiring reducts induced by Łukasiewicz and Gödel structures are locally finite, whereas this is not the case for the semiring reduct induced by the product structure. A *fuzzy language* r *over* Σ *and* $\mathcal{L}$ is a formal power series $r \in L\langle\langle \Sigma^* \rangle\rangle$. Then a *fuzzy automaton over* Σ *and* $\mathcal{L}$ is a weighted automaton over Σ and the semiring reduct $\mathcal{L}^*$ (see [60, 61]);[2] it is also an L-fuzzy automaton over Σ and $(L, \vee, \otimes)$. Fuzzy automata over the *product structure* occur in practical applications (see Sect. 5). As immediate consequences of Theorems 4.1 and 4.2, we obtain the following corollaries.

Corollary 4.3 ([33]). *Let* $\mathcal{L} = \langle L, \vee, \wedge, \otimes, \to, 0, 1\rangle$ *be a residuated lattice. Then for every fuzzy automaton over* $\mathcal{L}$ *there exists an equivalent deterministic fuzzy automaton over* $\mathcal{L}$ *iff the semiring reduct* $\mathcal{L}^*$ *is locally finite.*

Corollary 4.4. *Let* Σ *be an alphabet and* $\mathcal{L} = \langle L, \vee, \wedge, \otimes, \to, 0, 1\rangle$ *be a residuated lattice with locally finite semiring reduct* $\mathcal{L}^*$. *Then:*

- *For every fuzzy recognizable language* $r \in L\langle\langle \Sigma^* \rangle\rangle$, *we can effectively construct a minimum trim deterministic fuzzy automaton with behavior* r.
- *A fuzzy language* $r \in L\langle\langle \Sigma^* \rangle\rangle$ *is recognizable iff* r *has finite image and each of its cut languages is recognizable.*
- *A pumping lemma holds for fuzzy recognizable languages in* $L\langle\langle \Sigma^* \rangle\rangle$.
- *The equivalence problem is decidable for fuzzy automata over* Σ *and* $\mathcal{L}$.

Recently in [15], it has been proved that for every (non-deterministic) fuzzy automaton over a residuated lattice $\mathcal{L}$, a size (number of states) reduction algorithm exists, provided that $\mathcal{L}$ is a locally finite residuated lattice. The authors constructed a fuzzy automaton over the product structure (which is

[2] In [60, 61], the author considers also fuzzy automata over the semiring $\langle L, \vee, \wedge, 0, 1\rangle$ and for this requires $\mathcal{L}$ to be complete. For such automata, a type of pumping lemma is shown in [62]. The completeness axiom of $\mathcal{L}$ required in [15, 33, 60–62] is actually superfluous for fuzzy automata over the semiring reduct $\mathcal{L}^*$.

not locally finite) for which their reduction algorithm cannot be applied. The minimization problem for (either deterministic or non-deterministic) fuzzy automata over arbitrary residuated lattices remains open.

4.3 Fuzzy Automata with Outputs

Fuzzy automata with outputs have been mainly defined over the bounded distributive lattice $([0,1], \leq)$ (see [54, 17]). They are special cases of weighted transducers over the fuzzy semiring (for definitions on weighted transducers see [53]).

In [14, 49, 58, 69], size reduction algorithms have been developed for non-deterministic versions of fuzzy automata with outputs.

In [41], it is shown that a size reduction algorithm exists for complete L-fuzzy automata with outputs over a finite ℓ-monoid $(L, \vee, \cdot)$.

4.4 Fuzzy Abstract Families of Languages

In [2], a theory for full abstract families of fuzzy languages (full AFFLs) is presented. The underlying structure is a bounded ℓ-monoid $(L, \vee, \cdot)$ with $e = 1$ and its operation $\cdot$ being commutative. Furthermore, the lattice $(L, \leq)$ is complete satisfying

$$a \wedge \left(\bigvee_{i \in I} b_i \right) = \bigvee_{i \in I} (a \wedge b_i)$$

for every $a \in L$ and every family $(b_i \mid i \in I) \subseteq L$ of elements of L.

Rational operations between fuzzy languages over $(L, \vee, \cdot)$ are defined as the rational operations of formal power series over the semiring $(L, \vee, \cdot, 0, 1)$ (see [27]). For every two alphabets Σ and Δ, every homomorphism $h : \Sigma^* \to \Delta^*$ induces a *fuzzy homomorphism* $\overline{h} : \Sigma^* \to L\langle\langle \Delta^* \rangle\rangle$ mapping every word $w \in \Sigma^*$ to a fuzzy language with support $\{h(w)\}$. Then a family of fuzzy languages $\mathcal{R}$ is called a *full abstract family of fuzzy languages (full AFFL)* if it is closed under the rational operations, fuzzy and inverse fuzzy homomorphisms, and infimum with fuzzy recognizable languages. It is proved that the class of fuzzy recognizable languages over the ℓ-monoid $(L, \vee, \cdot)$ is a full AFFL. Furthermore, the concept of a fuzzy substitution is introduced, and the closure property of fuzzy recognizable languages under fuzzy substitutions is investigated.

4.5 Fuzzy Tree Languages

Recently, several authors have dealt with fuzzy tree languages. These are tree series over (complete or bounded) distributive lattices (for definitions on tree series, see [29]).

In [28], the authors study fuzzy tree languages over *completely distributive lattices*, i.e., complete lattices in which arbitrary suprema distribute over arbitrary infima and vice versa. They show that fuzzy recognizable and fuzzy rational tree languages coincide, i.e., a Kleene theorem, and moreover fuzzy recognizable tree languages have an equational characterization.

In [22], a multi-valued MSO logic over infinite trees is introduced and a Rabin-type theorem is proved for infinitary fuzzy tree languages over bounded distributive lattices.

5 Applications

In this section, we present two applications of fuzzy languages, with an effect to real world problems. First, we refer to an alternative method of syntactic pattern recognition using fuzzy languages. Then we define fuzzy discrete event systems which have successfully contributed to medicine.

A popular method for pattern recognition is the *syntactic pattern recognition*, where a pattern is classified by checking its syntax. Usually, patterns are represented by finite words over a finite alphabet. The letters of the alphabet, which are called *primitives*, are aimed to describe the features of the patterns. A pattern class is a language of patterns. The method of syntactic pattern recognition is the following. First, we construct finitely many regular grammars (with their terminal alphabet to be the set of primitives) taking into account the syntactic features according to which we wish to classify any pattern. The languages generated by these grammars (pattern classes) should be pairwise disjoint. Then for every constructed grammar, we consider the corresponding equivalent finite automaton. Now given a pattern to be classified, we check from which automaton it is accepted, and we classify the pattern in the pattern class represented by this automaton.

However, in many practical applications, the structural information of the patterns is inherently vague, i.e., the patterns are distorted or imperfectly formed. For instance, consider the case of recognizing handwritten characters, or determining the type of a geometrical pattern which is not perfectly sketched. In such situations, we consider the pattern classes to be fuzzy languages. Therefore, we define the pattern classes by using *fuzzy grammars*. These are weighted right-linear grammars over the fuzzy semiring (see [37, 27]). The corresponding equivalent weighted automata are actually multi-valued automata over the fuzzy semiring. It should be noted that now the supports of the fuzzy languages (pattern classes) are not required to be pairwise disjoint. Given a pattern to be classified, we compute its membership value to every pattern class, by using the constructed multi-valued automata. Then we look for the greatest value, and we classify the pattern into the corresponding class. Fuzzy syntactic pattern recognition has been applied in the identification of the skeletal maturity of children by using X-ray images of radius [55] (see also [35]). More precisely, the shapes of the radius of children were considered as patterns. Nine pattern classes were generated featuring the maturity of the radius. Then the skeletal maturity of a child was being identified by classifying an X-ray image of its radius.

Now we turn to fuzzy discrete event systems [44]. A *fuzzy discrete event system* (*FDES* for short) is a system $(\mathcal{A}_1, \ldots, \mathcal{A}_n)$ $(n > 0)$ of weighted automata over the semiring $\langle [0,1], \mathsf{max}, \cdot, 0, 1\rangle$ where $\cdot$ is the multiplication of the real numbers. Such automata are usually called fuzzy automata and they are actually $[0,1]$-fuzzy automata over the ℓ-monoid $([0,1], \mathsf{max}, \cdot)$. They can also be considered as fuzzy automata over the product structure (see Sect. 4). The input alphabet is not required to be the same for all the automata, thus we consider every automaton $\mathcal{A}_i$ $(1 \leq i \leq n)$ over an individual alphabet Σ_i. For the purposes of FDES theory, fuzzy automata lack their final distribution, i.e., they are of the form $\mathcal{A}_i = (Q_i, \mathrm{in}_i, \mathrm{wt}_i)$ for every $1 \leq i \leq n$. The elements of $\Sigma_1 \cup \cdots \cup \Sigma_n$ are called *events*. The FDES can be considered as a composite fuzzy automaton $\mathcal{A} = (Q, \mathrm{in}, \mathrm{wt})$ over $\Sigma_1 \cup \cdots \cup \Sigma_n$ and $[0,1]^n$, where:

- $Q = Q_1 \times \cdots \times Q_n$
- $\mathrm{in}((q_1, \ldots, q_n)) = (\mathrm{in}_1(q_1), \ldots, \mathrm{in}_n(q_n))$
- $\mathrm{wt}((q_1, \ldots, q_n), a, (q_1', \ldots, q_n')) = (r_1, \ldots, r_n)$ where $r_i = \mathrm{wt}_i(q_i, a, q_i')$ for every $1 \leq i \leq n$ with $a \in \Sigma_i$, and $r_i = 1$ for every $1 \leq i \leq n$ with $a \notin \Sigma_i$ provided that $q_i' = q_i$; in any other case, we let $\mathrm{wt}((q_1, \ldots, q_n), a, (q_1', \ldots, q_n')) = 0$

for every $(q_1, \ldots, q_n), (q_1', \ldots, q_n') \in Q, a \in \Sigma$.

In the following, we describe an important application of FDES to the implementation of a self-learning system for the selection of the suitable regimen for the HIV/AIDS (see [45, 76, 77]). The HIV/AIDS is among the most complex diseases to treat. One of the reasons for this complexity is that there is no cure for it. A treatment can only suppress the HIV virus and boost the immune system. Currently, there are only four classes of available anti-retroviral drugs and a regimen consists of a combination of two or more classes. Unfortunately, the HIV virus can easily develop resistance to the drugs. Thus, a decision for the suitable drug regimen for every particular patient turns to be a difficult task and can be successfully done only by experts. A wrong decision should be devastating since the patient may run out of options on available drugs. According to the experts, the following parameters must be considered for the choice of a suitable regimen:

- *Potency of the regimen*: Unlike other diseases, it is not reasonable to use the most potent regimen in the first stage of HIV/AIDS. In fact, initiating anti-retroviral therapy when the immune system is still intact does not prolong survival. The term "intact immune system" is already vague and this makes the HIV/AIDS treatment more complex than other diseases.
- *Adherence of the patient to the regimen*: This factor is very crucial unlike other diseases. The probability that a patient will benefit from the anti-retroviral therapy reduces dramatically if the patient skips even 5%

of doses. Moreover, this increases the risk that the virus will easier develop resistance to concrete drugs or even to a whole class of drugs. Unfortunately, statistics for HIV/AIDS and other chronic diseases show that the patients take only 50–70% of the required doses of long-term medications.

- *Adverse events*: These consist of side effects which may be mild to severe, and toxicity. Side effects like abdominal discomfort, loss of appetite, etc. are common especially in the first stages of HIV/AIDS treatment. On the other hand, toxicity usually causes liver problems, pancreatitis, etc. Unfortunately, in some cases, these problems turn out to be fatal.
- *Future drug options*: The HIV frequently develops resistance to the drugs. Thus, it is critical for a doctor, before concluding to any regimen, to consider the future drugs options after a potential occurrence of the resistance.

Clearly, the HIV/AIDS disease can be treated only by expert doctors. Moreover, the number of infected people increases all over the world. Actually, the number of experts is too small, especially in poor countries. Therefore, a computer program for the HIV/AIDS treatment regimen selection is desirable. Such a program has been built by using an FDES [45, 76, 77]. Here, we will briefly describe the contribution of the FDES to the program. The FDES is composed by four fuzzy automata $\mathcal{A}_1$, $\mathcal{A}_2$, $\mathcal{A}_3$, and $\mathcal{A}_4$ (over the semiring $\langle [0,1], \mathsf{max}, \cdot, 0, 1\rangle$), every one corresponding to one of the four aforementioned factors, respectively. The input alphabet is the same for all the automata. Every letter corresponds to a possible regimen which is a combination of two or more classes of drugs. The fuzzy automaton $\mathcal{A}_1$ has three states "initial", "medium", and "high" simulating the three instances of potency of a regimen. The states of the fuzzy automaton $\mathcal{A}_2$ are "initial", "challenging", "moderate", and "easy" modeling the possible values of the adherence. The states of $\mathcal{A}_3$ are "initial", "medium", "low", and "very low" simulating the level of the adverse events. Finally, the fuzzy automaton $\mathcal{A}_4$ has the states "initial", "medium", and "high" modeling the several options of future regimens. The initial distribution of every automaton assigns the value 1 to the "initial" state of every automaton, and the value 0 to every other state. The values of the weight assigning mappings for all the automata are determined by the expert doctors according to statistics and clinical experiments. Assume now that we have a particular patient and we ask the program to choose the optimal regimen. Initially, the system using a set of generic algorithms determines four vectors w_1, w_2, w_3, and w_4 with dimensions 3, 4, 4, and 3, respectively (these are the numbers of states of the automata $\mathcal{A}_1$, $\mathcal{A}_2$, $\mathcal{A}_3$, and $\mathcal{A}_4$, respectively). Then every one of the automata $\mathcal{A}_1$, $\mathcal{A}_2$, $\mathcal{A}_3$, and $\mathcal{A}_4$ takes as input a letter σ (i.e., a possible regimen) and produces a vector assigning a value to every one of its states. Let us denote these vectors by $q_{1\sigma}$, $q_{2\sigma}$, $q_{3\sigma}$, and $q_{4\sigma}$, respectively. Then the system computes the *performance index*

$$J(\sigma) = w_1^\mathsf{T} q_{1\sigma} + w_2^\mathsf{T} q_{2\sigma} + w_3^\mathsf{T} q_{3\sigma} + w_4^\mathsf{T} q_{4\sigma}$$

where w_i^{T} denotes the transpose of the vector w_i ($1 \leq i \leq 4$). The optimal treatment corresponds to the maximum $J(\sigma)$ for all regimens σ. For a second round treatment, the procedure is repeated. The vector states of the automata are now $q_{1\sigma}$, $q_{2\sigma}$, $q_{3\sigma}$, and $q_{4\sigma}$, but the system will compute new vectors w taking into account the current situation of the patient's health. In clinical experiments, this system matches the experts selection of regimen for 80% of the patients (see [45]).

Acknowledgments. I am deeply grateful to the editors for inviting me to write this chapter, and especially to Manfred Droste and Heiko Vogler for fruitful discussions and valuable comments on several versions of the chapter. I should also like to thank Symeon Bozapalidis for fruitful discussions. Peter Asveld, Andreas Maletti, Eleni Mandrali, Witold Pedrycz, and Arto Salomaa read previous versions of the chapter and made valuable remarks and suggestions. Andreas Maletti suggested the treatment of fuzzy recognizable languages with the "recognizable step" characterization. I am deeply grateful to all of them. I thank Arto Salomaa for providing me with reference [66] and indicating the seminal papers [48, 59]. I also thank Miroslav Ćirić for providing me with references [15, 33], Daowen Qiu for references [60, 61], and Feng Lin for clarifications on the papers [76, 77].

References

1. P.R.J. Asveld. A bibliography on fuzzy automata, grammars and languages. *Bulletin of the European Association for Theoretical Computer Science*, 58:187–196, 1996.
2. P.R.J. Asveld. Algebraic aspects of families of fuzzy languages. *Theoretical Computer Science*, 293:417–445, 2003.
3. R. Balbes and P. Dwinger. *Distributive Lattices*. University of Missouri Press, Columbia, 1974.
4. N.C. Basak and A. Gupta. On quotient machines of a fuzzy automaton and the minimal machine. *Fuzzy Sets and Systems*, 125:223–229, 2002.
5. R. Bělohlávek. Determinism and fuzzy automata. *Information Sciences*, 143:205–209, 2002.
6. G. Birkhoff. *Lattice Theory*, revised edition, volume XXV of *American Mathematical Society Colloquium Publications*. American Mathematical Society, Providence, 1961.
7. B. Borchardt. A pumping lemma and decidability problems for recognizable tree series. *Acta Cybernetica*, 16:509–544, 2004.
8. S. Bozapalidis and O. Louscou-Bozapalidou. On the recognizability of fuzzy languages I. *Fuzzy Sets and Systems*, 157:2394–2402, 2006.
9. S. Bozapalidis and O. Louscou-Bozapalidou. On the recognizability of fuzzy languages II. *Fuzzy Sets and Systems*, 159:107–113, 2008.
10. G. Bruns and P. Godefroid. Model checking with multi-valued logics. In *Proceedings of ICALP 2004*, volume 3142 of *Lecture Notes in Computer Science*, pages 281–293. Springer, Berlin, 2004.

11. J.R. Büchi. On a decision method in restricted second order arithmetic. In *Proceedings of International Congress for Logic, Methodology and Philosophy of Science*, pages 1–11, 1962
12. A.L. Buchsbaum, R. Giancarlo, and J.R. Westbrook. On the determinization of weighted finite automata. *SIAM Journal on Computing*, 5:1502–1531, 2002.
13. M. Chechik, B. Devereux, and A. Gurfinkel. Model-checking infinite state-space systems with fine-grained abstractions using SPIN. In *Proceedings of SPIN 2001*, volume 2057 of *Lecture Notes in Computer Science*, pages 16–36. Springer, Berlin, 2001.
14. W. Cheng and Z. Mo. Minimization algorithm of fuzzy finite automata. *Fuzzy Sets and Systems*, 141:439–448, 2004.
15. M. Ćirić, A. Stamenković, J. Ignjatović, and T. Petković. Factorization of fuzzy automata. In *Proceedings of FCT 2007*, volume 4639 of *Lecture Notes in Computer Science*, pages 213–225. Springer, Berlin, 2007.
16. A.B. Davey and H.A. Priestley. *Introduction to Lattices and Order*, 2nd edition. Cambridge University Press, Cambridge, 2002.
17. M. Doostfatemeh and S.C. Kremer. New directions in fuzzy automata. *International Journal of Approximate Reasoning*, 38:175–214, 2005.
18. M. Droste and P. Gastin. Weighted automata and weighted logics. *Theoretical Computer Science*, 380:69–86, 2007. Extended abstract in: *Proceedings of ICALP 2005*, volume 3580 of *Lecture Notes in Computer Science*, pages 513–525. Springer, Berlin, 2005.
19. M. Droste and P. Gastin. On aperiodic and star-free formal power series in partially commuting variables. *Theory of Computing Systems*, 42:608–631, 2008. Extended abstract in: D. Krob, A.A. Milchalev, and A.V. Michalev, editors, *Proceedings of the 12th International Conference on Formal Power Series and Algebraic Combinatorics, Moscow*, pages 158–169. Springer, Berlin, 2000.
20. M. Droste and P. Gastin. Weighted automata and weighted logics. In this *Handbook*. Chapter 5. Springer, Berlin, 2009.
21. M. Droste and W. Kuich. Semirings and formal power series. In this *Handbook*. Chapter 1. Springer, Berlin, 2009.
22. M. Droste, W. Kuich, and G. Rahonis. Multi-valued MSO logics over words and trees. *Fundamenta Informaticae*, 84:305–327, 2008.
23. M. Droste and G. Rahonis. Weighted automata and weighted logics on infinite words. *Russian Mathematics (Iz. VUZ)*, 2009, in press. Extended abstract in: *Proceedings of DLT'06*, volume 4036 of *Lecture Notes in Computer Science*, pages 49–58. Springer, Berlin, 2006.
24. S. Eilenberg. *Automata, Languages and Machines, volume A*. Academic Press, San Diego, 1974.
25. Z. Ésik and W. Kuich. A semiring-semimodule generalization of ω-regular languages I. *Journal of Automata, Languages and Combinatorics*, 10:203–242, 2005.

26. Z. Ésik and W. Kuich. A semiring-semimodule generalization of ω-regular languages II. *Journal of Automata, Languages and Combinatorics*, 10:243–264, 2005.
27. Z. Ésik and W. Kuich. Finite automata. In this *Handbook*. Chapter 3. Springer, Berlin, 2009.
28. Z. Ésik and G. Liu. Fuzzy tree automata. *Fuzzy Sets and Systems*, 158:1450–1460, 2007.
29. Z. Fülöp and H. Vogler. Weighted tree automata and tree transducers. In this *Handbook*. Chapter 9. Springer, Berlin, 2009.
30. S. Golan. *Semirings and Their Applications*. Kluwer Academic, Dordrecht, 1999.
31. A. Gurfinkel and M. Chechik. Multi-valued model checking via classical model checking. In *Proceedings of CONCUR 2003*, volume 2761 of *Lecture Notes in Computer Science*, pages 266–280. Springer, Berlin, 2003.
32. P. Hájek. *Metamathematics of Fuzzy Logic*. Kluwer Academic, Dordrecht, 1998.
33. J. Ignjatović, M. Ćirić, and S. Bogdanović. Determinization of fuzzy automata with membership values in complete residuated lattices. *Information Sciences*, 175:164–180, 2008.
34. D. Kirsten and I. Mäurer. On the determinization of weighted automata. *Journal of Automata, Languages and Combinatorics*, 10:287–312, 2005.
35. G. Klir and B. Yuan. *Fuzzy Sets and Fuzzy Logic*. Prentice Hall, Englewood Cliffs, 1995.
36. S. Konstandinidis, N. Sântean, and S. Yu. Fuzzification of rational and recognizable sets. *Fundamenta Informaticae*, 76:413–447, 2007.
37. W. Kuich and G. Rahonis. Fuzzy regular languages over finite and infinite words. *Fuzzy Sets and Systems*, 157:1532–1549, 2006.
38. W. Kuich and A. Salomaa. *Semirings, Automata, Languages*, volume 5 of *Monographs in Theoretical Computer Science. An EATCS Series*. Springer, Berlin, 1986.
39. O. Kupferman and Y. Lustig. Lattice automata. In *Proceedings of VMCAI 2007*, volume 4349 of *Lecture Notes in Computer Science*, pages 199–213. Springer, Berlin, 2007.
40. L. Lee and L.A. Zadeh. Note on fuzzy languages. *Information Sciences*, 1:421–434, 1969.
41. H. Lei and Y. Li. Minimization of states in automata theory based on finite lattice-ordered monoids. *Information Sciences*, 177:1413–1421, 2007.
42. Y. Li and W. Pedrycz. Fuzzy finite automata and fuzzy regular expressions with membership values in lattice-ordered monoids. *Fuzzy Sets and Systems*, 156:68–92, 2005.
43. Y. Li and W. Pedrycz. Minimization of lattice finite automata and its application to the decomposition of lattice languages. *Fuzzy Sets and Systems*, 158:1423–1436, 2007.

44. F. Lin and H. Ying. Modeling and control of fuzzy discrete event systems. *IEEE Transactions on Systems, Man and Cybernetics—Part B: Cybernetics*, 32:408–415, 2002.
45. F. Lin, H. Ying, R.D. MacArthur, J.A. Cohn, D.C. Barth-Jones, H. Ye, and L.R. Crane. Decision making in fuzzy discrete event systems. *Information Sciences*, 177:3749–3763, 2007.
46. O. Louskou-Bozapalidou. Fuzzy congruences on fuzzy algebras. *Applied Mathematical Sciences (Hikari)*, 1:815–819, 2007.
47. O. Louskou-Bozapalidou. Non-deterministic recognizability of fuzzy languages. *Applied Mathematical Sciences (Hikari)*, 1:821–826, 2007.
48. J. Łukasiewicz. O logice trojwartosciowej. *Ruch Filosoficzny*, 5:169–171, 1920.
49. D.S. Malik, J.N. Mordeson, and M.K. Sen. Minimization of fuzzy finite automata. *Information Sciences*, 113:323–330, 1999.
50. G. Malinowski. *Many-Valued Logics.* Oxford Science Publications, Oxford, 1993.
51. A. Mateescu, A. Salomaa, K. Salomaa, and S. Yu. Lexical analysis with a finite-fuzzy-automaton model. *Journal of Universal Computer Science*, 1:292–311, 1995.
52. M. Mohri. Finite-state transducers in language and speech processing. *Computational Linguistics*, 23:269–311, 1997.
53. M. Mohri. Weighted automata algorithms. In this *Handbook*. Chapter 6. Springer, Berlin, 2009.
54. J. Mordeson and D. Malik. *Fuzzy Automata and Languages: Theory and Applications.* Chapman & Hall/CRC, London, 2002.
55. A. Pathak and S.K. Pal. Fuzzy grammars in syntactic recognition of skeletal maturity from X-rays. *IEEE Transactions on Systems, Man and Cybernetics*, 16:657–667, 1986.
56. D. Perrin and E.J. Pin. *Infinite Words.* Elsevier, Amsterdam, 2004.
57. T. Petković. Varietes of fuzzy languages. In *Proceedings of the 1st International Conference on Algebraic Informatics*, pages 197–205. Aristotle University of Thessaloniki, Thessaloniki, 2005.
58. T. Petković. Congruences and homomorphisms of fuzzy automata. *Fuzzy Sets and Systems*, 157:444–458, 2006.
59. E.L. Post. Introduction to a general theory of elementary propositions. *American Journal of Mathematics*, 43:163–185, 1921.
60. D. Qiu. Automata theory based on complete residuated lattice-valued logic. *Science in China. Series F*, 44:419–429, 2001.
61. D. Qiu. Automata theory based on complete residuated lattice-valued logic (II). *Science in China. Series F*, 45:442–452, 2002.
62. D. Qiu. Pumping lemma in automata theory based on complete residuated lattice-valued logic: A note. *Fuzzy Sets and Systems*, 157:2128–2138, 2006.
63. G. Rahonis. Infinite fuzzy computations. *Fuzzy Sets and Systems*, 153:275–288, 2005.
64. J.B. Rosser and A.R. Turquette. *Many Valued Logics.* North-Holland, Amsterdam, 1952.

65. J. Sakarovitch. Rational and recognisable power series. In this *Handbook*. Chapter 4. Springer, Berlin, 2009.
66. A. Salomaa. Moisil and many-valued logic. In A. Iorgulescu, S. Marcus, S. Rudeanu, and D. Vaida, editors, *Grigore C. Moisil and his Followers in Theoretical Computer Science*, pages 29–39. Editura Academiei Romane, Bucharest, 2007.
67. A. Salomaa and M. Soittola. *Automata-Theoretic Aspects of Formal Power Series. Texts and Monographs in Computer Science.* Springer, Berlin, 1978.
68. E. Santos. Maximin automata. *Information and Control*, 13:363–377, 1968.
69. E. Santos. On reductions of maximin machines. *Journal of Mathematical Analysis and Applications*, 40:60–78, 1972.
70. B. Šešelja. Homomorphisms of poset-valued algebras. *Fuzzy Sets and Systems*, 121:333–340, 2001.
71. B. Šešelja and A. Tepavčević. On a generalization of fuzzy algebras and congruences. *Fuzzy Sets and Systems*, 65:85–94, 1994.
72. A. Tepavčević and G. Trajkovski. L-fuzzy lattices: An introduction. *Fuzzy Sets and Systems*, 223:209–216, 2001.
73. M. Ward and R.P. Dilworth. Residuated lattices. *Transactions of the American Mathematical Society*, 45:335–354, 1939.
74. W. Wechler. *The Concept of Fuzziness in Automata and Language Theory.* Akademie Verlag, Berlin, 1978.
75. W.G. Wee. On generalization of adaptive algorithm and applications of the fuzzy sets concept to pattern classification. PhD dissertation, Purdue University, Lafayette, 1967.
76. H. Ying, F. Lin, R.D. MacArthur, J.A. Cohn, D.C. Barth-Jones, H. Ye, and L.R. Crane. A fuzzy discrete event system approach determining optimal HIV/AIDS treatment regimens. *IEEE Transactions on Information Technology in Biomedicine*, 10:663–676, 2006.
77. H. Ying, F. Lin, R.D. MacArthur, J.A. Cohn, D.C. Barth-Jones, H. Ye, and L.R. Crane. A self-learning fuzzy discrete event system approach determining optimal HIV/AIDS treatment regimen selection. *IEEE Transactions on Systems, Man and Cybernetics—Part B: Cybernetics*, 37:966–979, 2007.
78. L.A. Zadeh. Fuzzy sets. *Information and Control*, 8:338–353, 1965.

Chapter 13: Model Checking Linear-Time Properties of Probabilistic Systems

Christel Baier, Marcus Größer, and Frank Ciesinski

Technische Universität Dresden, Fakultät Informatik,
Institut für Theoretische Informatik,
01062 Dresden, Germany
baier@tcs.inf.tu-dresden.de
groesser@tcs.inf.tu-dresden.de
ciesinsk@tcs.inf.tu-dresden.de

1 Introduction

This chapter is about the verification of Markov decision processes (MDPs) which are one of the fundamental models for reasoning about probabilistic phenomena of computer systems. MDPs have first been studied by Bellmann [14] and Howard [54] in the 1950s. While this early work on MDPs was motivated by optimization problems that appear in the context of operations research, nowadays MDPs are used in a variety of areas, including robotics, stochastic planning, control theory, reinforcement learning, economics, manufacturing, and semantics of randomized protocols. In the context of finite-state acceptors and transducers, MDPs served as basis for the introduction of probabilistic automata [81, 73], which again interact with the theory of weighted

M. Droste, W. Kuich, H. Vogler (eds.), *Handbook of Weighted Automata*,
Monographs in Theoretical Computer Science. An EATCS Series,
DOI 10.1007/978-3-642-01492-5_13, © Springer-Verlag Berlin Heidelberg 2009

automata by generalizing the concept of probabilities to weights in an arbitrary semiring.

In this chapter, Markov decision processes are used as an operational model for "probabilistic systems". Probabilism appears as a rather natural concept when providing the semantics of randomized algorithms or multi-agent systems with unreliable hardware components. In randomized algorithms, coin-tossing is used as an explicit probabilistic algorithmic concept. Examples for sequential randomized algorithms are the prominent primality tests by Miller and Rabin or by Solovay and Strassen or algorithms that operate with randomized data structures such as skip lists or universal hashing. For distributed systems, there is a wide range of coordination protocols to solve, e.g., mutual exclusion or leader election problems that utilize coin-tossing actions for symmetry breaking (see [44, 68]). For systems that operate with faulty components, such as communication channels that might corrupt or lose messages, or sensors that deliver wrong values in rare cases, probabilities can be used to specify the frequency of such exceptional behaviors. Probabilistic semantic models also play a central role in reasoning about systems that interact with an environment on which only partial information by means of stochastic assumptions about the I/O-operations of its interface is available.

The most popular operational models that support reasoning about probabilistic behaviors are Markovian models (Markov chains or Markov decision processes) where discrete probabilities are attached to the transitions. They enjoy the memoryless property stating that the future system evolution just depends on the current system state, but not on the specific steps that have been performed in the past. The memoryless property is inherent also in most non-stochastic automata models, such as labeled transition systems or weighted automata. In the stochastic setting, however, the memoryless property asserts that not only the enabled actions and the successor states are uniquely determined by the current state, but also the probabilities for the transitions. Markov chains are purely probabilistic, i.e., the possible behavior in each state is specified by a probabilistic distribution for the successor states. They can serve to formalize the stepwise behavior of sequential randomized algorithms. For modeling probabilistic parallel systems, Markov chains are not expressive enough to provide an *interleaving semantics*, which relies on the representation of concurrent (independent) actions α and β, executed by different processes, by a non-deterministic choice between the action sequences $\alpha\beta$ and $\beta\alpha$. Thus, models where probabilism and non-determinism co-exist are required to provide an operational semantics of randomized distributed algorithms. Stochastic models with nondeterminism are also needed for abstraction purposes (e.g., in the context of data abstraction, data-dependent conditional branching might be replaced with non-deterministic branching) or to model the potential interactions with an unpredictable environment (e.g., a human user). The operational semantics of such systems with probabilistic and non-deterministic behaviors can be described by a *Markov decision process* which is a stochastic model where the behavior in each state s is given

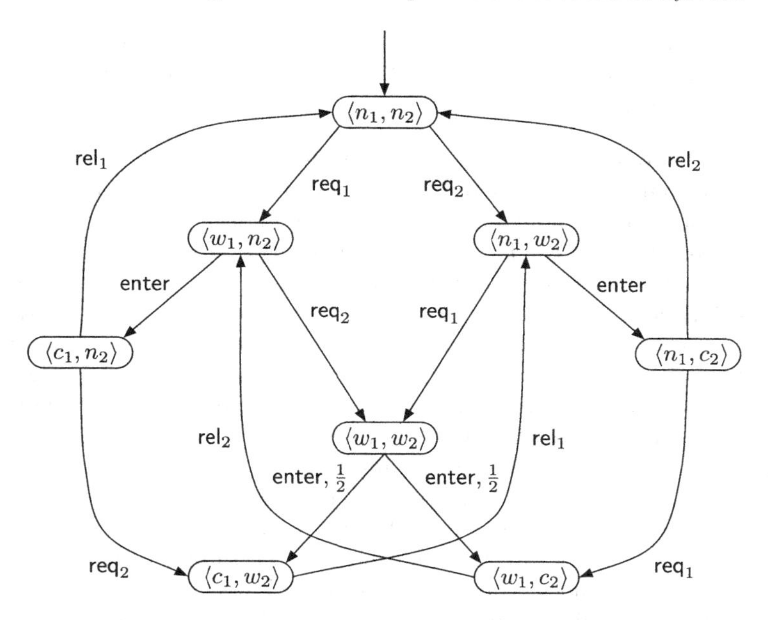

Fig. 1. MDP for a randomized mutual exclusion protocol

by a set of enabled actions which are augmented by distributions that specify the probabilistic effect of the execution of the enabled actions in state s. The idea is that when entering state s, first an enabled action α is chosen nondeterministically which is then executed and the associated distribution yields the probability for the next state. A Markov chain arises as a special case of an MDP where for each state the set of enabled actions is a singleton.

Example 1.1 (A randomized mutual exclusion protocol). We consider here a simple randomized mutual exclusion protocol for two concurrent processes $\mathcal{P}_1$ and $\mathcal{P}_2$. When process $\mathcal{P}_i$ is in its non-critical section (represented by location n_i), it can perform a request (via action req_i) and then move to a waiting section (location w_i). In the waiting section, the process waits for permission given by an arbiter to enter its critical section (location c_i) from where it can release and move to its non-critical section again (action rel_i). The arbiter that coordinates the access to the critical sections is randomized and permits process $\mathcal{P}_i$ to enter its critical section if the other process is in its non-critical section. If both processes request access to the critical section, the arbiter tosses a fair coin to decide which of the two processes has to wait and which process may enter the critical section. The composite behavior of the two concurrent processes and the arbiter can be modeled by the MDP depicted in Fig. 1. Note that the actions req, enter, and rel do not have a proper probabilistic effect, and thus yield a unique successor (the transition probabilities that equal 1 are omitted from the figure). Only in state $\langle w_1, w_2 \rangle$ there is a proper probabilistic choice performed by the arbiter to select the next process to enter the critical section. In all other states of the MDP where at least one of the processes is in its non-critical location, there is a nondeterministic choice between the enabled actions of the processes $\mathcal{P}_1$ and $\mathcal{P}_2$.

One can think of an MDP as a directed graph where the edges are augmented with action names and probabilities, i.e., positive real numbers ≤ 1, satisfying the side condition that for each state s and action name α either the probabilities attached to the outgoing α-transitions of s sum up to 1 (which ensures that they represent a probabilistic distribution) or s has no outgoing α-transitions at all (in which case α is not enabled in s). Thus, an MDP can be seen as a special instance of a *weighted automaton* where the weights are elements of the interval $]0,1]$. Applying the classical approach of weighted automata (see Chap. 3 of this handbook [35]) to an MDP (where we deal with the semiring $[0,1]$ with standard multiplication and where maximum serves as plus-operation), the weight of a finite path π is obtained by the product of the weights of the transitions on π and can be understood as the probability for π. To reason about the probabilities for properties over infinite paths (which are crucial for properties that impose conditions on the "long-run behaviors" such as liveness properties), the special interpretation of the weights as probabilities permits us to apply standard techniques of measure and probability theory. More precisely, the standard approach to define probabilities for events in an MDP relies on the sigma-algebra over infinite paths generated by the cylinder sets spanned by finite paths, and the probability measure is defined using Carathéodory's measure extension theorem. A summary of the relevant measure-theoretic concepts is presented in the appendix; see Sect. 8. This is an alternative approach to the interpretation of infinite words over weighted automata discussed in [35, 31] where special algebraic assumptions on the underlying semiring are made in order to define the weights of infinite paths and words. In particular, the semiring has to permit an infinite sum as well as a countably infinite product operation satisfying several commutativity, associativity, and distributivity laws. Another approach to interpret infinite words over weighted automata uses discounting, which is a well-known concept in mathematical economics as well as systems theory in which later events get less value than earlier ones [32–34].

The typical task for verifying a probabilistic system modeled by an MDP is to prove that certain temporal properties hold *almost surely*, i.e., with probability 1, or with some *high probability*, no matter how the non-determinism is resolved. The notion *qualitative property* is used when a certain event is required to hold almost surely (or dually with zero probability). Thus, qualitative properties assert that a certain path condition E holds for almost all or almost no paths. Depending on the type of condition E, the concept of qualitative properties is different from reasoning by means of purely functional properties that require a certain event E to hold for all paths or for no path. For example, if B is a set of states, then the qualitative reachability property asserting that "almost all paths will enter a state in B" is slightly weaker than the functional reachability property requiring that "all paths will enter a state in B". The notion *quantitative property* refers to a condition that requires a lower or upper bound in the open interval $]0,1[$ for the probability of a certain event or imposes some conditions on the mean value of a random function. For instance, typical requirements for a randomized mutual exclusion protocol are the qualitative property stating that "almost surely each waiting process will eventually get the grant to enter its critical section" and the quantitative property stating that "each waiting process has the chance of at least 99% to enter its critical section after having announced its demand five times". In the case of a communication protocol with a lossy channel, a natural quantitative requirement could be that "with probability ≥ 0.98, any message will be delivered correctly after at most three attempts to send it".

When speaking about probability bounds in the context of verification methods for MDPs, we range over all possible resolutions of the non-determinism. This corresponds to a *worst-case* analysis. Thus, if the given MDP is an interleaving model for a probabilistic multi-processor system, then the worst-case analysis ranges over all orders of concurrent actions and does not impose any restrictions on the relative speed of the processors. In cases where some non-deterministic choices in an MDP stand for the potential behaviors of an unknown environment, the worst-case analysis takes all possible activities of the environment into account. Similarly, when certain non-deterministic branches result from abstractions, then the worst-case analysis covers all concrete behaviors. By requiring that the probabilities for a given event E are 1 or sufficiently close to 1, the event E is supposed to characterize the "good" (desired) behaviors. Verification problems for MDPs can also be rephrased in the opposite way: if E describes the "bad" (undesired) behaviors, then the goal is to prove that E holds with probability 0 or some sufficiently small probability.

The notion *qualitative analysis* is used if the goal is to show that a certain event E appears with probability 0 or 1, while the notion *quantitative analysis* refers to the task of computing the maximal or minimal probabilities for E, when ranging over all schedulers, i.e., instances that resolve the non-determinism. Sometimes, some mild conditions on the resolution of non-deterministic choices along infinite paths are imposed, such as fairness assumptions. Other instances of a quantitative analysis are obtained when the goal is to establish lower or upper bounds for certain mean values, e.g., the average power consumption of a complex task or the expected number of rounds required to find a leader when imposing a leader election protocol. Such variants of the verification problem for MDPs will not be addressed in this chapter. Instead, we will concentrate on the quantitative reasoning by means of extremal probabilities under the full class of schedulers.

In the literature, many variants of classical temporal logics for non-probabilistic systems have been adapted to specify the requirements of a probabilistic system. One prominent example is the probabilistic variant of computation tree logic (PCTL) [47, 15, 12] which yields an elegant formalism to specify lower or upper probability bounds for reachability properties within a logical framework. While PCTL is a representative for logics that are based on the branching time view, one can also use purely path-based formalisms, such as linear temporal logic (LTL) or automata over infinite words, to specify the desired/good or undesired/bad event E, which is then the subject of a qualitative or quantitative analysis [79, 90, 91, 23]. For finite-state MDPs, the quantitative analysis against PCTL or LTL specifications mainly relies on a combination of graph algorithms, automata-based constructions and (numerical) algorithms for solving linear programs. Consequently, compared to the non-probabilistic case, there is the additional difficulty to solve linear programs, and also the required graph algorithms are more complex. This renders the state space explosion problem even more serious than in the non-probabilistic case and the feasibility of algorithms for the quantitative analysis crucially depends on good heuristics to increase efficiency. Among other features, the tool PRISM [61] contains a PCTL model checker for MDPs which has been successfully applied to, e.g., a series of randomized coordination algorithms, communication, and security protocols. To tackle the state space explosion problem, PRISM uses a tricky combination of data structures (multi-terminal binary decision diagrams and sparse matrices) for the internal representation of the MDP and the numerical computations required for a quantitative analysis [72]. Motivated by the success of the non-probabilistic model checker SPIN [51] where partial order reduction techniques [74, 52, 87, 38] for the verification of interleaving models are realized, the concept of partial order reduction has been adapted for the quantitative analysis of MDPs against linear-time specifications [9, 26] and implemented in the model checker LIQUOR [3, 5]. Several other techniques that attempt to speed up the verification algorithms for MDPs and/or to decrease the memory requirements have been proposed in the literature, such as symmetry reduction [62], iterative abstraction-refinement algorithms [25, 50], reduction techniques for linear programs [24, 5], and many others. Most of these techniques are orthogonal to the symbolic approach of PRISM and the partial order reduction approach of LIQUOR and can be applied in combination with them.

Somehow dual to verification problems are *controller synthesis problems* where one is typically interested in *best-case* scenarios (rather than the worst case) and attempts to construct a scheduler where the probabilities for the desired behaviors, say formalized by a linear-time property E, are maximal. Assuming that all non-deterministic choices are controllable and that the controller has complete knowledge of the computation leading to the current state, then the methods for computing maximal probabilities for the event E can easily be extended for constructing a scheduler that maximizes the probabilities for E. However, the assumption that complete knowledge about the history is available is unrealistic for multi-agent systems when controllers for a single agent or a coalition of agents are wanted. In this case, the adequate model are *partially observable* MDPs [84, 70, 71] that extend ordinary MDPs by an equivalence relation $\sim$ on the states. The idea is that equivalent states are not

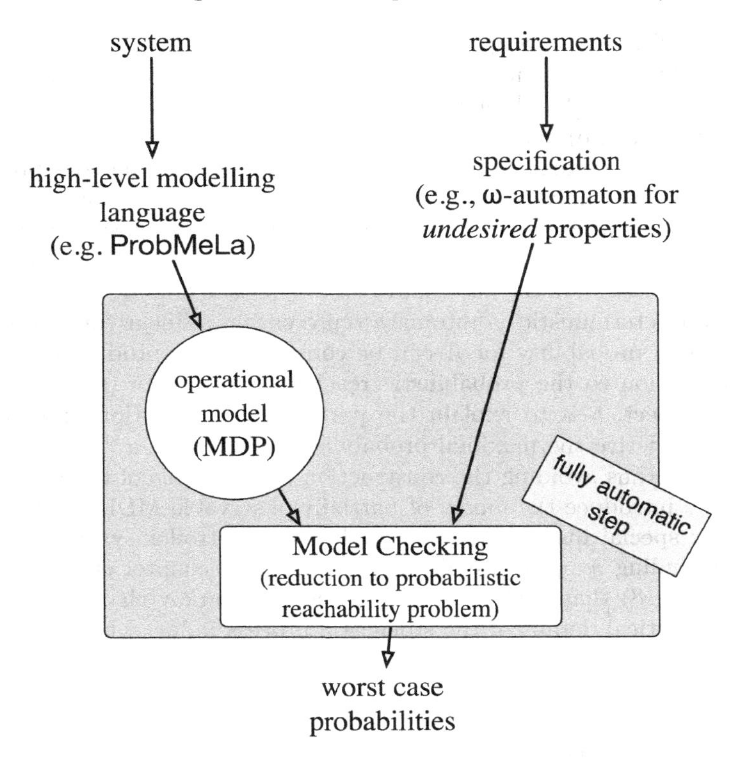

Fig. 2. Schema for the quantitative analysis

distinguishable by the controller and the task is to construct a controller (i.e., *observation-based scheduler*) for the partially observable MDP that maximizes the probabilities for E, where "observation-based" means that the scheduler can only observe the equivalence classes of the states that have been visited in the past, but not the states themselves. In general, the controller synthesis problem cannot be solved algorithmically as there is an undecidability result for the qualitative controller synthesis problem that asks for the existence of an observation-based scheduler where the probability to visit a certain set of states infinitely often is positive [2]. However, for simpler properties (e.g., safety properties) and special patterns of liveness properties, the qualitative controller synthesis problem is decidable.

About This Chapter

In the remaining sections of this chapter, we will present the main concepts of Markov decision processes as an operational model for probabilistic systems and present the basic steps for the (qualitative or quantitative) analysis against linear-time properties. Branching time properties will not be addressed here. For the basic steps to verify PCTL-like specifications for MDPs and the symbolic MTBDD-based approach, we refer to [82] and Chap. 10 in [11], as well as the literature mentioned there. We will start in Sect. 2 with the formal definition of an MDP and related notions (paths, schedulers and their induced probability measure) and illustrate the use of MDPs as an operational model for probabilistic systems by means of a few examples. The core problem of any quantitative analysis in an MDP is the problem of computing extremal reachability probabilities by means of optimization techniques. This will be explained in Sect. 3. The general case of ω-regular properties will be addressed in Sect. 4. We follow here the automata-based approach [90, 91, 23, 12] where we are given a deterministic ω-automata representing a linear-time property E. The maximal probability for E can be computed by a product construction and a reduction to the probabilistic reachability problem (see Fig. 2). The purpose of Sect. 5 is to explain the partial order reduction approach that attempts to derive the maximal probabilities for E from a "small" fragment of the MDP, thus avoiding the construction and analysis of the full MDP. In Sect. 6, we introduce the model of partially observable MDPs and report on results for special instances of the qualitative controller synthesis problem. Some concluding remarks are given in Sect. 7. The chapter ends with an appendix (Sect. 8) that contains the definition of Markov chains and explains the mathematical details of the stochastic process induced by a scheduler of an MDP.

2 Markov Decision Processes

Throughout this chapter, we will use Markov decision processes (MDPs) as an operational model for probabilistic systems. As in [80, 65, 27], the states of an MDP might have several enabled actions. Each of the actions that are enabled in state s is associated with a probability distribution which yields the probabilities for the successor states. This corresponds to the so-called reactive model in the classification of [89]. In addition, we assume here a labeling function that attaches to any state s a set of atomic propositions that are assumed to be fulfilled in state s. The atomic propositions will serve as atoms in the formal specifications for properties. For instance, to formalize deadlock freedom "processes $\mathcal{P}_1$ and $\mathcal{P}_2$ are never simultaneously in their critical sections" or starvation freedom "whenever $\mathcal{P}_1$ is in his waiting section, then $\mathcal{P}_1$ will eventually enter its critical section" for the randomized mutual exclusion protocol in Fig. 1, we can deal with temporal formulas that use the atomic propositions $\mathsf{wait}_i, \mathsf{crit}_i$ for $i = 1, 2$ which are attached to all states where the local state of process $\mathcal{P}_i$ is w_i or c_i, respectively. We will now give the formal definition of an MDP. For further basic definitions of, e.g., probability distribution, or Markov chain, we refer to the appendix (Sect. 8).

Definition 2.1 ((State-labeled) Markov decision process (MDP)). *A Markov decision process is a tuple*

$$\mathcal{M} = (S, \mathsf{Act}, \delta, \mu, \mathsf{AP}, L),$$

where:

- S *is a finite non-empty set of states.*
- Act *is a finite non-empty set of actions.*
- $\delta : S \times \mathsf{Act} \times S \to [0,1]$ *is a transition probability function such that for each* $s \in S$ *and* $\alpha \in \mathsf{Act}$, *either* $\delta(s, \alpha, .)$ *is a probability distribution on* S *or* $\delta(s, \alpha, .)$ *is the null-function (i.e.,* $\delta(s, \alpha, t) = 0$ *for any* $t \in S$*).*
- μ *is a probability distribution on* S *(called the initial distribution).*
- AP *is a finite set of atomic propositions.*
- $L : S \to 2^{\mathsf{AP}}$ *is a labeling function that labels a state* s *with those atomic propositions in* AP *that are supposed to hold in* s.

$\mathsf{Act}(s) = \{\alpha \in \mathsf{Act} \mid \exists t \in S : \delta(s, \alpha, t) > 0\}$ *denotes the set of actions that are enabled in state* s. *We require that* $\mathsf{Act}(s)$ *is non-empty for each state* $s \in S$.

The intuitive operational behavior of an MDP is the following. If s is the current state, then at first one of the actions $\alpha \in \mathsf{Act}(s)$ is chosen non-deterministically. Secondly, action α is executed leading to state t with probability $\delta(s, \alpha, t)$.

Action α is called a *probabilistic* action if it has a random effect, i.e., if there is at least one state s where α is enabled and that has two or more α-successors (an α-successor of state s is a state t such that $\delta(s, \alpha, t) > 0$). Otherwise, α is called *non-probabilistic.*

If all actions in Act are non-probabilistic and the initial distribution is a Dirac distribution, i.e., a probabilistic distribution that assigns probability 1 to some particular state, then our notion of an MDP reduces to an ordinary transition system with at most one outgoing α-transition per state and action α and exactly one initial state.

Example 2.2 (The Monty Hall problem). Before we proceed, let us have a look at a small example of an MDP. We consider here the Monty Hall problem: "Suppose you are a contestant on a game show, and you are given the choice of three doors: behind one door is a car, behind the others, goats. You choose a door, but you do not open it. Then the host, who knows what is behind the doors, has to open another door which has a goat behind it (if you initially picked the door with the car behind it, the host randomly chooses one of the other doors). He then asks whether you want to change your choice to the other unopened door. After either sticking to your first choice or switching to the other unopened door, you win what is behind the door that you have finally chosen. Considering that your goal is to win the car, is it to your advantage to switch your first choice?"

In Fig. 3, we depict an MDP for an abstraction of this problem where due to symmetry the information which exact door reveals the car is neglected. So, in the initial state s, the contestant chooses one of the three doors, each with equal probability $\frac{1}{3}$. State t_1 represents the case where the contestant has

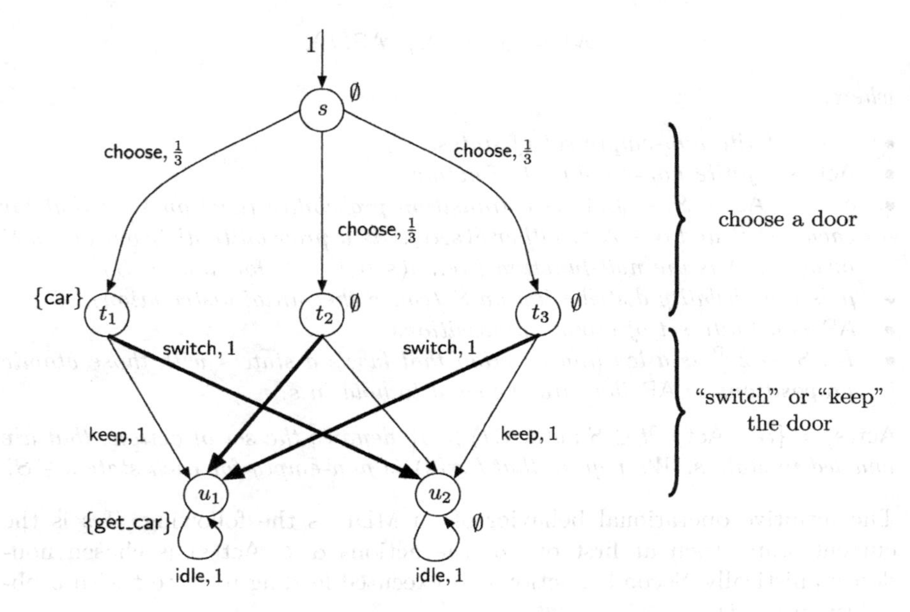

Fig. 3. The abstract "Monty Hall problem"

chosen the door with the car behind it, states t_2 and t_3 represent the other cases where the contestant has chosen a door with a goat behind it. Thus, state t_1 is labeled with the atomic proposition car, whereas t_2 and t_3 are labeled with $\emptyset$. After this first choice, the game show host opens a door which has a goat behind it. Note that now there are exactly two closed doors, one with a car behind it (represented by state u_1 and the labeling {get_car}) and one door with a goat behind it (represented by state u_2 and the labeling $\emptyset$). Now the contestant has the alternative to either stick to her/his chosen door or to switch to the other closed door. So, in states t_1, t_2, and t_3, there is a non-deterministic choice between the actions switch and keep, leading to u_1 or u_2. Note that the actions switch and keep are non-probabilistic as there are exactly two closed doors. After choosing switch or keep, the game is over. To complete the MDP, we added an idling self-loop to the states u_1 and u_2. For the sake of readability, we depict the transitions of the switch-action a little thicker and we omit the labeling of the actions of state t_2.

Paths and Schedulers of an MDP

Definition 2.3 (Path and corresponding notations). *An infinite path of an MDP is an infinite sequence* $\pi = ((s_0, \alpha_1), (s_1, \alpha_2), \ldots) \in (S \times \mathsf{Act})^\omega$ *such that* $\delta(s_i, \alpha_{i+1}, s_{i+1}) > 0$ *for* $i \in \mathbb{N}_{\geq 0}$. *We write paths in the form*

$$\pi = s_0 \xrightarrow{\alpha_1} s_1 \xrightarrow{\alpha_2} s_2 \xrightarrow{\alpha_3} \cdots$$

A finite path is a finite prefix of an infinite path that ends in a state. We use the notation $\mathsf{last}(\pi)$ *for the last state of a finite path* π *and* $|\pi|$ *for the length (number of actions) of a finite path. We denote by* $\mathsf{Path}_{\mathsf{fin}}^{\mathcal{M}}$*, resp.* $\mathsf{Path}_{\mathsf{inf}}^{\mathcal{M}}$*, the set of all finite, resp. infinite, paths of* $\mathcal{M}$*.*

In order to be able to talk about the probability, e.g., to get the car in the Monty Hall problem, we need another concept of the theory of MDPs, namely the concept of schedulers. Schedulers are a means to resolve the nondeterminism in the states, and thus yield a discrete Markov chain and a probability measure on the paths. Intuitively, a scheduler takes as input the "history" of a computation (formalized by a finite path π) and chooses the next action (resp. a distribution on actions).

Definition 2.4 (Scheduler). *For a given MDP* $\mathcal{M} = (S, \mathsf{Act}, \delta, \mu, \mathsf{AP}, L)$*, a history dependent randomized scheduler is a function*

$$\mathcal{U} : \mathsf{Path}_{\mathsf{fin}}^{\mathcal{M}} \rightarrow \mathsf{Distr}(\mathsf{Act}),$$

such that $\mathsf{supp}(\mathcal{U}(\pi)) \subseteq \mathsf{Act}(\mathsf{last}(\pi))$ *for each* $\pi \in \mathsf{Path}_{\mathsf{fin}}^{\mathcal{M}}$*. Here,* $\mathsf{Distr}(\mathsf{Act})$ *denotes the set of probability distributions on* Act *while* $\mathsf{supp}(\mathcal{U}(\pi))$ *denotes the support of* $\mathcal{U}(\pi)$*, i.e., the set of actions* $\alpha \in \mathsf{Act}$ *such that* $\mathcal{U}(\pi)(\alpha) > 0$*.*

A scheduler $\mathcal{U}$ is called *deterministic*, if $\mathcal{U}(\pi)$ is a Dirac distribution for each $\pi \in \mathsf{Path}_{\mathsf{fin}}$, i.e., $\mathcal{U}(\pi)(\alpha) = 1$ for some action α, while $\mathcal{U}(\pi)(\beta) = 0$ for every other action $\beta \neq \alpha$. Scheduler $\mathcal{U}$ is called *memoryless*, if $\mathcal{U}(\pi) = \mathcal{U}(\mathsf{last}(\pi))$ for each $\pi \in \mathsf{Path}_{\mathsf{fin}}$ (note that $\mathsf{last}(\pi)$ is a path of length 0). We denote by Sched, the set of all (history dependent, randomized) schedulers. We write $\mathsf{Sched}_{\mathsf{D}}$ to denote the set of deterministic schedulers, $\mathsf{Sched}_{\mathsf{M}}$ for the set of memoryless schedulers, and $\mathsf{Sched}_{\mathsf{MD}}$ for the set of memoryless deterministic schedulers. Note that the following inclusions hold.

- $\mathsf{Sched}_{\mathsf{M}} \subseteq \mathsf{Sched}$ and $\mathsf{Sched}_{\mathsf{D}} \subseteq \mathsf{Sched}$
- $\mathsf{Sched}_{\mathsf{MD}} \subseteq \mathsf{Sched}_{\mathsf{M}}$ and $\mathsf{Sched}_{\mathsf{MD}} \subseteq \mathsf{Sched}_{\mathsf{D}}$

A (finite or infinite) path $s_0 \xrightarrow{\alpha_1} s_1 \xrightarrow{\alpha_2} s_2 \xrightarrow{\alpha_3} \cdots$ is called a $\mathcal{U}$-path, if $\mathcal{U}(s_0 \xrightarrow{\alpha_1} \cdots \xrightarrow{\alpha_i} s_i)(\alpha_{i+1}) > 0$ for every $0 \leq i < |\pi|$.

Given an MDP $\mathcal{M}$ and a scheduler $\mathcal{U}$, the behavior of $\mathcal{M}$ under $\mathcal{U}$ can be formalized by a (possibly infinite-state) discrete Markov chain. By $\mathsf{Pr}^{\mathcal{M},\mathcal{U}}$, we denote the standard probability measure on the standard σ-algebra of the infinite paths of $\mathcal{M}$. Given a state s of $\mathcal{M}$, we denote by $\mathsf{Pr}_s^{\mathcal{M},\mathcal{U}}$ the probability measure that is obtained if $\mathcal{M}$ is equipped with the initial Dirac distribution μ_s, with $\mu_s(s) = 1$. A detailed definition of $\mathcal{M}$ and $\mathsf{Pr}^{\mathcal{M},\mathcal{U}}$ is provided in the appendix (Sect. 8). We also fix the following notation for convenience. Given an MDP $\mathcal{M}$, a scheduler $\mathcal{U}$, and a measurable path property E, we will write

$$\mathsf{Pr}^{\mathcal{M},\mathcal{U}}(E) \stackrel{\text{def}}{=} \mathsf{Pr}^{\mathcal{M},\mathcal{U}}\big(\{\pi \in \mathsf{Path}_{\mathsf{inf}}^{\mathcal{M}} \mid \pi \text{ satisfies } E\}\big)$$

for the probability that the property E holds in $\mathcal{M}$ under the scheduler $\mathcal{U}$.

Let us consider again the Monty Hall problem depicted in Fig. 3. It is easy to see that under the scheduler $\mathcal{U}$ with[1]

$$\mathcal{U}(s \xrightarrow{\mathsf{choose}} t_1)(\mathsf{keep}) = \mathcal{U}(s \xrightarrow{\mathsf{choose}} t_2)(\mathsf{switch}) = \mathcal{U}(s \xrightarrow{\mathsf{choose}} t_3)(\mathsf{switch}) = 1$$

the contestant will win the car with probability 1. This is, of course, an unrealistic scheduler as it uses the knowledge of whether the contestant has chosen the door with the car behind it or not. If the door with the car was chosen (state t_1), then the scheduler decides to keep the door, if a door with a goat was chosen (states t_2, t_3), then the scheduler decides to switch the door. As the contestant does not know whether he has chosen the door with the car behind it, the only realistic schedulers (that model a contestant's choice) are the two schedulers $\mathcal{U}_\mathsf{s}$ and $\mathcal{U}_\mathsf{k}$ with

$$\mathcal{U}_\mathsf{s}(s \xrightarrow{\mathsf{choose}} t_1)(\mathsf{switch}) = \mathcal{U}_\mathsf{s}(s \xrightarrow{\mathsf{choose}} t_2)(\mathsf{switch}) = \mathcal{U}_\mathsf{s}(s \xrightarrow{\mathsf{choose}} t_3)(\mathsf{switch}) = 1$$

and

$$\mathcal{U}_\mathsf{k}(s \xrightarrow{\mathsf{choose}} t_1)(\mathsf{keep}) = \mathcal{U}_\mathsf{k}(s \xrightarrow{\mathsf{choose}} t_2)(\mathsf{keep}) = \mathcal{U}_\mathsf{k}(s \xrightarrow{\mathsf{choose}} t_3)(\mathsf{keep}) = 1$$

where the contestant either decides to switch the door or to keep it. Simple computations show that with $\mathcal{M}$ being the MDP of Fig. 3

$$\begin{aligned}
\mathsf{Pr}^{\mathcal{M},\mathcal{U}_\mathsf{s}}(\Diamond\ \mathsf{get_car}) = \mathsf{Pr}^{\mathcal{M},\mathcal{U}_\mathsf{s}}(\{ & s \xrightarrow{\mathsf{choose}} t_1 \xrightarrow{\mathsf{keep}} u_1 \xrightarrow{\mathsf{idle}} \cdots, \\
& s \xrightarrow{\mathsf{choose}} t_2 \xrightarrow{\mathsf{switch}} u_1 \xrightarrow{\mathsf{idle}} \cdots, \\
& s \xrightarrow{\mathsf{choose}} t_3 \xrightarrow{\mathsf{switch}} u_1 \xrightarrow{\mathsf{idle}} \cdots \}) \\
= {} & 1 \cdot \tfrac{1}{3} \cdot 0 \cdot 1 + 1 \cdot \tfrac{1}{3} \cdot 1 \cdot 1 + 1 \cdot \tfrac{1}{3} \cdot 1 \cdot 1 \\
= {} & \tfrac{2}{3}
\end{aligned}$$

and similarly

$$\mathsf{Pr}^{\mathcal{M},\mathcal{U}_\mathsf{k}}(\Diamond\ \mathsf{get\_car}) = \frac{1}{3}.$$

This shows that it is an advantage to switch the door. Here, we used the LTL-like notation $\Diamond$ get_car to denote the reachability property stating that eventually a state where get_car holds will be reached. That is, given the atomic proposition get_car, a path $\pi = s_0 \xrightarrow{\alpha_1} s_1 \xrightarrow{\alpha_2} \cdots$ satisfies $\Diamond$ get_car if and only if there exists an index $i \in \mathbb{N}_{\geq 0}$ such that get_car $\in L(s_i)$.

Specifying Systems with an MDP-Semantics

In the literature, various modeling languages for probabilistic systems whose stepwise behavior can be formalized by an MDP have been proposed. Such languages can serve as a starting point for model checking tools that take as

[1] Note that this completely determines the scheduler as there are no other paths that end in a state with more than one enabled action.

```
msg = ...; // the data to be sent
msg_sent = false;
do:: msg_sent == false ->
        pif :0.9: -> msg_sent = true;
            :0.1: -> skip
        fip
od
```

Fig. 4. ProbMeLa-code for an unbounded retransmission protocol

input a description of the processes of a (parallel) system and a formalization of the property to be checked. Semantic rules are used to construct the MDP automatically. Then in further steps, model checking algorithms are applied. An example for such a modeling language is ProbMeLa [4], which is the input language of the model checker LIQUOR that supports the qualitative and quantitative analysis of MDPs against linear-time properties. ProbMeLa is a probabilistic version of the process meta-language ProMeLa which serves as modeling language in connection with the prominent model checker SPIN for non-probabilistic systems [51]. The core language of ProMeLa and ProbMeLa relies on Dijkstra's guarded commands `guard -> cmd` which can be used in loops (`do ... od`) and conditional commands (`if ... fi`). One of the additional probabilistic features of ProbMeLa is a probabilistic choice operator (`pif ... fip`) which can be seen as a probabilistic variant of conditional commands as probabilities are assigned to the commands rather than Boolean guards. Commands with the probabilistic choice operator have the form

```
pif :p1: -> cmd1;
      ⋮
    :pn: -> cmdn
fip
```

where the $\mathtt{p}_i$'s are values in $]0,1]$ that sum up to 1 and specify a probabilistic distribution for the commands $\mathtt{cmd}_i$.

For instance, the ProbMeLa-code in Fig. 4 might be a fragment of an unbounded retransmission protocol where some process sends data (variable `msg`) over a faulty medium such as a wireless radio connection or an otherwise noisy connection. This ProbMeLa-code specifies a process that iteratively attempts to send the message until it has been delivered correctly, where in each iteration of the loop the message will be delivered with probability 0.9 and will be lost with probability 0.1 (which is modeled here by the command `skip`). Termination of this protocol is guaranteed with probability 1 which means that almost surely the message will be delivered eventually. Note that there is an infinite computation where the message is lost in all iterations, but the probability for this to happen is 0. This simple example illustrates the difference between the functional and qualitative properties: the functional property requiring termination along all paths does not hold for the protocol,

```
do
 pif:0.5:-> first_fork = left; second_fork = right
    :0.5:-> first_fork = right; second_fork = left
 fip
 if ::forks[first_fork]==false -> forks[first_fork]=true;
      if
        ::forks[second_fork]==false -> forks[second_fork]=true;
          // philosopher is eating
          forks[second_fork]=false; forks[first_fork]=false
        ::forks[second_fork]==true -> forks[first_fork]=false
      fi
    ::forks[first_fork]==true -> skip
 fi
od
```

Fig. 5. ProbMeLa-code for each philosopher

while the qualitative property requiring termination along almost all paths (i.e., with probability 1) holds. Furthermore, we can establish the quantitative property stating that with probability 0.999 the message will be delivered within the first three iterations.

In the context of distributed systems, such as mutual exclusion protocols, leader election, or Byzantine agreement, coin tossing offers an elegant way to design coordination algorithms that treat all processes equally, but avoid deadlock situations since symmetry breaking is inherent in the random outcomes of the coin tossing experiment. A prominent example is the randomized solution of the dining philosophers problem suggested by Lehmann and Rabin [66]. The philosophers are sitting at a round table and neighboring philosophers have access to a shared resource (a fork). Each philosopher attempts to alternate infinitely often between a thinking and an eating phase, where the latter requires that the philosopher has picked up the fork to his left and the fork to his right. The Lehmann–Rabin protocol works as follows. As soon as a philosopher gets hungry, he decides to pick up his left or right fork by tossing a fair coin. If the selected fork is available, then he picks up this fork and checks whether the other fork is available, also. If so, then he takes it and starts to eat. Otherwise, he returns the taken fork, and repeats the whole procedure. This procedure can be described in ProbMeLa as shown in Fig. 5. If all philosophers operate on the basis of this protocol, then deadlock freedom is ensured on all paths. Starvation freedom holds almost surely under all schedulers, i.e., no matter how the steps of the philosophers are interleaved, with probability 1 each philosopher gets to eat eventually, if he intends to do so. Other classes of randomized algorithms, protocols, and scenarios are modeled likewise.

3 Maximal Reachability Probabilities

Computing maximal or minimal probabilities for reachability objectives is one of the core problems for a quantitative analysis of MDPs. In this section, we summarize the main concepts that rely on linear programming or an iterative approach which is known under the key word *value iteration*. Further details and other methods (e.g., policy iteration) can be found in any textbook on Markov decision processes; see, e.g., [80].

The notion "reachability" refers to the directed graph structure that is obtained from $\mathcal{M}$ by ignoring the labels of transitions and states. That is, a set B of states is said to be reachable from a state s if there exists a finite path in $\mathcal{M}$ that starts in state s and ends in some state $s' \in B$. In what follows, we will use the LTL-like notation $\Diamond B$ to denote the property "eventually reach B" where B is a set of states. That is, given a set $B \subseteq S$ of states, a path $\pi = s_0 \xrightarrow{\alpha_1} s_1 \xrightarrow{\alpha_2} \cdots$ satisfies $\Diamond B$ if and only if there exists an index $i \in \mathbb{N}_{\geq 0}$ such that $s_i \in B$. The quantitative analysis of an MDP $\mathcal{M}$ against a reachability specification amounts to establishing the best upper and/or lower probability bounds that can be guaranteed to reach a given set B of states, when ranging over all schedulers. That is, the goal is to compute

$$\sup_{\mathcal{U} \in \mathsf{Sched}} \mathsf{Pr}^{\mathcal{M},\mathcal{U}}(\Diamond B) \quad \text{and} \quad \inf_{\mathcal{U} \in \mathsf{Sched}} \mathsf{Pr}^{\mathcal{M},\mathcal{U}}(\Diamond B),$$

where the supremum and the infimum are taken over all schedulers $\mathcal{U}$ for $\mathcal{M}$. If $\mathcal{M}$ is clear from the context, we will omit the system $\mathcal{M}$ in the superscript of $\mathsf{Pr}^{\mathcal{M},\mathcal{U}}$.

For the rest of this chapter, we will restrict to the computation of the supremum because the results for the infimum are analogous. It is well known [80, 15] that the history of a scheduler is of no relevance when it comes to maximizing the probability of reaching a certain set of states and also that randomization in the schedulers is not needed. Thus, the supremum is attained by some memoryless deterministic scheduler (note that there are only finitely many such schedulers). Thus,

$$\sup_{\mathcal{U} \in \mathsf{Sched}} \mathsf{Pr}^{\mathcal{U}}(\Diamond B) = \sup_{\mathcal{U} \in \mathsf{Sched}_{\mathsf{M}}} \mathsf{Pr}^{\mathcal{U}}(\Diamond B) = \max_{\mathcal{U} \in \mathsf{Sched}_{\mathsf{MD}}} \mathsf{Pr}^{\mathcal{U}}(\Diamond B).$$

Note that the above chain of equality does not hold for arbitrary measurable path properties.

The standard method to compute these maxima is to compute $\mathsf{Pr}^{\mathcal{U}}_s(\Diamond B)$ for each state s in $\mathcal{M}$ via a recursive equation system. Formally, let an MDP $\mathcal{M} = (S, \mathsf{Act}, \delta, \mu, \mathsf{AP}, L)$ and a set $B \subseteq S$ of target states be given. The obligation is to compute

$$\mathsf{Pr}^{\max}_s(\Diamond B) \stackrel{\text{def}}{=} \max_{\mathcal{U} \in \mathsf{Sched}} \mathsf{Pr}^{\mathcal{U}}_s(\Diamond B) = \max_{\mathcal{U} \in \mathsf{Sched}_{\mathsf{MD}}} \mathsf{Pr}^{\mathcal{U}}_s(\Diamond B)$$

for each state $s \in S$. Let x_s denote this maximum for $s \in S$, that is,

$$x_s \stackrel{\text{def}}{=} \max_{\mathcal{U}\in\mathsf{Sched}_{\mathsf{MD}}} \mathsf{Pr}_s^{\mathcal{U}}(\Diamond B).$$

Then if $s \notin B$, it evidently holds that

$$x_s = \max\Big\{\sum_{t\in S} \delta(s,\alpha,t)\cdot x_t \;\Big|\; \alpha\in\mathsf{Act}(s)\Big\}.$$

Moreover, if $s \in B$, then obviously $x_s = 1$, and if there is no path from s to B, then $x_s = 0$. The following theorem [22, 80, 15] states that these characteristics are sufficient to specify the maximum values.

Theorem 3.1 (Equation system for maximal reachability probabilities). *Let $\mathcal{M}$ be an MDP with state space S and $B \subseteq S$. The vector $(x_s)_{s\in S}$ with $x_s = \mathsf{Pr}_s^{\max}(\Diamond B)$ is the unique solution of the following equation system:*

- *If $s \in B$, then $x_s = 1$.*
- *If B is not reachable from s, then $x_s = 0$.*
- *If $s \notin B$ and B is reachable from s, then*

$$x_s = \max\Big\{\sum_{t\in S} \delta(s,\alpha,t)\cdot x_t \;\Big|\; \alpha\in\mathsf{Act}(s)\Big\}.$$

Obviously, $x_s = \mathsf{Pr}_s^{\max}(\Diamond B)$ is a solution of the above equation system. The proof of its uniqueness is rather technical and omitted here.

Let us again consider the MDP of the Monty Hall problem depicted in Fig. 3. So, we are interested in the set of target states $B = \{u_1\}$. Then

$$\begin{array}{lll} x_{u_1} = 1 & x_{t_1} = \max\{x_{u_1}, x_{u_2}\} & x_s = \max\{\frac{1}{3}\cdot x_{t_1} + \frac{1}{3}\cdot x_{t_2} + \frac{1}{3}\cdot x_{t_3}\} \\ x_{u_2} = 0 & x_{t_2} = \max\{x_{u_1}, x_{u_2}\} & \\ & x_{t_3} = \max\{x_{u_1}, x_{u_2}\} & \end{array}$$

and the unique solution is $x_{u_1} = x_{t_1} = x_{t_2} = x_{t_3} = x_s = 1$ and $x_{u_2} = 0$.

To actually compute the values $\mathsf{Pr}_s^{\max}(\Diamond B)$ algorithmically, one can rewrite the equation system in Theorem 3.1 into the following linear program [22]:

- If $s \in B$, then $x_s = 1$.
- If B is not reachable from s, then $x_s = 0$.
- If $s \notin B$ and B is reachable from s, then $0 \le x_s \le 1$ and for each action $\alpha \in \mathsf{Act}(s)$:

$$x_s \ge \sum_{t\in S} \delta(s,\alpha,t)\cdot x_t.$$

With the objective to

$$\text{minimize} \sum_{s\in S} x_s,$$

the vector $(x_s)_{s\in S}$ with $x_s = \mathsf{Pr}_s^{\max}(\Diamond B)$ is the unique solution of this linear program. Identifying the states s such that the value of x_s is not fixed to 0 or 1 by the first two items of the linear program as $S_? = \{ s \in S \setminus B \mid B$ is reachable from $s \}$, one can rewrite the third item into

$$\big(1 - \delta(s,\alpha,s)\big) \cdot x_s - \sum_{t \in S_? \setminus \{ s \}} \delta(s,\alpha,t) \cdot x_t \geq \delta(s,\alpha,B)$$

where $\delta(s,\alpha,B) = \sum_{t \in B} \delta(s,\alpha,t)$. Thus, the third item in the above theorem can be read as a linear inequality $\mathbf{A} \cdot \mathbf{x} \geq \mathbf{b}$ where $\mathbf{x}$ is the vector $(x_s)_{s \in S_?}$ and $\mathbf{A}$ is a matrix with a row for each pair (s,α) with $s \in S_?$ and $\alpha \in \mathsf{Act}(s)$, two extra rows for each state $s \in S_?$ to represent the inequality $0 \leq x_s \leq 1$ and a column for each state $s \in S_?$. The precise values for $\mathsf{Pr}^{\mathsf{max}}_s(\Diamond B)$ can thus be computed by standard algorithms to solve linear programs, e.g., the simplex algorithm or polytime methods [83].

Corollary 3.2 (Complexity of computing maximal reachability probabilities). *For an MDP $\mathcal{M}$ with state space S, $B \subseteq S$ and $s \in S$, the values $\mathsf{Pr}^{\mathsf{max}}_s(\Diamond B)$ can be computed in time polynomial in the size of $\mathcal{M}$.*

This result, however, is more of theoretical interest than of practical relevance. In practice, one often uses a different approach to calculate the values $\mathsf{Pr}^{\mathsf{max}}_s(\Diamond B)$, namely an iterative approximation technique called *value iteration* (see, e.g., [80]) which is based on the following fact. The second item in Theorem 3.1 could be omitted and replaced by the requirement that the equations for x_s in the third item holds for every state $s \in S \setminus B$. However, the uniqueness of the solution vector $(x_s)_{s \in S} = (\mathsf{Pr}^{\mathsf{max}}_s(\Diamond B))_{s \in S}$ is then no longer guaranteed, but one can show that $(x_s)_{s \in S}$ is the least solution in $[0,1]^S$. For the value iteration, one fixes the value for $s \in B$ to $x_s = 1$ and starts with an initial value of $x_s = 0$ for all states $s \notin B$. One then iteratively recalculates the value according to item 3 of Theorem 3.1. That is,

$$\begin{aligned} x^i_s &= 1 && \text{for } s \in B \text{ and } i \in \mathbb{N}_{\geq 0} \\ x^0_s &= 0 && \text{for } s \notin B \\ x^{n+1}_s &= \mathsf{max}\Big\{ \sum_{t \in S} \delta(s,\alpha,t) \cdot x^n_t \mid \alpha \in \mathsf{Act}(s) \Big\} && \text{for } s \notin B. \end{aligned}$$

For the states $s \notin B$, it can be shown that

$$\lim_{n \to \infty} x^n_s = \mathsf{Pr}^{\mathsf{max}}_s(\Diamond B).$$

Note that $x^0_s \leq x^1_s \leq x^2_s \leq \cdots$. Thus, the values $\mathsf{Pr}^{\mathsf{max}}_s(\Diamond B)$ can be approximated by successively computing the vectors

$$\big(x^0_s\big)_{s \in S}, \big(x^1_s\big)_{s \in S}, \big(x^2_s\big)_{s \in S}, \ldots,$$

until $\mathsf{max}_{s \in S} |x^{n+1}_s - x^n_s|$ is below a termination threshold.

Let us once more consider the MDP of the Monty Hall problem depicted in Fig. 3 and the target set $B = \{u_1\}$. Thus, $x^i_{u_1} = 1$ for every $i \in \mathbb{N}_{\geq 0}$. Note that $x^i_{u_2} = x^{i-1}_{u_2}$, so $x^i_{u_2}$ will equal 0 for every $i \in \mathbb{N}_{\geq 0}$. With $x^i_{t_1} = x^i_{t_2} = x^i_{t_3} = \max\{1 \cdot x^{i-1}_{u_1}, 1 \cdot x^{i-1}_{u_2}\}$ we get

$$\begin{array}{llll}
x^0_{u_2} = 0 & x^0_{t_1} = x^0_{t_2} = x^0_{t_3} = 0 & x^0_s = 0 \\
x^1_{u_2} = 0 & x^1_{t_1} = x^1_{t_2} = x^1_{t_3} = 1 & x^1_s = \max\{\frac{1}{3} \cdot x^0_{t_1} + \frac{1}{3} \cdot x^0_{t_2} + \frac{1}{3} \cdot x^0_{t_3}\} = 0 \\
x^2_{u_2} = 0 & x^2_{t_1} = x^2_{t_2} = x^2_{t_3} = 1 & x^2_s = \max\{\frac{1}{3} \cdot x^1_{t_1} + \frac{1}{3} \cdot x^1_{t_2} + \frac{1}{3} \cdot x^1_{t_3}\} = 1 \\
x^3_{u_2} = 0 & x^3_{t_1} = x^3_{t_2} = x^3_{t_3} = 1 & x^3_s = \max\{\frac{1}{3} \cdot x^2_{t_1} + \frac{1}{3} \cdot x^2_{t_2} + \frac{1}{3} \cdot x^2_{t_3}\} = 1.
\end{array}$$

As all values did not change in the last iteration, we can conclude that the fixed point is reached.

Implementation Issues

It is obvious that for the set of states from which B is not reachable, the value iteration is not needed as $\mathsf{Pr}^{\mathsf{max}}_s(\Diamond B) = 0$ if and only if B is not reachable from s. Thus, it is advisable to first compute the set

$$S_0 \stackrel{\text{def}}{=} \{s \in S \mid B \text{ is not reachable from } s\}$$

with a standard backward reachability analysis and then set $x^i_s = 0$ for all states $s \in S_0$ and $i \in \mathbb{N}_{\geq 0}$. Note that this reduces the number of variables for which the value iteration has to be performed. Similarly, one can first compute the set

$$S_1 \stackrel{\text{def}}{=} \{s \in S \mid \mathsf{Pr}^{\mathsf{max}}_s(\Diamond B) = 1\}$$

of states s such that $\mathsf{Pr}^{\mathsf{max}}_s(\Diamond B) = 1$. For all states $s \in S_1$, we set $x^i_s = 1$ for all $i \in \mathbb{N}_{\geq 0}$ thus reducing again the number of variables for which the value iteration has to be performed. The computation of S_1 can be done efficiently by Algorithm 1 using a nested fixpoint computation.

In the formal description of the value iteration, each variable x_s with $s \notin B$ is updated in every iteration. This might of course lead to a great amount of unnecessary updates (consider for instance a unidirectional, very long chain where the last state forms the target set B). So, when implementing the value iteration one seeks to omit updating a value x_s if there is no successor t of s such that the value x_t has been changed during the last update of x_s. This idea is reflected in Algorithm 2 which iteratively propagates probabilities by means of a backward traversal of the MDP from B. The algorithm maintains a set T of states for which the x_t value has been changed. It then successively removes a state t from the set T and updates the value x_s for every state s that has t as a successor. If such a value x_s becomes altered, the state s is added to T.

There are several variants of Algorithm 2 that differ in the data structure used for the organization of the set T. For instance, the set T can be realized as a stack, a (priority) queue or using buckets to aggregate states. For the priority queue and the bucket structure, the sorting criterion is the value $\Delta(s)$. For more information on such implementation details and other heuristics, see [5].

Algorithm 1 Computation of S_1

$R := S$
$done := false$
while $done = false$ **do**
  $R' := B$
  $done' := false$
  while $done' = false$ **do**
    $R'' := R' \cup \{s \mid \exists \alpha \in \mathsf{Act}(s) : \mathsf{supp}(\delta(s,\alpha,.)) \subseteq R \wedge \mathsf{supp}(\delta(s,\alpha,.)) \cap R' \neq \emptyset\}$
    if $R'' = R'$ **then**
      $done' := true$
    end if
    $R' := R''$
  end while
  if $R' = R$ **then**
    $done := true$
  end if
  $R := R'$
end while
return R

Algorithm 2 Backward value iteration

compute S_0, S_1 and $S_? = S \setminus (S_0 \cup S_1)$
set $x_s := 1$ for all states $s \in S_1$ and $x_s := 0$ for all states $s \in S_0$
for all states $s \in S_?$ **do**
  $x_s := \max\{\sum_{t \in S_1} \delta(s,\alpha,t) \mid \alpha \in \mathsf{Act}(s)\}$
end for
$T := \{s \in S_? \mid x_s > 0\}$
repeat
  if $T \neq \emptyset$ **then**
    pick some state $t \in T$ and remove t from T
    for all states s such that $\exists$ action $\alpha \in \mathsf{Act}(s)$ with $\delta(s,\alpha,t) > 0$ **do**
      $x'_s := \max\{\sum_{u \notin S_0} \delta(s,\alpha,u) \cdot x_u \mid \alpha \in \mathsf{Act}(s)\}$
      if $x'_s > x_s$ **then**
        $\Delta(s) := x'_s - x_s$ and $x_s := x'_s$ and add s to T
      end if
    end for
  end if
until $T = \emptyset$ or termination threshold is matched

It should be noticed that if one is just interested in a qualitative reachability analysis (where the task is to check whether $\mathsf{Pr}^{\mathsf{max}}_s(\Diamond B)$ is 0 or 1), then the computation of the values $x_s = \mathsf{Pr}^{\mathsf{max}}(\Diamond B)$ is not necessary. In fact, algorithms to compute the sets S_0 and S_1 are sufficient.

4 Model Checking ω-Regular Properties

In the previous section, we have addressed the problem of computing extremal reachability probabilities in an MDP and have seen that this can be represented as a recursive equation system or a linear program which can be solved through fixpoint calculations or, e.g., the simplex algorithm. This approach can easily be extended to constrained reachability properties (that require to reach a certain target set B along finite paths that stay inside another set C) or the general class of regular safety properties (which impose conditions on the finite paths by means of a finite automaton). The quantitative analysis against reachability probabilities is also the core problem for quantitative reasoning about liveness properties or even arbitrary ω-regular linear-time properties. In this section, we explain how extremal probabilities for ω-regular properties represented by ω-automata can be calculated using a graph-based algorithm and a reduction to the problem of computing maximal reachability probabilities.

Before we proceed, we fix some notation on finite, resp. infinite words over a given alphabet. Following the standard notation, given an alphabet Σ, we write Σ^+ for the set of all non-empty finite words over Σ, Σ^* for the set of all finite words over Σ (including the empty word ε) and Σ^ω for the set of all infinite words over Σ. Given a finite non-empty word $\varsigma = \sigma_1\sigma_2\ldots\sigma_n$, the length $|\varsigma|$ of ς equals n. For an infinite word ς, the length is equal to ∞. Given a (non-empty finite or infinite) word $\varsigma = \sigma_1\sigma_2\sigma_3\ldots \in \Sigma^+ \cup \Sigma^\omega$ and $i \leq |\varsigma|$, we denote the ith letter of ς by ς_i (i.e., $\varsigma_i = \sigma_i$).

Recall from Definition 2.1 that in our approach each state of an MDP is labeled with a subset of a set AP of atomic propositions. Thus, each infinite path of such an MDP produces a trace which is an infinite word over the alphabet 2^{AP}.

Definition 4.1 (Trace of a path). *Given an MDP and an infinite path* $\pi = s_0 \xrightarrow{\alpha_1} s_1 \xrightarrow{\alpha_2} s_2 \xrightarrow{\alpha_3} \cdots$, *we define the infinite word*

$$\mathsf{trace}(\pi) \stackrel{\text{def}}{=} L(s_0)L(s_1)L(s_2)\ldots \in \left(2^{\mathsf{AP}}\right)^\omega$$

to be the trace of π.

Definition 4.2 (Linear-time property). *A linear-time property (LT property) over a given finite set* AP *of atomic propositions is a subset of* $(2^{\mathsf{AP}})^\omega$.

Given a state-labeled system, we say that a path π *satisfies* a given LT property E, if and only if $\mathsf{trace}(\pi) \in \mathsf{E}$. LT properties are used to specify the infinite behavior of a given system. For instance, the reachability property $\Diamond\,\mathsf{get\_car}$ is formally defined as the language

$$\mathsf{E} = \left\{\varsigma \in \left(2^{\mathsf{AP}}\right)^\omega \mid \exists i : \mathsf{get\_car} \in \varsigma_i\right\}.$$

There are various formalisms (logics, automata, algebraic expressions) to specify LT properties. A very prominent logical formalism in model checking is the linear temporal logic (LTL) [78]. In this chapter, however, we will utilize ω-automata to specify LT properties. While finite automata serve as acceptors for languages of finite words, the semantics of an ω-automaton refers to a language over infinite words. The syntax and operational semantics of ω-automata is roughly the same as for finite automata, except that the input of an ω-automata is an infinite word and the acceptance condition imposes constraints on infinite runs rather than on finite ones. As we assume that the reader of this handbook is already familiar with the concept of ω-automata, we just provide the definitions that are relevant for our purposes. For more information on ω-automata, we refer to, e.g., [85, 40].

Definition 4.3 (Deterministic ω-automaton). *We define a deterministic ω-automaton as a tuple*

$$\mathcal{A} = (Q, \Sigma, \delta_{\mathcal{A}}, q_0, \mathsf{Acc}),$$

where:

- Q *is a finite non-empty set of states.*
- Σ *is a finite non-empty input alphabet.*
- $\delta_{\mathcal{A}} : Q \times \Sigma \to Q$ *is a transition function.*
- $q_0 \in Q$ *is the initial state.*
- Acc *is an acceptance condition.*

In this chapter, we only consider acceptance conditions of the form

$$\mathsf{Acc} = \{(H_1, K_1), \ldots, (H_n, K_n)\}$$

and interpret them with a Rabin, resp. Streett semantics. Given a subset $T \subseteq Q$ of states, we call T:

- *Rabin-accepting*, if there exists $1 \leq i \leq n$ such that

$$T \cap H_i = \emptyset \quad \text{and} \quad T \cap K_i \neq \emptyset.$$

- *Streett-accepting*, if for every $1 \leq i \leq n$

$$T \cap H_i \neq \emptyset \quad \text{or} \quad T \cap K_i = \emptyset.$$

Thus, Rabin and Streett acceptance are complementary to each other.

Given a deterministic ω-automaton $\mathcal{A}$ with an acceptance condition $\mathsf{Acc} = \{(H_1, K_1), \ldots, (H_n, K_n)\}$ and an infinite word $\varsigma = \sigma_1\sigma_2\sigma_3\ldots$ over Σ, we define the *run* for ς in $\mathcal{A}$ to be the infinite state sequence $\rho = p_0, p_1, \ldots$ such that $p_0 = q_0$ and $p_i = \delta_{\mathcal{A}}(p_{i-1}, \sigma_i)$ for every $i \geq 1$. Given a run ρ, let

$$\mathsf{inf}(\rho) = \{p \in Q \mid \overset{\infty}{\exists} i \in \mathbb{N}_{\geq 0} : p_i = p\}$$

denote the set of states that occur infinitely often in ρ. We call a run ρ *Rabin-, resp. Streett-accepting*, if and only if $\mathsf{inf}(\rho)$ is Rabin-, resp. Streett-accepting. The *Z-accepted language* of $\mathcal{A}$ is defined as

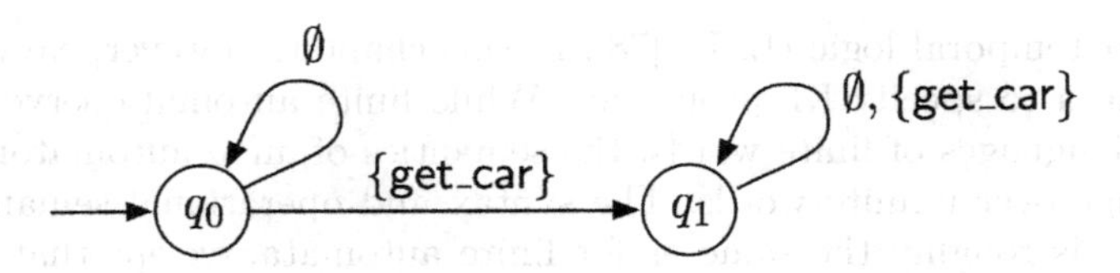

Fig. 6. DSA for $\Diamond$ get_car

$$\mathcal{L}_Z(\mathcal{A}) \stackrel{\text{def}}{=} \{\varsigma \in \Sigma^\omega \mid \text{the run for } \varsigma \text{ in } \mathcal{A} \text{ is Z-accepting}\}$$

for $Z \in \{Rabin, Streett\}$.

In order to argue more easily about the accepted language, we call the combination of a deterministic ω-automaton and Rabin acceptance (resp., Streett acceptance) a *deterministic Rabin automaton* (DRA) (resp., *deterministic Streett automaton* (DSA)). In the remainder of this section, we will drop the Z from Z-accepting, Z-accepted language, and $\mathcal{L}_Z(\mathcal{A})$, if it is clear from the context.

To use deterministic Rabin, resp. Streett automata as a formalism for representing LT properties, we will consider automata with the alphabet $\Sigma = 2^{\mathsf{AP}}$, where AP is the set of atomic propositions of the given system. For example, the reachability property

$$\mathsf{E} = \Diamond\ \mathsf{get\_car} = \left\{\varsigma \in \left(2^{\mathsf{AP}}\right)^\omega \mid \exists i : \mathsf{get\_car} \in \varsigma_i\right\}$$

over the set AP = {get_car} can be represented by the deterministic Streett automaton $\mathcal{A}$ shown in Fig. 6 where the acceptance condition is $\mathsf{Acc} = \{(\{q_1\}, \{q_0, q_1\})\}$. It is easy to see that $\mathcal{L}(\mathcal{A}) = \mathsf{E}$.

It is well known [85, 40] that the class of languages that are definable by deterministic Rabin (or Streett) automata coincides with the class of ω-regular languages and the class of MSO definable languages over infinite words. Thus, DRA and DSA are powerful enough to express many interesting specifications that arise in real-world scenarios. This includes simple temporal properties like "eventually" (reachability properties), "always" (safety properties), and liveness properties that result by combination of "eventually" and "always", such as "infinitely often" or "continuously from some moment on". But also more complex properties such as "each process will eventually enter its critical section" or "between two eating phases of a dining philosopher there is always a thinking phase" can be specified by DRA and DSA.

Natural requirements for MDPs and other types of probabilistic systems attach lower or upper probability bounds on such LT properties. That is, the typical task is to verify conditions such as "the probability that a waiting process is never allowed to enter its critical section is less than 0.005" for a randomized mutual exclusion protocol. But also qualitative properties play a crucial role where the goal is to establish conditions such as "with probability 1 the repeated attempt to deliver a message will eventually be successful" for a communication protocol or "with probability 1 a leader will eventually be elected" for a randomized leader election protocol.

In the remainder of this section, we explain the main steps for a quantitative analysis of an MDP $\mathcal{M}$ against an ω-regular LT property, specified by a DRA or DSA $\mathcal{A}$ (note that such properties are measurable). This requires us to compute the maximal or minimal probabilities

$$\mathsf{Pr}_s^{\max}(\mathcal{A}) \stackrel{\text{def}}{=} \sup_{\mathcal{U}\in\mathsf{Sched}} \mathsf{Pr}_s^{\mathcal{U}}(\mathcal{L}(\mathcal{A})) \quad \text{and} \quad \mathsf{Pr}_s^{\min}(\mathcal{A}) \stackrel{\text{def}}{=} \inf_{\mathcal{U}\in\mathsf{Sched}} \mathsf{Pr}_s^{\mathcal{U}}(\mathcal{L}(\mathcal{A}))$$

for the paths π of $\mathcal{M}$ that start in s and where π satisfies $\mathcal{L}(\mathcal{A})$, i.e., $\mathsf{trace}(\pi)$ belongs to $\mathcal{L}(\mathcal{A})$. As for the quantitative reachability analysis, the supremum and infimum are taken over all schedulers of the given MDP. In fact, there exist schedulers that maximize or minimize the probabilities for paths with a trace in $\mathcal{L}(\mathcal{A})$. Again, the supremum and infimum can be replaced with maximum and minimum. As in Sect. 3, we will focus here on explanations about the computation of maximal probabilities, i.e., the computation of the values $\mathsf{Pr}_s^{\max}(\mathcal{A})$. Analogous techniques are applicable to compute $\mathsf{Pr}_s^{\min}(\mathcal{A})$. However, since the class of ω-regular languages is closed under complementation, algorithms to compute $\mathsf{Pr}_s^{\max}(\mathcal{A})$ are even sufficient to reason about minimal probabilities. Given a deterministic ω-automaton $\mathcal{A}$ that specifies the desired behaviors, we may switch to a deterministic ω-automaton $\mathcal{B}$ for the complement language $\overline{\mathcal{L}(\mathcal{A})}$, i.e., $\mathcal{B}$ represents the undesired behaviors. (Here, we can exploit the duality of Rabin and Streett acceptance. That is, if $\mathcal{A}$ is a DSA, then we can use $\mathcal{B} = \mathcal{A}$, but treat $\mathcal{B}$ as a DRA, and vice versa.) We then apply the techniques for computing the maximal probabilities $\mathsf{Pr}_s^{\max}(\mathcal{B})$ for the "bad" event specified by $\mathcal{B}$. The greatest lower bound for the probabilities that can be guaranteed for the good behaviors is then obtained by the following equation:

$$\mathsf{Pr}_s^{\min}(\mathcal{A}) = 1 - \mathsf{Pr}_s^{\max}(\mathcal{B}).$$

The key for the quantitative analysis of MDPs against ω-regular properties lies in the concept of de Alfaro's end components [27, 28]. They can be seen as the MDP counterpart to terminal strongly connected components in Markov chains. Intuitively, an end component of an MDP is a non-empty strongly connected sub-MDP, that means an end component consists of a non-empty state set $T \subseteq S$ and a non-empty action set $A(t)$ for each state $t \in T$ such that, once T is entered and only actions in $A(t)$ are chosen, the set T will not be left and any state of T can be reached from any other state in T.

Definition 4.4 (End components). *Let $\mathcal{M} = (S, \mathsf{Act}, \delta, \mu, \mathsf{AP}, L)$ be an MDP. Then an end component of $\mathcal{M}$ is a pair (T, A) where $\emptyset \neq T \subseteq S$ and $A : T \to 2^{\mathsf{Act}}$ is a function such that the following three conditions are satisfied:*

- *$\emptyset \neq A(s) \subseteq \mathsf{Act}(s)$ for each state $s \in T$.*
- *If $s \in T$, $t \in S$ and $\alpha \in A(s)$ such that $\delta(s, \alpha, t) > 0$ then $t \in T$.*

- *The underlying directed graph* $(T, \rightarrow_A)$ *of* (T, A) *is strongly connected.*

Here, $\rightarrow_A$ *denotes the edge-relation induced by* A*, that is* $s \rightarrow_A t$ *if and only if* $\delta(s, \alpha, t) > 0$ *for some action* $\alpha \in A(s)$*.*

The importance of end components is due to the following observation about the limit behavior of paths that has been established by de Alfaro [27, 28]. Given an infinite path $\pi = s_0 \xrightarrow{\alpha_1} s_1 \xrightarrow{\alpha_2} s_2 \xrightarrow{\alpha_3} \cdots$ we denote by $\mathsf{Lim}(\pi)$ the pair (T, A) where $T = \mathsf{inf}(\pi)$ is the set of states in π that are visited infinitely often and $A : T \rightarrow 2^{\mathsf{Act}}$ is the function that assigns to any state $t \in T$ the set of actions $\alpha \in \mathsf{Act}$ such that $(s_i = t) \wedge (\alpha_{i+1} = \alpha)$ for infinitely many indices i. Given an MDP $\mathcal{M}$ and a (possibly history-dependent and randomized) scheduler $\mathcal{U}$, it holds that in the process induced by $\mathcal{U}$, almost all paths of $\mathcal{M}$ "end" in an end component, that is, their limit $\mathsf{Lim}(.)$ forms an end component.

Lemma 4.5 (Limiting behavior of MDPs and end components). *For any MDP* $\mathcal{M}$ *and scheduler* $\mathcal{U}$*,*

$$\mathsf{Pr}^{\mathcal{M},\mathcal{U}}\left(\{\pi \in \mathsf{Path}^{\mathcal{M}}_{\mathsf{inf}} \mid \mathsf{Lim}(\pi) \textit{ is an end component}\}\right) = 1.$$

This fundamental property of MDPs is one of the main features for computing $\mathsf{Pr}^{\max}_s(\mathcal{A})$ for a given DRA, resp. DSA $\mathcal{A}$ via a reduction to the problem of maximal reachability probabilities. Another feature is the product construction of the MDP and the automaton $\mathcal{A}$.

Definition 4.6 (Product-MDP). *As before, let* $\mathcal{M} = (S, \mathsf{Act}, \delta, \mu, \mathsf{AP}, L)$ *be an MDP and let* $\mathcal{A} = (Q, 2^{\mathsf{AP}}, \delta_{\mathcal{A}}, q_0, \mathsf{Acc})$ *be a DSA or DRA. The product of* $\mathcal{M}$ *and* $\mathcal{A}$ *is defined as the MDP*

$$\mathcal{M} \otimes \mathcal{A} = (S \times Q, \mathsf{Act}', \delta', \mu', Q, L')$$

where the transition function δ'*, the initial distribution* μ'*, and the labeling function* L' *are defined as follows:*

- $\mathsf{Act}' \stackrel{\text{def}}{=} \mathsf{Act}$
- $\delta'(\langle s, q\rangle, \alpha, \langle s', q'\rangle) \stackrel{\text{def}}{=} \begin{cases} \delta(s, \alpha, s') & \textit{if } q' = \delta_{\mathcal{A}}(q, L(s')) \\ 0 & \textit{otherwise} \end{cases}$
- $\mu'(\langle s, q\rangle) \stackrel{\text{def}}{=} \begin{cases} \mu(s) & \textit{if } q = \delta_{\mathcal{A}}(q_0, L(s)) \\ 0 & \textit{otherwise} \end{cases}$
- $L'(\langle s, q\rangle) \stackrel{\text{def}}{=} \{q\}$

Note that this construction requires a deterministic ω-automaton $\mathcal{A}$, as for a non-deterministic ω-automaton there is no straightforward way to define appropriate transition probabilities for the product. We may observe a one-to-one correspondence between the path

$$\pi = s_0 \xrightarrow{\alpha_1} s_1 \xrightarrow{\alpha_2} s_2 \xrightarrow{\alpha_3} \cdots$$

in the MDP $\mathcal{M}$ and the path

$$\pi^+ = \langle s_0, q_1 \rangle \xrightarrow{\alpha_1} \langle s_1, q_2 \rangle \xrightarrow{\alpha_2} \langle s_2, q_3 \rangle \xrightarrow{\alpha_3} \cdots$$

in $\mathcal{M} \otimes \mathcal{A}$ that starts in state $\langle s_0, q_1 \rangle$ where $q_1 = \delta_{\mathcal{A}}(q_0, L(s_0))$. Given a path π^+ in $\mathcal{M} \otimes \mathcal{A}$, the corresponding path in $\mathcal{M}$ is simply obtained by omitting all automata states q_i. Vice versa, given a path π as above, the corresponding path π^+ is obtained by adding the automaton states $q_{i+1} = \delta_{\mathcal{A}}(q_i, L(s_i))$ to π. Thus, π^+ emanates from π and the unique run for $\mathsf{trace}(\pi)$ in $\mathcal{A}$. Recall that the acceptance of a run imposes a condition on the set of states in the automaton that appear infinitely often in that run. Hence, whether or not $\mathsf{trace}(\pi)$ belongs to $\mathcal{L}(\mathcal{A})$ depends on the projection of $\mathsf{inf}(\pi^+)$ to the states in $\mathcal{A}$, i.e., the set $\{q \in Q \mid \exists s \in S : \langle s, q \rangle \in \mathsf{inf}(\pi^+)\}$. Since almost all paths in $\mathcal{M} \otimes \mathcal{A}$ constitute an end component (by Lemma 4.5), the algorithm to compute $\mathsf{Pr}_s^{\mathsf{max}}(\mathcal{A})$ relies on an analysis of the end components of the product where the acceptance condition of $\mathcal{A}$ holds.

Definition 4.7 (Accepting end components). *Given an MDP* $\mathcal{M} = (S, \mathsf{Act}, \delta, \mu, \mathsf{AP}, L)$ *and a DRA, resp. a DSA* $\mathcal{A} = (Q, 2^{\mathsf{AP}}, \delta_{\mathcal{A}}, q_0, \mathsf{Acc})$, *we call an end component* (T, A) *of* $\mathcal{M} \otimes \mathcal{A}$ *accepting if and only if the set*

$$T|_Q \stackrel{\text{def}}{=} \{q \in Q \mid \exists s \in S : \langle s, q \rangle \in T\}$$

is accepting in $\mathcal{A}$.

In the sequel, let AEC be the union of (the state-component of) all accepting end components in the product-MDP $\mathcal{M} \otimes \mathcal{A}$. That is, AEC consists of all states $\langle s, q \rangle \in S \times Q$ such that $\langle s, q \rangle \in T$ for at least one accepting end component (T, A) of $\mathcal{M} \otimes \mathcal{A}$. The probability for the paths in $\mathcal{M}$ under some scheduler $\mathcal{U}$ that have their trace in $\mathcal{L}(\mathcal{A})$ agrees with the probability for the paths in $\mathcal{M} \otimes \mathcal{A}$ whose limits yield an accepting end component under the corresponding scheduler $\mathcal{U}^+$ for the product. We use here the fact that each scheduler $\mathcal{U}$ for $\mathcal{M}$ can be mimicked by a scheduler $\mathcal{U}^+$ for $\mathcal{M} \otimes \mathcal{A}$. Formally, $\mathcal{U}^+$'s decision for a finite path π^+ agrees with $\mathcal{U}$'s decision for the path that results from π^+ by dropping the automaton component from the states. The maximal probabilities in $\mathcal{M} \otimes \mathcal{A}$ for paths π^+ where $\mathsf{Lim}(\pi^+)$ is an accepting end component agree with the maximal probability in $\mathcal{M} \otimes \mathcal{A}$ for reaching eventually the set AEC. Note that there is a scheduler that ensures that once AEC has been reached, the set AEC will never be left and almost surely the limiting behavior will constitute an accepting end component. This is the key property to provide a formal proof for the following theorem [27, 12].

Theorem 4.8 (Maximal probabilities for ω-regular properties). *Let* $\mathcal{M}$, $\mathcal{A}$, *and* AEC *be as above. Then for each state* $s \in S$:

$$\mathsf{Pr}_s^{\mathsf{max}}(\mathcal{A}) = \mathsf{Pr}^{\mathsf{max}}_{\langle s, \delta_{\mathcal{A}}(q_0, L(s)) \rangle}(\Diamond\, AEC).$$

On the basis of this theorem, an algorithm to compute $\mathsf{Pr}_s^{\mathsf{max}}(\mathcal{A})$ can proceed as follows. First, construct the product-MDP $\mathcal{M} \otimes \mathcal{A}$, then compute the accepting end components and finally apply the techniques explained in Sect. 4 to compute the maximal reachability probabilities for the target set AEC in the product. What remains is to explain the computation of the set AEC. The notion of an accepting end component just depends on the topological graph structure of the MDP $\mathcal{M} \otimes \mathcal{A}$ (the precise transition probabilities are irrelevant). Therefore, purely graph-based methods are sufficient to compute the set AEC. As the number of end components of an MDP $\mathcal{M}$ can grow exponentially in the size of $\mathcal{M}$, the naive approach that relies on an explicit computation of all accepting end components is hopelessly inefficient. Instead, one can use the concept of maximal end components to increase the efficiency. An end component (T, A) of an MDP $\mathcal{M}'$ is called *maximal* if there is no end component (T', A') of $\mathcal{M}'$ such that (i) $T \subseteq T'$, (ii) $A(t) \subseteq A'(t)$ for every $t \in T$, and (iii) $(T, A) \neq (T', A')$. Obviously, the state sets of maximal end components are pairwise disjoint and each end component is contained in some maximal end component. In particular, the total number of maximal end components is bounded by the number of states in $\mathcal{M}'$. We will explain first how the set of all maximal end components, which is denoted by MEC, can be computed in polynomial time.

Computing the Set of Maximal end Components

Here, we address the problem of computing the set MEC of all maximal end components in the product-MDP $\mathcal{M} \otimes \mathcal{A}$. Recall that the states contained in an end component (T, A) are strongly connected in the underlying directed graph $(T, \rightarrow_A)$ (see Definition 4.4). Furthermore, it is clear that a maximal end component in $\mathcal{M} \otimes \mathcal{A}$ is contained in some strongly connected component (SCC) of $(S \times Q, \rightarrow_{\mathcal{M} \otimes \mathcal{A}})$, the underlying directed graph of $\mathcal{M} \otimes \mathcal{A}$. Thus, an algorithmic scheme for calculating the set MEC of all maximal end components consists of the following steps [23, 27]:

1. In a first step, we compute a candidate set $\mathcal{C}$ of all possible maximal end components as follows. Compute the SCCs and define
$$\mathcal{C} = \{(T, A) \mid T \text{ is an SCC and } A(t) = \mathsf{Act}(t) \text{ for } t \in T\}.$$
2. Pick and remove a candidate (T, A) from $\mathcal{C}$.
3. For each state $t \in T$, remove all actions from $A(t)$ that violate the second condition of Definition 4.4, i.e., remove all $\alpha \in A(t)$ with $\sum_{u \in T}(t, \alpha, u) < 1$ and call that modified candidate (T, A').
4. Calculate all SCCs in the underlying directed graph $(T, \rightarrow_{A'})$ of (T, A') and insert them as new candidates into $\mathcal{C}$ (similar as in step 1, but with action sets restricted to A').
5. Repeat steps 2–4 until $\mathcal{C}$ reaches a fixpoint.

When the fixpoint is reached, the set $\mathcal{C}$ equals the set MEC. Since during each iteration for every candidate $(T, A) \in \mathcal{C}$ either there exist actions α in $A(t)$ that are removed, and thus causing a potential splitting of this candidate, or a fixpoint is reached for $\mathcal{C}$; the procedure obviously terminates. Furthermore, the number of iterations is bounded by the total number of pairs (t, α) where t is a state in $\mathcal{M} \otimes \mathcal{A}$ and α an action that is enabled in t. The costs per iteration are dominated by the calculation of the SCCs in steps 1 and 4. Thus, the complexity of the algorithm is polynomially bounded in the size of $\mathcal{M} \otimes \mathcal{A}$ (i.e., the total number of states and transitions in $\mathcal{M} \otimes \mathcal{A}$).

Remark 4.9. The algorithm to compute the maximal end components can also be used to increase the efficiency of the algorithms for computing the extremal probabilities for a reachability property $\Diamond B$. This is due to the fact that whenever s and t are states that belong to the same maximal end component, then $\mathsf{Pr}_s^{\mathsf{max}}(\Diamond B) = \mathsf{Pr}_t^{\mathsf{max}}(\Diamond B)$. Hence, the given MDP can be simplified by collapsing all states that belong to the same maximal end component into a single state. This reduces the number of variables in the equation system or linear program and can therefore lead to a major speed-up [5].

With the above algorithm for the computation of MEC, we can now explain how the set AEC can be computed for the product-MDP of a given MDP $\mathcal{M}$ and a given DRA, resp. DSA $\mathcal{A}$.

Case 1: $\mathcal{A}$ Is a Deterministic Rabin Automaton

Let $\mathsf{Acc} = \{(H_1, K_1), \ldots, (H_n, K_n)\}$ be the acceptance condition of $\mathcal{A}$ and let

$$H_i' \stackrel{\text{def}}{=} \{\langle s, q\rangle \mid s \in S,\ q \in H_i\} \quad \text{and}$$
$$K_i' \stackrel{\text{def}}{=} \{\langle s, q\rangle \mid s \in S,\ q \in K_i\}.$$

Thus, an end component (T, A) is accepting, if there is an index $1 \leq i \leq n$ such that

$$T \cap H_i' = \emptyset \quad \text{and} \quad T \cap K_i' \neq \emptyset.$$

Assume that (T, A) is accepting with respect to (H_i', K_i'). Let $\mathcal{M}_i'$ be the MDP that results from $\mathcal{M} \otimes \mathcal{A}$ by removing all states in H_i'. Then (T, A) is obviously an end component in $\mathcal{M}_i'$, and moreover the maximal end component of $\mathcal{M}_i'$ that contains (T, A) is also accepting with respect to (H_i', K_i'). Hence, the set AEC arises as the union of the sets AEC_i for $1 \leq i \leq n$ where AEC_i is the union of (the state-components of) all maximal end components (T, A) in $\mathcal{M}_i'$ where $T \cap K_i' \neq \emptyset$.

Case 2: $\mathcal{A}$ Is a Deterministic Streett Automaton

In this case, an efficient way to realize the quantitative analysis on the basis of Theorem 4.8 is obtained using the following lemma.

Lemma 4.10. *Given an MDP $\mathcal{M}'$ with state space S', a target set of states $B \subseteq S'$, and a set X with $B \subseteq X \subseteq \{t \in S' \mid \mathsf{Pr}_t^{\max}(\Diamond B) = 1\}$, then for every state $s \in \mathcal{M}'$:*

$$\mathsf{Pr}_s^{\max}(\Diamond B) = \mathsf{Pr}_s^{\max}(\Diamond X).$$

Let $\mathcal{M}' = \mathcal{M} \otimes \mathcal{A}$. We denote by *AMEC* the union of all maximal end components that have an accepting sub-component, i.e.,

$$AMEC \stackrel{\text{def}}{=} \{t \in S \times Q \mid \text{there exists } (T, A) \in \mathsf{MEC} \text{ such that } t \in T \text{ and there is some accepting end component } (T', A') \text{ with } T' \subseteq T\}.$$

It should be noticed that the correct term for *AMEC* is not "accepting maximal end components" but "maximal end components that contain at least one accepting sub-end component". We now show that with $B = AEC$ and $X = AMEC$ the condition of Lemma 4.10 holds, i.e.,

$$AEC \subseteq AMEC \subseteq \{t \mid \mathsf{Pr}_t^{\max}(\Diamond AEC) = 1\}.$$

This can be seen as follows. Let $t \in AMEC$ be included in the end component (T, A) that has the accepting sub-component (T', A'). As (T, A) is an end component, there is a scheduler $\mathcal{U}$ which ensures that starting in state t, almost surely each state in T will be visited infinitely often. Since $T' \subseteq T$, the set T' will be visited almost surely. As $T' \subseteq AEC$, we get that $\mathsf{Pr}_t^{\mathcal{U}}(\Diamond AEC) = 1$. Using Lemma 4.10, we thus can reformulate Theorem 4.8 as the following theorem.

Theorem 4.11 (Maximum probability for ω-regular properties, part II). *Let $\mathcal{M}$ be as before, $\mathcal{A}$ a DSA*[2]*, and AEC and AMEC be as above. Then for every state s of $\mathcal{M}$:*

$$\mathsf{Pr}_s^{\max}(\mathcal{A}) = \mathsf{Pr}^{\max}_{\langle s, \delta_{\mathcal{A}}(q_0, L(s))\rangle}(\Diamond AEC) = \mathsf{Pr}^{\max}_{\langle s, \delta_{\mathcal{A}}(q_0, L(s))\rangle}(\Diamond AMEC).$$

This observation allows us to switch from *AEC* to the larger set *AMEC* that arises by the union of certain maximal end components. In the following, we will describe how the set *AMEC* can be computed efficiently for a given MDP $\mathcal{M}$ and a given deterministic Streett automaton $\mathcal{A}$.

Calculating the Set AMEC

To compute the set *AMEC*, it remains to check for each maximal end component $(T, A) \in \mathsf{MEC}$ if it contains an accepting end component with respect to the given Streett acceptance condition Acc. In the sequel, let AMEC be the set of all maximal end components that contain an accepting end component. (Hence, *AMEC* is the set of all states that are contained in some $(T, A) \in$ AMEC.) For simplicity, we suppose that Acc consists of a single acceptance pair, say $\mathsf{Acc} = \{(H, K)\}$. Let

[2] Theorem 4.11 also holds if $\mathcal{A}$ is a DRA.

$$H' \stackrel{\text{def}}{=} \{\langle s, q\rangle \mid s \in S, q \in H\} \quad \text{and}$$
$$K' \stackrel{\text{def}}{=} \{\langle s, q\rangle \mid s \in S, q \in K\}.$$

Assume that a maximal end component (T, A) violates the given Streett acceptance condition. Then $T \cap H' = \emptyset$ and no sub-component can satisfy the acceptance condition by containing an H'-state. Hence, a sub-component can only satisfy the acceptance condition Acc if it does not contain a K'-state. These ideas lead to the following procedure. Consider a maximal end component $(T, A) \in \mathsf{MEC}$.

1. If $T \cap H' \neq \emptyset$ or $T \cap K' = \emptyset$, then $(T, A) \in \mathsf{AMEC}$.
2. Otherwise let $T_1 = T \setminus K'$ and A_1 such that $A_1(t) = \{\alpha \in A(t) \mid \sum_{t \in T_1}(s, \alpha, t) = 1\}$ for each state $t \in T_1$.
3. If an end component can be found in (T_1, A_1), then $(T, A) \in \mathsf{AMEC}$.
4. Otherwise, $(T, A) \notin \mathsf{AMEC}$.

5 Partial Order Reduction

In contrast to the previous sections, where advanced solution techniques for the value iteration have been discussed, we now focus on the state space explosion problem. There exist diverse methods for tackling the state space explosion problem for non-probabilistic as well as probabilistic systems. This includes symbolic model checking methods and various reduction techniques, see, e.g., [21] for an overview. The symbolic methods are mainly based on multi-terminal binary decision diagrams and focus on a compact internal representation of the (full) system [17, 45, 45, 6, 16, 72, 49, 69, 61]. So, these methods do not aim at avoiding the state space explosion, but at using a very compact representation of the given model. For instance, the PCTL-model checkers PRISM [60], ProbVERUS [48] and RAPTURE [56] are based on a symbolic representation of the system to be analyzed. In addition, hybrid approaches that combine the compact model representation of symbolic techniques with the good performance of numerical computations of explicit techniques have been developed [61]. Somewhat orthogonal to this approach are numerous reduction techniques, where the goal is to generate only a reduced sub-system which is "equivalent" (with respect to the properties to be verified) to the original system. Then model checking is applied to the reduced system, yielding the desired answer not only for the reduced system, but also for the original one. A large class of reduction techniques are bi-simulation–minimization techniques [55, 8, 77, 18, 58] that aim to aggregate bi-similar states and to construct an "equivalent" quotient of the original model. For models with non-trivial but interchangeable components, symmetry reduction techniques have been developed that use the inherent internal symmetries to reduce the state space.

Another class of reduction techniques are partial order reduction methods which have been thoroughly studied for non-probabilistic models [74, 52, 87, 38, 39, 76, 75] and have been extended to probabilistic systems in [9, 26, 7, 41, 42]. For partial order reduction, the starting point is usually a description of an asynchronous parallel system by a representation of the sub-systems that run in parallel, e.g., as in the language ProbMeLa that has been outlined in Sect. 2. The rough idea behind partial order reduction is to construct a reduced state graph by abolishing redundancies in the transition system that originate from the interleaving of independent activities that are executed in parallel. For independent actions α and β, the interleaving semantics represents their parallel execution by the nondeterministic choice between the action sequences $\alpha\beta$ and $\beta\alpha$. If $\alpha\beta$ and $\beta\alpha$ have the same effect to the control and program variables, and thus lead to the same state, the investigation of one order ($\alpha\beta$ or $\beta\alpha$) as a representative for both suffices under certain side conditions. More general, instead of constructing the full system $\mathcal{M}$, the goal of partial order reduction is to generate an "equivalent" sub-system $\mathcal{M}_{\mathsf{red}}$ of the full transition system $\mathcal{M}$. Here, "equivalence" is considered with respect to the type of property to be verified. Of course, the algorithmic construction and analysis of $\mathcal{M}_{\mathsf{red}}$ should be more efficient than model checking the full system $\mathcal{M}$. We give a small example to illustrate these ideas. Consider two processes $\mathcal{P}_1$ and $\mathcal{P}_2$ where $\mathcal{P}_1$ increments a variable x (action α) twice and $\mathcal{P}_2$ increments a variable y (action β) twice. Assume that we are only interested in the value of y, that is, each state is labeled with its y value. Then action α does not change the labeling, but action β does. Figure 7 shows the two processes and their parallel execution, where the shade of a state node represents its y value (the greater y is, the darker the node is). Now assume that we want to check whether the property

"The value of y never decreases."

holds on any path. For the system $\mathcal{P}_1|||\mathcal{P}_2$ of the parallel execution of $\mathcal{P}_1$ and $\mathcal{P}_2$ in Fig. 7, this means

"The shades of the nodes never get lighter."

along any path. Obviously, each path of the system satisfies this property. Now this property has a remarkable feature. In order to decide whether a path satisfies the property or not, it is only relevant what changes of the labeling occur along the path, but not how often a certain labeling is repeated before it changes. The property is so-called *stutter invariant*. It cannot distinguish between two paths that follow the same pattern of changes in the labeling (but may differ in the number of repetitions of a certain labeling). Such two paths are called *stutter equivalent*. Now consider the reduced system $(\mathcal{P}_1|||\mathcal{P}_2)_{\mathsf{red}}$ in Fig. 7. As any path of $\mathcal{P}_1|||\mathcal{P}_2$ has a stutter equivalent path in $(\mathcal{P}_1|||\mathcal{P}_2)_{\mathsf{red}}$ and the property under consideration cannot distinguish between such paths, it is sufficient to check whether all paths of the reduced system satisfy the

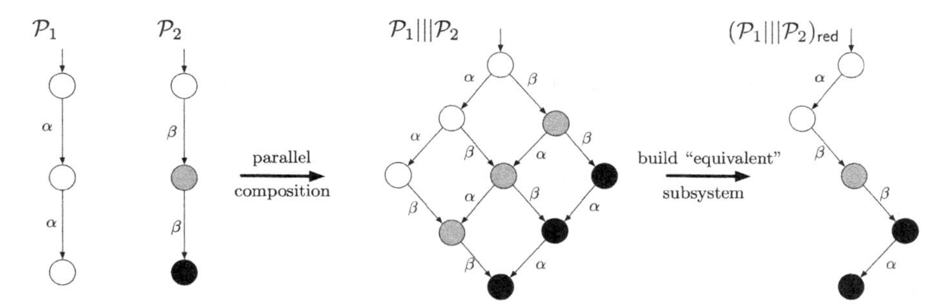

Fig. 7. The idea of partial order reduction

property. If all paths of the reduced system satisfy the property, so do all paths of the original system (and vice versa as the reduced system is a subsystem of the original one). Thus, the reduced system is "equivalent" to the original system with respect to the property.

The goal of partial order reduction is to give criteria with which an "equivalent" reduced system can be generated. These criteria heavily depend on the class of properties that one wants to preserve (e.g., LT properties, branching time properties) and on the kind of model. In the early 1990s, several partial order reduction techniques have been developed for non-probabilistic systems [86, 74, 87, 52, 38, 39, 75, 88, 76]. In the last few years, one instance of partial order reduction techniques, the so-called *ample set method* has been generalized to the probabilistic setting [9, 26, 7, 41, 42].

The interested reader might want to compare our notion of independent actions with the independence relation used in Sect. 2.1 on weighted distributed systems in Chap. 10 in this handbook [36].

The Ample Set Method for MDPs and LT Properties

The rough idea of the ample set method is to assign to any reachable state s of an MDP $\mathcal{M}$ an action-set $\mathsf{ample}(s) \subseteq \mathsf{Act}(s)$ and to construct a reduced system $\mathcal{M}_{\mathsf{red}}$ that results by using the action-sets $\mathsf{ample}(s)$ instead of $\mathsf{Act}(s)$. That is, starting from the initial states of $\mathcal{M}$, one builds up $\mathcal{M}_{\mathsf{red}}$ by only applying ample transitions. The reduced system should be equivalent to the original system in the desired sense, e.g., simulation equivalent or bisimulation equivalent, etc. Depending on the desired equivalence, the defined ample sets have to fulfill certain conditions to ensure the equivalence. These equivalences typically identify those paths whose traces (i.e., words obtained from the paths by projection on the state labels) agree up to stuttering. In this context, stuttering refers to the repetition of the same state labels.

Definition 5.1 (Stutter equivalence). *Two infinite words ς_1 and ς_2 over the alphabet Σ are called stutter equivalent, denoted by*

$$\varsigma_1 \equiv_{\mathrm{st}} \varsigma_2,$$

if and only if there is an infinite word $a_1 a_2 \ldots$ *over the alphabet* Σ *such that*

$$\varsigma_1 = a_1^{k_1} a_2^{k_2} \ldots \quad \textit{and} \quad \varsigma_2 = a_1^{n_1} a_2^{n_2} \ldots,$$

where $k_i, n_i \in \mathbb{N}_{\geq 1}$. *Two infinite paths* π_1 *and* π_2 *in a state-labeled MDP are called stutter equivalent, denoted by* $\pi_1 \equiv_{\mathrm{st}} \pi_2$, *if and only if their traces* $\mathsf{trace}(\pi_1)$ *and* $\mathsf{trace}(\pi_2)$ *over* 2^{AP} *are stutter equivalent.*

An LT property over AP is called stutter invariant, if it cannot distinguish between stutter equivalent paths.

Definition 5.2 (Stutter invariant LT properties). *An LT property* E *over* AP *is called stutter invariant if for all stutter equivalent words* $\varsigma_1, \varsigma_2 \in (2^{\mathsf{AP}})^\omega$ *we have that*

$$\varsigma_1 \in \mathsf{E} \quad \textit{if and only if} \quad \varsigma_2 \in \mathsf{E}.$$

We call two MDPs stutter equivalent if for each scheduler of one of the MDPs there exists a scheduler of the other MDP such that the schedulers yield the same probabilities for any stutter invariant LT property.

Definition 5.3 (Stutter equivalence for MDPs). *Given two MDPs* $\mathcal{M}_i = (S_i, \mathsf{Act}_i, \delta_i, \mu_i, \mathsf{AP}_i, L_i)$, *with* $i = 1, 2$, *we call* $\mathcal{M}_1$ *and* $\mathcal{M}_2$ *stutter equivalent, denoted by*

$$\mathcal{M}_1 \equiv_{\mathrm{st}} \mathcal{M}_2,$$

if and only if for each scheduler $\mathcal{U}_1$ *of* $\mathcal{M}_1$ *there exists a scheduler* $\mathcal{U}_2$ *of* $\mathcal{M}_2$ *such that,*

$$\mathsf{Pr}^{\mathcal{M}_1, \mathcal{U}_1}(\mathsf{E}) = \mathsf{Pr}^{\mathcal{M}_2, \mathcal{U}_2}(\mathsf{E})$$

for each stutter invariant measurable LT property $\mathsf{E} \subseteq (2^{\mathsf{AP}})^\omega$, *and vice versa.*

Before explaining the ample set method, we briefly illustrate its impact on probabilistic linear-time model checking. Assume that we are given two stutter equivalent MDPs $\mathcal{M}_1, \mathcal{M}_2$ and a stutter invariant measurable LT property E. Then

$$\sup_{\mathcal{U} \in \mathsf{Sched}^{\mathcal{M}_1}} \mathsf{Pr}^{\mathcal{M}_1, \mathcal{U}}(\mathsf{E}) = \sup_{\mathcal{U} \in \mathsf{Sched}^{\mathcal{M}_2}} \mathsf{Pr}^{\mathcal{M}_2, \mathcal{U}}(\mathsf{E}).$$

The corresponding equality with inf instead of sup certainly also holds. Hence, two stutter equivalent MDPs $\mathcal{M}_1$ and $\mathcal{M}_2$ are equivalent with respect to stutter invariant measurable linear-time specifications. A prominent class of stutter invariant measurable LT properties is the LTL fragment that does not use the "NextStep"-operator [90].

The following result (Theorem 5.8 below) has been established in [9]. It requires ample sets $\mathsf{ample}(s)$ for $s \in S$ that enjoy the properties (A0)–(A4) shown in Fig. 9 (these will be explained later) and asserts the stutter equivalence of the original MDP $\mathcal{M}$ and the reduced MDP $\mathcal{M}_{\mathsf{red}}$ that arises from $\mathcal{M}$ by removing all enabled actions of a state s that are not included in the ample set of s. The precise definition of $\mathcal{M}_{\mathsf{red}}$ is as follows.

Definition 5.4 (The reduced MDP). *Let $\mathcal{M} = (S, \mathsf{Act}, \delta, \mu, \mathsf{AP}, L)$ be an MDP, and suppose that for each state $s \in S$, $\mathsf{ample}(s)$ is a non-empty subset of $\mathsf{Act}(s)$. Then the reduced MDP $\mathcal{M}_{\mathsf{red}}$ is the MDP $(S_{\mathsf{red}}, \mathsf{Act}, \delta_{\mathsf{red}}, \mu, \mathsf{AP}, L_{\mathsf{red}})$ where the state space of $\mathcal{M}_{\mathsf{red}}$ is the smallest sub-set S_{red} of S such that:*

- $\{s \in S \mid \mu(s) > 0\} \subseteq S_{\mathsf{red}}$.
- *Whenever $s \in S_{\mathsf{red}}$ and $\alpha \in \mathsf{ample}(s)$ then $\{s' \in S \mid \delta(s, \alpha, s') > 0\} \subseteq S_{\mathsf{red}}$.*

The transition probability function $\delta_{\mathsf{red}} : S_{\mathsf{red}} \times \mathsf{Act} \times S_{\mathsf{red}} \to [0, 1]$ is given by

$$\delta_{\mathsf{red}}(s, \alpha, s') = \begin{cases} \delta(s, \alpha, s') & \textit{if } \alpha \in \mathsf{ample}(s) \\ 0 & \textit{otherwise.} \end{cases}$$

The labeling function $L_{\mathsf{red}} : S_{\mathsf{red}} \to 2^{\mathsf{AP}}$ is the restriction of $\mathcal{M}$'s labeling function to the state space of $\mathcal{M}_{\mathsf{red}}$, i.e., $L_{\mathsf{red}}(s) = L(s)$ for all $s \in S_{\mathsf{red}}$.

Thus, $\mathcal{M}_{\mathsf{red}}$ can be obtained by an on-the-fly algorithm which first generates all initial states of $\mathcal{M}$ and then successively expands each generated state s by considering all actions $\alpha \in \mathsf{ample}(s)$ and generating the α-successors of s that have not been generated before.

The following Theorem 5.8 [9, 43] ensures that given a deterministic ω-automaton $\mathcal{A}$ that accepts a stutter invariant language, it suffices to model check $\mathcal{M}_{\mathsf{red}}$ against $\mathcal{A}$ instead of $\mathcal{M}$, provided that the ample sets satisfy the conditions (A0)–(A4) in Fig. 9. Before presenting the theorem, we define *stutter actions*, resp. *independent actions* that are used in the condition (A1), resp. (A2). Stutter actions are actions that have no effect on the state labels, no matter in which state they are taken.

Definition 5.5 (Stutter action). *Given an MDP $\mathcal{M} = (S, \mathsf{Act}, \delta, \mu, \mathsf{AP}, L)$, we call an action $\alpha \in \mathsf{Act}$ a stutter action if and only if for all states $s, t \in S$,*

$$\delta(s, \alpha, t) > 0 \quad \textit{implies} \quad L(s) = L(t).$$

The main ingredient of any partial order reduction technique in the probabilistic or non-probabilistic setting is an adequate notion for independence of actions. The rough idea is a formalization of actions belonging to different processes that are executed in parallel and do not affect each other, e.g., as they only refer to local variables and do not require any kind of synchronization. In non-probabilistic systems, independence of two actions α and β means that, for any state s where both α and β are enabled, the execution of α does not affect the enabledness of β (i.e., the α-successor of s has an outgoing β-transition), and vice versa, and in addition the action sequences $\alpha\beta$ and $\beta\alpha$ lead to the same state. In the probabilistic setting, it is additionally required that $\alpha\beta$ and $\beta\alpha$ have the same probabilistic effect.

Definition 5.6 (Independence of actions, cf. [9, 26]). *Two actions α and β with $\alpha \neq \beta$ are called independent in an MDP $\mathcal{M}$ if and only if for each state $s \in S$ with $\{\alpha, \beta\} \subseteq \mathsf{Act}(s)$ it holds that:*

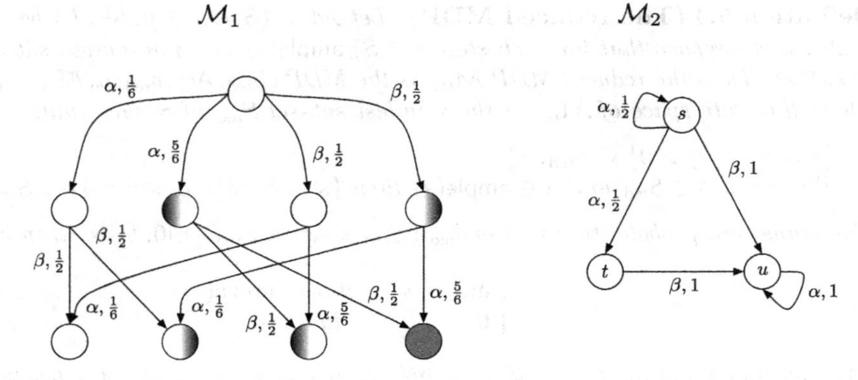

Fig. 8. Examples of independent actions

1. $\delta(s, \alpha, t) > 0$ *implies* $\beta \in \mathsf{Act}(t)$.
2. $\delta(s, \beta, u) > 0$ *implies* $\alpha \in \mathsf{Act}(u)$.
3. For each state $v \in S$:

$$\sum_{t \in S} \delta(s, \alpha, t) \cdot \delta(t, \beta, v) = \sum_{u \in S} \delta(s, \beta, u) \cdot \delta(u, \alpha, v).$$

Two different actions α *and* β *are called dependent if and only if* α *and* β *are not independent. If* $A \subseteq \mathsf{Act}$ *and* $\alpha \in \mathsf{Act} \setminus A$, *then* α *is called independent from* A *if and only if for each action* $\beta \in A$, α *and* β *are independent. Otherwise,* α *is called dependent on* A.

Example 5.7 (Independent actions). Figure 8 shows a fragment of an MDP $\mathcal{M}_1$ representing the parallel execution of independent actions α and β. For example, α might stand for the outcome of the experiment of tossing a "one" with a dice, while β stands for tossing a fair coin. In general, whenever α and β represent stochastic experiments that are independent in the classical sense, then α and β viewed as actions of an MDP are independent. However, there are also other situations where two actions can be independent that do not have a fixed probabilistic branching pattern. For instance, actions α and β in the MDP $\mathcal{M}_2$ in Fig. 8 are independent. To see this, first notice that only in state s both α and β are enabled. The α-successors t, s of s have a β-transition to state u, while the β-successor u has an α-transition to itself. The probabilistic effect under the action sequences $\alpha\beta$ and $\beta\alpha$ is the same as in either case state u is reached with probability 1.

Theorem 5.8 (Ample set method for MDPs). *Let* $\mathcal{M} = (S, \mathsf{Act}, \delta, \mu, \mathsf{AP}, L)$ *be an MDP and* $\mathsf{ample} : S \to 2^{\mathsf{Act}}$ *a function satisfying conditions (A0)–(A4) in Fig. 9. Then*

$$\mathcal{M} \equiv_{\mathsf{st}} \mathcal{M}_{\mathsf{red}},$$

where $\mathcal{M}_{\mathsf{red}}$ *denotes the reduced MDP that emanates from the MDP* $\mathcal{M}$ *and the ample sets defined by the function* ample *according to Definition 5.4.*

(A0) (Non-emptiness condition) For each state $s \in S$, it holds that $\emptyset \neq \mathsf{ample}(s) \subseteq \mathsf{Act}(s)$.

(A1) (Stutter condition) If $s \in S_{\mathsf{red}}$ and $\mathsf{ample}(s) \neq \mathsf{Act}(s)$, then all actions $\alpha \in \mathsf{ample}(s)$ are stutter actions.

(A2) (Dependence condition) For each path $\pi = s \xrightarrow{\alpha_1} \cdots \xrightarrow{\alpha_n} s_n \xrightarrow{\gamma} \cdots$ in $\mathcal{M}$ where $s \in S_{\mathsf{red}}$ and γ is dependent on $\mathsf{ample}(s)$ there exists an index $i \in \{1, \ldots, n\}$ such that $\alpha_i \in \mathsf{ample}(s)$.

(A3) (End component condition) For each end component (T, A) in $\mathcal{M}_{\mathsf{red}}$ we have that: $\alpha \in \bigcap_{t \in T} A(t)$ implies $\alpha \in \bigcup_{t \in T} \mathsf{ample}(t)$.

(A4) (Branching condition) If $\pi = s \xrightarrow{\alpha_1} s_1 \xrightarrow{\alpha_2} \cdots \xrightarrow{\alpha_n} s_n \xrightarrow{\alpha} \cdots$ is a path in $\mathcal{M}$ where $s \in S_{\mathsf{red}}, \alpha_1, \ldots, \alpha_n, \alpha \notin \mathsf{ample}(s)$ and α is probabilistic, then $|\mathsf{ample}(s)| = 1$.

Fig. 9. Conditions for the ample sets of MDPs

We now provide explanations why conditions (A0)–(A4) that have been proposed in [9] ensure the stutter equivalence of $\mathcal{M}$ and $\mathcal{M}_{\mathsf{red}}$. Condition (A0) simply assures that $\mathcal{M}_{\mathsf{red}}$ is a sub-MDP of $\mathcal{M}$ (recall that in Definition 2.1 we required that all states of an MDP are non-terminal). Thus, each scheduler of $\mathcal{M}_{\mathsf{red}}$ is also a scheduler of $\mathcal{M}$. So, the interesting part is the transformation of a given scheduler $\mathcal{U}$ of $\mathcal{M}$ into an "equivalent" scheduler $\mathcal{U}_{\mathsf{red}}$ of $\mathcal{M}_{\mathsf{red}}$ (where "equivalence" is understood with respect to the probabilities of stutter invariant measurable LT properties). The details of the scheduler transformation $\mathcal{U} \to \mathcal{U}_{\mathsf{red}}$ are rather technical and will not be explained here. The main idea is an iterative approach where an infinite sequence $\mathcal{U}_0 = \mathcal{U}, \mathcal{U}_1, \mathcal{U}_2, \ldots$ of schedulers for $\mathcal{M}$ is constructed such that:

- $\mathcal{U}_i, \mathcal{U}_{i+1}, \mathcal{U}_{i+2}, \ldots$ agree on all finite paths of length at most i.
- All finite $\mathcal{U}_i$-paths of length i are paths in $\mathcal{M}_{\mathsf{red}}$.
- $\Pr^{\mathcal{M},\mathcal{U}_i}(E) = \Pr^{\mathcal{M},\mathcal{U}}(E)$ for all stutter invariant measurable LT properties E.

The scheduler $\mathcal{U}_{\mathsf{red}}$ is then defined to be the limit of the schedulers $\mathcal{U}_i$, that is,

$$\mathcal{U}_{\mathsf{red}}(\pi) = \mathcal{U}_{i+1}(\pi)$$

if π is a path of length i in $\mathcal{M}_{\mathsf{red}}$.

The transformations $\mathcal{U}_i \mapsto \mathcal{U}_{i+1}$ all rely on the same schema. For simplicity, we just give a very rough sketch of the idea for the case $i = 0$ and assume that $\mathcal{U}_0 = \mathcal{U}$ is a deterministic scheduler. Suppose we are given a $\mathcal{U}$-path starting in state s that relies on the action sequence $\alpha_1\alpha_2\alpha_3\ldots$ and $\alpha_1 \notin \mathsf{ample}(s)$ (otherwise, $\mathcal{U}_{\mathsf{red}}$ just chooses α_1 with probability 1). If at least one of these actions belongs to $\mathsf{ample}(s)$, then we pick the smallest index i such that $\alpha_i \in \mathsf{ample}(s)$. Note that condition (A1) ensures that α_i is a stutter action as $\alpha_1 \notin \mathsf{ample}(s)$. Condition (A2) ensures that α_i is independent from $\alpha_1, \ldots, \alpha_{i-1}$.

Hence, we can switch from the action sequence $\alpha_1\alpha_2\ldots\alpha_{i-1}\alpha_i$ to the action sequence $\alpha_i\alpha_1\alpha_2\ldots\alpha_{i-1}$. Both action sequences can be executed from state s and yield the same distribution over the states that can be reached afterward. In addition, the action sequences $\alpha_1\alpha_2\ldots\alpha_{i-1}\alpha_i$ and $\alpha_i\alpha_1\alpha_2\ldots\alpha_{i-1}$ produce stutter equivalent paths that end in the same state (recall that α_i is a stutter action). These ideas are sketched in the following picture.

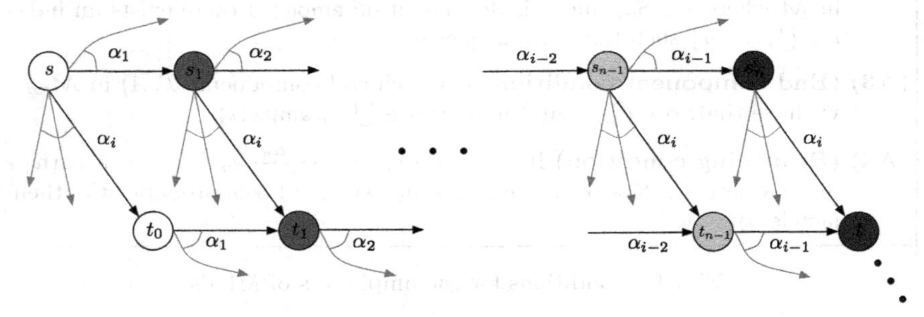

Since $\alpha_i \in \mathsf{ample}(s)$, scheduler $\mathcal{U}_1$ will choose α_i with some positive probability (the probabilities for the chosen actions rely on a rather complex formula that will not be discussed here). If the given $\mathcal{U}$-path does not contain an action in $\mathsf{ample}(s)$ then we pick an arbitrary action $\beta \in \mathsf{ample}(s)$ (this is possible by (A0)) and replace the action sequence $\alpha_0\alpha_1\ldots$ with $\beta\alpha_0\alpha_1\ldots$ (this is possible by (A2) as β is independent from each α_i). The scheduler $\mathcal{U}_1$ will then choose β with some positive probability. Note that this also yields some path that is stutter equivalent to the given $\mathcal{U}$-path. In summary, given a $\mathcal{U}$-path π starting in state s, the basic idea is to permute the first ample action of s that occurs along π to the beginning of the action sequence of π. If no such action exists, an arbitrary ample action of s is pre-pended to the action sequence of π. This step is then repeated ad infinitum to yield a scheduler $\mathcal{U}_{\mathsf{red}}$ of $\mathcal{M}_{\mathsf{red}}$. However, we cannot immediately conclude that $\mathcal{U}$ and $\mathcal{U}_{\mathsf{red}}$ yield the same probabilities for stutter invariant measurable LT properties because the generated $\mathcal{U}_{\mathsf{red}}$-paths might "delay" a certain action of a $\mathcal{U}$-path ad infinity as in the following example.

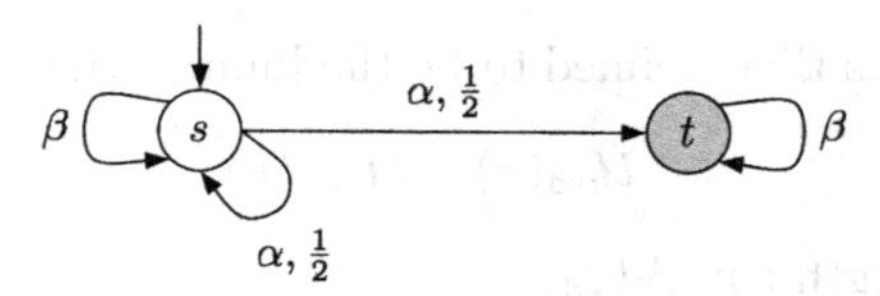

The state labeling is given by the shades of the states, thus β is a stutter action, while α is not. For $\mathsf{ample}(s) = \{\beta\}$ and scheduler $\mathcal{U}$ where $\mathcal{U}(\pi) = \alpha$ for all paths π with $\mathsf{last}(\pi) = s$, the construction sketched above (see [43] for the details) yields

$$\mathcal{U}_i(\underbrace{s \xrightarrow{\beta} s \xrightarrow{\beta} \cdots \xrightarrow{\beta} s}_{\text{length } j}) = \begin{cases} \beta & \text{for } j \leq i-1, \\ \alpha & \text{for } j = i. \end{cases}$$

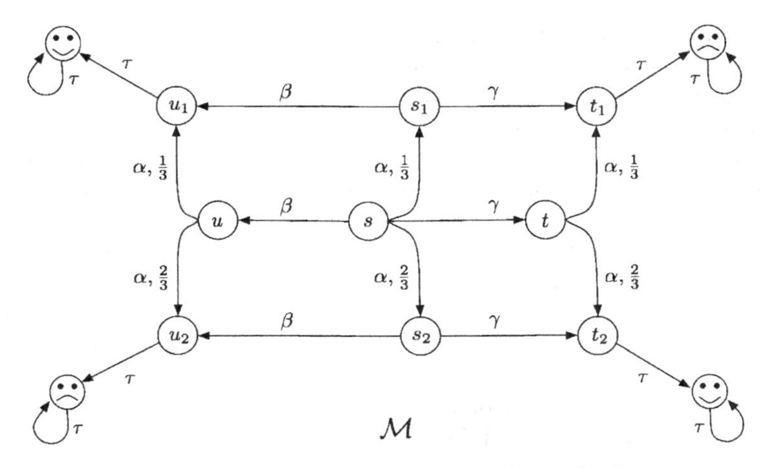

Fig. 10. Example to justify condition (A4)

Thus, scheduler $\mathcal{U}_{\mathsf{red}}$ always schedules β in the state s. In fact, $\mathcal{U}_{\mathsf{red}}$ is the only scheduler for $\mathcal{M}_{\mathsf{red}}$ as $\mathcal{M}_{\mathsf{red}}$ consists only of state s with the β-loop. Under $\mathcal{U}$ and each of the schedulers $\mathcal{U}_i$, we obtain probability 1 to reach the gray state t, while the probability to reach state t under $\mathcal{U}_{\mathsf{red}}$ is 0. However, in this example, conditions (A0), (A1), (A2), and (A4) hold, but the end component $(s, \{\beta\})$ of $\mathcal{M}_{\mathsf{red}}$ violates the end component condition (A3), which ensures that in the scheduler transformation almost surely there is no action of $\mathcal{M}$ that is postponed forever. Note that condition (A3) refers to end components in the reduced MDP $\mathcal{M}_{\mathsf{red}}$ rather than $\mathcal{M}$ (the definition of an end component has been provided in Definition 4.4).

It is worth noting that conditions (A0)–(A3) suffice in the non-probabilistic setting to ensure the equivalence between a transition system and its reduced system with respect to stutter invariant LT properties [74, 75]. However, for MDPs, we need the additional branching condition (A4). The intuitive reason for this is that the experiments

"first toss a coin, then decide between action β and γ" and
"first decide between action β and γ, then toss a coin"

are different. This becomes obvious in the example shown in Fig. 10. Starting in state s of $\mathcal{M}$, if first the coin is tossed (action α) and then, depending on its outcome, action β is chosen in state s_1 and action γ is chosen in state s_2, then this yields that a "smiling" state is reached with probability one. If, however, the choice between β and γ is resolved before the coin is tossed, that is the β-transition or the γ-transition is taken in state s, then taking α in state u, resp. state t, will not result in reaching the "smiling" states with probability one. Note that if we choose $\mathsf{ample}(s) = \{\beta, \gamma\}$ for the MDP $\mathcal{M}$ shown in Fig. 10, then conditions (A0)–(A3) are satisfied, whereas condition (A4) is violated. So, a scheduler of $\mathcal{M}$ might schedule a probabilistic non-ample action of the starting state s. Depending on the outcome (moving to state s_1 or s_2), the scheduler chooses different ample actions (of s). Thus, choosing α first, postpones the real non-deterministic decision between the ample actions β and γ. The reduced system $\mathcal{M}_{\mathsf{red}}$ is forced to decide immediately for a particular ample action β or γ of s (more precisely a distribution over the ample actions of s) in its first step before the outcome of α is known. This decision is fixed from then on. It is exactly this behavior that one has to forbid to gain stutter equivalence between the given system $\mathcal{M}$ and its reduced system. That means that if the system can branch probabilistically with non-ample actions (with respect to the starting state), then there should be only one ample action of the starting state. The additional branching condition (A4) ensures exactly this.

The above remarks only present rough explanations to justify conditions (A0)–(A4). For a full proof of Theorem 5.8, see [43].

Remark 5.9. Theorem 5.8 ensures that, given a deterministic ω-automaton that accepts a stutter invariant language, it suffices to model check $\mathcal{M}_{\mathsf{red}}$ instead of $\mathcal{M}$. As $\mathcal{M}_{\mathsf{red}}$ is in general smaller than $\mathcal{M}$, this yields a possible speed-up of the analysis. Of course, the algorithmic construction of appropriate ample sets together with the construction and the analysis of $\mathcal{M}_{\mathsf{red}}$ should be more efficient than model checking the full system $\mathcal{M}$. Note that even a reduction that eliminates only actions, but does not shrink the state space, might yield a speed-up of the analysis as the probabilistic model checking procedure relies on solving linear programs where the number of linear inequalities for any state s is given by the number of outgoing actions of s.

Experimental Results

The partial order approach for MDPs has been implemented in the model checker LIQUOR [3, 5] using heuristics for approximating the conditions (A2), (A3) and (A4) given in Fig. 9. These heuristics use a superset of the dependence relation and rely on a pre-analysis of the control flow graph induced by programs given in the specification language ProbMeLa (Sect. 2). Several case studies with LIQUOR have shown that the partial order reduction (POR) can lead to a major speed-up and can also decrease the space requirements. To give an impression on the dimension of the time and space requirements for realistic systems, the following table summarizes the results for a randomized leader election protocol (where variable N in the first column denotes the number of parallel processes in the model):

Randomized leader election

	without POR			with POR		
N	states	transitions	time	states	transitions	time
4	53621	156072	1.1 s	21063	78072	1.1 s
5	896231	$3.2 \cdot 10^6$	34 s	299670	$1.3 \cdot 10^6$	21 s
6	$1.1 \cdot 10^7$	$6.2 \cdot 10^7$	813 s	$4.1 \cdot 10^6$	$1.4 \cdot 10^7$	180 s

In other cases, for instance in models of parallel processes that share common synchronization points, the reduction can be even better. More results and more detailed information about applied heuristics and techniques can be found in [5].

6 Partially Observable MDPs

The analysis techniques of Sects. 3, 4, and 5 yield worst-case schedulers where the probability for a certain undesired event is maximal, or dually, where the probability for the desired behavior is minimal. To some extent, these techniques are also applicable to *controller synthesis problems* where the goal is to design a scheduler (i.e., a controller) that resolves the internal nondeterminism and optimizes the probabilities for a certain LT property. However, in this context, the general notion of a scheduler appears to be inadequate since it relies on the complete knowledge of the system history. Consider again the Monty Hall problem from Example 2.2 and the corresponding MDP $\mathcal{M}$ in Fig. 3. We saw in Example 2.2 that

$$\sup_{\mathcal{U}\in\mathsf{Sched}} \mathsf{Pr}^{\mathcal{M},\mathcal{U}}(\Diamond\,\mathsf{get\_car}) = 1,$$

where the supremum is attained by the scheduler $\mathcal{U}$ with

$$\mathcal{U}(s \xrightarrow{\mathsf{choose}} t_1)(\mathsf{keep}) = 1$$

and

$$\mathcal{U}(s \xrightarrow{\mathsf{choose}} t_2)(\mathsf{switch}) = \mathcal{U}(s \xrightarrow{\mathsf{choose}} t_3)(\mathsf{switch}) = 1.$$

As already pointed out in the example, this scheduler $\mathcal{U}$ does not reflect a realistic choice of the contestant, as the contestant does not know whether she/he has chosen the door with the car behind it, or not. So, the only realistic schedulers (that model a contestant's choice) are schedulers that make the same choice for each path that ends either in state t_1, t_2, or t_3. In this case, these are the two schedulers $\mathcal{U}_\mathsf{s}$ and $\mathcal{U}_\mathsf{k}$ with

$$\mathcal{U}_\mathsf{s}(s \xrightarrow{\mathsf{choose}} t_1)(\mathsf{switch}) = \mathcal{U}_\mathsf{s}(s \xrightarrow{\mathsf{choose}} t_2)(\mathsf{switch}) = \mathcal{U}_\mathsf{s}(s \xrightarrow{\mathsf{choose}} t_3)(\mathsf{switch}) = 1$$

and

$$\mathcal{U}_\mathsf{k}(s \xrightarrow{\mathsf{choose}} t_1)(\mathsf{keep}) = \mathcal{U}_\mathsf{k}(s \xrightarrow{\mathsf{choose}} t_2)(\mathsf{keep}) = \mathcal{U}_\mathsf{k}(s \xrightarrow{\mathsf{choose}} t_3)(\mathsf{keep}) = 1$$

where the contestant either decides to switch the door or to keep it. So, in this scenario, we are actually interested in computing the supremum, resp. infimum of $\mathsf{Pr}^{\mathcal{M},\mathcal{U}}(\Diamond\,\mathsf{get\_car})$ under all "realistic" schedulers. A model that allows us to express such requests is given by partially observable Markov decision processes (POMDPs) [84, 70, 71, 67].

Definition 6.1 (Partially observable Markov decision process). *A partially observable Markov decision process is a pair* $(\mathcal{M}, \sim)$, *where:*

- $\mathcal{M} = (S, \mathsf{Act}, \delta, \mu, \mathsf{AP}, L)$ *is a Markov decision process.*
- $\sim \subseteq S \times S$ *is an equivalence relation such that for all states* $s, t \in S$ *with* $s \sim t$ *we have* $\mathsf{Act}(s) = \mathsf{Act}(t)$.

If $s \in S$, *then* $[s]_\sim$ *denotes the equivalence class of state* s *with respect to* $\sim$.

Given a POMDP $(\mathcal{M}, \sim)$, an observation-based scheduler $\mathcal{U}$ is a scheduler for $\mathcal{M}$ that is consistent with $\sim$, i.e., $\mathcal{U}(s_0 \xrightarrow{\alpha_1} \cdots \xrightarrow{\alpha_n} s_n) = \mathcal{U}(t_0 \xrightarrow{\alpha_1} \cdots \xrightarrow{\alpha_n} t_n)$ if $s_i \sim t_i$ for $0 \leq i \leq n$. The set of observation-based schedulers is denoted by $\mathsf{Sched}^{(\mathcal{M},\sim)}$.

If we equip the MDP $\mathcal{M}$ for the Monty Hall problem with the equivalence relation $\sim$ given by $[s]_\sim = \{s\}$, $[u_1]_\sim = \{u_1\}$, $[u_2]_\sim = \{u_2\}$ and

$$[t_1]_\sim = [t_2]_\sim = [t_3]_\sim = \{t_1, t_2, t_3\},$$

then the deterministic observation-based schedulers of the POMDP $(\mathcal{M}, \sim)$ are the "realistic" schedulers that actually model a contestant's choice in the game. Thus, in the Monty Hall scenario, we are interested in computing

$$\sup_{\mathcal{U} \in \mathsf{Sched}_{\mathsf{D}}^{(\mathcal{M},\sim)}} \mathsf{Pr}^{\mathcal{M},\mathcal{U}}(\Diamond\, \mathsf{get\_car}),$$

resp. in computing the infimum. Here, $\mathsf{Sched}_{\mathsf{D}}^{(\mathcal{M},\sim)}$ denotes the set of deterministic observation-based schedulers of the POMDP $(\mathcal{M}, \sim)$. Unfortunately, we cannot expect to have algorithmic solutions for the task to compute extremal reachability probabilities, when ranging over observation-based schedulers. For a similar partial information model which uses distributed schedulers instead of observation-based schedulers, it has been shown that there is no algorithm that computes this supremum under all distributed schedulers. In fact, the supremum is not even approximable [37]. For the model of POMDPs, there even exist the following undecidability results for qualitative questions, which have recently been shown in [2, 43].

In what follows, we use LTL-notations to denote LT properties. The symbol $\Diamond$ stands for "eventually", $\Box$ for "always". Thus, the combination $\Box\Diamond$ denotes "infinitely often" and $\Diamond\Box$ means "continuously from some moment on".

Theorem 6.2 (Undecidability results for POMDPs). *The following problems are undecidable. Given a POMDP* $(\mathcal{M}, \sim)$ *and a set* B *of states in* $\mathcal{M}$, *is there a deterministic observation-based scheduler* $\mathcal{U}$ *for* $(\mathcal{M}, \sim)$ *such that:*

(a) $\mathsf{Pr}^{\mathcal{M},\mathcal{U}}(\Box\Diamond B) > 0$?
(b) $\mathsf{Pr}^{\mathcal{M},\mathcal{U}}(\Diamond\Box B) = 1$?

Those results as well as the undecidability result mentioned above on qualitative reachability from [37] are remarkable since the corresponding questions for (fully observable) MDPs are decidable in polynomial time (see Sect. 4). However, some other variants of qualitative verification problems for POMDPs have been shown to be decidable [29, 2, 43].

Theorem 6.3 (Decidable problems for POMDPs). *The following problems are decidable. Given a POMDP* $(\mathcal{M}, \sim)$ *and a set* B *of states in* $\mathcal{M}$*, does there exist* $\mathcal{U} \in \mathsf{Sched}^{(\mathcal{M},\sim)}$ *such that:*

(a) $\mathsf{Pr}^{\mathcal{M},\mathcal{U}}(\Box B) > 0$*?*
(b) $\mathsf{Pr}^{\mathcal{M},\mathcal{U}}(\Diamond B) > 0$*?*
(c) $\mathsf{Pr}^{\mathcal{M},\mathcal{U}}(\Box B) = 1$*?*
(d) $\mathsf{Pr}^{\mathcal{M},\mathcal{U}}(\Diamond B) = 1$*?*
(e) $\mathsf{Pr}^{\mathcal{M},\mathcal{U}}(\Box\Diamond B) = 1$*?*
(f) $\mathsf{Pr}^{\mathcal{M},\mathcal{U}}(\Diamond\Box B) > 0$*?*

In fact, in [2, 43], it has been shown that the problems (e) and (d) are reducible to each other and that the latter one can be reduced to the similar question for (fully observable) MDPs using an advanced powerset construction. The proof of (f) (see [43]) uses the interreducibility of (d) and (e), and (a) which has been shown in [29].

7 Conclusion

In this chapter, we have summarized the main features of Markov decision processes as an operational model for parallel probabilistic systems and model checking against ω-regular linear-time properties. We have supposed here that the properties are given by deterministic ω-automata. Instead of automata specifications, any logic that can be translated into automata can be used to provide a formalization of the requirements such as linear temporal logic, the mu-calculus, or monadic second-order logic. As quantitative reasoning about probabilistic systems relies on a combination of graph-based and numerical methods, heuristics that attack the state space explosion problem are even more important than in the non-probabilistic case. In this chapter, we have explained the partial order reduction approach. Several other reduction techniques to reduce the time and space requirements such as abstraction techniques, minimization with simulation-like relations, symmetry reduction, and symbolic approaches with variants of binary decision diagrams have been discussed in the literature and are topics of current research projects (see the references given in Sects. 1 and 5).

One of the key features of model checking tools is the concept of counterexamples that can be returned to the user if the checked property does not hold for the system. In the probabilistic setting, counterexamples are more complex, as single error traces are inadequate. First results on the generation of counterexamples for probabilistic systems and their use in abstraction-refinement approaches are presented in the recent papers [1, 46, 50]. Another current research trend is the investigation of alternating-time and game-based approaches that deal with MDP-like models representing the activities of several players. The concept of partially observable MDPs is one instance thereof (see Sect. 6), another instance are stochastic $2\frac{1}{2}$-player games (see, e.g., [20, 30, 19, 53]).

The classical model of Markov decision processes is adequate for the analysis against safety and liveness properties and other conditions on the temporal order of events, but not to reason about timing constraints within a dense time domain. The treatment of continuous-time Markov decision processes or other stochastic models where time-dependent distributions are attached to the transitions (e.g., [10]) or probabilistic variants of timed automata are examples for other very active research fields [64, 57, 63].

Many concepts for reasoning about MDPs viewed as acceptors for languages over finite words (Rabin's probabilistic finite automata [81]) can be generalized rather naturally for weighted automata. Such a generalization of concepts for MDPs to weighted automata is, however, less clear for the case of infinite words. It would also be interesting to see whether the measure-theoretic concepts that yield the basis to define the probabilities for ω-regular properties can be adapted to other classes of weighted automata to reason about the weights for (measurable) sets of infinite paths. This could yield an interesting alternative to the concept of discounting which is well known for MDPs augmented with a reward function that assigns rewards to states and/or actions (see, e.g., [80]) and has been discussed recently in [32–34] for weighted automata and to the approaches investigated in Chaps. 3 and 5 that enforce convergence of infinite series by imposing certain algebraic assumptions on the semiring of a weighted automaton.

8 Appendix

In this appendix, we give the formal definitions needed for the theory of MDPs that is used in the previous sections.

Markov Chains

We first start with the definition of a probability distribution.

Definition 8.1 (Probability distribution). *Let S be a countable set. A probability distribution on S is a function*

$$\mu : S \to [0,1] \quad \textit{such that} \quad \sum_{s\in S} \mu(s) = 1.$$

Given a probability distribution μ on S, $\mathsf{supp}(\mu)$ denotes the support *of μ, i.e., the set of states $s \in S$ with $\mu(s) > 0$. For each $s \in S$, μ_s denotes the unique*

Dirac distribution on S that satisfies $\mu_s(s) = 1$. By $\mathsf{Distr}(S)$, we denote the set of all probability distributions on S.

Next, we give the definition of a discrete Markov chain, which is basically a directed graph where the edges are labeled with a probability in $[0,1]$, such that in each state the probabilities of its outgoing edges sum up to one. Moreover, there is an initial probability distribution on the vertices of the graph.

Definition 8.2 (Discrete Markov chain). *A discrete Markov chain is a tuple*

$$\mathcal{M} = (S, \mathsf{p}, \mu),$$

where:

- *S is a countable non-empty set of states.*
- *$\mathsf{p} : S \times S \to [0,1]$ is the so-called transition probability function such that $\mathsf{p}(s,.)$ is a probability distribution on S for each $s \in S$.*
- *μ is a probability distribution on S (called the initial distribution).*

Let $T = \{(s,t) \mid \mathsf{p}(s,t) > 0, s,t \in S\}$ be the set of transitions with positive probability. We refer to the directed graph (S,T) as the *underlying graph* of $\mathcal{M}$. Note that (S,T) has no terminal nodes. A discrete Markov chain induces a stochastic process on the set S of its states in a natural way. The probability that the process starts in a certain state (the 0th step) is determined by the starting distribution. Moreover, being in state s in the $(n-1)$st step, the probability that the process is in state t in the nth step is equal to $p(s,t)$. The fact that those probabilities do not depend on the previous steps (*history-independent* or *memoryless*) is called the *Markov property.* For a detailed discussion on Markov chains, see, e.g., [59]. Before we go on, we fix some notation for paths of a discrete Markov chain.

Definition 8.3 (Path and corresponding notation). *An (in)finite path of a discrete Markov chain $\mathcal{M}$ is an (in)finite state sequence $\pi = s_0 s_1 \ldots$ such that $p(s_i, s_{i+1}) > 0$ for all i. Given a finite path $\pi = s_0 s_1 \ldots s_n$, the length $|\pi|$ of π equals n. For an infinite path π, the length is equal to ∞. Given a path $\pi = s_0 s_1 \ldots$ and $i \leq |\pi|$, we denote the ith state of π by π_i (i.e., $\pi_i = s_i$) and the ith prefix by $\pi\uparrow^i = s_0, s_1, \ldots, s_i$. We denote by $\mathsf{Path}_{\mathsf{fin}}$ (resp. $\mathsf{Path}_{\mathsf{inf}}$) the set of finite (resp. infinite) paths of a given discrete Markov chain and by $\mathsf{Path}_{\mathsf{fin}}(s)$ (resp. $\mathsf{Path}_{\mathsf{inf}}(s)$) the set of finite (resp. infinite) paths starting in the state s. The empty path is denoted by ϵ.*

If necessary, then we will index Path by the corresponding system, e.g., $\mathsf{Path}_{\mathsf{inf}}^{\mathcal{M}}$. We now define the probability space that formalizes the stochastic process induced by a discrete Markov chain.

Definition 8.4 (Basic cylinder). *Given a discrete Markov chain $\mathcal{M}$, we define, for every $\pi \in \mathsf{Path}_{\mathsf{fin}}^{\mathcal{M}}$, the* basic cylinder *of π as*

$$\Delta(\pi) = \left\{\rho \in \mathsf{Path}_{\mathsf{inf}}^{\mathcal{M}} : \rho\uparrow^{|\pi|} = \pi\right\}.$$

Definition 8.5 (Probability space of a discrete Markov chain). *Given a discrete Markov chain $\mathcal{M} = (S, \mathsf{p}, \mu)$, we define a probability space*

$$\Psi = \big(\mathsf{Path}_{\mathsf{inf}}^{\mathcal{M}}, \Delta, \mathsf{Pr}\big),$$

such that:

- *Δ is the σ-algebra generated by the empty set and the set of basic cylinders in $\mathsf{Path}_{\mathsf{inf}}^{\mathcal{M}}$.*
- *Pr is the uniquely induced probability measure which satisfies the following: $\mathsf{Pr}(\Delta(\epsilon)) = 1$ and for every basic cylinder $\Delta(s_0, s_1, \ldots, s_n)$ over S:*

$$\mathsf{Pr}\big(\Delta(s_0, s_1, \ldots, s_n)\big) = \mu(s_0) \cdot \mathsf{p}(s_0, s_1) \cdot \cdots \cdot \mathsf{p}(s_{n-1}, s_n).$$

Given a state $s \in S$, we denote by Pr_s the probability measure that is obtained if $\mathcal{M}$ is equipped with the starting distribution μ_s, thus $\mathsf{Pr}_s(\Delta(s)) = 1$. We call the a set $P \subseteq \mathsf{Path}_{\mathsf{inf}}$ of infinite paths measurable if and only if $P \in \Delta$.

The existence of the induced probability measure Pr follows from a well-known theorem in measure theory, which is known as Carathéodory's measure extension theorem. The uniqueness follows from the fact that the set of basic cylinders is intersection-stable. For more information on measure theory, see, e.g., [13].

Markov Decision Processes

We will now explain formally the probability space that emanates from a Markov decision process and a given scheduler. Let

$$\mathcal{M} = (S, \mathsf{Act}, \delta, \mu, \mathsf{AP}, L),$$

be an MDP and $\mathcal{U}$ a scheduler that resolves the nondeterminism in $\mathcal{M}$ (for the definition of an MDP and a scheduler see Sect. 2 of this chapter). The behavior of $\mathcal{M}$ under $\mathcal{U}$ can be formalized by an infinite-state discrete Markov chain $\mathcal{M}_{\mathcal{U}} = (\mathsf{Path}_{\mathsf{fin}}^{\mathcal{M}}, \mathsf{p}, \mu)$, where

$$\mathsf{p}(\pi, \pi') = \mathcal{U}(\pi)(\alpha) \cdot \delta\big(\mathsf{last}(\pi), \alpha, \mathsf{last}(\pi')\big),$$

for $\pi, \pi' \in \mathsf{Path}_{\mathsf{fin}}^{\mathcal{M}}$ with $|\pi'| = |\pi| + 1$, $\pi'\uparrow^{|\pi|} = \pi$ and α is the last action on the path π', i.e.,

$$\pi \xrightarrow{\alpha} \mathsf{last}(\pi') \;=\; \pi'.$$

As the states of $\mathcal{M}_{\mathcal{U}}$ are finite paths of $\mathcal{M}$, this notation is somewhat inconvenient. Consider $\Omega = (\mathsf{Path}_{\mathsf{inf}}^{\mathcal{M}_{\mathcal{U}}}, \Delta^{\mathcal{M}_{\mathcal{U}}})$ and $\Omega' = (\mathsf{Path}_{\mathsf{inf}}^{\mathcal{M}}, \Delta^{\mathcal{M}})$, where $\Delta^{\mathcal{M}_{\mathcal{U}}}$ (resp. $\Delta^{\mathcal{M}}$) is the σ-algebra generated by the empty set and the set of basic cylinders over $\mathcal{M}_{\mathcal{U}}$ (resp. $\mathcal{M}$). We define

$$\mathsf{f} : \mathsf{Path}_{\mathsf{inf}}^{\mathcal{M}_{\mathcal{U}}} \to \mathsf{Path}_{\mathsf{inf}}^{\mathcal{M}}$$

as $\mathsf{f}(\pi_0 \xrightarrow{\alpha_1} \pi_1 \xrightarrow{\alpha_2} \cdots) = \mathsf{last}(\pi_0) \xrightarrow{\alpha_1} \mathsf{last}(\pi_1) \xrightarrow{\alpha_2} \cdots$ (note that the π_i's are finite paths of $\mathcal{M}$). Then f is a measurable function and we define the following probability measure on $\Delta^{\mathcal{M}}$:

$$\mathsf{Pr}^{\mathcal{M},\mathcal{U}}(A') = \mathsf{Pr}^{\mathcal{M}_{\mathcal{U}}}(\mathsf{f}^{-1}(A')), \quad \text{for } A' \in \Delta^{\mathcal{M}}.$$

Then given a scheduler $\mathcal{U}$ for $\mathcal{M}$, the probability measure $\mathsf{Pr}^{\mathcal{M},\mathcal{U}}$ formalizes the behavior of $\mathcal{M}$ under $\mathcal{U}$, where we have the convenience to talk about measures of sets of infinite paths of $\mathcal{M}$. As for discrete Markov chains, given a state $s \in S$, we denote by $\mathsf{Pr}_s^{\mathcal{M},\mathcal{U}}$ the probability measure that is obtained if $\mathcal{M}$ is equipped with the starting distribution μ_s. For a detailed discussion on MDPs, see, e.g., [80].

We also fix the following notation for convenience. Given an MDP $\mathcal{M}$, a scheduler $\mathcal{U}$, and a path property E, we will write

$$\mathsf{Pr}^{\mathcal{M},\mathcal{U}}(E) = \mathsf{Pr}^{\mathcal{M},\mathcal{U}}(\{\pi \in \mathsf{Path}_{\mathsf{inf}}^{\mathcal{M}} \mid \pi \text{ satisfies } E\})$$

for the probability that the property E holds in $\mathcal{M}$ under the scheduler $\mathcal{U}$.

References

1. H. Aljazzar, H. Hermanns, and S. Leue. Counterexamples for timed probabilistic reachability. In *Proceedings of the 3rd International Workshop on Formal Modeling and Analysis of Timed Systems (FORMATS'05)*, volume 3829 of *Lecture Notes in Computer Science*, pages 177–195. Springer, Berlin, 2005.
2. C. Baier, N. Bertrand, and M. Grösser. On decision problems for probabilistic Büchi automata. In *Proceedings of the 11th International Conference on Foundations of Software Science and Computation Structures (FOSSACS'08)*, volume 4962 of *Lecture Notes in Computer Science*, pages 287–301. Springer, Berlin, 2008.
3. C. Baier and F. Ciesinski. Liquor: A tool for qualitative and quantitative linear time analysis of reactive systems. In *Proceedings of the 3rd International Conference on the Quantitative Evaluation of SysTems (QEST'06)*, pages 131–132. IEEE Computer Society Press, Los Alamitos, 2006.
4. C. Baier, F. Ciesinski, and M. Grösser. ProbMeLa: A modeling language for communicating probabilistic systems. In *Proceedings of the 2nd ACM–IEEE International Conference on Formal Methods and Models for Codesign (MEMOCODE'04)*, pages 57–66. IEEE Computer Society Press, Los Alamitos, 2006.
5. C. Baier, F. Ciesinski, M. Grösser, and J. Klein. Reduction techniques for model checking Markov decision processes. In *Proceedings of the 5th International Conference on Quantitative Evaluation of SysTems (QEST'08)*, pages 45–54. IEEE Computer Society Press, Los Alamitos, 2008.

6. C. Baier, E. Clarke, V. Hartonas-Garmhausen, M. Kwiatkowska, and M. Ryan. Symbolic model checking for probabilistic processes. In *Proceedings of the 24th International Colloquium on Automata, Languages and Programming (ICALP'97)*, volume 1256 of *Lecture Notes in Computer Science*, pages 430–440. Springer, Berlin, 1997.
7. C. Baier, P. d'Argenio, and M. Größer. Partial order reduction for probabilistic branching time. In *Proceedings of the 3rd Workshop on Quantitative Aspects of Programming Languages (QAPL'05)*, volume 153(2) of *Electronic Notes in Theoretical Computer Science*, pages 97–116. Springer, Berlin, 2006.
8. C. Baier, B. Engelen, and M. Majster-Cederbaum. Deciding bisimularity and similarity for probabilistic processes. *Journal of Computer and System Sciences*, 60:187–231, 2000.
9. C. Baier, M. Größer, and F. Ciesinski. Partial order reduction for probabilistic systems. In *Proceedings of the 1st International Conference on Quantitative Evaluation of SysTems (QEST'04)*, pages 230–239. IEEE Computer Society Press, Los Alamitos, 2004.
10. C. Baier, B. Haverkort, H. Hermanns, and J.P. Katoen. Efficient computation of time-bounded reachability probabilities in uniform continuous-time Markov decision processes. *Theoretical Computer Science*, 345(1):2–26, 2005.
11. C. Baier and J.-P. Katoen. *Principles of Model Checking.* MIT Press, Cambridge, 2008.
12. C. Baier and M. Kwiatkowska. Model checking for a probabilistic branching time logic with fairness. *Distributed Computing*, 11(3):125–155, 1998.
13. H. Bauer. *Wahrscheinlichkeitstheorie und Grundzüge der Maßtheorie.* de Gruyter, Berlin, 1978.
14. R. Bellmann. A Markovian decision process. *Journal of Mathematics and Mechanics*, 6(4):679–684, 1957.
15. A. Bianco and L. De Alfaro. Model checking of probabilistic and nondeterministic systems. In *Proceedings of the 15th Conference on Foundations of Software Technology and Theoretical Computer Science (FSTTCS'95)*, volume 1026 of *Lecture Notes in Computer Science*, pages 499–513. Springer, Berlin, 1995.
16. M. Bozga and O. Maler. On the representation of probabilities over structured domains. In *Proceedings of the 11th International Conference on Computer Aided Verification (CAV'99)*, volume 1633 of *Lecture Notes in Computer Science*, pages 261–273. Springer, Berlin, 1999.
17. J. Burch, E. Clarke, K. McMillan, D. Dill, and L. Hwang. Symbolic model checking: 10^{20} states and beyond. *Information and Computation*, 98(2):142–170, 1992.
18. S. Cattani and R. Segala. Decision algorithms for probabilistic bisimulation. In *Proceedings of the 13th International Conference on Concurrency Theory (CONCUR'02)*, volume 2421 of *Lecture Notes in Computer Science*, pages 371–385. Springer, Berlin, 2002.

19. K. Chatterjee. Stochastic ω-regular games. PhD thesis, University of California at Berkeley, 2007
20. K. Chatterjee, L. de Alfaro, and T. Henzinger. The complexity of stochastic Streett and Rabin games. In *Proceedings of the 32nd International Colloquium on Automata, Languages and Programming (ICALP'05)*, volume 3580 of *Lecture Notes in Computer Science*, pages 878–890. Springer, Berlin, 2005.
21. E. Clarke, O. Grumberg, and D. Peled. *Model Checking.* MIT Press, Cambridge, 1999.
22. C. Courcoubetis and M. Yannakakis. Markov decision processes and regular events (extended abstract). In *Proceedings of the 17th International Colloquium on Automata, Languages and Programming (ICALP'90)*, volume 443 of *Lecture Notes in Computer Science*, pages 336–349. Springer, Berlin, 1990.
23. C. Courcoubetis and M. Yannakakis. The complexity of probabilistic verification. *Journal of the ACM*, 42(4):857–907, 1995.
24. P.R. d'Argenio, B. Jeannet, H.E. Jensen, and K.G. Larsen. Reduction and refinement strategies for probabilistic analysis. In *Proceedings of the Joint International Workshop on Process Algebra and Performance Modeling and Probabilistic Methods in Verification (PAPM–PROBMIV'02)*, volume 2399 of *Lecture Notes in Computer Science*, pages 57–76. Springer, Berlin, 2002.
25. P.R. d'Argenio, B. Jeannet, H. Jensen, and K. Larsen. Reachability analysis of probabilistic systems by successive refinements. In *Proceedings of the 1st Joint International Workshop on Process Algebra and Performance Modeling and Probabilistic Methods in Verification (PAPM–PROBMIV'01)*, volume 2165 of *Lecture Notes in Computer Science*, pages 57–76. Springer, Berlin, 2001.
26. P.R. d'Argenio and P. Niebert. Partial order reduction on concurrent probabilistic programs. In *Proceedings of the 1st International Conference on Quantitative Evaluation of SysTems (QEST'04)*, pages 240–249. IEEE Computer Society Press, Los Alamitos, 2004.
27. L. de Alfaro. Formal verification of probabilistic systems. PhD thesis, Stanford University, 1997
28. L. de Alfaro. Stochastic transition systems. In *Proceedings of the 9th International Conference on Concurrency Theory (CONCUR'98)*, volume 1466 of *Lecture Notes in Computer Science*, pages 423–438. Springer, Berlin, 1998.
29. L. de Alfaro. The verification of probabilistic systems under memoryless partial-information policies is hard. In *Proceedings of the 2nd International Workshop on Probabilistic Methods in Verification (ProbMiV'99)*, pages 19–32. Birmingham University, Research Report CSR-99-9, 1999.
30. L. de Alfaro and R. Majumdar. Quantitative solution of omega-regular games. *Journal of Computer and System Sciences*, 68:374–397, 2004.

31. M. Droste and P. Gastin. Weighted automata and weighted logics. In this *Handbook*. Chapter 5. Springer, Berlin, 2009.
32. M. Droste and D. Kuske. Skew and infinitary formal power series. *Theoretical Computer Science*, 366:199–227, 2006.
33. M. Droste and G. Rahonis. Weighted automata and weighted logics with discounting. In *Proceedings of the 12th International Conference on Implementation and Applications of Automata (CIAA'07)*, volume 4783 of *Lecture Notes in Computer Science*, pages 73–84. Springer, Berlin, 2007.
34. M. Droste, J. Sakarovitch, and H. Vogler. Weighted automata with discounting. *Information Processing Letters*, 108(1):23–28, 2008.
35. Z. Esik and W. Kuich. Finite automata. In this *Handbook*. Chapter 3. Springer, Berlin, 2009.
36. I. Fichtner, D. Kuske, and I. Meinecke. Traces, series-parallel posets, and pictures: a weighted study. In this *Handbook*. Chapter 10. Springer, Berlin, 2009.
37. S. Giro and P.R. d'Argenio. Quantitative model checking revisited: neither decidable nor approximable. In *Proceedings of the 5th International Conference on Formal Modelling and Analysis of Timed Systems (FORMATS'07)*, volume 4763 of *Lecture Notes in Computer Science*, pages 179–194. Springer, Berlin, 2007.
38. P. Godefroid. *Partial Order Methods for the Verification of Concurrent Systems: an Approach to the State Explosion Problem*, volume 1032 of *Lecture Notes in Computer Science*. Springer, Berlin, 1996.
39. P. Godefroid, D. Peled, and M. Staskauskas. Using partial-order methods in the formal validation of industrial concurrent programs. In *Proceedings of the International Symposium on Software Testing and Analysis (ISSTA'96)*, pages 261–269. ACM, New York, 1996.
40. E. Grädel, W. Thomas, and T. Wilke, editors. *Outcome of the 2001 Dagstuhl Seminar on Automata, Logics, and Infinite Games: A Guide to Current Research*, volume 2500 of *Lecture Notes in Computer Science*. Springer, Berlin, 2002.
41. M. Größer and C. Baier. Partial order reduction for Markov decision processes: A survey. In *Proceedings of the 4th International Symposium on Formal Methods for Components and Objects (FMCO'05)*, volume 4111 of *Lecture Notes in Computer Science*, pages 408–427. Springer, Berlin, 2006.
42. M. Größer, G. Norman, C. Baier, F. Ciesinski, M. Kwiatkowska, and D. Parker. On reduction criteria for probabilistic reward models. In *Proceedings of the 26th Conference on Foundations of Software Technology and Theoretical Computer Science (FSTTCS'06)*, volume 4337 of *Lecture Notes in Computer Science*, pages 309–320. Springer, Berlin, 2006.
43. M. Größer. Reduction methods for probabilistic model checking. PhD thesis, Technische Universität, Dresden, 2008
44. R. Gupta, S. Smolka, and S. Bhaskar. On randomization in sequential and distributed algorithms. *ACM Computing Surveys*, 26(1):7–86, 1994.

45. G. Hachtel, E. Macii, A. Pardo, and F. Somenzi, Probabilistic analysis of large finite state machines. In *Proceedings of the 31st Design Automation Conference (DAC'94)*, pages 270–275. ACM, New York, 1994.
46. T. Han and J.-P. Katoen. Counterexamples in probabilistic model checking. In *Proceedings of the 13th International Conference on Tools and Algorithms for the Construction and Analysis of Systems (TACAS'07)*, volume 4424 of *Lecture Notes in Computer Science*, pages 60–75. Springer, Berlin, 2007.
47. H. Hansson and B. Jonsson. A logic for reasoning about time and reliability. *Formal Aspects of Computing*, 6(5):512–535, 1994.
48. V. Hartonas-Garmhausen, S. Campos, and E. Clarke. Probverus: Probabilistic symbolic model checking. In *Proceedings of the 5th International Workshop on Formal Methods for Real-Time and Probabilistic Systems (ARTS'99)*, volume 1601 of *Lecture Notes in Computer Science*, pages 96–110. Springer, Berlin, 1999.
49. H. Hermanns, M. Kwiatkowska, G. Norman, D. Parker, and M. Siegle. On the use of MTBDDs for performability analysis and verification of stochastic systems. *The Journal of Logic and Algebraic Programming: Special Issue on Probabilistic Techniques for the Design and Analysis of Systems*, 56:23–67, 2003.
50. H. Hermanns, B. Wachter, and L. Zhang. Probabilistic CEGAR. In *Proceedings of the 20th International Conference on Computer Aided Verification (CAV'08)*, volume 5123 of *Lecture Notes in Computer Science*, pages 162–175. Springer, Berlin, 2008.
51. G. Holzmann. *The SPIN Model Checker, Primer and Reference Manual.* Addison–Wesley, Reading, 2003.
52. G.J. Holzmann and D. Peled. An improvement in formal verification. In *Proceedings of the 7th International Conference on Formal Description Techniques (IFIP'94)*, pages 197–211. Chapman & Hall, London, 1995.
53. F. Horn. Random games. PhD thesis, RWTH Aachen and Université Paris 7, 2008
54. R. Howard. *Dynamic Programming and Markov Processes.* MIT Press, Cambridge, 1960.
55. T. Huynh and L. Tian. On some equivalence relations for probabilistic processes. *Fundamenta Informaticae*, 17:211–234, 1992.
56. B. Jeannet, P.R. d'Argenio, and K.G. Larsen. RAPTURE: A tool for verifying Markov decision processes. In *Proceedings of the International Conference on Concurrency Theory (CONCUR'02): Tools Day*, Technical Report, Faculty of Informatics, Masaryk University Brno, 2002.
57. M. Jurdzinski, F. Laroussinie, and J. Sproston. Model checking probabilistic timed automata with one or two clocks. *Logical Methods in Computer Science*, 4(3-1):4–20, 2008.
58. J.-P. Katoen, D. Klink, M. Leucker, and V. Wolf. Three-valued abstraction for continuous-time Markov chains. In *Proceedings of the 19th International Conference on Computer Aided Verification (CAV'07)*, vol-

ume 4590 of *Lecture Notes in Computer Science*, pages 316–329. Springer, Berlin, 2007.
59. J. Kemeny and J. Snell. *Denumerable Markov Chains*. Van Nostrand, Princeton, 1976.
60. M. Kwiatkowska, G. Norman, and D. Parker. PRISM: Probabilistic symbolic model checker. In *Proceedings of the 12th International Conference on Modelling Tools and Techniques for Computer and Communication System Performance Evaluation (TOOLS'02)*, volume 2324 of *Lecture Notes in Computer Science*, pages 113–140. Springer, Berlin, 2002.
61. M. Kwiatkowska, G. Norman, and D. Parker. Probabilistic symbolic model checking with PRISM: A hybrid approach. *International Journal on Software Tools for Technology Transfer (STTT)*, 6(2):128–142, 2004.
62. M. Kwiatkowska, G. Norman, and D. Parker. Symmetry reduction for probabilistic model checking. In *Proceedings of the 18th International Conference on Computer Aided Verification (CAV'06)*, volume 4144 of *Lecture Notes in Computer Science*, pages 238–248. Springer, Berlin, 2008.
63. M. Kwiatkowska, G. Norman, D. Parker, and J. Sproston. Verification of real-time probabilistic systems. In *Modeling and Verification of Real-Time Systems: Formalisms and Software Tools*, pages 249–288. Wiley, New York, 2008
64. M. Kwiatkowska, G. Norman, R. Segala, and J. Sproston. Automatic verification of real-time systems with discrete probability distributions. *Theoretical Computer Science*, 282:101–150, 2002.
65. K. Larsen and A. Skou. Bisimulation through probabilistic testing. *Information and Computation*, 94(1):1–28, 1991.
66. D. Lehmann and M.O. Rabin. On the advantage of free choice: A symmetric and fully distributed solution to the dining philosophers problem (extended abstract). In *Proceedings of the 8th Annual ACM Symposium on Principles of Programming Languages (POPL'81)*, pages 133–138. ACM, New York, 1981.
67. W. Lovejoy. A survey of algorithmic methods for partially observable Markov decision processes. *Annals of Operations Research*, 28(1):47–65, 1991.
68. N. Lynch. *Distributed Algorithms*. Morgan Kaufmann, San Francisco, 1996.
69. A. Miner and D. Parker. Symbolic representations and analysis of large probabilistic systems. In *Proceedings of the GI/Dagstuhl Research Seminar on Validation of Stochastic Systems*, volume 2925 of *Lecture Notes in Computer Science*. Springer, Berlin, 2003.
70. G. Monahan. A survey of partially observable Markov decision processes: Theory, models, and algorithms. *Management Science*, 28(1):1–16, 1982.
71. C. Papadimitriou and J. Tsitsiklis. The complexity of Markov decision processes. *Mathematics of Operations Research*, 12(3):441–450, 1987.

72. D. Parker. Implementation of symbolic model checking for probabilistic systems. PhD thesis, University of Birmingham, 2002.
73. A. Paz. *Introduction to Probabilistic Automata.* Academic Press, San Diego, 1971.
74. D. Peled. All from one, one for all: On model checking using representatives. In *Proceedings of the 5th International Conference on Computer-Aided Verification (CAV'93)*, volume 697 of *Lecture Notes in Computer Science*, pages 409–423. Springer, Berlin, 1993.
75. D. Peled. Partial order reduction: Linear and branching time logics and process algebras. In *Proceedings of the DIMACS Workshop on Partial Order Methods in Verification*, volume 29(10) of *DIMACS Series in Discrete Mathematics and Theoretical Computer Science*, pages 233–257. American Mathematical Society, Providence, 1997
76. D. Peled, V. Pratt, and G. Holzmann, editors. *Proceedings of the DIMACS Workshop on Partial Order Methods in Verification*, volume 29(10) of *DIMACS Series in Discrete Mathematics and Theoretical Computer Science.* American Mathematical Society, Providence, 1997.
77. A. Philippou, I. Lee, and O. Sokolsky. Weak bisimulation for probabilistic systems. In *Proceedings of the 11th International Conference on Concurrency Theory (CONCUR'00)*, volume 1877 of *Lecture Notes in Computer Science*, pages 334–349. Springer, Berlin, 2000.
78. A. Pnueli. The temporal logic of programs. In *Proceedings of the 18th Symposium on the Foundations of Computer Science (FOCS'77)*, pages 46–57. IEEE Computer Society Press, Los Alamitos, 1977.
79. A. Pnueli and L. Zuck. Probabilistic verification by tableaux. In *Proceedings of the Symposium on Logic in Computer Science (LICS'86)*, pages 322–331. IEEE Computer Society Press, Los Alamitos, 1986.
80. M.L. Puterman. *Markov Decision Processes: Discrete Stochastic Dynamic Programming.* Wiley, New York, 1994.
81. M.O. Rabin. Probabilistic automata. *Information and Control*, 6(3):230–245, 1963.
82. J.J.M.M. Rutten, M. Kwiatkowska, G. Norman, and D. Parker. *Mathematical Techniques for Analyzing Concurrent and Probabilistic Systems*, volume 23 of *CRM Monograph Series.* American Mathematical Society, Providence, 2004.
83. A. Schrijver. *Combinatorial Optimization: Polyhedra and Efficiency.* Springer, Berlin, 2003.
84. E.J. Sondik. The optimal control of partially observable Markov processes. PhD thesis, Stanford University, 1971.
85. W. Thomas. Automata on infinite objects. In *Handbook of Theoretical Computer Science, volume B*, Chapter 4, pages 133–191. Elsevier, Amsterdam, 1990.
86. A. Valmari. A stubborn attack on state explosion. *Formal Methods in System Design*, 1:297–322, 1992.

87. A. Valmari. State of the art report: Stubborn sets. *Petri-Net Newsletters*, 46:6–14, 1994.
88. A. Valmari. Stubborn set methods for process algebras. In *Proceedings of the DIMACS Workshop on Partial Order Methods in Verification*, volume 29(10) of *DIMACS Series in Discrete Mathematics and Theoretical Computer Science*, pages 213–231. American Mathematical Society, Providence, 1997.
89. R. van Glabbeek, S. Smolka, B. Steffen, and C. Tofts. Reactive, generative, and stratified models of probabilistic processes. In *Proceedings of the 5th IEEE Symposium on Logic in Computer Science (LICS'90)*, pages 130–141. IEEE Computer Society Press, Los Alamitos, 1990.
90. M.Y. Vardi. Automatic verification of probabilistic concurrent finite-state programs. In *Proceedings of the 26th Symposium on Foundations of Computer Science (FOCS'85)*, pages 327–338. IEEE Computer Society Press, Los Alamitos, 1985.
91. M.Y. Vardi and P. Wolper. An automata-theoretic approach to automatic program verification. In *Proceedings of the Symposium on Logic in Computer Science (LICS'86)*, pages 332–344. IEEE Computer Society Press, Los Alamitos, 1986.

Chapter 14: Applications of Weighted Automata in Natural Language Processing

Kevin Knight and Jonathan May

USC Information Sciences Institute,
4676 Admiralty Way, Suite 1001, Marina del Rey, CA 90292, USA
knight@isi.edu
jonmay@isi.edu

1 Background

Linguistics and automata theory were at one time tightly knit. Very early on, finite-state processes were used by Markov [40, 30] to predict sequences of vowels and consonants in novels by Pushkin. Shannon [53] extended this idea

M. Droste, W. Kuich, H. Vogler (eds.), *Handbook of Weighted Automata*,
Monographs in Theoretical Computer Science. An EATCS Series,
DOI 10.1007/978-3-642-01492-5_14,

to predict letter sequences of English words using Markov processes. While many theorems about finite-state acceptors (FSAs) and finite-state transducers (FSTs) were proven in the 1950s, Chomsky argued that such devices were too simple to adequately describe natural language [6]. Chomsky employed context-free grammars (CFGs) and then introduced the more powerful *transformational grammars* (TGs), loosely defined in [7]. In attempting to formalize TG, automata theorists like Rounds [51] and Thatcher [57] introduced the theory of tree transducers. Computational linguistics also got going in earnest with Woods' use of *augmented transition networks* (ATNs) for automatic natural language parsing [59]. In the final paragraph of his 1973 tree automata survey [58], Thatcher wrote:

> The number one priority in the area [of tree automata] is a careful assessment of the significant problems concerning natural language and programming language semantics and translation. If such problems can be found and formulated, I am convinced that the approach informally surveyed here can provide a unifying framework within which to study them.

At this point, however, mainstream work in automata theory, linguistics, and computational linguistics drifted apart. Automata theorists pursued a number of theory-driven generalizations [15, 20, 21], while linguists went the other way and eschewed formalism. Computational linguistics focused for a time on extensions to CFGs [11, 52], many of which were Turing equivalent. In the 1970s, speech recognition researchers returned to capturing natural language grammar with FSAs, this time employing transition weights that could be trained on machine-readable text corpora [29, 28, 2]. These formal devices had associated algorithms that were efficient enough for practical computers of that time, and they were remarkably successful at distinguishing correct from incorrect speech transcriptions. In the 1990s, this combination of finite-state string formalisms and large training corpora became the dominant paradigm in speech and text processing; generic software toolkits for weighted finite-state acceptors and transducers (WFSAs and WFSTs) were developed to support a wide variety of applications [23, 46].

The twenty-first century has seen a reawakened interest in tree automata among computational linguists [34, 54, 25], particularly for problems like automatic language translation, where transformations are sensitive to syntactic structure. Generic tree automata toolkits [42] have also been developed to support investigations. In the remainder of this chapter, we discuss how natural language applications use both string and tree automata.

2 WFST Techniques for Natural Language Processing

In this section, we use two sample applications to highlight ways in which finite-state transducers are used in natural language processing. The first application is transliteration of names and technical terms, and the second application is translation of natural language sentences. We conclude with a discussion of language modeling, an important part of both sample applications.

2.1 Example 1: Transliteration

Transliteration is the process by which names and technical terms are borrowed from one language to another. For some language pairs, this is a very simple or even trivial task—*Bill Gates* is written the same way in English and Spanish newspapers, and while the English word *conception* is changed to *concepción* in Spanish to preserve pronunciation; this is a regular and predictable pattern. However, the task becomes significantly more challenging when the language pairs employ different character sets and very different sound systems. For example, a Japanese newspaper may refer to アンジラナイト, using a sound-based character set called Katakana. If we know how Katakana encodes Japanese sounds, then we can sound out アンジラナイト as *anjiranaito*. The transformation from *anjiranaito* to some English term is still quite difficult, since among other constraints, Japanese words must end in vowels, and do not distinguish between l and r as is done in English.

After some thought, and perhaps the use of surrounding context, a bilingual speaker may realize *anjiranaito* was originally the English name *Angela Knight*. Here are more input/output samples:

```
マスターズトーナメント
masutaazutoonamento
Masters Tournament

アイスクリイーム
aisukuriimu
Ice Cream

ニューヨークタイムズ
nyuuyookutaimuzu
New York Times
```

Due to the large number of potential transliterations, this task is hard, even for humans. We can address the combinatorial explosion through the use of finite-state automata [33]. As a first attempt, we might contemplate a single finite-state transducer that converts a string of Katakana symbols K into strings of English letters from $\mathcal{E}$, the language of all English letter strings, with a corresponding probability of conversion for each English letter string E of $P(E|K)$, and chooses the most likely E in the language:

$$\operatorname*{argmax}_{E \in \mathcal{E}} P(E|K) \tag{1}$$

The corresponding transducer design looks like this:

$$\text{Katakana} \Rightarrow \boxed{\text{WFST}} \Rightarrow \text{English}$$

This would be a very complex transducer to design. For example, the Japanese *r* sound may turn into the English letter *R* or the English letter *L* (or some other letter sequence), and this decision depends on many other decisions. We also want to guarantee that the English output phrase is well formed. $P(E|K)$ represents both of these probabilities in one complicated step. By using Bayes' law, we can separate the probability associated with well-formed E from the probability of transformation between K and E:

$$\underset{E \in \mathcal{E}}{\operatorname{argmax}} P(E|K) = \underset{E \in \mathcal{E}}{\operatorname{argmax}} \frac{P(E)P(K|E)}{P(K)} \tag{2}$$

$$= \underset{E \in \mathcal{E}}{\operatorname{argmax}} P(E)P(K|E) \tag{3}$$

The corresponding transducer design now looks like this:

$$\boxed{\text{WFSA}} \Rightarrow \text{English} \Rightarrow \boxed{\text{WFST}} \Rightarrow \text{Katakana}$$

If we move from left to right, we can view this diagram as an explanation for Katakana strings. These explanations are often called "generative stories." According to this story, in order to produce a Katakana string, someone first generates a well-formed English phrase with probability $P(E)$ (according to the WFSA), and then someone converts that phrase into Katakana with probability $P(K|E)$ (according to the WFST). As generative stories go, this is actually a fairly realistic explanation of how Katakana words enter the Japanese vocabulary.

By contrast, if we move from right to left in the same diagram, we can convert a given Katakana string K into English by first sending it backward through the WFST, which will produce a multiplicity of English phrases that would transduce to K.[1] We can then intersect our multiplicity of phrases with the WFSA, in an effort to eliminate candidates that are not well formed.

This design, known as the *noisy-channel model* [53], has several advantages. First, the WFST can be greatly simplified, because it only models $P(K|E)$, the transformation of short English letter sequences into short Katakana sequences, and it does not need to pay attention to the global well formedness of the English. For example, an English T may nondeterministically transduce to a Katakana *to* or *ta*. The WFSA takes up the slack by enforcing global well formedness and assigns a $P(E)$ for any English string E, independent of any transformation. Also, we may have different resources for constructing the two devices—for example, we may have a large English dictionary to help us construct the WFSA.

[1] These English phrases can be represented as a finite-state acceptor, since the WFST preserves regularity in both directions.

The single WFST that represents $P(K|E)$ is still fairly complex. We would like to model the transformation in a series of small, easy-to understand steps. In this example, we break the initial transducer into a chain of three transducers, in the following design [33]:

$$\boxed{\text{WFSA A}} \Rightarrow \text{English} \Rightarrow \boxed{\text{WFST B}} \Rightarrow \text{English sounds}$$
$$\Rightarrow \boxed{\text{WFST C}} \Rightarrow \text{Japanese sounds} \Rightarrow \boxed{\text{WFST D}} \Rightarrow \text{Katakana}$$

According to this design, Katakana strings enter Japanese via the following path: (1) someone produces an English phrase, (2) that English phrase is converted into English sounds, (3) that English sound sequence is converted into Japanese sounds, and (4) those Japanese sounds are converted into Katakana symbols.

We justify this design in our probability model by using the conditional probability chain rule to break one probability distribution into a chain of independent distributions:

$$P(K|E) = \sum_{c \in C} P(K|c)P(c|E) \tag{4}$$

where C is any newly introduced parameter space that we can sum over.

This division can be repeated arbitrarily until we have the appropriate granularity of conditional probability, and hence WFST that we want for our model.

The probability model equation then becomes

$$\begin{aligned} &\operatorname*{argmax}_{E \in \mathcal{E}} P(E|K) \\ &= \operatorname*{argmax}_{E \in \mathcal{E}} \sum_{es} \sum_{js} \quad P(E) \cdot P(es|E) \\ &\cdot P(js|es) \cdot P(K|js) \end{aligned} \tag{5}$$

where es and js range over English and Japanese sound sequences, respectively.

Now that we have divided one complex automaton into a chain of automata, they are simple enough that we can build them—Fig. 1 shows fragments. WFSA A (Fig. 1a) nondeterministically generates English word sequences. WFST B (Fig. 1b) sounds out English word sequences. Note that this transducer can be used in either direction—given a sequence of words, forward application will output a sequence of sounds, and given a sequence of sounds, backward application will output a sequence of words. WFST C (Fig. 1c) converts English sounds into Japanese sounds. This is a highly nondeterministic process: an English consonant sound like T may produce a single Japanese sound t, or it may produce two Japanese sounds, such as $t\ o$ (as in the case of *Knight* transducing into *naito*). It is nondeterministic in the reverse direction as well, since a Japanese r sound may transduce to an English

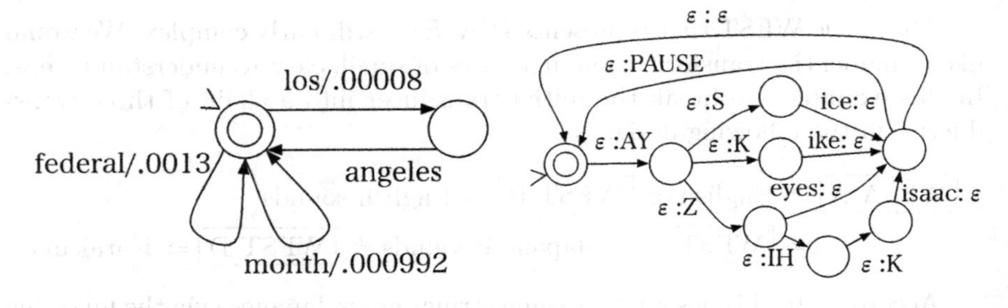

a WFSA A: Produce an English phrase.

b WFST B: Convert English phrase to English sounds.

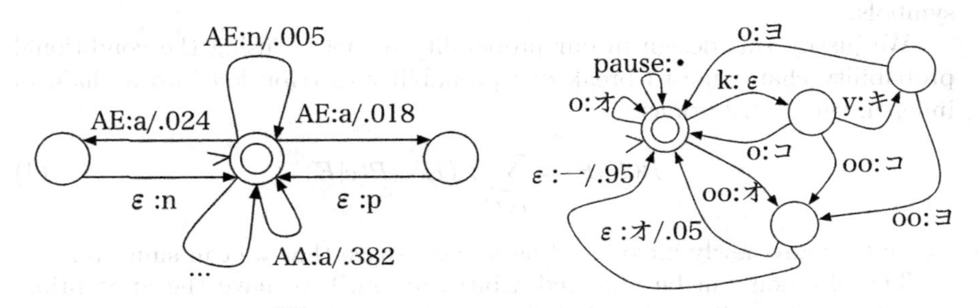

c WFST C: Convert English sounds to Japanese sounds.

d WFST D: Convert Japanese sounds to Katakana.

Fig. 1. Four automata fragments that transliterate English to Japanese [33]

R or *L*. However, the WFST can make its substitutions largely independent of context. Finally, WFST D (Fig. 1d) converts a Japanese sound sequence into Katakana writing. This is fairly deterministic, but requires some linguistic engineering to cover all the cases.

We can now translate a new Katakana string K by sending it backward through WFST D, then sending the result (which itself can be represented by a WFSA) backward through WFST C, and so on, finally intersecting it with WFSA A. In practice, this yields millions of English outputs, most of which consist of strange combinations of small (but legal) English words, e.g.,

```
Ann Gere Uh
Anne Jill Ahh
Angy Rugh
Ann Zillah
```

Here is where the probabilities from WFSA A and WFST C are important. If these are set to reflect what happens in the world (i.e., which English phrases are popular, and which sound substitutions are popular), then each potential

English output also comes with a score, and we can ask for the top-scoring ones (in order):

```
Angela Knight
Angela Nite
Andy Law Knight
Angela Nate
```

It turns out that passing our Katakana string through each transducer sequentially is only one of many possible search strategies. Another approach is to trivially transform A into a weighted transducer A' that is the partial weighted identity transducer with domain equal to the language accepted by A. We then compose transducers A', B, C, and D into a single weighted transducer, offline, then determinize and/or minimize it for deployment [45]; see Chap. 6. A third approach is to employ lazy composition [50, 63], which executes a parallel search through A', B, C, D, and K' (the identity transducer formed from the trivial automaton that accepts only K) by moving tokens from state to state in each. When the tokens all reach final states in their respective machines, an answer is generated; multiple answers can be created with backtracking or beam techniques.

What all of these strategies have in common is that they try to find the English word sequence(s) of highest probability, according to (5). Each of the factors in (5) is broken down further until we reach the probabilities actually stored on the transitions of our WFSAs and WFSTs. For example, consider $P(js|es)$. In Fig. 1c, we can see that WFST C converts English sounds into Japanese sounds via a one-to-many substitution process. Given an English sound sequence es and a Japanese sound sequence js, our WFST can convert es to js in several different ways, depending on how the individual English sounds take responsibility for subsequences of js. Each way can be represented by an alignment that specifies, for each Japanese sound, which English sound produced it. For example, there are four ways to align the sound sequences (L AE M P, r a n p u):

```
L  AE  M  P          L  AE  M  P          /* lamp */
|  |   |  |\         |  |     |\  \
r  a   n  p  u       r  a     n  p  u

L  AE  M  P          L  AE  M  P
|  |\   \   \        |\   \   \   \
r  a   n  p  u       r  a   n  p  u
```

Each alignment corresponds to a different transducing path through WFST C. While we may prefer the first alignment, the others may exist with some small probability. We therefore write the total probability of js given es as:

$$P(js|es) = \sum_{a} \prod_{i=1}^{|es|} P(j_{\mathrm{seq}_{es_i}}|es_i) \tag{6}$$

where alignment a maps each English sound es_i onto Japanese sound subsequence $j_{\text{seq}_{es_i}}$.

Where do transition probability values come from? It is hard for human designers to produce these numbers, so we typically learn them from online text corpora. In the case of WFSA A, we may gather English word frequencies and normalize them. For example, if we see the word *the* 1,200 times in a corpus of 12,000 words, we can assign $P(the) = 0.1$. In the case of WFST C, we can collect probabilities from manually aligned sequence pairs. For example, notice that the English sound M occurs twice in the following database:

```
L  AE  M  P              S  T  IY  M
|  |   |  |\             |\  \   \\ \
r  a   n  p u            s  u  t  i  i  m  u
```

From this, we may conclude that $P(n|M) = 0.5$ and $P(mu|M) = 0.5$. Of course, it is important to have thousands of such pairs, in order to get accurate probability estimates.

For both WFSA A and WFST C, what justifies this "count and divide" strategy for estimating probabilities? Here, we have followed the *maximum likelihood* principle [18], assigning those scores to our transitions that maximize the probability of the training corpus. This principle is especially handy when our training corpus is incomplete. For example, we may only have access to a plain (unaligned) bilingual dictionary:

```
L  AE  M  P              S  T  IY  M
r  a   n  p  u           s  u  t  i  i  m  u
```

and many other pairs.

Given any particular set of parameter values, such as $P(n|M) = 0.32$ (and so on), we can compute $P(js|es)$ for each example pair and multiply these together to get a corpus probability. Some sets of values will yield a high corpus probability, and others a low one. The *expectation–maximization* (EM) algorithm [12] can be used to search efficiently for a good set of values. In this case, highly accurate alignments can be generated automatically without human intervention. Other popular methods of parameter estimation include maximum entropy [27] and minimum error-rate [13].

2.2 Example 2: Translation

We now turn to our second sample application, automatic translation of sentences. This is more challenging for several reasons:

- There are hundreds of thousands of distinct words, versus dozens of distinct linguistic sounds.
- Each word may have many context-dependent meanings or translations.

- Translation often involves significant reordering. For example, in English, the verb comes in the middle of the sentence, while in Japanese, it comes at the end.
- Ensuring that our output is globally well formed requires capturing vast amounts of knowledge about the syntax of the target language, in addition to semantic understanding of how the world works.

While the automatic translation problem remains unsolved, substantial progress has been made in recent years. Much of this progress is due to automatic analysis of large manually-translated documents, such as are produced each year by the United Nations and the European Union.

We might start with the following design for translation:

Spanish ⇒ [WFST] ⇒ English

This design is again problematic, because each word must be translated in the context of all the other words. Therefore, we employ the noisy-channel model approach from Sect. 2.1:

[WFSA] ⇒ English ⇒ [WFST] ⇒ Spanish

In this scheme, the WFST can operate in a largely context-independent fashion. Sending a particular Spanish sentence backward through the WFST might yield many target hypotheses, e.g.,

```
John is in the table
John is on the table
John on is the table
      etc.
```

When we intersect this set with the English WFSA, grammatical hypotheses can be rewarded. Note that the WFSA helps out with both word choice and word ordering, and we can train this WFSA on vast amounts of monolingual English text.

How about the WFST? Brown et al. [4] proposed a particular model for $P(s|e)$ which would assign a conditional probability to any pair of Spanish and English strings. Knight and Al-Onaizan [32] cast this model as a sequence of finite-state automata. Figure 2 depicts the operation of these automata, and Fig. 3 shows automata fragments.

WFSA A (Fig. 3a) generates English word sequences according to some probability distribution that (we hope) assigns high probability to grammatical, sensible sequences. WFST B (Fig. 3b) decides, for each English word, whether to drop it, copy it, duplicate it, triplicate it, etc. The decision is

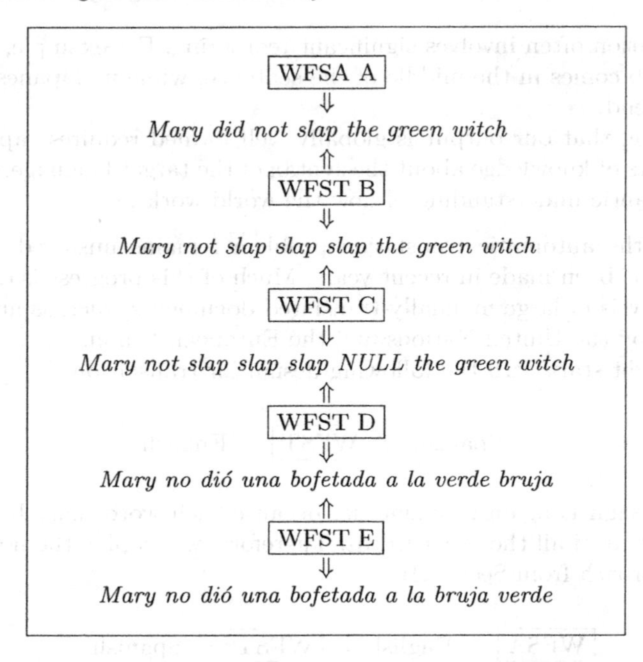

Fig. 2. The generative model of [4] as a cascade of automata

based only on the word itself, with no context information. After each resulting word, WFST C (Fig. 3c) inserts a NULL token[2] with probability 0.02. WFST D (Fig. 3d) then translates each word, one for one, into Spanish. Finally, WFST E (Fig. 3e) reorders the resulting Spanish words. Each transducer is simple enough to build; all of them are highly nondeterministic.

We do not use this transducer cascade in the forward direction, but rather in the reverse direction, to translate Spanish into English. We begin by sending our Spanish input backward through WFST E, to obtain various reorderings, including what we hope will be an English-like ordering. Ultimately, the results are intersected with WFSA A, which is designed to prefer well-formed English. Because of the scale of this problem, translating like this requires pruning the intermediate results. However, it is likely that we will accidentally prune out a good hypothesis before the WFSA A has had a chance to reward it. In practice, therefore, we must perform an integrated search in which all the automata weigh in simultaneously during the incremental construction of the English translation.

How are translation probabilities estimated? We first obtain quantities of manually-translated documents and process them into sentence pairs that are mutual translations. If we were provided with word alignments, e.g.,

[2] The NULL word is designed to generate Spanish function words that have no English equivalent.

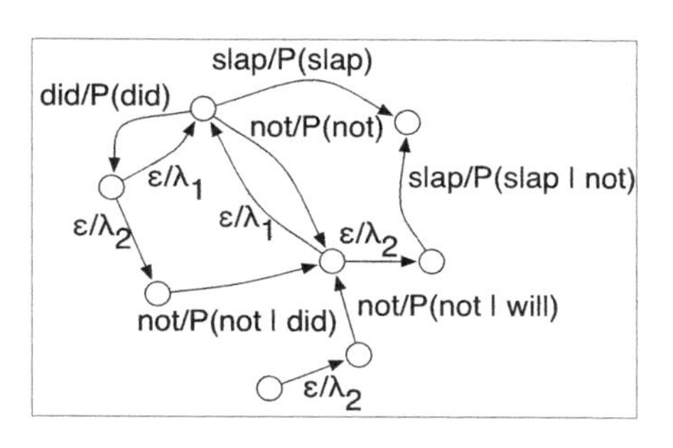

a WFSA A: Produce a phrase.

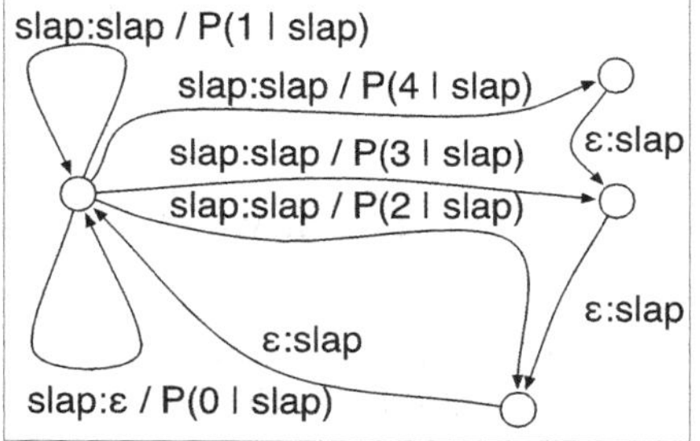

b WFST B: Delete or copy words.

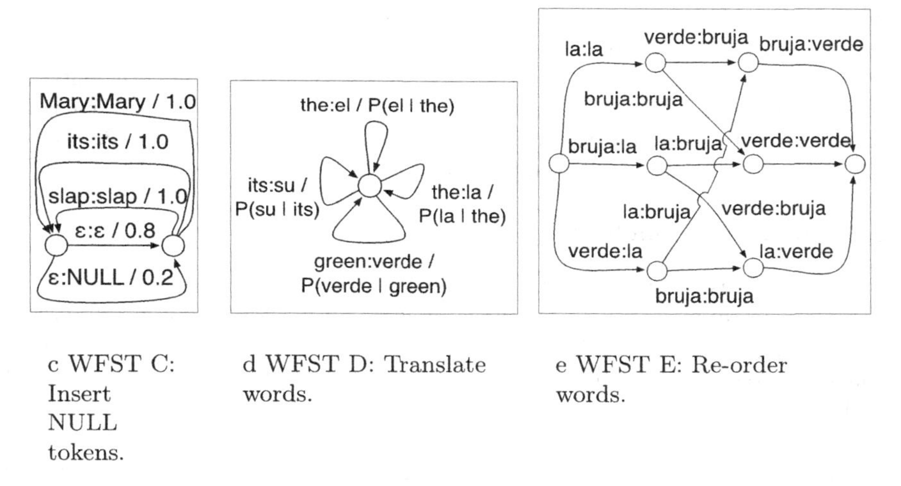

c WFST C: Insert NULL tokens.

d WFST D: Translate words.

e WFST E: Re-order words.

Fig. 3. Five automata fragments that translate Spanish to English. WFSA A produces an English phrase, and WFSTs B–E transform that phrase into Spanish

the	green	witch
la	bruja	verde

then we could estimate the parameters of WFSTs B, C, D, and E. For example, out of 1,000 alignment links connected to the word "green," perhaps 250 link to "verde," in which case $P(verde|green) = 0.25$. However, we are never provided with such alignment links, so again, we use the EM algorithm to guess both links and probability values.

The particular translation model of Fig. 3 was one of the first to be designed, and empirical experiments have revealed many weaknesses. One is that it translates word to word, instead of phrase to phrase. While it has the

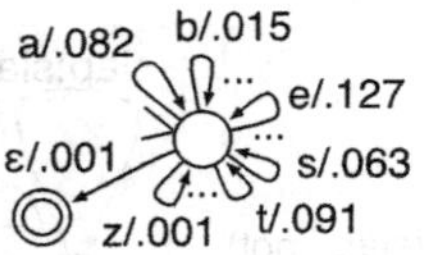

Fig. 4. WFSA for a 1-gram letter language model

capability for phrasal translation (as in *slap* ⇔ *dio una bofetada*), it does not execute such substitutions in a single step. More recent models, taking advantage of more computational power, remedy this. Other problems are more serious. For example, it is difficult to carry out large-scale reordering with finite-state machines, and it is difficult to make these reorderings sensitive to syntactic structure—e.g., the verb in English must somehow move to the end of the sentence when we translate Japanese. Furthermore, it is difficult to attain globally correct outputs, since the well formedness of English depends to some extent on hierarchical, nesting structure of syntactic constituents. For these reasons, some recent models of translation are appropriate for casting in terms of tree automata rather than string automata [60, 1, 61, 22, 16, 43, 19], and we investigate such models later in this chapter.

2.3 Language Modeling

In the previous sections, we focused on the transducers specific to each application. Here, we focus on *language modeling*, the problem of appropriately representing $P(E)$, a WFSA that models well-formed English sentences. Language models are used in any natural language application concerned with well-formed final output.

Shannon [53] observed that the generation of natural language text could be approximated to a reasonable measure by a WFSA that uses states to encode recently seen context. A simple example is a *1-gram* language model of characters, which simply encodes individual character frequencies. If in a corpus of 1,000,000 English characters, the letter e occurs 127,000 times, we estimate the probability $P(e)$ as 127,000/1,000,000, or 0.127. This model can be represented as a WFSA, as shown in Fig. 4.

A 2-gram model remembers the previous letter context—its WFSA has a state for each letter in the vocabulary. The transition between state r and state e outputs the letter e and has probability $P(e|r)$. We can train n-gram models in this way for any n. If we use such models to stochastically generate letter sequences, we observe the following results:

1-gram: thdo cetusar ii c ibt deg irn toihytrsen ...
2-gram: rt wo s acinth gallann prof burgaca ...
3-gram: restiche elp numarin cons dies rem ...
4-gram: what the legal troduce inortemphase ...
5-gram: we has decide in secuadoption on a ...
6-gram: thern is able to the bosnia around ...

While the 6-gram model generates more word-like items than the 1-gram model does, it still lacks sufficient knowledge of English grammar. For noisy-channel applications like translation and speech, a language model needs to know much more, in order to make decisions involving word choice and word order. Work in speech recognition in the 1970s and 1980s effectively trained and used *word* n-gram models, where the probability of a word depends on the previous $n-1$ words; since then, word n-gram models have been the dominant form of language model used in practical systems. This is somewhat surprising, given the work of Chomsky in the 1950s and 1960s which claimed that finite-state string processes were unsuitable for representing human grammars [6, 7, 44]. The largest language model built to date is a 7-gram model, built from one trillion words of English [3] and used for automatic language translation.

A language model should not assign zero probability to any string. For example, a 3-gram language model should accept a string even if it contains a word-triple that was never observed before in training. The process of *smoothing* reassigns some of the probability from seen events to unseen events. One simple technique is *interpolation* smoothing. For the 2-gram case, where we are calculating the likelihood of seeing word y given that the last recognized word was x, instead of estimating $P(y|x)$ as $\frac{\text{count}(xy)}{\text{count}(x)}$, which might be zero, we interpolate with the 1-gram probability of y:

$$P(y|x) = \lambda_1 \cdot \frac{\text{count}(xy)}{\text{count}(x)} + (1 - \lambda_1) \cdot \frac{\text{count}(y)}{N} \tag{7}$$

where N is the size of the training corpus. Likewise, $P(z|x,y)$ can be estimated as $\lambda_2 \frac{\text{count}(xyz)}{\text{count}(xy)} + (1 - \lambda_2)P(z|y)$. Once the counts have been collected from a training corpus, the λ_i values can be set to maximize the likelihood of a smaller (held-out) smoothing corpus, via the EM algorithm. Language models are often evaluated on the probability P they assign to a (further held-out) blind test corpus, or on the *perplexity*, which is $2^{\frac{-\log(P)}{N}}$.

Interpolation smoothing is not the best smoothing method available, but it can be implemented directly in a WFSA, as shown in Fig. 5. This formulation is space efficient, requiring only one transition per observed n-gram, rather than one transition per conceivable n-gram.

3 Applications of Weighted String Automata

In Sect. 2, we saw details of how WFSAs and WFSTs can be used to implement noisy channel models for two applications. In this section, we review recent work in other areas of natural language processing that uses similar techniques. In most cases, the structures and designs, though described in varied ways, are very similar and only differ in the data being modeled.

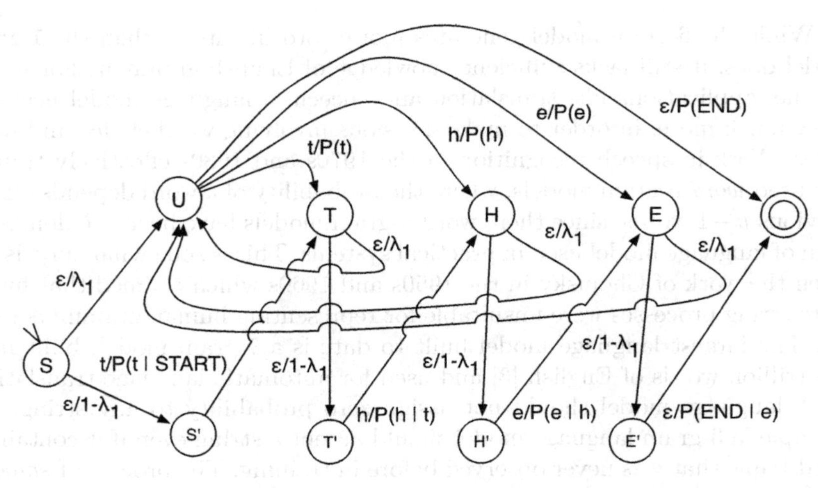

Fig. 5. Fragment of a WFSA for a 2-gram letter language model. At each state {S, T, H, E}, a decision is made to use 2-gram context by moving to states {S', T', H', E'}, respectively, or to use 1-gram context by moving to state U

3.1 Language Translation

We described a word-for-word model of language translation in Sect. 2. This model was implemented in a WFST framework by [32]. A phrase-for-phrase model was subsequently devised by [47] and implemented in a WFST framework by [37]. Translations from this model are much more accurate, and by using a WFST toolkit, Byrne et al. [37] are able to build a cascade of transducers and execute translations using generic finite-state procedures. The most problematic transducer is the one responsible for reordering—such a general transducer would be exceedingly large if built offline. In practice, given a particular source-language sentence, we can encode it and all of its local reorderings online as a temporary WFSA, which is then sent through the rest of the noisy-channel cascade.

3.2 Speech Recognition

Pereira et al. [49] apply the noisy-channel framework to the problem of speech recognition, i.e., recovering the sequence of spoken words that generated a given acoustic speech signal. A standard n-gram language model like that described in Sect. 2.3 is used. The noisy channel transducer, which generates $P(E|S)$ for a received acoustic speech signal S, is described as a chain of transducers as follows:

- For each word in S, a variety of phone sequences, i.e., individual units of speech, may be observed that can be interpreted as the word, with varying probabilities. For each word, a word-to-phone transducer is constructed, and the closure of these transducers over all words forms the complete word-to-phone transducer.

- Similar to the word-to-phone model, each phone can be expressed as a variety of audio signals. Again, the closure of phone-to-audio transducers for each phone is taken as the complete phone-to-audio transducer.

Once defined, the chain of transducers and the final language model are weighted with the method of maximum likelihood, directly observing probabilities from available training data, and possibly smoothing. Composition and decoding are handled entirely by generic automata operations as, for example, implemented in the AT&T FSM Toolkit [46].

3.3 Lexical Processing

In most natural language applications, it is necessary to cut an information stream into word units. This is especially hard in languages without whitespace, such as Chinese. Sproat et al. [56] show how to automatically break Chinese into words by constructing a series of WFSTs. Word-internal units must also be processed. We saw this in the case of transliteration (Sect. 2.1). Another problem is morphological analysis, in which a word is analyzed into *morphemes*, the smallest units of language that carry meaning. Languages like Turkish and Finnish are written with very long words that must often be broken into what would be equivalently represented by separate articles, prepositions, and nouns in other languages. For many other languages, simply finding the root form of an inflected word is a challenge. One of the most successful early introductions of finite-state processing into natural language processing was for morphology [31], and a weighted approach can be found in [9].

3.4 Tagging

A wide variety of natural language problems can be cast as *tagging* problems, in which each word of input is assigned a tag from some finite set. The classic example is part-of-speech tagging, which seeks to disambiguate the syntactic category of each word in a sentence. Given the sentence *The flag waves in the wind*, the tagger must realize that *flag* and *wind* are nouns, even though both can be verbs in other contexts (e.g., *wind a watch*). Finite-state methods are often applied to this task [8]; within the noisy channel framework, we can build an n-gram WFSA to model grammatical tag sequences, and a one-state WFST to model substitutions of words by tags. Another common tagging problem is to locate named entities (such as people, places, and organizations) in texts. Here, each word is tagged as either *B* (word begins a new entity), *I* (word is inside an entity), or *O* (word is outside an entity). This ternary tagging scheme covers cases where two entities may be adjacent in text. A sequence like *Japan gave Russia the Kuriles* would be tagged *B O B B I*.

3.5 Summarization

Text summarization is the shrinking of a document or set of documents into a short summary that contains a useful subset of the information. One application of summarization, headline generation, drops unnecessary words from an input text and performs limited transformation of the remaining words to form an appropriate news headline. The noisy-channel framework is followed to accomplish this task in [62], where the source is considered to be emitting a series of compressed sentences in "headlinese" which are then passed through a transducer that inserts extra words and transforms some words to form grammatical expanded sentences. Zajic et al. [62] tweak their results by introducing various penalties and feature weights onto the transducer arcs; these can be modeled by modifying weights accordingly or by introducing additional transducers that explicitly encode the transitions.

3.6 Optical Character Recognition

The automatic conversion of hard-copy printed material to electronic form is useful for preserving documents created before the digital age, as well as for digitizing writing that is still generated in a nondigital manner, e.g., converting handwritten notes. Scanner technology has progressed considerably in recent years thanks to probabilistic recognition techniques, which are representable in the noisy channel framework. Here, the noise metaphor is readily apparent; clear, uniformly represented characters are garbled by the noisiness of the printed page, incorrectly struck typewriter keys, or the human hand's inconsistency. Kolak et al. [36] use this approach, and built their final system with the AT&T FSM toolkit [46], thus using automata operations directly. The chain of transducers in this case first segments the words into characters, then groups the characters into subword sequences, and finally transforms the sequences into noise-filled sequences.

4 Applications of Weighted Tree Automata

String WFSTs are a good fit for natural language problems that can be characterized by left-to-right substitution. However, their expressiveness breaks down for more complex problems, such as language translation, where there is significant reordering of symbols, and where operations are sensitive to syntactic structure.

The usefulness of hierarchical tree structure was noticed early in linguistics, and automata theorists devised tree acceptors and transducers [51, 57] with the aim of generalizing string automata. Recently, natural language researchers have been constructing weighted syntax-based models for problems

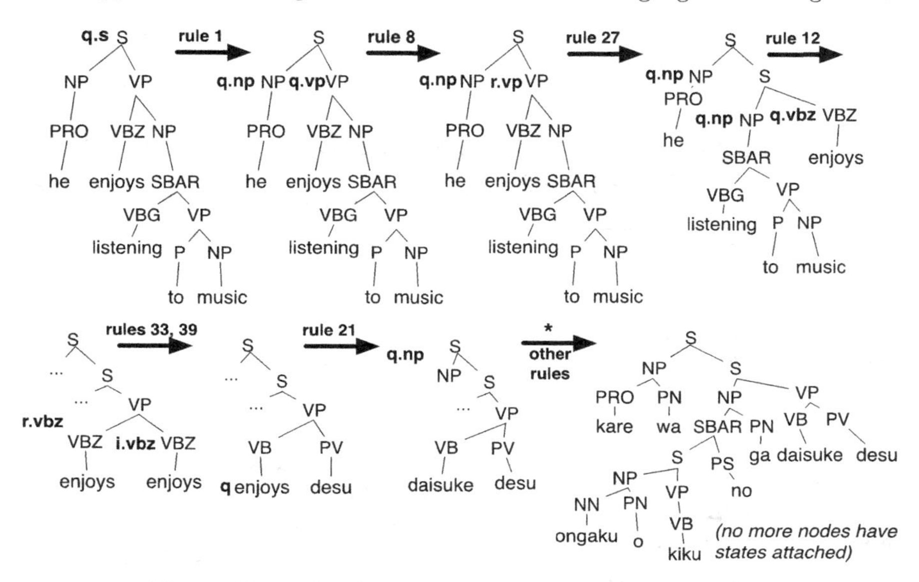

Fig. 6. Example of a syntax-based translation model [34]

such as language translation [61, 60, 1, 22, 16, 43], summarization [35], paraphrasing [48], question answering [14], and language modeling [5]. It has therefore become important to understand whether these natural language models can be captured by standard tree automata.

Figure 6 shows a syntax-based translation model that can be contrasted with the string-based model depicted in Figs. 2 and 3. In the upper left of the figure is an English tree, and in the lower right is a Japanese tree. In between, we see a top-down model of transformation in which pieces of English syntax are matched and replaced with pieces of Japanese syntax. Ultimately, individual English words and phrases are replaced with Japanese ones. This transformation can be carried out by a top-down tree transducer with ε-transitions, as defined by [51, 57], a fragment of which is shown in Fig. 8. This type of transducer is theoretically quite powerful, employing rules that copy unbounded pieces of input (as in Rule 4) and rules that delete pieces of input without processing them (as in Rule 34). It is well known that copying and deleting complicate matters—for example, the class of transformations induced by copying transducers is not closed under composition, which is a significant departure from the string case.

Figure 7 shows some natural transformations that arise in translating one human language to another. In the first example, an English noun-phrase (NP1) must be moved after the verb when we translate to Arabic. A standard noncopying tree transducer cannot handle this case, because it is necessary to "grab and reorder" structures that are deep in the input tree (such as VB and NP2), while the standard transducer can only get hold of the direct children

a Non-local re-ordering (*English/Arabic*).

b Non-constituent Phrasal Translation (*English/Spanish*).

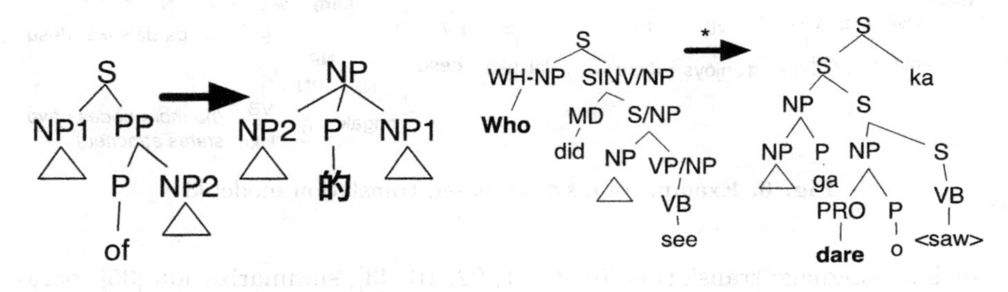

c Lexicalized re-ordering (*English/Chinese*).

d Long distance re-ordering (*English/Japanese*).

Fig. 7. Examples of reordering made possible with a syntax-based translation model

of an input node. For this reason, [24] introduces a class of top-down tree transducers whose rules have extended left-hand sides. An example of such a rule is

```
q S(x0:NP VP(x1:VB x2:NP)) -> S(q x1, r x0, s x2)
```

In [19], Galley et al. give algorithms for acquiring such tree transducers from bilingual data. The English side of this data must be automatically parsed; this is typically done with statistical techniques such as in [10]. At the time of this writing, the largest such transducer has 500 million rules, and the empirical performance of the associated translation system compares favorably with string-based methods. Currently, work at the intersection of tree automata and natural language processing is active:

- On the empirical side, researchers aim to improve tree-based translation by building better models of translation and better rule-extraction algorithms. To further those goals, the availability of a toolkit for manipulating tree automata and tree transducers, such as [42], is important. Similar toolkits for string automata and transducers [46, 23] have enabled better

`/* translate */`

1. q.s S(x0, x1)	$\rightarrow^{0.9}$	S(q.np x0, q.vp x1)
2. q.s S(x0, x1)	$\rightarrow^{0.1}$	S(q.vp x1, q.np x0)
3. q.np x	$\rightarrow^{0.1}$	r.np x
4. q.np x	$\rightarrow^{0.8}$	NP(r.np x, i x)
5. q.np x	$\rightarrow^{0.1}$	NP(i x, r.np x)
6. q.pro PRO(x0)	$\rightarrow^{1.0}$	PRO(q x0)
7. q.nn NN(x0)	$\rightarrow^{1.0}$	NN(q x0)
8. q.vp x	$\rightarrow^{0.8}$	r.vp x
9. q.vp x	$\rightarrow^{0.1}$	S(r.vp x, i x)
10. q.vp x	$\rightarrow^{0.1}$	S(i x, r.vp x)
11. q.vbz x	$\rightarrow^{0.4}$	r.vbz x
12. q.vbz x	$\rightarrow^{0.5}$	VP(r.vbz x, i x)
13. q.vbz x	$\rightarrow^{0.1}$	VP(i x, r.vbz x)
14. q.sbar x	$\rightarrow^{0.3}$	r.sbar x
15. q.sbar x	$\rightarrow^{0.6}$	SBAR(r.sbar x, i x)
16. q.sbar x	$\rightarrow^{0.1}$	SBAR(i x, r.sbar x)
17. q.vbg VBG(x0)	$\rightarrow^{1.0}$	VP(VB(q x0))
18. q.pp PP(x0, x1)	$\rightarrow^{1.0}$	NP(q.np x1, q.p x0)
19. q.p P(x0)	$\rightarrow^{1.0}$	PN(q x0)
20. q he	$\rightarrow^{1.0}$	kare
21. q enjoys	$\rightarrow^{0.1}$	daisuki
22. q listening	$\rightarrow^{0.2}$	kiku
23. q to	$\rightarrow^{0.1}$	o
24. q to	$\rightarrow^{0.7}$	ni
25. q music	$\rightarrow^{0.8}$	ongaku
26. r.vp VP(x0, x1)	$\rightarrow^{0.9}$	S(q.vbz x0, q.np x1)
27. r.vp VP(x0, x1)	$\rightarrow^{0.1}$	S(q.np x1, q.vbz x0)
28. r.sbar SBAR(x0, x1)	$\rightarrow^{0.1}$	S(q.vbg x0, q.pp x1)
29. r.sbar SBAR(x0, x1)	$\rightarrow^{0.9}$	S(q.pp x1, q.vbg x0)
30. r.np NP(x0)	$\rightarrow^{0.1}$	q.pro x0
31. r.np NP(x0)	$\rightarrow^{0.8}$	q.nn x0
32. r.np NP(x0)	$\rightarrow^{0.1}$	q.sbar x0
33. r.vbz VBZ(x0)	$\rightarrow^{0.7}$	VB(q x0)

`/* insert */`

34. i NP(x0)	$\rightarrow^{0.3}$	PN(wa)
35. i NP(x0)	$\rightarrow^{0.3}$	PN(ga)
36. i NP(x0)	$\rightarrow^{0.2}$	PN(o)
37. i NP(x0)	$\rightarrow^{0.1}$	PN(ni)
38. i SBAR(x0, x1)	$\rightarrow^{0.7}$	PS(no)
39. i VBZ(x0)	$\rightarrow^{0.2}$	PV(desu)

Fig. 8. Fragment of a top-down tree transducer with ε-transitions implementing a syntax-based translation model [34]

model development where the domain of strings is involved. Similarly, tree automata toolkits allow the reenvisioning of previous models in a clean transducer framework [25] as well as the rapid development of new models [41].

- On the algorithms side, researchers create more efficient procedures and data structures for executing tree-based inferences. For example, in [26], Huang and Chiang present efficient algorithms for extracting the k most probable trees from a context-free grammar, which is useful for extracting the k-best translations from a large set of hypotheses encoded as a grammar. In [24], Graehl and Knight give EM algorithms for training tree transducers. In [41], May and Knight show improvements from determinizing weighted tree automata. In [38], Maletti gives an $O(mn^4)$ algorithm for minimizing a weighted tree automaton with m rules and n states.
- On the theory side, researchers investigate which automata models both (1) fit natural language phenomena, and (2) possess good theoretical properties. There is still much work to be done—for example, transformations induced by extended left-hand side transducers are not closed under composition, even in the noncopying, nondeleting case. Researchers have also been investigating connections between tree automata and synchronous grammars [55]; the latter of which have been developed independently in the natural language community.

Another area of promise is syntax-based language modeling [5]. Here, we build a probability distribution over all English trees, rather than all English strings.[3] We hope to concentrate probability on objects that are grammatically well formed. Returning to our noisy-channel framework, we can then envision a language model represented by a regular tree grammar [20] and a channel model consisting of a cascade of tree transducers.

4.1 Open Problems

Knight and Graehl, in [34], presented a list of open problems pertinent to the use of weighted tree automata in natural language processing applications. Although some of those problems have since been at least partially solved (for example, in Tiburon [42] we have an instantiation of a useful, generic tree transducer toolkit, and the properties of extended tree transducers were studied in [39]) some still remain and new problems often arise, such as:

- Eppstein [17] presents an algorithm for finding the k best paths through a WFSA which runs in $O(m + n \log n + k)$ time, while [26] present an algorithm for finding the k best derivations of a weighted tree automaton which runs in $O(m + nk \log k)$ time. It is unknown whether the separation of k and n can be achieved in the tree case.

[3] We can still get the probability of a string by summing over all the trees who have that yield.

- Is there an algorithm for minimizing deterministic weighted tree automata that improves on the $O(mn^4)$ runtime of [38]?
- Is there an algorithm to determine whether two instances of a class of weighted tree transducers may be composed to form a single instance of that class that captures the same sequential transformation, even though the class itself is known to be not closed under composition?
- Is there a class of weighted tree transducers that (a) is expressive enough to capture natural language translation phenomena, (b) is closed under composition, (c) allows for unbounded output, (d) admits an algorithm for efficient weight training from input/output examples, and (e) preserves regularity?
- Can we cast other existing NLP models in the language of tree machinery, as was done in [25] for the model of [61], and can we extend these models in interesting ways?

5 Conclusion

In this chapter, we have surveyed some of the natural language applications in which weighted automata play a role. We expect the number of applications to grow over the coming years as automata theorists, linguists, and engineers collaborate to solve difficult problems, and as computational power grows to support new research avenues. Thatcher's vision, that we may use automata to solve problems, model behavior, and make advances in the natural language domain seems realized in current research efforts, and we have high hopes for the future.

References

1. H. Alshawi, S. Douglas, and S. Bangalore. Learning dependency translation models as collections of finite-state head transducers. *Computational Linguistics*, 26(1):45–60, 2000.
2. J.K. Baker. The DRAGON system—An overview. *IEEE Transactions on Acoustics, Speech, and Signal Processing*, ASSP-23(1):24–29, 1975.
3. T. Brants, A.C. Popat, P. Xu, F.J. Och, and J. Dean. Large language models in machine translation. In *Proceedings of the 2007 Joint Conference on Empirical Methods in Natural Language Processing and Computational Natural Language Learning, Prague, Czech Republic, June 2007*, pages 858–867. Association for Computational Linguistics, Stroudsburg, 2007.
4. P.F. Brown, S.A.D. Pietra, V.J.D. Pietra, and R.L. Mercer. The mathematics of statistical machine translation: Parameter estimation. *Computational Linguistics*, 19(2):263–312, 1993.

5. E. Charniak. Immediate-head parsing for language models. In *Proceedings of the 39th Annual Meeting of the Association for Computational Linguistics, Toulouse, France, July 2001*, pages 116–123. Association for Computational Linguistics, Stroudsburg, 2001.
6. N. Chomsky. Three models for the description of language. *IRE Transactions on Information Theory*, 2(3):113–124, 1956.
7. N. Chomsky. *Syntactic Structures*. Mouton, The Hague, 1957.
8. K.W. Church. A stochastic parts program and noun phrase parser for unrestricted text. In *Second Conference on Applied Natural Language Processing Proceedings, Austin, TX, February 1988*, pages 136–143. Association for Computational Linguistics, Stroudsburg, 1988.
9. A. Clark. Memory-based learning of morphology with stochastic transducers. In *Proceedings of the 40th Annual Meeting of the Association for Computational Linguistics, Philadelphia, PA, July 2002*, pages 513–520. Association for Computational Linguistics, Stroudsburg, 2002.
10. M. Collins. Head-driven statistical models for natural language parsing. PhD thesis, University of Pennsylvania, Philadelphia, PA, 1999.
11. M. Dalrymple. *Lexical Functional Grammar*. Academic Press, New York, 2001.
12. A.P. Dempster, N.M. Laird, and D.B. Rubin. Maximum likelihood from incomplete data via the EM algorithm. *Journal of the Royal Statistical Society, Series B*, 39(1):1–38, 1977.
13. R.O. Duda and P.E. Hart. *Pattern Classification and Scene Analysis*. Wiley, New York, 1973.
14. A. Echihabi and D. Marcu. A noisy-channel approach to question answering. In *Proceedings of the 41st Annual Meeting of the Association for Computational Linguistics, Sapporo, Japan, July 2003*, pages 16–23. Association for Computational Linguistics, Stroudsburg, 2003.
15. S. Eilenberg. *Automata, Languages, and Machines*. Academic Press, New York, 1974.
16. J. Eisner. Learning non-isomorphic tree mappings for machine translation. In *The Companion Volume to the Proceedings of 41st Annual Meeting of the Association for Computational Linguistics, Sapporo, Japan, July 2003*, pages 205–208. Association for Computational Linguistics, Stroudsburg, 2003.
17. D. Eppstein. Finding the k shortest paths. *SIAM Journal on Computing*, 28(2):652–673, 1998.
18. R.A. Fisher. On the "probable error" of a coefficient of correlation deduced from a small sample. *Metron. International Journal of Statistics*, 1:3–32, 1921.
19. M. Galley, M. Hopkins, K. Knight, and D. Marcu. What's in a translation rule? In *Proceedings of the Human Language Technology Conference of the North American Chapter of the Association for Computational Linguistics: HLT–NAACL 2004, Boston, MA, May 2004*, pages 273–280. Association for Computational Linguistics, Stroudsburg, 2004.

20. F. Gécseg and M. Steinby. *Tree Automata.* Akadémiai Kiadó, Budapest, 1984.
21. F. Gécseg and M. Steinby. Tree languages. In G. Rozenberg and A. Salomaa, editors, *Handbook of Formal Languages, volume 3*, Chapter 1, pages 1–68. Springer, Berlin, 1997.
22. D. Gildea. Loosely tree-based alignment for machine translation. In *Proceedings of the 41st Annual Meeting of the Association for Computational Linguistics, Sapporo, Japan, July 2003*, pages 80–87. Association for Computational Linguistics, Stroudsburg, 2003.
23. J. Graehl. Carmel finite-state toolkit. http://www.isi.edu/licensed-sw/carmel, 1997.
24. J. Graehl and K. Knight. Training tree transducers. In *Proceedings of the Human Language Technology Conference of the North American Chapter of the Association for Computational Linguistics: HLT–NAACL 2004, Boston, MA, May 2004*, pages 105–112. Association for Computational Linguistics, Stroudsburg, 2004.
25. J. Graehl, K. Knight, and J. May. Training tree transducers. *Computational Linguistics*, 34(3):391–427, 2008.
26. L. Huang and D. Chiang. Better k-best parsing. In *Proceedings of the Ninth International Workshop on Parsing Technology, Vancouver, Canada, October 2005*, pages 53–64. Association for Computational Linguistics, Stroudsburg, 2005.
27. E.T. Jaynes. Information theory and statistical mechanics. *Physical Review (Series II)*, 106(4):620–630, 1957.
28. F. Jelinek. Continuous speech recognition by statistical methods. *Proceedings of the IEEE*, 64(4):532–556, 1976.
29. F. Jelinek, L.R. Bahl, and R.L. Mercer. Design of a linguistic statistical decoder for the recognition of continuous speech. *IEEE Transactions on Information Theory*, IT-21(3):250–256, 1975.
30. D. Jurafsky and J.H. Martin. *Speech and Language Processing: An Introduction to Natural Language Processing, Computational Linguistics, and Speech Recognition*, 2nd edition. Chapter 4: N-grams. Prentice Hall, Englewood Cliffs, 2009.
31. R. Kaplan and M. Kay. Regular models of phonological rule systems. *Computational Linguistics*, 20(3):331–378, 1994.
32. K. Knight and Y. Al-Onaizan. Translation with finite-state devices. In *Machine Translation and the Information Soup: Third Conference of the Association for Machine Translation in the Americas, AMTA'98, Langhorne, PA, October 1998*, volume 1529 of *Lecture Notes in Computer Science*, pages 421–437. Springer, Berlin, 1998.
33. K. Knight and J. Graehl. Machine transliteration. *Computational Linguistics*, 24(4):599–612, 1998.
34. K. Knight and J. Graehl. An overview of probabilistic tree transducers for natural language processing. In *Computational Linguistics and Intelligent Text Processing 6th International Conference, CICLing 2005, Mexico*

City, Mexico, February 2005, volume 3406 of *Lecture Notes in Computer Science*, pages 1–24. Springer, Berlin, 2005.

35. K. Knight and D. Marcu. Summarization beyond sentence extraction: A probabilistic approach. *Artificial Intelligence*, 139(1):91–107, 2002.
36. O. Kolak, W. Byrne, and P. Resnik. A generative probabilistic OCR model for NLP applications. In *Proceedings of the 2003 Human Language Technology Conference of the North American Chapter of the Association for Computational Linguistics, Edmonton, Canada, May–June 2003*, pages 55–62. Association for Computational Linguistics, Stroudsburg, 2003.
37. S. Kumar and W. Byrne. A weighted finite state transducer implementation of the alignment template model for statistical machine translation. In *Proceedings of the 2003 Human Language Technology Conference of the North American Chapter of the Association for Computational Linguistics, Edmonton, Canada, May–June 2003*, pages 63–70. Association for Computational Linguistics, Stroudsburg, 2003.
38. A. Maletti. Minimizing deterministic weighted tree automata. In *Proceedings of the 2nd International Conference on Language and Automata Theory and Applications*, pages 371–382. Universitat Rovira I Virgili, Tarragona, 2008.
39. A. Maletti, J. Graehl, M. Hopkins, and K. Knight. The power of extended top-down tree transducers. *SIAM Journal on Computing*, 39(2):410–430, 2009.
40. A.A. Markov. Essai d'une recherche statistique sur le texte du roman "Eugene Onegin" illustrant la liaison des epreuve en chain (Example of a statistical investigation of the text of "Eugene Onegin" illustrating the dependence between samples in chain). *Izvistia Imperatorskoi Akademii Nauk (Bulletin de l'Académie Impériale des Sciences de St.-Pétersbourg)*, 7:153–162, 1913. English translation by Morris Halle, 1956.
41. J. May and K. Knight. A better n-best list: Practical determinization of weighted finite tree automata. In *Proceedings of the Human Language Technology Conference of the NAACL, Main Conference, New York, NY, June 2006*, pages 351–358. Association for Computational Linguistics, Stroudsburg, 2006.
42. J. May and K. Knight. Tiburon: A weighted tree automata toolkit. In O.H. Ibarra and H.-C. Yen, editors, *Proceedings of the 11th International Conference of Implementation and Application of Automata, CIAA 2006, Taipei, Taiwan, August 2006*. volume 4094 of *Lecture Notes in Computer Science*, pages 102–113. Springer, Berlin, 2006.
43. I.D. Melamed. Multitext grammars and synchronous parsers. In *Proceedings of the 2003 Human Language Technology Conference of the North American Chapter of the Association for Computational Linguistics, Edmonton, Canada, May–June 2003*, pages 79–86. Association for Computational Linguistics, Stroudsburg, 2003.

44. G.A. Miller and N. Chomsky. Finitary models of language users. In R.D. Luce, R.R. Bush, and E. Galanter, editors, *Handbook of Mathematical Psychology, volume II*, pages 419–491. Wiley, New York, 1963.
45. M. Mohri. Finite-state transducers in language and speech processing. *Computational Linguistics*, 23(2):269–312, 1997.
46. M. Mohri, F.C.N. Pereira, and M.D. Riley. AT&T FSM library. http://www.research.att.com/~fsmtools/fsm, 1998. AT&T Labs—Research.
47. F. Och, C. Tillmann, and H. Ney. Improved alignment models for statistical machine translation. In *Proceedings of the 1999 Joint SIGDAT Conference of Empirical Methods in Natural Language Processing and Very Large Corpora, College Park, MD, June 1999*, pages 20–28. Association for Computational Linguistics, Stroudsburg, 1999.
48. B. Pang, K. Knight, and D. Marcu. Syntax-based alignment of multiple translations: Extracting paraphrases and generating new sentences. In *Proceedings of the 2003 Human Language Technology Conference of the North American Chapter of the Association for Computational Linguistics, Edmonton, Canada, May–June 2003*, pages 102–109. Association for Computational Linguistics, Stroudsburg, 2003.
49. F. Pereira, M. Riley, and R. Sproat. Weighted rational transductions and their application to human language processing. In *Human Language Technology, Plainsboro, NJ, March 1994*, pages 262–267. Morgan Kaufmann, San Mateo, 1994.
50. M. Riley, F. Pereira, and E. Chun. Lazy transducer composition: A flexible method for on-the-fly expansion of context-dependent grammar network. In *Proceedings, IEEE Automatic Speech Recognition Workshop, Snowbird, UT, December 1995*, pages 139–140.
51. W.C. Rounds. Mappings and grammars on trees. *Theory of Computing Systems*, 4:257–287, 1970.
52. I.A. Sag, T. Wasow, and E.M. Bender. *Syntactic Theory*, 2nd edition. CSLI Publications, Stanford, 2003.
53. C. Shannon. A mathematical theory of communication. *Bell System Technical Journal*, 27:379–423, 1948. 623–656
54. S.M. Shieber. Synchronous grammars as tree transducers. In *Proceedings of the Seventh International Workshop on Tree Adjoining Grammar and Related Formalisms (TAG+ 7), Vancouver, Canada, May 2004*, pages 88–95.
55. S.M. Shieber. Unifying synchronous tree adjoining grammars and tree transducers via bimorphisms. In *11th Conference of the European Chapter of the Association for Computational Linguistics, Trento, Italy, April 2006*, pages 377–384. Association for Computational Linguistics, Stroudsburg, 2006.
56. R. Sproat, W. Gales, C. Shih, and N. Chang. A stochastic finite-state word-segmentation algorithm for Chinese. *Computational Linguistics*, 22(3):377–404, 1996.

57. J.W. Thatcher. Generalized^2 sequential machine maps. *Journal of Computer and System Sciences*, 4(4):339–367, 1970.
58. J.W. Thatcher. Tree automata: An informal survey. In A.V. Aho, editor, *Currents in the Theory of Computing*, pages 143–172. Prentice Hall, Englewood Cliffs, 1973.
59. W.A. Woods. Transition network grammars for natural language analysis. *Communications of the Association for Computing Machinery*, 13(10):591–606, 1970.
60. D. Wu. Stochastic inversion transduction grammars and bilingual parsing of parallel corpora. *Computational Linguistics*, 23(3):377–404, 1997.
61. K. Yamada and K. Knight. A syntax-based statistical translation model. In *Proceedings of the 39th Annual Meeting of the Association for Computational Linguistics, Toulouse, France, July 2001*, pages 523–530. Association for Computational Linguistics, Stroudsburg, 2001.
62. D. Zajic, B. Dorr, and R. Schwartz. Automatic headline generation for newspaper stories. In *Proceedings of the ACL-02 Workshop on Text Summarization (DUC 2002), Philadelphia, PA, July 2002*, pages 78–85. Association for Computational Linguistics, Stroudsburg, 2002.
63. B. Zhou, S.F. Chen, and Y. Gao. Folsom: A fast and memory-efficient phrase-based approach to statistical machine translation. In *Proceedings of the IEEE/ACL 2006 Workshop on Spoken Language Technology, Palm Beach, Aruba, December 2006*, pages 226–229, IEEE Press, New York, 2006.

Index

M. Droste, W. Kuich, H. Vogler (eds.), *Handbook of Weighted Automata*,
Monographs in Theoretical Computer Science. An EATCS Series,
DOI 10.1007/978-3-642-01492-5,

9783642260490